Advances in Enhancing Unconventional Oil/Gas Recovery

Advances in Enhancing Unconventional Oil/Gas Recovery

Editors

Tao Zhang
Zheng Sun
Dong Feng
Wen Zhao

MDPI

Basel • Beijing • Wuhan • Barcelona • Belgrade • Novi Sad • Cluj • Manchester

Editors
Tao Zhang
Southwest Petroleum University
Chengdu
China

Zheng Sun
China University of Mining and Technology
Xuzhou
China

Dong Feng
China University of Geoscience (Beijing)
Beijing
China

Wen Zhao
Research Institute of Petroleum Exploration and Development
Beijing
China

Editorial Office
MDPI AG
Grosspeteranlage 5
4052 Basel, Switzerland

This is a reprint of articles from the Special Issue published online in the open access journal *Processes* (ISSN 2227-9717) (available at: https://www.mdpi.com/journal/processes/special_issues/523CUK5BFD).

For citation purposes, cite each article independently as indicated on the article page online and as indicated below:

Lastname, A.A.; Lastname, B.B. Article Title. *Journal Name* **Year**, *Volume Number*, Page Range.

ISBN 978-3-7258-1687-3 (Hbk)
ISBN 978-3-7258-1688-0 (PDF)
doi.org/10.3390/books978-3-7258-1688-0

Contents

About the Editors

Tao Zhang

Tao Zhang works at the Department of Petroleum and Natural Gas Engineering at Southwest Petroleum University. His primary focus is on teaching and research in microscopic seepage theory and simulation (LBM, PNM, MD, etc.), gas reservoir engineering theory (shale gas, coalbed methane, tight gas, etc.), and numerical simulation of oil/gas reservoirs. He has led over 10 projects funded by the National Natural Science Foundation of China, the Sichuan Provincial Natural Science Fund, the China Postdoctoral Science Foundation, and various oil and gas companies. Tao Zhang has published nearly 40 papers as the first or corresponding author in journals such as *SPE Journal*, *Water Resources Research*, and *Physical Review E*. His work has been cited more than 3100 times, with an h-index of 30 (as of July 2024). He has been invited to speak at renowned conferences, such as SPE ATCE seven times, and has been awarded five invention patents and two second prizes at the provincial and ministerial levels. Additionally, he serves as an Editorial Board Member for Cambridge Scholars Publishing and the journal *Processes* and is a youth Editorial Board Member of the journal *Petroleum Science*.

Zheng Sun

Zheng Sun earned a Ph.D. in Development of Oil and Gas Reservoirs from the China University of Petroleum in Beijing in 2020. He is an Associate Professor at the State Key Laboratory for Fine Exploration and Intelligent Development of Coal Resources, China University of Mining and Technology. His research interests include microscopic fluid behavior characterization, geological CO2/H2 storage and utilization, and advanced numerical simulation. Dr. Zheng Sun has published over 100 scientific journal works, all focusing on the mentioned research topics. Recently, according to the report jointly released by Elsevier and Stanford University, he has been selected as one of the World's Top 2% Scientists. Also, he has supervised 10+ national projects on fundamental research, financially supported by the National Natural Science Foundation and Ministry of Science and Technology, and has participated in 20+ social-service projects coordinating with oil/gas companies. Given the research efforts contributing to the oil/gas reservoir development, Dr. Zheng Sun won the First Prize for Science and Technology Advances awarded by Hubei Province, and several industrial prizes. He is the Assistant Editor for the *International Journal of Hydrogen Energy*, Associate Editor for the *Journal of Petroleum Exploration and Production*, and a reviewer for multiple well-recognized scientific journals.

Dong Feng

Dong Feng earned a Ph.D. in the Development of Oil and Gas Reservoirs from the China University of Petroleum, Beijing, in 2021. He lectures at the School of Energy Resources, China University of Geosciences, Beijing. His research interests include micro- and nanoscale flow in nanoporous media, phase behavior in conventional reservoirs, enhanced oil/gas recovery, and CCUS. As a first /corresponding author, He has published over 30 papers on oil/gas reservoir development. Dr. Dong Feng won Beijing's Second Prize for Natural Science and several industrial prizes. He was also selected for the National Postdoctoral Innovation Talent Support Program and the Beijing Association for Science and Technology Youth Talent Support Program. He was a young Editorial Board Member and reviewer for multiple well-recognized scientific Journals.

Wen Zhao

Wen Zhao earned a Ph.D. in Geology from the China University of Petroleum, Beijing, in 2022. Currently, he is a researcher at the Research Institute of Petroleum Exploration and Development, PetroChina. His research interests include underground fluid migration and accumulation, such as tight gas accumulation, CO2 storage process with physical experiment (CT Scan, Fluid Charging), and numerical simulation method (Lattice Boltzmann Method, LBM; Pore Network Method, PNM).

processes

MDPI

Editorial

Special Issue: Advances in Enhancing Unconventional Oil/Gas Recovery

Tao Zhang [1,*], Zheng Sun [2], Dong Feng [3] and Wen Zhao [4]

1 Department of Petroleum and Natural Gas Engineering, Southwest Petroleum University, Chengdu 610500, China
2 State Key Laboratory of Coal Resources and Safe Mining, China University of Mining and Technology, Xuzhou 221116, China; sunzheng@cumt.edu.cn
3 School of Energy Resources, China University of Geosciences, Beijing 100083, China; fdcup404@cup.edu.cn
4 Research Institute of Petroleum Exploration and Development, CNPC, Beijing 100083, China; zhaow625@126.com
* Correspondence: tobiascheuing@163.com

In recent years, unconventional reservoirs such as tight gas/oil reservoirs [1], shale gas/oil reservoirs [2], and coalbed methane [3] have garnered significant attention and have played a crucial role in meeting the world's growing energy demands. Characterized by low porosity and low permeability, these reservoirs differ substantially from conventional ones, with pore sizes often falling into the micro- or even nanoscale. These unique features render traditional theories, approaches, and technologies less effective for unconventional reservoirs. Microscopic fluid distribution modes, fluid transport mechanisms, and fluid phase behavior evolve with pore size, presenting numerous challenges in accurately describing these relationships. Consequently, the development of unconventional reservoirs demands many novel solutions to enhance oil and gas recovery efficiencies.

This Special Issue of *Processes* collects together original research and reviews focused on the latest advances in enhancing unconventional oil and gas recovery. The contributions span laboratory measurements and modeling, field case studies, multi-scale simulation studies, mathematical modeling, and more. The studies presented here provide new insights into the complex dynamics of unconventional reservoirs and propose innovative solutions to the challenges they present.

Citation: Zhang, T.; Sun, Z.; Feng, D.; Zhao, W. Special Issue: Advances in Enhancing Unconventional Oil/Gas Recovery. *Processes* **2024**, *12*, 1386. https://doi.org/10.3390/pr12071386

Received: 29 June 2024
Accepted: 1 July 2024
Published: 3 July 2024

1. Reservoir Characterization

The accurate characterization of unconventional reservoirs is critical for effective exploration and production. This section includes a range of studies focusing on different aspects of reservoir characterization. Ali et al. [4] present multiple rock physics diagnostic models to quantitatively characterize shallow marine sediments in tight gas fields of the Middle Indus Basin; they found that sorting the grain and coating cement onto the grain's surface both affected the cementation process. Xu et al. established an interpretation model that integrates response characteristics of logging curves, including shale content, porosity, permeability, and saturation, into tight sandstone reservoirs. Jia et al. predicted longitudinal superimposed "sweet spots" in tight gas reservoirs through a case study of Block G in Canada. The findings indicate that the key engineering parameters for horizontal wells are lateral length, number of stages, fluid volume, and proppant quantity, with lateral length emerging as a crucial independent parameter. Gao and Lei employed a combination of site surveys and numerical simulations to address the gas over-run issue in the upper corner of the "U-type ventilation" end-mining working face, and developed a gas control scheme for it involving "U+ omni-directional large-diameter high-level boreholes along the roof strike". Zhang and Li et al. offer a dynamic permeability model for shale matrix post-hydraulic fracturing, considering mineral and pore size distribution, dynamic gas entrapment, and poromechanics variations.

2. Fluids in Porous Media

Understanding fluids' properties and behaviors in the porous media of unconventional reservoirs is essential for optimizing extraction techniques. (1) In order to determine fluid properties, Wan and Wang et al. optimized sulfur solubility testing methods for high-sulfur natural gas; the enhanced filtration system can measure sulfur dissolution directly in real gas samples. Compared to the previous method, this approach results in less residual sulfur during the testing process, and the measured sulfur solubility is 2.13% higher. Zhao et al. introduced oil removal technology for water injection in low-permeability reservoirs using a micro-vortex flow approach, which demonstrated a 9.4% increase in oil removal efficiency while reducing the coagulant dosage by 30.0%. (2) For nanoscale effects, Zhou and Wang et al. provided molecular insights into the displacement resistance of deep oil with associated gas methane during water flooding in nano-pore throats, and they observed that oil and methane displacement efficiency was enhanced as the methane content increased. Wu et al. [5] examined methane transport behavior in nanoscale porous media through pore network modeling, highlighting the dramatic improvement in apparent gas permeability when pressure is low (1 MPa) and pore sizes are at nanoscale (5–80 nm). Wan and Niu et al. incorporated wettability-dependent viscosity into a pore network modeling framework to understand the nanoconfined effects for single water-phase flow. (3) For multiphase flow, Chen et al. and Wei et al. utilized digital core methods to study gas and water flow behavior in fractured tight gas reservoirs and shale oil reservoirs, focusing on matrix imbibition and gas invasion behavior. Both of the studies were able to provide a visualization of the fluids' distribution in unconventional rocks. Ye and Gong et al. used the volume of fluid (VOF) method to explore gas invasion behavior under different conditions, and the effects of different drilling fluid properties and fracture morphology on gas invasion were analyzed. Li et al. investigated the effects of pore water content on stress sensitivity in tight sandstone oil reservoirs in the Mahu Block, Xinjiang Province, China. They noticed that the stress sensitivity coefficient is not a constant value but decreases as water saturation increases, therefore optimizing a power-law model for stress-sensitive damage.

3. Productivity Evaluation

Because of the low porosity and low permeability of unconventional reservoirs, a hydraulic fracturing operation is commonly adopted to enhance productivity, causing the fluid flows in these reservoirs to exhibit highly nonlinear characteristics. Therefore, novel methods are established to accurately evaluate productivity, which is vital for predicting reservoir performance and optimizing recovery strategies. (1) In terms of analytical or semi-analytical models, Wang and Xie et al. introduced a new capacity evaluation method for both vertical and horizontal wells in offshore low-permeability reservoirs with natural fractures. The method incorporated vertical gradation and fractal analysis to effectively represent the fracture's complexity and connectivity, and demonstrated a reduction in the average prediction error to less than 15%. Using a dual-medium model to characterize the stress sensitivity, Chen and Sun developed semi-analytical models for gas–water two-phase productivity prediction in carbonate gas reservoirs. In addition, Wang and Li et al. established a semi-analytical gas–water two-phase productivity prediction model for deep CBM reservoirs. Although analytical methods can evaluate productivity quickly, the evaluation accuracy is not very high in highly heterogeneous reservoirs. (2) In terms of numerical simulations, Liu and Zhou et al. investigated fluid–solid coupling mechanisms around open-hole wellbores under transient seepage conditions based on the Biot consolidation theory and the finite difference method, attempting to simulate the pore pressure field and effective stress field around the wellbore. To investigate the impact of various drainage methods on the production performance of coalbed gas wells, Wang and Liu et al. developed a fluid–solid coupling mathematical model by accounting for the effects of coal and shale matrix shrinkage, effective stress, and interlayer fluid flow on reservoir properties. Zhou and Pu et al. [6] focused on connectivity analysis and injection–production optimization in strong, heterogeneous sandstone reservoirs, and the practical application results show that the method can increase the annual output of oil by 1.3%.

4. Enhanced Oil/Gas Recovery

The oil/gas recovery rate for unconventional reservoirs is very low compared with conventional ones. Enhancing oil and gas recovery in unconventional reservoirs is a primary goal which poses large challenges. Xiao et al. [7] reviewed the development and influencing factors of microbial consortiums for enhanced oil recovery after polymer flooding. The process of updating microbial-enhanced oil recovery includes microbial products, mechanisms, merits and demerits, displacement fluid, and effects of the conditions on microorganisms. Zhang and Bai et al. investigated CO_2 injection for enhanced gas recovery and geo-storage in complex tight sandstone gas reservoirs through indoor experiments, and they examined the effects of displacement rate, fracture dip angle, core permeability, and core wetness and dryness on CO_2 gas displacement efficiency and storage efficiency. To avoid rapid water-cut rise and low water-flooding efficiency, Lu et al. proposed a new method, a combination of a rock property test, a brittleness evaluation, and a productivity numerical model to optimize the fracture parameters of low-permeability sandstone reservoirs under fracturing flooding conditions. Additionally, novel products have been developed to enhance oil/gas recovery: Liu and Huang et al. developed a novel water-based drilling fluid (PLUS/KCL) for complex pressure system formations in the South China Sea, Kong et al. optimized foam acid systems for plugging removal in low-pressure oil and gas reservoirs, and Li et al. examined the adsorption behavior and dynamics of micrometer-sized polymer microspheres on quartz sand surfaces.

5. Machine Learning and Data Science Applications

Machine learning and data science are increasingly utilized for parameter optimization, trend prediction, and decision-making in unconventional reservoirs. Ye and Li et al. [8] predicted single-well production rates after hydraulic fracturing in unconventional gas reservoirs using ensemble learning models. Five machine learning algorithms were applied alongside reservoir geology, engineering parameters, and production data to develop the foundational models, indicating that the ensemble model offers higher predictive accuracy and reliability than any single algorithm. Li utilized the Bayesian Adaptive Direct Searching Algorithm for the fast-assisted history matching of fractured vertical wells in coalbed methane reservoirs. The application showed that the proposed method is capable of deriving reasonable estimates of key reservoir properties within 50 numerical simulation runs, which is far more efficient than existing methods. Gao et al. present the intelligent optimization of gas flooding based on a multi-objective approach for a well pattern in the Middle East. Over 400 days, the cumulative oil production increased 25.3% compared to the average distribution method and 12.7% compared to the reservoir engineering method. Based on large data resources, it is possible to predict the evolution of target parameters quickly, while saving measurement costs. To give examples, Su et al. explored the scaling mechanism and countermeasures in tight sandstone gas reservoirs using a BP neural network model, and Sun et al. applied XGBoost for bottomhole pressure prediction in carbonate reservoirs.

6. Conclusions

This Special Issue presents a comprehensive collection of research addressing critical areas in unconventional oil and gas recovery. The integration of multidisciplinary research, multiscale simulations, advanced laboratory tests, and cutting-edge computational tools highlights the forward-thinking nature of the studies, each aiming to overcome the unique challenges posed by unconventional reservoirs and improve overall recovery efficiencies.

We hope that the innovative approaches and findings presented here can provide valuable insights and practical solutions for industry professionals and researchers alike.

Author Contributions: Investigation, T.Z.; writing—original draft preparation, T.Z., Z.S., D.F. and W.Z.; writing—review and editing, T.Z., Z.S., D.F. and W.Z. All authors have read and agreed to the published version of the manuscript.

Conflicts of Interest: The authors declare no conflicts of interest.

List of Contributions

1. Xu, B.; Wang, Z.; Song, T.; Zhang, S.; Peng, J.; Wang, T.; Chen, Y. Modeling of Quantitative Characterization Parameters and Identification of Fluid Properties in Tight Sandstone Reservoirs of the Ordos Basin. *Processes* **2024**, *12*, 278.
2. Jia, Y.; Huang, W.; Wang, P.; Su, P.; Kong, X.; Liu, L.; Shan, Y. Prediction of Longitudinal Superimposed "Sweet Spot" of Tight Gas Reservoir: A Case Study of Block G, Canada. *Processes* **2023**, *11*, 666.
3. Gao, H.; Lei, Y. Research on Gas Control Technology of "U+ Omni-Directional Roof to Large-Diameter High-Level Drilling Hole" at the End Mining Face of Multi-Source Goaf. *Processes* **2023**, *11*, 320.
4. Zhang, Q.; Li, H.; Li, Y.; Wang, H.; Lu, K. A Dynamic Permeability Model in Shale Matrix after Hydraulic Fracturing: Considering Mineral and Pore Size Distribution, Dynamic Gas Entrapment and Variation in Poromechanics. *Processes* **2024**, *12*, 117.
5. Wan, Y.; Wang, L.; Yang, Y.; Ding, Z.; Tang, D.; Zhang, D.; Zhang, L. Research on Optimization of Sulfur Solubility Testing Method for High-Sulfur Natural Gas. *Processes* **2024**, *12*, 1210
6. Zhao, D.; Xie, W.; Zhu, J.; Li, B.; Wang, L.; Chen, T.; Sheng, Y.; Huang, X. Oil Removal Technology for Water Injection in Low-Permeability Reservoirs: A Micro-Vortex Flow Approach. *Processes* **2024**, *12*, 1092.
7. Zhou, L.; Wang, W.; Yan, Y.; Zhao, C.; Zhong, J.; Liu, Y. Molecular Insight into the Occurrence Characteristics of Deep Oil with Associated Gas Methane and the Displacement Resistance in Water Flooding in Nano-Pore Throat. *Processes* **2023**, *11*, 2529.
8. Wan, Y.; Niu, N.; Lu, W.; Zhou, Y.; Wang, B.; Lu, S. Unusual Water Flow in Ultra-Tight Porous Media: Integration of Profession and Innovation. *Processes* **2023**, *11*, 1245.
9. Chen, F.; Duan, Y.; Wang, K. Study on the Flow Behavior of Gas and Water in Fractured Tight Gas Reservoirs Considering Matrix Imbibition Using the Digital Core Method. *Processes* **2024**, *12*, 709.
10. Wei, J.; Li, J.; Yang, Y.; Zhang, A.; Wang, A.; Zhou, X.; Zeng, Q.; Shang, D. Digital-Rock Construction of Shale Oil Reservoir and Microscopic Flow Behavior Characterization. *Processes* **2023**, *11*, 697.
11. Ye, C.; Gong, J.; Liu, K.; Pei, J.; Xu, S.; Xu, P. Study on Gas Invasion Behavior of Gas–Liquid Displacement in Fractured Reservoirs. *Processes* **2022**, *10*, 2533.
12. Li, X.; Gu, K.; Xu, W.; Song, J.; Pan, H.; Dong, Y.; Yang, X.; You, H.; Wang, L.; Fu, Z.; et al. Effects of Pore Water Content on Stress Sensitivity of Tight Sandstone Oil Reservoirs: A Study of the Mahu Block (Xinjiang Province, China). *Processes* **2023**, *11*, 3153.
13. Wang, K.; Xie, M.; Liu, W.; Li, L.; Liu, S.; Huang, R.; Feng, S.; Liu, G.; Li, M. New Method for Capacity Evaluation of Offshore Low-Permeability Reservoirs with Natural Fractures. *Processes* **2024**, *12*, 347.
14. Chen, D.; Sun, Z. A Semi-Analytical Model for Gas–Water Two-Phase Productivity Prediction of Carbonate Gas Reservoirs. *Processes* **2023**, *11*, 591.
15. Wang, S.; Li, D.; Li, W. A Semi-Analytical Model for Production Prediction of Deep CBM Wells Considering Gas-Water Two-Phase Flow. *Processes* **2023**, *11*, 3022.
16. Liu, E.; Zhou, D.; Su, X.; Wang, H.; Liu, X.; Xu, J. Research on Fluid–Solid Coupling Mechanism around Openhole Wellbore under Transient Seepage Conditions. *Processes* **2024**, *12*, 412.
17. Wang, W.; Liu, S.; Sang, S.; Du, R.; Liu, Y. A Study on the Production Simulation of Coal–Shale Interbedded Coal Measure Superimposed Gas Reservoirs under Different Drainage Methods. *Processes* **2023**, *11*, 3424.
18. Zhang, L.; Bai, T.; Zhao, Q.; Zhang, X.; Cheng, H.; Li, Z. CO_2 Injection for Enhanced Gas Recovery and Geo-Storage in Complex Tight Sandstone Gas Reservoirs. *Processes* **2023**, *11*, 2059.

19. Lu, C.; Ma, L.; Guo, J.; Zhao, L.; Xu, S.; Chen, B.; Zhou, Y.; Yuan, H.; Tang, Z. Fracture Parameters Optimization and Field Application in Low-Permeability Sandstone Reservoirs under Fracturing Flooding Conditions. *Processes* **2023**, *11*, 285.
20. Liu, H.; Huang, Y.; Liu, Z.; Dong, L.; Wang, M. The Development and Application of Novel Water-Based Drilling Fluid for Complex Pressure System Formation in the South China Sea. *Processes* **2023**, *11*, 1323.
21. Kong, X.; Liu, B.; Xu, H.; Shen, J.; Li, S. Optimization and Performance Evaluation of Foam Acid Systems for Plugging Removal in Low Pressure Oil and Gas Reservoirs. *Processes* **2023**, *11*, 649.
22. Li, J.; Luo, T.; Yan, W.; Cheng, T.; Cheng, K.; Yu, L.; Cao, J.; Yang, Z. Adsorption Behavior and Adsorption Dynamics of Micrometer-Sized Polymer Microspheres on the Surface of Quartz Sand. *Processes* **2023**, *11*, 1432.
23. Li, Z. Fast Assisted History Matching of Fractured Vertical Well in Coalbed Methane Reservoirs Using the Bayesian Adaptive Direct Searching Algorithm. *Processes* **2023**, *11*, 2239.
24. Gao, M.; Wei, C.; Zhao, X.; Huang, R.; Li, B.; Yang, J.; Gao, Y.; Liu, S.; Xiong, L. Intelligent Optimization of Gas Flooding Based on Multi-Objective Approach for Efficient Reservoir Management. *Processes* **2023**, *11*, 2226.
25. Su, X.; Zhou, D.; Wang, H.; Xu, J. Research on the Scaling Mechanism and Countermeasures of Tight Sandstone Gas Reservoirs Based on Machine Learning. *Processes* **2024**, *12*, 527.
26. Sun, H.; Luo, Q.; Xia, Z.; Li, Y.; Yu, Y. Bottomhole Pressure Prediction of Carbonate Reservoirs Using XGBoost. *Processes* **2024**, *12*, 125.

References

1. Zhang, T.; Wang, B.; Zhao, Y.; Zhang, L.; Qiao, X.; Zhang, L.; Guo, J.; Vo Thanh, H. Inter-layer interference for multi-layered tight gas reservoir in the absence and presence of movable water. *Pet. Sci.* **2024**, *21*, 1751–1764. [CrossRef]
2. Tang, H.; Luo, S.; Liang, H.; Zeng, B.; Zhang, L.; Zhao, Y.; Song, Y. Integrated Numerical Simulation of Hydraulic Fracturing and Production in Shale Gas Well Considering Gas-Water Two-Phase Flow. *Pet. Explor. Dev.* **2024**, *51*, 684–696. [CrossRef]
3. Da, L.; Pan, Y.; Xiao, Y.; Wang, A.; Wang, Q.; Wei, C.; Fan, D. Parameter Design Method for Destressing Boreholes to Mitigate Roadway Coal Bursts: Theory and Verification. *Rock Mech. Rock Eng.* **2024**. [CrossRef]
4. Ali, M.; Ashraf, U.; Zhu, P.; Ma, H.; Jiang, R.; Lei, G.; Ullah, J.; Ali, J.; Vo Thanh, H.; Anees, A. Quantitative Characterization of Shallow Marine Sediments in Tight Gas Fields of Middle Indus Basin: A Rational Approach of Multiple Rock Physics Diagnostic Models. *Processes* **2023**, *11*, 323. [CrossRef]
5. Wu, S.; Fang, S.; Ji, L.; Wen, F.; Sun, Z.; Yan, S.; Li, Y. Transport Behavior of Methane Confined in Nanoscale Porous Media: Impact of Pore Evolution Characteristics. *Processes* **2022**, *10*, 2746. [CrossRef]
6. Zhou, Y.; Pu, L.; Dang, S.; He, J.; Pu, S. Study on Connectivity Analysis and Injection–Production Optimization of Strong Heterogeneous Sandstone Reservoir Based on Connectivity Method. *Processes* **2023**, *11*, 2816. [CrossRef]
7. Xiao, H.; Amir, Z.; Junaidi, M. Development of Microbial Consortium and Its Influencing Factors for Enhanced Oil Recovery after Polymer Flooding: A Review. *Processes* **2023**, *11*, 2853. [CrossRef]
8. Ye, F.; Li, X.; Zhang, N.; Xu, F. Prediction of Single-Well Production Rate after Hydraulic Fracturing in Unconventional Gas Reservoirs Based on Ensemble Learning Model. *Processes* **2024**, *12*, 1194. [CrossRef]

processes

Article

Quantitative Characterization of Shallow Marine Sediments in Tight Gas Fields of Middle Indus Basin: A Rational Approach of Multiple Rock Physics Diagnostic Models

Muhammad Ali [1,†], Umar Ashraf [2,3,†], Peimin Zhu [1], Huolin Ma [1], Ren Jiang [4], Guo Lei [5], Jar Ullah [1], Jawad Ali [6], Hung Vo Thanh [7,8,*] and Aqsa Anees [2,3,*]

1 Institute of Geophysics and Geomatics, China University of Geosciences, Wuhan 430074, China
2 Institute of International Rivers and Eco-Security, Yunnan University, Kunming 650500, China
3 Institute for Ecological Research and Pollution Control of Plateau Lakes, School of Ecology and Environmental Science, Yunnan University, Kunming 650500, China
4 Research Institute of Petroleum Exploration and Development, PetroChina, Beijing 100083, China
5 SuperTech Energy Co., Ltd., Building 30, Chuangxin Park, Changping District, Beijing 102299, China
6 School of Environmental Science and Engineering, Shaanxi University of Science and Technology, Xi'an 710021, China
7 Laboratory for Computational Mechanics, Institute for Computational Science and Artificial Intelligence, Van Lang University, Ho Chi Minh City 700000, Vietnam
8 Faculty of Mechanical—Electrical and Computer Engineering, Van Lang University, Ho Chi Minh City 700000, Vietnam
* Correspondence: hung.vothanh@vlu.edu.vn (H.V.T.); aqsaanees01@outlook.com (A.A.)
† These authors contributed equally to this work.

Abstract: For the successful discovery and development of tight sand gas reserves, it is necessary to locate sand with certain features. These features must largely include a significant accumulation of hydrocarbons, rock physics models, and mechanical properties. However, the effective representation of such reservoir properties using applicable parameters is challenging due to the complicated heterogeneous structural characteristics of hydrocarbon sand. Rock physics modeling of sandstone reservoirs from the Lower Goru Basin gas fields represents the link between reservoir parameters and seismic properties. Rock physics diagnostic models have been utilized to describe the reservoir sands of two wells inside this Middle Indus Basin, including contact cement, constant cement, and friable sand. The results showed that sorting the grain and coating cement on the grain's surface both affected the cementation process. According to the models, the cementation levels in the reservoir sands of the two wells ranged from 2% to more than 6%. The rock physics models established in the study would improve the understanding of characteristics for the relatively high V_p/V_s unconsolidated reservoir sands under study. Integrating rock physics models would improve the prediction of reservoir properties from the elastic properties estimated from seismic data. The velocity–porosity and elastic moduli-porosity patterns for the reservoir zones of the two wells are distinct. To generate a rock physics template (RPT) for the Lower Goru sand from the Early Cretaceous period, an approach based on fluid replacement modeling has been chosen. The ratio of P-wave velocity to S-wave velocity (V_p/V_s) and the P-impedance template can detect cap shale, brine sand, and gas-saturated sand with varying water saturation and porosity from wells in the Rehmat and Miano gas fields, both of which have the same shallow marine depositional characteristics. Conventional neutron-density cross-plot analysis matches up quite well with this RPT's expected detection of water and gas sands.

Keywords: rock physics template; rock physics diagnostics; tight gas sand reservoir; shallow marine sediments; Lower Goru formation

Citation: Ali, M.; Ashraf, U.; Zhu, P.; Ma, H.; Jiang, R.; Lei, G.; Ullah, J.; Ali, J.; Vo Thanh, H.; Anees, A. Quantitative Characterization of Shallow Marine Sediments in Tight Gas Fields of Middle Indus Basin: A Rational Approach of Multiple Rock Physics Diagnostic Models. *Processes* **2023**, *11*, 323. https://doi.org/10.3390/pr11020323

Academic Editor: Yidong Cai

Received: 6 December 2022
Revised: 15 January 2023
Accepted: 17 January 2023
Published: 18 January 2023

1. Introduction

Seismic reflections of rock properties play an imperative role in seismic reservoir characterization and require integrating knowledge and concepts from different disciplines [1–5]. The seismic reflection method is a cost-effective and efficient way to characterize the reservoirs representing an indirect measurement acquired by the propagation of sound waves through various subsurface formations [6–8]. Understanding the seismic response of various folded structures is crucial to interpret seismic profiles in terms of tectonics and geology [9–11]. This type of interpretation of the seismic profiles is called structural interpretation, which implies the mapping of the geologic features of the subsurface to configurations such as folds, faults, and fractures [12,13]. The difference in rock elastic properties in the subsurface determines seismic reflections [14]. Elastic properties of rocks, that is, P- and S-wave velocities and bulk density, depend in turn on the porosity, lithology, rock texture, and pore fluid [15,16]. Acoustic impedance, the product of bulk density and velocity, is a common result of seismic inversion [17]. By itself, acoustic impedance is meaningless to the geologist and engineer. However, once it is interpreted in terms of reservoir rock properties, namely pore fluid, porosity, and lithology, it transforms into vital parameters for reservoir evaluation and drilling decisions [18].

Due to this link between geological properties and the seismic response, it is possible to transform the information from wells using seismic amplitudes through a method or a function [19]. This integration implies a transformation based on inverse and forward modeling methods. The rock physics model (RPM) is a transformation used to interpret a seismic response into reservoir rock properties and vice versa using inverse and forward modeling [15,20]. In other words, inverse modeling aims to reconstruct a geologic model using seismic data. In contrast, forward modeling is applied when core and well data are available to derive seismic signatures of rock properties for a particular geologic model. This type of interpretation of seismic reflections is called a quantitative seismic interpretation that quantifies the petrophysical parameters of the subsurface for each rock unit within the geometrical interpretation framework.

The only way to establish relationships between various rock attributes and rock physics models is via controlled experiments by simple cross-plotting between velocity and porosity [21,22]. Controlled experiments are conducted at data factories which can be categorized into three types of data: laboratory, well (wireline), and computational rock physics. Rock physics diagnostics (RPD) is a methodology for establishing RPMs that describe the relationship between seismic velocity and porosity [23]. To perform RPD, reservoir rock properties should be examined from the wireline logging dataset or laboratory core data analysis [24]. Certainly, the former approach has its advantages, such as in situ measurements of rock properties, a large number of data points, and high vertical resolution. However, one faces difficulties interpreting lateral changes in reservoir rock properties [25]. RPD is useful for understanding the relationships between seismic velocity and porosity and quantifying the cement content from well log data, which is an important aspect of the mechanical strengths of reservoir rocks and fluid flow [19,26]. Moreover, it describes the texture of the rocks, particularly grain size sorting, the position of diagenetic cement, the effect of clay, etc. All these parameters control the response of the elastic properties of pure minerals and their mixtures [14,22].

A number of rock physics models were developed, and relevant studies were conducted to establish and understand the relationship between various rock attributes using laboratory, wireline, and digital rock physics data [24]. According to published literature, several RPD studies have been conducted to predict rock properties quantitatively using the RPM, which matched the well data under examination [27–32]. The authors [33] performed RPD to estimate the amount of contact and noncontact cement in sandstone formations from Saudi Arabia. This study combined stiff-sand, soft-sand, and intermediate stiff-sand models to quantify the cement content present in sedimentary rock. Thin section analysis was used to determine the amount of quartz cement. It was found that the thin section was well correlated with the results from RPD. In another study [34], the author presented a RPD study of an offshore gas field to examine wireline data from three wells. They established an RPM as a constant-cement model. A correction of measured shear-wave velocity was also conducted as it appeared unreasonably low. Moreover, they performed perturbational forward modeling to predict the seismic response of various geological scenarios away from well control. RPD was also used in a study for the prediction of the missing S-wave velocity curve [35]. The authors in [36] performed RPD to evaluate the amount of contact cement. Utilizing RPD, we can quantitatively explain the well data.

Numerous retrospective research has been conducted to address the methods for determining and estimating these geological parameters in the Lower Goru sand reservoir. Therefore, it is important to develop the finest RPM and template to identify a previously unidentified, insufficiently unconsolidated reservoir. We illustrate the deterministic methodology to establish RPMs for the well using well data intervals from a clastic depositional environment: tight gas sandstone (TGS) in the Lower Goru fields. Later, the derived reservoir properties can be extrapolated to the prospects away from well control to investigate specific what-if scenarios, such as plausible lithology or fluid variations, thereby building catalogues of RPMs and templates of rock properties. We propose that our rock physics model will provide further insights for accurate reservoir characterization of the Middle and Lower Goru clastic reservoir, Pakistan, and other sites around the world that have the same geology.

2. Geological Setting and Stratigraphy

The study area lies within the N-E pattern dominating the Middle Indus Basin, also known as the Lower Indus Basin. The Middle Indus Basin is situated in a manner that the Indian Shield is abutted to the east; the Sulaiman Fold, Thrust Belt, and Kirther Ranges in the west; Jacobabad-Khairpur High in the south; with Sargodha High in the north [37]. The research area lies close to Jacobabad-Khairpur High (Figure 1). Khairpur High is located in the northwest direction of the study area, which is distinguished by the highest geothermal gradient considered to be effective up to 40.68 °F/328 ft. Khairpur High is dominant in forming structural traps in Miano, Kadanwari, and neighboring areas [13,38,39]. Since the Upper, Middle, and Lower Indus Basins are quite abundant in hydrocarbons and cover rare complete petroleum systems. Many wells in the Lower Indus basin have been drilled [37,40,41]. The strongest rock source is known for being organically abundant, black-colored shales deposited in a quite oxygen-depleted marine environment [42].

In the field of study, the Sembar Formation Lower Cretaceous shale is recognized for the foundation of oil and gas contained in the region due to its thermal maturity and organic richness [43,44]. The Sembar formation is overlaid by the Goru Formation, divided into two distinct sections, the Lower Goru member and the Upper Goru member [44]. The Goru Formation was deposited in the Cretaceous age [37]. The Goru Formation comprises shale, limestone, sandstone, and siltstone [45]. The layered limestone is thin as well as fine-grained within the reservoir rocks. Goru Formation has been deposited on the shelf of the shallow marine environment [12,46]. Limestone and shale have occupied the upper section. Basal and massive sands dominate the lower section of the Goru Formation, and these are the main reservoir rocks in the Middle Indus basin (Dolan, 1990). The lower section of the Lower Goru formation was additionally divided (from bottom to top) into four separate sand intervals: A, B, C, and D (Figure 2) [47,48]. The B, C, and D sand intervals behave as a potential hydrocarbon reservoir in the study area. Two wells, namely, Miano-07 and Rehmat-02, from two different gas fields were examined for rock physics analysis.

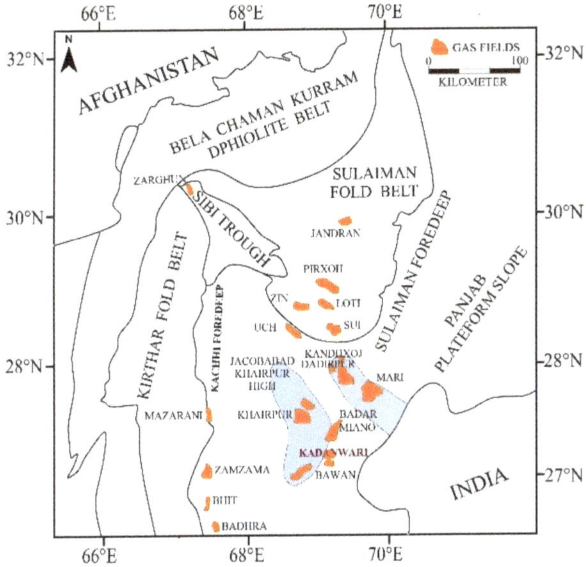

Figure 1. A regional geological map of Pakistan gas fields within the Middle Indus Basin, complete with the structural settings, tectonics, and local multi-sedimentary basins that are close to the Jacobabad-Khairpur High.

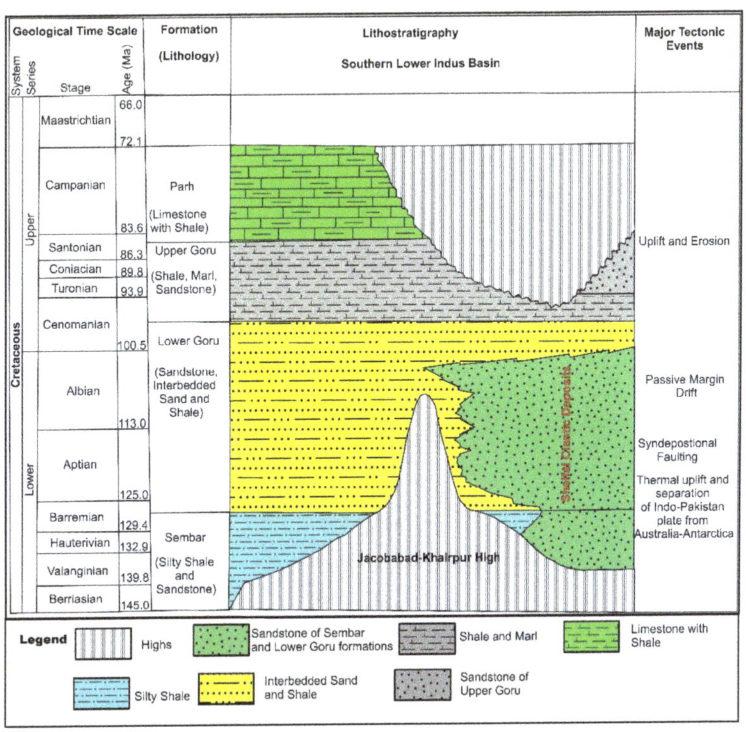

Figure 2. Stratigraphic chart of the Lower Indus Basin.

3. Materials and Methodology

The study was carried out using well data from Miano-07 and Rehmat-02. In the research, the conventional geophysical logs such as caliper (CAL), gamma-ray (GR), spontaneous potential (SP), neutron porosity (NPHI), density (RHOB), shallow resistivity (LLS), deep resistivity (LLD), sonic (DT), and shear sonic (DTS) were available. The workflow of the technique first applies the conditioning of the well logging data [49–51]. Then petrophysical analysis was conducted across the zone of interest to enhance the log data quality and estimate attributes of the reservoir such as porosity (∅), the volume of shale (V_{sh}), and water saturation (S_W) [52,53]. In the second stage, the RHOB log was corrected. The elastic parameters are computed in a subsequent step. To fine-tune the Gassmann fluid substitution model, we will need to identify its petrophysical characteristics in Step 3. Log tracking was then used to compare the degrees of agreement between the predicted and measured elastic properties. Many diagnostic models of rock physics are contrasted and assessed to characterize the velocity–porosity behavior of a clastic reservoir.

The following steps were followed to validate the rock physics model.

3.1. Density Log Conditioning and Curve Synthesis

After uploading and inspecting the well log data, it was discovered that most log responses are aligned in-depth and did not require a shift in depth. However, when testing the raw log data, it was found that they have CAL log readings, which can affect the log quality. In particular, pad-type devices (e.g., micro-resistivity and density tools) can suffer from inadequate pad contact and may give a substandard quality result. Figure 3a demonstrates the measured RHOB curve response that needs log conditioning at various depths.

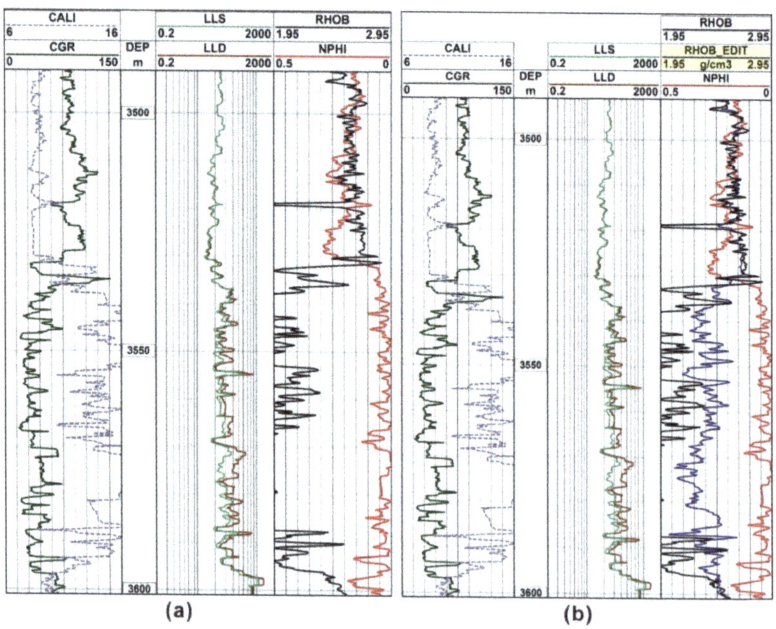

Figure 3. (**a**) Measured RHOB curve on track-4 (black) for Remat-02 well requires conditioning. (**b**) Synthesized RHOB curve (blue) on top of measured RHOB curve. The yellow box on track-4 shows the corrected RHOB curve.

The RHOB log plays the most important role in providing a good source of porosity and clay volume data, and the rock physics study is reliant on the RHOB log. Various methods are available for creating a synthetic RHOB; Gardner's empirical relationship is one of the most famous methods [54]. The main aim of curve synthesis and reconstruction is to shape the logging data at break-out/wash-out intervals on an absent section or where the data is incorrect. Similarly, the RHOB curve was designed at the bad hole interval by curve synthesis, rebuilding the synthetic curve, and applying data from the nearest interval with fine borehole conditions. Gardner et al. [54] investigated primary sedimentary rocks of different lithologies through a wide range of basins, geological ages, and depths and determined relationships between the P-wave velocity and density in rocks of different lithologies. The relationship between the P-wave velocity and density for shale has long been used in seismic analyses:

$$\rho_{syn} = \alpha \times \left(v_p\right)^{\beta} \quad (1)$$

where V_p is the compressional wave velocity in m/s, ρ_{syn} is the synthetic RHOB in g/cc, and $\alpha = 0.31$ when V_p is in m/s or $\alpha = 0.23$ when V_p is in ft/s and $\beta = 0.25$. Major sedimentary rocks generally define a narrow stripe along this trend line, except for coals and evaporate.

In this study, we used the default Gardner parameters to transform velocity data to density. To get a visual feel for the differences between the default Gardner method and the measured RHOB log, Figure 3b is plotted as a typical comparison for the methods derived from velocity logs.

In Figure 3b, the curve in blue color is the evaluated RHOB data, where a few parts of inadequate quality data have been observed against the breakout/wash-out portion. Similarly, for this interval, the synthetic RHOB curves were generated using Gardner's equation and V_p data in the interval 3525–3580 m. Subsequently, Gardner's equation is mentioned below as a mathematical expression (Gardner's Equation (1)).

Figure 3b shows that the log plot verifies the reliability of the synthetic RHOB curve. This also indicates a good match with the measured RHOB data; thus, where the whole condition is fine, it can replace the RHOB data with the portion of the bad hole where the measured data is low. Thus, the calculated RHOB was replaced by the synthesized RHOB curve at 3525–3580 m.

Finally, the synthetic RHOB curve was generated where the measured RHOB curve was not acceptable. Before starting the petrophysical interpretation on Schlumberger Techlog software, all these changes were made. Determining fluids and reservoir parameters were carried out with emphasis on the reservoir, and the closest non-reservoir zones were plotted. V_p, V_s, and ρ were obtained from these zones, which were used later in the rock physics analysis.

3.2. Well Log Response for Identifying Hydrocarbon Reservoir

An alternative procedure for identifying hydrocarbon-bearing zones is mostly based on conventional well log data obtained by the exploratory well-run sondes [55]. The Miano-07 has penetrated Lower Goru sand at a depth of 3420 m, followed by Upper Goru shale at a depth interval of 3320–3410 m (Figure 4). Another well, Rehmat-02, has intruded the Lower Goru at the depth interval of 3420–3650 m. Figure 5 shows the well log response of Rehmat-02 at the targeted sand interval within the Lower Goru Formation. Wireline log reflexes, for example, GR, LLS, LLD, NPHI, and RHOB logs, were used to estimate petrophysical parameters for selected depth intervals in the gas-bearing zones. Figures 4 and 5 show the log responses of P-wave velocity (V_p), S-wave velocity (V_s), P-impedance, and V_p/V_s from each of these wells in the hydrocarbon-bearing regions. The Rehmat-02 well revealed a clean sand reservoir with low RHOB and GR values and a low V_p/V_s ratio. Miano-07 well showed a shaly sand reservoir with alternate high-low GR, low V_p/V_s ratio, and low RHOB responses. Shale and silt lamina produces a relatively high GR response, implying a shaly sand reservoir. The sand reservoir in Miano-07 and Rehmat-02 is part of the Lower Goru Formation of the Early Cretaceous age, deposited in a shelf in the shallow marine environment [12]. Water-saturated reservoirs ranging from 15 to 23% are cemented reservoirs with an effective porosity ($\varnothing_e$) of 15 to 25% and V_{sh} ranging from 2.5% in the clean sand reservoir to 17% in the shallow reservoir. Table 1 shows the petrophysical parameters of Rehmat-02 and Miano-07 at the targeted depths.

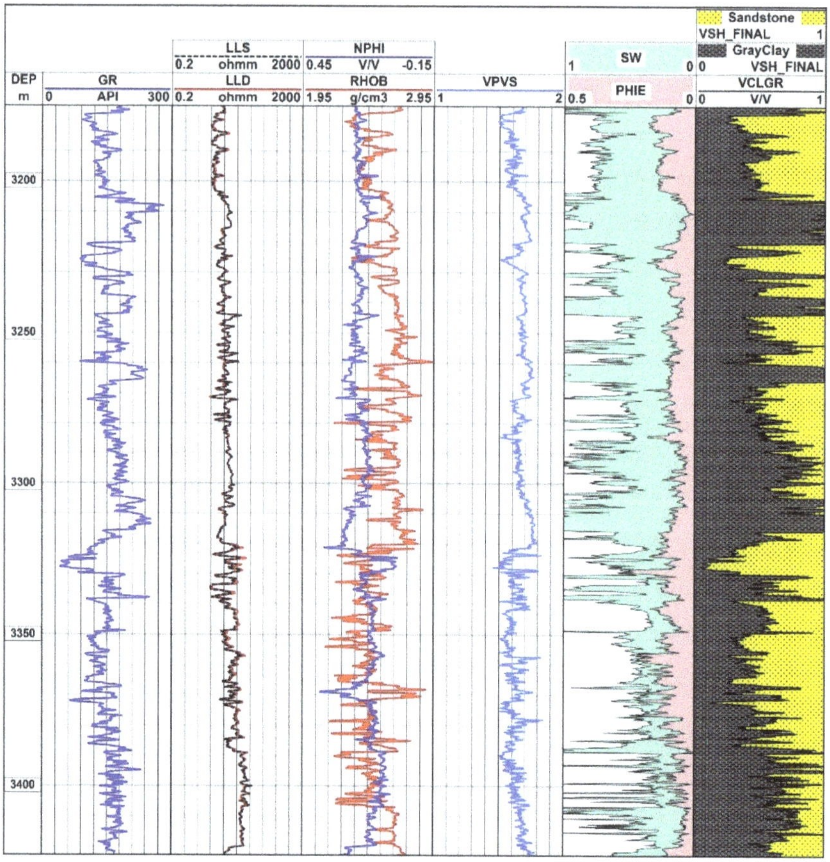

Figure 4. Well log data from Miano-07 well at targeted gas zone. The reservoir zones are within the Lower Goru formation.

Table 1. Prediction of petrophysical parameters from chosen depth intervals of two wells in the Middle Indus Basin.

Wells	Gas Zones		Petrophysical Parameters		
	Depth Interval (m)	Thickness (m)	V_{sh} (%)	$\varnothing_e$	S_w
Miano-07	3325	1.5	0.45	10	0.55
Rehmat-02	3540	5	0.4	15	0.45

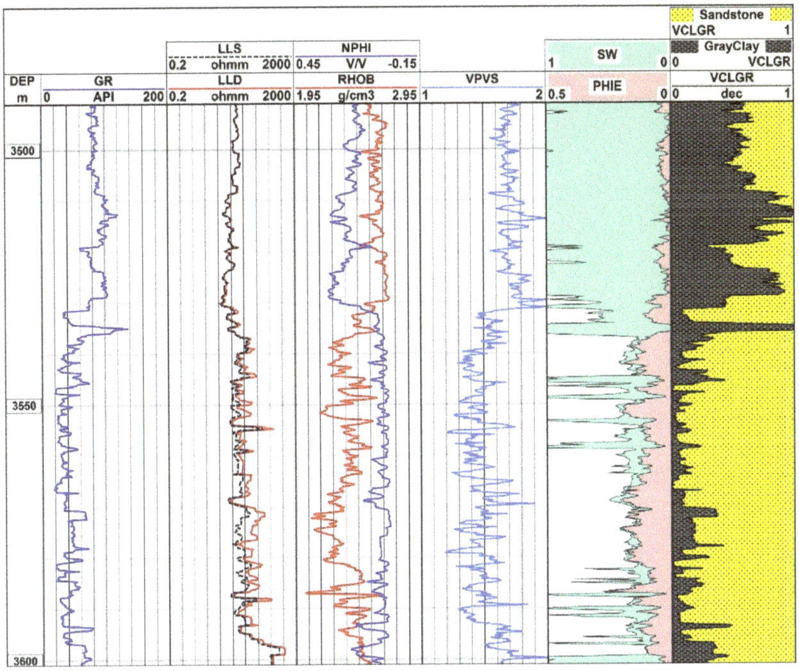

Figure 5. Well log data from Rehmat-02 well at gas zone. The reservoir zone is within the Lower Goru sand formation.

3.3. Rock Mechanical Properties from Well Logs

The significant goal of petrophysical analyses was to obtain proper physical parameters to differentiate fluid and lithology from borehole (log) data. The three fundamental independent elastic parameters that might result from well logs data are V_p, V_s, and ρ. These three fundamental parameters can be used to extract several helpful elastic parameters and have different abilities to discriminate between lithologies and fluids. The elastic properties of rocks can be classified into compressional-type and shear-type parameters [56]. Shear-type parameters are not sensitive to fluid because shear stress cannot transmit in fluids. Examples of shear-type parameters include V_s, SI, and G. Parameters such as V_p, acoustic impedance (AI), Bulk Modulus (K), and Lame's first parameter (λ), which is sensitive to the fluid and are usually known as compressional-type parameters. Some of the compressional-type and shear-type parameters and their hybrids will be used for the well analysis to describe fluid saturation zones and indicate fluid and lithological changes.

The elastic parameters are estimated by using equations are:

Poisson's ratio (PR) $$\sigma = \frac{3K-2G}{2(K+G)} \quad (2)$$

Bulk modulus (K) $$K = \rho\left(v_P^2 - \frac{4}{3}v_S^2\right) \quad (3)$$

Shear modulus (G) $$G = \rho \times v_S^2 \quad (4)$$

Young's modulus (E) $$E = \frac{9K \times G}{3K+G} \quad (5)$$

Compressional modulus (μ) $$\mu = \rho \times v_P^2 \quad (6)$$

Lame's parameter (λ) $$\lambda = K - \frac{2\mu}{3} \quad (7)$$

In the Lower Goru sand, the values of G, E, K, and UCS (unconfined compressive strength) in the Upper Goru shale are increased, followed by an increase in sediments. The value of the Poisson ratio (PR) does differ from 0.29 to 0.31 in the Upper Goru shale, with a decrease to 0.13 in the Lower Goru sand formation. In the Lower Goru sand formation, the gas-producing zone is determined by the low PR values between depths of 3533 m and 3597 m in Rehmat-02 (Figure 6) and similarly, depths of 3333–3339 m in the Miano-07 well (Figure 7).

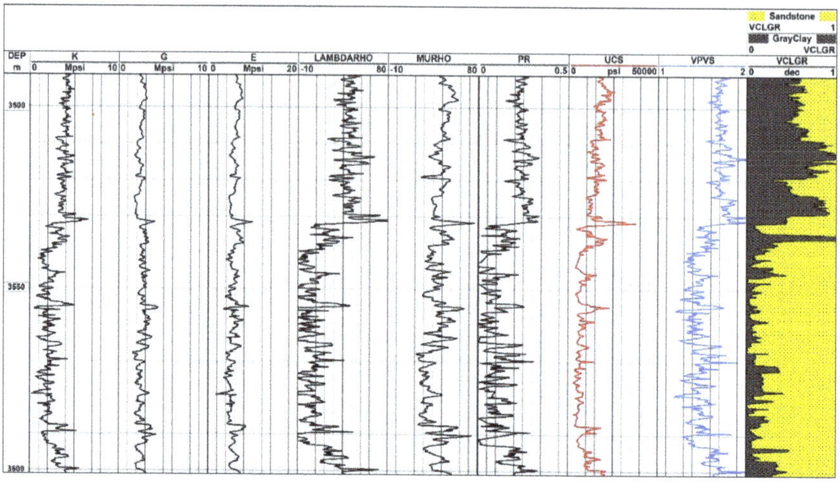

Figure 6. Mechanical rock parameters calculated for available depth interval in Rehmat-02 well. The lithology is also displayed in the colored column next to each plot.

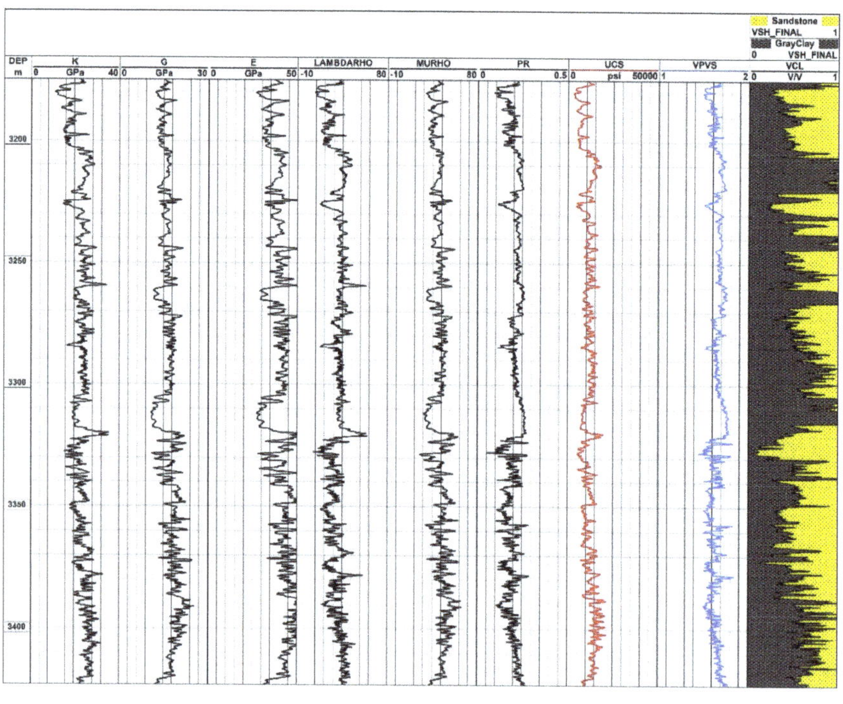

Figure 7. Mechanical rock parameters calculated for available depth interval in Miano-07 well. The lithology is also displayed in the colored column next to each plot.

3.4. Rock Physics Model

We focused on building reliable RPD modeling and RPT analyses for the Lower Goru sand reservoir in the Middle Indus Basin. The model's friable sand (FS), contact cement (CC), and constant cement (CC) for specific areas of the Miano-07 and Rehmat-02 wells have been implemented to associate seismic and elastic rock properties with geological properties and to estimate the degree to which these geological directions may impact the fluid effect of the seismic parameters and the lithology. We have also employed the Hertz–Mindlin theory (HMT) to operate the modified Hashin–Shtrikman (HS) limits. An effort was made to drive models for clean sands, shaly sands, and shales since those are the lithologies found under the study zones in the Miano-07 and Rehmat-02 wells.

3.5. Rock Physics Diagnostics (RPD)

A standard procedure in the rock physical model is establishing specific trends in porosity–seismic velocity to be determined from the entire data volume and assigning these individual tendencies to correct depth intervals and deposition series. This process is usually referred to as an RPD. The RPD is normally performed on the core and wireline log data [26]. This procedure explains the rock texture: the position of diagenetic cemented and uncemented, grain size sorting, and the influence of shale or clay, etc. The V_p and V_s are usually not well-thought out to be the correct fluid indicator device because of the coupling effect among V_p and V_s via K and G.

In comparison, rock physics trends show up separately in the modulus–porosity or impedance–porosity plane than in the velocity–porosity plane. K is extremely sensitive to pore filling (S_w) deformation due to seismic wave changes resulting from various pore volumes. Any fluid typically does not impact the G. The RPD functional place is the plane of rock physics, which can be (a) modulus–porosity; (b) impedance–porosity; and (c) velocity–porosity plane. Compared to rock diagnostic velocity–porosity planes, we favored elastic moduli porosity or impedance–porosity planes primarily to change the correlated fluid effects in K. In the current research, V_p, K, and G act as a function of porosity for the Miano-07 and Rehmat-02 wells and are overlaid for the three different combined models CC, CC, and FS. These RPD models, with previous knowledge of the quantification of cement, are beneficial to predict the constant cement model with a percentage of cement that will be utilized for searching coordination numbers in HMT to build the dry modulus of rock under the crucial porosity developed by the rock physics template (RPT). Since we do not have the amount of quartz cement in the study in the form of thin sections, the model reported in [57] was applied to quantify cement for RPT generation in the basin.

The cement quantification model is designed to fix log data into a rock modulus–porosity and velocity–porosity system. To define the FS model, HMT and the lower HS bound were implemented. The upper bound HS was applied for cement sand. As a general rule, lithology regulates the elastic properties of rock (texture and composition), porosity (type and quantity), pore fluid, frequency, anisotropy, and depth that include different parameters such as temperature, differential pressure, age, and lithology. In a complex local geological setting, these control parameters are usually different. HMT is utilized to build the dry modulus of high porous rocks, commonly called critical porosity ($\varnothing_e$). The vital porosity of the sand is endorsed (~40%). The authors in [58] reported that the dry elastic moduli at very high porosity could be extracted using the following relationships:

$$K_{HM} = \left[\frac{n^2(1-\varnothing_c)G^2}{18\pi^2(1-PR)^2}P\right]^{\frac{1}{3}} \tag{8}$$

$$M_{HM} = \frac{(5-4PR)}{5(1-2PR)}\left[\frac{3n^2(1-\varnothing_c)G^2}{2\pi^2(1-PR)^2}P\right]^{\frac{1}{3}} \tag{9}$$

where M_{HM}, K_{HM} = dry rock shear and bulk moduli, respectively; n = coordination number; P = net effective pressure, at $\varnothing_c$; PR = Poisson's ratio for solid phase; and

M = shear modulus (mineral modulus) for solid phase. The coordination number (average number of contacts per grain) increases with reduced porosity, resulting from more well-organized packing under increased efficient pressure [59]. High porosity shales have quite a limited number of coordinates and inverses. Cement coordination numbers are estimated to be 8.6 and 4.6 (at crucial porosity) for the reservoir sand and shale, respectively.

The model for rock physics concerns the elastic properties of rock physics. Rocks consist of two fundamental components: pore fluid and mineral frame. The mineral frame consists of several minerals. The conventional way of dealing with this condition is to systematically construct a solitary or efficient mineral whose elastic properties rely on the elastic properties of the mineral components that might be used to measure the dry mineral frame's elastic properties. The "effective solid" hypothesis is that the effective mineral is a complex consisting of several elastic components of pure minerals with known volumetric fractions. The elastic composite's K and G have upper and lower boundaries [59]. Established on the HS bounds between zero porosity and crucial porosity, the elastic dry rock modulus can be calculated. For the measurement of dry elastic rock modules with HS bounds for the Miano-07 and Rehmat-02 wells in the Middle Indus Basin, the RHOB and elastic modulus of the various dry solid matrix, for example, feldspar, quartz, and clay are set out in Table 2.

Table 2. The expected elastic moduli and density of minerals for RPT [59].

Minerals	Bulk Modulus (GPa)	Shear Modulus (GPa)	Density (g/cc)
Quartz	36.6	45.0	2.65
Feldspar	75.6	25.6	2.60
Clay	20.9	6.9	2.58

The HS bounds deliver the thinnest variety of elastic moduli without indicating the geometries of components [59,60]:

$$K_{HS\pm} = K_1 + \frac{f_2}{(K_2 - K_1)^{-1} + f_1\left(K_1 + \frac{4}{3}M_1\right)^{-1}} \tag{10}$$

$$M_{HS\pm} = M_1 + \frac{f_2}{(M_2 - M_1)^{-1} + \frac{2f_1(K_1+2M_1)}{5M_1\left(K_1+\frac{4}{3}M_1\right)}} \tag{11}$$

where M_{HS}, K_{HS} = shear and bulk modules, respectively, are calculated via HS bounds; M, K = shear and bulk mineral moduli, respectively, of various components (index mentions to separate phase 1 or 2); and f = fraction of volume of separate phases.

Computing the upper and lower bounds is usually based on material stiffness and softness. The upper bound means the material is rigid, and parameters in (Equations (10) and (11)) will be subscribed with 1. In the current work, the goal of modeling can be achieved using the modified HS bound in which if the material is soft, the bound might be lower, and the parameters can be subscribed with 1. This attaches to end members in the porosity–moduli plane (crucial porosity and zero porosity), i.e., the mineral point and fundamental point are linked by unconsolidated lines, which are modified by lower HS boundaries [58]. CC and CC models with porosity due to cementation take the temperament of V_p and elastic rock modules (K and G). A thick sand interval (Figure 5) was determined in the Rehmat-02 well by low and different GR values along with high V_p (approximately 3 to 3.5 km/s). This sand body is surrounded by shale, contrasting with the sandstone reservoir's GR and Vp. For Miano-07, we noted the GR log response varying from shale to clean sand in the reservoir zone (Figure 4). In both wells, Miano-07 and Rehmat-02, the shaly sand and the clean sand reservoir zone reflect the same stratigraphic structures, although they are located at various depths and in diverse gas fields. Lower Goru sand in Miano-07 and Rehmat-02 wells consists of shale, cemented sandstones, and friable sands. The hardness of rock depends not only on mineralogy and porosity but also on the rock's texture (microstructure), i.e., the arrangements on the pore scale of solid phase components. The V_p is graphed in contradiction of porosity for sands that appreciate the microstructure effects. The cement model presupposes that the porosity reduces the unchanging deposition of cement layers at the grain surface in proportion to the initial critical porosity value. Diagenetic cement significantly increases the rigidity of the sand by improving the grain contacts [21]. For diagnostic purposes, the elastic properties are cross-plotted from similar depth intervals of two wells (3533 m and 3597 m for Rehmat-02 and 3335–3339 m for Miano-07) in moduli–porosity and velocity–porosity planes overlaying hypothetical models of rock physics (Figures 8a, 9a and 10a). The cross-plots show that increasing depth of clay content is related to increasing V_p and elastic values. The sorted well data of Miano-07 through the Lower Goru sand was divided into two sections. One is the sand that contains more quartz cement at the contact of grains, and the second is the sand gained from Lower Goru sand with 2–6% cement. A great portion of the scatter is detected in elastic moduli–porosity and velocity–porosity cross-plots for Rehmat-02. This dispersed impact is due to the presence of clay in the sand of the Lower Goru.

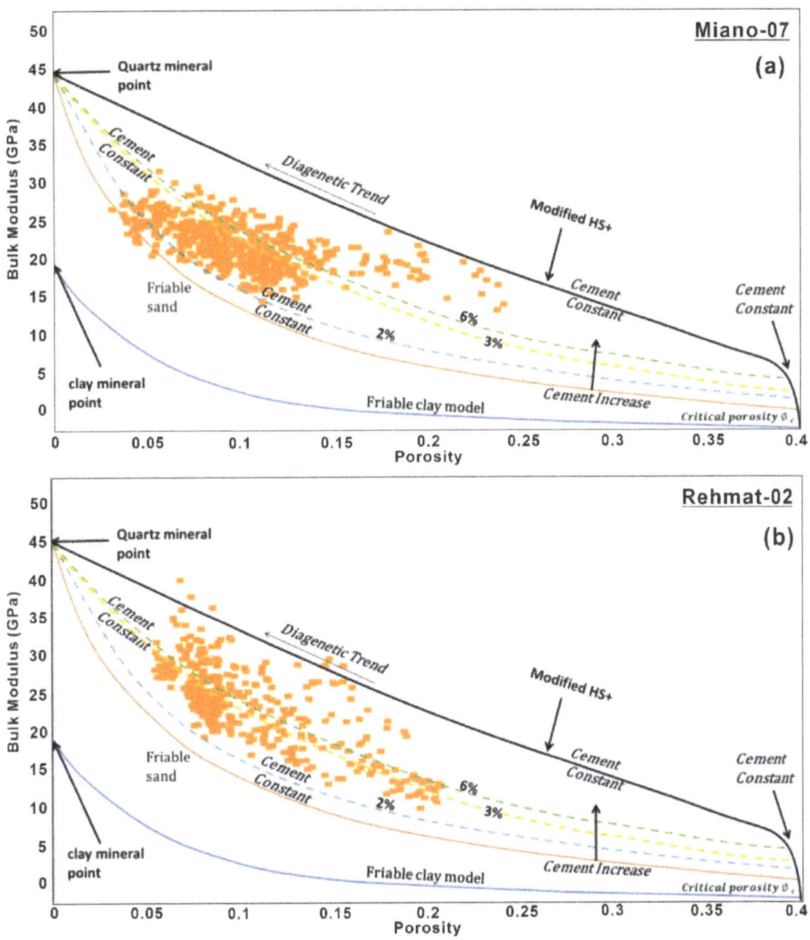

Figure 8. RPD cross-plots of (**a**) Miano-07, and (**b**) Rehmat-02, between Bulk Modulus (K) and porosity of reservoir sands utilizing the CC, CC and FS (uncemented sand) models.

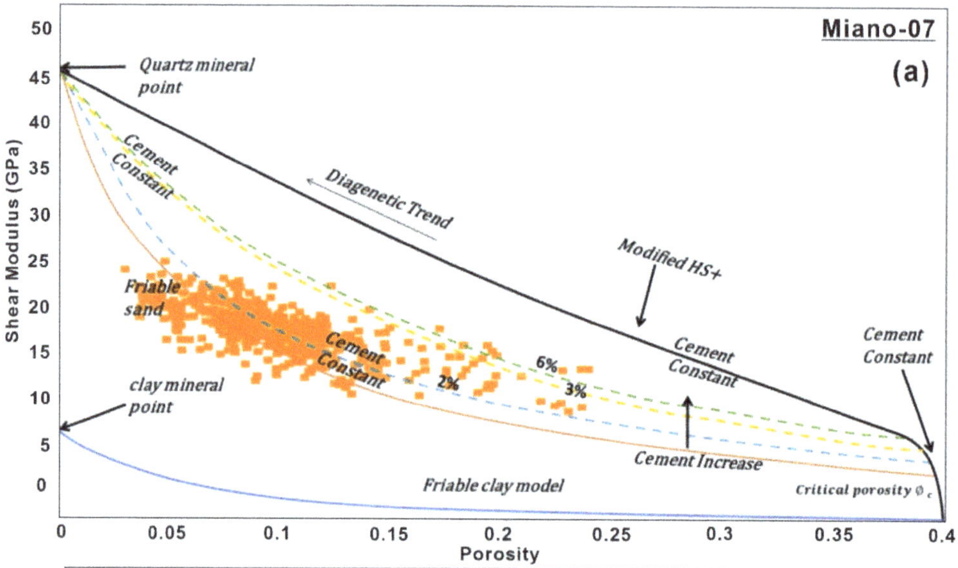

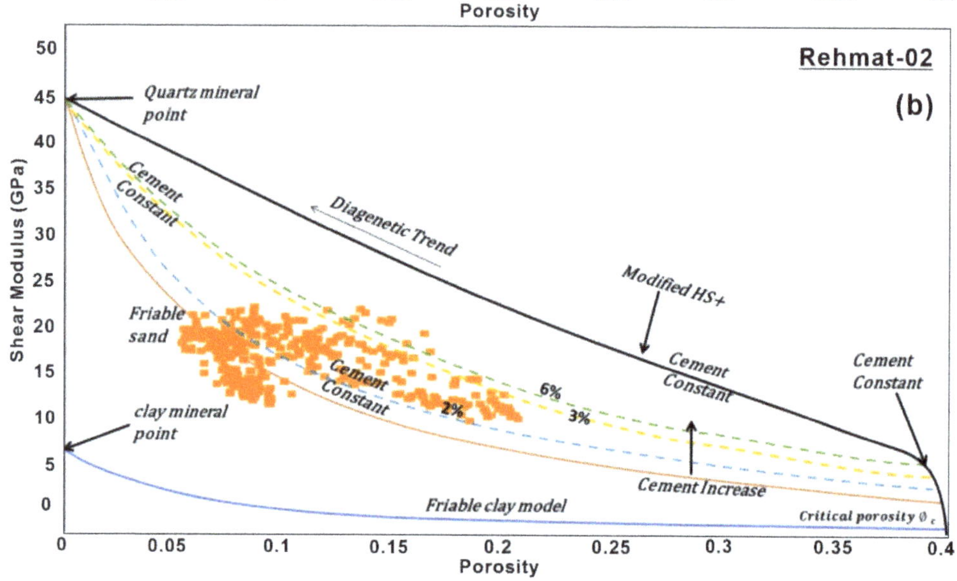

Figure 9. RPD cross-plots from (**a**) Miano-07 and (**b**) Rehmat-02, of Shear Modulus (G) versus porosity of reservoir sands utilizing the CC, CC, and FS (uncemented sand) models.

The data from the Lower Goru sand of Rehmat-02 was sorted and separated into three parts: friable sand (uncemented sand), sands with large volumes of quartz cement at grain contacts, and sands with 2–6% cement. The upper curve refers to pure quartz cases, whereas the lower curve applies to uncemented rock (Figures 8b, 9b and 10b). We observed that the maximum data value falls on the upper line of the velocity–porosity plane.

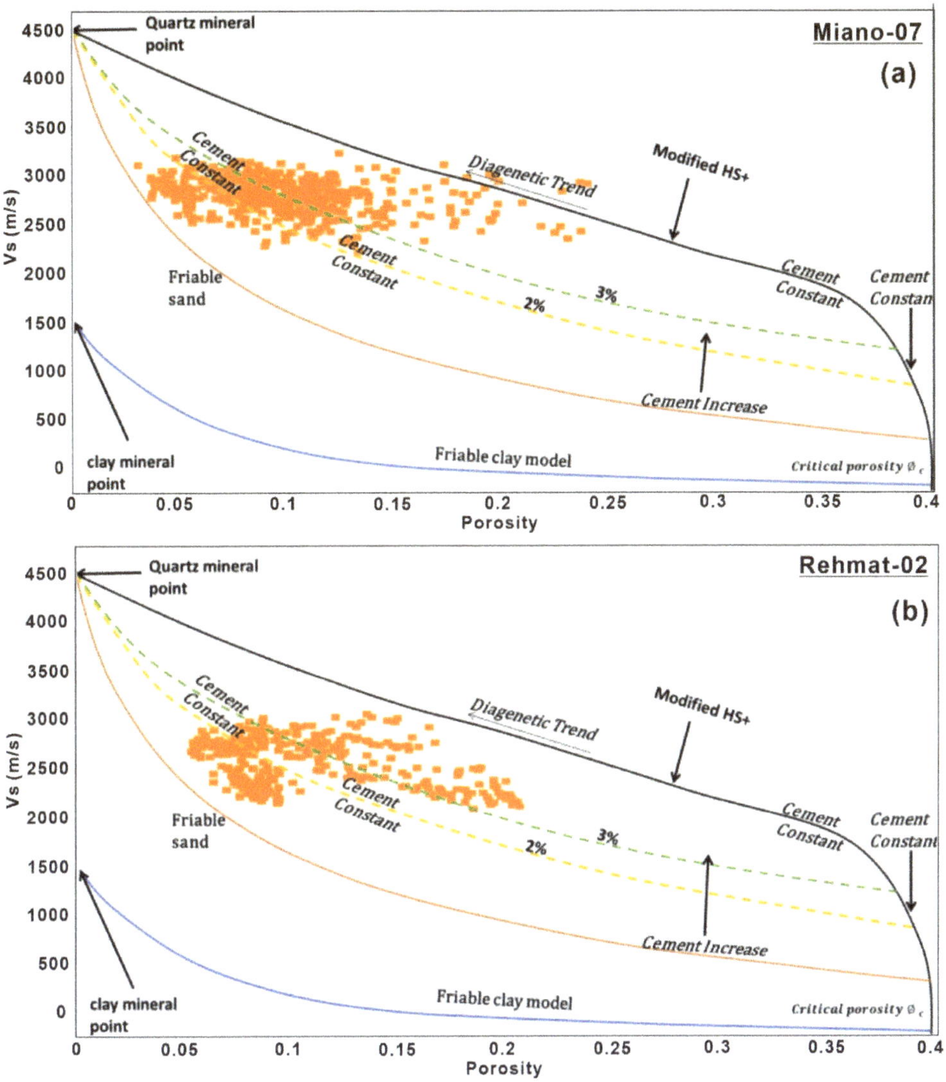

Figure 10. RPD cross-plots from (**a**) Miano-07 and (**b**) Rehmat-02, of V_S versus porosity of reservoir sands utilizing the CC, CC, and FS (uncemented sand) models.

The plane reflected all cement deposited exactly on grain contacts with a flat slope for Miano-07 and steep slope for Rehmat-2. The data value drops along constant cement lines, which resembles cement's uniform deposition on the grain's surface. For both the Miano-07 and Rehmat-02 wells, the data value dropped at grain contacts and had a steep slope on the upper link of the bulk modulus–porosity plane. The elastic moduli–porosity plane values decreased among the constant cement line and showed a gentle slope towards the steep slope. Microcrack variability is a possible key explanation for the disparity or shift in slope between the porosity of the velocity plane and the porosity of the modulus. Recall that the highest data value is on a constant cement line (6%) that reflects uniformly deposited cement on the grain surface. As observed in Figures 8–10, a CC model shows the reservoir zones recognized in the depth intervals: 3533 m and 3597 m for Rehmat-02 and 3335–3339 m for Miano-07. In these RPD cross-plots, the value of the distributions indicates complex geological trends showing lower to higher porosity values. Gas-saturated pay zone grew distinctive and specific elastic moduli–porosity and velocity–porosity patterns for the Miano-07 and Rehmat-02 wells. In the Lower Goru sand reservoir in the Miano-07 well, strong cementation in excess of 6% was noted. In the Rehmat-02, well, the uncemented sand and cemented sand with diverse cementation from 2 to over 6% were noted. Together with the three models of RPD, the CC-CC models were well-thought out to examine the reservoir rock texture and dry rock modulus of these two wells.

3.6. Gassmann Fluid Substitution

Gassmann fluid substitution is usually used for elastic estimation modulus of fluid-saturated rocks such as hydrocarbon or brine. In a RPD and elastic modulus, Gassmann's equation was used to measure the effect of fluid substitution on the desired saturation. Utilizing the fluid substitution model [61], the elastic modulus of saturated porous rocks was computed, which provided the association between Ks of saturated porous rock, dry rock modulus, pore fluid, and solid matrix.

$$K_{Sat} = K_{Dry} + \left[\frac{\left(1 - \frac{K_{Dry}}{K_{mineral}}\right)}{\frac{\varnothing}{K_{fluid}} + \frac{1-\varnothing}{K_{mineral}} + \frac{K_{Dry}}{K_{mineral}}} \right] \tag{12}$$

$$\mu_{Sat} = \mu_{Dry} \tag{13}$$

where, K_{Dry} and μ_{Dry} are derived from the rock physics model, the K and G of the dry modulus (because fluid G is nil, the G of the fluid-saturated rock is equal to μ_{Dry}); and $\varnothing$ is the porosity. The saturated water rocks density is

$$\rho = \rho_b(1 - \varnothing) + \rho_{fluid}\varnothing \tag{14}$$

The solid-phase modulus can be computed as the "effective" fluid phase K. The elements can incorporate water, oil, and gas. The immiscible system of the effective K ($K_{fluid-av}$) is expressed by the authors in [62] if all the individual liquid phases are perfectly connected hydraulically, i.e., the gas pressure is similar to that in the oil and in the water.

$$\frac{1}{K_{fluid-av}} = \frac{S_{br}}{K_{br}} + \frac{S_{oil}}{K_{oil}} + \frac{S_{gas}}{K_{gas}} \tag{15}$$

$$\rho_{fluid} = S_w \rho_w + (1 - S_w)\rho_{hc} \tag{16}$$

where $K_{fluid-av}$= effective fluid bulk modulus, K_{br} = brine bulk modulus, S_{br} = brine saturation, K_{oil}= oil bulk modulus, S_{oil} = oil saturation, K_{gas} = gas bulk modulus, S_{gas} = gas saturation, S_w= water saturation, ρ_w = water density, and ρ_{hc} = hydrocarbon density.

The authors in [63] reported the fluid modulus (K_{fluid})

$$K_{fluid} = \left(\frac{S_w}{K_w} + \frac{(1 - S_w)}{K_{hc}} \right) \tag{17}$$

where K_{hc} = hydrocarbon modulus and K_w = water bulk modulus.

K_{Sat} is needed to inquire about fluid substitution, and the V_p, and V_s are then lastly calculated by using these equations:

$$V_p = \sqrt{\left(\frac{K_w + \mu}{\rho_b} \right)} \tag{18}$$

$$V_s = \sqrt{\left(\frac{\mu}{\rho_b} \right)} \tag{19}$$

4. Results and Discussion

In the initial phase, we estimated the RPM and RPT parameters, which were utilized to discriminate lithologies and fluid behavior using multiple cross-plots. In the last phase of the result and discussion, we utilized the results of RMP and RPT with mechanical parameters to discriminate the rock behavior by log plot and cross-plot analyses.

4.1. Rock Physics Template (RPT) Analysis

The RPT technology has grown into a lithology and fluid prediction instrument. A cross-plot between V_p/V_s ratio vs. AI, V_p/V_s ratio vs. PR, and SI vs. AI are the most popular types of RPT. RPT analysis of well log data can be an effective, independent exercise for analyzing and monitoring the quality of well data and assessing seismic detectability of various fluid and lithology scenarios [57]. RPT models must consider regional geological factors. In RPMs, the geological restrictions contain lithology, mineralogy, burial depth, diagenesis, pressure, and temperature. Table 2 shows reservoir parameters and computational fluid properties of the Miano-07 well under the specified pressure and temperature. Gassmann fluid substitution modeling employs the density and K of the various fluids, as described in Table 3.

Table 3. Reservoir parameters for RPT model.

Parameters	Symbols	Numerical Values	Unit
Bulk modulus of sand	k_s	3.7×10^{10}	Pa
Shear modulus of sand	μ_s	4.4×10^{10}	Pa
Density of sand	ρ_s	2.65	g/cc
Bulk modulus of clay	k_c	2.15×10^{10}	Pa
Shear modulus of clay	μ_c	1.18×10^{10}	Pa
Density of clay	ρ_c	2.59	g/cc
Reservoir temperature	T	174.85	°C
Reservoir pressure	P	37.14	MPa
Salinity of brine	S_b	20,000	Ppm
Specific gravity of the gas	G_{gas}	0.6	API
Bulk modulus of gas	k_{gas}	7.78×10^7	Pa
Density of gas	ρ_{gas}	1.7123	g/cc
Bulk modulus of brine	k_b	3.18×10^9	Pa
Density of brine	ρ_b	1	g/cc
An aspect ratio of sand	α_s	0.12	Unitless
An aspect ratio of shale	α_c	0.035	Unitless

The outcome of RPD for sands and shales, with log data from 3335–3339 m from Miano-07 and 3533–3597 m from Rehmat-02, were overlaid in the V_p/V_s vs. AI, V_p/V_s vs. PR, and SI vs. AI, in shading by GR values (Figures 11–13). The shale and sand plainly showed up in the line of two trends. Each line of porosity sand had its saturation line starting with 100% brine-saturated sands and ending with gas-saturated. Figure 11a,b provides the context of the upper shale-trend line demonstrating pure shale, although the line below displays clean compacted quartz sand filled with brine. In the V_p/V_s vs. AI cross-plot field, there is a line reflecting that gas saturations are nearly perpendicular to the sand trend line on a RPT. This template classifies seal rock (i.e., reservoirs of gas sand), brine-saturated sand, and cap shale for the Miano-07 (Figure 11a) and Rehmat-02 (Figure 11b) wells.

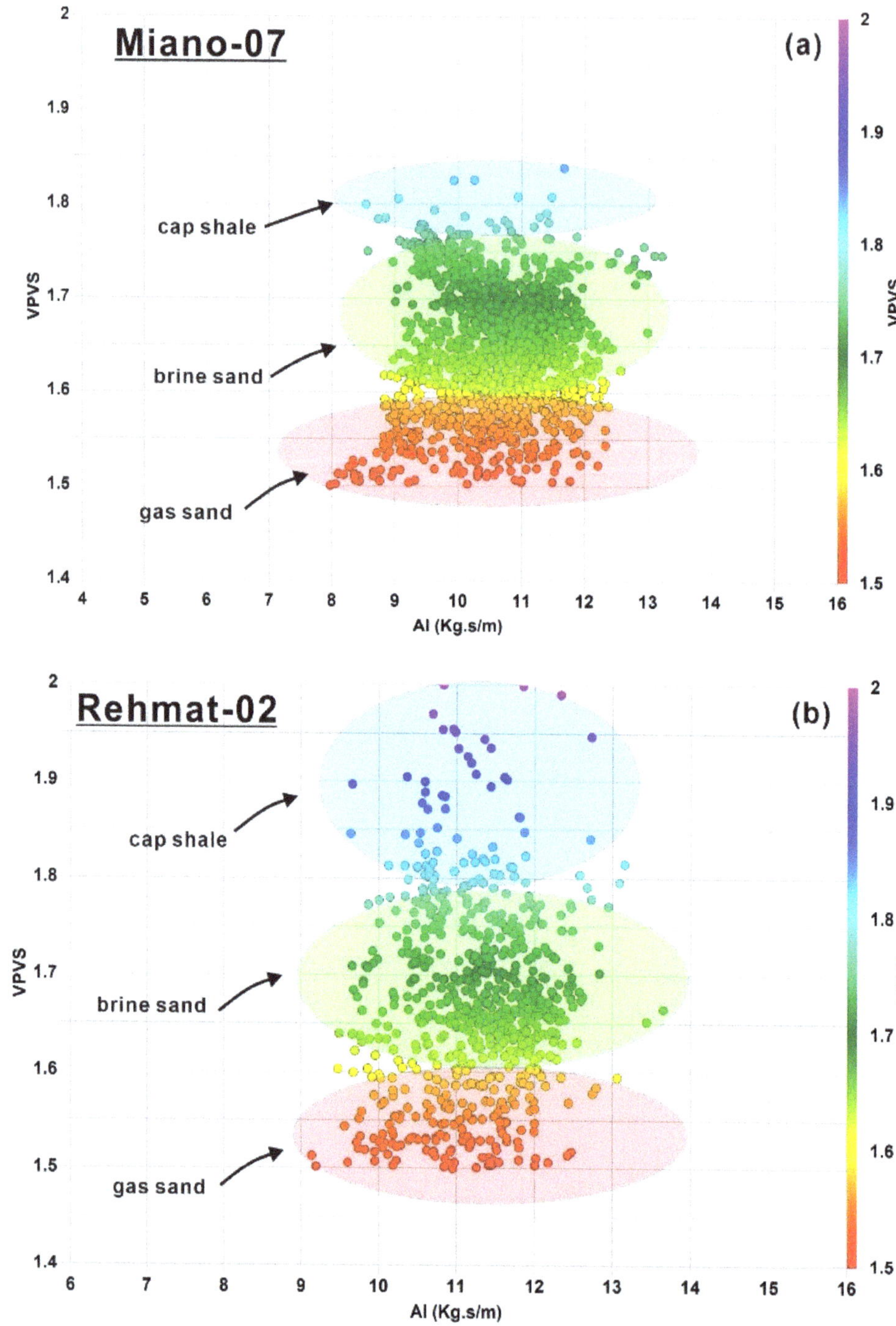

Figure 11. The RPT across V_p/V_s versus AI of (**a**) Miano-07 and (**b**) Rehmat-02 wells at the targeted intervals to showcase the gas-saturated (red box), brine sand (green box), and cap shale (blue box) zones.

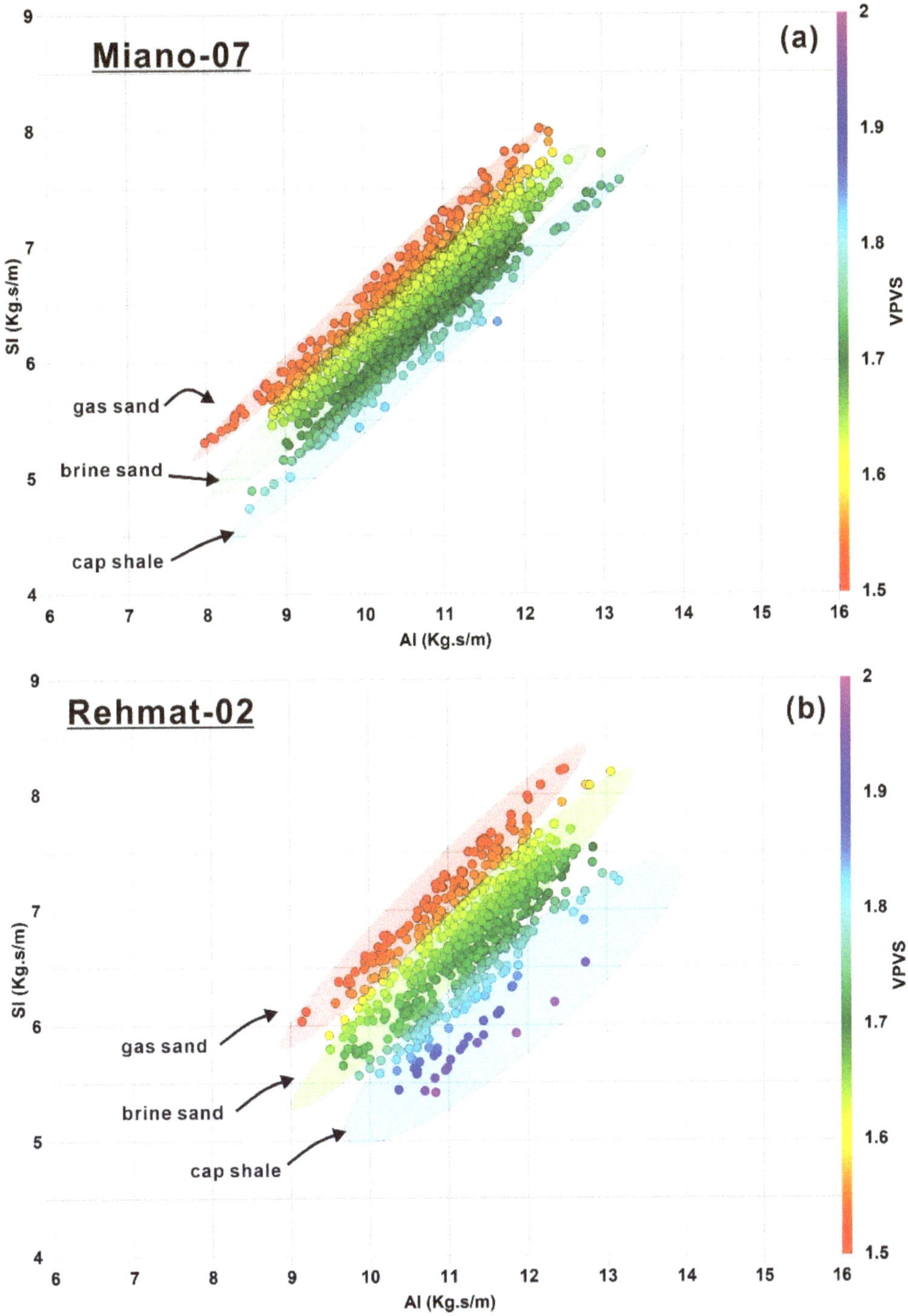

Figure 12. The RPT across SI versus AI of (**a**) Miano-01 and (**b**) Rehmat-02 wells at the targeted intervals to showcase gas-saturated (red box), brine sand (green box), and cap shale (blue box) zones.

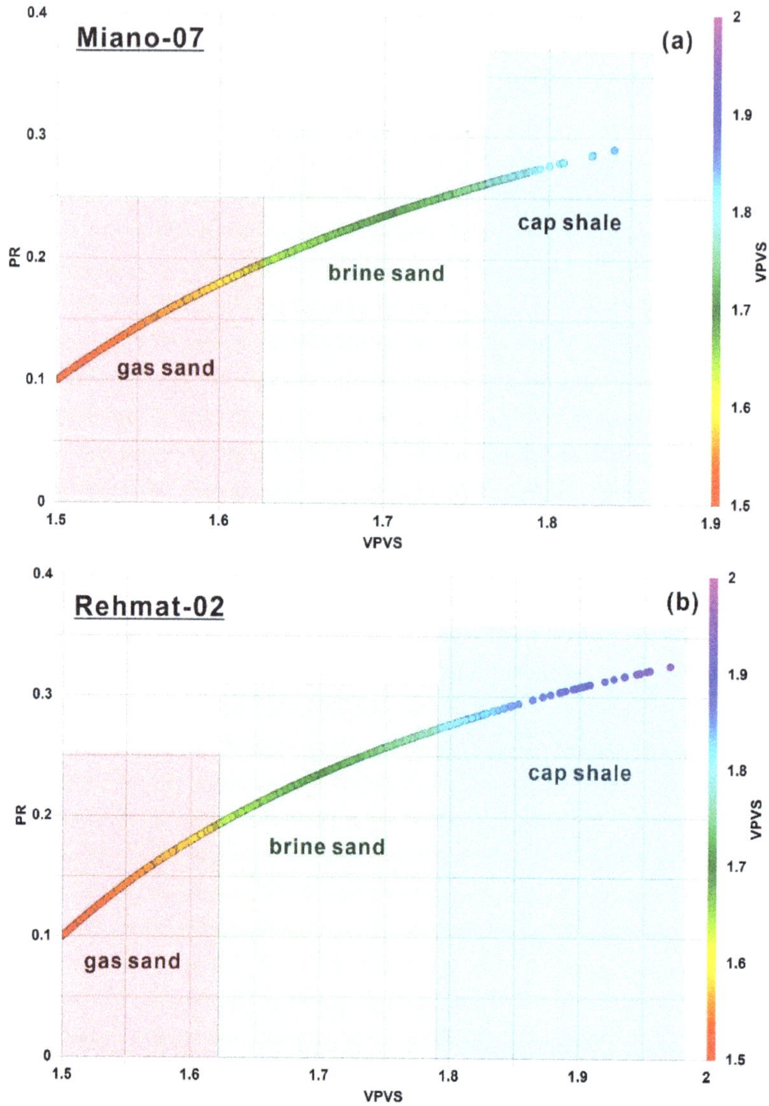

Figure 13. The RPT across PR versus AI of (**a**) Miano-01 and (**b**) Rehmat-02 wells at the targeted intervals to showcase the gas-saturated (red box), brine sand (green box), and cap shale (blue box) zones.

The RPT is shown in Figure 12 as the cross-plots of AI in competition with S-impedance (SI) measured from the well log data of the Miano-07 (Figure 12a) and Rehmat-02 (Figure 12b) wells. A clear distinction between gas reservoir, brine sand, and shale cluster points shows lithology discrimination, with full concentration of the sand clusters at an average porosity of 24% to 26% for both Miano-07 and Rehmat-02 wells. The points with minor AI exhibited high porosity and gas saturation for the gas reservoir plots. The points between sand and shale porosity patterns demonstrate the shaly sand and brine sand. The slight increase in AI at these porosities (24–26%) for the two wells shows that the reservoirs are composed of shaly sand.

Figure 13a,b shows the RPT deployed as the cross-plot of V_p/V_s with PR computed from well log data, focused on available V_p and V_s velocities and the RHOB logging data at the Miano-07 and Rehmat-02 wells. Moreover, the ratio of compressional V_p and V_s velocities and PR could also be calculated. The PR is a diagnostic lithological indicator. It measures the change of a body of material when it is affected by compression or tension [25]. Any change in PR could indicate a change in pore fluid. The gas-saturated zones have a PR (0.23 to 0.4) comparable to the surrounding zones. The shale lithology was plotted at a PR greater than the gas-saturated and brine sand zones.

The identification of gas zones can be established by examining the relationship between V_p/V_s and PR. The gas-saturated (red box), brine sand (green box), and cap shale (blue box) in the research area can be classified using the guideline plot based on extra log information using the V_p/V_s versus PR cross-plot. The area on the cross-plot shows a good fit with our preceding observations of cross-plots about the location of the gas reservoir.

4.2. RPM and RPT-Based Discrimination Characteristics of Reservoir

Both the Miano-07 and Rehmat-02 wells showed a dominant sequence of a sand–shale pattern. These two wells were studied in terms of their lithology and fluid forms. The high GR values represents the lithology of shale. Due to its high conductivity, shale lithology causes the resistivity to shift to the far-left side. Sand lithology with high resistivity and high AI is a highly gas-saturated area. The overall porosity of the rocks (sand–shale lithology) depends on grain size, accumulation, shape, sorting, intergranular matrix, and cement. Rock physics modeling and RPT studies have provided an approach to realize reservoir microstructures, cementation, fluid types, and lithology under specific temperature and pressure conditions. For the chosen intervals of 3325–3335 m from Miano-07 (Figure 14a), the NPHI and RHOB cross-plots showed the existence of gas-bearing Lower Goru sand which is highlighted by a red box. The cross-plot in Figure 14b demonstrates that the presence of laminated shale in the Lower Goru formation affects the Rehmat-02 well at the interval of 3533–3597 m. The log examination, along with the NPHI and RHOB cross-plot (Figure 14), identified clean sand gas-bearing zones (red colored boxes) and shaly sand gas-bearing zones that lie below the SS (sandstone) line.

Rock physics modeling and RPT were used to characterize the active seismic sands away from well regulations. For the Miano-07 (Figure 15) and Rehmat-02 (Figure 16) wells, conventional log curves along with the mechanical and elastic properties agree with each other and reliably indicate the existence of shaly sand and sand reservoirs. Overall, low GR, low V_p/V_s, low PR ratio value, higher resistivity, and a cross-over between NPHI and RHOB logs were observed. As a result, log data displayed on the developed RPT fit well with conventional wireline log data, as seen in the intersection between the NPHI and RHOB logs. The application of the RPT is a more effective tool for characterizing the reservoir in any geological setting. It can be known as reservoir identification, local geological prediction, and risk minimization in undeveloped areas. In addition to applying RPT to the whole Lower Goru tight sand reservoir in Pakistan and elsewhere with similar geological trends and reservoir distribution, the anticipated model may be used to estimate porosity from seismically determined impedance anywhere in the globe.

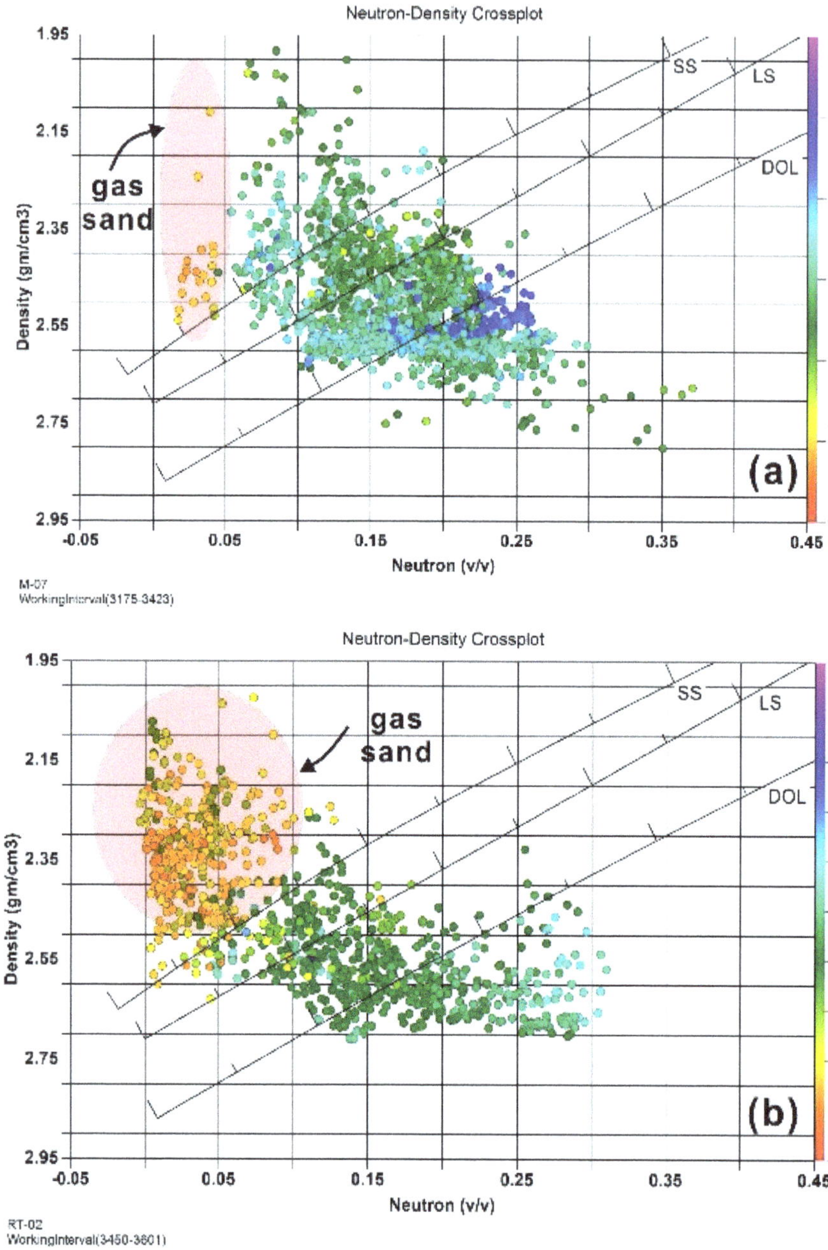

Figure 14. The RPT across RHOB versus NPHI of (**a**) Miano-01 and (**b**) Rehmat-02 wells at the targeted intervals to showcase the gas-saturated (red box), brine sand (green box), and cap shale (blue box) zones.

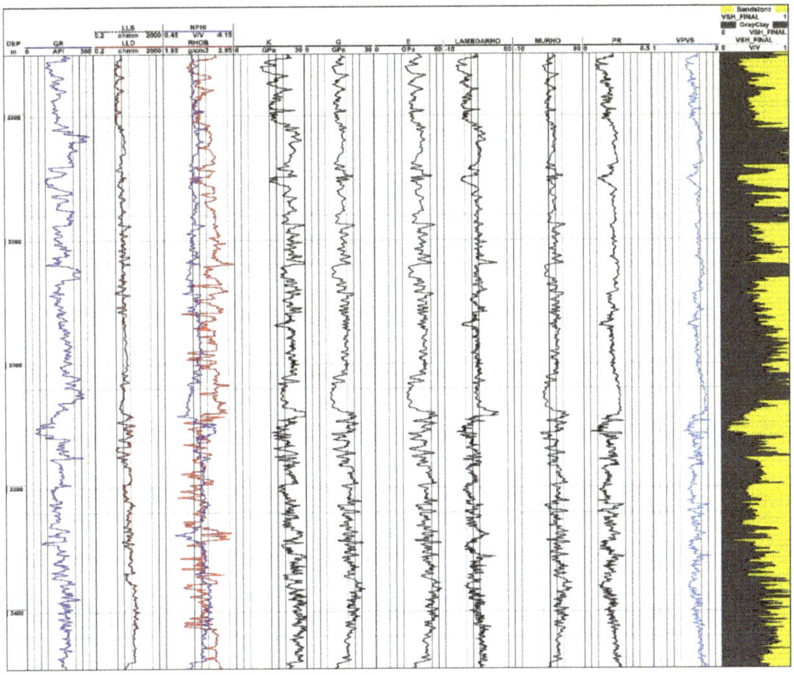

Figure 15. Standard petrophysical logs (tracks-2–4), mechanical properties (tracks-5–9), and elastic/seismic logs (tracks-10–11) provide a reliable and consistent interpretation of lithology (track-12) of Miano-07.

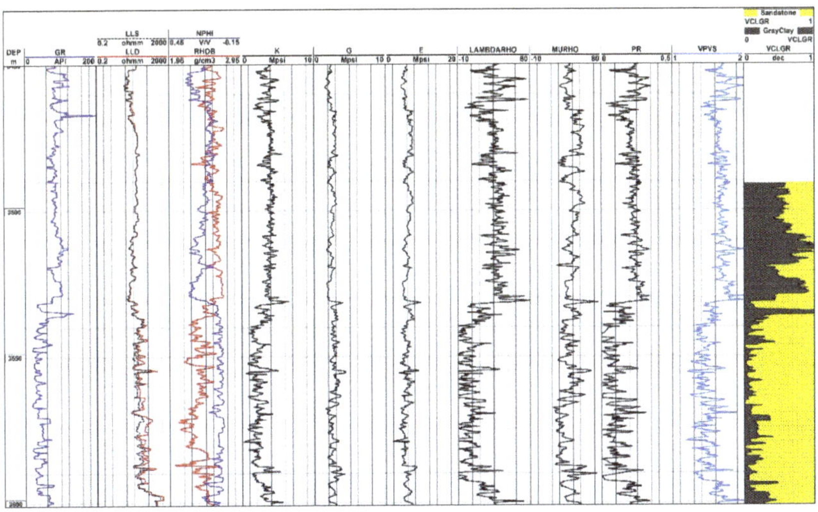

Figure 16. Standard petrophysical logs (tracks-2–4), mechanical properties (tracks-5–9), and elastic/seismic logs (tracks-10–11) provide a reliable and consistent interpretation of lithology (track-12) of Rehmat-02.

5. Conclusions

A model was designed for an unconsolidated reservoir in a Lower Goru gas field on the supposition that the rock's internal structure is identical to the idealized structure anticipated by the theoretical model if well log data points drop close to a theoretical line in the velocity–porosity plane. Several models were compared: friable sand, contact cement, and constant cement with varying degrees of cementation. The gas sands water saturation in the range of 21–60%, density in the range of 2.25–2.35 g/cc, with P-velocity and S-velocity in the range of 2800–2950 m/s and 2251–2372 m/s, respectively, inferred that the reservoir rocks are unconsolidated.

The RPM established in the study would improve the understanding of characteristics for the relatively high V_p/V_s unconsolidated reservoir sands under the study. The integration of RPMs would improve the prediction of reservoir properties from the elastic properties.

Through extensive cross-plot analysis of various rock properties, lithology and fluid discrimination were achieved with shear velocity versus porosity, shear impedance, and V_p/V_s versus acoustic impedance cross-plots, in agreement with the estimated water saturation and oil sand porosity. The shear velocity versus porosity cross-plots estimated oil sand porosities ranges from 10–15%, with a 2−3% degree of diagenetic process of constant cement model, alongside clear lithology distinction between oil sands and shales. These templates could be correlated with the seismic inverted results from the same field or with deposition environment to predict the lithology of undrilled areas and their fluid content.

Future development will include close work with real seismic and well log data to further calibrate and validate the proposed methods of reservoir characterization as well as to chart the areas of their applicability.

Author Contributions: M.A. and U.A. conceived this study. P.Z. supervised this project. J.U. and J.A. undertook the responsibility of arranging the data for this project. M.A. and U.A. wrote the manuscript. H.M., G.L. and R.J. helped in formal analysis and software. H.V.T. and A.A. discussed the data. P.Z. reviewed the manuscript and provided the necessary funding. All authors have read and agreed to the published version of the manuscript.

Funding: This research was supported by the National Natural Science Foundation of China (No. 41774145). We also extend our gratitude towards Yunnan Provincial Government Leading Scientist Program, No. 2015HA024.

Data Availability Statement: The original contributions presented in the study are included in the article Further inquiries can be directed to the corresponding authors.

Acknowledgments: The authors would like to thank the Directorate General of Petroleum Concessions (DGPC), Pakistan, for releasing the well data.

Conflicts of Interest: The authors declare no conflict of interest.

References

1. Anees, A.; Zhang, H.; Ashraf, U.; Wang, R.; Thanh, H.V.; Radwan, A.E.; Ullah, J.; Abbasi, G.R.; Iqbal, I.; Ali, N. Sand-Ratio Distribution in an Unconventional Tight Sandstone Reservoir of Hangjinqi Area, Ordos Basin: Acoustic Impedance Inversion-Based Reservoir Quality Prediction. 2022. Available online: https://repository.kaust.edu.sa/bitstream/handle/10754/685280/feart-10-1018105.pdf?sequence=1&isAllowed=y (accessed on 20 November 2022).
2. Ashraf, U.; Anees, A.; Shi, W.; Wang, R.; Ali, M.; Jiang, R.; Vo Thanh, H.; Iqbal, I.; Zhang, X.; Zhang, H. Estimation of porosity and facies distribution through seismic inversion in an unconventional tight sandstone reservoir of Hangjinqi area, Ordos basin. *Front. Earth Sci.* **2022**, *10*, 1014052. [CrossRef]
3. Bashir, Y.; Faisal, M.A.; Biswas, A.; Ali, S.H.; Imran, Q.S.; Siddiqui, N.A.; Ehsan, M. Seismic expression of miocene carbonate platform and reservoir characterization through geophysical approach: Application in central Luconia, offshore Malaysia. *J. Pet. Explor. Prod.* **2021**, *11*, 1533–1544. [CrossRef]
4. Iltaf, K.H.; Yue, D.; Wang, W.; Wan, X.; Li, S.; Wu, S.; Liu, R.; Weijia, Z.; Mehboob, S.; Shah, S.A. Facies and petrophysical modeling of Triassic Chang 6 tight sandstone reservoir, Heshui oil field, Ordos basin, China. *Lithosphere* **2021**, *2021*, 9230422.
5. Manzoor, U.; Ehsan, M.; Radwan, A.E.; Hussain, M.; Iftikhar, M.K.; Arshad, F. Seismic driven reservoir classification using advanced machine learning algorithms: A case study from the lower Ranikot/Khadro sandstone gas reservoir, Kirthar fold belt, lower Indus Basin, Pakistan. *Geoenergy Sci. Eng.* **2023**, 211451. [CrossRef]
6. Abbas, A.; Zhu, H.; Anees, A.; Ashraf, U.; Akhtar, N. Integrated seismic interpretation, 2D modeling along with petrophysical and seismic attribute analysis to decipher the hydrocarbon potential of Missakeswal area. *Pakistan. J. Geol. Geophys.* **2019**, *7*, 1–12.
7. Anees, A.; Zhong, S.W.; Ashraf, U.; Abbas, A. Development of a computer program for zoeppritz energy partition equations and their various approximations to affirm presence of hydrocarbon in missakeswal area. *Geosciences* **2017**, *7*, 55–67.

8. Ashraf, U.; Zhu, P.; Anees, A.; Abbas, A.; Talib, A. Analysis of Balkassar area using velocity modeling and interpolation to affirm seismic interpretation, Upper Indus Basin. *Geosciences* **2016**, *6*, 78–91.
9. Alalimi, A.; AlRassas, A.M.; Vo Thanh, H.; Al-qaness, M.A.; Pan, L.; Ashraf, U.; AL-Alimi, D.; Moharam, S. Developing the efficiency-modeling framework to explore the potential of CO2 storage capacity of S3 reservoir, Tahe oilfield, China. *Geomech. Geophys. Geo-Energy Geo-Resour.* **2022**, *8*, 128.
10. Riaz, M.S.; Bin, S.; Naeem, S.; Kai, W.; Xie, Z.; Gilani, S.M.M.; Ashraf, U. Over 100 years of faults interaction, stress accumulation, and creeping implications, on Chaman Fault System, Pakistan. *Int. J. Earth Sci.* **2019**, *108*, 1351–1359. [CrossRef]
11. Vo Thanh, H.; Lee, K.-K. 3D geo-cellular modeling for Oligocene reservoirs: A marginal field in offshore Vietnam. *J. Pet. Explor. Prod. Technol.* **2022**, *12*, 1–19.
12. Ashraf, U.; Zhang, H.; Anees, A.; Mangi, H.N.; Ali, M.; Zhang, X.; Imraz, M.; Abbasi, S.S.; Abbas, A.; Ullah, Z. A core logging, machine learning and geostatistical modeling interactive approach for subsurface imaging of lenticular geobodies in a clastic depositional system, SE Pakistan. *Nat. Resour. Res.* **2021**, *30*, 2807–2830. [CrossRef]
13. Ashraf, U.; Zhu, P.; Yasin, Q.; Anees, A.; Imraz, M.; Mangi, H.N.; Shakeel, S. Classification of reservoir facies using well log and 3D seismic attributes for prospect evaluation and field development: A case study of Sawan gas field, Pakistan. *J. Pet. Sci. Eng.* **2019**, *175*, 338–351. [CrossRef]
14. Dvorkin, J.; Nur, A.; Yin, H. Effective properties of cemented granular materials. *Mech. Mater.* **1994**, *18*, 351–366. [CrossRef]
15. Ali, M.; Ma, H.; Pan, H.; Ashraf, U.; Jiang, R. Building a rock physics model for the formation evaluation of the Lower Goru sand reservoir of the Southern Indus Basin in Pakistan. *J. Pet. Sci. Eng.* **2020**, *194*, 107461. [CrossRef]
16. Ellis, D.V.; Singer, J.M. *Well Logging for Earth Scientists*; Springer: Berlin/Heidelberg, Germany, 2007; Volume 692.
17. Ashraf, U.; Zhang, H.; Anees, A.; Ali, M.; Zhang, X.; Shakeel Abbasi, S.; Nasir Mangi, H. Controls on reservoir heterogeneity of a shallow-marine reservoir in Sawan Gas Field, SE Pakistan: Implications for reservoir quality prediction using acoustic impedance inversion. *Water* **2020**, *12*, 2972. [CrossRef]
18. Avseth, P.; Mukerji, T.; Mavko, G. *Quantitative Seismic Interpretation: Applying Rock Physics Tools to Reduce Interpretation Risk*; Cambridge University Press: Cambridge, UK, 2010.
19. Datta Gupta, S.; Chatterjee, R.; Farooqui, M. Rock physics template (RPT) analysis of well logs and seismic data for lithology and fluid classification in Cambay Basin. *Int. J. Earth Sci.* **2012**, *101*, 1407–1426. [CrossRef]
20. Durrani, M.Z.A.; Talib, M.; Sarosh, B. Rock physics-driven quantitative seismic reservoir characterization of a tight gas reservoir: A case study from the Lower Indus Basin in Pakistan. *First Break* **2020**, *38*, 43–53. [CrossRef]
21. Durrani, M.Z.; Willson, K.; Chen, J.; Tapp, B.; Akram, J. Rational rock physics for improved velocity prediction and reservoir properties estimation for granite wash (tight sands) in Anadarko Basin, Texas. *Int. J. Geophys.* **2014**, *2014*, 209351. [CrossRef]
22. Gray, D.; Day, S.; Schapper, S. Rock physics driven seismic data processing for the Athabasca oil sands, Northeastern Alberta. *CSEG Rec. March* **2015**, 32–40.
23. Guo, Z.; Lv, X.; Liu, C.; Chen, H.; Cai, Z. Characterizing gas hydrate–bearing marine sediments using elastic properties—Part 1: Rock physical modeling and inversion from well logs. *J. Mar. Sci. Eng.* **2022**, *10*, 1379. [CrossRef]
24. Mavko, G.; Mukerji, T.; Dvorkin, J. *The Rock Physics Handbook*; Cambridge University Press: Cambridge, UK, 2020.
25. David, C.; Le Ravalec-Dupin, M. Rock physics and geomechanics in the study of reservoirs and repositories. *Geol. Soc. Lond. Spec. Publ.* **2007**, *284*, 1–14. [CrossRef]
26. Avseth, P.A. *Combining Rock Physics and Sedimentology for Seismic Reservoir Characterization of North Sea Turbidite Systems*; Stanford University: Stanford, CA, USA, 2000.
27. Arévalo-López, H.S.; Dvorkin, J.P. Porosity, mineralogy, and pore fluid from simultaneous impedance inversion. *Lead. Edge* **2016**, *35*, 423–429. [CrossRef]
28. Ganguli, S.S.; Vedanti, N.; Dimri, V. 4D reservoir characterization using well log data for feasible CO2-enhanced oil recovery at Ankleshwar, Cambay Basin-A rock physics diagnostic and modeling approach. *J. Appl. Geophys.* **2016**, *135*, 111–121. [CrossRef]
29. Hossain, Z.; Mukerji, T.; Dvorkin, J.; Fabricius, I.L. Rock physics model of glauconitic greensand from the North Sea. *Geophysics* **2011**, *76*, E199–E209. [CrossRef]
30. Ogebule, O.Y.; Jahren, J.; Mondol, N.H. Compaction, rock physics and rock properties of sandstones of the Stø Formation: Case study of five wells from the south-western Barents Sea, Norway. *Mar. Pet. Geol.* **2020**, *119*, 104448. [CrossRef]
31. Pan, H.; Li, H.; Grana, D.; Zhang, Y.; Liu, T.; Geng, C. Quantitative characterization of gas hydrate bearing sediment using elastic-electrical rock physics models. *Mar. Pet. Geol.* **2019**, *105*, 273–283. [CrossRef]
32. Wollner, U.; Dvorkin, J.P. Elastic mineral facies: Selecting site-specific elastic moduli of clay. *Geophysics* **2017**, *82*, MR111–MR119. [CrossRef]
33. Hossain, Z.; de Newton, P.V. Advanced Rock Physics Diagnostic: A New Method For Cement Quantification. In Proceedings of the 2013 SEG Annual Meeting, Beijing, China, 22–27 September 2013.
34. Wollner, U.; Yang, Y.; Dvorkin, J.P. Rock-physics diagnostics of an offshore gas field. *Geophysics* **2017**, *82*, MR121–MR132. [CrossRef]
35. Mavko, G.; Dvorkin, J.; Walls, J. A Rock Physics and Attenuation Analysis of a Well from the Gulf of Mexico. In *SEG Technical Program Expanded Abstracts 2005*; Society of Exploration Geophysicists: Houston, TX, USA, 2005; pp. 1585–1588.
36. Dræge, A. A diagenetic rock physics approach for siliciclastics. *Lead. Edge* **2011**, *30*, 1368–1375. [CrossRef]
37. Kadri, I.B. *Petroleum Geology of Pakistan*; Pakistan Petroleum Limited: Karachi, Pakistan, 1995.

38. Ahmad, N.; Chaudhry, S. Kadanwari gas field, Pakistan: A disappointment turns into an attractive development opportunity. *Pet. Geosci.* **2002**, *8*, 307–316. [CrossRef]
39. Berger, A.; Gier, S.; Krois, P. Porosity-preserving chlorite cements in shallow-marine volcaniclastic sandstones: Evidence from Cretaceous sandstones of the Sawan gas field, Pakistan. *AAPG Bull.* **2009**, *93*, 595–615. [CrossRef]
40. Ehsan, M.; Gu, H. An integrated approach for the identification of lithofacies and clay mineralogy through Neuro-Fuzzy, cross plot, and statistical analyses, from well log data. *J. Earth Syst. Sci.* **2020**, *129*, 1–13. [CrossRef]
41. Mahmood, M.F.; Ahmad, Z.; Ehsan, M. Total organic carbon content and total porosity estimation in unconventional resource play using integrated approach through seismic inversion and well logs analysis within the Talhar Shale, Pakistan. *J. Nat. Gas Sci. Eng.* **2018**, *52*, 13–24. [CrossRef]
42. Brink, G.I.; Logan, A.M. Sequence stratigraphic approach to the Goru Petroleum system in the middle Indus basin, Pakistan. *AAPG Bull.* **1997**, *81*, 1364–1365.
43. Ehsan, M.; Gu, H.; Ali, A.; Akhtar, M.M.; Abbasi, S.S.; Miraj, M.A.F.; Shah, M. An integrated approach to evaluate the unconventional hydrocarbon generation potential of the Lower Goru Formation (Cretaceous) in Southern Lower Indus basin, Pakistan. *J. Earth Syst. Sci.* **2021**, *130*, 1–16. [CrossRef]
44. Kazmi, A.H.; Jan, M.Q. *Geology and Tectonics of Pakistan*; Graphic Publishers: Santa Ana, CA, USA, 1997.
45. Wandrey, C.J.; Law, B.; Shah, H.A. *Patala-Nammal Composite Total Petroleum System, Kohat-Potwar Geologic Province, Pakistan*; US Department of the Interior, US Geological Survey Reston: Reston, VA, USA, 2004.
46. Ahmad, N.; Fink, P.; Sturrock, S.; Mahmood, T.; Ibrahim, M. Sequence stratigraphy as predictive tool in lower goru fairway, lower and middle Indus platform, Pakistan. *PAPG ATC* **2004**, *1*, 85–104.
47. Dar, Q.U.; Pu, R.; Baiyegunhi, C.; Shabeer, G.; Ali, R.I.; Ashraf, U.; Sajid, Z.; Mehmood, M. The impact of diagenesis on the reservoir quality of the early Cretaceous Lower Goru sandstones in the Lower Indus Basin, Pakistan. *J. Pet. Explor. Prod. Technol.* **2022**, *12*, 1437–1452. [CrossRef]
48. Krois, P.; Mahmood, T.; Milan, G. Miano Field, Pakistan, A Case History of Model Driven Exploration. In Proceedings of the Pakistan Petroleum Convention, Islamabad, Pakistan, 25–26 November 1998; pp. 112–131.
49. Ali, M.; Jiang, R.; Ma, H.; Pan, H.; Abbas, K.; Ashraf, U.; Ullah, J. Machine learning-A novel approach of well logs similarity based on synchronization measures to predict shear sonic logs. *J. Pet. Sci. Eng.* **2021**, *203*, 108602. [CrossRef]
50. Hussain, M.; Liu, S.; Ashraf, U.; Ali, M.; Hussain, W.; Ali, N.; Anees, A. Application of machine learning for lithofacies prediction and cluster analysis approach to identify rock type. *Energies* **2022**, *15*, 4501. [CrossRef]
51. Jiang, R.; Ji, Z.; Mo, W.; Wang, S.; Zhang, M.; Yin, W.; Wang, Z.; Lin, Y.; Wang, X.; Ashraf, U. A Novel Method of Deep Learning for Shear Velocity Prediction in a Tight Sandstone Reservoir. *Energies* **2022**, *15*, 7016. [CrossRef]
52. Rider, M. *The Geological Interpretation of Well Logs*; Rider-French Consulting Limited: Houston, TX, USA, 1996.
53. Sarasty, J.J.; Stewart, R.R. Analysis of well-log data from the White Rose oilfield, offshore Newfoundland. *CREWES Res. Rep.* **2003**, *15*, 1–16.
54. Gardner, G.; Gardner, L.; Gregory, A. Formation velocity and density—The diagnostic basics for stratigraphic traps. *Geophysics* **1974**, *39*, 770–780. [CrossRef]
55. Masoudi, R.; Karkooti, H.; Othman, M.B.; Darman, N.H. Reliable Characterization and Modeling of the Capillary Transition Zone and Flow Dynamics in the Oil Rim Reservoirs. In Proceedings of the SPE Enhanced Oil Recovery Conference, Kuala Lumpur, Malaysia, 19–21 July 2011.
56. Yuedong, Q.; Yunfeng, G.; Hongwei, A. Study and application of optimal technique of shear-wave ve-locity prediction based on Xu-White model. *J. Oil Gas Technol.* **2007**, *29*, 100–105.
57. Avseth, P.; Bachrac, R. Seismic Properties of Unconsolidated Sands: Tangential Stiffness, Vp/Vs Ratios and Diagenesis. In Proceedings of the 2005 SEG Annual Meeting, Houston, TX, USA, 6–11 November 2005.
58. Dvorkin, J.; Nur, A. Elasticity of high-porosity sandstones: Theory for two North Sea data sets. *Geophysics* **1996**, *61*, 1363–1370. [CrossRef]
59. Mavko, G.; Mukerji, T.; Dvorkin, J. *The Rock Physics Handbook: Tools for Seismic Analysis of Porous Media*; Cambridge University Press: Cambridge, UK, 2009.
60. Mavko, G.; Mukerji, T. Bounds on low-frequency seismic velocities in partially saturated rocks. *Geophysics* **1998**, *63*, 918–924. [CrossRef]
61. Gassmann, F. On the elasticiy of porous media. *Mitt. Aus Dem Inst. Fur Geophys.* **1951**, *96*, 1–21. [CrossRef]
62. Domenico, S. Effect of brine-gas mixture on velocity in an unconsolidated sand reservoir. *Log Anal.* **1977**, *18*, 882–894. [CrossRef]
63. Wood, A.B. *Textbook of Sound Being an Account of the Physics of Vibrations with Special Reference to Recent Theoretical and Technical Developments*, 2nd ed.; G. Bell and Sons Ltd.: London, UK, 1941; p. 578.

processes

Article

Modeling of Quantitative Characterization Parameters and Identification of Fluid Properties in Tight Sandstone Reservoirs of the Ordos Basin

Bo Xu [1,2], Zhenhua Wang [3], Ting Song [4], Shuxia Zhang [3], Jiao Peng [1,2], Tong Wang [1,2,*] and Yatong Chen [1,2]

1 School of Petroleum Engineering, Xi'an Shiyou University, Xi'an 710065, China; bxu@xsyu.edu.cn (B.X.); jp@xsyu.edu.cn (J.P.); c12370904@126.com (Y.C.)
2 Key Laboratory of Special Stimulation Technology for Oil and Gas Fields in Shaanxi Province, Xi'an 710065, China
3 Research Institute of Shannxi Yanchang Petroleum (Group) Co., Ltd., Xi'an 710065, China; 13571843652@163.com (Z.W.); 18710563168@163.com (S.Z.)
4 No.11 Oil Production Plant, Changqing Oilfield Company, PetroChina, Qingyang 745000, China; songtin_cq@petrochina.com.cn
* Correspondence: wangtong214@126.com

Abstract: The Ordos Basin has abundant resources in its tight sandstone reservoirs, and the use of well logging technology stands out as a critical element in the exploration and development of these reservoirs. Unlike conventional reservoirs, the commonly used interpretation models are not ideal for evaluating tight sandstone reservoirs through logging. In order to accurately evaluate parameters and identify fluid properties in the tight sandstone reservoirs of the Ordos Basin, we propose the adaption of conventional logging curves. This involves establishing an interpretation model that integrates the response characteristics of logging curves to tight sandstone reservoirs in accordance with the principles of logging. In this approach, we create interpretation models specifically for shale content, porosity, permeability, and saturation within the tight sandstone reservoir. Using the characteristics of the logging curves and their responses, we apply a mathematical relationship to link these parameters and create a template for identifying fluid properties within tight sandstone reservoirs. The average absolute errors of the new multi-parameter weighting method shale content interpretation model and porosity classification saturation interpretation model for quantitative evaluation of reservoir shale content and oil saturation are small, and the accuracy meets the production requirements. In this paper, the four-step method is used to identify the fluid properties of tight sandstone reservoirs step by step, which is to use the interrelationship between curves, eliminate the useless information, enhance the useful information, and finally solve the problem of identifying the fluid properties of tight sandstone reservoirs, which is difficult to identify, and realize the linear discrimination of the interpretation standard, which improves the accuracy of interpretation. The proven multi-information, four-step, step-by-step fluid property identification template has an accuracy of more than 90%. The interpretation model has been applied to 20 wells on the block with a compliance rate of 95.23%, providing the basis for accurately establishing the tight sandstone interpretation standard. The newly introduced log evaluation approach for tight sandstone reservoirs effectively overcomes the technical hurdles that have previously hindered the evaluation of such reservoirs in the Ordos Basin. This method is suitable for wide application and can be used for quantitative evaluation of tight sandstone reservoirs in different regions.

Keywords: tight sandstone; logging interpretation; model; fluid property identification; siliciclastic reservoirs; unconventional petroleum resources

Citation: Xu, B.; Wang, Z.; Song, T.; Zhang, S.; Peng, J.; Wang, T.; Chen, Y. Modeling of Quantitative Characterization Parameters and Identification of Fluid Properties in Tight Sandstone Reservoirs of the Ordos Basin. *Processes* **2024**, *12*, 278. https://doi.org/10.3390/pr12020278

Academic Editors: Dicho Stratiev and Qingbang Meng

Received: 15 December 2023
Revised: 20 January 2024
Accepted: 25 January 2024
Published: 26 January 2024

1. Introduction

As global unconventional oil and gas exploration and development rapidly progresses, scientists are increasingly focusing on the development of tight sandstone reservoirs.

Effective identification of fluids in these reservoirs is considered to be the first and crucial step for their efficient development [1,2]. Accurate identification of fluids in tight sandstone reservoirs is of great importance to the overall development of such reservoirs [3]. China's land-phase tight sandstones are generally characterized by tight lithology and strong heterogeneity, making it difficult for conventional logging techniques to be useful for fluid identification [4,5]. Currently, logging curve identification for tight sandstones mainly includes lithology logging curves (natural potential SP, natural gamma GR), porosity logging curves (acoustic logging AC, density DEN, neutron CN), and resistivity logging curves (microlateral MLL, octolateral LL8, etc.) [6,7]. Based on the above logging curves for pore fluid identification, the selection of logging curves needs to be optimized and standardized [8,9].

The Ordos Basin is located at the intersection of the stable zone in eastern China and the active zone in western China. It is surrounded by multiple rifts as shown in Figure 1 [10]. Internally, the basin has a generally smooth structure characterized by a dip angle of less than 1°. The tectonic structure is simple, with gentle tectonics, stable subsidence, minimal fracturing and low activity [11,12]. The basin can be divided into six primary tectonic units: the northern Yimeng uplift, the western thrust belt, the western Tianhuan depression, the central Yishan slope, the southern Weibei uplift, and the eastern Jinxi fault fold belt [13].

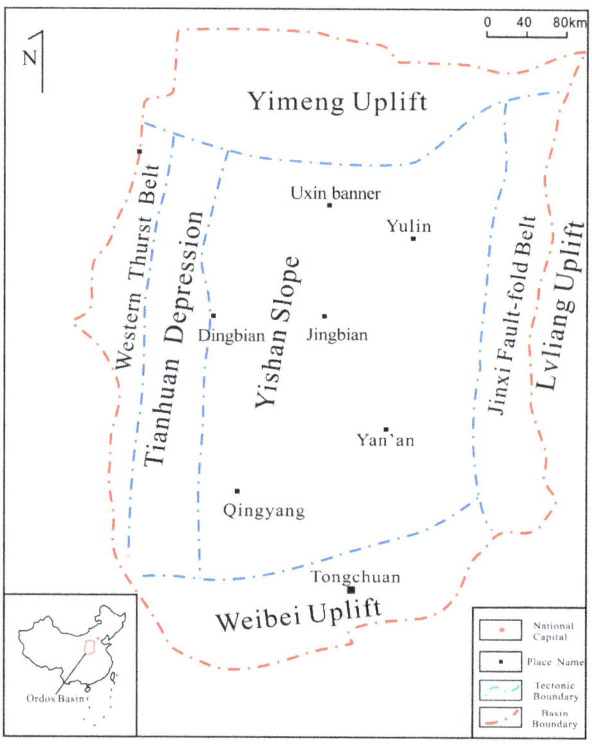

Figure 1. Highly generalized tectonic map of the Ordos Basin study area.

Tight sandstone reservoirs in the Ordos Basin exist mainly in the lower assemblage, especially in the Chang 7–Chang 10 section [14]. These reservoirs exhibit poor physical properties, characterized by complex pore throat structure, pronounced heterogeneity, and complicated rock–electric relationships [15,16]. There are the following difficulties in logging interpretation: the interpretation model based on the relationship study of the "four properties" (lithological characteristics, physical characteristics, electrical characteristics and oil-bearing characteristics) of the conventional reservoir cannot effectively meet the evaluation of logging in tight sandstone reservoirs [17–19]. Due to the influence of the sandstone skeleton, the contribution of fluid properties to well logging information in tight sandstone reservoirs is much smaller than that of the rock skeleton, making it difficult to discriminate fluid properties [20]. Therefore, the results of well logging interpretation in tight sandstones are not satisfactory [21,22]. Aiming at the problems of conventional logging curves on tight sandstone interpretation and evaluation, this paper starts from tight sandstone reservoirs in Ordos Basin and establishes or selects the interpretation model of the lithology, porosity, permeability and saturation of tight sandstone reservoirs according to the four properties of reservoirs and the logging principle. We then enhance the relevant signals within the log curves using the reflection of reservoir characteristics in the log curves, supplemented by oil test and recovery data. This process eliminates extraneous information to provide a standardized interpretation for distinguishing between oil and water reservoirs.

2. Characterization of Tight Sandstone Reservoirs

2.1. Lithological Characteristics

According to the thin section electron micrographs and core photographs of the lower assemblage core of the Yanchang Formation in the Ordos Basin (Figures 2 and 3), it can be seen that the pore type is dominated by intergranular pores, which are well developed. There are more fillers, mainly colluvium, and the colluvium is mainly illite and quartz, with better storage capacity. Combined with the petrographic description data of the logged wells, the main rock type of the lower assemblage of tight sandstone reservoirs is fine sandstone, with a small amount of medium sandstone and siltstone (Figure 4a). The mineralogical composition is dominated by feldspar sandstone (Figure 4b), with feldspar ranging from 55.13% to 83.54% of the fractions, with an average of 56.53%, quartz content ranging from 16.05% to 35.06%, with an average of 22.06%, and lithic content ranging from 3.66% to 10.71%, with an average of 5.61%. The variation in the lithology of the tight sandstone in the longitudinal direction is detailed in Figure 5. In the logging map, it can be seen that the lithology of the lower assemblage group is better, which is dominated by shaly sandstone and sandstone, and has better mining potential. Lower assemblage group tight sandstone casting thin section statistics are summarized in Table 1.

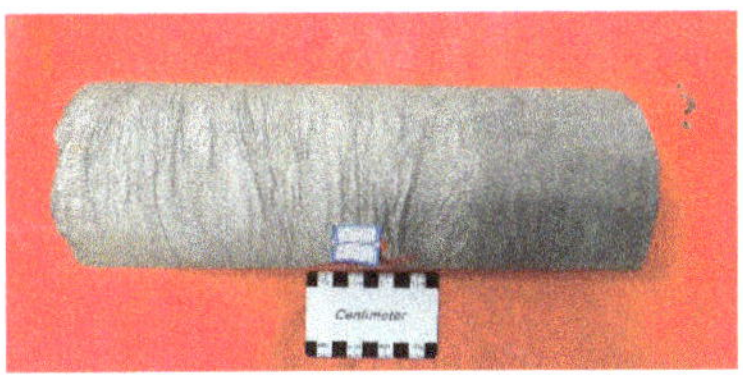

(a)

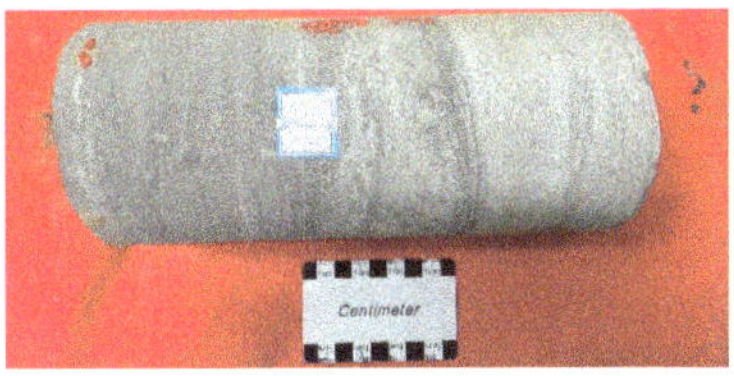

(b)

Figure 2. Photographs of tight sandstones of the lower assemblage of the Yanchang Formation. (**a**) Coring well L96, depth 1635 m, convolute bedding. (**b**) Coring well L108, depth 1508 m, convolute bedding.

(a)

(b)

Figure 3. Pore characteristics of the lower assemblage tight sandstone of the Yanchang Formation. (**a**) Core electron microscope picture. (**b**) Picture of a thin section of a core cast.

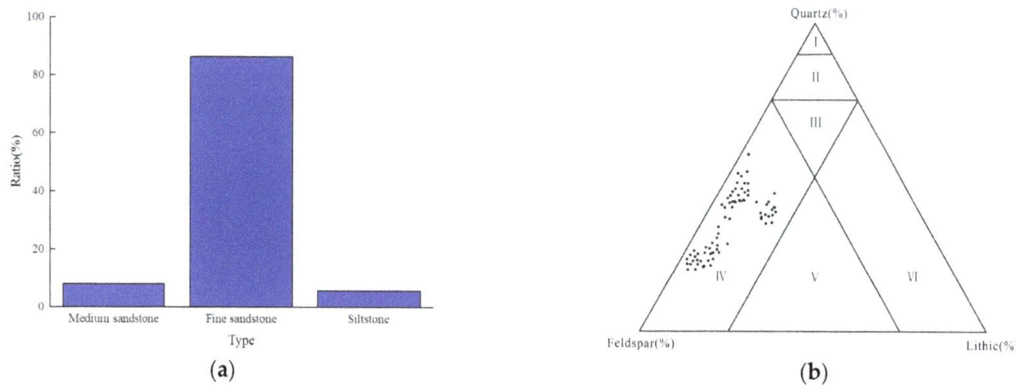

Figure 4. Sandstone lithology analysis diagram. (**a**) Sandstone lithology statistical histogram. (**b**) Sandstone compositional triangulation. I—Pure quartz sandstone; II—quartz sandstone; III—secondary lithic feldspar sandstone or secondary feldspar lithic sandstone; IV—feldspar sandstone; V—lithic feldspar sandstone or feldspar lithic sandstone; VI—lithic sandstone.

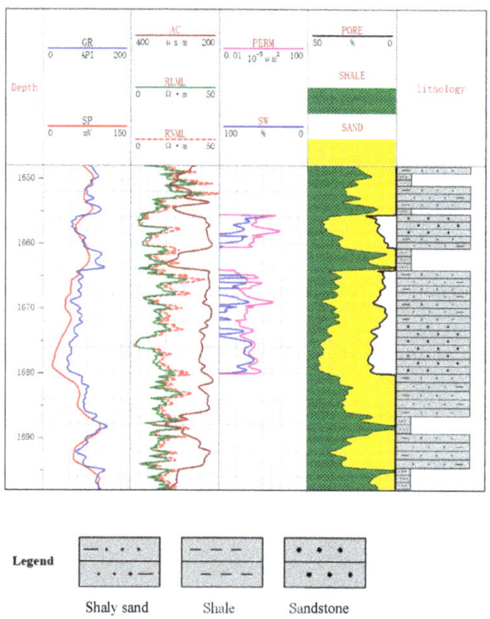

Figure 5. L110 well logging composite histogram.

2.2. Physical Characteristics

In logging, the physical properties of the reservoir are typically characterized using the porosity and permeability parameters. The analysis of porosity and permeability variation patterns is primarily based on data from 252 sandstone samples. According to the core analysis data, the porosity distribution interval from Chang 7 to Chang 9 ranges from 1.1% to 18.81% (Figure 6a), with an average of 8.33%, and the main distribution range is between 2% and 12%, accounting for 84.52% of the porosity samples in this interval. The distribution range of permeability is 0.01–37.71 $\times 10^{-3}$ μm^2 with a mean of 0.33 $\times 10^{-3}$ μm^2 (Figure 6b). The main distribution range of permeability is between 0.07 $\times 10^{-3}$ μm^2 and 0.4 $\times 10^{-3}$ μm^2, which accounted for 61.5% of the total number of samples, while samples between 0.01 × 10–3 μm^2 and 0.1 $\times 10^{-3}$ μm^2 accounted for 48% of the samples, samples between 0.1 and 1×10^{-3} μm^2 accounted for 50.4% of the samples, and samples larger than 1×10^{-3} μm^2 accounted for only 1.6% of the total number of samples.

Table 1. Statistical table of cast thin section identification of tight sandstones of some lower assemblage group.

Core Samples	Well	Depth	Terrigenous Debris			Thin-Section Porosity (%)
			Quartz (%)	Feldspar (%)	Lithic (%)	
1	L96	1634.5	33.0	58.0	8.0	4
2	L108	1507.3	31.0	57.0	6.0	8
3	L108	1513.5	28.0	67.0	5.0	3
4	L71	1479.7	25.0	66.0	8.0	2
5	L120	1312.6	26.0	65.0	7.0	1
6	L92	1677.7	28.0	67.0	4.0	2
7	L92	1671.9	28.0	66.0	6.0	3
8	L92	1751.4	20.0	72.0	8.0	1
9	P198	1418.1	21.0	67.0	11.0	5
10	P198	1531.5	29.0	61.0	10.0	4
11	U113	1767	25.0	67.0	7.95	3
12	P200	1817.4	27.8	63.3	8.89	1
13	P200	1820.2	28.6	64.8	6.59	2
14	U129	1745.7	25.6	64.4	10.0	5
15	Q1	832.9	31.4	52.3	16.28	2
16	X105	1081.3	27.2	64.1	8.7	5
17	X105	1193.6	31.5	55.4	13.04	7
18	S1040	1080.2	28.2	64.7	7.06	2
19	S32	1471.5	21.7	72.3	6.0	8
20	S32	1553	23.3	66.3	10.4	3

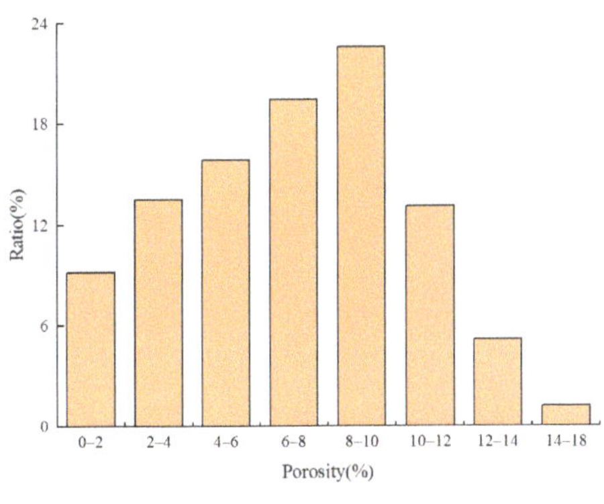

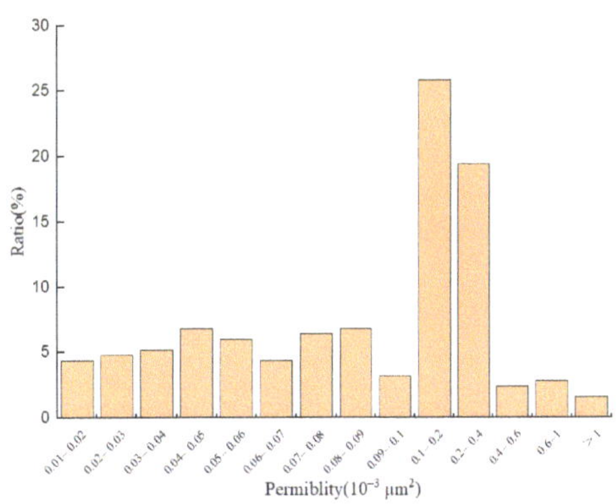

Figure 6. Physical characteristics distribution histogram. (**a**) Porosity distribution histogram. (**b**) Permeability distribution histogram.

2.3. Electrical Characteristics

Oil-bearing reservoirs in tight sandstones are characterized by low natural gamma, negative anomalies in natural potential, and high resistivity, and the resistivity of oil-bearing systems is generally greater than 20 Ω·m [23,24]. The layers with high shale content are all characterized by high natural gamma, small natural potential amplitude differences, relatively low resistivity, and high acoustic time difference values. As a whole, the resistivity of the tight sandstone reservoirs, on the other hand, is relatively high, ranging from 20 Ω·m to 50 Ω·m (Figure 7).

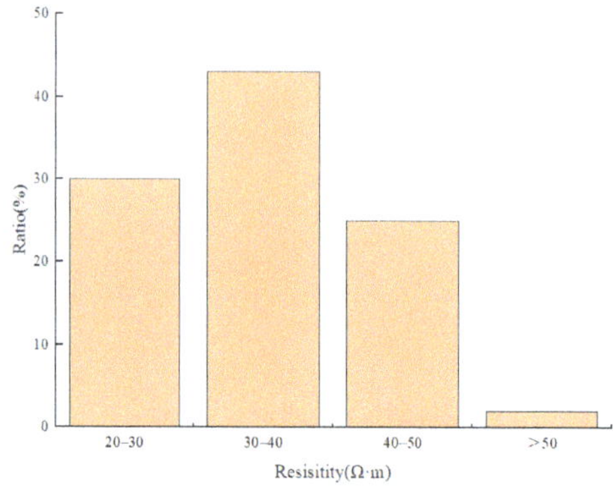

Figure 7. Resistivity distribution histogram.

2.4. Oil-Bearing Characteristics

The oil-bearing grades of tight sandstone reservoirs mainly have three grades: oil stains, oil traces and fluorescence, with oil stains accounting for 14.76%, oil traces accounting for 28.01% and fluorescence accounting for 26.36%. According to the statistics, the logging level of oil-producing reservoirs is generally above the oil trace, and the logging level above the oil trace accounts for 42.77% of the total number of wells (Figure 8a). The oil saturation is the calculated value from the logging, the distribution ranges from 2.52% to 40.81% and the average value is 20.81% (Figure 8b). From the distribution histogram, it can be seen that the distribution of oil saturation is mainly concentrated between 10% and 30%, accounting for 82.39% of the total number of samples, indicating that the tight sandstone reservoirs of the lower assemblage are not full of oil [25,26]. Higher resistivity values do not reflect the oil content but are more influenced by the rock skeleton. When calculating oil saturation in tight sandstone reservoirs, the original Archie model needs to be improved or a new oil saturation interpretation model needs to be established.

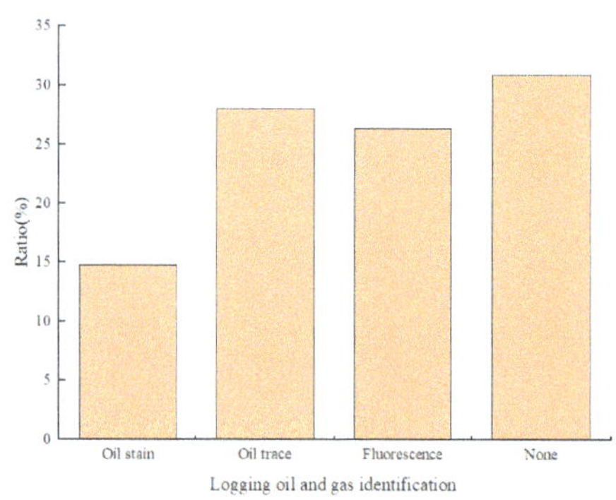

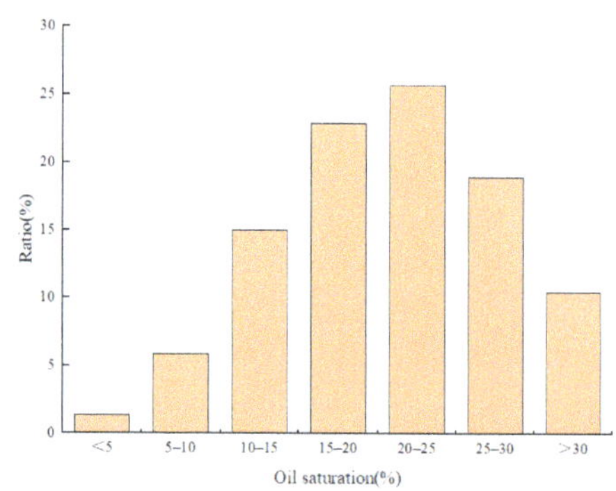

Figure 8. Oil-bearing characteristics distribution figure. (**a**) Logging grade distribution histogram. (**b**) Oil saturation distribution histogram.

3. Research on Evaluation Methods of Logging Interpretation

For fluid identification of tight sandstone reservoirs, it is necessary to first select the logging parameters of special areas in the reservoir and take this parameter as a standard [27]. Subsequently, by highlighting specific parameters in distinct areas, the reservoir characteristics become discernible and the influence of extraneous elements such as the rock skeleton can be minimized [28,29]. This approach facilitates rapid and accurate identification of fluid properties in tight sandstone reservoirs. In addition, variations in pore structure, which affect the oil content and oil-bearing properties in tight sandstone reservoirs, also affect the fluid response characteristics to some extent [30]. According to the actual situation of tight sandstone reservoirs, the reservoir itself has a variety of characteristics, and using only a single parameter or a certain logging method will make it difficult to identify the fluid properties of tight sandstone reservoirs [31]. Based on this, when analyzing the fluid parameters in the reservoir, it is necessary to combine various factors in the reservoir with a comprehensive analysis; this optimization process allows refinement of the reservoir geological characteristics [32]. In the case of tight sandstones, there are numerous factors that affect the identification of the reservoir fluid, including shale content, porosity, permeability, and water saturation. The establishment of a logging interpretation model is especially important in the Lower Assemblage reservoir of the Ordos Basin, which is tight and whose fluid distribution is complicated [33,34].

3.1. Calculation of Shale Content

Usually, the interpretation of shale content in sandstone reservoirs is sought by GR and SP. However, the lithology of the tight sandstone reservoir is dominated by fine sandstone with small mean grain size and strong adsorption, which adsorbs certain radioactive materials, so the GR logging value is high [35]. In contrast, the shale content interpreted in terms of GR and SP is on the high side due to the poor physical properties of the tight sandstone reservoir and the reduced SP amplitude difference. To eliminate the influence of non-formation factors on the shale content evaluation, all logging curves are analyzed for shale content reflection in the reservoir and the logging principles are analyzed [36]. The acoustic propagation in the tight sandstone layer is affected by the lithology and the contact mode of the rock particles, the propagation mode is nonlinear, the acoustic time difference value decreases and the calculated shale content is small. Therefore, in order to reduce the error of the logging curve in calculating the shale content of tight sandstone reservoirs, a compensating acoustic curve is introduced to weight the shale index calculated by *GR* and *AC* (or *SP* and *AC*) to explain the shale content of tight sandstone reservoirs [37].

$$\Delta GR = \frac{GR - GR_{min}}{GR_{max} - GR_{min}} \tag{1}$$

ΔGR: natural gamma calculated shale index; GR: natural gamma logging value, API; GR_{min}: gamma value for pure sandstone, API; GR_{max}: gamma value for pure mudstone, API.

$$\Delta AC = \frac{AC - AC_{min}}{AC_{max} - AC_{min}} \tag{2}$$

ΔAC: compensated acoustic calculation of shale index; AC: compensated acoustic logging values, μs/m; AC_{min}: compensated acoustic value for pure sandstone, μs/m; AC_{max}: compensated acoustic value for pure mudstone, μs/m.

$$\Delta SP = 1 - \frac{\Delta SP_m}{\Delta SP_{sa}} \tag{3}$$

$$\Delta SP_m = SP_m - SP_{sh} \tag{4}$$

$$\Delta SP_{sa} = SP_{sa} - SP_{sh} \tag{5}$$

ΔSP: natural potential calculated shale index; SP_m: measured actual natural potential of sandstone, mV; SP_{sh}: measured natural potential of mudstone, mV; SP_{sa}: natural potential value of water-bearing pure sandstone, mV.

$$SH = (1 - K) \times \Delta GR + K \times \Delta AC \tag{6}$$

$$SH = (1 - K) \times \Delta SP + K \times \Delta AC \tag{7}$$

$$V_{sh} = \frac{2^{GCUR \times SH} - 1}{2^{GCUR} - 1} \tag{8}$$

SH: shale index, radix; *K*: weighting factor; *GCUR*: the lithology coefficient, for older formations is 2 and for newer formations is 3.7.

3.2. Calculation of Porosity

For the interpretation of porosity, the Wiley model derived from the time-averaged formula is generally used, but tight sandstone reservoirs, with complex pore structures and nonlinear propagation of acoustic waves in the formation, have higher error in the porosity calculated by the Wiley model based on the time-averaged formula [38]. In 1986, three Raymer–Hunt–Gardner logging analysts of the French TOTAL Petroleum Company, after a thorough study of the work of their predecessors, took into account the influence of pore structure on acoustic wave propagation and proposed the formula of the acoustic formation factor, which is found to be modeled by Raymer–Hunt–Gardner through a comparative analysis [39,40]. The accuracy of porosity interpretation in tight sandstone reservoirs is significantly superior to that of the Wylie model. Therefore, the Raymer–Hunt–Gardner model is selected for porosity calculation in tight sandstone reservoirs:

$$\Delta AC_{CC} = \Delta AC - V_{sh} \times (\Delta AC_{sh} - \Delta AC_{ma}) \tag{9}$$

$$C = \frac{\Delta AC_{ma}}{(2 \times \Delta AC_f)} \tag{10}$$

$$\Phi_e = 1 - C - \sqrt{C^2 - \frac{\Delta AC_{ma}}{\Delta AC_f} + \frac{\Delta AC_{ma}}{\Delta AC_{CC}}} \tag{11}$$

ΔAC_{cc}: corrected acoustic time difference value, μs/m; C is a constant and is the reciprocal of the coefficient of skeletal lithology; ΔAC_{sh}: mudstone acoustic time difference, μs/m; ΔAC_{ma}: rock skeleton acoustic time difference, μs/m; ΔAC_f: pore fluid acoustic time difference, μs/m; Φ_e: effective porosity of rock, *f*.

3.3. Calculation of Permeability

Permeability determines the capacity of the reservoir and is a very important parameter in logging evaluation, but it is also the most difficult geological parameter to calculate accurately [41]. At present, the logging calculation of permeability generally adopts the empirical formula proposed by Timur, and in different blocks, the corresponding coefficients and indices of the empirical formula are determined [42,43]. In tight sandstone reservoirs, the correlation between permeability and porosity may decrease, but, overall, it remains positively correlated with porosity and negatively correlated with bound water saturation. The empirical formula for permeability calculation is still followed here, using the porosity of the tight sandstone in the lower assemblage of the study area. Permeability and bound water saturation data and the coefficients and exponents of the empirical formula are determined by the fitting method, and bound water saturation can be derived from conventional logging curves.

$$K = 0.126 \times \frac{\Phi^{0.08}}{{S_{wi}}^{1.11}} \tag{12}$$

K: permeability, 10^{-3} μm^2; Φ: porosity, f; S_{wi}: bound water saturation, f.

3.4. Calculation of Oil Saturation

Oil saturation is the core of logging interpretation, and the commonly used oil saturation formula is Archie's formula and its improved type [44]. To improve the accuracy of saturation interpretation, it is necessary to have accurate cementation index m, saturation index n, and saturation constants a and b. According to the petrographic experimental data, the appropriate way to correct the petrographic parameters in Archie's formula was elucidated, which can better control the accuracy of saturation calculation [45]. Through the analysis of the porosity–formation factor cross plot, it is found that all the data become two trends with a porosity of 7.1% as the demarcation; thus, here, according to the size of porosity, the rock electrical data are categorized, and the values of a, b, m and n are obtained, respectively, in order to improve the interpretation accuracy of oil saturation in tight sandstone reservoirs.

$$S_w = \sqrt[n]{\frac{a \times b \times R_w}{\Phi \times R_t^m}} \tag{13}$$

$$S_o = 1 - S_w \tag{14}$$

S_w: water saturation, f; S_o: oil saturation, f; a, b: lithological constants, dimensionless; m: lithological index, dimensionless; n: saturation index, dimensionless; R_w: formation water resistivity, Ω·m; R_t: formation true resistivity, Ω·m; Φ: porosity, f.

3.4.1. Porosity Classification Method for Calculating Oil Saturation

Firstly, the sandstone samples of the lower assemblage are selected, the resistivity value of the rock samples is measured, combined with the formation water resistivity Rw, the formation factor F is found, and the m-value and a-value are obtained using regression analysis; the centrifugal method is used to measure the saturation and resistivity under different centrifugal speeds, and the power function relationship is established by the cross plot of the resistivity coefficients with the water-bearing saturation degree, so that the n-value and b-value can be determined.

Based on the porosity–formation factor cross plot, the data are categorized according to the porosity of 7.1%. When the porosity $\Phi \geq 7.1\%$, the corresponding rock electrical parameters a = 1.320, b = 1.0705, m = 1.736, n = 1.629 are calculated based on the relationship between porosity and formation factors and the relationship between water saturation and the resistivity index (Figure 9).

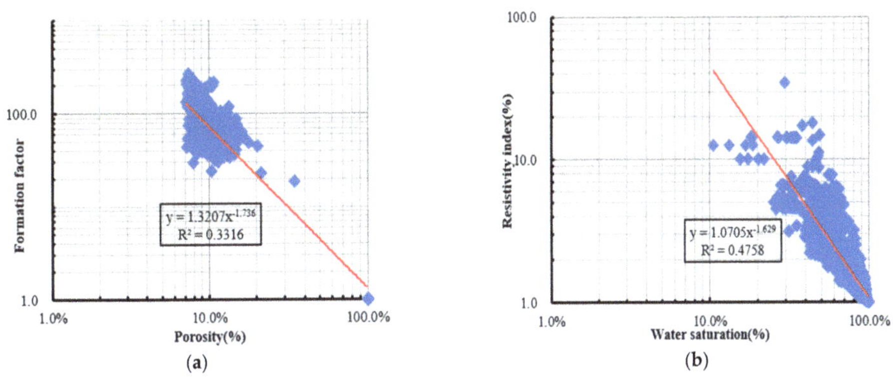

Figure 9. The relevant parameter cross plot when the porosity ≥7.1%. (**a**) Porosity and formation factor cross plot. (**b**) Water saturation and resistivity index cross plot.

When the porosity $\Phi < 7.1\%$, the rock electrical parameters a = 1.8751, b = 1.1749, m = 1.297, n = 1.872 are determined in the same way (Figure 10).

3.4.2. Calculation of Oil Saturation by Acoustic Time Difference Logging

When the interplay between the "four properties" of the reservoir is examined, it becomes clear that in tight sandstone reservoirs, resistivity does not accurately reflect the oil-bearing nature of the reservoir. Consequently, using the resistivity curve to calculate reservoir oil saturation results in significant errors [46]. The logging curve is a comprehensive reflection of formation information, and the acoustic time-difference logging curve value contains formation skeleton information and porosity information, which in turn contains water-bearing porosity and oil-bearing porosity, and according to the definition of oil-bearing saturation, the ratio of oil-bearing porosity to total porosity is oil-bearing saturation [47]. Therefore, it is possible to build a volumetric model from the logging principle of compensated acoustic waves, convert it into a mathematical model, remove invalid information, extract the required information, and build a calculation model for oil bearing saturation (Figure 11).

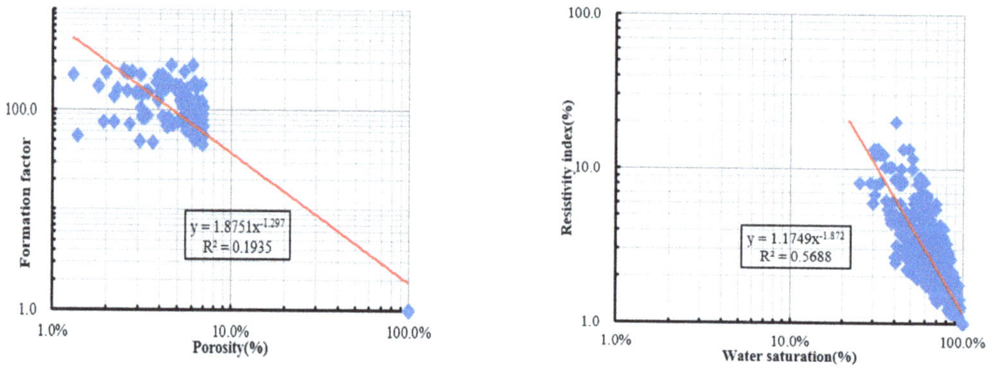

Figure 10. The relevant parameter cross plot when the porosity <7.1%. (**a**) Porosity and formation factor cross plot. (**b**) Water saturation and resistivity index cross plot.

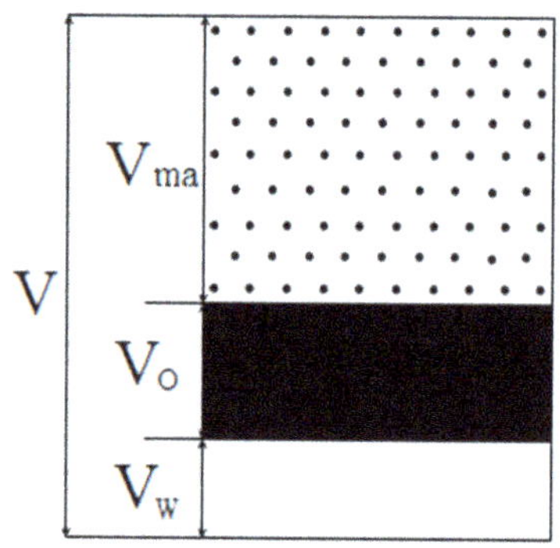

Figure 11. Volumetric modeling for calculating oil bearing saturation.

The acoustic logging volume model is converted into an equivalent model to establish the relationship between each parameter, and the mathematical relationship equation between porosity and acoustic time difference is as follows:

$$AC = AC_{ma} \times (1 - \Phi_w - \Phi_o) + AC_o \times \Phi_o + AC_w \times \Phi_w \tag{15}$$

It is known from the definition of porosity that the total porosity is the sum of oil-bearing porosity and water-bearing porosity of the formation; therefore:

$$\Phi = \Phi_o + \Phi_w \tag{16}$$

This is then finalized according to the definition of oil-bearing saturation:

$$S_o = \frac{(AC - AC_{ma})}{(AC_o - AC_w) \times \Phi} - \frac{(AC_w - AC_{ma})}{(AC_o - AC_w)} \tag{17}$$

AC: acoustic time difference logging value, μs/m; AC_w: sonic value of formation water at formation condition, μs/m; AC_o: acoustic wave value of crude oil under formation conditions, μs/m; AC_{ma}: acoustic value of sandstone skeleton, μs/m; Φ_w: water-bearing porosity, f; Φ_o: oil bearing porosity, f; Φ: total porosity of formation, f.

4. Fluid Property Identification

Reservoir lithology, physical properties and oil-bearing aspects are inherently interrelated and mutually constraining. The logging curve serves as a comprehensive representation of the lithological, physical, and petrophysical properties. In tight sandstone reservoirs, the proportion of fluid in the logging response is reduced, making it difficult for the resistivity curve, which is optimal for reflecting fluid properties, to comprehensively and accurately represent the oil-bearing situation of the formation [48].

The cross-plot method is to select the pairs of logging parameters and draw a cross plot to classify the fluid properties. As mentioned above, the logging parameters such as GR, RT and AC can distinguish oil and water layers and can be used to construct a cross plot to semi-quantitatively identify the fluid properties of the lower assemblage tight reservoir, targeting the geological and logging characteristics of tight sandstone reservoirs, focusing on extracting oil-bearing information from logging signals, synthesizing and enhancing useful information, and eliminating factors that affect the identification of oil-bearing properties. The reservoir fluid properties are progressively recognized through four steps. Using the oil test data of the study area (Table 2), combined with the GR, AC, and RILD values of the test oil test formation, AC/GR-RILD, AC-RILD, GR-AC*RILD/100, and AC-GR*RILD/100 cross plots are made, respectively (Figure 12). The above steps lead to the final identification of fluid properties.

Table 2. Fluid identification data for some lower assemblage group tight sandstones.

Well	GR/ API	AC/ μs/m	AC/GR	RILD /Ω·m	RILM /Ω·m	LL8 /Ω·m	Explanation of Conclusions
L77	94.45	243.41	2.58	43.29	43.48	56.28	O/W
L76	88.81	234.33	2.64	30.08	35.43	44.16	O/W
L63	99.32	282.68	2.85	63.56	64.96	130.25	O/W
L63	100.19	263.09	2.63	28.61	26.54	66.24	O/W
L89	95.2	269.14	2.83	42.18	57.61	166	O/W
L78	78.85	236.09	2.99	32.29	32.95	58.19	O/W
L128	50.23	238.46	4.75	28.2	27.45	34.1	O/W
L72	63.35	240.66	3.80	30.43	30.69	31.74	O/W
L109	77.93	238	3.05	28.06	29.98	35.41	O/W
L139	82.03	245	2.99	35.33	33.4	58.08	O/W
L183	73.64	241.75	3.28	70.29	68.74	63.1	O/W
L232	77.76	242.25	3.12	40.78	34.42	43.9	O/W
L233	84.02	235.14	2.80	99.03	65.62	103.89	WWO
L252	84.78	241.46	2.85	75.92	55.29	96.33	WWO
L251	76.41	222	2.91	41.02	49.63	68.6	WWO
L261	73.66	220.25	2.99	45.89	53.03	119.85	WWO
L46	82.55	230.06	2.79	20.13	22.99	24.6	WWO
L37	67.63	247.32	3.66	27.53	30.39	38.09	WWO
L43	81.19	223.08	2.75	44.9	54.08	55.15	WWO
L47	68.87	231.65	3.36	21.5	22.72	21.79	WWO
L55	79.76	233.46	2.93	27.79	26.94	31.81	W
L41	83.01	232.4	2.80	16.44	19.85	21.66	W
L76	59.38	262.63	4.42	13.73	13.74	15.83	W
L40	81.65	237	2.90	22.85	23.49	29.95	W
L75	87.62	235.58	2.69	26.26	30.21	56.54	W
L63	83.44	227.09	2.72	14.84	14.44	11.91	W
L74	98.25	255.24	2.60	67.58	60.05	47.88	W
L50	85.44	211.16	2.69	34.47	38.31	75.8	D
L51	108.62	255.48	2.35	46.86	59.97	60.05	D
P23	104.9	228.34	2.18	45.69	37.3	86.7	D
L80	99	231.5	2.34	20.98	25.26	32.35	D

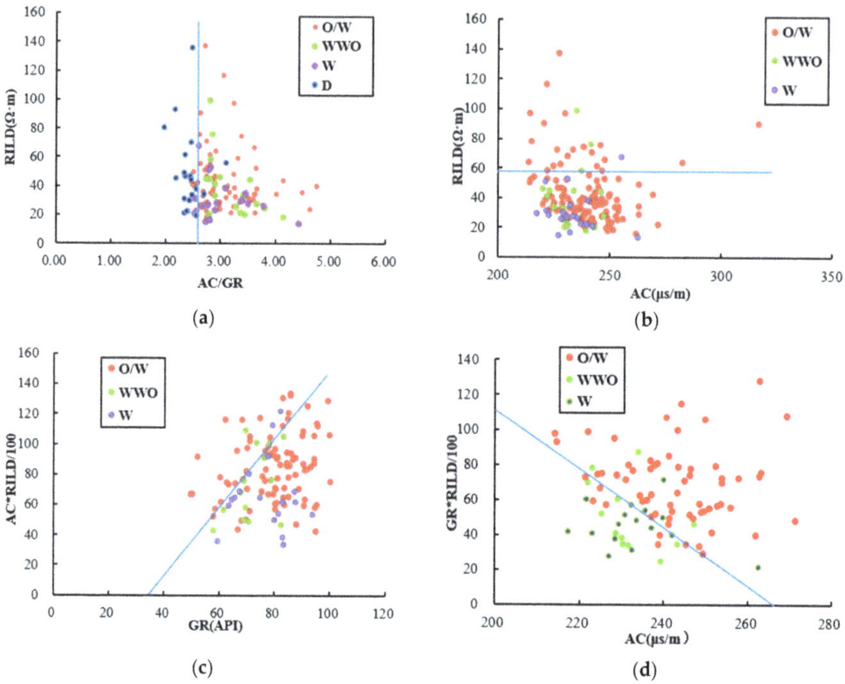

Figure 12. Logging fluid identification cross plot. (**a**) AC/GR-RILD cross plot. (**b**) AC-RILD cross plot. (**c**) GR-AC*RILD/100 cross plot. (**d**) AC-GR*RILD/100 cross plot.

In the first step, the AC/GR-RILD cross plot is generated, and if AC/GR < 2.54, the formation is dry, the dry layer of the tight sandstone reservoir is effectively identified, and the identified dry layer(D) data are removed.

In the second step, the remaining data are used to produce AC-RILD cross plot; if RILD ≥ 58 Ω·m, the formation is an oil–water layer (O/W), the fluid properties of some layers can be identified, and the identified oil–water layer data are removed.

In the third step, the remaining data are used to generate the GR-AC*RILD/100 cross plot; if AC*RILD/100 > 2.6*GR-91.31, the reservoir is an oil–water layer, and the reservoir data with identified fluid properties are removed again.

In the fourth step, the AC-GR*RILD/100 cross plot is made with the remaining data; if GR*RILD/100 ≥ 110.5–0.41*AC, the reservoir is an oil–water layer, and the remaining reservoirs are water layers (W) and water with an oil layer (WWO).

The oil–water layer identification template established in four steps is used to finalize a linear discrimination criterion for oil–water layers:

When AC/GR < 2.54, the reservoir is a dry layer.

Condition 1: RILD ≥ 58 Ω·m;

Condition 2: AC*RILD/100 ≥ 2.6*GR-91.31;

Condition 3: GR*RILD/100 ≥ 110.5–0.41*GR.

If any one of conditions 1–3 is satisfied, the reservoir is an oil–water layer; otherwise, the reservoir is a water layer.

According to the statistical statistics, it is concluded that the accuracy of the fluid property identification template or linear discriminating criterion established by the four-step method to discriminate the fluid properties of tight sandstone is over 93.29%.

5. Examples of Logging Interpretation Model Applications

Using the newly constructed lithology, porosity, permeability and saturation interpretation model and interpretation standard, 20 wells in the study area are secondarily interpreted, and the interpretation error of each parameter is less than 5%, with 95.23% agreement between the interpretation conclusion and the oil test conclusion (Table 3).

Table 3. Comparison of model interpretation results with actual interpretation findings.

Well	GR/ API	AC/ μs/m	RILD /Ω·m	POR /%	PERM /10^{-3} μm²	SW /%	Oil Test (m³)		Explanation of Conclusions	Explanatory Model			
							Oil	Water		Condition 1	Condition 2	Condition 3	Result
L108	62.29	238.09	22.04	13.37	0.36	58.96	2.04	19.86	O/W			yes	O/W
L34	78.12	232.58	38.03	10.35	0.52	55.32	1	2.6	O/W			yes	O/W
L80	75.29	246.61	58.46	14.03	1.47	30.72	1.83	1.84	O/W	yes			O/W
L96	70.42	251.82	42.41	15.63	1.99	33.35	1.58	0.73	O/W		yes		O/W
L110	78.22	253.54	33.64	15.95	0.95	37.64	0.87	1.78	O/W			yes	O/W
L121	78.08	253.00	21.89	15.37	1.66	50.96	3	0.4	O/W			yes	O/W
L128	53.95	244.83	32.30	15.60	1.19	39.50	2.7	5.63	O/W		yes		O/W
L71	80.61	245.25	36.15	12.83	0.78	45.17	1	5	O/W			yes	O/W
L92	64.25	231.17	13.27	11.36	0.16	95.10	/	/	W				W
L120	84.34	222.03	116.27	6.73	8.53	22.55	0.49	1.83	O/W	yes			O/W
P198	67.30	211.74	30.43	6.34	0.07	45.52	2	1	O/W			yes	O/W
P200	89.89	217.87	19.29	7.07	0.77	56.00	1.6	7	O/W			yes	O/W
P201	83.45	259.12	29.56	16.52	1.26	39.12	1.1	5.62	O/W			yes	O/W
U113	77.12	253.07	43.14	15.83	1.76	32.56	5	8	O/W			yes	O/W
U127	82.73	225.17	30.36	7.81	0.20	85.75	0	31.55	W			yes	O/W
U129	62.46	221.44	52.30	9.01	1.08	52.78	1.3	1.8	O/W		yes		O/W
Q1	85.89	216.10	25.89	6.66	0.28	60.15	1.4	2.88	O/W			yes	O/W
X105	62.22	233.50	44.04	12.12	0.74	42.76	5.9	9	O/W		yes		O/W
S1040	85.03	230.08	40.22	9.67	0.40	48.18	0.94	2.231	O/W			yes	O/W
S32	80.59	243.50	20.69	13.27	0.36	51.72	3.67	5.403	O/W			yes	O/W

Figure 13 shows a graph of the interpreted results for the L110 well. The interpreted porosity, permeability and oil-bearing saturation are close to the core analysis values. The logging depth is 1672–1680 m. The natural potential and natural gamma curve characteristics are consistent with the lithological characteristics of the sandy mudstone profile and the RLL8, RILM and RILD logging curve characteristics are consistent with the oil and water formations identified by the logging interpretation. Resistivity averages 32.7 Ω·m, porosity averages 9.96 PU, permeability is 0.87×10^{-3} μm² and oil saturation is 51.3%. It can be seen from the shot hole layer that after the layer was put into production, the initial production of liquid was 8.33 m³, oil production was 2.70 t, and water bearing was 67.59%, which is an oil and water layer, which is consistent with the interpretation results.

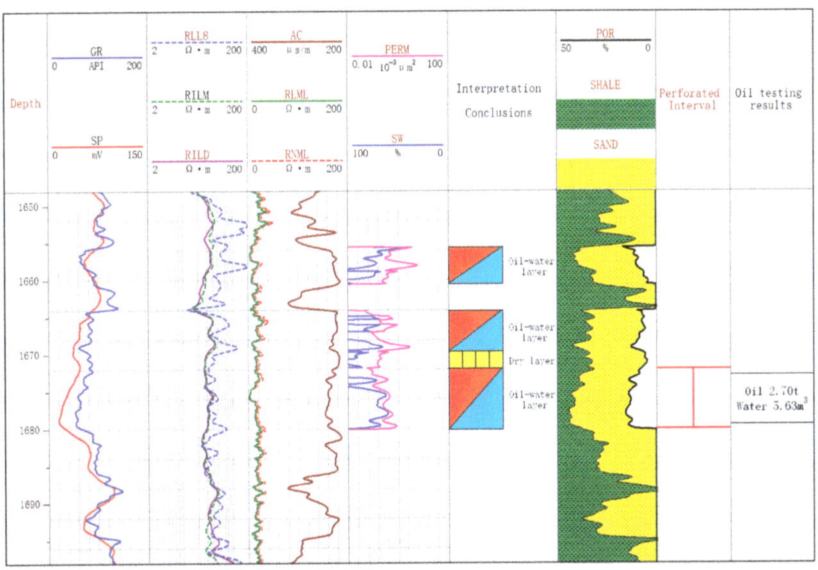

Figure 13. Comprehensive histogram of L110 logging interpretation.

6. Conclusions

(1) The investigation of the evaluation method for tight sandstone reservoirs in the lower assemblage of the south-central Ordos involves the exploration of the interrelationships among the "four properties". Based on this research, specialized models focusing on parameters such as the shale content, porosity, permeability, and saturation degree of tight sandstone reservoirs are developed or selected. These models demonstrate effective applicability to tight sandstones.

(2) For the problem of high shale content in tight sandstone reservoirs, GR and AC (SP and AC) are used to calculate the shale index, and the weighting method is effective.

(3) The established porosity classification method and acoustic time difference method for calculating oil saturation in tight sandstone reservoirs overcame the difficulty of resistivity reflecting the weakening of oil bearing and improved the accuracy of interpretation of oil saturation in reservoirs.

(4) The multi-information four-step method gradually recognizes the fluid characteristics of tight sandstone reservoirs and improves the compliance rate of log interpretation, which is applied to 20 wells in the block with a compliance rate of 95.23%, and lays the foundation for accurately establishing the interpretation standard of tight sandstone. This method is not only important for the development of tight sandstone reservoirs in the lower assemblage of the Ordos Basin but also for the identification of fluid properties of tight sandstone reservoirs in other blocks.

Author Contributions: Writing—original draft: B.X.; writing—review and editing: Z.W. and T.W.; methodology: B.X. and T.S.; validation: S.Z. and J.P.; formal analysis: T.S. and Y.C. All authors have read and agreed to the published version of the manuscript.

Funding: This research received no external funding.

Data Availability Statement: We state that the data are unavailable due to privacy or ethical restrictions of the company and university.

Conflicts of Interest: Authors Zhenhua Wang, Shuxia Zhang were employed by the company Research Institute of Shannxi Yanchang Petroleum (Group) Co., Ltd. Author Ting Song was employed by the Changqing Oilfield Company. The remaining authors declare that the research was conducted in the absence of any commercial or financial relationships that could be construed as a potential conflict of interest.

References

1. Zhou, L.H.; Chen, C.W.; Han, G.M.; Yang, F. Difference characteristics between continental shale and tight oil and exploration practice: A case from Huanghua Depression, Bohai Bay Basin. *Earth Sci.* **2021**, *46*, 555–571.
2. Huang, W.; Zhang, X.L.; Zhao, J.Y. Fluild Identification of Tight Reservoir in Chang 9 Formation, Ordos Basin. *J. Northwest Univ.* **2015**, *45*, 811–818.
3. Wen, B.; Zhou, Y. Study on fluid identification of low porosity and low permeability reservoir. *Petrochem. Ind. Technol.* **2017**, *24*, 126.
4. Albrecht, D.; Reitenbach, V. Laboratory measurements of fluid transport properties on tight gas sandstones and applications. *Energy Procedia* **2014**, *59*, 366–373. [CrossRef]
5. Bi, M.W.; Chen, S.Y.; Zhou, Z.H. Genesis of secondary pores of tight sandstone reservoir in He 8th Member in Su 6 area of Sulige gas field. *J. China Univ. Petrol.* **2015**, *39*, 8–16.
6. Guo, J.; Ling, Z.; Xu, X.; Zhao, Y.; Yang, C.; Wei, B.; Zhang, Z.; Zhang, C.; Tang, X.; Chen, T.; et al. Saturation Determination and Fluid Identification in Carbonate Rocks Based on Well Logging Data: A Middle Eastern Case Study. *Processes* **2023**, *11*, 1282. [CrossRef]
7. Bi, L.R.; Mao, Z.Q.; Xiao, C.W.; Liu, X.L. Application of normal distribution method in identifying oil (gas) and water zones. *J. Oil Gas Technol.* **2006**, *28*, 76–78.
8. Cheng, X.Z.; Zhou, C.C.; Fan, Y.R.; Wang, Z.H.; Huang, S.X. Genesis and identification of low contrast pay zones in fresh-water reservoirs. *J. Oil Gas Technol.* **2007**, *29*, 62–68.
9. Li, L.; Yuan, Z.; Hui, K.; Liu, S. Accumulation regularity of Upper Paleozoic gas in north Ordos Basin. *Oil Gas Geol.* **2000**, *21*, 268–271.

10. He, X.; Chen, S.; Hu, C.; Zhang, H.; Mou, F.; Dai, L.; Lu, Y.; Fu, X.; Han, M. Pore Structure Change in the Continental Shale Oil Reservoir and Its Main Influencing Factors: A Case Study of the Chang 7 Member in the Ordos Basin. *Processes* **2023**, *11*, 2314. [CrossRef]
11. Zhu, H.; Chen, K.; Liu, K.; He, S. A sequence stratigraphic model for reservoir sand-body distribution in the Lower Permian Shanxi Formation in the Ordos Basin, northern China. *Mar. Pet. Geol.* **2008**, *25*, 731–743. [CrossRef]
12. Chen, Z.; Li, X.; Chen, H.; Duan, Z.; Qiu, Z.; Zhou, X.; Hou, Y. The Characteristics of Lithofacies and Depositional Model of Fine-Grained Sedimentary Rocks in the Ordos Basin, China. *Energies* **2023**, *16*, 2390. [CrossRef]
13. Zheng, W.; Hu, X.; Chen, S.; Liu, J.; Jia, C. Characteristics of sedimentary evolution in the Upper Paleozoic Daniudi Gasfield, Ordos Basin. *Acta Sedimentol. Sin.* **2015**, *33*, 306–313.
14. Mao, K.Y. Logs fluid typing methods and adaptive analysis of tight sandstone reservoir of Yingcheng formation in Lishu Fault. *Adv. Earth Sci.* **2016**, *31*, 1056–1066.
15. Jamil, M.; Siddiqui, N.A.; Rahman, A.H.B.A.; Ibrahim, N.A.; Ismail, M.S.B.; Ahmed, N.; Usman, M.; Gul, Z.; Imran, Q.S. Facies Heterogeneity and Lobe Facies Multiscale Analysis of Deep-Marine Sand-Shale Complexity in the West Crocker Formation of Sabah Basin, NW Borneo. *Appl. Sci.* **2021**, *11*, 5513. [CrossRef]
16. Jamil, M.; Siddiqui, N.A.; Usman, M.; Wahid, A.; Umar, M.; Ahmed, N.; Haq, I.U.; El-Ghali, M.A.K.; Imran, Q.S.; Rahman, A.H.A.; et al. Facies analysis and distribution of Late Palaeogene deep-water massive sandstones in submarine-fan lobes, NW Borneo. *Geol. J.* **2022**, *57*, 4489–4507. [CrossRef]
17. Sun, J.M.; Wei, X.H.; Chen, X.L. Fluid identification in tight sandstone reservoirs based on a new rock physics model. *J. Geophys. Eng.* **2016**, *13*, 526–535. [CrossRef]
18. Wang, F.; Yang, X.M.; Zhang, Y.H.; Bian, H.Y. Application of multi-algorithmic fusion methods to fluid identity in tight reservoir. *Prog. Geophys.* **2015**, *30*, 2785–2792.
19. He, M.; Gu, H.M.; Wan, H. Log interpretation for lithology and fluid identification using deep neural network combined with MAHAKIL in a tight sandstone reservoir. *J. Pet. Sci. Eng.* **2020**, *194*, 107498. [CrossRef]
20. Li, H.; Tang, H.M.; Qin, Q.R.; Zhou, J.L.; Qin, Z.J.; Fan, C.H.; Su, P.; Wang, Q.; Zhong, C. Characteristics, formation periods and genetic mechanisms of tectonic fractures in the tight gas sandstones reservoir: A case study of xujiahe formation in YB area, sichuan basin, China. *J. Pet. Sci. Eng.* **2019**, *178*, 723–735. [CrossRef]
21. Li, B.; Tan, M.J.; Zhang, H.T. Interpretation method of nuclear magnetic resonance dual-TW logging in oil-wet tight sandstone reservoirs. *Appl. Geophys.* **2020**, *5*, 796–808. [CrossRef]
22. Chen, J.; Xie, R.C.; Liu, C.C.; Duan, Y.; Li, H.; Xiang, Y. Logging fluid identification and quantitative evaluation of Jurassic tight sandstone gas reservoir in Zhongjiang gas field. *Nat. Gas Ind.* **2019**, *39*, 136–141.
23. Fu, L.; Qin, Z.; Xie, A.; Chen, L.; Li, J.; Wang, N.; Qin, Q.; Mao, K. The relation of the "four properties" and fluid identification of the carboniferous weathering crust volcanic reservoir in the Shixi Oilfield, Junggar Basin, China. *Front. Earth Sci.* **2022**, *10*, 983572. [CrossRef]
24. Zou, C.; Zhu, R.; Wu, S.; Yang, Z.; Tao, S.; Yuan, X. Types, characteristics, genesis and prospects of conventional and unconventional hydrocarbon accumulations: Taking tight oil and tight gas in China as an instance. *Acta Pet. Sin.* **2012**, *33*, 173–187.
25. Dai, J.; Ni, Y.; Wu, X. Tight gas in China and its significance in exploration and exploitation. *Pet. Explor. Dev.* **2012**, *39*, 277–284. [CrossRef]
26. Guo, X.; Liu, K.; Jia, C.; Song, Y.; Zhao, M.; Zhuo, Q.; Lu, X. Hydrocarbon accumulation processes in the Dabei tight-gas reservoirs, Kuqa Subbasin, Tarim Basin, northwest China. *AAPG Bull.* **2016**, *100*, 1501–1521. [CrossRef]
27. Wu, H.; Zhao, J.; Wu, W.; Li, J.; Huang, Y.; Chen, M. Formation and diagenetic characteristics of tight sandstones in closed to semi-closed systems: Typical example from the Permian Sulige gas field. *J. Pet. Sci. Eng.* **2020**, *199*, 108248. [CrossRef]
28. Shaogui, D.; Yiren, F.; Guoxin, L. Influence of formation water salinity on physical properties of argillaceous sandstone. *Well Logging Technol.* **2006**, *30*, 113–115.
29. Haiyong, M.; Lifa, Z.; Xiuqin, D. Hydrochemical characteristics and geological significance of the Chang 8 formation of the Yanchang Formation in Jiyuan area, Ordos Basin. *J. Northwest Univ.* **2013**, *43*, 253–257.
30. Fan, R.; Zhou, L.; Wu, J.; Zeng, T.; Zhou, X.F. Research on tight sandstone reservoir fluids identification in Xujiahe formation, northeastern Sichuan Basin. *Petrol. Geol. Recov. Effic.* **2015**, *22*, 67–71.
31. Mode, A.W.; Anyiam, O.A.; Aghara, I.K. Identification and petrophysical evaluation of thinly bedded low-resistivity pay reservoir in the Niger Delta. *Arab. J. Geosci.* **2014**, *8*, 2217–2225. [CrossRef]
32. Xu, Z.; Jiang, S.; Liu, L.; Wu, K.; Li, R.; Liu, Z.; Shao, M.; Jia, K.; Feng, Y. Natural gas accumulation processes of tight sandstone reservoirs in deep formations of Songliao Basin, NE China. *J. Nat. Gas Sci. Eng.* **2020**, *83*, 103610. [CrossRef]
33. Wang, Q.; Chen, D.; Wang, F.; Gao, X.; Zou, Y.; Tian, Z.; Li, S.; Chang, S.; Yao, D. Origin and distribution of an under-pressured tight sandstone reservoir: The Shaximiao Formation, Central Sichuan Basin. *Mar. Petrol. Geol.* **2021**, *132*, 121–128. [CrossRef]
34. Zhou, X.; Xu, G.; Cui, H.; Zhang, W. Fracture development and hydrocarbon accumulation in tight sandstone reservoirs of the Paleogene Huagang Formation in the central reversal tectonic belt of the Xihu Sag, East China Sea. *Pet. Explor. Dev.* **2020**, *47*, 499–512. [CrossRef]
35. Jia, J.; Yin, W.; Qiu, N.; Wang, G.; Ma, L.; Liu, Y.; Liu, N. Migration and accumulation of crude oil in Upper Triassic tight sand reservoirs on the southwest margin of Ordos Basin, Central China: A case study of the Honghe Oilfield. *Geol. J.* **2017**, *53*, 2280–2300. [CrossRef]

36. Yang, H.; Fu, J.; Wei, X.; Liu, X. Sulige field in the Ordos Basin: Geological setting, field discovery and tight gas reservoirs. *Mar. Pet. Geol.* **2008**, *25*, 387–400. [CrossRef]
37. He, Y.H.; Yang, X.; Wang, X.J.; Zhong, Z.Q.; Hou, D.J. Fluid identification in low permeability reservoir in Jiyuan area, Ordos Basin. *China Petrol. Explor.* **2016**, *21*, 110–115.
38. Wyllie, M.R.J.; Gregory, A.R.; Gardner, L.W. Elastic wave velocities in heterogeneous and porous media. *Geo-Physics* **1956**, *21*, 41–70. [CrossRef]
39. Raymer, L.L.; Hunt, E.R.; Gardner, J.S. An improved sonic transit time-to-porosity transform. In Proceedings of the SPWLA 21st Annual Logging Symposium, Lafayette, LA, USA, 8 July 1980; SPWLA-1980-P. OnePetro: Richardson, TA, USA, 1980.
40. Martin, J.P.; Nicoletis, S.; Raiga-Clemenceau, J. The concept of acoustic formation factor for more accurateporosity determination from sonic transit time data. *Log Anal.* **1988**, *29*, SPWIA-1988-v29n1a4.
41. Gao, J.; Ma, B.; Lu, Y.; Zhang, W.; Cao, Q. Origin of authigenic kaolinite with implications for Permian tight gas sandstone reservoirs in the northern Ordos Basin, central China. *J. Nat. Gas Sci. Eng.* **2022**, *99*, 104429. [CrossRef]
42. Wang, G.; Chang, X.; Yin, W.; Li, Y.; Song, T. Impact of diagenesis on reservoir quality and heterogeneity of the Upper Triassic Chang 8 tight oil sandstones in the Zhenjing area, Ordos Basin, China. *Mar. Pet. Geol.* **2017**, *83*, 84–96. [CrossRef]
43. Zhao, G.P. Characterization of fluid inclusions and timing of gas accumulation in Upper Paleozoic reservoirs of Hangjinqi area, Ordos Basin. *Oil Gas Geol.* **2017**, *38*, 905–912.
44. Archie, G.E. The electrical resistivity log as an aid in determining some reservoir characteristics. *Trans. Am.* **1942**, *146*, 41–56. [CrossRef]
45. Liu, Y.; Ye, J.; Zong, J.; Wang, D.; Cao, Q.; Yang, B.; Li, W.; Zhao, J. Analysis of forces during tight oil charging and implications for the oiliness of the tight reservoir: A case study of the third member of the Palaeogene Shahejie Formation, Qibei slope, Qikou sag. *Mar. Pet. Geol.* **2022**, *144*, 105819. [CrossRef]
46. He, Y.; Guo, H.; Lan, H.; Ling, C.; Fu, M. The Effect of Single Sandstone Stacking Pattern on the Sandstone Reservoir Physical Properties—A Case Study from the Shanxi Formation in the Daniudi Area, Northeastern Ordos Basin. *Energies* **2022**, *15*, 4740. [CrossRef]
47. Zhao, G.; Li, X.; Liu, M.; Dong, C.; Chen, D.; Zhang, J. Reservoir Characteristics of Tight Sandstone and Sweet Spot Prediction of Dibei Gas Field in Eastern Kuqa Depression, Northwest China. *Energies* **2022**, *15*, 3135. [CrossRef]
48. Wang, R.; Sun, W. A Study on Micro Cracks in Super Low Permeability Sandstone Reservoir of the Upper Triassic Yanchang Formation in the Ordos Basin. *Geol. Rev.* **2009**, *55*, 444–448.

processes MDPI

Article

Prediction of Longitudinal Superimposed "Sweet Spot" of Tight Gas Reservoir: A Case Study of Block G, Canada

Yuepeng Jia *, Wensong Huang, Ping Wang, Penghui Su, Xiangwen Kong , Li Liu and Yunpeng Shan

PetroChina Research Institute of Petroleum Exploration and Development, Beijing 100083, China
* Correspondence: jia1996@petrochina.com.cn; Tel.: +86-178-1316-5875

Abstract: In this paper, taking Block G in Canada as an example, combined with the data of the working area, the Pearson–MIC comprehensive evaluation method was adopted to optimize the key parameters of productivity. Based on the analytic hierarchy process, the weight of each parameter was calculated, the grade of evaluation index of the "sweet spot" was divided, the standard of the sweet spot was established, and the distribution of the superimposed sweet spot was finally depicted. The results show that lateral length, number of stages, volume of fluid, and amount of proppant are the key engineering parameters of horizontal well, and lateral length is an independent key engineering parameter. The cumulative gas production in the first two years was normalized on the lateral length to eliminate the engineering influence, and the total organic carbon (TOC) was finally determined as the key geological parameter, whereas porosity and water saturation were the secondary key parameters. The area of Type I sweet spots accounts for 24.2% in the Series Upper and 23.1% in the Series Lower. This study proposed a new sweet spot prediction idea based on the influence of geological factors on productivity, and its results also laid a foundation for the subsequent placement of horizontal wells in Block G.

Keywords: longitudinal superimposed sweet spot; key parameters; Pearson correlation coefficient; MIC; tight gas reservoir

Citation: Jia, Y.; Huang, W.; Wang, P.; Su, P.; Kong, X.; Liu, L.; Shan, Y. Prediction of Longitudinal Superimposed "Sweet Spot" of Tight Gas Reservoir: A Case Study of Block G, Canada. *Processes* **2023**, *11*, 666. https://doi.org/10.3390/pr11030666

Academic Editors: Tao Zhang, Zheng Sun, Dong Feng, Hung Vo Thanh and Wen Zhao

Received: 2 January 2023
Revised: 11 February 2023
Accepted: 21 February 2023
Published: 22 February 2023

1. Introduction

The term "sweet spot" refers to an oil and gas enrichment area that can be preferentially explored and developed under existing economic and technical conditions [1–3]. There are three prediction methods: (1) identification of favorable reservoirs through seismic inversion technology; (2) comprehensive analysis and prediction of a sweet spot based on geological research of reservoir formation, diagenesis, sedimentation, and sequence; and (3) prediction of a sweet spot with mathematical analysis methods by analyzing the influence of reservoir physical property, reserve abundance, reservoir thickness, and other formation parameters on productivity based on existing data. With the gradual accumulation of production materials and the deepening of reservoir understanding, the "static–dynamic method" is widely adopted, which use dynamic data to screen the main control parameters and combining geological models to predict the sweet spot.

Sweet spots are divided into geological sweet spots and engineering sweet spots. The geological sweet spot manifests higher yield under the same fracturing technology, and its main parameters include porosity, permeability, gas saturation, reservoir thickness, total organic carbon, vitrinite reflectance, etc. The engineering sweet spot refers to the area conducive to fracturing construction, and its main parameters include brittle mineral content, brittleness index, fracture pressure, pore pressure, crustal stress, etc. [4–8]. Scholars have done a lot of research on the evaluation and prediction of the sweet spot. Pan et al. (2016) elaborated on the geological conditions of the sweet spot of shale gas reservoirs based on a large amount of data [9]. Chen et al. (2018) proposed a set of multivariate statistical methods to identify potential sweet spots based on the relationship between the production of tight oil production wells and geological factors [10]. Wei et al. (2021) quantitatively evaluated the comprehensive sweet spot of shale oil by establishing the evaluation function of sweet spot characterization parameters and calculating the comprehensive weight factor based on the BP neural network, Eaton method, and other methods [11]. Chen et al. (2016) raised the concept of shale oil and gas vertical sweet spots by comprehensively considering various factors [12]. Zhou et al. (2020) analyzed the key geological parameters of tight sandstone reservoir, and proposed the sweet spot model for qualitative identification of factor analysis and quantitative evaluation criteria [13]. Li et al. (2020) established the horizontal well section optimization model by using the combination weight method in the form of normalized index value and weight product [14]. Han et al. (2019) used RGB color fusion method to get the prediction result of sweet spot for five key attributes [15]. Liao et al. (2019) qualitatively analyzed the influence of geological and engineering sweet spot parameters on gas production, and quantitatively characterized the sweet spot coefficient by using the radar area model and weight model [16,17]. Montgomery et al. (2005) uses seven parameters for evaluation: total organic carbon, original oil and gas potential, vitrinite reflectance, content and composition of natural gas, formation temperature, and hydrocarbon conversion ratio [18]. Chalmers and Bustin (2007) used total organic carbon, organic matter type, vitrinite reflectance, mineral composition, water saturation, pore diameter, gas adsorption capacity, porosity, and permeability when evaluating the Cretaceous shale of British Columbia in the Western Canada Basin [19]. Curtis (2002) selects reservoir thickness, mineral, total organic carbon, organic matter type, vitrinite reflectance, permeability, pressure, brittle mineral content, and other parameters to optimize favorable areas in Antrim, Ohio, New Albany, Barnett, and Lewis and other blocks [20]. Pejman et al. (2017) analyzed log data through two methods of data mining and machine learning technology to identify areas containing high total organic carbon (TOC) and brittle rocks that are prone to fracture [21]. Heege et al. (2015) evaluated the underexplored shale and mapped out sweet spots across shale basins based on the properties of North American shale (Marcellus, Bakken, Haynesville, and Barnett), i.e., vitrinite reflectance, porosity, depth, thickness, absorbed gas content, and brittleness index [22]. Munther et al. (2019) simplified the complexity of unconventional resource evaluation and evaluated the oil and gas resource prospects of Goldwyer by considering the parameters of total organic carbon, total porosity, water saturation and brittleness index, and depicting the average map of different attributes in a weighted way [23]. Sahoo et al. (2020) integrated data analytics and fluid geochemical data (mainly Ph/nC18 and Pr/nC17) to map the sweet spots within a producing shale play of North America under the limited conditions of core data, logging data, and seismic data [24]. Tang et al. (2020) illustrated a new tool that uses gradient-boosting decision trees (GBDTs) and ensemble models to classify reservoir quality and identify logging curves more effectively, which further helps search and identify sweet spots automatically [25]. Yuan et al. (2022) applied advanced well logging and slim-compact microimager (SCMI) technologies to conduct high-resolution quantitative analysis of the Goldwyer Formation in the Canning Basin, Western Australia, from in situ macroscopic scale to laboratory microscale, and believed that the most favorable lithofacies in the sweet spot develop numerous organic matter/mineral nanopores for hydrocarbon storage from a microscopic view [26]. Nelson et al. (2022) proposed a workflow to quantify completion quality (CQ) and reservoir quality (RQ), and offered a method by which to compare the predicted sweet spots against available production data via their correlation coefficient, and finally resulted reasonable formation property maps, and the successful identification of sweet spots [27]. At present, the prediction of sweet spot is generally based on simple geological research without considering the factors of productivity. This paper takes the Upper Montney Formation of Block G in Canada as an example to analyze the impact of various geological parameters on the productivity of horizontal wells through Pearson correlation coefficient and maximal information coefficient (MIC) method, optimizes the key parameters, and finally determines the weight of each parameter and predicts the sweet spot by using the analytic hierarchy process. This study can not only provide technical support for efficient development of Block G in Canada, but also provide reference for similar gas fields.

2. Geology

Block G in Canada is located in the western edge of the Western Canada Basin. Three sets of strata were adopted from bottom to top, namely the Triassic, Jurassic, and Cretaceous, in which the oil-bearing strata mainly include the Triassic Montney Formation, Doig Phosphate Formation, and Cretaceous Cadomin Formation. The Triassic Montney Formation is divided into Upper Montney and Lower Montney, wherein the Upper Montney Formation is the target layer of this study and belongs to the "marine" dense siltstone deposits on the passive continental margin. The sedimentation center of sandstone migrates from northeast to southwest, and the sandstone is stacked forward on the plane, belonging to thick massive tight sandstone. Its thickness is about 150~200 m, and the altitude is −1950 m~−2880 m. The stratum shows the characteristics of multicycle sandstone progradation filling mode, which is subdivided into seven layers, AA and A~F, from bottom to top [28]. At present, two sets of development layer series are adopted: Series Upper includes three layers (F, E, D), and Series Lower includes four layers (C, B, A and AA) (Figure 1).

The content of brittle minerals such as quartz and calcite in the Upper Montney Formation is relatively high, accounting for approximately 40% of the total mineral components. The pores of rock particles are small and the degree of diagenesis is high. The mudstone particles are silt with 20~50 μm particles relatively concentrated according to Udden–Wentworth classification (Table 1) [29], and the pore types are mainly intergranular pore and intercrystalline pore. The extremely fine sandstone has relatively high content of argillaceous and organic matter between grains. The gamma ray (GR) curve shows a certain sawtooth shape, indicating that it has obvious characteristics of "self-source and self-reservoir", strong hydrocarbon generation capacity, and adsorbed gas, similar to shale [30]. The hydrogen index (HI) sample value of the reservoir ranges from 25 to 50, and the type of organic matter belongs to type II with high organic matter abundance. The TOC is generally between 2% and 4%.

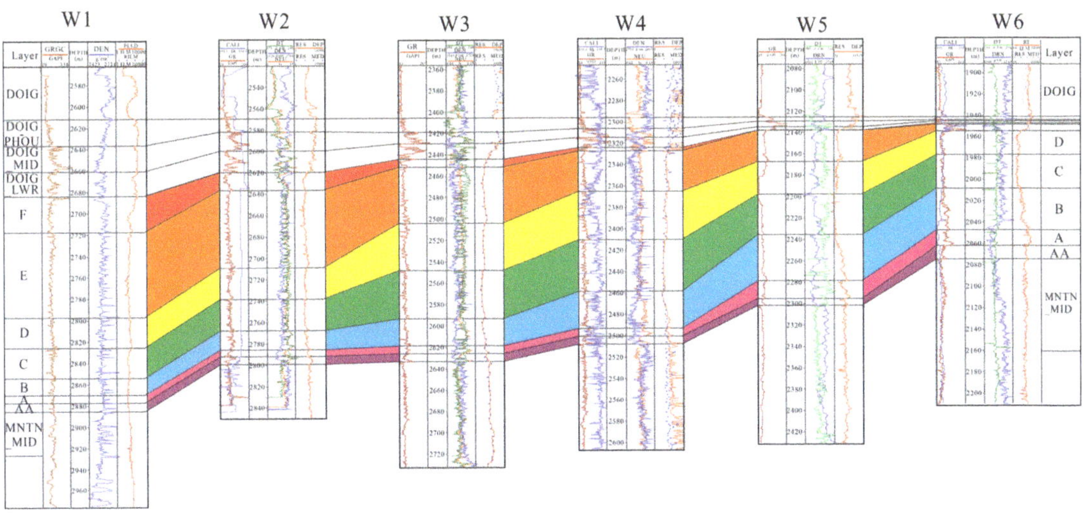

Figure 1. Cross section of Upper Montney Formation in the Western Canada Basin.

Table 1. Udden–Wentworth Scale.

Size (mm)	Particels	Rocks
>256	Bowlder gravel	Gravelstone
5~256	Cobble gravel	
4~64	Pebble gravel	
2~4	Granule gravel	
1~2	Very coarse sand	Sandstone
1/2~1	Coarse sand	
1/4~1/2	Medium sand	
1/8~1/4	Fine sand	
1/16~1/8	Very fine sand	
1/256~1/16	Silt	Mudstone
<1/256	Clay	

3. Methodology

3.1. Pearson–MIC Comprehensive Evaluation Method

Pearson correlation coefficient is used to measure the degree of correlation between two variables. Its value is between −1 and 1, reflecting the linear relationship between variables and the direction of correlation. The greater the absolute value of correlation coefficient, the stronger the correlation. Positive values represent positive correlation and negative values represent negative correlation. It is generally believed that the relationship between the absolute value of the obtained correlation coefficient and the correlation strength is as follows: [0.0, 0.2), uncorrelated or very weak correlation; [0.2, 0.4), weak correlation; [0.4, 0.6), medium intensity correlation; [0.6, 0.8), strong correlation; and [0.8, 1.0], extremely strong correlation.

When two groups of variables with linear relationships are continuous and have normal distribution or unimodal symmetric distribution close to normal distribution, the correlation between these two groups of variables can be expressed by product difference correlation coefficient. The formula is

$$r = \frac{\sum_{i=1}^{n}(X_i - \overline{X})(Y_i - \overline{Y})}{\sqrt{\sum_{i=1}^{n}(X_i - \overline{X})^2}\sqrt{\sum_{i=1}^{n}(Y_i - \overline{Y})^2}} \tag{1}$$

In the formula, r is the Pearson correlation coefficient, $X_i, \overline{X}$ is a parameter value of group X variable and its average value, and $Y_i, \overline{Y}$ is a parameter value of group Y variable and its average value.

The maximal information coefficient (MIC), is used to measure the linear or nonlinear correlation between two variables X and Y. The basic mathematical principle is: for any two variables X and Y in the sample data set D, the sample size in the sample data set is N, and the two variables have i and j possible values respectively, so that the correlation map of the two variables can be divided into $i \times j$, and then calculate the maximum information coefficient MIC by the following formula:

$$MIC(X;Y|D) = \max_{ij<B(N)} \frac{I^*(X,Y,D,i,j)}{\log[\min(i,j)]} \tag{2}$$

$$I^*(X,Y,D,i,j) = \max[I(X,Y,D|_G,i,j)] \tag{3}$$

$$B(N) = N^{0.6} \tag{4}$$

In the formula, I^* is the MIC of X, Y in the $i \times j$ grid, and N is the number of samples.

The Fisher-z transformation is a method used in statistics to test hypotheses of correlation coefficient, and its formula is

$$z = \frac{1}{2} In\left(\frac{1+r}{1-r}\right) = \operatorname{arctan}h(r) \tag{5}$$

After getting z, calculate the standard deviation of z (SDz) and select the significance threshold α, and check the standard deviation of the corresponding mean value. This article takes α as 0.95, and the corresponding inspection value is 1.95. From this, the upper limit Max and lower limit Min are calculated. If the upper and lower limits are on the same side of zero, the sample is statistically significant. We have

$$Max = z + (1.96 \times SDz) \tag{6}$$

$$Min = z - (1.96 \times SDz) \tag{7}$$

$$SDz = \frac{1}{(N-3)^2} \tag{8}$$

3.2. Analytic Hierarchy Process

The analytic hierarchy process (AHP) refers to a systematic method that takes a complex multiobjective decision-making problem as a system, decomposes the objective into multiple objectives or criteria, and then into several levels of multiple indicators, and calculates the hierarchical single ranking and total ranking through the qualitative index fuzzy quantitative method to serve as the objective and multiple scheme optimization decision-making [31]. The specific steps include establishing the hierarchical structure model, constructing the pairwise matrix, hierarchical single ranking and its consistency check, hierarchical total ranking and its consistency check.

According to the elements included in the analysis system and the relationship between the elements, a hierarchical structure model is established, which is generally divided into target layer, scheme layer and criterion layer. Then a pairwise matrix is constructed: suppose there are n factors in a layer, compare their influence on a criterion in the previous layer, and determine the proportion of the layer to a criterion. The above comparison is between two factors, i.e., $a_{ij} = b_i/b_j$. We have

$$A = (a_{ij})_{n \times n} = \begin{pmatrix} a_{11} & \cdots & a_{1n} \\ \vdots & \ddots & \vdots \\ a_{n1} & \cdots & a_{nn} \end{pmatrix} \tag{9}$$

Hierarchical single ranking is the process of determining the influence degree of each factor in the next layer on a factor in the previous layer, that is, the process of solving the pairwise matrix constructed above, generally including iterative method, sum product method, and square root method. The square root method is introduced here.

(1) Multiply the elements of A by rows: $\mu_{ij} = \prod b_{ij}$.
(2) Take the n root of the resulting product: $\mu_i = \sqrt[n]{\mu_{ij}}$.
(3) Normalize the root vector to obtain the eigenvector W: $W = \mu_{ij}/\sum \mu_i$.
(4) Calculate the maximum eigenroot of the pairwise matrix:

$$\lambda_{\max} = \sum [(AW)_i/(nW)_i]$$

In practice, the pairwise matrix is required to meet the general consistency, and a consistency check is required. Only through inspection can it be explained that the pairwise matrix is logically reasonable, and the analysis of the results can continue. The steps of consistency inspection are as follows.

(1) Calculate consistency index CI: $CI = (\lambda_{\max} - 1)/(n-1)$.
(2) Look up the table to determine the corresponding average random consistency index RI (RI is the average of the maximum eigenvalue of a certain number of n-order random matrices and its value is only related to the order (n) of the matrix).

(3) Calculate the consistency ratio CR and make judgment: $CR = CI/RI$.

When $CR < 0.1$, it is considered that the consistency of the pairwise matrix is acceptable. When $CR > 0.1$, it is considered that the pairwise matrix does not meet the consistency requirements, and the pairwise matrix needs to be revised.

4. Analysis on Key Parameters of Horizontal Well Productivity in Tight Gas Reservoir

4.1. Interfering Factor of Horizontal Well Productivity in Tight Gas Reservoir

There are many factors affecting the productivity of horizontal wells in tight gas reservoirs, and these are mainly divided into geological factors and engineering factors. Geological factors include permeability, reservoir thickness, porosity, water saturation, total organic carbon (TOC), brittle mineral content, crustal stress, etc. Engineering factors include lateral length, fracturing technology, number of stages, volume of fracturing fluid, amount of proppant, etc.

Unconventional reservoir development requires large-scale hydraulic fracturing, and the role of permeability is limited, and the data scattering about it and the cumulative production of horizontal wells in the first two years is disordered, without an obvious linear relationship, which has no obvious correlation with the productivity of horizontal wells. Reservoir thickness is an important condition for sandstone to have sufficient reservoir space, but the data scattering about it and horizontal well productivity is triangular, indicating that there is no correlation between them. Porosity is the key index to determine the abundance, and the overall scattering about it and production in the first two years is positively correlated, which has a certain impact on the productivity of horizontal wells. Water saturation is negatively correlated with horizontal well productivity. Total organic carbon (TOC) refers to the carbon content in the residual organic matter after oil and gas spill in the source rock, which is the material basis for generation of oil and gas, and the data scattering between it and production in the first two years is positively correlated, indicating that it has a certain correlation with the productivity of horizontal wells (Figure 2).

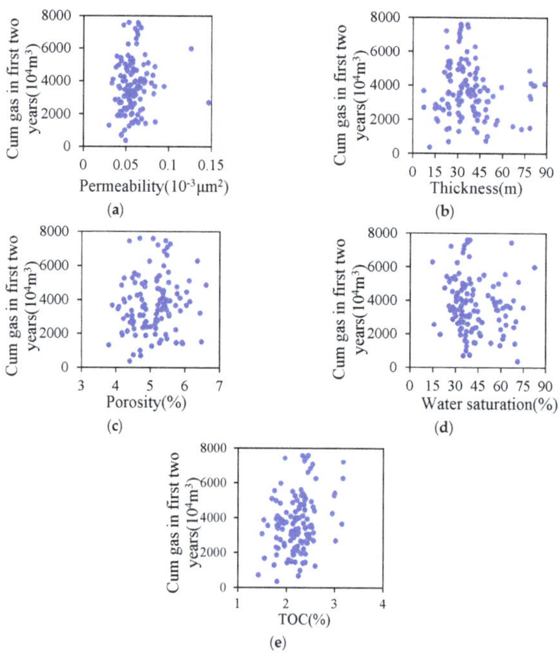

Figure 2. Relationship between cumulative gas production of horizontal well in the first two years and geological parameters. (**a**) Permeability; (**b**) thickness; (**c**) porosity; (**d**) water saturation; (**e**) total organic carbon (TOC).

Under the same conditions, the longer the lateral length of a horizontal well is, the more well controlled reserves are, and the higher the production is. The data scattering of lateral length of horizontal wells and cumulative production in the first two years shows a good positive linear correlation. Similarly, the greater the number of stages, the number of clusters, volume of fracturing fluid and amount of proppant, the larger the reconstruction volume formed by fracturing, the stronger the conductivity of artificial fractures, and the higher the production of wells. These parameters have obvious positive correlation with cumulative production in the first two years. The relationship between flowback rate and productivity is complex. In matrix, flowback rate and productivity are negatively correlated. In fractures, they are positively correlated. In addition, slippage effect, stress sensitivity, production system, etc. will also affect their correlation. The data scattering of flowback rate and productivity is disorder, indicating that there is no single correlation between them (Figure 3).

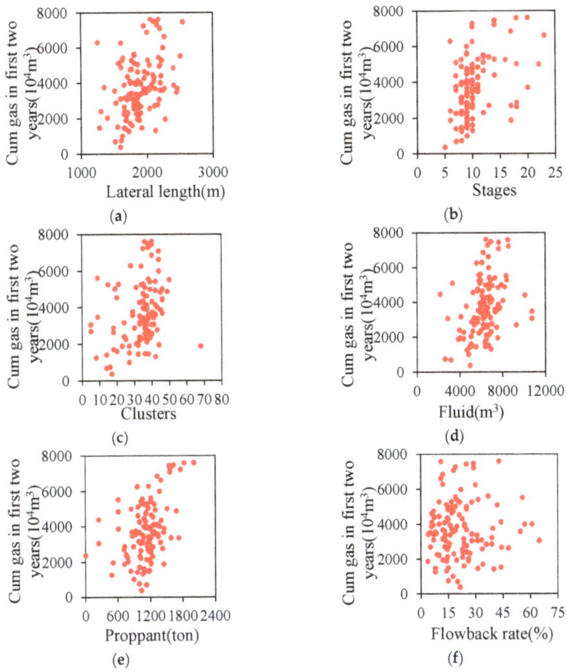

Figure 3. Relationship between cumulative gas production of horizontal well in the first two years and engineering parameters. (**a**) Lateral length; (**b**) number of stages; (**c**) number of clusters; (**d**) volume of fracturing fluid; (**e**) amount of proppant; (**f**): flowback rate.

4.2. *Analysis on Engineering Parameters of Horizontal Wells in Tight Gas Reservoirs*

The physical property distribution of the same layer in Block G is relatively stable. In order to reduce the influence of geological factors, the correlation between the cumulative gas production in the first two years of horizontal wells and various engineering parameters is calculated by layers (the purpose of calculation for each layer is only to eliminate the influence of geological factors on engineering factors, and then analyze the influence of engineering parameters on horizontal well productivity. The results of each layer are consistent in theory.). In order to eliminate the impact caused by the difference of fracturing technology, horizontal wells with similar fracturing technology are selected that were put into production from 2010 to 2013. Fisher's z-transform shows that there are few wells in Layer AA, A, B, C, D, and F, and the samples do not obey the normal distribution and have no statistical significance, so they are not considered. There are most wells in Layer E, which is statistically significant. Moreover, the impact of absolute engineering factors on productivity is not affected by geological factors, so take Layer E as an example to analyze the impact of engineering parameters on productivity (Table 2).

Table 2. Correlation between cumulative gas production in the first two years of horizontal wells in Layer E and engineering parameters.

Engineering Parameters	Min	Max	Mean	Pearson	MIC
Lateral length (m)	1400.15	2268.06	1853.65	0.6338	0.6466
Stages	7	23	11	0.5190	0.5205
Clusters	14	68	34	0.3259	0.4873
Fluid (m^3)	2606.70	8583.36	6167.47	0.6217	0.6503
Proppant (ton)	791.00	2008.00	1225.96	0.5996	0.6149
Flowback rate (%)	3.78	61.00	22.80	−0.0674	0.1455

These engineering parameters are selected as the key engineering parameters whose Pearson correlation coefficient and MIC are greater than 0.5 with the cumulative gas production in the first two years of the horizontal wells. Finally, the optimized key engineering parameters are lateral length, number of stages, volume of fracturing fluid, and amount of proppant.

4.3. *Analysis on Geological Parameters of Horizontal Wells in Tight Gas Reservoirs*

In order to analyze the impact of geological parameters on the productivity of horizontal wells, it is necessary to eliminate the impact of engineering factors on the productivity of horizontal wells. Therefore, the cumulative gas production in the first two years of horizontal wells is normalized to the optimized engineering parameters.

The engineering parameters are mutually independent and mutual influence. For example, the longer the lateral length of horizontal well is, the more the number of stages may be. The greater the volume of fracturing fluid is, the greater the amount of proppant may be. Through cluster analysis, it is found that the correlation coefficient between the lateral length and the number of stages, volume of fracturing fluid, and amount of proppant is greater than 0.5, so the lateral length is determined as an independent key engineering parameter (Table 3).

Table 3. Correlation between each key engineering parameter.

	Lateral Length	Stages	Fluid	Proppant
Lateral length	-			
Stages	0.5160	-		
Fluid	0.5045	0.3657	-	
Proppant	0.5018	0.4461	0.5407	-

The cumulative gas production in the first two years of the horizontal well is normalized on the lateral length, and the Pearson correlation coefficient and MIC between normalized production and various geological parameters are calculated. It is found that the correlation between the normalized cumulative gas production in the first two years and various geological parameters is weak (Table 4). Therefore, the requirements for correlation are reduced. The key geological parameter is the parameter whose Pearson correlation coefficient and MIC are greater than 0.3, and the secondary key parameter is the parameter whose Pearson correlation coefficient or MIC is greater than 0.3. Finally, TOC is determined as the key geological parameter, and porosity and water saturation are the secondary key parameters.

Table 4. Correlation between geological parameters and normalized cumulative gas production in the first two years.

	Permeability	Thickness	Porosity	Water Saturation	TOC
Pearson	0.1225	−0.0834	0.2387	−0.2034	0.3068
MIC	0.1701	0.1207	0.3261	0.3053	0.3928

5. Prediction of Longitudinal Superimposed Sweet Spot of Tight Gas Reservoir

The steps to predict the longitudinal superimposed sweet spot based on the analytic hierarchy process include: establishing hierarchical structure model, constructing a pairwise matrix, hierarchical ranking and consistency check, grading of evaluation indicators, and comprehensive evaluation and index division of longitudinal superimposed sweet spot.

(1) Establishing hierarchical structure model: Block G has been effectively developed with clear geological understanding. This prediction of sweet spot further determines the favorable development area in the determined geological area. The target layer is the sweet spot. The criterion layer is the TOC, porosity, water saturation. The scheme layer is as follows: Type I sweet spot, Type II sweet spot, Type III sweet spot and nonsweet spot.

(2) Constructing pairwise matrix: According to the results of Pearson-MIC comprehensive evaluation, the scale of TOC is set as 1, and the scale of porosity and water saturation is set as 2, so as to construct a pairwise matrix (Table 4).

(3) Hierarchical ranking and consistency check: Calculate the weight of each evaluation parameter according to the principle of analytic hierarchy process. The calculation meets the consistency check and is consistent with the actual production (Table 5).

Table 5. Pairwise matrix and weight of Block G.

Z	TOC	Porosity	Water Saturation	Weight
TOC	1	2	2	0.50
Porosity	1/2	1	1	0.25
Water saturation	1/2	1	1	0.25

(4) Grading of evaluation indicators: This paper involves the evaluation of three parameters, and the value range of each parameter is quite different. For the convenience of comparison, the evaluation index of the sweet spot is divided into three grades: excellent, medium, and poor (Table 6). Each cell in the Petrel work area of Block G corresponds to the index parameter value. Judge the range of evaluation indicator parameter values of each cell, divide the grade of evaluation indicators, and assign values.

Table 6. Pairwise matrix and weight of Block G.

Parameter	Excellent	Medium	Poor
	1	**0.5**	**0**
TOC (%)	>2.15	2~2.15	<2
Porosity (%)	>6	5~6	<5
Water saturation (%)	<30	30~50	>50

(5) Comprehensive evaluation and index division of longitudinal superimposed sweet spot: The comprehensive score M of each cell in the Petrel work area of Block G is calculated by summing up the weights of each parameter. Based on the weighted average of the thickness of each layer in corresponding series, the plane distribution of the cell comprehensive score S of the Series Upper and the Series Lower is depicted. On this basis, referring to the existing sweet spot standard, the comprehensive score

is divided into four grades to determine the longitudinal superimposed sweet spot standard of Block G (Table 7). According to the standards in Table 6, the longitudinal superimposed sweet spot of Series E and Series Lower in Block G is depicted (Figure 4). We have

$$M = \sum_{i=1}^{n} W_i Q_i \tag{10}$$

$$S = \sum_{i=1}^{l} Z_i M_i \tag{11}$$

In the formula, M is the comprehensive score of each cell, W_i is the weight of evaluation index, Q_i is the score of evaluation index, S is the comprehensive score of each cell in the series, and Z_i is the thickness weight of each layer in the series.

Table 7. Comprehensive evaluation grade table of sweet spot in Block G.

Grade of Sweet Spot	Type I Sweet Spot	Type II Sweet Spot	Type III Sweet Spot	Non-Sweet Spot
Comprehensive score	>0.7	0.6~0.7	0.5~0.6	<0.5

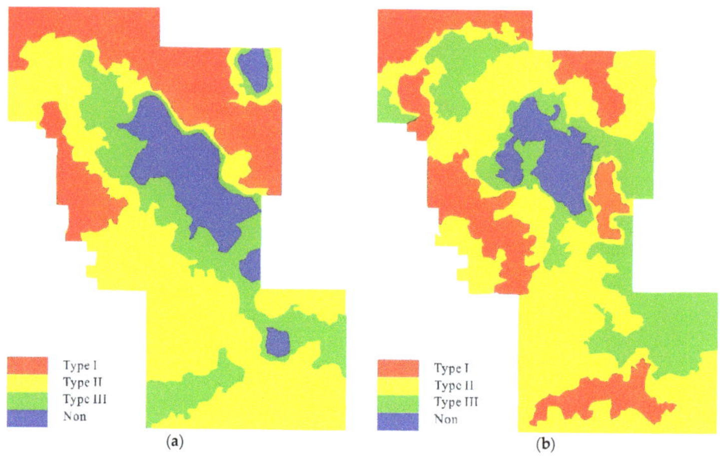

Figure 4. Plane distribution maps of longitudinal superimposed sweet spot in Block G. (**a**) Series Upper. (**b**) Series Lower.

In the Series Upper, Layer E has the best physical properties and the largest thickness, making the most outstanding contribution to the sweet spot. The Type I sweet spot accounts for 24.2% in the Series Upper, which is mainly distributed in the north and northeast of the reservoir with good physical properties. The Type II sweet spot accounts for 42.4%, mainly distributed in the south where the physical property is slightly inferior but the reservoir thickness is relatively thick. The Type III sweet spot accounts for 22.8%, and the non-sweet spot accounts for 10.6%. These two areas are mainly concentrated in the middle of the reservoir with ordinary physical properties and thin thickness.

In the Series Lower, Type I sweet spot accounts for 23.1%, which is relatively scattered, and the southern part is greatly affected by thin Layer AA and Layer A with good reservoir physical properties. The Type II sweet spot accounts for 48.0%, and its distribution is relatively scattered. The Type III sweet spot accounts for 21.7%, mainly distributed in the southeast and northwest. The non-sweet spot accounts for 7.2%, concentrated in the middle.

6. Conclusions

(1) The influence of engineering parameters on productivity is more significant. According to Pearson–MIC comprehensive evaluation method, the lateral length of horizontal well, number of stages, volume of fracturing fluid, and amount of proppant are determined as the key engineering parameters.
(2) Through cluster analysis, the lateral length is selected as an independent key engineering parameter, and the cumulative gas production in the first two years is normalized on the lateral length to eliminate the impact of the engineering. After calculating the correlation between the normalized cumulative gas production in the first two years and geological parameters, the TOC is determined as the key geological parameter, and porosity and water saturation are the secondary key parameters.
(3) The sweet spot standard of Block G is formulated based on the weight of each parameter obtained by AHP and the sweet spot evaluation index grade. The distribution of longitudinal superimposed sweet spot in Block G of Series Upper and Series Lower is obtained by thickness weighted average, laying a foundation for the subsequent placement of new wells.

Author Contributions: Writing—review and editing, Y.J.; supervision, W.H.; methodology, P.S.; project administration, P.W.; data curation, X.K. and L.L.; visualization, Y.S. All authors have read and agreed to the published version of the manuscript.

Funding: This research was funded by Major National Science and Technology Projects (No. 2016ZX05029-005), Major Science and Technology Projects of CNPC (No. 2019D-4412), and Special Science and Technology Project of CNPC (No. 2021DJ3303).

Institutional Review Board Statement: Not applicable.

Informed Consent Statement: Not applicable.

Data Availability Statement: Our data presented is original and not inappropriately selected, manipulated, enhanced, or fabricated.

Conflicts of Interest: The authors declare no conflict of interest.

References

1. Yang, S.; Zhang, J.; Huang, W.; Zhang, Y.; Tang, X. "Sweet spot" types of reservoirs and genesis of tight sandstone gas in Kekeya area, Turpan-Hami Basin. *Acta Pet. Sin.* **2013**, *34*, 272–282. [CrossRef]
2. Ma, Y.; Cai, X.; Zhao, P. China's shale gas exploration and development: Understanding and practice. *Pet. Explor. Dev.* **2018**, *45*, 561–574. [CrossRef]
3. Kumar, A.; Seth, P.; Shrivastava, K.; Manchanda, R.; Sharma, M.M. Integrated Analysis of Tracer and Pressure-Interference Tests to Identify Well Interference. *SPE J.* **2020**, *25*, 1623–1635. [CrossRef]
4. Zhao, Z.; Du, J. *Tight Oil and Gas*; Petroleum Industry Press: Beijing, China, 2012; pp. 100–128.
5. Zou, C.; Tao, S.; Hou, L. *Unconventional Oil and Gas Geology*; Geological Publishing House: Beijing, China, 2014.
6. Sun, L.; Zou, C.; Jia, A.; Wei, Y.; Zhu, R.; Wu, S.; Guo, Z. Development characteristics and orientation of tight oil and gas in China. *Pet. Explor. Dev.* **2019**, *46*, 1015–1026. [CrossRef]
7. Xu, M. *Research on the Geological and Engineering Sweet Spot Index of Shale Gas*; China University of Petroleum: Beijing, China, 2016.
8. Zhao, W.; Jia, C.; Jiang, L.; Zhang, T.; He, M.; Zhang, F.; Jiang, Z.; Li, X.; Wu, K. Fluid charging and hydrocarbon accumulation in the sweet spot, Ordos Basin, China. *J. Pet. Sci. Eng.* **2021**, *200*, 108391. [CrossRef]
9. Pan, R.; Gong, Q.; Yan, J.; Jin, J. Elements and gas enrichment laws of sweet spots in shale gas reservoir: A case study of the Longmaxi Fm in Changning block, Sichuan Basin. *Nat. Gas Ind.* **2016**, *36*, 7–13. [CrossRef]
10. Chen, Z.; Yang, C.; Jiang, C.; Dan, K.; Kezhen, H.; Liu, X.; Melinda, Y. Production characteristics and sweet-spots mapping of the Upper Devonian Lower Mississippian Bakken Formation tight oil in southeastern Saskatchewan, Canada. *Pet. Explor. Dev.* **2018**, *45*, 626–635. [CrossRef]
11. Wei, Y.; Li, J.; Lu, S.; Song, Z.; Zhao, R.; Zhang, Y.; Wang, J.; Liu, X. Comprehensive evaluation method of sweet spot zone in Lacustrine shale oil reservoir and its application: A case study of shale oil in lower 1st member of the Shahejie formation in the Raoyang sag. *J. China Univ. Min. Technol.* **2021**, 1–12. [CrossRef]
12. Chen, G.; Bai, Y.; Chen, X.; Xu, B.; Zhu, Y.; Feng, R.; Chen, L. A new identification method for the longitudinal integrated shale oil/gas sweet spot and its quantitative evaluation. *Acta Pet. Sin.* **2016**, *37*, 1337–1342+1360. [CrossRef]
13. Zhou, L.; Liu, H.; Zhou, K.; Hao, J.; Wang, X. "Sweet spot" identification and evaluation of tight sandstone reservoir. *Bull. Geol. Sci. Technol.* **2020**, *39*, 165–173. [CrossRef]

14. Li, M.; Sun, J.; Hu, J.; Liu, D.; Yuan, B. Sweet spot selection model for horizontal shale gas well section. *Sci. Technol. Eng.* **2020**, *20*, 9844–9850.
15. Han, F.; Song, G.; Wu, Z.; Shen, X.; Han, F. Multi-attribute Fusion Prediction Technique for Tight Oil "Sweets". *Spec. Oil Gas Reserv.* **2019**, *26*, 94–99. [CrossRef]
16. Liao, D.; Lu, B. An evaluation method of engineering sweet spots of shale gas reservoir development: A case study from the Jiaoshiba Gas Field, Sichuan Basin. *Nat. Gas Ind.* **2018**, *38*, 43–50. [CrossRef]
17. Liao, D.; Lu, B.; Chen, Y. An evaluation method of geological sweet spots of shale gas reservoir development: A case study from the Jiaoshiba Gas Field, Sichuan Basin. *Acta Pet. Sin.* **2019**, *40*, 144–151. [CrossRef]
18. Montgomery, S.L.; Jarvie, D.M.; Bowker, K.A.; Pollastro, R.M. Mississippian Barnett Shale, Fort Worth basin, north-central Texas: Gas-shale play with multi-trillion cubic foot potential. *Am. Assoc. Pet. Geol.* **2005**, *89*, 155–175. [CrossRef]
19. Chalmers, G.R.L.; Bustin, R.M. The organic matter distribution and methane capacity of the Lower Cretaceous strata of northeastern British Columbia, Canada. *Int. J. Coal Geol.* **2007**, *70*, 223–239. [CrossRef]
20. Curtis, J.B. Fractured shale gas systems. *AAPG* **2002**, *86*, 1921–1938. [CrossRef]
21. Pejman, T.; Farzam, J.; Muhammad, S. Data Mining and Machine Learning for Identifying Sweet Spots in Shale Reservoirs. *Expert Syst. Appl.* **2017**, *88*, 435–477. [CrossRef]
22. Heege, J.T.; Zijp, M.; Nelskamp, S.; Douma, L.; Verreussel, R.; Veen, J.T.; Bruin, G.; Peters, R. Sweet spot identification in underexplored shales using multidisciplinary reservoir characterization and key performance indicators: Example of the Posidonia Shale Formation in the Netherlands. *J. Nat. Gas Sci. Eng.* **2015**, *27*, 558–577. [CrossRef]
23. Munther, A.; Reza, R. Sweet-spot mapping through formation evaluation and property modelling using data from the Goldwyer Formation of the Barbwire Terrace, Canning Basin. *Petroleum* **2019**, *5*, 13–29. [CrossRef]
24. Sahoo, A.K.; Basu, S.; Vishal, V.; Srivastava, M. Hierarchical clustering of Pristane and Phytane data to identify the sweet spots in the shale reservoirs. *Energy Clim. Chang.* **2020**, *1*, 100011. [CrossRef]
25. Tang, J.; Fan, B.; Xiao, L.; Weitz, D. A New Ensemble Machine-Learning Framework for Searching Sweet Spots in Shale Reservoirs. *SPE J.* **2020**, *26*, 482–497. [CrossRef]
26. Yuan, Y.; Rezaee, R.; Gu, J.; Wu, S.; Al-Khdheeawi,, E.A.; Wang, J.; Pan, B. High-resolution coupling of stratigraphic sweet spot lithofacies and petrophysical properties: A multiscale study of Ordovician Goldwyer Formation, Western Australia. *Pet. Sci.* **2022**, *11*, 011. [CrossRef]
27. Tatsipie, N.R.K.; Sheng, J.J. Statistical distribution of geomechanical properties and 'Sweet Spots' identification in part of the upper Bakken. *Pet. Res.* **2022**, *10*, 005. [CrossRef]
28. Mu, L.; Zhang, M.; Xia, Z.; Qu, L.; Jiang, P. Quantitative characterization technique for tight gas reservoir in Groundbirch, Canada. *Acta Pet. Sin.* **2017**, *38*, 363–374. [CrossRef]
29. Haldar, S.K. *Introduction to Mineralogy and Petrology*; Elsevier: Amsterdam, The Netherlands, 2020.
30. Jiang, P.; Mu, L.; Zhang, M.; Zhao, W. Difference of reservoir characteristics between domestic and oversea tight gas of CNPC and its developing trends. *Nat. Gas Geosci.* **2015**, *26*, 1095–1105. [CrossRef]
31. Xu, S. *Practical Decision-Making Method: The Principle of Analytic Hierarchy Process*; Tianjin University Press: Tianjin, China, 1989.

processes

Article

Research on Gas Control Technology of "U+ Omni-Directional Roof to Large-Diameter High-Level Drilling Hole" at the End Mining Face of Multi-Source Goaf

Hong Gao [1,2,3,*] and Yun Lei [2,3]

[1] School of Safety Science and Engineering, Anhui University of Science & Technology, Huainan 232001, China
[2] CCTEG Shenyang Research Institute, Fushun 113122, China
[3] State Key Laboratory of Coal Mine Safety Technology, Fushun 113122, China
* Correspondence: g.h.f.w@163.com; Tel.: +86-15998150563

Abstract: Aiming at the gas over-run problem in the upper corner of the "U-type ventilation" end-mining working face in the multi-source goaf of the #15 coal seam in the Phoenix Mountain Mine, site survey and numerical simulation methods were adopted, which showed that the maximum caving zone height of the #15 coal seam is 14.87 m, and the maximum height of the fissure zone is 51.63 m. On this basis, the gas control scheme of the "U+ omni-directional large-diameter high-level borehole along roof strike" in the end-mining working face was formulated. After adopting this scheme, the extracted gas concentration of each borehole will reach 5–20%, the gas extraction flow rate will reach 1 m^3/min–2.5 m^3/min, the gas concentration at the upper corner of the working face will be controlled below 0.54%, and the gas concentration in the return airway will be controlled below 0.35%, achieving the expected effect of gas control.

Keywords: multi-source goaf; U-type ventilation; omni-directional large-diameter high-level borehole; gas control; effect evaluation

Citation: Gao, H.; Lei, Y. Research on Gas Control Technology of "U+ Omni-Directional Roof to Large-Diameter High-Level Drilling Hole" at the End Mining Face of Multi-Source Goaf. *Processes* **2023**, *11*, 320. https://doi.org/10.3390/pr11020320

Academic Editors: Tao Zhang, Zheng Sun, Dong Feng, Hung Vo Thanh and Wen Zhao

Received: 9 December 2022
Revised: 15 January 2023
Accepted: 16 January 2023
Published: 18 January 2023

1. Introduction

The problem of gas control has always been one of the important factors that seriously restrict the safe and efficient production of coal mines [1]. In particular, some mines with multi-seam coal mining are greatly affected by gas gushing from adjacent coal seams or goafs at the stope face [2].

Aiming at the problem of gas control in the mining process of the working face, Han Chengqiang [3] proposed a method combining the carbon isotope method and layered measurement and obtained the proportion of gas in each coal seam during the process of gas drainage from the coal seam, which provides a basis for the calculation of the residual gas content of the coal seam. Aiming at the problem of gas control in the U-type ventilation working face of high-gas mines, Gao Hong et al. [4–6] proposed a gas drainage technique combining a high-level suction roadway and a bottom-level suction roadway, which reduced the gas gushing to the working face. Li Jianwei, Gu Wangxin, et al. [7,8] studied the distribution rules of plastic zones around boreholes with different apertures, which provided a basis for the selection of borehole diameters for gas drainage. Thanh et al. [9–14] used different models to predict the CO_2 trapping index of deep saline aquifers and found that the developed models obtained excellent predictive performance. Wang Yilei et al. [15,16] determined the design parameters of high-level drilling in Tangkou Coal Mine in accordance with the theory of the vertical "tri-zone" and "O" ring and achieved significant results. Gao Yukun et al. [17] solved the problem of gas control in the upper corners of the Pingshan Coal Mine by adopting high-level borehole drainage measures, and they optimized the layout of the boreholes by adopting numerical simulation methods. Aiming at the problem of gas control in irregular working faces, Jiao Zhenhua et al. [18] studied the law of gas migration in the goaf and adopted comprehensive gas control measures for the sealing of gas-gushing channels, upper-corner drainage and high-level borehole drainage in the goaf. On the basis of analyzing the source of gas emissions in the working face and its influencing factors, Gao Liang et al. [19] formulated gas control measures in a goaf in a directional roof with a long drilling segmental fissure, joint roadway intubation and large-diameter coal pillar drilling bridging the goaf, which achieved good results. Xu Xingfu, Zou Yinxian, et al. [20–23] used numerical simulation methods to simulate and analyze different high-level borehole parameters and optimized the layout of high-level boreholes. Based on the development law of the overlying rock fissures, Hu Jinlin et al. [24] used theoretical analysis, numerical simulation and other methods to study the fissure evolution law in the overburden strata of the working face with the mining process and provided a basis for the selection of high-level drilling layers. Guo Xinhong, Shi Kun, et al. [25,26] used the carbon isotope method to obtain the proportion of gas emissions in each coal seam and determined the optimal layout position of the high-level borehole. Li Quanxin, Tao Yunqi, et al. [27,28] proposed a drainage mode for the large-area gas drainage of medium-hard coal seams, regional gas drainage for broken and soft coal seams and mining pressure relief gas for "replacing the roadway with boreholes". The accurate coverage of the whole area was realized, and continuous sampling was carried out throughout the period so as to extract as much as possible. Zhang Haoquan [29] adopted numerical simulation methods to study the near-field gas drainage area in the goaf of Shoushan No. 1 Coal Mine and proposed near-field gas drainage technology in the goaf based on a low-level roof and long boreholes, which effectively controlled the gas concentration in the upper corner.

The Phoenix Mountain Coal Mine in the Jincheng Mining Area adopts a "U"-type ventilation method, and the gas content in some areas is relatively high, especially in the adjacent seam, where a large amount of gas gushes out to the goaf, which seriously affects the safe production of the coal mine. This research analyzed the gas source of the working face and the heights of the "three zones" of the overburden of the working face, and on this basis, it conducted a "U+ omni-directional large-diameter high-level borehole along Roof strike" gas extraction test in the goaf, which effectively solved the problem of gas over-run in this area.

2. Project Overview

Located 5 km north of Jincheng City, the Phoenix Mountain Coal Mine belongs to Shanxi Jincheng Anthracite Coal Mining Group Co., Ltd. The main coal-bearing strata in the minefield are the Shanxi Formation and Taiyuan Formation, with a total thickness of 147.48 m. There are 14 coal seams, with a total thickness of 14.25 m. The minable coal seams are No. 3, No. 9 and No. 15 coal seams. As of 2017, No. 3 and No. 9 coal seams have been mined out, and the mined coal seam is No. 15, with a production capacity of 3.0 Mt/a.

The XV1306 working face is located in the three-level 151-panel area, the coal seam to be mined is #15, and the elevation of the working face is 524.0–598.1 m. In terms of the specific position of the working face in the mining area, the relationship with adjacent working faces, the mining of adjacent working faces, etc., the north of this working face is the XV1305 working face (mined), the south is the XV1307 working face (excavated), the west is the mine boundary, and the east is the track lane, belt lane and return airway of the 151 panels (all have been excavated). The average thickness of the overlying #3 coal seam on the working face is 6.1 m, from which all coal has been extracted. The overlying #9 coal is the goaf, and the area was damaged by a small coal mine of the 92,308, 92,310, and 91,324 working faces. The thickness of the coal seam in the working face is 1.90–3.10 m, with an average thickness of 2.55 m. The method of one-time full-height mining was adopted, and the thickness of the coal is 2.0–3.0 m. The coal seam dip angle of the working face is 0°–14°, with an average of 4°.

The gas content of the No. 15 coal seam is relatively large, while the average distance between the No. 15 coal seam and the upper part of the No. 9 coal seam is 35 m. In addition to the gas gushing out of this coal seam, the amount of gas gushing out of the adjacent

No. 3 coal seam and No. 9 coal seam is also relatively large during the mining process. Therefore, influenced by multi-source goafs in the No. 3 coal seam and No. 9 coal seam, the problem of gas control in the upper corner of the U-type ventilation coal mining face is particularly important.

3. Field Detection of Fissure's Developing Height

3.1. Time Selection and Principle Analysis of Probe Hole

In order to study the developing height of the fissure zone in the XV1306 working face, the YTJ20 borehole peering instrument was used. By detecting the position of the fissure in the borehole and marking its position, a basis was provided for determining the developing position and height of the fissure zone.

The TJ20 rock stratum detection recorder (Figure 1) was developed by Xuzhou Zhongkuang Huatai Technology Development Co., Ltd. (Xuzhou, China), which is convenient and practical. Before detection, firstly, holes are drilled into the coal wall or rock stratum, and a micro-camera is put in the hole to realize real-time monitoring. The image is transmitted to the display of the recorder through a cable. When there are damaged areas, fissure zones, separation areas or shear fracture zones in the surrounding rock, they can be seen intuitively, and the positions of abnormal areas can be recorded.

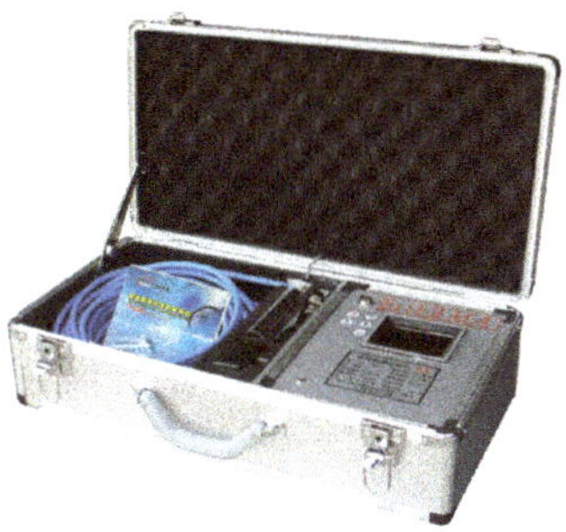

Figure 1. YTJ20 rock stratum detection recorder.

In this exploration, the No. 1 drill hole in the No. 8 drilling field of the XV1306 stope face and the No. 5 drill hole in the eighth roof strike drilling field were selected. The drilling construction parameters are shown in Table 1. When the distance from the stope face is 29 m, during pre-mining detection, the fissure zone has not yet fully developed. When the distance from the stope face is 15 m, the fissure zone is affected by mining, and fissures have begun to develop. This period is used as post-mining data.

Table 1. Construction parameters of probe hole.

Hole Number	Hole Depth (m)	Angle with Roadway (°)	Angle of Inclination (°)
1	55	4	35
5	65	11	45

Finally, the image before mining and the image after mining were compared and analyzed, and the developing height of the fissure zone was determined through field measurements, which provided a basis for the design of the high-level boreholes.

3.2. Processing and Analysis of Detection Images

Through the borehole peeping instrument, the pre-mining and post-mining rock stratum of the No. 1 drill hole in the No. 8 drill field of the XV1306 working face and the No. 5 hole along the eighth roof strike drill field were peeked at, and the following abnormal pictures were detected and analyzed (Figures 2–5).

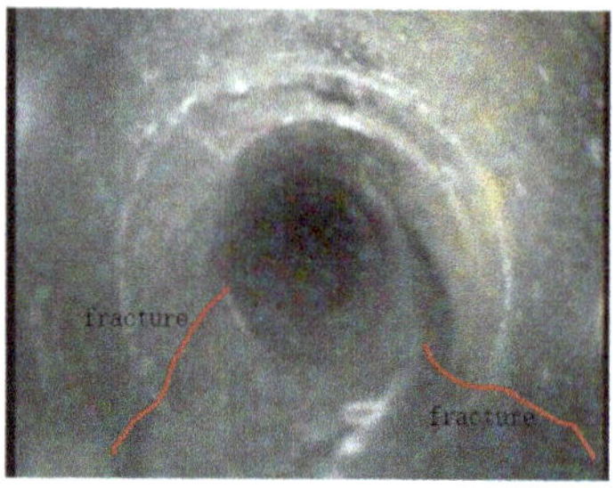

Longitudinal crack at 14.37 m of measuring hole

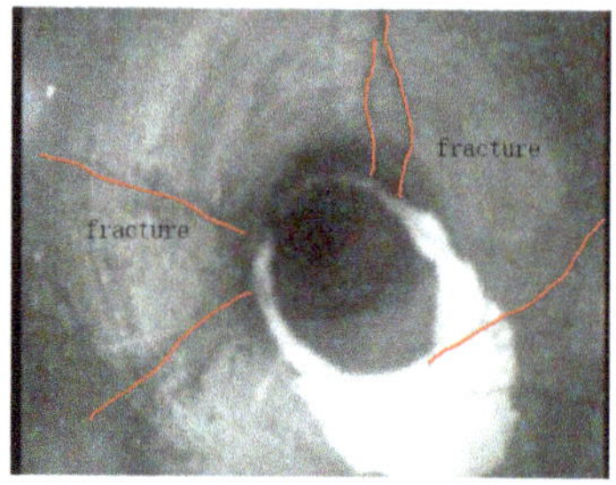

Longitudinal crack at 21.37 m of measuring hole

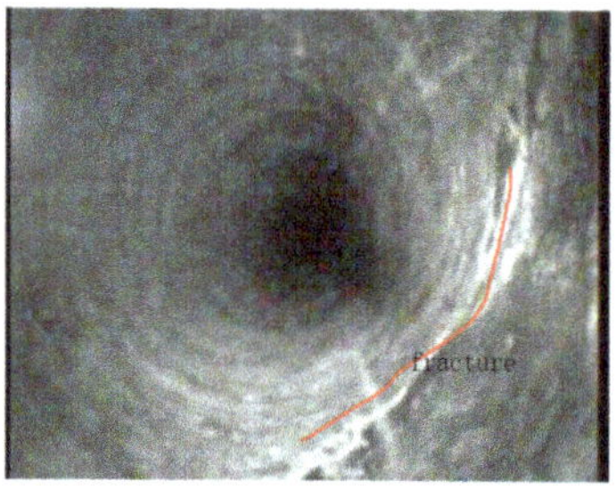

Transverse crack at 19.66 m of measuring hole

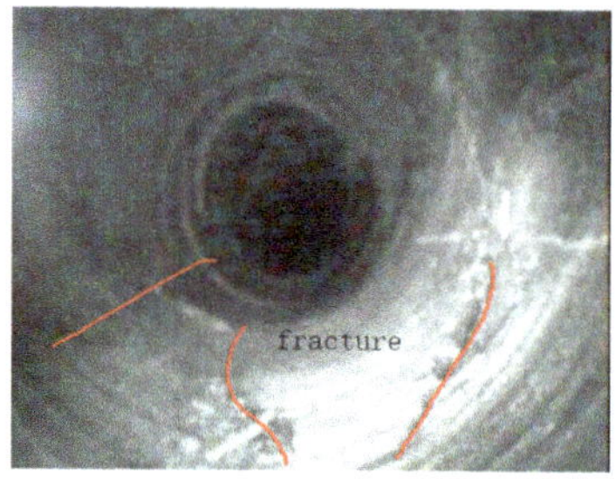

Longitudinal crack at 22.68 m of measuring hole

Figure 2. Peeping results of strata before mining in No. 1 hole 29 m away from the working face.

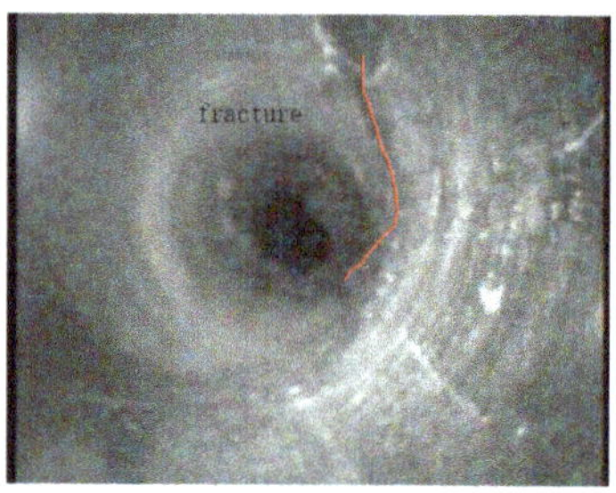

Longitudinal crack at 34.58 m of measuring hole

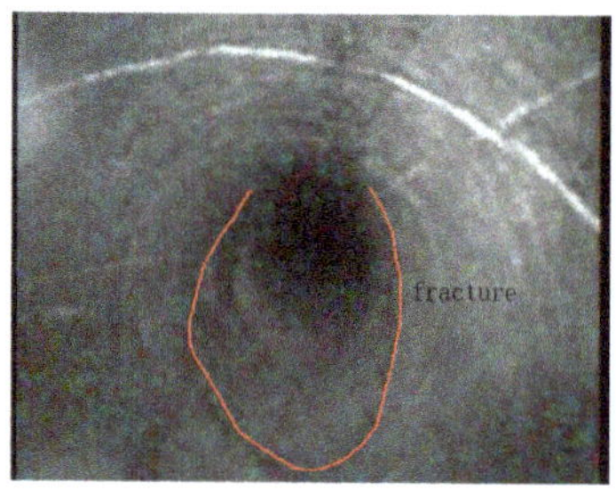

Transverse crack at 36.43 m of measuring hole

Figure 3. *Cont.*

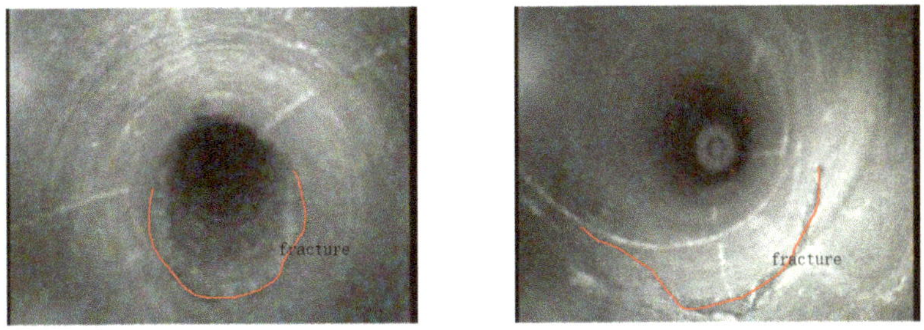

Transverse crack at 39.55 m of measuring hole

Transverse fissure at 46.43 m of measuring hole

Figure 3. Peeping results of strata after mining in No. 1 hole 15 m away from the working face.

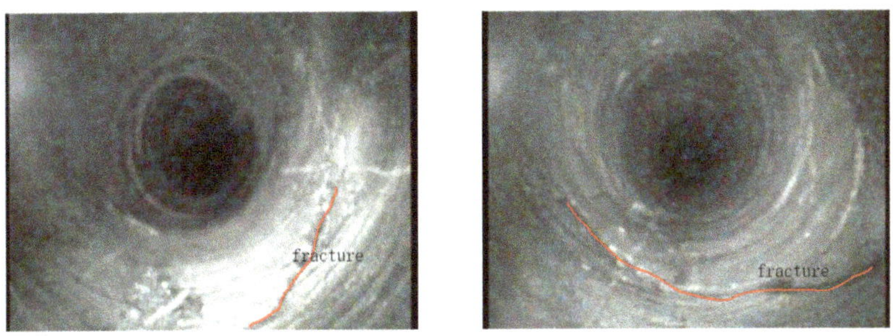

Transverse crack at 14.86 m of the measuring hole

Transverse crack at 18.35 m of measuring hole

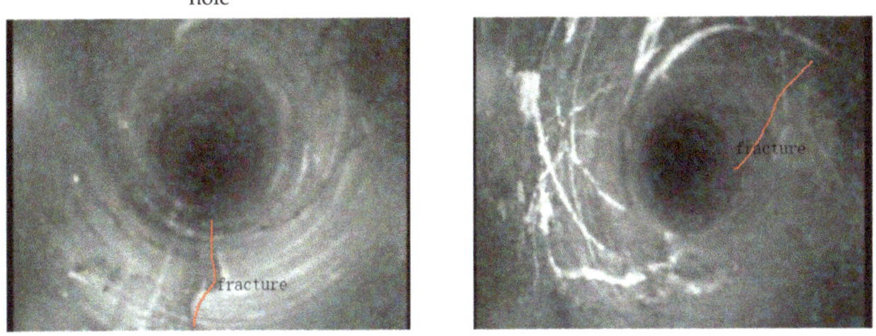

Longitudinal crack at 19.08 m of measuring hole

Longitudinal crack at 15.83 m of measuring hole

Figure 4. Peeping results of strata before mining in No. 5 hole 29 m away from the working face.

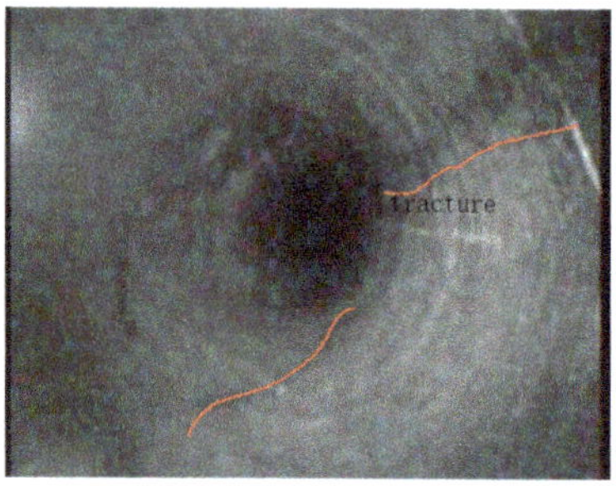

Longitudinal crack at 35.54 m of measuring hole

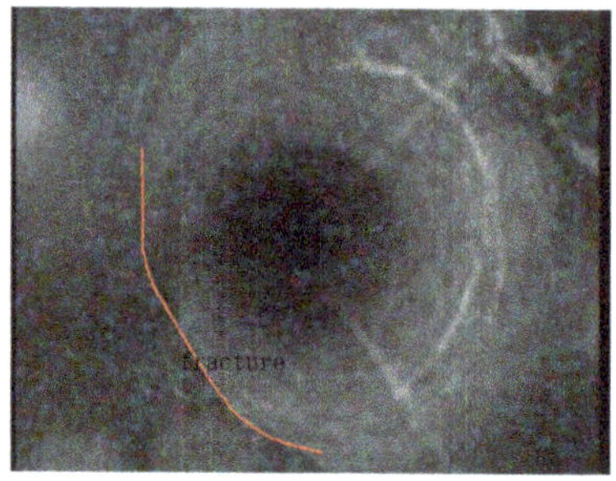

Transverse crack at 28.95 m of measuring hole

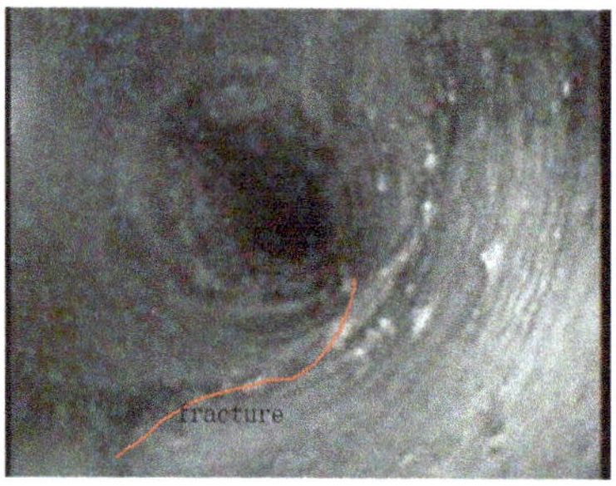

Longitudinal crack at 27.15 m of measuring hole

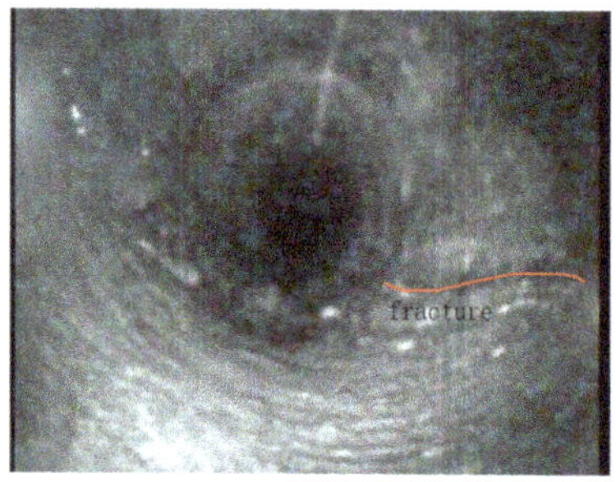

Longitudinal crack at 38.85 m of measuring hole

Figure 5. Peeping results of strata after mining in No. 5 hole 15 m away from the working face.

By collecting and analyzing the fissure positions of the pre-mining and post-mining boreholes, the stratum where the fissures are located can be calculated according to the elevation angle and depth of the boreholes. Usually affected by mining, the fissure zone will form a fracture surface and fracture space at a layer of a certain height. After the actual measurement, it is concluded that the height of the fissure surface in the caving zone of the #15 coal seam is 8.24~13.49 m, and the height of the fissure surface in the fissure zone is 19.19~27.46 m (see Table 2).

Table 2. Fracture height calculation table.

Hole Number	Depth of Borehole Fracture Location (m)	Angle of Inclination (°)	Height of Fracture (m)	Notes
1	14.37	35	8.24	Before mining
1	21.37	35	12.25	Before mining
1	19.66	35	11.27	Before mining
1	22.68	35	13.00	Before mining
5	14.86	45	10.50	Before mining
5	18.35	45	12.97	Before mining
5	19.08	45	13.49	Before mining
5	15.83	45	11.19	Before mining
1	34.58	35	19.83	After mining
1	36.43	35	20.89	After mining
1	39.55	35	22.67	After mining
1	46.43	35	26.62	After mining
5	35.54	45	25.12	After mining
5	28.95	45	20.46	After mining
5	27.15	45	19.19	After mining
5	38.85	45	27.46	After mining

4. Numerical Simulation of Fissure's Developing Height in Mining Overburden

4.1. Numerical Calculation Model and Boundary Conditions

4.1.1. Model Establishment

The numerical simulation was performed using Flac^{3D} software, which is simulation calculation software developed by ITASCA Company in the United States, and the generalized finite difference method was used to simulate the nonlinear continuum mechanical behavior.

The numerical calculation model takes the geological conditions of the #15 coal seam of the Phoenix Mountain Mine as the prototype of the model and establishes a numerical simulation mechanical model according to the distribution and lithology of the rock stratum (Figure 6).

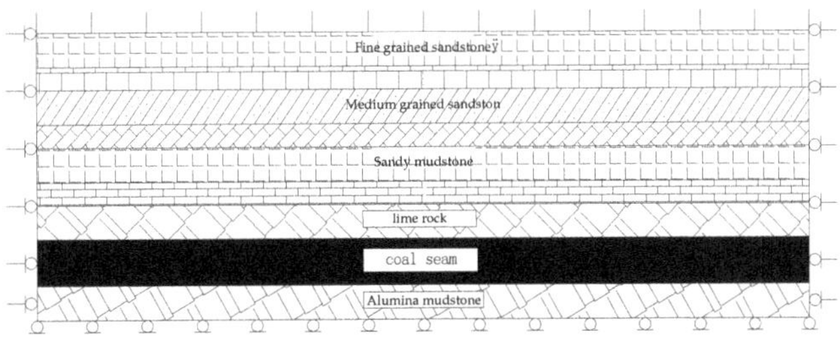

Figure 6. Mechanical model of numerical simulation.

In the actual mine, the #3 and #9 coal seams on the #15 coal seam have been mined. Considering the compaction of the goaf and the conduction of forces in the model during the simulation, according to experience, the cohesive force strength and tensile strength of the lithology of the #3 and #9 coal seams are taken as 1/10 of the original lithology strength.

The mechanical property parameters of each rock stratum in the calculation model are shown in Table 3.

According to the actual conditions and the needs of this calculation, the geometric dimensions of the established model are 160 m × 120 m × 120 m in length × width × height, and the established numerical calculation model is shown in Figure 7.

Table 3. Rock mechanical parameter table.

Serial Number	Rock Character	Density (Kg/m^3)	Elastic Modulus (GPa)	Cohesion (MPa)	Tensile Strength (MPa)	Internal Friction angle (°)	Poisson's Ratio
1	Coal	1400	2.7	1.6	0.8	28	0.35
2	Fine-grained sandstone	2626	9.9	2.6	3.0	43.6	0.23
3	Medium-grained sandstone	2644	6.5	2.4	2.7	42.3	0.24
4	Sandy mudstone	2649	5.4	2.1	0.1	38.8	0.11
5	Lime rock	2625	52.5	2.8	6.0	77.5	0.21
6	Alumina mudstone	2646	6.0	2.3	1.4	40.6	0.18

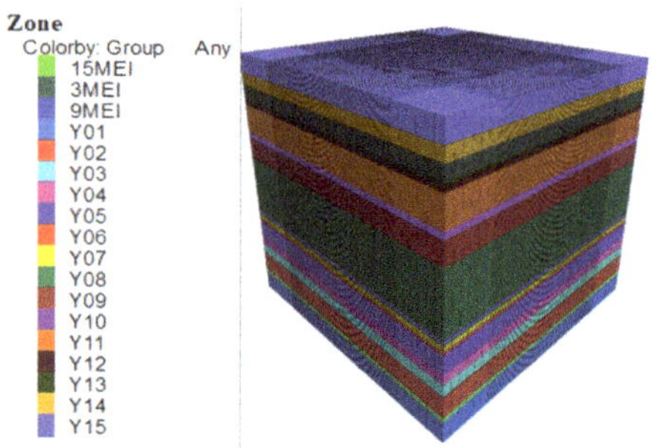

Figure 7. Numerical calculation model.

4.1.2. Model Boundary and Load Conditions

The calculation model boundary conditions are as follows:

1. The boundaries of the four sides of the model are fixed; that is, the horizontal displacement is zero, and horizontal stress is applied.
2. The bottom boundary of the model is fixed; that is, the horizontal displacement and vertical displacement are both zero.
3. The top of the model is a free boundary, and vertical stress is applied. The load conditions of the model were determined: the #15 coal seam is about 370 m away from the surface, and the upper boundary of the model is used as the equivalent stress by applying the gravity of the overlying strata. In the model, the #15 coal seam is 100 m away from the upper part of the model, and the upper boundary of the model is about 270 m away from the surface.

$$\begin{cases} P=\gamma H \\ P'=\lambda H \end{cases} \tag{1}$$

In the formula, P is principal stress, MPa; P' is horizontal stress, MPa; γ is bulk density, KN/m^3; H is vertical depth, m; and λ is the coefficient of lateral pressure.

γ is taken as 25, and H is taken as 270. Referring to the data of the Phoenix Mountain Mine, the ground stress of the mine is dominated by tectonic stress, and its coefficient of lateral pressure λ is taken as 1.12.

According to Formula (1), the equivalent stress load applied on the upper part of the model is 6.75 MPa, and a load of 7.56 MPa is applied in the horizontal direction. The initial balance diagram of the model is shown in Figure 8.

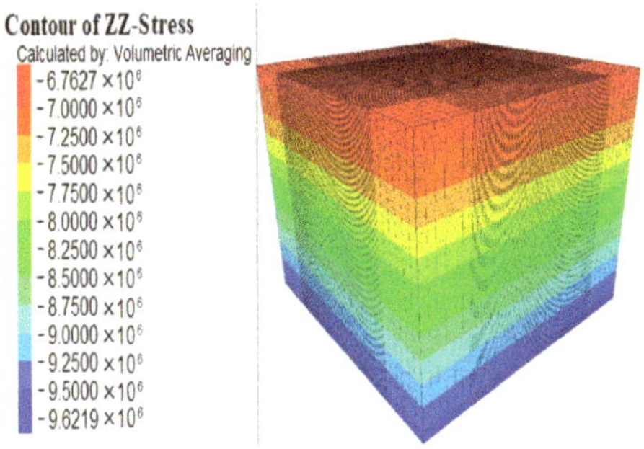

Figure 8. Initial equilibrium diagram of model.

4.2. Numerical Simulation Scheme

The numerical simulation calculation mainly analyzed the various changing characteristics of the surrounding rock mass in the goaf under the existing mining conditions of the working face. The calculation mainly included:

1. The stress distribution characteristics of the surrounding rock in the goaf;
2. The distribution characteristics of the plastic zone of the surrounding rock in the goaf.

The stratum attitude of the Phoenix Mountain Mine is nearly horizontal. According to the mining data of the Phoenix Mountain Mine, when its XV1306 working face is mined, the first weighting step distance is about 30 m, and the periodic weighting step distance is 11–24 m. Simulations were run to monitor the changes in the surrounding rock of the working face when the working face advances 50 m, 60 m, 80 m and 100 m; that is, the stress situation and damage characteristics of the surrounding rock mass of the #15 coal seam working face during the first and periodic weighting of the working face were calculated using simulation theory.

4.3. Analysis of Numerical Simulation Results

When the whole caving method is adopted, the caving zone, fracture zone and bending subsidence zone will generally appear in the roof strata of the goaf from the bottom to the top. The formation and development of these three zones not only directly affect the evolution of the stress field around the mining face but also directly affect the evolution of the stress field. It is of great significance to accurately determine the heights of the collapse zone and fracture zone for the arrangement of strata in the high-level borehole. The strata in the caving zone are deposited in the goaf in a loose state after caving, while the strata in the fissure zone are still in a stress state. The height of the fracture zone can be determined by analyzing the stress state of the rock stratum using numerical simulations, and the height of the collapse zone can be determined by analyzing the change in the plastic zone.

3. The stress distribution characteristics of the surrounding rock in the goaf with different advances of the working face. The distribution of the stress field in the stope is also continuously changing, while the stope goes through the process from the first weighting to periodic weighting. Figure 9 shows the vertical principal stress profiles of the central part of the goaf with different advance distances.

As shown in the figure, when the working face advances 50 m, the release height of the principal stress of the roof is about 28.00 m (Figure 9a); when the working face advances 60 m, the release height of the principal stress of the roof is about 33.45 m (Figure 9b); when the working face advances 80 m, the release height of the principal stress of the roof is about 44.80 m (Figure 9c); and when the working face advances 100 m, the release height of the principal stress of the roof is about 51.63 m (Figure 9d). It can be seen in Figure 9 that with the advancement of the working face, after the stope is fully mined, the maximum height of the pressure release on the roof in the goaf is about 51.63 m; that is, the height of the loose circle of the roof-surrounding rock is about 51.63 m.

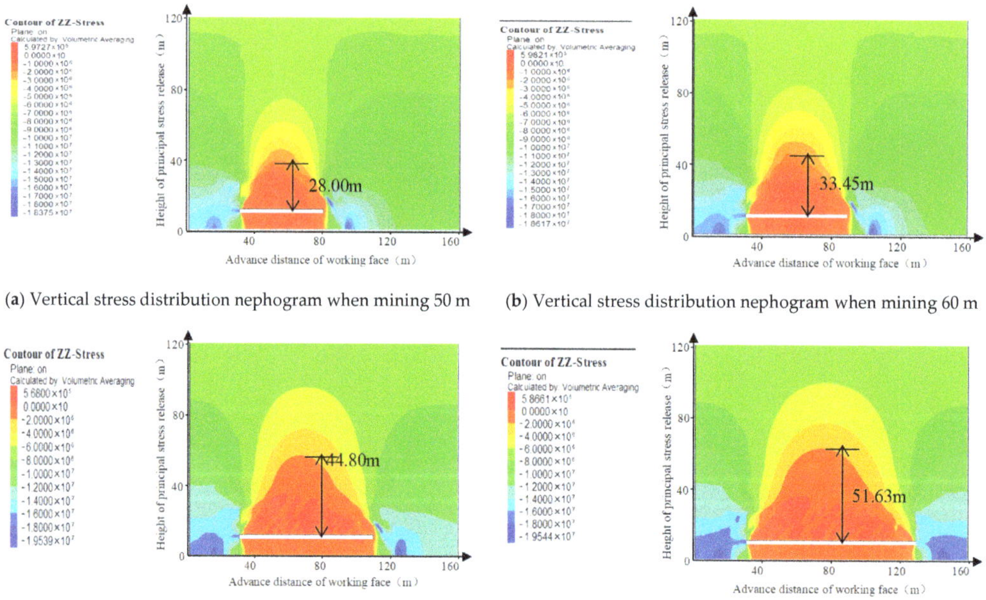

(**a**) Vertical stress distribution nephogram when mining 50 m (**b**) Vertical stress distribution nephogram when mining 60 m

(**c**) Vertical stress distribution nephogram when mining 80 m (**d**) Vertical stress distribution nephogram when mining 100 m

Figure 9. Distribution nephogram of vertical principal stress in goaf.

4. The distribution characteristics of the plastic zone of the surrounding rock in the goaf with different advance distances. Figure 10 shows the distribution nephogram of the plastic zone of the surrounding rock in the goaf with different advance distances.

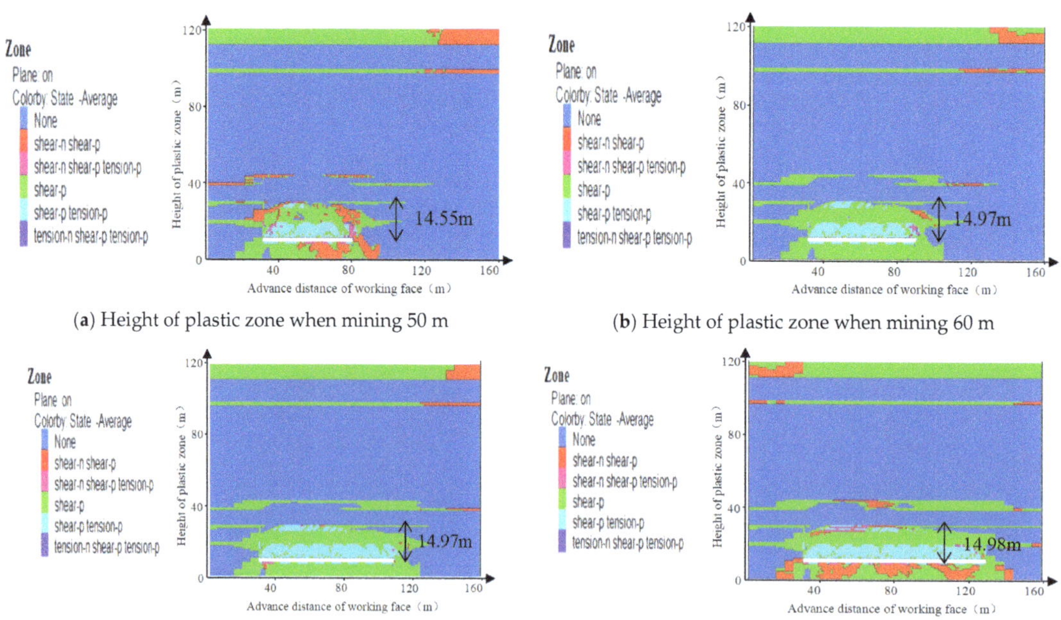

(**a**) Height of plastic zone when mining 50 m (**b**) Height of plastic zone when mining 60 m

(**c**) Height of plastic zone when mining 80 m (**d**)Height of plastic zone when mining 100 m

Figure 10. Distribution characteristics of the plastic zone of the surrounding rock in the goaf.

As shown in the figure, when the working face advances 50 m, the height of the plastic zone of the roof-surrounding rock is about 14.55 m (Figure 10a); when the working face advances 60 m, the height of the plastic zone of the roof-surrounding rock is about 14.97 m (Figure 10b); when the working face advances 80 m, the height of the plastic zone of the roof-surrounding rock is about 14.97 m (Figure 10c); and when the working face advances 100 m, the height of the plastic zone of the roof-surrounding rock is about 14.98 m (Figure 10d). It can be seen in Figure 10 that with the advancement of the working face, the crushing height of the surrounding rock of the roof is about 14.87 m; that is, the height of the roof-caving zone is about 14.87 m.

5. Reasonable Arrangement of Parameters of Omni-directional Roof High-Level Boreholes

Affected by the roof caving of the coal seam, the high-level boreholes constructed in the low-level drilling field all have a certain effective drainage range. With the advancement of the stope face, some boreholes will lose their drainage function. Affected by the drilling angle and the roof-caving shape of the goaf, blind drainage areas are easy to find in drainage boreholes between low-level drilling fields [30,31]. Under such conditions, an omni-directional high-level borehole can be constructed, which can reduce drilling field construction on the one hand and effectively solve the problem of blind drainage areas on the other hand.

The high-level drilling field is arranged at a position 5–8 m above the roadway roof. An omni-directional drainage borehole is a 180° borehole constructed in a high-level drilling field. The drilling covers a large area and has a wide radiation range. According to the lithology of the roof of the #15 coal seam and the occurrence of the coal seam, the location with the better drilling effect was designed. Combined with the weighting step distance of the old roof of the #15 coal seam roof, the caving fissure zone is determined to be 5 to 8 times the mining height, and in terms of the characteristics of gas occurrence in the goaf, the zone 10–30 m away from the upper corner is the O-shaped gas enrichment zone, and the position of the final borehole is finally determined. The design of a high-level drilling field includes the high-level suction roadway and all boreholes in the drilling field. The omni-directional drainage borehole can drain the gas in the adjacent seam and the goaf at the same time, and the high-level borehole constructed in the opposite direction can advance the recovery position of the working face, drain the gas in the goaf of the upper adjacent seam and control the gas in the goaf of the adjacent seam in advance to reduce the impact on the gas gushing out of the mining seam.

The Phoenix Mountain Mine mainly comprises a high-level drilling field constructed next to the return airway of the XV1306 working face. The high-level drainage boreholes constructed in the drilling field are arranged in a fan shape. There are a total of 28 boreholes with a depth of 48–160 m and a diameter of 94 mm. The spacing of the drilling holes is 0.5 m, the spacing between the left and right sides of the drilling field is 5–10 m, and the spacing of the drilling field's final boreholes is 10–20 m. The specific parameters are shown in Table 4.

The omni-directional high-level drainage borehole layout of the XV1306 working face is shown in Figure 11.

Table 4. XV1306 working face high-level drilling field drainage borehole layout parameters.

Hole No.	Hole Depth (m)	Angle with XXV 1212 Lane (°)	Elevation Angle (°)	Height Difference between Terminal Hole and Roof (m)	Hole No.	Hole Depth (m)	Angle with XXV 1212 Lane (°)	Elevation Angle (°)	Height Difference between Terminal Hole and Roof of Drilling Site (m)
1	156.5	7.8	0.3	15.5	15	48.3	81	6.5	13.5
2	157.2	9.5	0.2	16	16	55.1	59.8	7.8	14
3	158	11.1	0.4	16.5	17	67.4	45.3	8.7	14.5
4	159	12.8	0.3	17	18	82.5	35.3	8.3	15
5	160.1	14.4	0.6	17.5	19	99.4	28.8	8.6	15.5
6	161	16	0.4	18	20	117.1	24.1	8.3	16
7	162.7	17.6	0.7	18.5	21	135.6	21.2	8.4	16.5
8	138.9	20.7	0.4	18.5	22	154.6	18.5	8.1	16.5
9	120.2	23.8	0.5	18	23	153.2	16.9	8.2	16
10	102.1	28.3	0.4	17.5	24	151.8	15.1	8	15.5
11	85.5	33.7	1.2	17	25	150.7	13.5	8.1	15
12	70	42.8	1.8	16.5	26	149.6	11.6	7.9	14.5
13	57.1	56.1	2.1	14	27	148.7	10	8.1	14
14	49	76	3.6	13.5	28	147.9	8.1	7.8	13.5

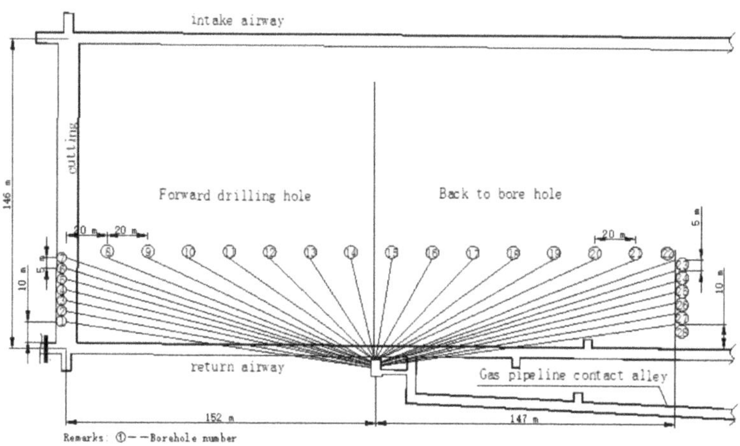

Figure 11. Omni-directional high-level drainage borehole layout.

The high-level boreholes are constructed by ordinary drilling; the drilling rig model is ZDY-3200S, the drill pipe length is 1.5 m, with a diameter of 73 mm, and the drill bit specification is 94 mm. The "Two Sealing and One Grouting" sealing method with pressure is adopted. After the hole is sealed, the borehole is connected to the drainage system, and the drainage parameters are promptly inspected according to regulations. The investigation data on the gas drainage parameters shall be added to the on-site management layout in a timely manner, and according to the investigation results of the drainage parameters, invalid boreholes shall be promptly removed to optimize the gas drainage system.

6. Analysis of Gas Drainage Effect of High-Level Boreholes in the Final Mining Face and the Influence of Air Distribution Volume on High-Level Boreholes

1. Analysis of the effect of gas drainage from head-on high-level boreholes

From the time-varying curves of the flow rate and the concentration of some high-level boreholes in the No. 8 drilling field of the XV1306 working face, it can be seen that the gas flow rate in the boreholes is mostly between 1 and 2.5 m^3/min, and the average gas flow rate in the boreholes is about 1.5 m^3/min. Most of the extracted gas has a concentration between 5% and 20%, and the maximum reaches 23%. The scalar quantity of drilling single-hole drainage reaches between 0.05 m^3/min and 0.5 m^3/min, with an average of 0.15 m^3/min, achieving a good gas extraction effect.

Affected by the mining stress, the coal seam roof produces a large number of fissures, and a large amount of gas gradually flows into the mining face through the fissures. When the roof is subjected to pressure for the first time, with the caving of the roof, a large amount of gas flows into the goaf and gradually stabilizes in the later period. The gas flow increases gradually with the proximity of periodic pressure in the later period.

The gas flow and concentration curves of some boreholes in the No. 8 drilling field of Lane XV1306 are shown in Figures 12 and 13.

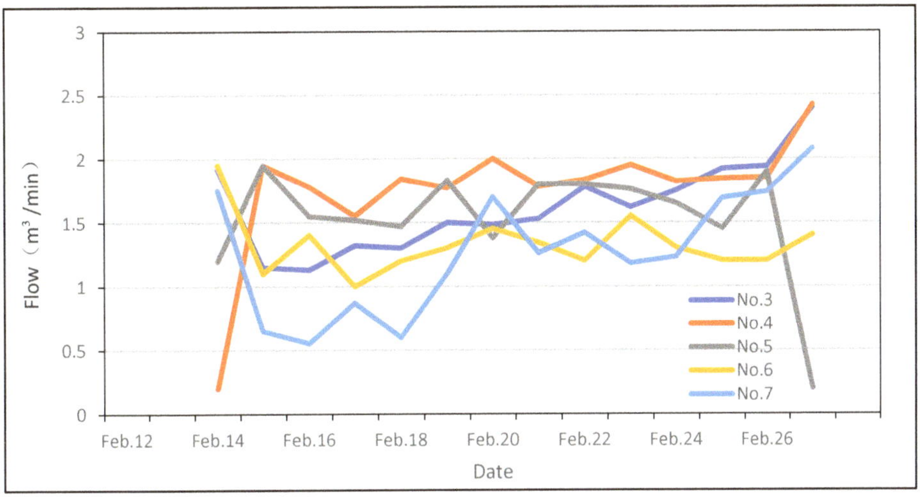

Figure 12. Gas flow curve of some boreholes in No. 8 drilling field of Lane XV1306.

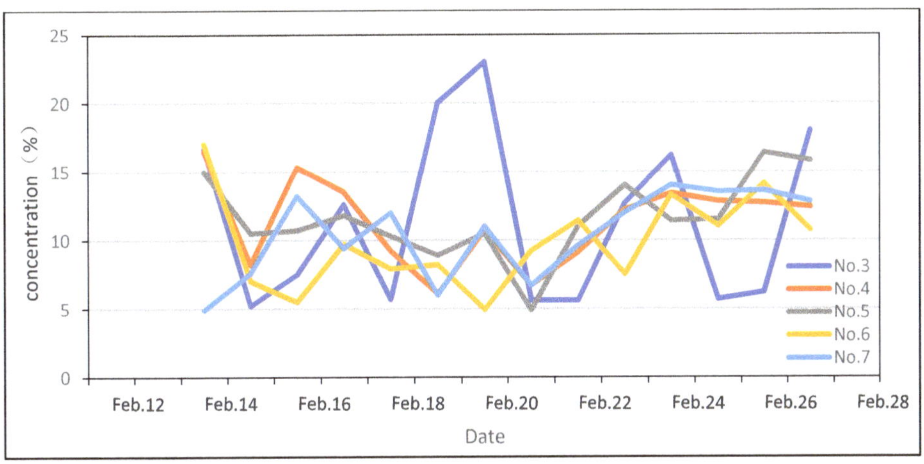

Figure 13. Gas concentration curve of some boreholes in No. 8 drilling field of Lane XV1306.

Practice has proved that when the negative drainage pressure increases to a certain level, the flow rate cannot be increased. On the contrary, the flow rate can be increased when the negative pressure is reduced to an appropriate level. The drainage effect is optimal when the negative pressure is 15–22.4 KPa in the high drilling field of the XV1306 working face.

2. Analysis of gas drainage effect of reverse high-level boreholes

Since reverse boreholes are less affected by mining stress, the gas drainage effect is relatively stable during the mining of the working face. From the time-varying curves of the flow and concentration of the No. 8 drilling field of the XV1306 working face facing away from the high position, it can be seen that the gas flow rate in the boreholes is mostly between 0.5 and 0.6 m^3/min; the gas drainage concentration is mostly between 14 and 20% and up to 25% at most, achieving a relatively good gas extraction effect.

Figures 14 and 15 show the gas flow and gas concentration curves of the reverse boreholes in the No. 8 drilling field of Lane XV1306. After adopting the reverse high-level borehole drainage method, the gas existing in the overlying goaf on the XV1306 working face is controlled in advance. Since it is less affected by the mining of the working face, a stable drainage effect is achieved, and the working face is reduced. The amount of gas gushing from the working face after mining to the reverse drainage position of the borehole reduced the pressure of gas in the goaf of the working face and achieved remarkable results.

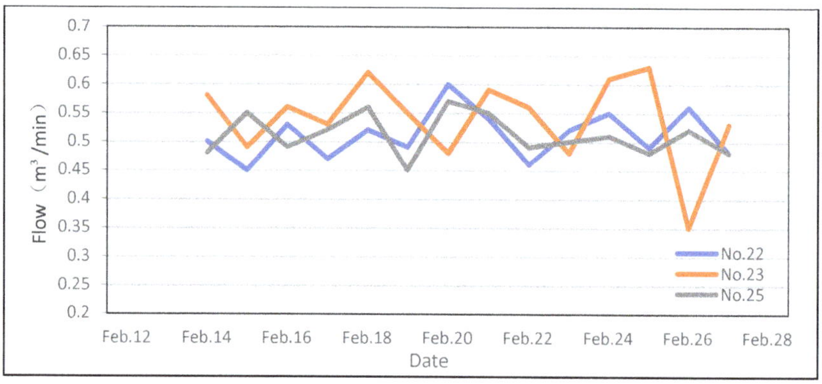

Figure 14. Gas drainage flow curve of some reverse boreholes.

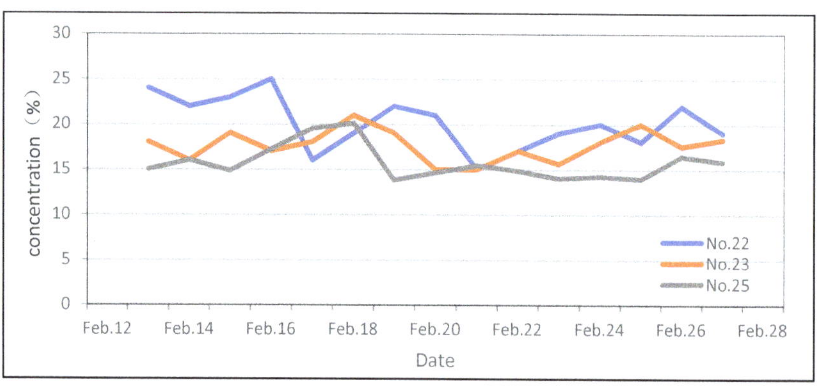

Figure 15. Gas drainage concentration curve of some reverse boreholes.

After adopting omni-directional high-level borehole drainage on the roof, the gas concentration in the upper corner is controlled below 0.54%, the gas concentration in the return airway is controlled below 0.35%, and the gas drainage effect has been significantly improved.

3. Site Management of High-Level Drilling Field

Since the high-level drilling site is 5–8 m higher than the roadway roof and is located on the outside of the roadway, the ventilation route of the high-level drilling field is cut off after the recovery progress of the working face pushes past the drilling field, and the gas pipeline connection roadway becomes a blind alley. Therefore, it is necessary to seal the connection roadway of the gas pipeline connected to the high-level drilling field at the mouth of the blind alley and retain the original drainage system to ensure that the omni-directional high-level borehole on the roof is always in a state of drainage. By monitoring the measuring device on the gas pipeline outside the airtight area, the negative drainage pressure is adjusted to optimize the drainage system.

4. Influence of Air Distribution Volume on High-level Boreholes

Combined with the gas emission characteristics of the XV1306 coal mining face of the Phoenix Mountain Mine, the relationship between the air distribution volume of the coal mining face and the scalar volume of the high-level borehole is analyzed and studied.

The air volume of the coal mining face is calculated according to the gas emissions using the following formula:

$$Q_{\text{mining}} = \frac{Q_{CH_4}}{c} K \tag{2}$$

In the formula, Q_{mining} is the air volume required by the coal mining face, m^3/min; Q_{CH_4} is the Ventilation Air Methane (VAM), m^3/min; c is the maximum allowable gas content in the return air flow of the coal mining face, 0.8%; and K is the ventilation coefficient of the coal mining face and taken as 1.8.

The gas emission volume of the XV1306 working face is composed of two parts: Ventilation Air Methane (VAM) and the gas drainage volume. When the overlying layer on the XV1306 working face is coal pillars, the gas emission rate is 11.48 m^3/min. Calculated according to a daily output of 3769 t/d and an air distribution volume of 1008 m^3/min, the maximum Ventilation Air Methane (VAM) is 4.48 m^3/min. The gas scalar volume extracted is 7 m^3/min.

It can be seen from Table 5 that, taking the average daily output of 3769 t/d as an example, the maximum Ventilation Air Methane (VAM) is 4.48 m^3/min, and the drainage volume is 7 m^3/min. One ZWY105/132 drainage pump can provide a drainage flow of 80 m^3/min, and the corresponding drainage volume is 3.73 m^3/min, which cannot meet the drainage demand of the working face. Therefore, two ZWY105/132 drainage pumps were selected in the later stage; the drainage volume can reach 160 m^3/min, and the corresponding drainage volume is 7.46 m^3/min, which can meet the gas drainage requirements of the working face.

Table 5. Relation between daily output, gas content, air distribution volume and high-level borehole drainage volume of XV1306 working face.

Daily Output (t)	Working Face Emission Quantity (m^3/min)	Working Face Air Distribution (m^3/min)	Maximum Gas Discharge in Strong Wind (m^3/min)	Extraction Quantity (m^3/min)
2500	7.1	719	3.2	3.9
3000	8.44	818	3.64	4.8
3500	10.27	916	4.07	6.2
3769	11.48	1008	4.48	7
4000	11.91	1015	4.51	7.4

Lane XV1214 is the return airway of the XV1306 working face, with a designed net section of 10.75 m^2. To ensure that the wind speed does not exceed 4 m/s, the air distribution volume should be less than 2580 m^3/min. When the amount of gas gushing out from the working face is large, it is possible to appropriately increase the air distribution volume of the return airway within the wind speed limit. At the same time, combined with high-level boreholes for gas drainage, the gas concentration in the return airway corner can be effectively controlled to prevent gas over-run.

7. Conclusions

1. Through on-site detection and numerical simulation, it is concluded that the caving zone of the #15 coal seam has a maximum height of 14.87 m, and the maximum height of the fissure zone is 59.29 m, which provides a basis for the selection of the high-level borehole position in the working face.
2. According to the calculation results of the heights of the fissure zone and caving zone of the coal seam roof, the gas control scheme of the "U+ omni-directional large-diameter high-level borehole along roof strike" in the final mining face was formulated. After adopting this scheme, the drainage concentration of each borehole will reach 5–20%, with a gas drainage flow rate of 1 mm^3/min–2.5 m^3/min, and the average gas drainage volume will reach 0.15 m^3/min. The maximum gas concentration in the upper corner is 0.54%, and the maximum gas concentration in the return airway is 0.35%, which achieved the expected gas control effect.
3. The reasonable air distribution volume of the 1307 stope face should be less than 2580 m^3/min. When the amount of gas gushing out from the working face is large, the air distribution volume of the return airway can be appropriately increased within the wind speed limit, combined with high-level boreholes for gas drainage, which can effectively prevent gas over-run in the upper corner.

Author Contributions: H.G. led the research, completed the numerical simulation and drafted the manuscript; Y.L. conducted the detection, prepared the pictures and tables and critically revised the article. All authors have read and agreed to the published version of the manuscript.

Funding: This work was financially supported by the National Key R&D Program of China (No. 2021YFC31008) and the Liaoning Province Doctoral Research Launch Fund (No. 2022-BS-365).

Data Availability Statement: Not applicable.

Acknowledgments: The authors would like to thank all editors and anonymous reviewers for their comments and suggestions.

Conflicts of Interest: The funders had no role in the design of the study; in the collection, analyses or interpretation of data; in the writing of the manuscript; or in the decision to publish the results.

References

1. Huang, H. Application of Comprehensive Gas Control Technology at the First Mining Face of Short-distance Outburst Coal Seam Groups. *China Coal* **2020**, *46*, 42–46.
2. Wang, X.J.; Liu, Y.F. Study on comprehensive control of gas in close coal seam mining. *Coal Technol.* **2021**, *40*, 135–138.
3. Han, C.Q. Research on gas tracing technology of joint drainage in close-range outburst coal seam group. *Coal Mine Saf.* **2022**, *53*, 68–73, 80.
4. Gao, H.; Yang, H.W. Study on gas extraction technology of high and bottom drainage roadway in high gas mine. *Ind. Min. Autom.* **2021**, *47*, 100–106.
5. Gao, H. Layer Layout of High Drainage Roadway in Zhaozhuang Mine and Analysis of Gas Drainage Effect. *Coal Eng.* **2018**, *50*, 8–11.
6. Li, H.; Liu, M.J.; Yang, H.W.; Gao, H.; Xia, T.Q. Influence Range Simulation of Loose Blasting Borehole in the Coal-Rock Mass. *Therm. Sci.* **2019**, *23*, 1457–1464. [CrossRef]
7. Li, J.W.; Liang, W.X.; Fu, W. Simulation and application of plastic zone of borehole based on gas extraction effect. *Coal Mine Saf.* **2022**, *53*, 170–174, 180.
8. Gu, W.; Lu, Y.; Shi, S.L.; Wu, K.; Yin, H.Q.; Huang, L.X.; Peng, S.M. Optimization and Application of Borehole Plastic Zone Range Based on Gas Drainage Effect. *China Saf. Prod. Sci. Technol.* **2021**, *17*, 25–30.
9. Thanh, H.V.; Yasin, Q.; Al-Mudhafar, W.J.; Lee, K.K. Knowledge-based machine learning techniques for accurate prediction of CO_2 storage performance in underground saline aquifers. *Appl. Energy* **2022**, *314*, 118985. [CrossRef]
10. Thanh, H.V.; Lee, K.K. Application of machine learning to predict CO_2 trapping performance in deep saline aquifers. *Energy* **2022**, *239*, 122457. [CrossRef]
11. Al-qaness, M.A.A.; Ewees, A.A.; Thanh, H.V.; AlRassas, A.M.; Dahou, A.; Elaziz, M.A. Predicting CO_2 trapping in deep saline aquifers using optimized long short-term memory. *Environ. Sci. Pollut. Res.* **2022**, *137*, 1–15. [CrossRef] [PubMed]

12. Majid, S.-F.; Hung, V.T.; Zhenxue, D.; Abolfazl, M.; Mohammad, R.; Umar, A.; Ahmed, E.R. Exploring the power of machine learning to predict carbon dioxide trapping efficiency in saline aquifers for carbon geological storage project. *J. Clean. Prod.* **2022**, *372*, 133778. [CrossRef]
13. Hung, V.T.; Aiyoub, Z.; Majid, S.-F.; Umar, A.; Zhang, H.M. Application of hybrid artificial intelligent models to predict deliverability of underground natural gas storage sites. *Renew. Energy* **2022**, *200*, 169–184. [CrossRef]
14. Thanh, H.V.; Binh, D.V.; Kantoush, S.A.; Nourani, V.; Saber, M.; Lee, K.-K.; Sumi, T. Reconstructing daily discharge in a megadelta using machine learning techniques. *Water Resour. Res.* **2022**, *58*, e2021WR031048. [CrossRef]
15. Wang, Y.L. Application of High-level Drilling Technology in Gas Control in Extra-thick Coal Seams. *Coal Mine Saf.* **2017**, *48*, 120–122.
16. Wang, Y.L.; Guo, Z.J. Optimization and application of high-level directional drilling technology in Tangkou Coal Mine. *Coal Mine Saf.* **2018**, *49*, 152–155. [CrossRef]
17. Gao, Y.K.; Xiao, S.L.; Yan, J.H.; Huang, Z.A.; Wang, C.J.; Fu, M.M. Numerical simulation of high-level borehole gas drainage in goaf of Pingshan Coal Mine. *Coal Mine Saf.* **2017**, *48*, 160–163. [CrossRef]
18. Jiao, Z.H.; Lin, W.G.; Huang, H.T.; Guo, K.B.; Yuan, Q.P. Study on gas control in different regions along irregular mining face in goaf. *Coal Eng.* **2021**, *53*, 89–93. [CrossRef]
19. Gao, L.; Li, X.J.; Pan, J.C. Main control factors and control measures of gas emission in mining face of Buertai Mine. *China Saf. Prod. Sci. Technol.* **2019**, *15*, 130–135. [CrossRef]
20. Xu, X.F.; Xue, W.C.; Li, Y.Z.; Wang, Y.L.; Wang, T.Y.; Wang, H.B. Study on optimization of high-level borehole drainage parameters in rock burst coal seam. *Coal Mine Saf.* **2017**, *48*, 157–159. [CrossRef]
21. Feng, P.Y.; Zhang, J.Y.; Wu, J.G.; Guan, L.H. Research on Optimization of Gas Drainage Parameters in High Boreholes Based on Principal Stress Judgment. *Ind. Saf. Environ. Prot.* **2022**, *48*, 13–17.
22. Zou, Y.X. Numerical Simulation Research on Optimization of Gas Parameters in High-level Borehole Gas Drainage in Fusheng Coal Mine. *Min. Saf. Environ. Prot.* **2014**, *4*, 15–17, 21.
23. Guo, D. Research on Optimization of High-level Borehole Parameters for Gas Drainage. *Coal Eng.* **2013**, *45*, 60–62. [CrossRef]
24. Hu, J.L.; Zhao, J.; Wang, S.L.; Zhang, Z.R. Research on Reasonable Layout of High-level Boreholes Based on the Development of Fissures in Overburden Rock. *Coal Eng.* **2021**, *53*, 90–93.
25. Guo, X.H.; Ren, Z.J. Research on Layer Optimization of High-level Borehole Gas Drainage under the Condition of Multiple Coal Seams. *Coal Mine Saf.* **2021**, *52*, 194–198.
26. Shi, K. Research on Coal and Gas Co-mining Technology of Protective Seam on Close-distance Coal Seam Group. *Coal Eng.* **2018**, *50*, 64–66.
27. Li, Q.X.; Xu, C.; Liu, J.L.; Fang, J.; Zhang, J.; Liu, F. Key Technology and Engineering Practice of Directional Drilling for Global Gas Drainage in Coal Mines. *J. Coal Sci.* **2022**, *47*, 3108–3116.
28. Tao, Y.Q.; Zhang, J.Z.; Guo, M.G.; Yu, H.; Pan, R.J. Research and Engineering Practice of Mining Pressure Relief Gas Drainage Replacing Borehole for Roadway. *Min. Saf. Environ. Prot.* **2022**, *49*, 43–48.
29. Zhang, H.Q. Research on Near-field Gas Drainage Technology in Goaf of Hard Overlying Rock Working Face. *Coal Eng.* **2022**, *54*, 74–78.
30. Thanh, H.V.; Sugai, Y.; Sasaki, K. Impact of a new geological modelling method on the enhancement of the CO_2 storage assessment of E sequence of Nam Vang field, offshore Vietnam. *Energy Sources Part A Recovery Util. Environ. Eff.* **2019**, *42*, 1499–1512. [CrossRef]
31. Thanh, H.V.; Menad, N.A.; Lee, K.K. A Robust machine learning models of carbon dioxide trapping indexes at geological storage sites. *Fuel* **2022**, *316*, 123391. [CrossRef]

processes

MDPI

Article

A Dynamic Permeability Model in Shale Matrix after Hydraulic Fracturing: Considering Mineral and Pore Size Distribution, Dynamic Gas Entrapment and Variation in Poromechanics

Qihui Zhang [1], Haitao Li [1,*], Ying Li [1,*], Haiguang Wang [2] and Kuan Lu [3]

1 State Key Laboratory of Oil and Gas Reservoir Geology and Exploitation, Southwest Petroleum University, Chengdu 610500, China; zhang.qihuii@outlook.com
2 Central Sichuan Division, Petro China Southwest Oil & Gas Field Company, Suining 629000, China; whg2021_01@petrochina.com.cn
3 China Petroleum Engineering & Construction Corp. Beijing Company, Beijing 100085, China; lukuan@cnpc.com.cn
* Correspondence: lihaitao@swpu.edu.cn (H.L.); yingingli@163.com (Y.L.)

Abstract: Traditional research on apparent permeability in shale reservoirs has mainly focussed on effects such as poromechanics and porosity-assisted adsorption layers. However, for a more realistic representation of field conditions, a comprehensive multi-scale and multi-flowing mechanism model, considering the fracturing process, has not been thoroughly explored. To address this research gap, this study introduces an innovative workflow for dynamic permeability assessment. Initially, an accurate description of the pore size distribution (PSD) within three major mineral types in shale is developed using focussed ion beam-scanning electron microscopy (FIB-SEM) and nuclear magnetic resonance (NMR) data. Subsequently, an apparent permeability model is established by combining the PSD data, leading to the derivation of dynamic permeability. Finally, the PSD-related dynamic permeability model is refined by incorporating the effects of imbibition resulting from the fracturing process preceding shale gas production. The developed dynamic permeability model varies with pore and fracture pressures in the shale reservoir. The fracturing process induces water blockage, water-film formation, and water-bridging phenomena in shale, requiring additional pressure inputs to counteract capillary effects in hydrophilic minerals in shale, But also increases the overall permeability from increasing permeability at larger scale pores. Unlike traditional reservoirs, the production process commences when the fracture is depleted to 1–2 MPa exceeds the pore pressure, facilitated by the high concentration of hydrophobic organic matter pores in shale, this phenomenon explains the gas production at the intial production stage. The reduction in adsorption-layer thickness resulting from fracturing impacts permeability on a nano-scale by diminishing surface diffusion and the corresponding slip flow of gas. this phenomenon increases viscous-flow permeability from enlarged flow spacing, but the increased viscous flow does not fully offset the reduction caused by adsorbed-gas diffusion and slip flow. In addition to the phenomena arising from various field conditions, PSD in shale emerges as a crucial factor in determining dynamic permeability. Furthermore, considering the same PSD in shale, under identical pore spacing, the shape factor of slit-like clay minerals significantly influences overall permeability characteristics, much more slit-shaped pores(higher shape factor) reduce the overall permeability. The dynamic permeability-assisted embedded discrete fracture model (EDFM) showed higher accuracy in predicting shale gas production compared to the original model.

Keywords: apparent permeability; shale gas; imbibition; flow back; Knudsen flow; adsorpted gas; free gas

Citation: Zhang, Q.; Li, H.; Li, Y.; Wang, H.; Lu, K. A Dynamic Permeability Model in Shale Matrix after Hydraulic Fracturing: Considering Mineral and Pore Size Distribution, Dynamic Gas Entrapment and Variation in Poromechanics. *Processes* **2024**, *12*, 117. https://doi.org/10.3390/pr12010117

Academic Editors: Guining Lu

Received: 5 December 2023
Revised: 21 December 2023
Accepted: 26 December 2023
Published: 2 January 2024

1. Introduction

Shale, as an unconventional reservoir, necessitates distinct exploitation techniques. The production of shale gas has played a pivotal role in global energy supply [1]. The sedimentary structure of shale amalgamates various minerals, including organic matter, clay, carbonates, feldspars, quartz, and pyrite [2–4]. Pores within the shale matrix exist across multiple scales, from nano- to micro-scale, rendering shale formation difficult to characterise [5–7]. Additionally, shale gas exploration using the fracturing technique has had a significant impact on the poro-mechanism of shale gas reservoirs (SGR). To enhance the ultimate recovery of SGR, a comprehensive understanding of reservoir characterisation, flowing mechanism, and dynamic petrophysics play key roles in shale gas production [3,8–11].

For shale characterisation, researchers have systematically classified the minerals found within shale into three predominant categories based on their petrophysical properties: brittle minerals, clay, and organic matter (Figure 1). Studies suggest that the pores within these three minerals contribute to shale permeability [12–14]. These include inter-granular pores formed between clay minerals, round pores caused by erosion and grains overlapping in brittle minerals, and intra-granular pores within organic matter [3,15–20]. For quantifying the pore structure and distribution within different minerals, researchers have applied different routines to quantitatively analyse the pore size distribution (PSD). The numerical description of PSD addresses the fractural theory and the log–Gaussian distribution. The fractural theory is more commonly used to describe the micro-fracture while the log–Gaussian distribution describes PSD directly from experiment results [10,21–25]. The experimental invasion approach includes mercury intrusion porosimetry (MIP), low-temperature nitrogen adsorption (LTNA), and carbon dioxide adsorption. The non-invasive methodologies include nuclear magnetic resonance (NMR) and optical measurements methods include SEM and focussed ion beam-scanning electron microscopy (FIB-SEM) [5,8,25,26] (Figure 1). However, the single experimental methodology data do not reflect whole scale PSD owing to the range limitations inherent to each method, and there is no whole scale methodology that can both reflect PSD and shape factor concurrently [17,27,28]. Recent research has attempted to describe whole-scale PSD using the integration of nano- and micro-scale approaches [23,29]. However, a simple integration method will engender an accuracy issue at the integration boundary. In this article, to describe PSD, we devised a novel coupling method to integrate the NMR and SEM data based on the mechanism of nuclear relaxivity.

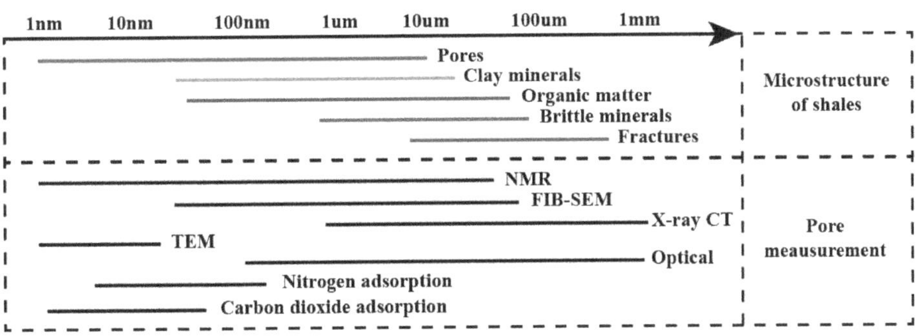

Figure 1. Different structures and measurement scales in shale.

As previously noted, SRG development requires a fracturing technique. Based on observation, a significant amount of hydrophobic organic matter surrounds the brittle minerals. Thus, substantial nano- and micro-pores escalate the capillary phenomenon during fracture, causing substantial fluid leakage during fracturing [13,14,30–33]. The water imbibition process caused by the fluid leakage compressed the original gas in place and displaced it into a deeper location within the pores under the fracturing pressure. the imbibed fracturing fluid will race to control the adsorption sites of the adsorbed gas owing to the stronger adsorption capacity of water molecules, resulting in a reduction in the adsorption layer thickness and water-blockage. As the pressure of the fracturing process decreases, the entrapped gas expands and drains the water at the pore entrance [4,13,34]. The imbibition shows a difference based on mineral petrophysics, and fracturing fluid leakage results in water blockage in pores, drastically impacting the production of shale gas [7,26,35,36]. Quantitative analysis of fracturing fluid leakage in pores and its corresponding impact on gas production has been explained through experiments conducted on capillary tubes with varying radii. Additionally, scholars have utilized molecular simulations to describe the impact on the reduction of adsorbed gas due to fracturing fluid leakage, a factor proven to greatly influence the gas's apparent permeability [2,17,18,37].

Water and gas transportation within shale formations covers multiple scales and follows different flow regimes owing to the petrophysics of SGR [12]. Confinement in nano-micro pores can alter the phase behaviour of fluids in the pores, and, consequently, the flow regime owing to the distortion of vapour–liquid equilibrium conditions [19,21,38,39]. Based on the Knudsen number, the flow regime can be classified as adsorption gas diffusion, slip flow, and viscous flow. Recent research [35,40] involving accurate explanation has focussed on free gas adsorption. Further explanations related to the flow regime in SGR will be presented later in the article.

Conventional numerical description for dynamic permeability in SGR integrated stress-dependent pore radii variation and poro-elastic poromechanics. Figure 2 illustrates the experimental results from multiple studies. It is well-documented [17,41] that geological characteristics during SGR depletion and production are a deterministic index of shale gas capability in hydrocarbon generation and production process. Numerous researchers have investigated the impact on permeability from dynamic radius at varying stress levels [42–46]. Based on numerical simulations, researchers have concluded that effective stress on pores will change the pore formation, and, thus, significantly impact the apparent permeability by changing the flow mechanism. Additionally, the matrix undergoes a slight increase early during production and becomes larger at the later stage because of slip flow and Knudsen diffusion [30,38,47–49].

This study aimed to comprehensively address all key attributes of dynamic permeability in Shale Gas Reservoirs (SGRs) through a detailed workflow (Figure 3). Firstly, complex PSD characterization is conducted based on the stitched FIB-SEM data and NMR data, including imaging, segmentation, and statistical analysis to extract equivalent pore radius at different scales. The whole-scale PSD of equivalent granular-shaped pores is characterized by coupling both experimental results at this stage. After that, the PSD in shale is separated by three minerals with the help of Bayesian statistics for Log-Gaussian characterization. The characterized PSD includes probability density at various radii and the shape factor of clay pores. This Log-Gaussian-based PSD description in different minerals is then utilized for radius correlation considering water film thickness under field conditions and to extract the critical liquid bridging parameters. After this, the correlated PSD becomes a crucial factor influencing the fracturing-induced imbibition process. During this stage, the force balance at different imbibed conditions under varying pressures and their corresponding impact on the adsorbed gas in pores are characterized. Finally, with the probability density at each correlated pore scale, critical water blockage, and liquid bridging condition at different pressures under varying radii, the apparent permeability was calculated at each correlated equivalent radius in the capillary bundle model. Combined with probability density, liquid bridging, and water blockage, the dynamic permeability encompasses all the aforementioned flowing mechanisms and water–gas conditions to reflect real-world field conditions.

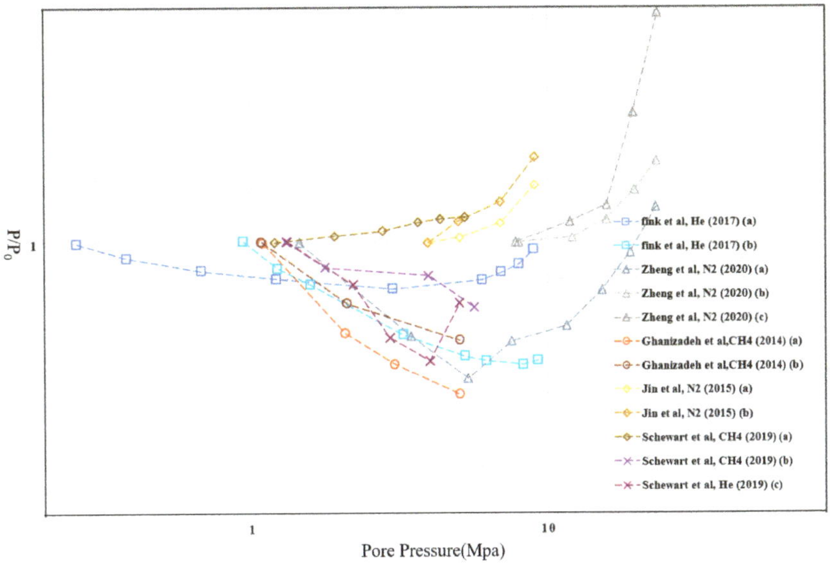

Figure 2. Dynamic permeability from the experimental approach [17,42].

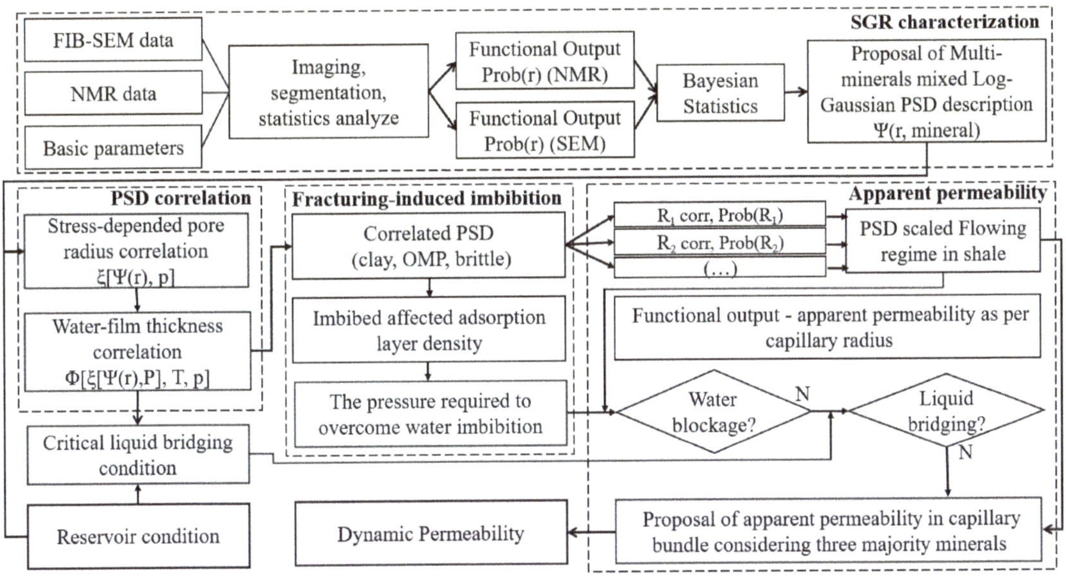

Figure 3. The proposed workflow of extracting dynamic permeability in shale matrix.

To validate the dynamic permeability model, it was integrated into the EDFM simulator as an amendment permeability module. The proposed model demonstrated an increased precision in reservoir production prediction compared to the original simulated results.

2. Complex Pore Characterisation

2.1. Description of PSD within Different Minerals

For shale characterisation, numerous studies [21,48,49] have examined dynamic petrophysics from pores within different minerals. Drawing insights from visualisation technology, our proposed mineral-pore system is divided into three different pores: granular-shaped inter-pores between brittle minerals; inter-pores between tetrahedral sheets-shaped clay minerals; and intra-pores within organic matter for characterisation (see Figure 4).

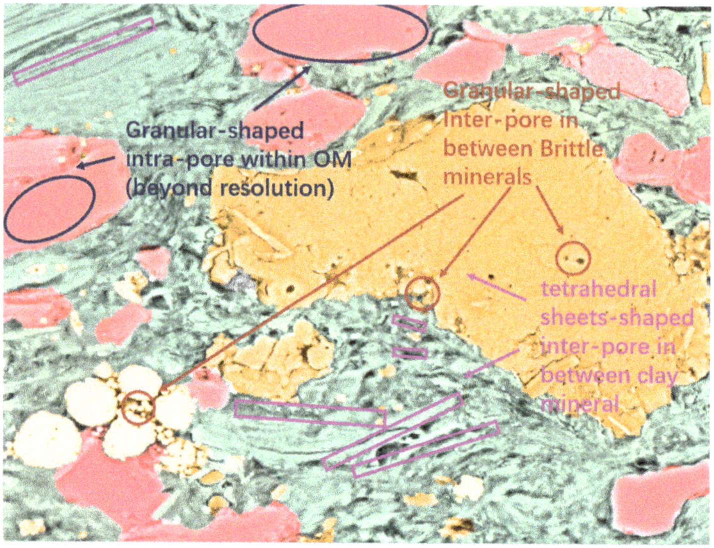

Figure 4. Different types of pores in shale FIB-SEM image (filter: green-clay; red-organic matter; orange-brittle).

Quantifying PSD involves a diverse set of implementations across the nano- and micro-scales. Numerous studies contribute to the realm of accurate PSD description by utilising cutting-edge visualisation techniques and characterising interactive systems within three types of minerals [8,13,50,51]. However, those pore size characterised methodologies with merits and limitations, the visualisation methods such as X-ray CT, FIB-SEM, and TEM are capable of describing the micro-nano pore and mineral shape from a small 2D planer or 3D structure, whereas the adsorption testing and NMR can reflect the PSD based on a shale sample scale. Additionally, different implementations reflect varied results at different scales because of the theory behind the methodology. In this study, for complex shale PSD characterisation, a coupled NMR and FIB-SEM methodology is used to accurately reflect the valid PSD based on the NMR methodology and the characterisation of pore geometry based on FIB-SEM.

In our study, nano-micro PSD is extracted from FIB-SEM. Sup-resolution image from the WY formation is created utilising the FIB-SEM line-by-line scanning image stitching technology [52]. To characterise the PSD in shale, we stitched 16,000 FIB-SEM images with 1024×1024 pixels (50 nm per pixel) via python codes, pores within sup-resolution images are extracted from the greyscale threshold segmentation software, the surface pore area and related probability density, shape factor statistics are extracted form the stiched FIB-SEM image. combined with the image segmentation software (Figure 5), the equivalent pore distribution with radius is shown in Figure 6. Based on the box-plot statistics, we conclude that there are two spikes in the boxplot statics above 0.1 um. Consequently, referring to different nano-micro test results [3,53] and different PSD models [10] (Figure 7), we assume the nano-micro scaled PSD of the WY formation follows the log–Gaussian distribution.

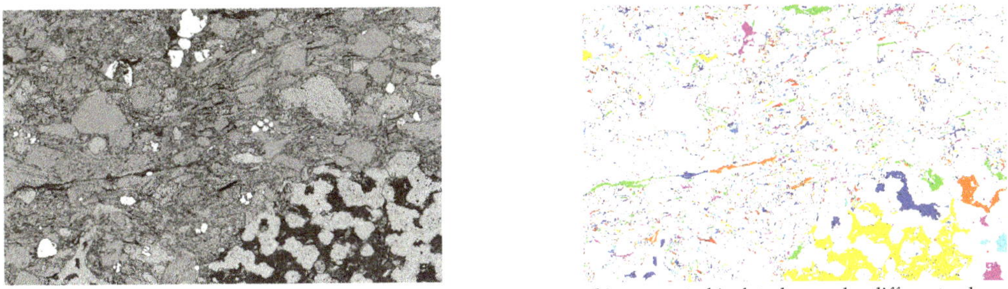

(**a**) stitched FIB-SEM image (**b**) segmented isolated pores by different colour

Figure 5. Sup-resolution image and corresponding segmented image.

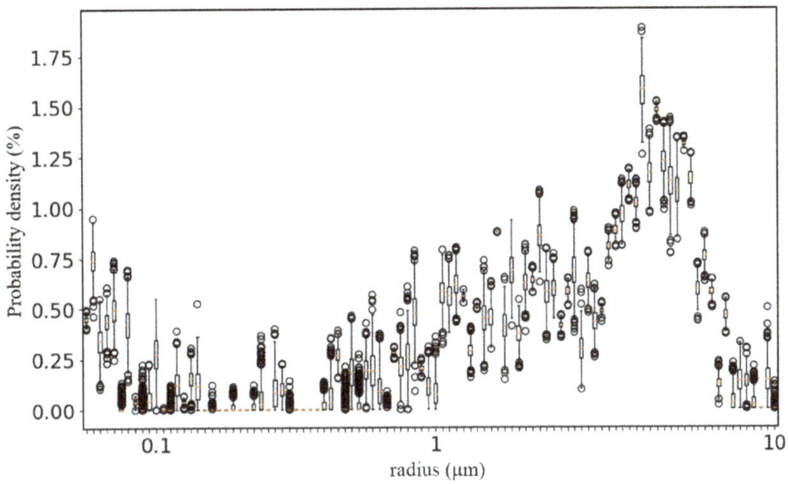

Figure 6. Statistics of Micro-nano scaled PSD from stitched FIB-SEM image (100 sets).

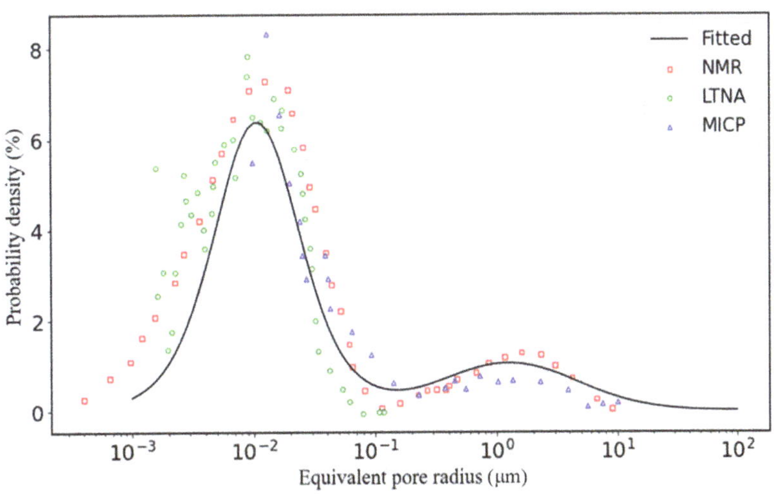

Figure 7. Different PSD testing results.

2.2. Coupling PSD by Comparing NMR and FIB-SEM Data

To overcome the limitations of the coupled PSD distribution in previous studies, this study utilised both FIB-SEM and NMR data to comprehensively clarify pore structure and PSD [2,48,54]. FIB-SEM can directly observe the PSD within different minerals in the shale matrix, while NMR transfers the signal decay to obtain the volume of pores. The transformation between pore volume and NMR relaxivity data was described based on the fundamental principles of the porous media NMR theory [55].

$$\frac{1}{T_{2,S}} = \rho_2 S_V \tag{1}$$

where, S_V is the ratio of surface area to pore volume; T_2 is the surface relaxivity—a value that characterises the strength of relaxation induced by the solid/fluid interface.

As mentioned earlier in this study, pores in shale are considered to have granular and slit geometries. The NMR interpretation alone cannot directly illustrate the geometry of pores, and FIB-SEM has limitations in describing nano-scaled pores in shale due to resolution constraints. To comprehensively characterise the PSD and the corresponding shape factor, we integrated NMR data and FIB-SEM data using information from larger granular-shaped organic matter pores (OMPs) in the range of approximately 100 nm to 1 μm, where pore scale can be characterised by both methodologies.

PSD in larger OMPs is directly described from visualized FIB-SEM data, while the original T_2 data interpretation relies on the pore structure. By adjusting the S_v, the PSD curve from interpreted NMR data aligns with FIB-SEM results, and the corresponding S_v ratio reflects the equivalent granular pore capillary bundle PSD. Consequently, an equivalent granular-shaped capillary bundle model is generated from NMR data with FIB-SEM coupled S_v to demonstrate the full-scaled distribution in the shale matrix.

Thus, the PSD at varying equivalent radii in shale is considered as:

$$P(r) = \sum \frac{1}{rlg\sqrt{2\pi}\sigma} EXP\left[-\frac{(lgr-\mu)^2}{2\sigma^2}\right] \tag{2}$$

where σ and μ are standard derivations and mean; r is the equivalent radii.

2.3. Log–Gaussian Mixed Capillary Model for Describing Overlapped Pores Distribution in Clay and Brittle Minerals Based on Coupled Data

Based on multiple scholarly studies, it is evident that the PSD in various shale formations cannot be simplistically characterised as a mere combination of Gaussian distributions arising from organic matter and inorganic pores [23]. Notably, certain PSD tests conducted on the Barnett shale did not exhibit the conventional bimodal distribution with two distinct peaks in the curve. This underscores the complexity and unique characteristics of shale formations, challenging the application of a one-size-fits-all approach [27] in describing their PSD [25,28]. Fortunately, research into distinct pore types has revealed that the pore distribution at various equivalent radii can still be effectively described based on the probability density of the three predominant minerals. The portrayal of PSD within the shale matrix is described as follows:

$$P(r) = \sum^{K} c_k P(r, \mu_k, \sigma_k)(k = b, c, om) \tag{3}$$

$$P(r|\mu_k, \sigma_k) = \frac{1}{\sqrt{2\pi}\sigma_k} e^{-\frac{(r-\mu_k)^2}{2\sigma_k}} \tag{4}$$

where, r is the equivalent pore radius; σ_k and μ_k are the standard deviation and mean, respectively, of the corresponding pore distribution of brittle mineral b, clay mineral c, and organic matter om. c_k reflects the volume concertation of brittle, organic matter and clay, respectively.

As mentioned, the micro-scale distribution is considered a Gaussian mixed model (GMM) of three clusters. To accurately interpret the slit-like PSD, the Gaussian distribution of equivalent clay pores can be extracted, thanks to the Expectation Maximisation Algorithm. A latent variable, z is added to the algorithm to describe probability of categorisation. Considering the full range probability of categorising a random pore includes two results: clay mineral pores and brittle mineral pores. The latent variable, z_k = 1 reflects the correct categorisation. The full range probability can then be described with the help of latent variable $z \in \{(0, 1)\ (1, 0)\}$ by describing the random categorisation based on Bayesian interpretation. The joint probability distribution of z, as follows:

$$p(z) = p(z_1)p(z_2)\cdots p(z_K) = \prod_{k=1}^{K} \pi_k^{z_k} \tag{5}$$

$$P(r|z) = \prod^{K} P(r|\mu_k, \sigma_k) \tag{6}$$

Because, $z_k \in (0, 1)$. Thus, the conditional probability of categorising pores are described as:

$$P(r|z) = \sum^{K} c_k P(r|\mu_k, \sigma_k)(k = b, c, om) \tag{7}$$

c_k is the probability that certain mineral pores will be selected, which are described via volume concertation. Based on the algorithm mentioned above, we already know the prior probability $P(z)$ and likelihood $P(r|z)$. The posterior probability $P(z|r)$ is described as follows:

$$P(z|r) = P(z_k = 1|r) = \frac{P(z_k = 1)P(r|z_k = 1)}{\sum_{j=1}^{K} P(r|\mu_k, \sigma_k)} \tag{8}$$

with prior probability, likelihood, and posterior probability from the Bayes decision theory, the perimeter μ_k and σ_k of the GMM, which represents the two majority micro-scale pores, can be accurately predicted using the converged expectation–maximization algorithm with the help of a Python-coded iteration function. The schematic workflow is shown in Figure 8. Thus, the equivalent capillary bundle model reflecting the brittle mineral and entrapped clay pores will be characterised.

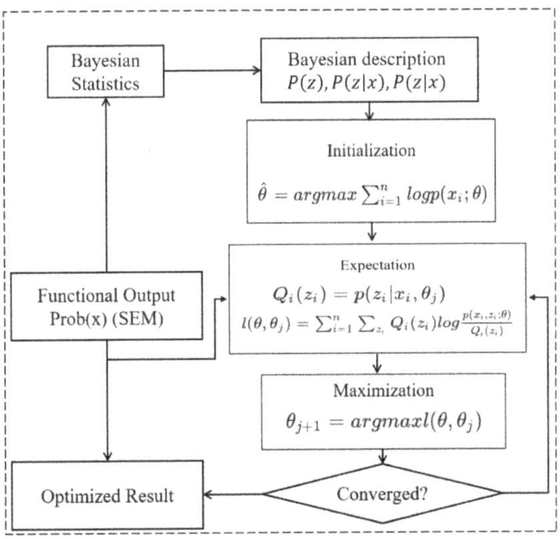

Figure 8. The schematic workflow of Bayesian-based expectation–maximization algorithm for PSD separation.

2.4. Capillary Bundle Model for Describing Shale Matrix

The process of constructing a shale matrix model based on PSD extraction is illustrated in Figure 9. Initially, the FIB-SEM image was analysed using a segmentation algorithm to extract pores ranging from micro- to meso-scale, including slit-like pores between clay minerals. Thereafter, the data were processed to eliminate the impact of micro fractures. By combining the nano-scale NMR data, an equivalent capillary tube could be established, facilitating a comprehensive characterisation of the shale matrix.

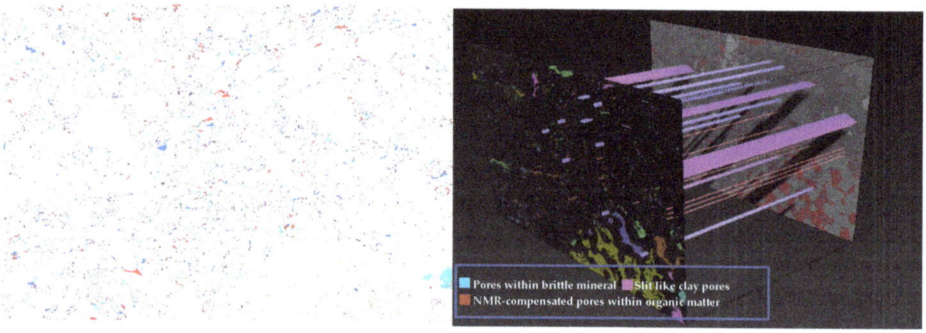

Figure 9. The conceptual model of capillary bundle for describing shale matrix.

2.5. Tortuosity of Capillary Tube Model

However, the GMM-based capillary tube model cannot directly reflect the flow through porous media and the equivalent capillary length cannot be simply explained as the length of shale matrix [45]. The dimensionless parameter, τ, which is called hydraulic tortuosity, is used to describe the average length of all particle path lines that pass through a cross-section—in this case, the shale matrix. Hydraulic tortuosity was first proposed by Kozeny via the well-known Kozeny–Carman equation [56]:

$$K = \frac{\phi^3}{\beta\, \tau_h^2 S^2} \tag{9}$$

where, ϕ is the porosity defined as the fraction of pore space in the porous medium; S is the specific surface area equal to the ratio of interstitial surface area to bulk volume; and β is the shape factor (a constant depending on the type of granular material).

However, this method is not a perfect fit for the capillary bundle model in shale owing to its high heterogeneity. Another representative theoretical model was derived from a fixed bed of randomly packed identical particles [39]:

$$\tau = 1.23 \frac{(1-\phi)^{4/3}}{\xi^2 \phi} \tag{10}$$

where, ξ is the shape factor (sphericity) of the particle; ξ = 1 for sphere particle and ξ < 1 for non-spherical particles. In this study, the length of the capillary bundle model considering hydraulic tortuosity for each mineral pore is described as follows [57]:

$$\Omega(L) = 1.23 \frac{L(1-\phi_i c_i)^{4/3}}{{\xi_i}^2 \phi_i} \tag{11}$$

2.6. Stress Dependent Pore Radius Correlation

2.6.1. Pores with Brittle Mineral and Clay Mineral

Shale is viewed as a naturally fractured reservoir sensitive to changes in stress conditions and increased effective stress decreases the fracture aperture. Based on laboratory research, the size changes for non-organic matter (brittle minerals and clay) are not as strong as for organic matter [58,59]. In this study, the equation applied for describing brittle minerals and clay minerals is as follows:

$$\phi = \phi_r + (\phi_i - \phi_r)e^{(a\sigma'_m)} \tag{12}$$

where, ϕ represents the porosity of the medium, ϕ_r signifies the residual porosity under high-stress conditions, ϕ_i denotes the initial porosity, a is a media property determined through experimental methods, and σ'_m prime stands for the mean effective stress. The numerical values for parameters a have been extracted from an earlier investigation [60].

Thus, assuming the length in the equivalent capillary bundle model in the shale matrix remains constant during production, the radius in the equivalent capillary bundle model is described as follows:

$$\varnothing \propto R^2\Omega(L) \tag{13}$$

Considering the tortuosity-compensated capillary length remains constant during reservoir depletion, substituting (13) into (12) the radius changes in the circular-shaped equivalent capillary bundle model could be described as follows:

$$R_{iom}{}^2\Omega(L) = R_r{}^2\Omega(L) + \left(R_i{}^2\Omega(L) - R_r{}^2\Omega(L)\right)e^{(a\sigma'_m)} \tag{14}$$

$$\Lambda(r)_{iom} = \sqrt{R_r{}^2(1 - e^{a\sigma'_m}) + r^2e^{a\sigma'_m}} \tag{15}$$

2.6.2. Water Film Thickness Considering Adsorption

The stability and thickness of adsorbed films on substrates are primarily influenced by force interactions between the thin film and the solid surface, which can be measured by the disjoining pressure. Many studies have investigated the thickness of water films in geological media [1,61,62]. In the thermodynamic system of water film adsorption on a flat substrate, the relationship between disjoining pressure and relative humidity can be described [62]. To consider the structural force caused by OMP, Li [62] proposed a novel model that takes into account both hydrophobic and hydrophilic minerals. They consider the water film with slit-like pores as the combination of surface force on both slit-like pores. Hence, the water films adhering to slit surfaces maintain a flat configuration, and due consideration has been given to the capillary pressure arising from the gas–water meniscus interface.The disjoining pressure for capillary tube and slit-like pores can be described as follows [16,63]:

For equivalent sphere pores:

$$\Pi_{capillary}(h) = \frac{r}{r-h}\cdot\Pi_{flat}(h) + P_c = \frac{r}{r-h}\cdot\Pi_{flat}(h) + \frac{\gamma}{r-h} \tag{16}$$

$$\Pi_{flat}(h) = \Pi_m(h)+\Pi_e(h) + \Pi_s(h) \tag{17}$$

$$\Pi_m(h) = -\frac{A_{gws}\left(15.96\frac{h}{l} + 2\right)}{12\pi h^3\left(1 + 5.32\frac{h}{l}\right)^2} \tag{18}$$

$$\Pi_e(h) = 2\varepsilon_r\varepsilon_0(k_BT/ze)^2\cdot K^2 \tag{19}$$

$$\Pi_s = ke^{-\frac{h}{\lambda}} \tag{20}$$

where, $\Pi_1(h)$ denotes the force between the water film and the surface on the same side, accounting for both short- and long-range interactions, MPa; Similarly, $\Pi_2(h)$ represents the force between the water film and the opposing surface; $\Pi_3(h)$ signifies the force between two adsorbed water films, considering only molecular action, Mpa. It's worth noting that Awgw in $\Pi_3(h)$ represents the Hamaker constant for liquid-gas-liquid interactions caused by two adsorbed water films.

The relationship between water film thickness h, r described by disjoining pressure $\Pi(h)$ and P_v/P_0 in the case of water film inside granular and slit-liked capillary can be described via the implicit function, the corresponding relationship between:

$$\Pi_{capillary}(h) = -\frac{RT}{V_m}\ln\frac{P_v}{P_0} \tag{21}$$

The water film thickness in the capillary bundle model at reservoir condition (23 Mpa) is shown in Figure 10. The designated gas flow radius based on the capillary bundle radius is described as follows:

$$\Phi(r) = r - h(T, V_m, P_v/P_0) \tag{22}$$

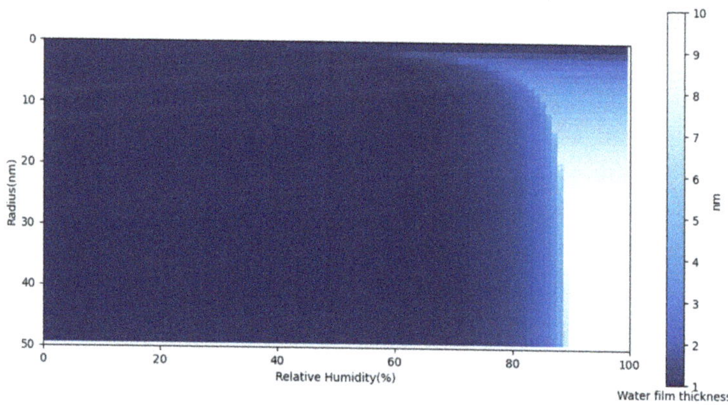

Figure 10. The corresponding water film thickness at varying radii and RH.

2.6.3. Pores in Organic Matter

Organic matter in shale reservoirs is considered as both source rock and porous media for gas. A high total organic carbon (TOC) implies greater adsorption and desorption capacity. As shale reservoir depletes, adsorbed gas begins to desorb from the pore surface. The matrix shrinkage process is relatively similar to that of coalbed methane with coal; recently, studies have referred to the model of coal to quantitatively analyse this phenomenon [64,65]. In this study, a well-known mathematical model [47] is applied to calculate the variation of permeability for the organic matter. Additionally, many studies regarding coalbed methane have revealed the relationship between porosity and permeability, based on which, researchers have explored the relationship between porosity and permeability of organic matter [24]. For shale reservoir under specific stress, the porosity change and porosity–permeability relationship before and after depletion are described as follows:

$$k = k_i e^{-3c_f \Delta\sigma} \tag{23}$$

$$\Delta\sigma = -\frac{v}{1-v}(p - p_i) + \frac{E}{3(1-v)}\varepsilon_l\left(\frac{p}{p+p_\varepsilon} - \frac{p_i}{p_i+p_\varepsilon}\right) \tag{24}$$

$$\frac{k}{k_i} = \left(\frac{\phi}{\phi_i}\right)^3 \tag{25}$$

where, k represents media permeability; k_i stands for the initial media permeability; c_f denotes media compressibility, typically determined by fitting the correlation curve to laboratory test data; $\Delta\sigma$ represents the change in effective stress; ν is Poisson's ratio; p signifies media pressure; p_i is the initial media pressure; E stands for Young's modulus;

and ε_l and $p\varepsilon$ denote the Langmuir-type organic matter shrinkage constants. Thus, the pore radius described by the pressure sensitivity are shown in:

$$\Lambda(r)_{om} = R_i \sqrt[6]{e^{-3c_f \frac{v}{1-v}(p-p_i)+\frac{E}{3(1-v)}\varepsilon_l(\frac{p}{p+p\varepsilon}-\frac{p_i}{p_i+p\varepsilon})}} \tag{26}$$

3. Dynamic Permeability Model

3.1. Flowing Mechanism Considering Hydraulic Fracturing Process

It is conceptualised that the micro-fracture system comprises three parts: the granular-shaped organic matter pores; brittle mineral pores; and slit-like entrapped pores within clay. The conceptual model for gas transportation after gas production is illustrated in Figure 11. Smaller pores in hydrophilic minerals (brittle, clay) in the shale with higher capillary forces are comparable easily to drain the water into the pore space, resulting in water blockage, while larger hydrophilic minerals with smaller capillary force are required smaller pressure differences to overcome water imbibition caused from hydraulic fracturing. organic matter pores suffered the least impact from water imbibition processes due to their hydrophobic characteristics. As the reservoir is depleted, the original gas in the pore will displace the imbibed water outside the pore, water film will form around the pore wall. Thus, the gas flow of pores within the shale matrix can be recognized in two categories, fully water-saturated pores due to imbibition, and water-film-affected pores. Brittle minerals and clay minerals pore with different shape factors displaced different shapes in water film.

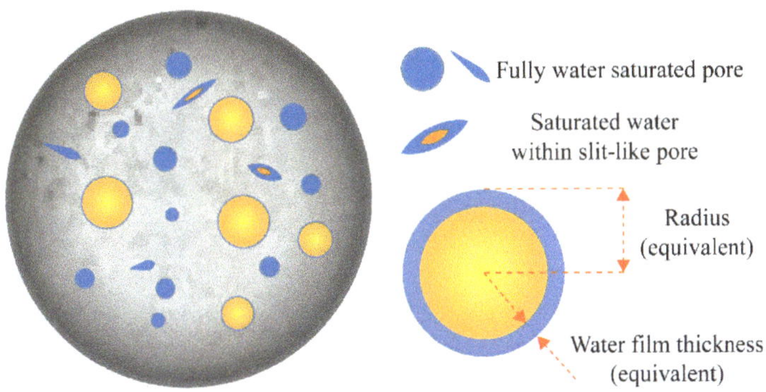

Figure 11. The conceptual model of micro-nano pores in fracturing process.

To quantitatively analyse the pore, during the fracturing process, substantial fracturing fluid was pumped into the deeper shale matrix throughout the capillary tunnel [14]. Owing to differences in petrophysics, the flow mechanism for the capillary tube in different minerals varies [66]. If the water is leaking off into open system throats, gas will be displaced into nearby pores or throats, the closed system takes account of the gas entrapment in a single throat, recent studies have explored the imbibition process as in a closed system (Figure 11), based on its ultra-low permeability [13,26] and experiment results [14]. The concepts of open and closed systems are illustrated in Figure 12. In this article, to express the fracturing process, we assume the capillary bundle model in a closed system. However, the capillary bundle model is not applicable for super heterogeneity shale reservoir because the model considered the capillary bundle as in a single mineralogy, corresponding water-blockage phenomenon and in the proposed model will be drastically influenced at the muti-mineralogy capillary model.

Figure 12. The schematic diagram of water–gas interaction in open (**left**) and closed (**right**) systems.

For the interpretation of water saturation and the water-blockage phenomenon in the proposed model, we consider the force at both sides of the capillary tube (Figure 13). Fluid and gas saturation in the conceptualised model during production are analysed via the force balance equation at the water–gas interface in pores. The force balance can be described during the fracking and flow-back periods for the water–gas characterisation. Eventually, the water-blockage phenomenon at different pressures can be recognised. Following the force-balanced equation considering compressed OGIP, capillary pressure, fracturing liquid pressure, and fracturing pressure are investigated. The water blockage phenomenon at different pores and fracturing pressures can be characterised during the fracturing period in the equivalent capillary bundle model [38]. The force balance equation at different minerals is shown as follows.

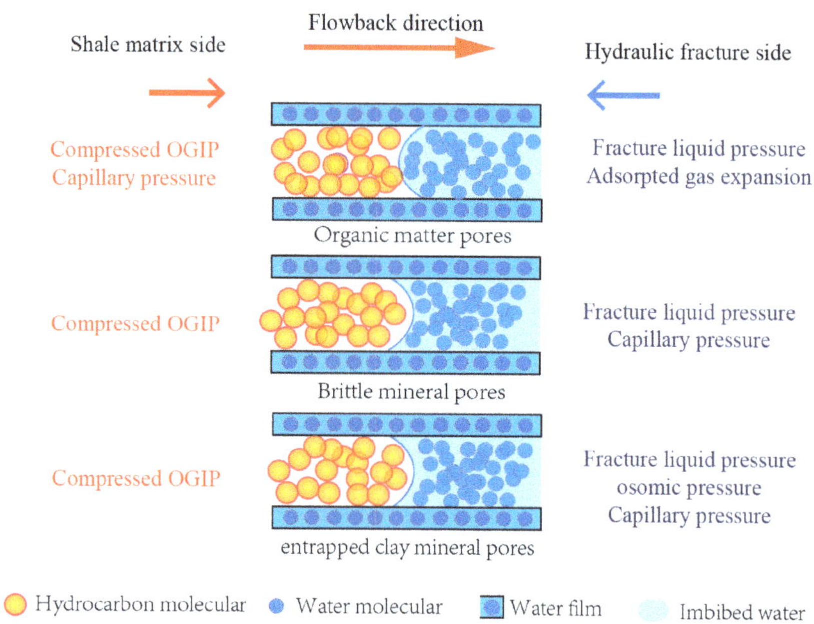

Figure 13. The conceptual model of water–mineral interaction force balance.

For OMP:

$$\frac{P_{g0}L_0}{L_0 - x_0} + \frac{4\sigma cos\theta_{omp}}{d} - P_f + \frac{4ZRTn_{omp}x(p)}{d(L_0 - x(p))} = 0 \tag{27}$$

For brittle minerals pores:

$$\frac{P_{g0}L_0}{L_0 - x_0} + \frac{4\sigma cos\theta_b}{d} - P_f + \frac{4ZRTn_b x(p)}{d(L_0 - x(p))} = 0 \tag{28}$$

For entrapped clay minerals pores:

$$\frac{P_{g0}L_0}{L_0 - x_0} + \frac{4\sigma cos\theta_c}{d} - P_f + \frac{4ZRTn_c x(p)}{d(L_0 - x(p))} + nE_\pi RT\left(C_0 - C_f\right) = 0 \quad (29)$$

where, x_0 is the equivalent initial water saturated capillary length; σ is the interfacial tension of water N/m; and L_0 reflects the tortuosity-considered equivalent capillary length. In this article, they are described as follows:

$$L_0 = \Omega(L)L = 1.23\frac{L(1-\phi_i c_i)^{4/3}}{\xi_i{}^2\phi_i} \quad (30)$$

3.2. Water Bridging Mechanism during Production

As previously mentioned, a water film is created because of force interactions. During the flow-back production period, as water is displaced from the capillaries through imbibition, hydrocarbon molecules begin to get adsorbed in the organic matter pores (OMP). Nevertheless, a thick water film persists within the capillary bundle, regardless of wettability. Surface forces become dominant in stabilising the liquid film. As the water film thickens, a liquid bridge forms, eventually spontaneously filling the entire capillary. This phenomenon of transition from the vapour phase to the liquid phase, known as 'capillary condensation', significantly impacts the apparent permeability.

Researchers [3,66] have discussed the characteristics of water-film thickness and stability within nano-pores and developed an approach considering fluid/pore-wall interactions to describe the phase behaviour of thin water-film transition into liquid condensation (Figure 14). They have identified that the instability mechanisms of adsorbed films differ inside slits and capillaries, with capillaries having higher surface interactions and easier condensation. The phase behaviour of adsorbed water film within nano-porous montmorillonite and shale was investigated, with water condensation found in hydrophilic clay samples, but only partial condensation in shale samples owing to the hydrophobic repulsion of organic minerals. This may impact the storage and transportation of gas in these nano-pores.

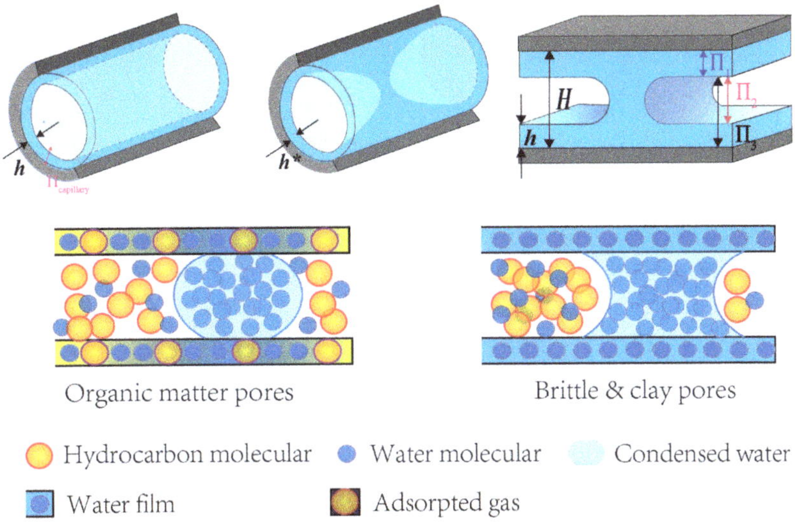

Figure 14. Water bridging mechanism at different water film thickness (h) and critical water film thickness (h*) at different mineralogies.

Generally, the instability condition of the thin wetting film could be determined by disjoining pressure isotherms, which was formulated as $\partial\Pi(h)/\partial h > 0$ [67]. Thus, the critical film thickness, h^* can be characterised as follows:

$$\left.\frac{\partial\Pi(h)}{\partial h}\right|_{h=h^*} = 0 \tag{31}$$

As per the previously derived water-film thickness equation, the corresponding critical water-film thickness within the capillary bundle model can be described as follows:

$$\Pi\left(h^*\right) = -\frac{RT}{V_m}\cdot\ln\frac{P_v^*}{P_0} \tag{32}$$

The corresponding vapor pressure, P_v^* and saturated vapor pressure, P_0 are described as follows [68]:

$$P_0 = 7.0746 - (1657.46/(T - 273.15 + 227.02)) \tag{33}$$

$$P_v^* = c_{water}P \tag{34}$$

3.3. Mixed Flow Regime

The capillary bundle model is considered stacked-up flow channels, each flow channel can be described as independent, such flows within different pore sizes show different flow regimes, have higher importance in engineering and are easy to get analytical solutions. The equivalent capillary bundle model adheres to the flow regimes within porous media, accommodating the coexistence of various flow regimes. Considering the pore-scale distribution in different minerals within the shale matrix, we categorise the flow regimes within a single capillary tube during shale gas flow-back production. This includes molecular flow from the adsorption layer and the Knudsen layer-affected viscous flow (Figure 15). Researchers have demonstrated that these two dominant flow regimes within nano-micro structures could be investigated independently [69–71]. For permeability calculations, both flow regimes could be considered independently. Subsequently, permeability could be calculated by adding free gas molecular flow and slip-correlated viscous flow.

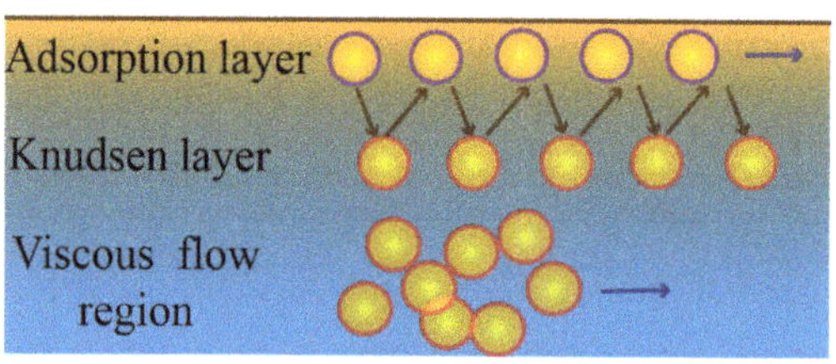

Figure 15. Mixed-flow region in equivalent capillary tube.

Gas-flow regimes in unconventional reservoirs are divided by the K_n (ratio of a molecular mean free path (MFP) to the average channel size). Considering a dynamic pore aperture, the effective K_n is described as follows:

$$K_n = \frac{\lambda}{b_{eff}} \tag{35}$$

$$\lambda = \frac{\mu_g}{P_p}\sqrt{\frac{\pi ZRT}{2M}} \tag{36}$$

where,

$$\mu_g = \mu_{g0}\left[1 + \frac{Y_1}{T_r^5}\left(\frac{P_r^4}{T_r^{20} + P_r^4}\right) + Y_2\left(\frac{P_r}{T_r}\right)^2 + Y_3\left(\frac{P_r}{T_r}\right)\right] \tag{37}$$

$$Z = \left(0.702e^{-2.5T_r}\right)P_r^2 - \left(5.524e^{-2.5T_r}\right)P_r + \left(0.044T_r^2 - 0.164T_r + 1.15\right) \quad (38)$$

$$P_r = P_p/P_g\,,\ T_r = T/T_g \quad (39)$$

M is the molecular mass, dimensionless; b_{eff} is the effect pore radius, m; μ_{g0}, is the viscosity at standard condition, Pa-s; and P_r and T_r are the relative pressure and temperature, respectively. MPa; Y_1 , Y_2 and Y_3 are the fitting curve coefficient, dimensionless; corresponding data will be displayed in Tables 1 and 2.

Table 1. Basic parameters of the dynamic permeability model for shale gas reservoir.

Parameters	Value
The maximum pore spacing, R_{fmax} (μm)	10 [54]
The minimum pore spacing, R_{fmin} (μm)	0.001
Porosity for fracture networks, ϕ	0.05 [57]
TMAC, f	0.79 [24]
Temperature, T (K)	393
Pore pressure, p (MPa)	23
Confining pressure, Pc (MPa)	100.56
Critical temperature, T_c, K	190.6
Critical pressure, P_c, Pa	4.599×10^6
Fitting constant, Y_1	7.9
Fitting constant, Y_2	-9×10^{-6}
Fitting constant, Y_3	0.28
The Langmuir pressure, P (MPa)	6.72×10^6
Adapted coefficient (Organic matter), a	5×10^{-8}
Media coefficient (in-organic matter), C_f (1/Mpa)	22
Molecular weight, M (kg/mol)	16×10^{-3}
Universal gas constant, R (J/(mol·K))	8.314
Interfacial tension (water-gas), σ, Nm^{-1}	25 [38]
Rarefaction coefficient, α	1.19 [24]
Diffusion coefficient, D_a^0	1×10^{-12} [24]
Adsorption isothermals, ΔH, $Jmol^{-1}$	14×10^3
Electrical conductivity, e, (Fm^{-1})	8.85×10^{-12}
Potential difference (solid-liquid), mV	50
CH_4 molecular radius, m	0.38×10^{-9}
CH_4 viscosity, MPa·s	0.0184
Water concentration in CH_4	0.001

Table 2. The parameters of different minerals.

Parameters	OMP	Brittle	Clay
Porosity concentration	0.9	0.03	0.07
STD derivation	0.32	0.17	0.23
Mean	0.97×10^{-6}	0.03×10^{-6}	0.062×10^{-6}
Contact angle (°)	116	10	12
Adsorbed gas concentration (mol/m^2)	7×10^{-6}	1.92×10^{-6}	1.8×10^{-6}

The viscosity correlation factor considering the rarefaction effect in capillary structure is as follows:

$$u_g = \frac{u_{g0}}{1 + \alpha Kn} \quad (40)$$

3.3.1. Free Gas Viscous Flow after Fracturing Process

In smaller Knudsen-numbered equivalent capillaries, viscous flow predominates compared to the previously mentioned flow regime. Unlike the earlier-discussed flow regimes, viscous flow is highly sensitive to the geometry of the flow region. Within the equivalent capillary bundle context mentioned earlier, viscous flow in the mixed-flow regimes is distinguished between rectangular and round shapes. For entrapped clay minerals pores, the velocity distribution within a constant wide-height ratio slit-like clay pore is described using Fourier's equation:

$$v(x,y) = \frac{\nabla P}{u_g}\frac{h^2}{12}\sum_{n=1,3,5\ldots}^{\infty}\frac{48}{\pi^3 n^3}\left(1-\frac{cosh(n\pi x/h)}{cosh(n\pi w/2)}\right)sin(n\pi y/h) \tag{41}$$

For the inter-pore within the shale matrix such as brittle minerals pores and larger-scaled OMP pores, the viscous flow velocity along the radius could be expressed as follows:

$$v(r) = \frac{{R_f}^2}{4\mu_g}\left(1-r^2\right)\nabla P \tag{42}$$

3.3.2. Adsorption Gas Diffusion with Impact of Fracturing Process

As the reservoir depletes, adsorbed gas in the organic matter pore surface diffuses along the flow-back direction. This phenomenon is explained by the Max-Stefan theory, given that the gas molecules are adsorbed as a single layer within entrapped organic matter pores and the diffusion process is powered by the electrochemical potential gradient. Assuming the desorption-diffusion process follows the Langmuir iso-thermal adsorption curve and the spherical shape of organic matter pore, the shale gas diffusion velocity in the adsorption layer driven by electrochemical gradient is given as follows [45]:

$$v_d = D_d C_a \frac{ZRT}{p^2}\nabla P \tag{43}$$

where, D_d is the diffusion coefficient, m^2/s, C_a is the molar concentration of adsorbed gas mol/m^3.

Based on the molecular simulation results, during the fracturing process, water molecules will displace the originally adsorbed methane (CH_4). Substantially, CH_4 molecular concertation, C_a at different pressures is described as follows:

$$C_a = \frac{4px}{\pi N_A d_m^3 (P_L + P)\Omega(L)} \tag{44}$$

where, N_A is the Avogadro constant, $6.02 \times 10^{23}/mol$, P_L is the Langmuir pressure, Pa.

Corresponding diffusion coefficient with fracturing impact is described as follows:

$$D_a = {D_a}^0 \frac{px}{(P_L + P)\Omega(L)} \tag{45}$$

where, ${D_a}^0$ is the diffusion coefficient from experiment, x and $\Omega(L)$ are the imbibition length in capillary and tortuosity-considered capillary length, respectively.

3.3.3. Correlation of Free Gas Flux from Slip Flow and Knudsen Diffusion

The movement of free gas within a capillary tube involves continuous flow because of molecular collisions and Knudsen flow caused by molecular-wall collisions. At nanoscale, the flow regime is significantly influenced by the Knudsen diffusion phenomenon; researchers consider this among the first boundary conditions in fluid dynamics. The viscous flow of free gas is impacted by the corresponding Knudsen diffusion boundary [18]. The slip phenomenon within the Knudsen layer will indeed impact the viscous flow of free gas.

In this scenario, the additional free gas viscous flow influenced by the slip condition in an equivalent granular-shaped capillary bundle, acting as a substitute for OMP and brittle minerals pores, could be expressed as follows [72]:

$$N_S = \frac{p}{ZRT}\frac{FK_v}{\mu_g}\nabla P \tag{46}$$

$$F = 4\left(\frac{2}{f} - 1\right)K_n = \sqrt{\frac{9\pi ZRT}{M}}\frac{u_g}{pR_f}\left(\frac{2}{f} - 1\right) \tag{47}$$

where, M is the molecular mass, g/mol; f is the tangential momentum accommodation coefficient (TMAC), dimensionless; and R_f is the pore spacing, m.

Additional free-gas flux velocities from rectangular equivalent slit-shaped clay minerals from X, Y axes directions are expressed as:

$$v(x) = \frac{\nabla P}{u_g}\frac{h^2}{12}\frac{2-f}{f}\frac{6K_n}{1-bK_n}\sum_{n=1,3,5\ldots}^{\infty}\frac{8}{\pi^2 n^2}\left(1 - \frac{cosh(n\pi x/h)}{cosh(n\pi\varepsilon/2)}\right) \tag{48}$$

$$v(y) = \frac{\nabla P}{u_g}\frac{h^2}{12}\frac{2-f}{f}\frac{6K_n}{1-bK_n}\sum_{n=1,3,5\ldots}^{\infty}\frac{8tanh(n\pi\varepsilon/2)sin(n\pi y/h)}{\pi^2 n^2} \tag{49}$$

where, ε is the shape factor of clay minerals, width/height, dimensionless.

The Knudsen diffusion refers to the collision between the CH_4 molecules and the pore wall. The diffusion process is recognised as the secondary boundary, which will also affect the free-gas flux. To characterise this phenomenon, studies have used the diffusion coefficient, D_k from a linear algorithm [45].

The Knudsen diffusion-induced free-gas flux is expressed as follows:

$$N_k = \frac{1}{ZRT}D_k = \frac{1}{ZRT}\frac{2R_f}{3}\sqrt{\frac{8ZRT}{\pi M}}\nabla P \tag{50}$$

where, N_k is the gas flux from the Knudsen diffusion.

The additional flux due to the Knudsen layer affecting slip-gas flow could be expressed as:

$$N = \frac{p}{Z\mu_{g0}RT}\frac{{R_f}^2}{8}\left(1 + \frac{2-f}{f}\frac{4K_n}{1-bK_n}\right)\nabla P \tag{51}$$

3.4. Development of Dynamic Apparent Permeability Model

Above mentioned work in this study have already delineated the equivalent capillary bundle model for describing PSD, the water-film thickness and corresponding water-bridging mechanism influenced by fracturing, dynamic poro-mechanism, and the flow-velocity model for each capillary. Subsequent work will consolidate these mechanisms and models to extract dynamic permeability at varying pressures.

Combining Equations (42), (46) and (50), the free-gas flux considering slip flow at different granular pores is described as follows:

$$\begin{aligned} N_f &= \int \pi v(r)dr + N_k + N_s \\ &= \pi\frac{p(1+\alpha K_n)}{Z\mu_{g0}RT}\left[\left(1 + \frac{2-f}{f}\frac{4K_n}{1-bK_n} + D_a C_a\frac{16M}{\pi * \mu_{g0}K_n^2}\right)\frac{R_f^2}{8}\right]\nabla P\frac{A}{L} \end{aligned} \tag{52}$$

$$N_{ads} = v_d\frac{A}{L} = D_d C_a\frac{ZRT}{p^2}\nabla P\frac{A}{L} \tag{53}$$

Based on the Darcy equation, the permeability model for a single capillary is described as follows:

$$Q = \frac{p}{ZRT}\frac{K}{\mu_{g0}}A\frac{\Delta P}{L} \tag{54}$$

Thus, the apparent permeability at the corresponding radius is described as follows:

$$K_{app} = (1+\alpha K_n)\left[\left(1+\frac{2-f}{f}\frac{4K_n}{1-bK_n}+D_aC_a\frac{16M}{\pi * \mu_{g0}K_n^2}\right)\frac{R_f^2}{8}\right] \tag{55}$$

where, R_f is the equivalent pore spacing. Combining Equation (12) on stress depends correlation and Equation (26) on water-film thickness, R_f is express as follows:

$$\begin{gathered} R_f(brittle) = \Lambda(r)_{iom}\Phi(r) \\ = \sqrt{[r-h(T,V_m,P_v/P_0)]^2(1-e^{a\sigma_m'}) + [r-h(T,\mathrm{V_m},P_v/P_0)]^2 e^{a\sigma_m'}} \end{gathered} \tag{56}$$

$$\begin{gathered} R_f(OMP) = \Lambda(r)_{om}\Phi(r) \\ = [r-h(T,V_m,P_v/P_0)]\sqrt[6]{e^{-3c_f\frac{v}{1-v}(p-p_i)+\frac{E}{3(1-v)}\varepsilon_l\left(\frac{p}{p+p_\varepsilon}-\frac{p_i}{p_i+p_\varepsilon}\right)}} \end{gathered} \tag{57}$$

Combining Equations (31) and (32), the water-bridging mechanism at varying pressures is expressed as follows:

$$\left.\frac{\partial\left(\frac{r}{r-h}\cdot\Pi_{flat}(h)+\frac{\gamma}{r-h}\right)}{\partial h}\right|_{h=h^*} = 0 \tag{58}$$

The water-bridging mechanism at different pore pressures is described as follows:

$$\Gamma(r,\ P) = \begin{cases} 1\ P > P_{thw} \\ 0\ P < P_{thw} \end{cases} \tag{59}$$

where, combining Equations (32), (33) and (34), the $P_{threshold}$ is described as follows:

$$P_{thw} = \frac{1}{c}7.0746 - (1657.46/(T-273.15+227.02))e^{\left(\frac{r}{r-h*}\cdot\Pi_{flat}(h*)+\frac{\gamma}{r-h*}\right)} \tag{60}$$

Based on Equations (27), (28) and (29), the water blockage from the fracturing process is expressed as follows:

$$T(r,\ P) = \begin{cases} 1\ P > P_{thb} \\ 0\ P < P_{thb} \end{cases} \tag{61}$$

$$P_{thb} = \begin{cases} \frac{P_{g0}L_0}{L_0-x_0} + \frac{4\sigma cos\theta_{omp}}{d} + \frac{4ZRTn_{omp}x(p)}{d(L_0-x(p))} \\ \frac{P_{g0}L_0}{L_0-x_0} + \frac{4\sigma cos\theta_b}{d} + \frac{4ZRTn_b x(p)}{d(L_0-x(p))} \\ \frac{P_{g0}L_0}{L_0-x_0} + \frac{4\sigma cos\theta_c}{d} + \frac{4ZRTn_c x(p)}{d(L_0-x(p))} + nE_\pi RT\left(C_0-C_f\right) \end{cases} \tag{62}$$

Thus, the dynamic permeability model of brittle minerals and OMP at varying pressures is derived as follows:

$$\begin{gathered} K(P) = \sum_{r=1,2\ldots}\frac{1}{rlg\sqrt{2\pi}\sigma}EXP\left[-\frac{(lgr-\mu)^2}{2\sigma^2}\right] \\ (1+\alpha K_n)\left[\left(1+\frac{2-f}{f}\frac{4K_n}{1-bK_n}+D_aC_a\frac{16M}{\pi\mu_{g0}K_n^2}\right)\frac{R_f^2}{8}\right]\Gamma(r,\ P)T(r,\ P) \end{gathered} \tag{63}$$

Combining Equations (41), (48) and (49), the free-gas flux considering the slip flow at clay pores is described as follows:

$$N_{freegas} = \oint v(x,y) + \int v(x) + \int v(y) + N_{ads} \tag{64}$$

$$\begin{aligned} N_{freegas} = &\oint \frac{\nabla P}{u_g}\frac{h^2}{12}\sum_{n=1,3,5\ldots}^{\infty}\frac{48}{\pi^3 n^3}\left(1-\frac{cosh(n\pi x/h)}{cosh(n\pi w/2)}\right)sin(n\pi y/h) \\ &+\int \frac{\nabla P}{u_g}\frac{h^2}{12}\frac{2-f}{f}\frac{6K_n}{1-bK_n}\sum_{n=1,3,5\ldots}^{\infty}\frac{8}{\pi^2 n^2}\left(1-\frac{cosh(n\pi x/h)}{cosh(n\pi\varepsilon/2)}\right) \\ &+\int \frac{\nabla P}{u_g}\frac{h^2}{12}\frac{2-f}{f}\frac{6K_n}{1-bK_n}\sum_{n=1,3,5\ldots}^{\infty}\frac{8tanh(n\pi\varepsilon/2)sin(n\pi y/h)}{\pi^2 n^2} + D_d C_a \frac{ZRT}{P^2}\nabla P \end{aligned} \tag{65}$$

$$N_{freegas} = wh\frac{\nabla P h^2}{12\mu_g}(1+\alpha K_n)F \tag{66}$$

$$\begin{aligned} F = 1 - &\frac{192}{\pi^5\varepsilon}\sum_{n=1,3,5\ldots}^{\infty}\frac{tanh(n\pi\varepsilon/2)}{n^5} + \frac{2-f}{f}\frac{6K_n}{1-bK_n}\sum_{n=1,3,5\ldots}^{\infty}\frac{8}{\pi^2 n^2}\left[\left(1-\frac{2tanh(n\pi\varepsilon/2)}{n\pi\varepsilon}\right)\right] \\ &+\frac{2-f}{f}\frac{3K_n}{1-bK_n}\sum_{n=1,3,5\ldots}^{\infty}\frac{32tanh(n\pi\varepsilon/2)}{\pi^3 n^3} + D_a C_a \frac{48M}{\pi\mu_g}K_n^2 \end{aligned} \tag{67}$$

Corresponding permeability on clay minerals pores is described as follows:

$$\begin{aligned} K(P) = &\sum_{r=1,2\ldots}\frac{1}{rlg\sqrt{2\pi}\sigma}EXP\left[-\frac{(lgr-\mu)^2}{2\sigma^2}\right]\frac{h^2}{12}(1+\alpha K_n)\Bigg\{1 \\ &-\frac{192}{\pi^5\varepsilon}\sum_{n=1,3,5\ldots}^{\infty}\frac{tanh(n\pi\varepsilon/2)}{n^5} \\ &+\frac{2-f}{f}\frac{6K_n}{1-bK_n}\sum_{n=1,3,5\ldots}^{\infty}\frac{8}{\pi^2 n^2}\left[\left(1-\frac{2tanh(n\pi\varepsilon/2)}{n\pi\varepsilon}\right)\right] \\ &+\frac{2-f}{f}\frac{3K_n}{1-bK_n}\sum_{n=1,3,5\ldots}^{\infty}\frac{32tanh(n\pi\varepsilon/2)}{\pi^3 n^3} \\ &+D_a C_a\frac{48M}{\pi\mu_g}K_n^2\Bigg\}\Gamma(r,\ P)T(r,\ P) \end{aligned} \tag{68}$$

4. Model Validation

It is difficult to validate the proposed model with the imbibition process via experiments because of the significant pressure difference during hydraulic fracturing or the nano-scaled pore structure in shale, in this study, the proposed model is separated into two parts for validation. First, we conducted a comparison with the conventional dynamic permeability model and a porous-meter experiment to examine the righteousness of the PSD-coupled dynamic permeability model. The proposed model was then applied in MRST-shalemaster—an open-sourced EDFM simulator, to assess its accuracy while using conventional intrinsic permeability and the proposed dynamic permeability model.

4.1. Validation of Apparent Permeability Model for Single Capillary

To validate the permeability model in an experiment-viable condition, this study compared (Figure 16a) the permeability against a single capillary tube based on Sheng (2019) and Wang (2022) [24,66]. These two methods explain the permeability at a single capillary tube and originate from Javadpour's [47] study and the Poiseuille theory, respectively. Figure 16 illustrates that the proposed capillary model is in good agreement with both models. Sheng's work [24] focused on the influence on the shape factor, ignored the wettability on the Knudsen flow while Wang's work [66] consider the constant film thickness. Javadepour-based models, compared to the Poiseuille-based AP model showed a relatively higher number in permeability, both AP models showed that the shape factor and water film thickness can be ignored at larger-scaled pores. In addition, our model uses equivalent round pore radius to describe the clay minerals while Sheng [24] uses height and shape factors to describe PSD in ellipse pores, making the curve of eclipse mineral pores align with granular pores and ignoring the PSD in the real situation, this interpretation will underestimate the influenced of shape factor, from our work, using equivalent radius interpretation by NMR data for describing slit-liked clay pores, limited the total flow area, shape factor is then becoming a dominating parameters in AP model.

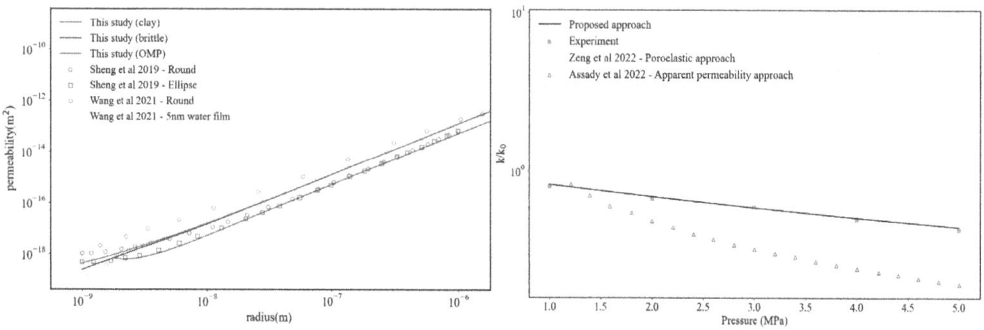

(**a**) Comparison between AP models (**b**) Dynamic permeability at varying pressures

Figure 16. Validation of the proposed dynamic permeability model [24,66,73,74].

4.2. Validation of Dynamic Permeability with Experiment Data

Experimental validation of the PSD-integrated dynamic permeability is conducted under the same shale sample extracted from the WY formation after the NMR tests. The cylindrical core sample is 2.5 cm in diameter with 5 cm in length. Permeability was tested under the pulse gradient porous meter with CH_4 as injecting media to maintain pressure from 1.2 Mpa to 5 Mpa. Owing to the limitation of the experiment setup, this part of the numerical model neglected the imbibition-induced water blockage and desorption phenomenon; the input parameters of this part are provided in Table 1. The comparison of the model's calculation and experimental data is shown in Figure 16b.

As previously noted (Figure 2), studies from different samples showed different trends as the pressure increased. Zeng [73] suggested that the two main contributions to the change in permeability include the stress-induced matrix shrinkage and gas-slippage weakening, samples with nature fractures trends to reflect a "U-shape" in the dynamic permeability while others showed a declining trend as the pressure increases [74]. To validate the proposed dynamic permeability model, this part takes the comparison with the stress-dependence poroelastic approach and the gas-slippage dominated AP model, which is recognised as the upper boundary and lower boundary of the dynamic permeability model, respectively [73]. Our numerical model is situated in between both models because the proposed model takes into account both dynamic flow regime and poroelastic properties. Comparison with numerical model and experiment data also matches well.

4.3. Validation of Dynamic Permeability under Field Conditions: Numerical Simulation

It is challenging to reflect fracturing-influenced field conditions. To validate the proposed dynamic permeability during production, we made amendments to the permeability module in an open-source EDFM simulator based on the Matlab Reservoir Simulation Toolkit (MRST) [75]. The results of the amended EDFM with the original EDFM using the WY1 field production data are used to validate the overall dynamic permeability, and the relevant SGR parameters are presented in Table 1.

The WY 1 platform employs a long-section, multi-cluster fracturing with segmented perforation mode. The main segments have lengths ranging from 110 m to 135 m, averaging 120 m. Following the principle of segmenting without crossing small layers and choosing segments with similar physical properties, a single segment consists of 10 perforation clusters, each cluster having a length of 0.35. For longer segments (exceeding 130 m), a single segment consists of 11 clusters. In segments with a risk of fractures, a single segment consists of 12 clusters, there are a total of 15 fracturing segments, with cluster spacing in the range of 4.75 to 10.65 m. According to the fracturing design, each segment can be recognised as isolated. Thus, we set the EDFM simulation into three cases, Case 1 includes 12 hydraulic fractures with 100 m fracture half-length; Case 2 includes 11 hydraulic fractures and 105 m fracture half-length; and Case 3 includes 10 hydraulic fractures with 130 m fracture half-length, related SGR parameters are shown in Table 3. The simulation takes into account the randomly generated natural fracture mapping from MRST toolbox, for comparison purposes, the random seed was set as 61,320 to get the extract same natural fracture pattern

Table 3. The main parameters from EDFM and Amended EDFM model.

Parameters	EDFM	Amended EDFM
Matrix permeability (*nD*)	600	Dynamic
Fracturing pressure (Mpa)	-	50
Porosity	0.03	
Fracture permeability (*D*)	0.1	
Natural fracture perm (*D*)	0.01	
Bottom hole pressure (MPa)	15	
Initial pore pressure (MPa)	23	
Case1 contribution	0.2	
Case2 contribution	0.7	
Case3 contribution	0.1	

Figure 17 illustrates the pressure distribution in three cases, to reflect 15 segments in the field, we use the contribution factor from three cases to reflect the cumulative production (Table 3). The results, in terms of gas production rate and cumulative production, are plotted in Figure 18, including field data that incorporates the shut-in period. The amended EDFM result initially underestimated the cumulative gas production, followed by an overestimation. However, when compared with the EDFM result, it demonstrated higher accuracy in predicting cumulative gas production over time.

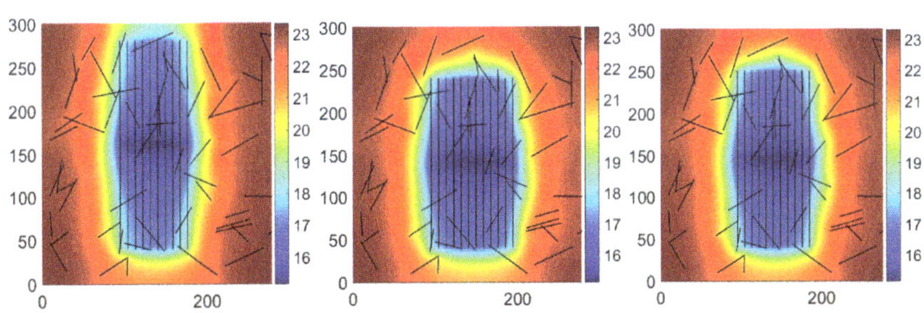

Figure 17. The pressure distribution in three types of cases.

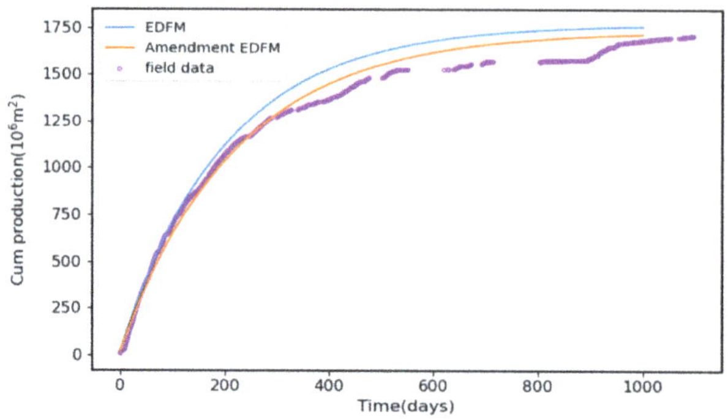

Figure 18. The cumulated production curve along the production time.

5. Result and Discussion

5.1. Dynamic Permeability under Different Reservoir Conditions

Figure 19 illustrates the apparent permeability at varying fracture and pore pressure. Figure 19a presents the dynamic permeability when pore pressure is smaller than the fracture pressure. Figure 19b demonstrates the dynamic permeability when the pore pressure is bigger than the fracture pressure, owing to the capillary phenomenon. First, the graph shows that dynamic permeability decreased gradually, and unlike the experimental result, the trend at high pore pressure did not reach a plateau because the water-bridging phenomenon resulted in water blockage at high pore pressure. Second, at the same pore pressure, the effect of the imbibition process is mostly on the adsorption layer distribution on the pore wall and the corresponding pore-size shrinkage. Third, dynamic porous parameters also play a key role in dynamic permeability, including PSD distribution, water-film thickness, pressure, and temperature during production. The above-mentioned phenomenon will be discussed in detail in the following section to address the field suggestions based on dynamic permeability.

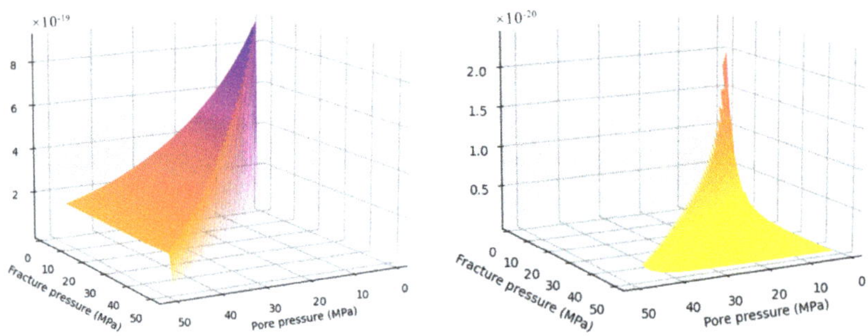

(**a**) Pressure in pore is larger than in fracture (**b**) Pressure in pore is smaller than in fracture

Figure 19. The dynamic permeability model under different pressures.

5.2. Fracturing-Induced Water Blockage and Water-Film Thickness during Production

Reservoir parameters used in this section are listed in Table 1 to describe the contribution along the pore radius. A schematic demonstration of permeability contribution at varying pore radii is presented in Figure 20, which is, the PSD ratio times the corresponding permeability. Four reservoir pressure conditions (29 Mpa, 23 Mpa, 22 Mpa, and 21 MPa) were compared to describe the fracturing-induced water-blockage phenomenon. We can conclude that, after the hydraulic fracturing process, as the reservoir depleted, hydrophobic OMP started to exhaust and contribute to the permeability owing to its ultra-strong capillary forces. As for hydrophilic minerals, extra capillary forces are required for water-blockage removal because of its wettability. Additionally, the water film mechanism inside pores will escalate the capillary force by changing the equivalent pore radius and substantially impact the permeability owing to its high concertation in OMP. However, the water-bridging mechanism will occur only under 50 nm pores under reservoir conditions.

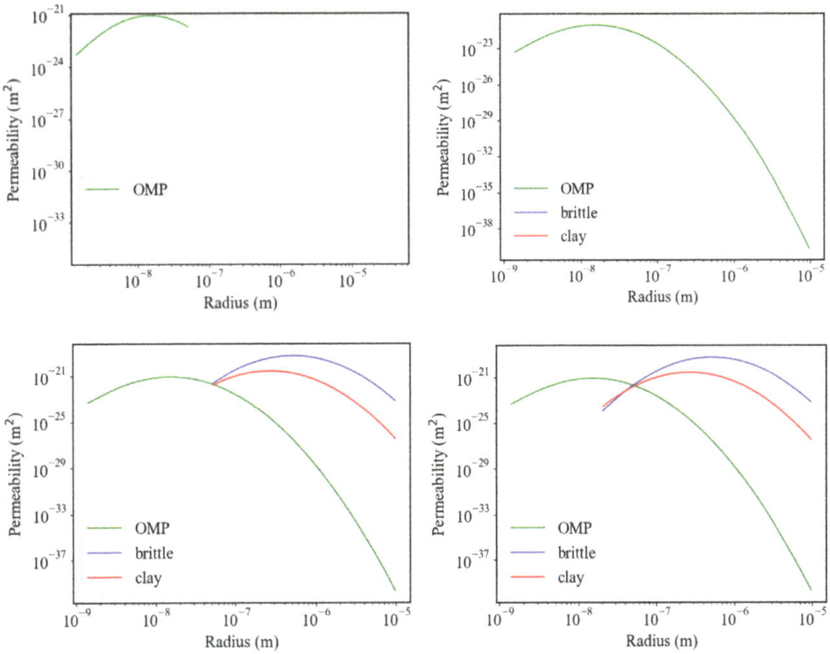

Figure 20. Permeability contributions as per equivalent radii under different pressure.

Figure 20 demonstrates the relationship of three different minerals in shale at varying pressures and temperatures, within three different types of minerals. Note that in the permeability system, smaller pores (1 nm–200 nm) are more affected by the water-film phenomenon. Compared to brittle minerals and OMP, pores within clay minerals are more affected because of their slit-shaped structure. Additionally, Figure 21 illustrates the influences on pore pressure and reservoir temperature. Comparison of the permeability model conducted under 20 MPa and 23 Mpa at 383 K showed that greater pore pressure will have a heavier impact on permeability. The comparison between 440 K and 383 K at 23 Mpa pore pressure concluded that higher temperature will have a greater impact on permeability. Also, for a different type of shale rocks, the overall wettability of brittle minerals, clay and OMP at different TOC will have an influence on the waterblockage phenomenon at different pressure, smaller contact angles will escalate the waterblockage phenomenon, consequentially reducing the overall permeability.

5.3. Fracturing Effected Adsorption Layer Thickness and Consequentially Impacted Permeability

The fracturing response has a significant impact on the poro-mechanical of the pore. Figure 22a demonstrates the change of diffusion coefficient under different fracturing pressures. Figure 22a utilizes the 1 minus diffusion coefficient to at different fracturing pressure to illustrate the impact on the diffusion phenomenon. From the curve, we can see that the fracturing process will significantly impact the diffusion coefficient. The fracturing impact ratio increases as the fracturing pressure increases. For hydrophilic brittle and clay minerals, under identical fracturing pressure, small pores undergo greater diffusion coefficient damage than bigger pores owing to their wettability. For hydrophobic OMPs, smaller pores undergo less diffusion coefficient damage owing to strong capillary at smaller pores.

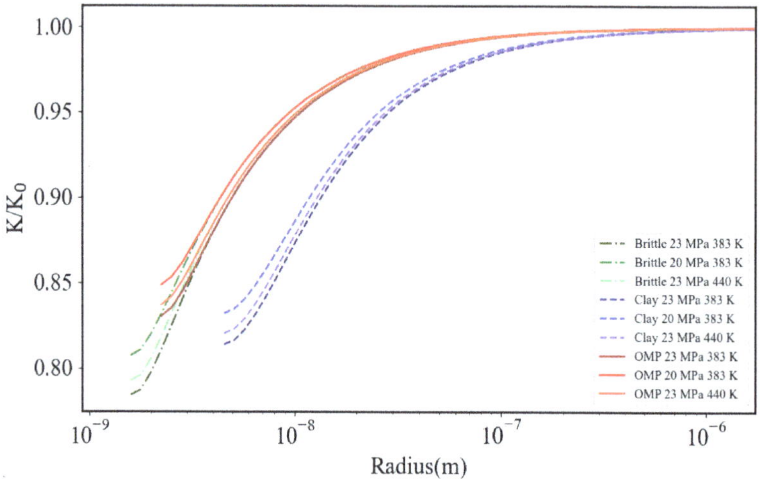

Figure 21. Permeability changes under different field conditions.

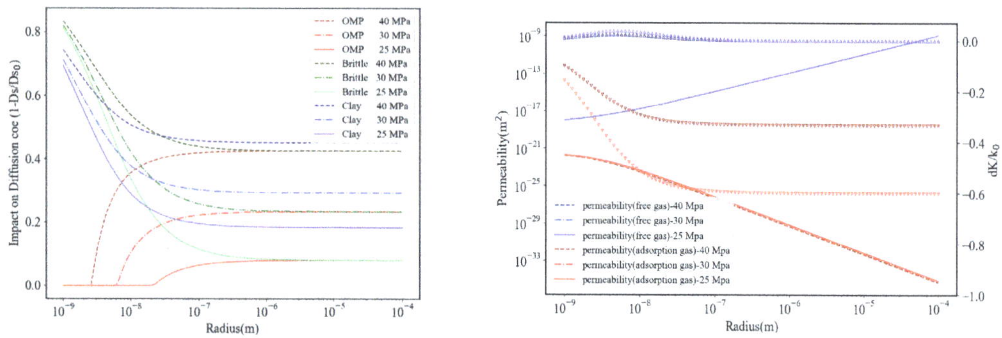

(**a**) Impact on Diffusion coefficient at varying radius (**b**) Impact on permeability at varying radius

Figure 22. The impact on the diffusion coefficient at different fracturing pressures and corresponding permeability dynamics.

Apart from the impact on the diffusion coefficient, fracturing-induced thinning of the desorption portion layer also contributes to the desorption-induced matrix shrinkage, directly increasing the permeability contributed by free gas [43,67,72]. Figure 22b illustrates the permeability changes with different fracturing pressures for both free gas and adsorbed gas. The triangle marks with the corresponding colours depict the derived permeability at different radii. Higher fracturing pressure resulted in a reduction of adsorbed gas permeability by around 30–60%, while the incremental in free gas permeability was only at 0.02%.

5.4. *Influence of Clay-Shaped Factors*

The dynamics of the proposed model are significantly influenced by clay minerals due to their high concentration at a larger pore scale. Besides the concentration of clay minerals, the shape factor (SF) also plays a vital role in overall permeability.

The FIB-SEM description can reflect the shape factor distribution at a larger scale in the matrix (Figure 23). Neglecting the granular-shaped brittle mineral pores contributing at a small SF, from observations, the real SF distribution in slit-like pores also follows a Gaussian distribution. Thus, we present three cases with smaller, medium, and higher

SF (7, 20, 43) dominated distributions with the same porosity to discuss the impact on permeability (Figure 24a).

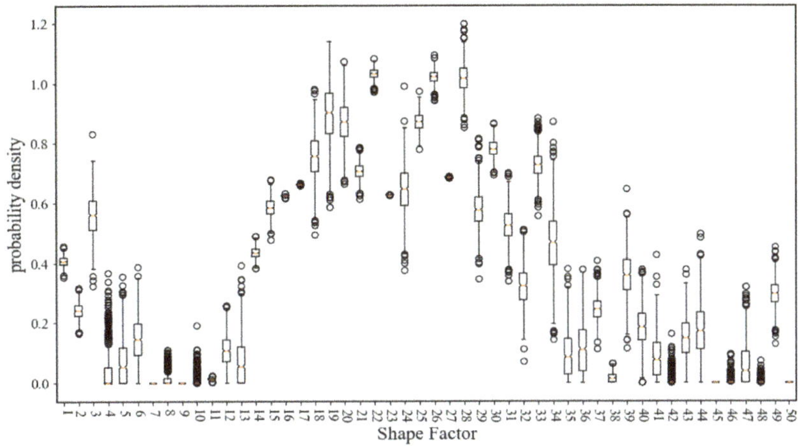

Figure 23. The probability density from different shape factors on clay mineral pores.

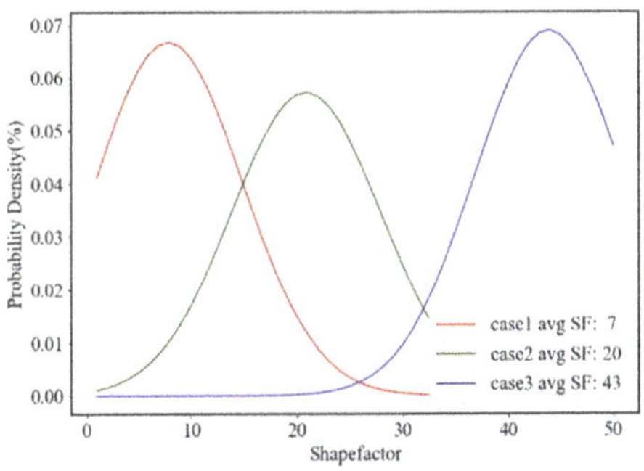

(**a**) Probability density of SF in three cases

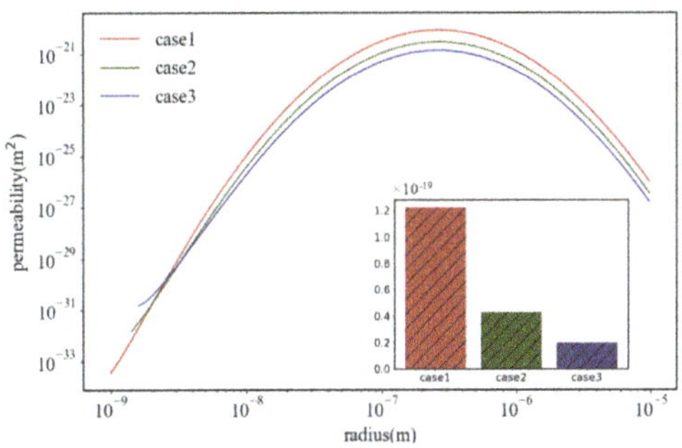

(**b**) Permeability at varying radii from three cases

Figure 24. The three cases of probability density at varying shape factors and corresponding permeability.

Figure 24b describes the contributed permeability at different radii and overall permeability. From the contributed permeability curve, the permeability at different radii with a small shape factor has a larger number. Overall permeability is reduced by 70% when comparing Case 1 (mean SF = 7) and Case 2 (mean SF = 20), but reduced by 83% when compared with Case 3 (mean SF = 43). The increment of SF and the corresponding overall permeability follows a logarithmic relationship.

6. Conclusions

In this study, we developed the entire workflow to extract dynamic permeability under the field conditions impacted by the fracturing process, The dynamic permeability model explains the water-gas interaction in the shale matrix, which is perfect for explaining the gas flow during the reservoir depletion and well-testing period to prevent water blockage with the influence of hydraulic fracturing. For field application, this methodology could be used to assist in setting a development plan for enhancing EUR during the early period by controlling the bottom-hole pressure to optimise the matrix permeability. Also, this model demonstrated dynamic permeability in the shale matrix at different reservoir pressures. This article also demonstrated that the proposed dynamic permeability model, combining hydraulic fracture and natural fracture with the EDFM showed higher accuracy in predicting the production rate.

Key findings in this study include:

1. Bayesian-assisted Gaussian description for the three majority minerals derived from coupled FIB-SEM and NMR data proves to be viable for accurately describing the PSD in shale. The corresponding dynamic permeability model demonstrates an intimate association with experimental data.
2. The fracturing-induced imbibition process results in water blockage and water bridging mechanisms during shale gas production, impacting the dynamic permeability in the matrix. The water blockage phenomenon significantly reduces permeability in nano-scaled brittle minerals and clay. Substantial water blockage requires a larger pressure gradient to overcome. The impact on total permeability, however, depends on the subsequent PSD in the micro-scale. Nano-scaled dominated OMP and micro-scaled dominated brittle and clay pores reduce the impact on permeability from water blockage. Water bridging occurs only in nano-scale OMP below 50nm at high pressure and temperatures. Due to the high concentration of OMP, the permeability contribution cannot be neglected.
3. Reservoir depletion has a substantial impact on permeability, showing a declining trend as pore pressure increases. In addition to poromechanics, the fracturing-induced imbibition phenomenon reduces the thickness of the water film inside pores, significantly impacting adsorbed gas permeability, and only marginally boosting the contribution of free gas to permeability.
4. Due to the high permeability of large-scale clay mineral pores, the SF of clay minerals significantly influences dynamic permeability. With constant pore spacing, higher SF clay mineral pores reduce permeability in the shale matrix.

Author Contributions: Q.Z.: Conceptualization, Methodology, Software, Formal Analysis, Writing—Original Draft; H.L.: Investigation, Data Curation, Reviewing and Editing (equal); Supervision; Y.L.: Visualization, Reviewing and Editing (equal); H.W.: Investigating; K.L.: Validation. All authors have read and agreed to the published version of the manuscript.

Funding: This study was supported by the Mechanism of SNPs to increase the sealing efficiency of mudstone caprock for CO_2 geological storage (the National Natural Science Foundation of China, Project no. 42272176), Mechanisms of flow-chemistry-biology coupling and enhanced caprock integrity in CO_2 geological sequestration based on the synergistic effect of nanoparticles and methanogens (Sino-German mobility programme, Project no: M0469), Chinese Postdoctoral ScienceFoundatio (2022M722636).

Data Availability Statement: Data are contained within the article.

Conflicts of Interest: Author Wang Haiguang was employed by the company Central Sichuan Division, Petro China Southwest Oil & Gas Field Company; Author Lu Kuan was employed by the company China Petroleum Engineering & Construction Corp. Beijing Company. The remaining authors declare that the research was conducted in the absence of any commercial or financial relationships that could be construed as a potential conflict of interest.

Nomenclature

total organic carbon	TOC
surface relaxivity	$T_{2,S}$
ratio of surface area to pore volume	S_v
standard derivations of Log-gaussian distribution	σ
mean of Log-gaussian distribution	μ
volume concertation of brittle, organic matter and clay, respectively	c_k
hydraulic tortuosity	τ
porosity	ϕ

surface area	S
shape factor	β
sphericity of particle	ξ
residual porosity at high stress	ϕ_r
initial porosity	ϕ_i
mean effective stress	σ'_m
force between water film and surface	$\Pi_1(h)$
force between water film and opposite surface	$\Pi_2(h)$
force between two adsorbed water films	$\Pi_3(h)$
water film thickness	h
equivalent proe radius	r
idea gas constant	R
media permeability	k
initial media permeability	k_i
media compressibility	c_f
change of effective stress	$\Delta\sigma$
Poisson's ratio	ν
media pressure	p
Langmuir-type organic matter shrinkage constants	$\varepsilon_l, p_\varepsilon$
initial water saturated capillary length	x_0
interfacial tension of water	σ
tortuosity-considered equivalent capillary length	L_0
critical film thickness	h^*
saturated vapor pressure	P_v^*
fitting curve coefficient	Y_1, Y_2, Y_3
viscous of gas	u_g
Knudsen number	Kn
Rarefaction coefficient	α
diffusion coefficient	D_d
molar concentration of adsorbed gas	C_a
Avogadro constant	N_A
langmuir pressure	P_L
tortuosity-considered capillary length	$\Omega(L)$
molecular mass	M
shape factor of clay minerals	ε
Young's Modulus	E
Adsorption isothermals	ΔH
Electrical conductivity	e

References

1. Li, M.; Pang, X.; Xiong, L.; Hu, T.; Chen, D.; Zhao, Z.; Hui, S. Application of mathematical statistics to shale gas-bearing property evaluation and main controlling factor analysis. *Sci. Rep.* **2022**, *12*, 9859. [CrossRef] [PubMed]
2. Chai, D. Comprehensive Characterization of Shale Gas Seepage in Nanoscale Organic-Rich Shales. Ph.D. Thesis, University of Houston, Ann Arbor, MI, USA, 2020. Available online: https://www.proquest.com/dissertations-theses/comprehensive-characterization-shale-gas-seepage/docview/2494342993/se-2?accountid=132643 (accessed on 13 March 2022).
3. Li, F.; Wang, M.; Liu, S.; Hao, Y. Pore characteristics and influencing factors of different types of shales. *Mar. Pet. Geol.* **2019**, *102*, 391–401. [CrossRef]
4. Saif, T.M.A. Multi-Scale Multi-Dimensional Imaging and Characterization of Oil Shale Pyrolysis. AGU Fall Meeting Abstracts. 2017. Available online: https://ui.adsabs.harvard.edu/abs/2017AGUFMEP21H..08G (accessed on 20 December 2023).
5. Chandra, D.; Vishal, V. A critical review on pore to continuum scale imaging techniques for enhanced shale gas recovery. *Earth-Sci. Rev.* **2021**, *217*, 103638. [CrossRef]
6. Goral, J. Digital Rock Physics for Multiscale Characterization of Heterogeneous Petroleum Geomaterials. Master's Thesis, The University of Utah, Salt Lake City, UT, USA, 2022. Available online: https://www.proquest.com/pqdtglobal/docvie-w/1983971354/abstract/B0790A58BFF94CC7PQ/1 (accessed on 19 May 2022).
7. Zhang, D.; Meegoda, J.N.; da Silva, B.M.G.; Hu, L. Impact of de-ionized water on changes in porosity and permeability of shales mineralogy due to clay-swelling. *Sci. Rep.* **2021**, *11*, 20049. [CrossRef] [PubMed]
8. Cai, J.; Lin, D.; Singh, H.; Wei, W.; Zhou, S. Shale gas transport model in 3D fractal porous media with variable pore sizes. *Mar. Pet. Geol.* **2018**, *98*, 437–447. [CrossRef]

9. Clarkson, C.R.; Williams-Kovacs, J.D. Modeling Two-Phase Flowback of Multifractured Horizontal Wells Completed in Shale. *Spe J.* **2013**, *18*, 795–812. [CrossRef]
10. Cui, J.; Cheng, L. A theoretical study of the occurrence state of shale oil based on the pore sizes of mixed Gaussian distribution. *Fuel* **2017**, *206*, 564–571. [CrossRef]
11. Liu, Y.Y.; Ma, X.H.; Zhang, X.W.; Guo, W.; Kang, L.X.; Yu, R.Z.; Sun, Y.P. A deep-learning-based prediction method of the estimated ultimate recovery (EUR) of shale gas wells. *Pet. Sci.* **2021**, *18*, 1450–1464. [CrossRef]
12. Guckenheimer, J.; Holmes, P. *Micro and Nano Flow*, 5th ed.; Applied Mathematical Sciences; Springer: New York, NY, USA, 1997; Volume 42.
13. Wang, J.; Rahman, S.S. Investigation of Water Leakoff Considering the Component Variation and Gas Entrapment in Shale During Hydraulic-Fracturing Stimulation. *SPE Reserv. Eval. Eng.* **2016**, *19*, 511–519. [CrossRef]
14. Wang, K.; Jiang, B.; Ye, K.; Li, H.; Tan, Y. Spontaneous imbibition model for micro–nano–scale pores in shale gas reservoirs considering gas–water interaction. *J. Pet. Sci. Eng.* **2022**, *209*, 109893. [CrossRef]
15. Afagwu, C.; Abubakar, I.; Kalam, S.; Al-Afnan, S.F.; Awotunde, A.A. Pressure-transient analysis in shale gas reservoirs: A review. *J. Nat. Gas Sci. Eng.* **2020**, *78*, 103319. [CrossRef]
16. Fu, J.; Su, Y.; Chen, Z.; Li, L.; Wang, W.; Zhan, S. Distribution of a water film confined in inorganic nanopores in real shale gas reservoirs. *J. Pet. Sci. Eng.* **2022**, *209*, 109831. [CrossRef]
17. Gao, Q.; Han, S.; Cheng, Y.; Li, Y.; Yan, C.; Han, Z. Apparent permeability model for gas transport through micropores and microfractures in shale reservoirs. *Fuel* **2021**, *285*, 119086. [CrossRef]
18. Wang, H.; Wang, W.; Su, Y.; Jin, Z. Lattice Boltzmann Model for Oil/Water Two-Phase Flow in Nanoporous Media Considering Heterogeneous Viscosity, Liquid/Solid, and Liquid/Liquid Slip. *SPE J.* **2022**, *27*, 3508–3524. [CrossRef]
19. Wang, K.; Ye, K.; Jiang, B.; Li, H.; Tan, Y. The mechanism of gas-water extraction in micro- and nanoscale pores in shale gas reservoirs: Based on gas-water interactions. *Chem. Eng. Sci.* **2022**, *248*, 117259. [CrossRef]
20. Xu, J.; Wu, K.; Li, R.; Li, Z.; Li, J.; Xu, Q.; Li, L.; Chen, Z. Nanoscale pore size distribution effects on gas production from fractal shale rocks. *Fractals* **2019**, *27*, 1950142. [CrossRef]
21. Alharthy, N.S.; Nguyen, T.N.; Teklu, T.W.; Kazemi, H.; Graves, R.M. Multiphase Compositional Modeling in Small-Scale Pores of Unconventional Shale Reservoirs. In *SPE Annual Technical Conference and Exhibition*; SPE: Abu Dhabi, United Arab Emirates, 2013; p. D031S052R008. [CrossRef]
22. Cao, P.; Liu, J.; Leong, Y.-K. A fully coupled multiscale shale deformation-gas transport model for the evaluation of shale gas extraction. *Fuel* **2016**, *178*, 103–117. [CrossRef]
23. Shen, R.; Zhang, X.; Ke, Y.; Xiong, W.; Guo, H.; Liu, G.; Zhou, H. An integrated pore size distribution measurement method of small angle neutron scattering and mercury intrusion capillary pressure. *Sci. Rep.* **2021**, *11*, 17458. [CrossRef]
24. Sheng, G.; Javadpour, F.; Su, Y. Dynamic porosity and apparent permeability in porous organic matter of shale gas reservoirs. *Fuel* **2019**, *251*, 341–351. [CrossRef]
25. Xiao, D.; Lu, Z.; Jiang, S.; Lu, S. Comparison and integration of experimental methods to characterize the full-range pore features of tight gas sandstone—A case study in Songliao Basin of China. *J. Nat. Gas Sci. Eng.* **2016**, *34*, 1412–1421. [CrossRef]
26. Sakhaee-Pour, A.; Bryant, S.L. Pore structure of shale. *Fuel* **2015**, *143*, 467–475. [CrossRef]
27. Wang, D.; Yao, J.; Chen, Z.; Song, W.; Cai, M.; Tian, M.; Zhang, J.; Xu, W. Image-based model for dynamic apparent gas permeability in Organic-rich shales. *Fuel* **2022**, *318*, 123588. [CrossRef]
28. Pang, Y.; Wang, S.; Yao, X.; Hu, X.; Chen, S. Evaluation of Gas Adsorption in Nanoporous Shale by Simplified Local Density Model Integrated with Pore Structure and Pore Size Distribution. *Langmuir* **2022**, *38*, 3641–3655. [CrossRef] [PubMed]
29. Blauch, M.E. Developing Effective and Environmentally Suitable Fracturing Fluids Using Hydraulic Fracturing Flowback Waters. In *SPE Unconventional Resources Conference/Gas Technology Symposium*; SPE: Abu Dhabi, United Arab Emirates, 2010; p. SPE-131784-MS. [CrossRef]
30. Duong, A.N. Rate-Decline Analysis For Fracture-Dominated Shale Reservoirs: Part 2. In *SPE/CSUR Unconventional Resources Conference–Canada*; OnePetro: Richardson, TX, USA, 2014; p. D021S008R001. [CrossRef]
31. Inman, M. Natural gas: The fracking fallacy. *Nature* **2014**, *516*, 28. [CrossRef] [PubMed]
32. Sun, J.Q.; Ma, T.H.; Zhang, L.Y.; Huang, B.; Li, A.S. Study on Hydraulic Fracturing Affected by Natural Fractures. *Appl. Mech. Mater.* **2014**, *501–504*, 2056–2059. [CrossRef]
33. Tian, Z.; Wei, W.; Zhou, S.; Sun, C.; Rezaee, R.; Cai, J. Impacts of gas properties and transport mechanisms on the permeability of shale at pore and core scale. *Energy* **2022**, *244*, 122707. [CrossRef]
34. Fu, J.; Su, Y.; Li, L.; Wang, W.; Wang, C.; Li, D. Productivity model with mechanisms of multiple seepage in tight gas reservoir. *J. Pet. Sci. Eng.* **2022**, *209*, 109825. [CrossRef]
35. Dongari, N.; Sharma Iitk, A.; Durst, F. Pressure-driven diffusive gas flows in micro-channels: From the Knudsen to the continuum regimes. *Microfluid. Nanofluidics* **2009**, *6*, 679–692. [CrossRef]
36. Xu, R. Pore Scale Study of Gas Sorption and Transport in Shale Nanopore Systems. Ph.D. Thesis, The University of Texas at Austin, Austin, TX, USA, 2020. [CrossRef]
37. Shi, K.Y.; Chen, J.Q.; Pang, X.Q.; Jiang, F.J.; Hui, S.S.; Zhao, Z.C.; Chen, D.; Cong, Q.; Wang, T.; Xiao, H.-Y.; et al. Wettability of different clay mineral surfaces in shale: Implications from molecular dynamics simulations. *Pet. Sci.* **2023**, *20*, 689–704. [CrossRef]

38. Wang, K.; Ma, L.; Taylor, K.G. Microstructure changes as a response to CO2 storage in sedimentary rocks: Recent developments and future challenges. *Fuel* **2023**, *333*, 126403. [CrossRef]
39. Zhao, Z.; Zhou, X.-P. Pore-scale diffusivity and permeability evaluations in porous geomaterials using multi-types pore-structure analysis and X-μCT imaging. *J. Hydrol.* **2022**, *615*, 128704. [CrossRef]
40. Alkouh, A.; Wattenbarger, R.A. Estimation of Effective-Fracture Volume Using Water-Flowback and Production Data for Shale-Gas Wells. *J. Can. Pet. Technol.* **2014**, *53*, 290–303. [CrossRef]
41. An, C.; Fang, Y.; Liu, S.; Alfi, M.; Yan, B.; Wang, Y.; Killough, J. Impacts of Matrix Shrinkage and Stress Changes on Permeability and Gas Production of Organic-Rich Shale Reservoirs. In *SPE Reservoir Characterisation and Simulation Conference and Exhibition*; OnePetro: Richardson, TX, USA, 2017; p. D021S009R002. [CrossRef]
42. Cao, P.; Liu, J.; Leong, Y.-K. Combined impact of flow regimes and effective stress on the evolution of shale apparent permeability. *J. Unconv. Oil Gas Resour.* **2016**, *14*, 32–43. [CrossRef]
43. Yang, D.; Wang, W.; Chen, W.; Wang, S.; Wang, X. Experimental investigation on the coupled effect of effective stress and gas slippage on the permeability of shale. *Sci. Rep.* **2017**, *7*, 44696. [CrossRef] [PubMed]
44. Zhu, W.; Ma, D. Effective stress characteristics in shale and its effect on shale gas productivity. *J. Nat. Gas Geosci.* **2018**, *3*, 339–346. [CrossRef]
45. Javadpour, F. Nanopores and Apparent Permeability of Gas Flow in Mudrocks (Shales and Siltstone). *J. Can. Pet. Technol.* **2009**, *48*, 16–21. [CrossRef]
46. Li, Y.; Dong, P.; Zhou, D. A dynamic apparent permeability model for shale microfractures: Coupling poromechanics, fluid dynamics, and sorption-induced strain. *J. Nat. Gas Sci. Eng.* **2020**, *74*, 103104. [CrossRef]
47. Shi, J.Q.; Durucan, S. Drawdown Induced Changes in Permeability of Coalbeds: A New Interpretation of the Reservoir Response to Primary Recovery. *Transp. Porous Media* **2004**, *56*, 1–16. [CrossRef]
48. Afagwu, C.; Alafnan, S.; Mahmoud, M.A.; Patil, S. Permeability model for shale and ultra-tight gas formations: Critical insights into the impact of dynamic adsorption. *Energy Rep.* **2021**, *7*, 3302–3316. [CrossRef]
49. Balczár, I.; Korim, T.; Dobrádi, A. Correlation of strength to apparent porosity of geopolymers–Understanding through variations of setting time. *Constr. Build. Mater.* **2015**, *93*, 983–988. [CrossRef]
50. Gu, X.; Mildner, D.F.R.; Cole, D.R.; Rother, G.; Slingerland, R.; Brantley, S.L. Quantification of Organic Porosity and Water Accessibility in Marcellus Shale Using Neutron Scattering. *Energy Fuels* **2016**, *30*, 4438–4449. [CrossRef]
51. Zhang, Q.; Wu, X.; Meng, Q.; Wang, Y.; Cai, J. Fractal models for gas–water transport in shale porous media considering wetting characteristics. *Fractals* **2020**, *28*, 2050138. [CrossRef]
52. Hagita, K.; Higuchi, T.; Jinnai, H. Super-resolution for asymmetric resolution of FIB-SEM 3D imaging using AI with deep learning. *Sci. Rep.* **2018**, *8*, 5877. [CrossRef] [PubMed]
53. Cai, M.; Su, Y.; Hao, Y.; Guo, Y.; Elsworth, D.; Li, L.; Li, D.; Li, X. Monitoring oil displacement and CO_2 trapping in low-permeability media using NMR: A comparison of miscible and immiscible flooding. *Fuel* **2021**, *305*, 121606. [CrossRef]
54. Andersen, P.Ø. Steady-State Gas Flow from Tight Shale Matrix Subject to Water Blocking. *SPE J.* **2021**, *26*, 3970–3985. [CrossRef]
55. Jiang, Y.; Xu, G.; Bi, H.; Shi, Y.; Gao, Y.; Han, X.; Zeng, X. A new method to determine surface relaxivity of tight sandstone cores based on LF-NMR and high-speed centrifugation measurements. *J. Pet. Sci. Eng.* **2021**, *196*, 108096. [CrossRef]
56. Taghavinejad, A.; Sharifi, M.; Heidaryan, E.; Liu, K.; Ostadhassan, M. Flow modeling in shale gas reservoirs: A comprehensive review. *J. Nat. Gas Sci. Eng.* **2020**, *83*, 103535. [CrossRef]
57. Ahmadi, M.; Mohammadi, S.; Nemati Hayati, A. Analytical derivation of tortuosity and permeability of monosized spheres: A volume averaging approach. *Phys. Rev. E Stat. Nonlinear Soft Matter Phys.* **2011**, *83*, 026312. [CrossRef]
58. Ashrafi Moghadam, A.; Chalaturnyk, R. Expansion of the Klinkenberg's slippage equation to low permeability porous media. *Int. J. Coal Geol.* **2014**, *123*, 2–9. [CrossRef]
59. Miguel, A.F.; Serrenho, A. On the experimental evaluation of permeability in porous media using a gas flow method. *J. Phys. D Appl. Phys.* **2007**, *40*, 6824–6828. [CrossRef]
60. Rutqvist, J.; Wu, Y.-S.; Tsang, C.-F.; Bodvarsson, G. A modeling approach for analysis of coupled multiphase fluid flow, heat transfer, and deformation in fractured porous rock. *Int. J. Rock Mech. Min. Sci.* **2002**, *39*, 429–442. [CrossRef]
61. Tuller, M.; Or, D.; Dudley, L.M. Adsorption and capillary condensation in porous media: Liquid retention and interfacial configurations in angular pores. *Water Resour. Res.* **1999**, *35*, 1949–1964. [CrossRef]
62. Li, J.; Li, X.; Wu, K.; Feng, D.; Zhang, T.; Zhang, Y. Thickness and stability of water film confined inside nanoslits and nanocapillaries of shale and clay. *Int. J. Coal Geol.* **2017**, *179*, 253–268. [CrossRef]
63. Raviv, U.; Laurat, P.; Klein, J. Fluidity of water confined to subnanometre films. *Nature* **2001**, *413*, 51–54. [CrossRef]
64. Li, J.; Li, X.; Wang, X.; Li, Y.; Wu, K.; Shi, J.; Yang, L.; Feng, D.; Zhang, T.; Yu, P. Water distribution characteristic and effect on methane adsorption capacity in shale clay. *Int. J. Coal Geol.* **2016**, *159*, 135–154. [CrossRef]
65. Teng, T.; Wang, J.G.; Gao, F.; Ju, Y.; Jiang, C. A thermally sensitive permeability model for coal-gas interactions including thermal fracturing and volatilization. *J. Nat. Gas Sci. Eng.* **2016**, *32*, 319–333. [CrossRef]
66. Wang, H.; Su, Y.; Qiao, R.; Wang, J.; Wang, W. Investigate Effects of Microstructures on Nanoconfined Water Flow Behaviors from Viscous Dissipation Perspectives. *Transp. Porous Med.* **2021**, *140*, 815–836. [CrossRef]
67. Li, Q.; Han, Y.; Liu, X.; Ansari, U.; Cheng, Y.; Yan, C. Hydrate as a by-product in CO_2 leakage during the long-term sub-seabed sequestration and its role in preventing further leakage. *Environ. Sci. Pollut. Res.* **2022**, *29*, 77737–77754. [CrossRef]

68. de la Calle-Arroyo, C.; López-Fidalgo, J.; Rodríguez-Aragón, L.J. Optimal designs for Antoine Equation. *Chemom. Intell. Lab. Syst.* **2021**, *214*, 104334. [CrossRef]
69. Song, H.; Li, B.; Li, J.; Ye, P.; Duan, S.; Ding, Y. An Apparent Permeability Model in Organic Shales: Coupling Multiple Flow Mechanisms and Factors. *Langmuir* **2023**, *39*, 3951–3966. [CrossRef]
70. Zhan, J.; Soo, E.; Fogwill, A.; Cheng, S.; Cai, H.; He, R.; Chen, Z. An Integrated Multi-Scale Numerical Simulation of Transient Gas Flow in Shale Matrix. In *Offshore Technology Conference Asia*; OTC: Springfield, MO, USA, 2018; p. D021S007R002. [CrossRef]
71. Zhang, F.; Emami-Meybodi, H. Flowback Fracture Closure of Multifractured Horizontal Wells in Shale Gas Reservoirs. In *SPE/AAPG Eastern Regional Meeting*; OnePetro: Richardson, TX, USA, 2018; p. D043S007R003. [CrossRef]
72. Yu, H.; Xu, H.; Fan, J.; Zhu, Y.; Wang, F.; Wu, H.-A. Transport of Shale Gas in Microporous/Nanoporous Media: Molecular to Pore-Scale Simulations. *Energy Fuels* **2020**, *35*, 911–943. [CrossRef]
73. Zeng, J.; Liu, J.; Guo, J. Characterization of gas transport in shale: A multi-mechanism permeability modeling approach. *Chem. Eng. J.* **2022**, *438*, 135604. [CrossRef]
74. Assady, A. The Impact of Stress Dependent Permeability Alteration on Gas Based EOR in the Bakken Formation. Ph.D. Thesis, The University of North Dakota, Ann Arbor, MI, USA, 2022. Available online: https://www.proquest.com/dissertations-theses/impact-stress-dependent-permeability-alteration/docview/2721678992/se-2?accountid=132643 (accessed on 13 March 2022).
75. Wang, B.; Fidelibus, C. An Open-Source Code for Fluid Flow Simulations in Unconventional Fractured Reservoirs. *Geosciences* **2021**, *11*, 106. [CrossRef]

processes MDPI

Article

Research on Optimization of Sulfur Solubility Testing Method for High-Sulfur Natural Gas

Ying Wan [1], Li Wang [1,2,*], Yan Yang [1], Zhao Ding [1], Daqing Tang [1], Dihong Zhang [1] and Linling Zhang [1]

1 Exploration and Development Research Institute of PetroChina Southwest Oil and Gas Field Company, Chengdu 610213, China; wan_y@petrochina.com.cn (Y.W.); yangy2020@petrochina.com.cn (Y.Y.); dingzhao@petrochina.com.cn (Z.D.); tangdq@petrochina.com.cn (D.T.); zhangdihong@petrochina.com.cn (D.Z.); zhanglinling@petrochina.com.cn (L.Z.)
2 National Energy R&D Center of High Sulfur Gas Exploitation, Chengdu 610051, China
* Correspondence: wli-111@petrochina.com.cn

Abstract: At present, the methods for sulfur solubility testing of high-sulfur natural gas generally use laboratory proportioning gas samples and then connect equipment to test the sulfur solubility of the gas samples based on the adsorption deposition mechanism. However, these testing methods generally have the following problems: (1) The equipment used in these test methods has safety hazards such as leakage at pipe and valve connections. (2) The sulfur solubility of real gas samples cannot be measured directly. (3) The equipment is difficult to clean, and it is especially difficult to clean the sulfur deposited at pipe elbows and valve connections. This will lead to low sulfur solubility test results. (4) The thermal insulation performance during the test process is not good, and temperature changes have a great impact on gas volume measurement. In order to solve the above problems, a testing method and comprehensive experimental device for the solubility of elemental sulfur in high-sulfur natural gas were established. This test method wraps the entire experimental device with a metal shell, which has good safety and thermal insulation performance, and it uses customized pipeline connections, which have high flushing efficiency, less sulfur deposition, and more accurate experimental results. The upgraded filtration system can directly measure the sulfur dissolution of real gas samples, and a CS_2 solution recovery process is added to reduce the risk of leakage and environmental pollution. This method and equipment were used to test the elemental sulfur solubility determination of real gas samples from a high-sulfur gas well. The research results show that the solubility of elemental sulfur is related to temperature, pressure, and H_2S concentration and increases with the increase in temperature, pressure, and H_2S concentration. Compared with the previous method, this method has less residual sulfur during the test process, the measured sulfur solubility is 2.13% greater, and the test results are more accurate and reliable. This research result provides important basic data support for accurately measuring the elemental sulfur solubility of real gas samples in high-sulfur gas reservoirs and dealing with sulfur deposition problems.

Keywords: high-sulfur natural gas; sulfur solubility; experimental test

Citation: Wan, Y.; Wang, L.; Yang, Y.; Ding, Z.; Tang, D.; Zhang, D.; Zhang, L. Research on Optimization of Sulfur Solubility Testing Method for High-Sulfur Natural Gas. *Processes* **2024**, *12*, 1210. https://doi.org/10.3390/pr12061210

Academic Editor: Qingbang Meng

Received: 13 May 2024
Revised: 9 June 2024
Accepted: 11 June 2024
Published: 13 June 2024

1. Introduction

High-sulfur gas reservoirs in China contain abundant resources, with cumulative proven high-sulfur gas reserves exceeding 1 trillion cubic meters [1]. Among them, high-sulfur gas reservoirs in the Sichuan Basin have large reserves, high output, and high development value [2,3]. However, during the development process, elemental sulfur will undergo phase transitions. Hydrogen polysulfide is formed by the reaction of elemental sulfur with H2S. The reaction is a reversible chemical reaction. As temperature and pressure decrease, hydrogen polysulfide decomposes to form sulfur and hydrogen sulfide. When sulfur reaches the critical saturation state in the gas, the decrease in temperature and pressure will lead to the precipitation of sulfur from the gas phase. When the precipitated sulfur cannot be carried by fluid hydrodynamics, sulfur accumulates and deposits in formation pores, wells, and surface pipelines. Sulfur deposition will lead to pipeline corrosion and formation and wellbore blockage, affecting the smooth production of gas wells [4–7]. It will also affect the fracture conductivity in the production process of a gas well, and the decrease in the conductivity in production will significantly reduce the recovery rate of the gas well [8].

Elemental sulfur generally exists in the form of hydrogen polysulfide and other compounds under initial reservoir conditions. However, during the production of high-sulfur gas reservoirs, the temperature and pressure are continuously reduced. When sulfur in gases reaches critical saturation levels, elemental sulfur accumulates and deposits in pore space, wellbores, and shaft linings. It affects two-phase seepage, reduces conductivity, blocks fracture channels, and causes gas well production reduction or shutdown [9–11]. At the same time, for high-production gas wells, due to high production and high flow rate, sulfur elements are easily carried out, and it is difficult for deposition to occur. However, the higher the production allocation, the faster the formation pressure drops, and the solubility of sulfur decreases rapidly, causing more sulfur elements to precipitate, affecting production capacity [12,13]. Therefore, it is necessary to accurately measure the elemental sulfur solubility in high-sulfur gas wells to optimize the gas well working system and solve the problem of sulfur deposition. For accurate experimental testing of sulfur solubility, scholars have proposed different methods.

Kennedy et al. [14] measured the sulfur solubility in methane, carbon dioxide, hydrogen sulfide, and mixtures of these gases and proved that sulfur solubility is related not only to temperature and pressure, but also to gas components. The method involves using a hard glass tube filled with glass wool to adsorb the elemental sulfur precipitated from the gas to be tested, then passing oxygen through it and igniting it, followed by absorbing the combustion products with a weak solution of sodium carbonate, and finally calculating the sulfur content through back titration. The sulfur dissolution measured by this method is low and poses safety risks. Roof [15] modified Kennedy's test method and measured sulfur solubility in pure hydrogen sulfide using carbon disulfide to absorb the gas during the test. Compared with Kennedy's experimental method, the safety is greatly improved, but a hand pump is used to control the back pressure. The back pressure control accuracy is low, which affects the full adsorption of elemental sulfur, and the conclusions drawn are limited to low-temperature and low-pressure conditions. Based on the Roof test method, Swift et al. [16] extended the tested temperature and pressure conditions to formation conditions and concluded that under a certain pressure, as the temperature increases, the solubility of elemental sulfur increases. The disadvantage is that the gas used in the test is pure gas, and the results are quite different from the solubility of elemental sulfur in the actual natural gas in the formation. Brunner and Woll [17] improved the test methods of Roof and Swift. Their test principle is also to absorb the precipitated sulfur with a CS_2 solution and then evaporate and measure the sulfur mass. The difference is that they indirectly obtained the volume of the gas by condensing the pure hydrogen sulfide gas and then weighing it, which made up for the accuracy problem of the gas metering device at that time. However, weighing solid hydrogen sulfide gas had greater safety risks. Brunner et al. [18] extended the study of Kennedy et al. and measured the solubility of sulfur in a number of different acid gas mixtures. The results show that in multi-component mixed gases, the H_2S content has the greatest impact on elemental sulfur, and the solubility of elemental sulfur increases with the increase in H_2S content. Gu et al. [19,20] discovered that the solubility of elemental sulfur in CH_4 and CO_2 was poorly studied experimentally. They used a static method device to measure the solubility of sulfur in pure gases of CH_4, CO_2, and H_2S and their mixtures. It was concluded that the content of H_2S in the gas component is the main factor affecting the solubility of elemental sulfur. Fu et al. [21] established a method for calculating the sulfur saturation of formations and found that physical deposition is the main method of elemental sulfur deposition. In order to accurately obtain the saturated elemental sulfur content in high-sulfur gas reservoirs, Yang et al. [22] measured the solubility of elemental sulfur in high-sulfur gas at certain temperatures and different pressures. Experiments have shown that the solubility of elemental sulfur in high-sulfur gases increases as the pressure increases. The experimental process was optimized based on the static method experimental device, which was equipped with an upgraded filter so that the experimental device could directly measure the sulfur solubility of real gas samples. However, it did not guarantee a constant temperature throughout the experimental process. Temperature changes have an impact on gas measurement. Serin et al. [23] used chromatographic analysis to study the solubility of sulfur in supercritical CO_2, supplementing the research on the solubility of sulfur in CO_2 at lower temperatures. The advantage of this method is that the volume of the sulfur-containing gas sample is known and the amount is small. However, the disadvantage is that the mass spectrometer is sensitive to detection. The residual sulfur in the device will significantly reduce the measured sulfur solubility. And real gas samples cannot be measured. Cloarec et al. [24], based on the test method of Serin et al., improved the gas expansion equipment of the equipment and added a set of sulfur capture devices. This resulted in less residual sulfur in the device and more accurate experimental results. Zhang [25] tried to measure the sulfur solubility of real gas samples and added a condensing device based on the Brunner test method. It connects high-sulfur gas saturated with elemental sulfur to a condensing device and calculates the solubility of elemental sulfur in high-sulfur gas reservoirs by precipitating elemental sulfur in the condensing device. However, he did not consider that elemental sulfur easily deposits at pipeline elbows and valves, and the device was not equipped with corresponding reinjection purge equipment, resulting in a low measured sulfur solubility. Zeng et al. [26] designed a downhole sampler suitable for ultra-deep high-sulfur gas fields to provide more accurate data samples for the dynamic detection of gas fields, considering the characteristics of the large buried depth, highly heterogeneous reservoir, and complex gas–water relationship of the Puguang gas field. Based on the test method of Cloarec et al., Wu et al. [27] added a gas metering device to measure the sulfur solubility of real gas samples by controlling the gas volume entering the mass spectrometer. However, sulfur deposition in pipelines still has a great impact on test results.

Based on the research of the above scholars, it is found that when the solubility of elemental sulfur in high-sulfur natural gas is measured by mass spectrometer, although smaller sulfur-containing gas samples are required and the test is less dangerous, the mass spectrometer analysis is sensitive, and the standard sample configuration and the deposition of sulfur content in the pipeline will significantly affect the test results, so the traditional adsorption deposition method is selected as the theoretical basis for the optimization method. However, the previous adsorption deposition method has the following shortcomings: 1. The sulfur solubility of real gas samples is rarely measured directly. 2. The deposited sulfur at pipeline elbows and valve connections cannot be cleaned well, making the measured sulfur solubility low. 3. There is a lack of transition area between the reaction, condensation, and metering systems for reducing the impact of temperature changes on the metering results. 4. The back pressure control system has low accuracy, and there are fluctuations in the airflow when measuring real gas samples, which affects the full adsorption of elemental sulfur. 5. A CS_2 solution needs to be added multiple times during the test, which poses risks of leakage and environmental pollution. In order to develop high-sulfur gas reservoirs more safely and rationally, it is necessary to design a safer and more accurate sulfur solubility testing method based on the shortcomings of the previous condensation adsorption method.

2. Integrated Equipment for Sulfur Solubility Testing

The test method is based on the condensation adsorption mechanism and uses a solvent (CS_2) that can quickly dissolve elemental sulfur in the gas and does not react with other components as the absorption liquid. It involves calculating the solubility of elemental sulfur in high-sulfur natural gas by measuring the elemental sulfur content in the absorption liquid and the gas volume passing through the meter. By first measuring the sulfur solubility of several pressure- and temperature-sensitive points of the experimental gas sample and then using the interpolation method to calculate the elemental sulfur content of the experimental gas sample at different temperatures and pressures, the supporting design experimental device has the following advantages: 1. The entire set of equipment is wrapped in a metal shell, which improves the safety of the test and ensures constant temperature throughout the process, reducing the interference of temperature changes on the test results. 2. It uses two sets of customized filters to filter formation water, drilling fluid, and other impurities in real gas samples and directly tests the elemental sulfur solubility of real gas samples. 3. The use of customized pipelines can effectively flush the deposited sulfur at pipeline elbows and valves, and the sulfur solubility measurement results are more accurate. 4. The back pressure control system is more accurate and can adjust the back pressure according to the flow rate at the outlet of the metering system to ensure full adsorption of elemental sulfur. 5. A CS_2 recovery system is added, which can condense and recover CS_2 steam during the test, reducing the risk of leakage and environmental pollution.

2.1. Test Device

The sulfur dissolution test uses an independently developed and designed elemental sulfur solubility meter (Figure 1). The device has a maximum working pressure of 100 MPa and a maximum working temperature of 150 °C. The entire set of equipment is made of Hastelloy alloy, which can meet the test requirements of temperature resistance, pressure resistance, and corrosion resistance. Compared with the previous test device for measuring elemental sulfur solubility, this device has improved and optimized the test process and upgraded many key components. The optimized test device mainly consists of six parts: sample transfer and sample preparation system, back pressure control system, condensation adsorption system, flow metering control system, off-gas treatment system, and carbon disulfide reinjection system.

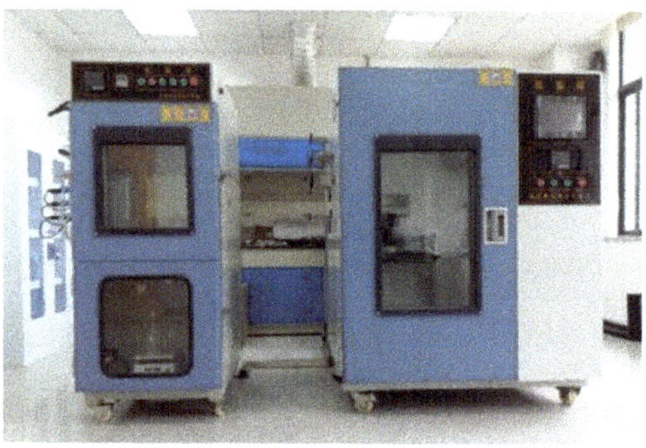

Figure 1. Integrated equipment for sulfur solubility testing.

2.1.1. Sample Preparation System

The sample preparation system is shown in Figure 2a. The system mainly consists of a constant pressure and speed Duplex pump, a sample preparation device, an air bath thermostat, a filter, and a swing device. Its main function is to simulate the depletion exploitation method by controlling temperature and pressure, gradually reducing the test pressure and temperature and thereby directly measuring the content of elemental sulfur in the sample. At the same time, excess sulfur powder can also be added before introducing the gas sample to make the elemental sulfur in the gas sample reach a saturated state, thereby measuring the solubility of elemental sulfur under different temperature and pressure conditions. In order to test sulfur solubility more accurately and safely, the testing process and equipment have been optimized as follows: (1) In order to improve the thermal insulation performance, air bath heating is used instead of electric heating, which makes heating safer and more uniform. (2) In order to ensure the sealing safety of the mixing chamber, the use of valves and the occurrence of leak points are minimized. Therefore, there is only one temperature probe in the mixing chamber, and the pressure is automatically controlled by a constant pressure and speed Duplex pump. (3) In order to directly measure real gas samples, two sets of filters are equipped to filter out formation water, drilling fluid, and other impurities that may be present in real gas samples. When one group becomes clogged, another group can be switched.

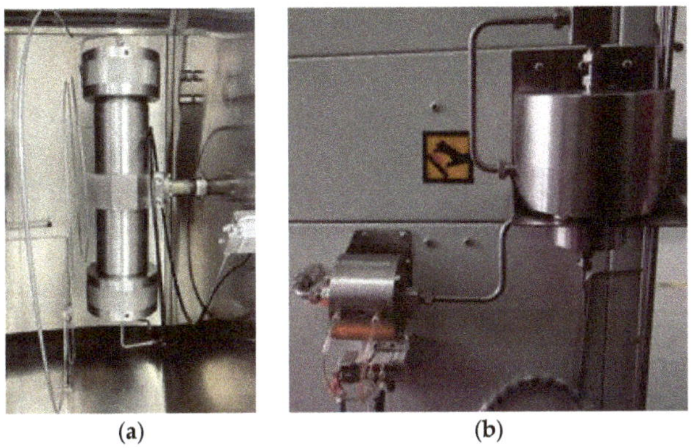

Figure 2. (**a**) Sample preparation system; (**b**) back pressure control system.

2.1.2. Back Pressure Control System

The back pressure control system is shown in Figure 2b. The system mainly consists of an emergency brake valve, a back pressure tracking pump, and a back pressure valve. Its main function is to slowly balance and reduce the pressure of the high-pressure, high-sulfur gas sample in the mixing chamber to normal pressure (0.1 MPa). The critical point of elemental sulfur solubility is reached through the reduction in pressure, causing elemental sulfur in the gas to precipitate. In order to test sulfur solubility more accurately and safely, the testing process and equipment have been optimized as follows: In order to directly measure real gas samples, a high-precision back pressure tracking pump and a piston needle back pressure valve were selected to improve the stability of the back pressure system, stabilize the gas flow rate, and allow the elemental sulfur in the gas to be fully absorbed by the condensation adsorption system.

2.1.3. Condensation Adsorption System

The condensation adsorption system is shown in Figure 3a. The system mainly consists of an air bath thermostat, a coil, and an adsorption tank. Its main function is to lower the temperature of the gas sample and use CS_2 solution to adsorb elemental sulfur in the gas. In the previous test process, the condensation adsorption system mostly used several glass bottles in series, and the bottle mouths were sealed with rubber stoppers. However, during the actual test process, due to the instability of the airflow, it is easy for the rubber stopper of the last glass bottle to collapse when the gas flow suddenly increases, posing a safety hazard. Moreover, when the high-temperature airflow in the mixing chamber directly enters the CS_2 solution, it will cause part of the CS_2 solution to volatilize, making the gas measurement result larger. To test sulfur solubility more accurately and safely, the testing process and equipment have been optimized as follows: (1) To improve the accuracy of measurement, a section of coil pipe is added before and after the adsorption tank and an umbrella-shaped gas–liquid separation device is installed in front of the flow meter to reduce the impact of temperature changes and CS_2 vapor on the measurement results. (2) To enhance the safety of the sulfur solubility testing, the adsorption tank is constructed with a whole digging approach, carving out six interconnected tanks on a single piece of

Hastelloy alloy material. The top of the tanks is also sealed with Hastelloy alloy material, ensuring both the safety of the equipment and the thorough adsorption of elemental sulfur.

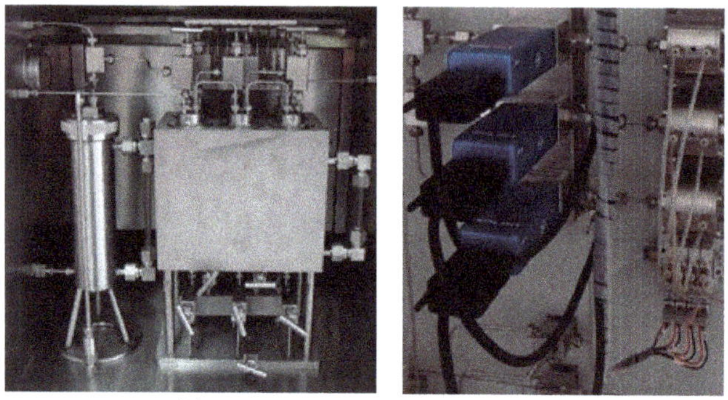

(a) (b)

Figure 3. (**a**) Condensation adsorption system; (**b**) metering system.

2.1.4. Metering System

The metering system is shown in Figure 3b. The system mainly consists of pneumatic valves, flow meters, and pressure sensors. Its main function is to measure the gas volume after reducing the pressure to normal pressure. To test sulfur solubility more accurately and safely, the testing process and equipment have been optimized as follows: (1) Because the minimum flow rate measured by the flow meter is 5% of the maximum range, to improve the accuracy of measurement, three anti-sulfur flow meters with different ranges were selected and automatically switched according to the flow rate during the test. (2) For real gas samples, even if the gas pressure is reduced to normal pressure, the gas flow will still be unstable. To directly measure the real gas sample, a flow control system is designed to control the back pressure according to the flow rate at the outlet end of the flowmeter, to avoid the flow rate at the outlet end suddenly reaching 0 or increasing suddenly.

2.1.5. Off-Gas Treatment System

The off-gas treatment system is shown in Figure 4a. The system consists of three glass cylinders with a length of 1 m containing a desulfurizer. Its main function is to absorb waste gas during testing to avoid leakage and environmental pollution. Previous sulfur dissolution test methods mostly used solutions to absorb exhaust gas, but this method requires frequent replacement of the absorption solution and has safety hazards. To test sulfur solubility more safely, the testing process and equipment have been optimized as follows: Three exhaust gas recovery tanks are used. The tanks are independent of each other and can be switched freely. When all the desulfurizer in one tank turns black, it can be quickly switched to another one. The inlet and outlet of the used tail gas recovery tank can be sealed with dead-end plugs to prevent any leakage and environmental contamination.

2.1.6. Carbon Disulfide Reinjection System

The carbon disulfide reinjection system is shown in Figure 4b. The system is mainly composed of a constant speed and constant pressure and speed Duplex pump and a pneumatic valve. Its main function is to clean the pipeline after the test is completed and collect elemental sulfur that may precipitate outside the adsorption system. To test sulfur solubility more safely, the testing process and equipment have been optimized as follows: (1) To enhance the flushing efficiency, the entire set of equipment employs custom pipelines, which can effectively flush elemental sulfur from difficult-to-clean areas such as valve connections, thus improving measurement accuracy. (2) To improve the safety of the test, a CS_2 recovery system was added. After the test, the toxic and harmful CS_2

liquid is condensed and recovered for reuse, thereby reducing environmental pollution and safety hazards.

(**a**) (**b**)

Figure 4. (**a**) Off-gas treatment system; (**b**) carbon disulfide reinjection system.

2.2. Test Process

(1) After checking the air tightness of the equipment, put excess sulfur powder (50 g) into the sample preparation room. Adjust the temperature and pressure of the equipment to the set values, then transport the gas in the sampling chamber to the sample preparation chamber, and rock and mix for more than 24 h, so that the elemental sulfur solubility of the gas in the system reaches saturation.

(2) As a result of the back pressure being controlled, the gas in the sample preparation is slowly (5 mL/min~10 mL/min) passed through the elemental sulfur adsorption system, and the gas volume passing through is measured with a gas metering device.

(3) When the gas pressure in the sample preparation room drops to 0.1 MPa, clean all pipelines with CS_2 solution and add the cleaning solution to the adsorption tank. The solid material obtained by heating the CS_2 solution in the adsorption system is weighed with a precision balance to obtain the mass of elemental sulfur. At the same time, CS_2 steam is condensed and recovered.

To simulate the depletion exploitation method, during measurement, a back pressure system is used to gradually reduce the formation pressure to the wellhead pressure at a certain temperature. The elemental sulfur solubility was calculated for each pressure point. Each pressure point is determined using the same method used to determine elemental sulfur content.

2.3. Test Data Processing Methods

According to the gas state equation, the sulfur-containing gas volume measured during the test process can be converted into the gas volume V_0 under standard conditions. The calculation formula used is

$$V_0 = \frac{P_1 \times V_1 \times 293.15}{101.325 \times (273.15 + t_1)}$$

where V_0 is the volume of gas under standard conditions (101.325 kPa, 20.0 °C), m^3; P_1 is the atmospheric pressure during testing, MPa; V_1 is the gas volume under the pressure of P_1, m^3; t_1 is the test temperature, °C.

According to the volume of sulfur-containing gas in the standard state and the mass of elemental sulfur measured by a precision balance, the solubility of elemental sulfur in sulfur-containing gas is calculated. The calculation formula used is

$$C = \frac{m_1}{V_0}$$

where C is the elemental sulfur content in sulfur-containing gas, g/m^3; V_0 is the volume of gas under standard conditions (101.325 kPa, 20.0 °C), m^3; m_1 is the mass of elemental sulfur, g.

3. Test Results

The samples used in the test came from two production wells in a gas field in the Sichuan Basin. Samples 1 and 2 are taken from the same gas sample of production well 1, and samples 3 and 4 are taken from the same gas sample of production well 2. For samples 1 and 3, the new method was used to measure sulfur solubility. For samples 2 and 4, a traditional method was used to measure sulfur solubility. The natural gas components of the gas samples from production well 1 are as follows: H_2S content is 6.75%, CH_4 content is 84.31%, C_2H_6 content is 0.07%, CO_2 content is 8.5%, N_2 content is 0.36%, H_e content is 0.01%, H_2 content is 0.003%. For production well 2, the natural gas components of the gas samples are as follows: H_2S content is 16.78%, CH_4 content is 79.1%, C_2H_6 content is 0.06%, CO_2 content is 5.12%, and N_2 content is 0.31%. The test results are as follows Tables 1 and 2.

Table 1. Solubility of gas samples from Well 1 at different temperatures and pressures.

Pressure/(MPa)	Solubility at Different Temperatures/($g \cdot m^{-3}$)					
	Sample 1			Sample 2		
	110 °C	90 °C	70 °C	110 °C	90 °C	70 °C
10	0.4589	0.3369	0.1976	0.4578	0.3338	0.1974
20	0.5142	0.3927	0.2834	0.5001	0.3855	0.2793
30	0.5838	0.4259	0.3429	0.5665	0.4215	0.3375
40	0.6491	0.5384	0.4158	0.6268	0.5198	0.4029
50	0.7958	0.6479	0.4428	0.7609	0.6206	0.4256

Table 2. Solubility of gas samples from Well 2 at different temperatures and pressures.

Pressure/(MPa)	Solubility at Different Temperatures/($g \cdot m^{-3}$)			
	Sample 3		Sample 4	
	90 °C	70 °C	90 °C	70 °C
10	0.5762	0.2301	0.5736	0.2296
20	0.8437	0.3643	0.8315	0.3593
30	1.1726	0.5842	1.1541	0.5784
40	1.2733	0.9017	1.2444	0.8821

4. Results

4.1. *Effect of Temperature and Pressure on Sulfur Solubility*

According to the test results, it is found that when the formation pressure is 50 MPa and the formation temperature is 110 °C, the elemental sulfur solubility of sample 1 is 0.7958 g/m^3. When the formation pressure remains constant and the formation temperature decreases to 70 °C, the elemental sulfur solubility decreases to 0.4428 g/m^3. Similarly, when the formation temperature remains constant and the formation pressure decreases to 10 MPa, the elemental sulfur solubility decreases to 0.4589 g/m^3. Similarly, the solubility of elemental sulfur in other samples follows a similar trend. That is, at the same temperature, the solubility of elemental sulfur increases with increasing air pressure. At the same pressure, the solubility of elemental sulfur increases with the increase in gas temperature. This may be because as temperature and pressure increase, chemical dissolution processes strengthen gradually, accelerating the formation of polysulfide compounds from elemental sulfur in the gas, thereby increasing sulfur solubility in natural gas.

4.2. Effect of Hydrogen Sulfide Content on Sulfur Solubility

The elemental sulfur solubility of sample 1 (H_2S content 6.75%) and sample 2 (H_2S content 16.78%) was compared at the same temperature and pressure. The results are shown in the figure below.

It can be seen from Figure 5 that at the same formation temperature and pressure, the elemental sulfur solubility of sample 2 with high H_2S content is greater than that of sample 1 with low H_2S content, and the difference in elemental sulfur solubility between the two increases with the increase in formation temperature and pressure. It can be concluded that under the same temperature and pressure, as the hydrogen sulfide content increases, the solubility of elemental sulfur shows an upward trend.

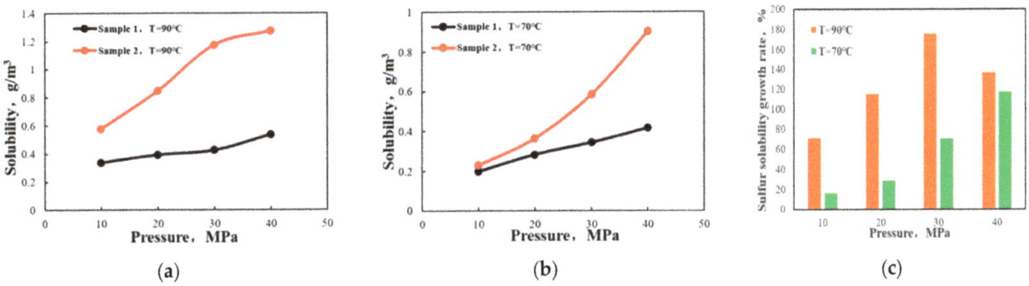

Figure 5. This figure shows experimental results of the influence of H_2S content on sulfur solubility. (**a**) T = 90 °C, comparison of elemental sulfur solubility; (**b**) T = 70 °C, comparison of elemental sulfur solubility; (**c**) T = 90 °C, increase in elemental sulfur solubility.

4.3. Comparison of Results from Different Testing Methods

For Sample 2 and Sample 4, the traditional sulfur concentration testing method was used to test the solubility of elemental sulfur, and it was compared with the results measured by the new method. The results are shown in Figure 6.

According to Figure 6, it can be seen that the solubility of elemental sulfur measured by the new method is generally greater than that measured by the traditional method, and the gap is even greater under high temperature and high pressure. For samples from well 1, when the formation pressure is 50 MPa and the formation temperature is 110 °C, the solubility difference measured by the two experimental devices is 4.59%, while when the formation pressure is 10 MPa and the formation temperature is 70 °C, the solubility difference measured by the two experimental devices is only 0.14%. Excluding the data of the highest and lowest pressure at each temperature-sensitive point, the solubility of elemental sulfur measured by the new method is on average 2.13% greater than that measured by the traditional method.

An analysis shows that, on the one hand, chemical dissolution plays a dominant role under high temperature and pressure. The equipment supporting the new testing method can better maintain the temperature and pressure required by the experiment. However, for the traditional testing method, the temperature and pressure will decrease slightly over time, the dissolved elemental sulfur in the gas sample will be relatively small, and the final elemental sulfur precipitated will decrease. On the other hand, the supporting equipment of the new test method uses customized pipeline connections and adds a CS_2 reinjection process. After the test, the pipeline was cleaned repeatedly. There was less residual sulfur in the pipeline, more elemental sulfur was collected, and the measured solubility of elemental sulfur was relatively large.

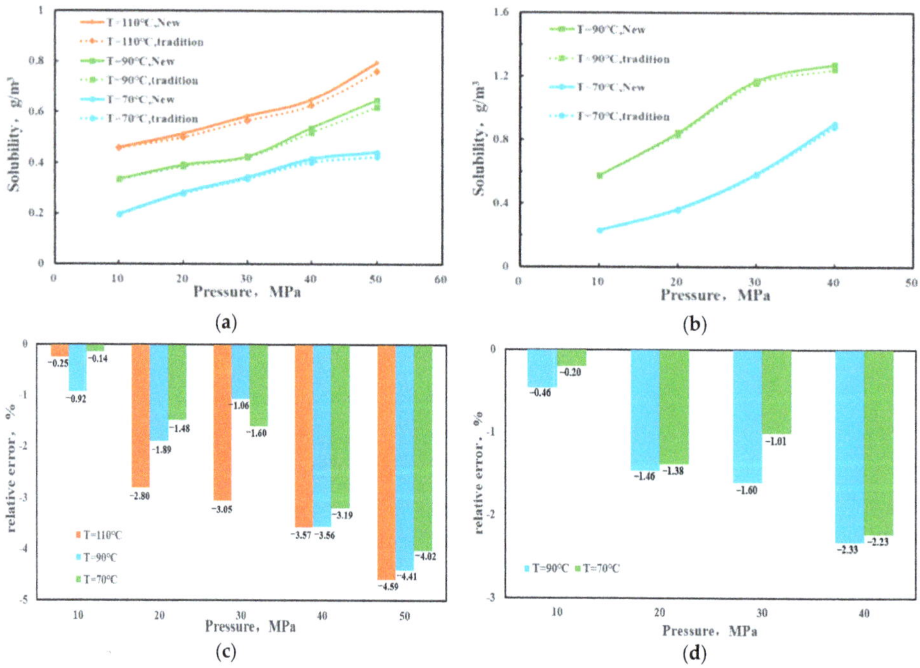

Figure 6. This figure shows a comparison of results from different testing methods. (**a**) Comparative results of elemental sulfur solubility of well 1 samples; (**b**) comparative results of elemental sulfur solubility of well 2 samples; (**c**) difference in solubility of elemental sulfur in well 1 samples; (**d**) difference in solubility of elemental sulfur in well 2 samples.

5. Conclusions

(1) Based on the fact that sulfur solubility measurement methods have low accuracy and can rarely directly measure real gas samples, a testing method and integrated testing equipment for elemental sulfur solubility in high-sulfur gas reservoirs were established. This test method wraps the entire test device with a metal shell, which has good safety and insulation. Using customized pipeline connections, the flushing efficiency is high, there is less sulfur deposition, and the test results are more accurate. The upgraded filtration system can directly measure the sulfur dissolution of real gas samples. The CS_2 recycling process is increased to reduce the risk of leakage and environmental pollution.

(2) A new method was used to test the solubility of elemental sulfur in real sulfur-containing gas samples from a gas well. The results show that the solubility of elemental sulfur is positively correlated with temperature, pressure, and H_2S content and increases with the increase in temperature, pressure, and H_2S concentration.

(3) The same gas samples were tested using new methods and traditional methods. The results show that the new method obtains more elemental sulfur; the measured sulfur solubility is correspondingly greater, with an average difference of 2.13%; and the test results are more accurate and reliable.

Author Contributions: Conceptualization, Y.W. and L.W.; methodology, Y.Y.; validation, Z.D., D.T. and D.Z.; investigation, L.Z.; resources, L.W.; data curation, Y.Y.; writing—original draft preparation, Y.W.; writing—review and editing, L.W.; supervision, Y.Y.; project administration, L.W. All authors have read and agreed to the published version of the manuscript.

Funding: This research received no external funding.

Data Availability Statement: The data presented in this study are available on request from the corresponding author. The data is unavailable due to privacy.

Conflicts of Interest: Authors Ying Wan, Li Wang, Yan Yang, Zhao Ding, Daqing Tang, Dihong Zhang and Linling Zhang were employed by the Exploration and Development Research Institute of PetroChina Southwest Oil and Gas Field Company. The authors declare that the research was conducted in the absence of any commercial or financial relationships that could be construed as a potential conflict of interest. The Exploration and Development Research Institute of PetroChina Southwest Oil and Gas Field Company had no role in the design of the study; in the collection, analyses, or interpretation of data; in the writing of the manuscript, or in the decision to publish the results.

References

1. Huang, L. New progresses in safe, clean and efficient development technologies for high-sulfur gas reservoirs. *Nat. Gas Ind. B* **2015**, *2*, 360–367. [CrossRef]
2. Hu, Y.; Peng, X.; Li, Q.; Li, L.; Hu, D. Progress and development direction of technologies for deep marine carbonate gas reservoirs in the Sichuan Basin. *Nat. Gas Ind. B* **2020**, *7*, 149–159. [CrossRef]
3. Ma, X. A golden era for natural gas development in the Sichuan Basin. *Nat. Gas Ind. B* **2017**, *4*, 163–173. [CrossRef]
4. Hu, Y.; Peng, Y.; Zhang, W.; Song, M. Corrosion Inhibition Performance on Tubing Steel in Oil Wells Based on Schiff Base Surfactants. *Int. J. Electrochem. Sci.* **2017**, *12*, 4360–4369. [CrossRef]
5. Hu, Y.; Peng, Y.; Hu, F.; He, F.; Guo, P. Anti-corrosion Performance of Chromium-coated Steel in a Carbon Dioxide-saturated Simulated Oilfield Brine. *Int. J. Electrochem. Sci.* **2017**, *12*, 5628–5635. [CrossRef]
6. Hu, Y.; Peng, Y.; Zhao, J.; Guo, P.; Long, S. Electrochemical Analysis of the Corrosion Behavior of Drill Pipe Steel under Oil/Water Emulsion Condition. *Int. J. Electrochem. Sci.* **2017**, *12*, 8526–8534. [CrossRef]
7. Yong, H.; Dong, H.; Xian, P. Key development technologies for high-sulfur gas reservoirs: New progress, challenges and key research directions. *Nat. Gas Ind.* **2022**, *42*, 23–31.
8. Peng, Y.; Luo, A.; Li, Y.; Wu, Y.; Xu, W.; Sepehrnoori, K. Fractional model for simulating Long-Term fracture conductivity decay of shale gas and its influences on the well production. *Fuel* **2023**, *351*, 129052. [CrossRef]
9. Li, J.; Cao, K.; Duan, Y. Seepage mechanism and productivity decline of shale-gas well. *Nat. Gas Explor. Dev.* **2011**, *34*, 10.
10. Peng, H.; Zhang, M.; Liang, X. Establishment of material balance equation for a gas reservoir considering sulfur deposition. *Nat. Gas Explor. Dev.* **2011**, *34*, 9.
11. Peng, Y.; Ma, H.; Li, Z.; Zhou, L.; Zhao, J.; Wen, X. Material balance evaluation method for water-bearing gas reservoirs considering the influence of relative permeability. *Pet. Sci. Technol.* **2024**, 1–16. [CrossRef]
12. Li, H.; Liu, P.; Liu, C.; Wu, L.; Chen, J. Technologic bottleneck of effectively developing high-sour gas fields in eastern Sichuan basin and its countermeasure effect. *Nat. Gas Explor. Dev.* **2015**, *38*, 10.
13. Zhou, M.; Xiang, Y.; Zhang, W.; Zhang, N.; Wang, Y.; Chen, Z.; Wang, H. Develop features of Chuangxing bioreef gas reservoirs in eastern Sichuan Basin and technical countermeasures. *Nat. Gas Explor. Dev.* **2020**, *43*, 7.
14. Kennedy, H.T.; Wieland, D.R. Equilibrium in the Methane-Carbon Dioxide-Hydrogen Sulfide-Sulfur System. *Trans. Aime* **1960**, *219*, 166–169. [CrossRef]
15. Roof, J.G. Solubility of Sulfur in Hydrogen Sulfide and in Carbon Disulfide at Elevated Temperature and Pressure. *Soc. Pet. Eng. J.* **1971**, *11*, 272–276. [CrossRef]
16. Swift, S.C.; Manning, F.S.; Thompson, R.E. Sulfur-Bearing Capacity of Hydrogen Sulfide Gas. *Soc. Pet. Eng. J.* **1976**, *16*, 57–64. [CrossRef]
17. Brunner, E.; Woll, W. Solubility of Sulfur in Hydrogen Sulfide and Sour Gases. *Soc. Pet. Eng. J.* **1980**, *20*, 377–384. [CrossRef]
18. Brunner, E.; Place, M.C.; Woll, W. Sulfur Solubility in Sour Gas. *J. Pet. Technol.* **1988**, *40*, 1587–1592. [CrossRef]
19. Gu, M.; Li, Q.; Chen, W.; Guo, T. Solubility of solid sulfur in super/near-critical H_2s-containing sour fluid mixtures (II) The thermodynamic model. *J. Chem. Ind. Eng.* **1993**, *3*, 321–327.
20. Gu, M.; Li, Q.; Chu, X.; Chen, W.; Guo, T. Solubility of solid sulfur in super/near-critical H_2s-containing sour fluid mixtures (I) The thermodynamic model. *J. Chem. Ind. Eng.* **1993**, *3*, 315–320.
21. Fu, D.; Guo, X.; Du, Z. Sulfur deposition mechanism of the gas reservoir swith high sulfur content. *J. Southwest Pet. Univ. (Sci. Technol. Ed.)* **2009**, *31*, 109–111.
22. Yang, X.F.; Huang, X.P.; Zhong, B. Experimental Test and Calculation Methods of Elemental Sulfur Solubility in High Sulfur Content Gas. *Nat. Gas Geosci.* **2009**, *20*, 416–419.
23. Serin, J.; Jay, S.; Cézac, P.; Contamine, F.; Mercadier, J.; Arrabie, C.; Legros-Adrian, J. Experimental studies of solubility of elemental sulphur in supercritical carbon dioxide. *J. Supercrit. Fluids* **2010**, *53*, 12–16. [CrossRef]
24. Cloarec, E.; Serin, J.; Cézac, P.; Contamine, F.; Mercadier, J.; Louvat, A.; Lopez, A.C.; Caneghem, P.V.; Forster, R.; Kim, U. Experimental Studies of Solubility of Elemental Sulfur in Methane at 363.15 K for Pressure Ranging from (4 to 25) MPa. *J. Chem. Eng. Data* **2012**, *57*, 1222–1225. [CrossRef]

25. Zhang, G. *Study on Feature of Phase Behaviors and Seepage Mechanism in High Sour Gas Reservoir-Take Samples from Changxing Gas Reservoir in Yuanba as Study Object*; Chengdu University of Technology: Chengdu, China, 2014.
26. Zeng, D.; Peng, X.; Fu, D.; Hu, J.; Wu, X.; Zhang, J. Development dynamic monitoring technologies used in the Puguang high-sulfur gas field. *Nat. Gas Ind. B* **2019**, *6*, 191–197. [CrossRef]
27. Wu, Y.; Wang, N.; Zhang, G.; Zhang, M.; Mo, C. Testing Method and Prediction Model of Elemental Sulfur Solubility in Sour Gas Reservoirs. *Spec. Oil Gas Reserv.* **2022**, *29*, 104–109. [CrossRef]

processes MDPI

Article

Oil Removal Technology for Water Injection in Low-Permeability Reservoirs: A Micro-Vortex Flow Approach

Dawei Zhao [1,*], Weihong Xie [1], Jingyi Zhu [1], Bing Li [1], Lirong Wang [1], Tao Chen [2], Yuxin Sheng [3] and Xiujie Huang [4]

1 PetroChina Planning & Engineering Institute, Beijing 100083, China; xieweihong@petrochina.com.cn (W.X.); jyzhu@petrochina.com.cn (J.Z.); libing@petrochina.com.cn (B.L.); wanglirong.66@petrochina.com.cn (L.W.)
2 School of Environment & Energy Engineering, Beijing University of Civil Engineering and Architecture, Beijing 100032, China; chentao@bucea.edu.cn
3 PetroChina Jidong Oilfield Company, Tangshan 063004, China; shengyuxin@petrochina.com.cn
4 Sinochem Holdings Corporation Ltd., Beijing 100081, China; huangxiujie@sinochem.com
* Correspondence: zhaodawei@petrochina.com.cn

Abstract: Gravity settling is a widely employed technology that removes oil from produced water in oilfields. However, with the transition of reservoir development to low-permeability reservoirs, conventional produced water settling tanks face limitations in the treatment efficiency and coagulant dosage. This study presents an innovative approach that optimizes sedimentation tank structures and integrates micro-vortex flow technology to enhance coagulation and flocculation. Through chemical dosage experiments, comparative experiments, and long-term observation, the micro-vortex flow reactor demonstrates a 9.4% increase in oil removal efficiency while reducing the coagulant dosage by 30.0%. The MOR equipment achieved a 20.5% higher oil removal efficiency than conventional methods while maintaining effluent oil and suspended solids below 20 mg/L. The long-term observation experiment of MOR equipment further highlights oil removal efficiency of 94.2% and the micro-vortex reactor's excellent anti-pollution performance. The MOR equipment significantly reduces the land occupancy area by over 50% compared to conventional methods, thanks to the implementation of micro-vortex flow technology that effectively addresses the limitations associated with traditional settling tanks. This study contributes to advancing efficient and sustainable practices in waterflooding reservoirs, particularly for meeting stringent standards of water injection in low-permeability oilfields.

Keywords: produced water; micro-vortex flow; oil removal; enhanced coagulation; multifunctional

Citation: Zhao, D.; Xie, W.; Zhu, J.; Li, B.; Wang, L.; Chen, T.; Sheng, Y.; Huang, X. Oil Removal Technology for Water Injection in Low-Permeability Reservoirs: A Micro-Vortex Flow Approach. *Processes* **2024**, *12*, 1092. https://doi.org/10.3390/pr12061092

Academic Editor: Qingbang Meng

Received: 7 March 2024
Revised: 30 April 2024
Accepted: 1 May 2024
Published: 27 May 2024

1. Introduction

In the petroleum and natural gas industry, the largest quantity of a byproduct resulting from oil and gas extraction is produced water, primarily generated through two processes. Firstly, a mixture of formation water and crude oil is produced in the extraction process, typically sourced from underground water surrounding oil wells. Secondly, water injection into reservoirs to maintain pressure, known as waterflooding, results in the injected water eventually becoming part of the produced water or wastewater [1]. After crude oil collection and initial processing, the water separated from the oil–water mixture is the target of oilfield water treatment. Its properties are influenced by factors such as the geological conditions of the crude oil, the quality of injected water, and crude oil gathering and transportation. Produced water mainly contains the following substances [2,3]:

- Suspended Solids: Particle diameters range from 1 to 100 μm, with solids larger than 100 μm easily removed by settling. Suspended solids include various types of mud and sand (clay, silt, fine sand, etc.), corrosion products and scale (Fe_2O_3, CaO, FeS, $CaCO_3$, etc.), bacteria (sulfate-reducing bacteria, saprophytic bacteria, etc.), organic matter, etc.

- Colloidal Particles: Particle sizes range from 0.001 to 1 μm, with a composition similar to suspended solids but smaller in size and harder to separate.
- Floating Oil and Dispersed Oil: Oil droplets larger than 100 μm are dispersed oil, while those between 10 and 100 μm are floating oil, both of which can be removed through natural settling over time or simple physical processes.
- Emulsified Oil: Oil droplets have a diameter of 0.001 to 10 μm, and emulsified oil is difficult to remove through natural settling and requires chemical methods, air flotation, etc., making it a focal point in water treatment.
- Dissolved Substances: Primarily includes dissolved inorganic salts smaller than 0.001 μm and gases dissolved in the 0. 3 to 0.5 nm range.

Produced water poses a detrimental impact on the environment due to its complex composition and costly disposal methods [4]. Produced water injection is a cost-effective approach for disposing of it. In numerous onshore wells, the practice of waterflooding to improve recovery is widely employed and considered as the preferred option [5,6]. Typically, the primary control parameters' concentration for water injection includes oil, suspended solids, and bacteria. Failure to meet the designated water quality standards during water injection activities can damage the subterranean reservoir and result in a persistent reduction in its permeability, thereby impacting the productivity of the oilfield [6–8]. It is noteworthy that, unlike discharging produced water into the surface environment, although produced water typically has high Total Dissolved Solids (TDSs), which equal 50,000 to 250,000 mg/L, there are no requirements for TDS levels when treating and re-injecting formation water. This is because water injection with reduced TDS back into the underground formations may induce water-sensitive reactions [9], causing clay swelling, which could lead to formation blockages instead.

Reinforcing the continuous exploration and development of low-permeability and ultra-low-permeability oil reservoirs is crucial to enhancing its crude oil reserves and production in the future [10]; water drive development is still an effective way to develop low-permeability reservoirs [11]. The importance of treating produced water continues to grow. In oilfields, cost-effective and operationally simple physical oil removal methods are commonly employed [12,13]. Various physical methods are employed in oilfields, and the main principles and challenges are as follows:

- Natural oil removal method [14–16]: The proposed approach leverages the density disparity between oil and water to effectively separate a substantial quantity of floating oil. Although it requires extended hydraulic retention time and occupies significant space, it demonstrates strong resilience to variations in water volume and quality. However, it has limited effectiveness in separating dispersed and emulsified oils.
- Inclined plates (tubes) technology [17,18]: Based on the principles of shallow pools, integrating inclined plates into the oil removal system increases the surface area for separation, reduces the separation elevation, extends the wetted perimeter of water flow, reduces the hydraulic radius of water flow, and stabilizes flow dynamics, thereby facilitating the segregation of oil and water. When combined with flocculants, the effectiveness is further improved.
- Coalescence technology [19–21]: The produced water passes through coalescence fillers. As the water flows through the fillers, oil droplets collide and adhere to them, gradually increasing and hastening the separation of oil and water. The selection of filler material has a significant impact on the efficiency of oil removal. When water contains significant sediments and suspended solids, the fillers are prone to clogging, making equipment maintenance challenging.
- Air flotation separation technology [22–24]: Employing dispersed micro-bubbles as a medium, this method facilitates the adsorption of suspended solids and oil droplets in water. Typically, it necessitates the concurrent utilization of flocculants for optimal efficacy. While proficient in eliminating oil and suspended solids, it demands stringent bubble size criteria. Otherwise, the treatment outcome may fall short of expectations.

- Hydraulic cyclone technology [25–27]: Exploiting the density variance between oil and water, this approach employs cyclones or eddy currents to engender centrifugal force for oil–water segregation, which is an efficient oil removal technology. Nonetheless, its susceptibility to impact loads is inadequate, its capacity for suspended solids removal is moderate, it cannot separate emulsified oil, and treating recovery water from backwashing poses challenges.

According to the development of international low-permeability oil and gas reservoirs, low-permeability reservoirs can be divided into general low-permeability (K = 1.0 to 10.0 mD), special low-permeability (K = 0.5 to 1.0 mD), and ultra low-permeability (K = <0.5 mD) based on the permeability rate (K) [28]. Oilfield enterprises determine the parameters of the water injection index through core experiments or on-site statistical experiments [29]. Industry standards also provide recommended water injection index parameters based on permeability rates, as shown in Table 1.

Table 1. Recommended water injection parameters based on permeability rates [30].

Number	Permeability Rate (K) (mD)	Oil (mg/L)	Suspended Solids (mg/L)
1	0.5 to 2.0	≤30.0	≤25.0
2	0.05 to 0.5	≤15.0	≤20.0
3	0.01 to 0.05	≤10.0	≤15.0
4	<0.01	≤5.0	≤8.0

The oil and suspended solids requirements of the water injection determine the depth of the oilfield-produced-water treatment process; the higher the water injection requirement, the more emulsified oil and colloids need to be removed. Low-permeability reservoirs usually have the problems of a poor water injection effect and low injection-production ratio. They are faced with the challenge that the water quality of injection is high, and the conventional process cannot meet the treatment demand [31].

For produced water with high emulsification levels and a high content of colloids, conventional physical methods alone may not be achieved through treatment. Typically, chemical methods need to be combined, with the most common approach being the addition of coagulants and flocculants [32]. These chemical agents compress the double electric layers of colloidal particles in water, reduce colloidal stability, and promote collisions between destabilized colloidal ions, forming larger flocs. During this process, oil droplets are also adsorbed, resulting in the effective removal of both oil and suspended solids. A coagulation sedimentation tank is widely used in oilfields. Compared to a natural settling tank, a central reaction barrel incorporates swirling or turbulent flow to uniformly mix coagulants or flocculants with produced water, forming large aggregates, as shown in Figure 1. The produced water coming out of the central reaction barrel undergoes the separation of oil and suspended solids in the settling zone. The selection of suitable chemicals and dosages depends on the emulsification level and water quality characteristics (cations, anions, salinity) and often requires chemical tests [33]. The central reaction barrel may not achieve sufficient chemical reaction intensity when facing highly emulsified produced water, resulting in uneven agent diffusion. Inadequate contact efficiency with oil droplets and suspended solids ultimately fails to meet the expected water quality standards for oil removal.

To meet the standards for water injection, a filtration process is typically implemented after physical or chemical oil removal treatments. The filtration process can involve the use of a filter made of one or a combination of porous materials such as walnut shells, quartz sand, anthracite coal, magnetite, and garnet [34,35]. Water passes through the filtration bed of porous granular materials, where impurities are trapped in the voids and adsorbed onto the filtration media, resulting in water purification. If the influent oil content before the filtration stage is high, a large amount of oil will gradually adsorb onto the filtration media, obstructing the pore channels and causing media fouling, a reduced filtration rate, and deterioration in effluent quality. Therefore, it is essential to use treated water or gas for backwashing to reduce media contamination. When the filtration pressure continues to rise until the filtration system fails, and backwashing cannot restore filtration performance, replacing the media becomes necessary [36]; common backwashing methods include (1) water washing alone at fluidization velocity; (2) air scouring followed by fluidized water washing; (3) combining simultaneous air scouring and sub-fluidized water washing [37]. The design values for influent oil and suspended solids in the filtration process need to be comprehensively considered based on the settings of the oil removal and filtration processes. The recommended influent standards for filtration processes given by the Petroleum Industry Standardization Technical Committee in the petroleum industry standard are oil $\leq$ 100 mg/L and suspended solids $\leq$ 50 mg/L [38]. If the effluent water quality requirements are not high or the oil removal section performs well, only one filtration stage may be required. Depending on actual operating conditions, two- or three-stage filters can also be set up. An effective oil removal process not only enables the filtration system to demonstrate its outstanding capability in removing suspended solids and ensuring water quality compliance but also reduces energy consumption and media replacement costs by minimizing backwashing [39]. Therefore, the effluent water quality from the oil removal section significantly influences the parameter settings (filtration stages, filtration rate, backwash frequency, etc.) and operational performance (effluent quality, energy consumption levels, etc.) of the filtration section. Membrane filters can also be used in the filtration process [40], but they require higher influent water quality, and the cost of membrane replacement after membrane fouling is relatively high [41]. Oilfield companies are cautious when applying membrane filters. However, membrane technology with strong anti-fouling performance has also been widely used in oilfields, such as ceramic membranes [42].

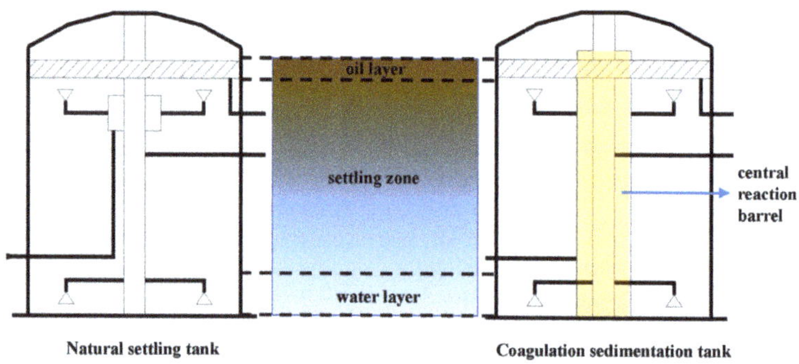

Figure 1. Schematic diagram of natural settling tank and coagulation sedimentation tank.

The oil removal process has a natural settling tank and a coagulation settling tank, and the filtration process has one or two filtration stages; this combination of produced water treatment processes is widely used in conventional oilfields, as shown in Figure 2. To meet more stringent water injection standards for low-permeability fields, the advancement in water treatment technology aims to achieve integration, high efficiency, and cost-effectiveness. Gravity separation is currently the most commonly employed method for treating produced water from oilfields due to its cost-effectiveness and simplicity [13]. The produced water is stored in a system consisting of two vertically settling tanks. After a retention time of 6 to 8 h, the upper layer of oil is skimmed off, and the water is discharged from the bottom of the tank. In cases of highly emulsified or high-pour-point oils, chemical agents like coagulants and flocculants are added to aid in the separation process. The dosage of these agents needs to be determined through experimentation [43]. These agents are mixed in a conventional coagulation tank using turbulent agitation. Their performance depends on their diffusion and contact effects within the tank. The performance of the agents depends on their diffusion and contact effects within the tank. However, the traditional process of using two large sedimentation tanks for oil removal has some shortcomings, such as a large footprint, prolonged retention time for oil–water separation, low efficiency of coagulant reactions, and insufficient removal efficiency of emulsified oil [44,45]. Yang has proposed a challenge in the Changqing Oilfield where the expected design water injection volume for water flooding development cannot be achieved due to the changing formation water absorption capacity [46]. Following the decrease in formation water absorption capacity, elevating the water injection pressure is a method to facilitate injecting a larger volume of water. Nonetheless, an increase in water injection pressure signifies that higher-power water injection pumps will necessitate more electricity consumption [11]. One of the crucial factors affecting the formation of water absorption capacity is the quality of the injected water [8,47]. These challenges underscore the significant demand for high-quality water injection in low-permeability oilfields.

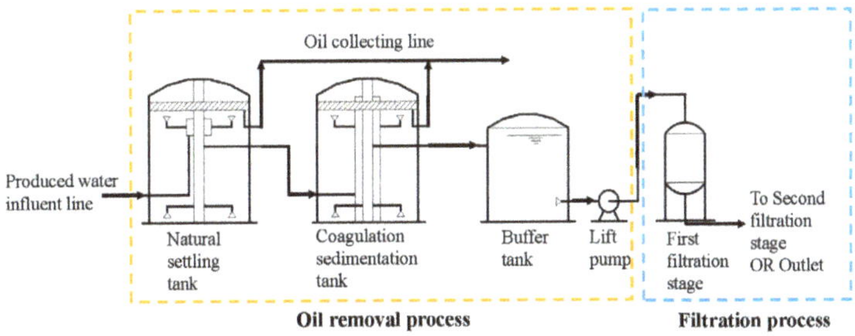

Figure 2. Typical oil removal process and filtration process of oilfield-produced-water treatment.

To mitigate the drawbacks of the existing two-stage settling tank oil removal process, optimization can be pursued in two principles: (1) re-optimizing the design structure to incorporate both natural settling and coagulation sedimentation, thereby reducing the time wasted on inefficient natural settling and achieving improved efficiency and integration to buffer fluctuations in water quality and quantity; (2) improving the efficiency of contact between the coagulants and oil droplets, and achieving enhanced oil removal using a reduced amount of agents during the coagulation sedimentation stage.

Conventional physical and chemical water treatment methods have been widely utilized in the past due to their mature application and low operational costs. However, as water quality requirements become more stringent, the treatment efficiency and effectiveness of these methods may need to be further improved to meet the new standards. Consequently, recent research efforts have been focused on enhancing conventional technologies and integrating them with new techniques to develop more efficient water treatment technologies. For instance, El-Sayed et al. developed a new carbon thin film (ACTF) by hydrolyzing wood sawdust and treating the residue. They tested its effectiveness in removing oil from synthetic produced water. The ACTF showed high oil adsorption capacity, reaching a maximum of 700 mg of oil per g of ACTF at a bed height of 5 mm and flow rate of 0.5 mL/min. The Yoon–Nelson model accurately describes the breakthrough curve for oil adsorption. This study highlights the promising potential of ACTF as an efficient technique for removing oil from produced water [48]. El-Maghrabi focused on the use of mesoporous silica (MCM-41) to remove oil from produced water in the oil and gas industry. A continuous fixed-bed experiment was conducted to examine how the flow rate and bed height affect the breakthrough characteristics of the adsorption system. The results showed a maximum oil removal efficiency of 70.26% at a flow rate of 0.5 mL/min and a bed height of 1.5 mm. The Thomas model provided a good fit, indicating the suitability of MCM-41 as an effective adsorbent for oil removal [49]. Hollanda et al. conducted a study to characterize oilfield-produced water and evaluate photochemical systems and combined processes for its treatment, contributing to a deeper understanding of the properties and treatment options for oilfield-produced water [50]. In another study, Das et al. evaluated the application of various coagulants and their associated costs for treating oil- and gas-produced water. This evaluation helps identify the most efficient and cost-effective coagulant for water treatment [51]. Similarly, Khor et al. analyzed the performance, energy consumption, and cost of chemical and electrochemical coagulation for treating produced water, providing valuable insights into the feasibility and cost-effectiveness of various coagulation methods for oil- and gas-produced water treatment [52]. Al Hawli et al. developed a hybrid electro-coagulation/forward osmosis system to treat produced water, combining electrocoagulation and forward osmosis to enhance water treatment efficiency and effectiveness [53]. Additionally, Shah et al. proposed a novel settling tank for the treatment of produced water, utilizing computational fluid dynamics simulations and particle image velocimetry experiments. This innovative approach improves the settling process and enhances overall treatment performance [54]. Zeliff et al. investigated the combination of hydrocyclone separation with other established separation techniques for treating produced water, aiming to optimize the separation process and achieve improved water treatment outcomes [55]. Research has demonstrated significant progress in enhancing the efficiency of produced water treatment by improving existing technologies and incorporating new methods [35]. Nevertheless, the development of new chemical agents, the use of electrochemical technology, or the use of membrane technology [56] increases the cost of produced water treatment. It is more economical to improve the conventional physical processing technology, but the stability of the technology needs to be observed for a long time.

In response to the current inefficiency of physical oil removal methods in removing emulsified oil, the problem of incomplete chemical reactions in chemical oil removal methods, and the requirement for certain oil content in the influent for filtration processes, our research has revealed that micro-vortex flow technology is a promising technique that can alleviate the limitations of these traditional methods. A micro-vortex refers to a small-scale vortex or swirling motion that occurs in a fluid, such as water, at a miniature level. It is characterized by its compact size and typically forms in localized areas, often near orifices or other flow disturbances. Micro-vortex flow plays a crucial role in fluid dynamics, influencing processes like mixing, particle interaction, and energy transfer within the fluid [57]. Micro-vortex flow technology is a straightforward and efficient auxiliary method used in water treatment. It involves the utilization of small-scale vortices or swirling motions to enhance various processes during water treatment. When water passes through orifices, it generates a locally uniform and isotropic turbulence, leading to the formation of numerous vortices of different sizes. The larger vortices transfer energy to the smaller ones, resulting in the relative motion of particles and the experience of centrifugal inertial forces within the vortex flow layer; particles of a similar vortex scale collide with each other in the vortex flow layer [58]. It can be used in flocculation and sedimentation units to improve the efficiency of coagulation and sedimentation processes [59]. Ban et al.'s study proposed a novel microporous flocculation magnetic fluidized bed (MFMFB) reactor for treating low-concentration lead-polluted groundwater. The reactor promoted enhanced flocculation through microporous flocculation, anisotropy, micro-vortex, and effective energy consumption [60]. Wang designed a novel vortex flocculation reactor, featuring a column–cone–column structure and incorporating vortex generators to create micro-vortex flow and enhance flocculation efficiency [57]. Zhengong's study found that micro-vortex coagulation plays a crucial role in vortex clarification technology. By using micro-vortex flow technology to modify standard clarification tanks, the treatment capacity of the vortex clarification tank has been enhanced [61]. Harnessing exceptional diffusion capabilities of micro-vortex flow not only facilitates the uniform dispersion and contact of coagulants but also promotes the coalescence of minute oil droplets into larger ones [62], thereby facilitating their removal. This dual effect of micro-vortex flow enhances the overall coagulation process and augments oil removal efficiency.

In this study, we have developed multifunctional oil removal (MOR) equipment that integrates regulation buffering, natural settling, chemical mixing, and coagulation sedimentation processes. Incorporating micro-vortex flow reactors aims to improve the efficiency of chemical mixing and coagulation/flocculation. Through experiments evaluating the enhancement effect of micro-vortex flow technology in MOR equipment, our objective is to streamline on-site processes while ensuring qualified inlet water quality for the filter process. Ultimately, our goal is to improve the quality of water injection in low-permeability oil reservoirs and minimize environmental impacts [63].

2. Materials and Methods

2.1. Chemicals

The chemical agents utilized in the experiment underwent laboratory screening at an early stage. Coagulant screening encompassed polymeric aluminum ferric sulfate, polymeric ferric sulfate, and polymeric aluminum chloride. Flocculant screening included cationic polyacrylamide (with degrees of ionization of 40, 50, 60, and 90) as well as anionic polyacrylamide. On-site water sample tests were conducted by the laboratory to select the coagulant and flocculant with optimal performance in removing oil and suspended solids as pilot-scale chemical agents (Tables A1 and A2).

The chemicals employed in the pilot test consisted of a coagulant, specifically yellow powder-form poly-aluminum chloride produced by Gongyi Dongfang Purification Materials Factory with an active content of 26%. Each bag weighed 25 kg. The recommended dosage for its application in water treatment by chemical manufacturers ranges from 10 to 30 mg/L. Furthermore, a white powder-form flocculant was used, which was cationic polyacrylamide manufactured by Daqing Tenghui Limited Company. It had a cationic degree of 90 and an effective substance content of 99%. Each bag also weighed 25 kg. The recommended dosage for its application in water treatment by chemical manufacturers is between 0.1 and 1.0 mg/L.

2.2. Field Water

The pilot equipment was installed adjacent to the existing oil removal tanks at a union station in an oilfield, as shown in Figure 3. This union station receives water from multiple blocks, resulting in a complex composition and significant fluctuations in quantity. The incoming water exhibits varying levels of oil content, ranging from 60 to 300 mg/L, as well as suspended solids content ranging from 30 to 60 mg/L. After undergoing thorough oil removal and filtration processes, the treated water is primarily utilized for reinjection into low-permeability formations.

Figure 3. The MOR (multifunctional oil removal) equipment at the oilfield union station.

2.3. Experimental Equipment Design and Principles

2.3.1. Main Components of Equipment

The MOR equipment is designed based on conventional vertical settling tanks but with innovative modifications. It is divided into three interconnected chambers: the natural oil removal (NR) chamber, the chemical reaction (CR) chamber, and the coagulation sedimentation (CS) chamber. The design of the inlet and outlet components in each chamber has been optimized, incorporating trumpet-shaped inlets and outlets along with circular weirs. This design ensures uniform water distribution and minimizes the impact of water flow disturbances on treatment efficiency. Additionally, the equipment includes oil collection and sludge discharge components to prevent the accumulation of pollutants, such as oil and mud, during prolonged operation. The main components of this equipment are illustrated in Figure 4a: (1) the NR Chamber Inlet; (2) NR chamber outlet; (3) chemical-dosing device; (4) micro-vortex flow reactors (installed in CR chamber); (5) CS chamber distribution outlet; (6) inclined tube device; (7) NR chamber oil collection weir; (8) CS chamber oil collection pipe; and (9) CS chamber outlet.

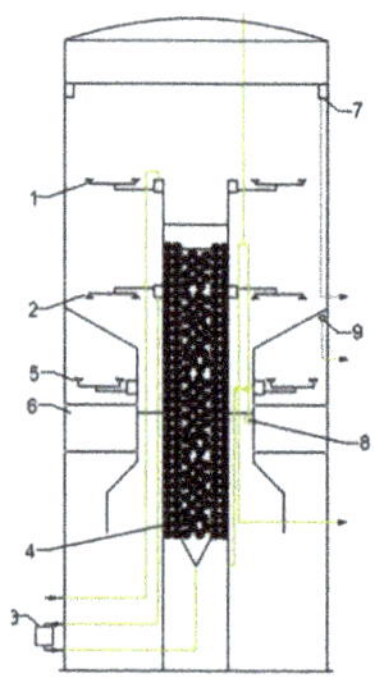

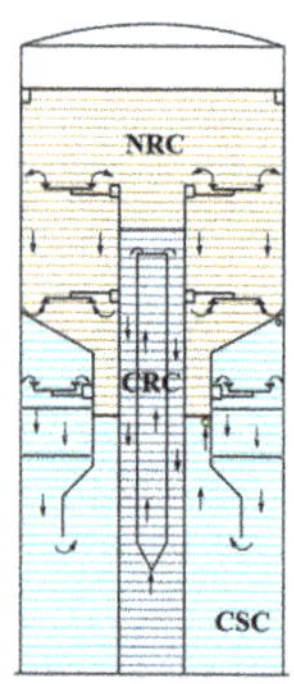

Figure 4. (**a**) The main components of the equipment. (**b**) Three main chambers of the equipment: NR (natural oil removal) chamber; CR (chemical reaction) chamber; and CS (coagulation sedimentation) chamber. The arrows in the figure indicate the direction of the water flow.

The positional relationship between the chambers and the direction of water flow is depicted in Figure 4b. The retention time of the equipment is determined by calculating the unit area load; values as outlined in the national standards [64] were taken into consideration. Upon entering the MOR equipment, the incoming water undergoes a stabilization and settling process in the NR chamber. The unit area load ranges from 0.98 to 1.48 $m^3/m^2{\cdot}h$ (producing a water descent rate of 0.27 mm/s to 0.41 mm/s) with a hydraulic retention time of 2.0 to 2.5 h. Chemicals are added before the water enters the CR chamber, equipped with a double-layered barrel containing micro-vortex flow reactors. During the process, water flows upward from the inner layer, reaches the top, and then flows downward in the outer layer. The double-layered barrel design ensures prolonged contact time between the chemicals and the produced water. Furthermore, this process promotes the occurrence of micro-vortex flow. Within the CR chamber, diffusion, coagulation, and flocculation reactions are enhanced by the micro-vortex flow reactors, taking approximately 15 to 20 min. Finally, the treated effluent from the CR chamber flows into the CS chamber for sedimentation, where inclined tube device is installed. The hydraulic retention time within the CS chamber is approximately 1.5 to 2.0 h, and the unit area load ranges from 1.17 to 1.76 $m^3/m^2{\cdot}h$ (produced water descent rate of 0.33 mm/s to 0.49 mm/s). The equipment processing flow rate corresponding to the above parameters is 20 to 30 m^3/h. The MOR equipment is designed to handle inlet water with oil content $\leq$ 1000 mg/L and suspended solids content $\leq$ 300 mg/L [64]. It is also equipped with a liquid-level control device, allowing direct entry of water into the filter section without the need for a buffer tank.

2.3.2. Micro-Vortex Flow Reactor

The micro-vortex flow reactors are essential for the CR chamber, where they play a crucial role in enhancing the coagulation process throughout the equipment. Micro-vortex flow pertains to the creation of localized, uniformly isotropic turbulence as water passes through the apertures of the micro-vortex flow reactors, as shown in Figure 5a. Consequently, the shape of the micro-vortex flow reactors may vary, encompassing polyhedral hollow spheres, pall rings, or saddle rings, among others. Pall rings and saddle rings have advantages such as high surface area and effective liquid distribution. However, they also have the following disadvantages: (1) There is a lack of rolling motion hinders the self-cleaning ability of pall rings and saddle rings, unlike spherical shapes. This can result in particle accumulation or fouling on the surface, reducing reactor efficiency over time. (2) The shape and structure of pall rings and saddle rings make them susceptible to contamination. Gaps or recessed areas allow for the accumulation of sediments, particles, or biofilms, which can impact water treatment quality and efficiency.

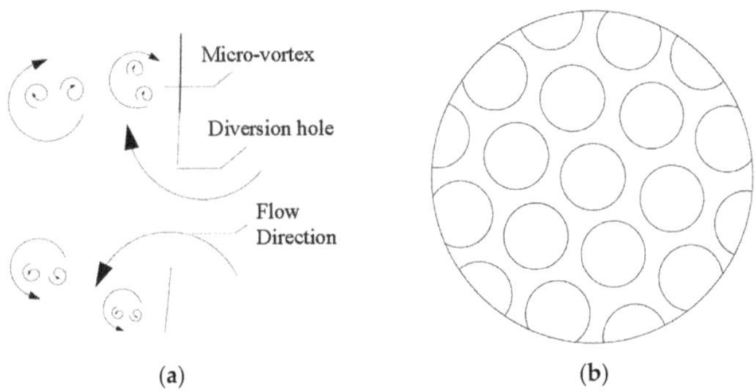

Figure 5. (**a**) The micro-vortex flow reaction principle. (**b**) The polyhedral hollow spheres micro-vortex flow reactor.

In this equipment, polyhedral hollow spheres made of ABS (Acrylonitrile Butadiene Styrene) have been selected for their ability to achieve a specific cleaning effect through the rolling friction between the spheres, thereby preventing blockage and fouling within the chamber. As shown in Figure 5b, several diversion holes of the same size are uniformly distributed on the spherical shell to realize the generation of water flow and a vortex. Our selection of the micro-vortex flow reactors and their specifications, including a diameter of 200 mm and diversion holes with a diameter of 30 mm, was informed by the research findings of other scholars [62,65] in the field of drinking water purification.

2.3.3. Principles and Process of Oil Removal in the Chamber

- The NR (natural oil removal) chamber

The NR chamber uses a physical method that takes advantage of the density difference between oil and water to effectively remove oil from the water. The ascent of oil droplets follows the Stokes formula, which describes particle motion in a viscous fluid.

The main goal in this chamber is to remove oil larger than 50 μm, both free-floating and dispersed particles. After that, the effluent goes to the NR chamber outlet where it is directed to an external dosing device. This device introduces coagulants and flocculants to enhance the treatment process before reintroducing the water into the CR chamber.

- The CR (chemical reaction) chamber

Three mechanisms are employed to effectively remove oil from the CR chamber: coagulation, demulsification, and three-dimensional contact flocculation.

Coagulation involves the gradual aggregation of oil droplets into larger globules. Micro-vortex flow reactors facilitate collision and coagulation among dispersed and partially emulsified oil particles, resulting in the formation of larger oil globules. These reactors offer advantages such as easy installation, clog resistance, and convenient maintenance.

The demulsification process starts by adding coagulants and flocculants before the CR chamber in the MOR equipment. These chemicals undergo hydrolysis in water, leading to processes like adsorption and electrostatic neutralization. This destabilizes emulsified oil and colloidal particles, which then enter the micro-vortex flow generated by the reactors. This facilitates the coagulation and enlargement of dispersed oil droplets. The combination of micro-vortex flow and chemical treatment enhances aggregation and settling of oil particles, improving efficiency in removing oil.

In the CR chamber, controlled flow dynamics create an environment that promotes three-dimensional contact flocculation, resulting in a strong adsorption and continuous absorption of fine particles. This interaction facilitates the gradual growth and aggregation of flocs. The CS chamber receives these enriched flocs containing captured particles. The CR chamber optimizes the coagulation and enlargement of dispersed oil and partially emulsified particles, specifically in the size range of 10 μm to 50 μm. Additionally, it enhances chemical mixing, creating ideal conditions for effective oil–water separation in the subsequent chamber.

- The CS (coagulation sedimentation) chamber

The CS chamber uses an inclined tube area based on shallow pool and coalescence principles for efficient oil removal. Flocs and oil droplets enter the settling zone of the tubes, where oil droplets accumulate on the upper wall and suspended solids gather on the lower wall, enlarging their size for effective removal.

The inclined tube oil removal is based on the "shallow pool theory," which states that the efficiency of gravity separation equipment depends on its horizontal cross-sectional area rather than its depth, assuming a constant effective volume in the sedimentation tank. Therefore, ideal gravity separation equipment should have a large surface area and shallow depth.

By incorporating honeycomb inclined tubes (as shown in Figure 6) into the CS chamber, the effective separation area increases and the separation height decreases. The presence of inclined tubes (or plates) also enhances water flow's wet perimeter, reduces the hydraulic radius, and promotes laminar flow conditions by lowering the Reynolds number (Re). Moreover, a higher Froude number (Fr) stabilizes water flow, facilitating oil–water separation.

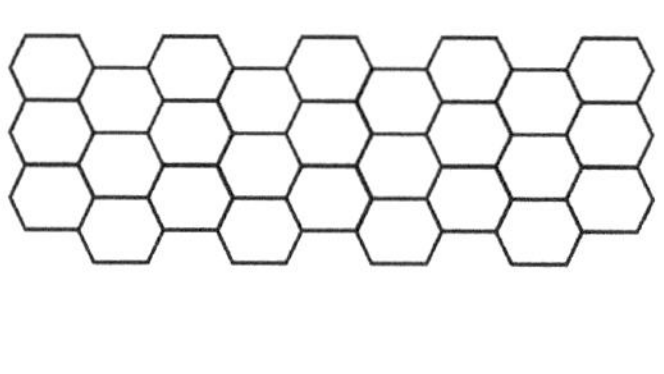

(a)

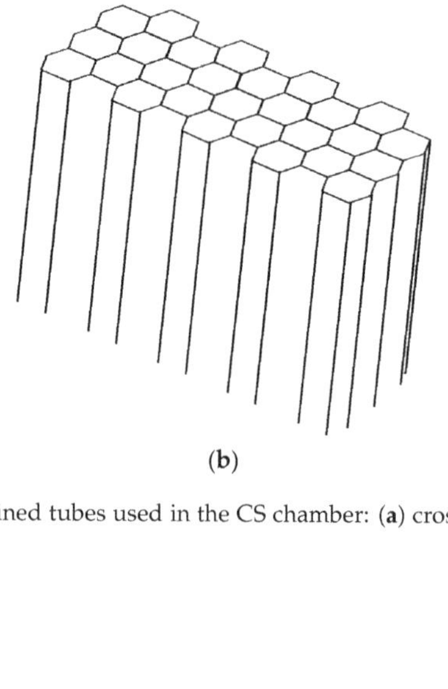

(b)

Figure 6. Schematic of the honeycomb inclined tubes used in the CS chamber: (**a**) cross-sectional view, (**b**) 3D view.

2.4. Pilot Experiment Design

2.4.1. Flocculant Dosage Experiment for Chemical Addition

The amount of chemicals added during produced water treatment is crucial as it directly affects both the cost and effectiveness. Cationic polyacrylamide is used as a flocculant with a recommended dosage range of 0.5 to 1 mg/L. To evaluate the impact of different dosages, we conducted experiments using four combinations of coagulants and flocculants, where polyaluminum chloride was used as the coagulant at levels of 16 mg/L and 22 mg/L.

The experiment involved combining two types of chemicals to establish four distinct dosage combinations, aiming to determine the optimal coagulant dosage by assessing the efficiency of oil removal and suspended solids removal for each combination. The detailed design of the dosage combinations for the experiment is provided in Table 2.

Table 2. Dosage Combinations for Chemical Addition Experiment.

Number	Coagulant (mg/L)	Flocculant (mg/L)	Oil Removal Rate (%)	Suspended Solids Removal Rate (%)
1	16.0	0.5	A1	B1
2	16.0	1.0	A2	B2
3	22.0	0.5	A3	B3
4	22.0	1.0	A4	B4

Ai and Bi represent the oil and suspended solids removal rates, respectively, where "i" denotes the experiment combination number. Based on the experimental results, the average oil removal rate (AOR) and the average suspended solids removal rate (ASR) for different chemical concentrations in each combination are calculated to evaluate the impact of the chemical dosage on pollutant removal. For example, the AOR for the coagulant dosage of 16 mg/L is calculated as (A1 + A2)/2, while the ASR is calculated as (B1 + B2)/2. Similarly, the AOR and ASR for coagulant dosages of 22 mg/L and flocculant dosages of 0.5 mg/L and 1.0 mg/L are determined.

2.4.2. Coagulant Dose Optimization and Micro-Vortex Flow Controlled Experiment

The aim of this experiment section was to determine the optimal coagulant dosage range for the MOR equipment and assess the effectiveness of coagulation enhancement by micro-vortex flow reactors. The pilot-scale MOR equipment operation was divided into two stages: without micro-vortex flow reactors in the first stage, and with micro-vortex flow reactors in the second stage. By comparing treatment efficiency between these stages, we evaluated the enhanced oil removal effect of micro-vortex flow reactors and determined an appropriate coagulant dosage range during the treatment process.

2.4.3. Comparative Study of Oil Removal Processes

We selected the MOR equipment equipped with micro-vortex flow reactors to compare its efficiency in removing oil with the oil removal process currently used in the union station for produced water treatment, both using a coagulant (polyaluminum chloride) and flocculant (cationic polyacrylamide) at optimized dosages. The oil removal process employed in the union station primarily consists of one natural settling tank (with a detention time of 7.5 h) and one coagulation sedimentation tank (with a detention time of 6.5 h), both with a volume of 3000 m^3. To ensure that the MOR equipment operates under the same conditions as the natural settling tank, the unit area load in the settling tank is calculated based on its effective volume. By adjusting the flow rate of the MOR equipment, the unit area load can be made equivalent to that of the settling tank. The natural settling tank with a diameter of 15.8 m and a treatment flow rate of 291 m^3/h has a unit area load of $Q/A = 1.48\ m^3/m^2{\cdot}h$. Similarly, the coagulation sedimentation tank with a diameter of 15.3 m and a treatment flow rate of 291 m^3/h has a unit area load of $Q/A = 1.59\ m^3/m^2{\cdot}h$.

In this case, a unit load index of 1.48 $m^3/m^2 \cdot h$ is chosen for the MOR equipment, which has a diameter of 2.5 m. Calculations reveal that the flow rate of the MOR equipment should be 30 m^3/h.

2.4.4. Long-Term Performance Observation

The majority of studies have focused on short-term evaluations of treatment technologies, but there is a need for further investigation into their long-term performance, stability, and reliability. It is essential to conduct long-term monitoring and evaluation to ensure sustainable and effective treatment of produced water.

Due to the complexity of the integrated equipment structure, issues such as clogging and material contamination are prone to occur over the long run [2,66]. Consequently, the initially achieved water quality standards may gradually become less stable. To assess the long-term stability and reliability of the MOR equipment, we conducted an 80-day pilot study at the union station oilfield site. Water quality tests were performed three times daily, with samples collected from the equipment's inlet, outlet of the NR chamber, and equipment outlet. The daily data were averaged for each sampling point. The oil removal efficiency and suspended solids removal efficiency of the equipment were then calculated using stage averages and variances, providing valuable insights into its performance effectiveness and stability.

2.5. *Methods for Data Collection and Analysis*

During water sample collection, it is crucial to obtain samples from the designated sampling port of the equipment. Prior to sampling, open the sampling tap and let it run smoothly for 3 min before collecting the sample. The oil content in the water samples is measured using the petroleum ether extraction method. First, extract the water samples and determine the oil content (in mg) in the extracted samples using a standard curve. Measure the volume of the extracted water sample (in L) and then calculate its oil content (in mg/L).

The measurement method for suspended solids content utilizes the membrane filtration weighing technique. After collecting the water samples, carefully filter them through a membrane. By considering both the volume of filtered water (in L) and weight increase in the filter membrane (in mg), accurately calculate suspended solids content in water (in mg/L).

3. Results and Discussion

3.1. *Flocculant Dose Optimization*

Based on the experimental findings, this study calculated the average oil removal rates (AORs) and average suspended solids removal rates (ASRs) for a specific reagent under various combinations, with the results presented in Tables 3 and 4. When increasing the flocculant dosage from 0.5 mg/L to 1 mg/L, the ASR decreased by 3.5% and the AOR has not changed. The decrease with the increased flocculant dosage can be attributed to several factors, including an excessive flocculant causing colloids to acquire a positive charge and an excessive flocculant dosage can deplete the alkalinity in the water, hindering the complete hydrolysis of the flocculant, resulting in residual amounts of the flocculant exceeding permissible limits in treated water. Conversely, the AOR and ASR increased by 11.2% and 18.1%, respectively, when the coagulant dosage was raised from 16 mg/L to 22 mg/L. The higher dosage of the coagulant improves oil and suspended solids removal efficiency by enhancing particle coagulation and destabilization. Increasing the coagulant dosage promotes the formation of larger, denser flocs, facilitating their separation from water. Considering cost implications, a flocculant dosage of 0.5 mg/L is considered suitable for MOR equipment.

Table 3. Chemical combination test results.

Number	Coagulant (mg/L)	Flocculant (mg/L)	Influent Oil (mg/L)	Effluent Oil (mg/L)	Oil Removal (%)	Influent Suspended Solids (mg/L)	Effluent Suspended Solids (mg/L)	Suspended Solids Removal (%)
1	16.0	0.5	82.9	19.3	76.7	32.2	17.8	44.7
2	16.0	1.0	113.2	18.9	83.3	36.2	22.0	39.2
3	22.0	0.5	243.3	13.4	94.5	48.0	18.8	60.8
4	22.0	1.0	108.1	13.1	87.9	38.8	15.8	59.3

Table 4. Coagulant and flocculant combination test AOR and ASR results.

Number	Dosage (mg/L)	AOR (%)	ASR (%)
1	0.5 (F *)	85.6	52.8
2	1.0 (F *)	85.6	49.3
3	16.0 (C *)	80.0	42.0
4	22.0 (C *)	91.2	60.1

* "F" means the flocculant; "C" means the coagulant.

3.2. *Coagulant Dose Optimization and Micro-Vortex Flow Controlled Experiment*

The experiment was conducted in two stages to test the enhanced coagulation/flocculation effect of micro-vortex flow reactors. The dosage of the coagulant was tested under two equipment states. In the first stage, the CR chamber of the MOR equipment had no micro-vortex flow reactors and the inlet/outlet water flow rate was 30 m^3/h. In the second stage, the CR chamber of multifunctional oil removal equipment was filled with micro-vortex flow reactors, and the inlet water conditions and reagent dosage were set to be identical to those in the first stage. The flocculant dosage remained at 0.5 mg/L while different dosages (0 mg/L, 4 mg/L, 10 mg/L, 16 mg/L, 22 mg/L, 28 mg/L, and 34 mg/L) of the inorganic coagulant (polyaluminum chloride) were used successively. It is important to note that due to on-site testing being experimental in nature, there were fluctuations in influent water quality. The findings from both stages are presented in Tables 5 and 6, respectively, while Figure 7 illustrates the results.

Table 5. MOR equipment operation without micro-vortex flow reactors under different coagulant concentrations.

Number	Coagulant (mg/L)	Flocculant (mg/L)	Influent Oil (mg/L)	Effluent Oil (mg/L)	Oil Removal (%)	Influent Suspended Solids (mg/L)	Effluent Suspended Solids (mg/L)	Suspended Solids Removal (%)
1	0.0	0.0	282.9	64.4	77.2	38.2	22.2	41.9
2	4.0	0.5	152.6	25.5	83.3	60.4	22	63.6
3	10.0	0.5	156.2	19.1	87.8	28	26.1	6.8
4	16.0	0.5	142.9	19.3	86.5	32.2	17.8	44.7
5	22.0	0.5	243.3	13.4	94.5	48	18.8	60.8
6	28.0	0.5	160.1	14.5	90.9	56.6	17.6	68.9
7	34.0	0.5	137.4	62	54.9	64.4	61.8	4.0

Based on the experimental results, the oil removal rate for MOR equipment without micro-vortex flow reactors is over 90% within a coagulant dosage range of 22 to 28 mg/L. For MOR equipment with micro-vortex flow reactors, the oil removal rate is over 95% within a coagulant dosage range of 10 to 22 mg/L. However, when the coagulant dosage exceeds 34 mg/L, both oil and suspended solids removal rates decrease significantly due to excessively high chemical concentrations causing colloidal stability in water and affecting sedimentation effectiveness.

Table 6. MOR equipment with micro-vortex flow reactors under different coagulant concentrations.

Number	Coagulant (mg/L)	Flocculant (mg/L)	Influent Oil (mg/L)	Effluent Oil (mg/L)	Oil Removal (%)	Influent Suspended Solids (mg/L)	Effluent Suspended Solids (mg/L)	Suspended Solids Removal (%)
1	0.0	0.0	284.3	37.1	87.0	52.1	21.8	58.2
2	4.0	0.5	121.3	11.9	90.2	34.6	19.8	42.8
3	10.0	0.5	67.5	3.1	95.4	50	8.5	83.0
4	16.0	0.5	255.6	6.5	97.5	28.2	12.1	57.1
5	22.0	0.5	145.4	6.3	95.7	27.8	18.3	34.2
6	28.0	0.5	68.3	3.6	94.7	23.8	11.9	50.0
7	34.0	0.5	83.3	53.8	35.4	58.1	45.7	21.3

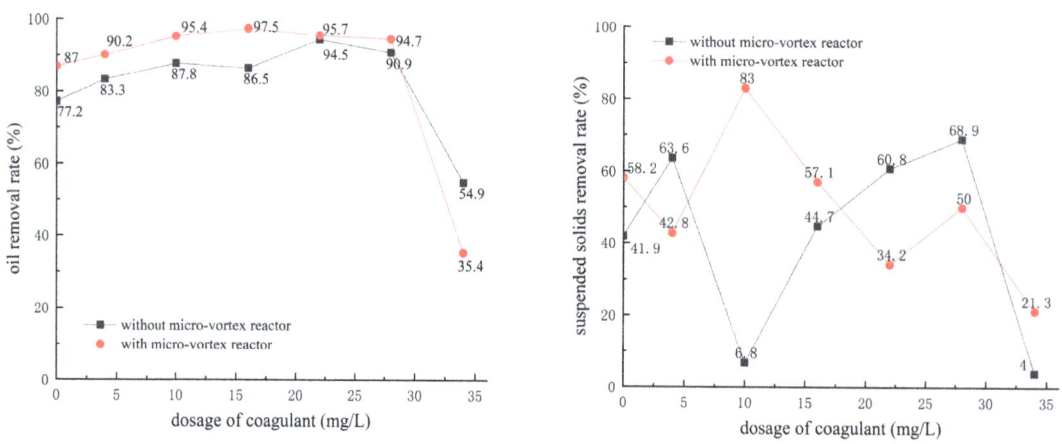

Figure 7. A comparison of (**a**) oil and (**b**) suspended solids removal efficiency with/without micro-vortex flow reactors.

Under the same experimental conditions, MOR equipment equipped with micro-vortex flow reactors exhibits an average oil removal rate that is 9.4% higher than that of MOR equipment without micro-vortex flow reactors. This suggests that the micro-vortex flow reactors enhance oil droplet coalescence and chemical dispersion, leading to improved oil removal performance. This finding aligns with previous studies conducted by other researchers [61,62]. While the average suspended solids removal rate also increases by 7.9%, there is significant fluctuation in suspended solids removal efficiency attributed to relatively low influent suspended solids content.

3.3. *Comparative Study with Union Station Oil Removal Process*

3.3.1. Union Station Oil Removal Process

The oil removal process at the union station consists of a natural settling tank and a coagulation sedimentation tank. The inorganic coagulant dosage in the coagulation (polyaluminum chloride) sedimentation tank is 22 mg/L, while the MOR equipment uses a dosage of 14 mg/L. Both processes employ an organic flocculant dosage of 0.5 mg/L.

The results of the 7-day operation of the combined natural settling tank and coagulation sedimentation tank process (referred to as the "dual-tank" process) are shown in Figure 8. The total influent oil content fluctuates between 55 and 262 mg/L, with an average of 172.7 mg/L. The average effluent oil content after treatment in the natural settling tank is 70.1 mg/L, while it is 42.8 mg/L after treatment in the coagulation sedimentation tank. The average oil removal rate of the "dual-tank" process is 71.7%.

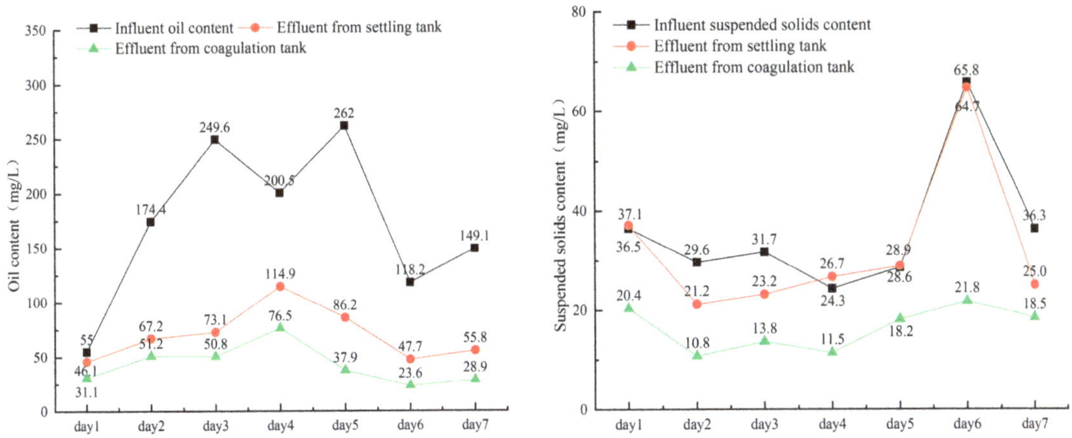

(a) (b)

Figure 8. Oil content (**a**) and suspended solids content (**b**) of the "natural settling tank + coagulation sedimentation tank" process were measured at the union station.

The influent water has an average suspended solids content of 36.1 mg/L, while the effluent after natural settling tank treatment has an average suspended solids content of 32.4 mg/L, indicating a low removal efficiency in the natural settling tank. According to the Stokes' Law analysis, the settling velocity of suspended solids is directly proportional to the square of their particle size. When colloidal particles (d = 0.001 to 1 μm) are present in the suspension, relying solely on gravity separation would require a lengthy retention time. Additionally, the natural settling chamber is the first chamber of the equipment, which needs to withstand the impact of changes in water quality and quantity, serving as a buffer and homogenization mechanism to ensure a smooth water flow into the second chamber. The average suspended solids content of the effluent after the coagulation sedimentation tank treatment is 16.4 mg/L, indicating that coagulation has effectively removed suspended solids. The average suspended solids removal rate of the "dual-tank" process is 52.7%.

The "dual-tank" process moderately removed oil and suspended solids. However, the effluent's relatively high oil content indicates the need for further optimization to meet stricter quality standards. This may involve additional treatment steps like filtration or advanced processes to improve oil removal efficiency and prevent filter media fouling. Additionally, monitoring and adjusting the dosage of an inorganic coagulant and organic flocculant in the sedimentation tank can optimize removal performance and ensure efficient operation.

3.3.2. MOR Equipment

The influent water quality for the MOR equipment and the "dual-tank" oil removal process at the union station is the same. The 7-day experiment results for the MOR equipment, as shown in Figure 9, indicate that the average oil content of the influent water is 172.7 mg/L. Following treatment in the MOR equipment's NR chamber, the average oil content in the effluent water is reduced to 88.2 mg/L. After further treatment in the CS chamber, the average oil content in the effluent water is reduced to 7.5 mg/L. The MOR equipment achieves an average oil removal rate of 95.7%, demonstrating a significant oil removal effect.

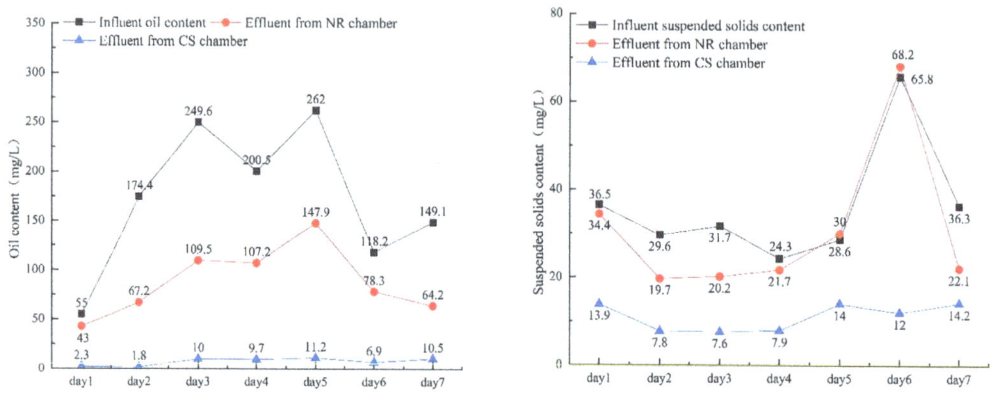

(a) (b)

Figure 9. Oil content (**a**) and suspended solids content (**b**) of the MOR equipment were measured.

The influent water has an average suspended solids content of 36.1 mg/L, which decreases to 30.9 mg/L after treatment in the NR chamber of the MOR equipment. Further treatment in the CS chamber reduces the suspended solids content to 11.1 mg/L. The MOR equipment achieves a remarkable removal rate of 69.3% for suspended solids, indicating that the equipment also performs well in the removal of suspended solids.

According to Table 7, the treatment process of the MOR equipment and the "dual-tank" oil removal process at the union station can be divided into two stages: natural settling and coagulation sedimentation. Under identical influent water quality, both processes exhibit similar treatment effects during the natural settling stage. However, after natural settling, the oil removal rate at the union station is approximately 10% higher than that of the MOR equipment's NR chamber. This disparity can be attributed to a longer retention time of 7.5 h in the natural settling tank at the union station compared to 2.5 h in the MOR equipment's NR chamber. A longer retention time facilitates better oil removal efficiency by leveraging gravity separation based on density differences between oil and water.

Table 7. Comparative Experiment of MOR Equipment and "dual-tank" Oil Removal Process.

Processes	Sampling Point	Retention Time (h)	Average Oil (mg/L)	Average Suspended Solids (mg/L)	Oil Removal (%)	Suspended Solids Removal (%)
Inflow	Inlet line		172.7	36.1	-	-
Natural settling	Effluent of the union station's natural settling tank	7.5	70.1	32.4	59.4	10.2
	Effluent of NR chamber	2.5	88.2	30.9	48.9	14.4
Coagulation sedimentation	Effluent of union station's coagulation sedimentation tank	6.5	42.8	16.4	75.2	54.6
	Effluent of CS chamber	2.0	7.5	11.1	95.7	69.3

During the coagulation–sedimentation stage, the MOR equipment exhibits a remarkable increase in the oil removal rate, from 48.9% to 95.7%. In contrast, the "dual-tank" oil removal process at the union station only shows an increase from 59.4% to 75.2%. The superior performance of the MOR equipment can be attributed to several factors:

- In the preceding chamber (CR) of the MOR equipment, effluent water undergoes coalescence through micro-vortex flow, increasing the particle size of small oil droplets.
- The micro-vortex flow in the CR enhances reagent diffusion, ensuring uniform contact with oil droplets and facilitating electrical neutralization and adsorption effects.
- The micro-vortex flow strengthens reagent diffusion, enabling the formation of a high-quality three-dimensional contact coagulation layer that effectively removes suspended solids and oil.
- Efficient oil removal conditions are already established within the influent entering the CS chamber, including larger coalesced droplets, the demulsification of reagents, and stable floc formation. These conditions significantly increase overall oil removal efficiency in the shallow pool section.

The MOR equipment outperformed the "dual-tank" process in comparative experiments with identical influent conditions and retention times, showing superior performance in removing oil and suspended solids from the produced water. It achieved a 20.5% higher oil removal rate within a shorter hydraulic retention time. Both effluent oil and suspended solids concentrations were below 20 mg/L, meeting the requirements for filtration stage entry. Additionally, the MOR equipment eliminates the need for a buffer tank, effectively fulfilling the functions of three tanks simultaneously as depicted in Figure 10.

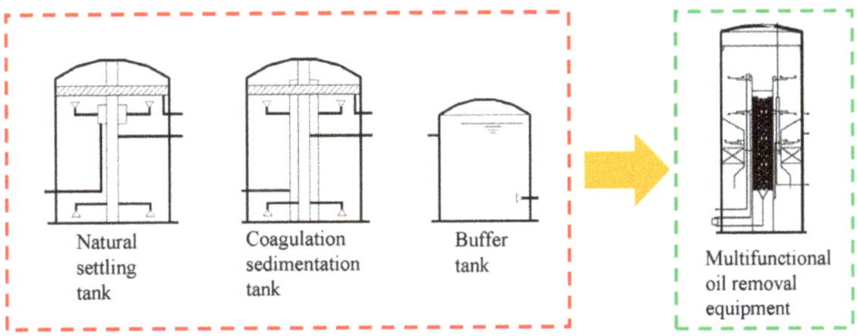

Figure 10. The MOR equipment achieves the functions of a natural settling tank, coagulation sedimentation tank, and buffer tank.

3.4. Long-Term Observations of MOR Equipment Performance

The MOR equipment underwent operational trials at the Daqing Oilfield Union Station for approximately three months, maintaining an inlet flow rate of 25 to 30 m^3/h. An organic flocculant, cationic polyacrylamide, was applied at a dosage of 0.5 mg/L along with an inorganic coagulant, polyaluminum chloride, at a dosage of 16 mg/L.

Water quality evaluations were conducted over a period of 80 days during the experimental phase, with daily sampling at the MOR equipment's inflow point, NR chamber's output point, and CS chamber's outlet point. It is important to note that operational produced water was introduced into the influent on day seven, and acidized fracturing water was added on days nineteen and twenty. Furthermore, untreated operational wastewater was introduced into the influent on days 70 to 72 and 77 to 79.

Experiment results are shown in Figure 11 and Table 8. This study was divided into three stages based on the experimental duration: the early stage (1 to 30 days), the middle stage (31 to 60 days), and the late stage (61 to 80 days). The average oil content in the influent of the MOR equipment during the first two stages was 155.4 mg/L. Some fluctuations were observed in treatment efficiency, resulting in an average oil removal rate of 93.6%. However, there was a significant increase in the average influent oil content to approximately 371.2 mg/L during days 61 to 80 of the experiment. Despite this change, the variance decreased to 4 while maintaining an average removal rate of around 96%. This suggests that with continued operation, the equipment demonstrated improved treatment efficiency and stability. The fitting line for the oil removal rate also supports this conclusion (the red line in Figure 11a). Overall, over a span of 80 days, the MOR equipment showed an impressive average oil removal efficiency of 94.2%, resulting in an effluent with an oil concentration of approximately 9 mg/L. Importantly, despite significant variations in influent quality during extended operation, consistent and effective oil removal performance was consistently demonstrated by the MOR equipment.

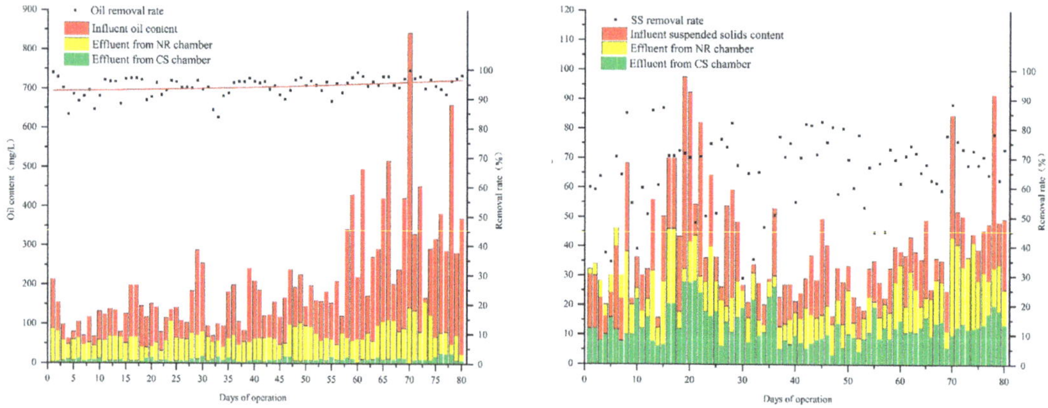

(a) (b)

Figure 11. (**a**) Observing the long-term performance of oil removal by applying MOR equipment; (**b**) observing the long-term performance of suspended solids removal by applying MOR equipment.

Table 8. The Variance and Mean of Produced Water Treatment Results by the MOR Equipment.

Days of Operation	Oil Removal Rate Average (%)	Oil Removal Rate Variance	Suspended Solids Removal Rate Average (%)	Suspended Solids Removal Rate Variance
Days 1 to 30	93.3	12.3	63.1	233.6
Days 31 to 60	93.6	11.4	64.4	222.9
Days 61 to 80	96.0	4.1	70.2	47.6
Whole period	94.2	10.8	65.4	187.3

In terms of suspended solids removal, the performance of the MOR equipment exhibits significant fluctuations. From days 1 to 60, the average influent concentration of suspended solids is 37.5 mg/L, with a removal efficiency ranging between 21.4% and 86.9%, and a large variance exceeding 200 in mean values. The average removal rate is 63.7%. However, from days 61 to 80, the equipment's treatment effect stabilizes gradually. The average influent concentration of suspended solids increases to 45.7 mg/L, with a decrease in variance to approximately 50. The average removal rate for suspended solids increases to 70.2%. The MOR equipment demonstrates certain efficiency in removing suspended solids, resulting in an effluent concentration averaging at around 12.7 mg/L; however, the removal effect fluctuates due to relatively low influent concentrations.

In the 80-day operational test, MOR equipment demonstrated excellent resistance to shocks and pollution. The analysis indicates that the spherical structure of the micro-vortex flow reactor significantly enhances its self-cleaning performance. The collisions and rolling between these spheres prevent oil adhesion while effectively promoting the condensation and diffusion of chemicals. The self-cleaning effect of the micro-vortex reactor is rarely addressed in other studies, which may be due to less blockage and adhesion occurring in drinking water purification processes compared to higher oil content in produced water [57,61]. Therefore, our research emphasizes the superior sustainability of spherical reactors in treating produced water, enhancing coagulation efficiency, and outstandingly removing oil throughout the entire 80-day operation.

4. Conclusions

This study employed multifunctional oil removal (MOR) equipment to enhance the efficacy of oil removal in produced water from low-permeability oilfields and address treatment challenges. The investigation was conducted at an oilfield union station, yielding the subsequent key findings:

(1) The micro-vortex flow action promotes a uniform diffusion and contact of chemicals, resulting in a 30% reduction in the coagulant dosage compared to MOR equipment without micro-vortex flow reactors.
(2) Comparative experiments were conducted to evaluate the performance of MOR equipment versus the "natural settling tank + coagulation sedimentation tank" process. The MOR equipment demonstrated superior removal rates, surpassing the conventional process by 20.5% for oil and 14.7% for suspended solids. It also reduced the coagulant dosage by 30%, while maintaining a hydraulic retention time of 5 h, resulting in an effluent with oil and suspended solids content below 20 mg/L.
(3) Long-term operational observations have revealed the inherent self-cleaning effect of micro-vortex flow reactors, ensuring consistent and stable equipment performance over an extended duration. The MOR equipment consistently achieved oil removal rates exceeding 93% and suspended solids removal rates surpassing 63% during a continuous operation period of 80 days. The average concentrations in the effluent were measured at approximately 9 mg/L for oil and around 12.7 mg/L for suspended solids. As the operation time increased, higher removal rates and stability were observed during the third period.

Despite promising experimental results, micro-vortex flow technology and MOR equipment still face future challenges:

(1) It is crucial to assess whether the oil removal efficiency can be maintained when dealing with different qualities of produced water and higher flow rates in larger treatment systems.
(2) When dealing with high-viscosity organic compounds, such as polymers in chemically enhanced recovery produced water, the micro-vortex reactor's resistance to contamination needs re-evaluation, and regular maintenance may be necessary.
(3) The shape and material of the micro-vortex reactor can be further studied to achieve a better enhanced coagulation effect.

In conclusion, the MOR equipment demonstrated superior efficiency in oil removal using a single sedimentation tank. Furthermore, incorporating micro-vortex flow technology offers significant advantages such as improved water treatment efficiency, reduced chemical consumption, energy-efficient practice promotion, and minimized sludge generation through self-cleaning capabilities for long-term utilization at low costs. Integrating MOR equipment with micro-vortex flow technology enhances water injection quality in low-permeability reservoirs and facilitates sustainable development. Additionally, it can effectively be used in offshore oilfields, remote and small-scale stations, and stations facing challenges in meeting water treatment standards.

5. Patents

This research has led to the development of several patents, including a patented system and process for treating produced water [67], and two patented degreasing units [68,69].

Author Contributions: Conceptualization, D.Z. and W.X.; Data curation, Y.S. and X.H.; Formal analysis, L.W.; Investigation, L.W.; Methodology, J.Z. and B.L.; Project administration, B.L.; Resources, Y.S.; Supervision, J.Z.; Validation, L.W.; Visualization, D.Z.; Writing—original draft, D.Z.; Writing—review and editing, W.X. and T.C. All authors have read and agreed to the published version of the manuscript.

Funding: This research received no external funding.

Data Availability Statement: The data that support the findings of this study are available from the corresponding author, D. Z., upon reasonable request.

Acknowledgments: The authors would like to express their deep appreciation to the staff and technicians of the oil fields for granting us access to necessary data and facilities. We also extend our gratitude to our colleagues and fellow researchers for their intellectual contributions and stimulating discussions throughout this study. Our heartfelt thanks go out to everyone who played a role in successfully completing this research project.

Conflicts of Interest: Author Yuxin Sheng was employed by the PetroChina Jidong Oilfield Company, Xiujie Huang was employed by the Sinochem Holdings Corporation Ltd. The remaining authors declare that the research was conducted in the absence of any commercial or financial relationships that could be construed as a potential conflict of interest.

Appendix A

Results of screening tests for coagulant and flocculant type and concentration used in the test are shown in Tables A1 and A2.

Table A1. Aggregation effect of different concentrations of commonly used coagulants in oilfield.

Chemical Composition	Manufacturer	Concentration/ $(mg \cdot L^{-1})$	Turbidity before Dosing/ NTU	Oil Content before Dosing $/(mg \cdot L^{-1})$	Turbidity after Dosing/ NTU	Oil Content after Dosing $/(mg \cdot L^{-1})$	Oil Removal Efficiency/%	Turbidity Removal Rate/%
polyaluminum ferric sulfate	GY2	1	60.0	120.3	20.3	29.65	75.29	66.17
	GY2	5	60.2	120.0	15.07	12.44	89.63	74.88
	GY2	10	59.5	120.3	12.5	5.47	95.45	79.17
	GY2	20	59.8	120.0	11.8	5.23	95.64	80.27
	GY2	30	60.2	119.5	9.21	5.15	95.7	84.7
	GY2	40	60.2	120.0	7.74	4.53	96.23	87.1
	GY2	50	60.2	120.0	2.28	1.45	98.79	96.2
polyaluminum chloride (active component 26%)	GYT *	1	60.0	120.3	18.2	20.7	82.79	69.67
	GYT *	5	60.2	120.0	14.2	11.3	90.58	76.41
	GYT *	10	59.5	120.3	10.3	8.34	93.07	82.69
	GYT *	20	60.2	119.5	9.24	7.15	94.17	84.65
	GYT *	30	60.2	119.5	8.45	4.24	96.45	85.96
	GYT *	40	60.2	120.0	6.74	3.23	97.31	88.80
	GYT *	50	60.2	120.0	2.1	1.5	98.75	96.51
polyferric sulfate	BS	1	59.8	120.3	22.3	19.05	84.13	62.83
	BS	5	60.3	120.3	21.3	17.33	85.56	59.5
	BS	10	60.2	120.0	16.6	5.47	95.45	72.33
	BS	20	60.1	120.0	14.3	4.53	96.22	76.17
	BS	30	60.2	120.0	9.7	3.67	96.94	83.89
	BS	40	60.0	119.8	6.72	2.67	97.78	88.8
	BS	50	60.2	120.0	5.3	2.1	98.25	91.20
polyaluminum chloride active component 30%	HNLQ	1	60.3	120.3	13.2	12.21	89.83	78
	HNLQ	5	60.2	120.0	11.6	7.33	93.90	80.67
	HNLQ	10	60.1	120.0	9.08	5.35	95.54	84.87
	HNLQ	20	60.1	120.3	7.18	4.77	96.03	88.03
	HNLQ	30	60.0	120.0	5.87	3.89	96.76	90.22
	HNLQ	40	60.0	120.0	5.47	1.42	98.82	90.88
	HNLQ	50	60.0	120.0	1.35	1.1	99.08	97.75
polyaluminum ferric sulfate	JC	1	60.3	120.2	22.1	74.53	37.89	63.17
	JC	5	60.2	120.0	20.1	52.44	56.30	66.5
	JC	10	60.1	120.0	16.9	50.79	57.67	71.83
	JC	20	60.1	120.3	15.5	44.77	62.69	74.17
	JC	30	60.0	120.0	14.9	28.3	76.42	75.17
	JC	40	60.0	120.0	14.4	12	90	76
	JC	50	60.1	120.0	10.5	8.2	93.17	82.53
polyaluminum chloride (active component 22%)	MT	1	60.3	120.2	16.9	19.21	83.99	71.83
	MT	5	60.2	120.0	15.8	18.84	84.3	73.67
	MT	10	60.1	120.2	13.5	12.91	89.24	77.5
	MT	20	60.0	120.0	12.8	9	92.5	78.67
	MT	30	60.0	120.0	11.8	8.3	93.08	80.33
	MT	40	60.0	120.2	11.2	7.53	93.74	81.33
	MT	50	60.0	120.0	8.3	6.9	94.25	86.17

* Chemical used in field pilot evaluation.

Table A2. Aggregation effect of different concentrations of commonly used flocculants in oilfield.

Chemical Composition	Manufacturer	Concentration/ (mg·L^{-1})	Turbidity before Dosing/ NTU	Oil Content before Dosing /(mg·L^{-1})	Turbidity after Dosing/ NTU	Oil Content after Dosing /(mg·L^{-1})	Oil Removal Efficiency/%	Turbidity Removal Rate/%
cationic polyacrylamide (ionic degree 50)	HNLQ	0.1	60.0	120.3	17.3	9.88	91.77	71.17
	HNLQ	0.2	60.0	120.0	16.2	8.5	92.92	73
	HNLQ	0.5	60.2	120.0	15.7	8.4	93	73.83
	HNLQ	1	59.5	120.3	14.6	7.8	93.5	75.67
	HNLQ	2	59.6	120.0	13.7	6.2	94.83	77.17
	HNLQ	4	60.2	120.0	12.2	5	95.83	79.67
cationic polyacrylamide (ionic degree 60)	YLZ	0.1	59.8	120.3	22.7	12.95	89.21	62.17
	YLZ	0.2	60.0	120.0	22.1	11.9	90.08	63.17
	YLZ	0.5	60.3	120.3	21.9	11.5	90.42	63.5
	YLZ	1	60.2	120.0	21.6	10.6	91.17	64
	YLZ	2	60.1	120.0	20.8	10.2	91.5	65.33
	YLZ	4	60.0	119.8	19.0	9.42	92.15	68.33
cationic polyacrylamide (ionic degree 90)	MT *	0.1	60.3	120.3	15.7	25	79.17	73.83
	MT *	0.2	60.0	120.0	15.4	18	85	74.33
	MT *	0.5	60.2	120.0	15.3	14.23	88.14	74.5
	MT *	1	60.1	120.0	15.2	10.58	91.18	74.67
	MT *	2	60.1	120.3	14.3	7.56	93.7	76.17
	MT *	4	60.0	120.0	13.8	3	97.5	77
cationic polyacrylamide (ionic degree 50)	JC	0.1	60.3	120.2	19.1	73.37	38.86	68.17
	JC	0.2	60.0	120.0	18.9	62.1	48.25	68.5
	JC	0.5	60.2	120.0	18.6	61.2	49	69
	JC	1	60.1	120.0	18	51.74	56.88	70
	JC	2	60.1	120.3	17	38.02	68.32	71.67
	JC	4	60.0	120.0	15.8	31.51	73.74	73.67
anionic polyacrylamide	YLZ-2	0.1	60.3	120.2	18.6	25.23	78.98	69
	YLZ-2	0.2	60.0	120.0	18.3	20.9	82.58	69.5
	YLZ-2	0.5	60.2	120.0	18.1	19.02	84.15	69.83
	YLZ-2	1	60.1	120.2	17.5	18.26	84.78	70.83
	YLZ-2	2	60.0	120.0	17.3	12.91	89.24	71.17
	YLZ-2	4	60.0	120.2	17	9.09	92.43	71.67
cationic polyacrylamide (ionic degree 60)	KP	0.1	60.2	120.0	26.2	25.23	78.98	56.33
	KP	0.2	60.0	120.0	23.5	24.1	79.92	60.83
	KP	0.5	60.1	120.2	22.61	23.2	80.67	62.32
	KP	1	60.0	120.0	21.4	19.5	83.75	64.33
	KP	2	60.0	120.2	20.3	16.6	86.17	66.17
	KP	4	60.1	120.2	15.6	10.3	91.42	74

* Chemical used in field pilot evaluation.

References

1. Igunnu, E.T.; Chen, G.Z. Produced Water Treatment Technologies. *Int. J. Low-Carbon Technol.* **2014**, *9*, 157–177. [CrossRef]
2. Fakhru'l-Razi, A.; Pendashteh, A.; Abdullah, L.C.; Biak, D.R.A.; Madaeni, S.S.; Abidin, Z.Z. Review of Technologies for Oil and Gas Produced Water Treatment. *J. Hazard. Mater.* **2009**, *170*, 530–551. [CrossRef] [PubMed]
3. Nasiri, M.; Jafari, I.; Parniankhoy, B. Oil and Gas Produced Water Management: A Review of Treatment Technologies, Challenges, and Opportunities. *Chem. Eng. Commun.* **2017**, *204*, 990–1005. [CrossRef]
4. Venkatesan, A.; Wankat, P.C. Produced Water Desalination: An Exploratory Study. *Desalination* **2017**, *404*, 328–340. [CrossRef]
5. Zhong, W.J.; Fu, Y.C.; Tian, S.; Huang, Y.; Wang, X. A New Oilfield Injection Water Quality Decision-Making Method. *Pet. Sci. Technol.* **2022**, *41*, 123–138. [CrossRef]
6. Kassab, M.A.; Abbas, A.E.; Elgamal, I.; Shawky, B.M.; Mubarak, M.F.; Hosny, R. Review on the Estimating the Effective Way for Managing the Produced Water: Case Study. *Open J. Mod. Hydrol.* **2021**, *11*, 19–37. [CrossRef]
7. Bennion, D.B.; Thomas, F.B.; Imer, D.; Ma, T.; Schulmeister, B. Water Quality Considerations Resulting in the Impaired Injectivity of Water Injection and Disposal Wells. *J. Can. Pet. Technol.* **2001**, *40*, 53–61. [CrossRef]
8. Wang, J.; Fu, Y.-C.; Huang, Y.; Wang, X.-T.; Tian, S. A Study of the Effect of Oilfield Water Quality on Reservoir Injury. In *Proceedings of the International Field Exploration and Development Conference*; Springer: Berlin/Heidelberg, Germany, 2021.
9. Kong, B.; Wang, S.; Chen, S. Minimize Formation Damage in Water-Sensitive Montney Formation with Energized Fracturing Fluid. *SPE Reserv. Eval. Eng.* **2017**, *20*, 562–571. [CrossRef]
10. Li, X.; Yang, Z.; Li, S.; Huang, W.; Zhan, J.; Lin, W. Reservoir Characteristics and Effective Development Technology in Typical Low-Permeability to Ultralow-Permeability Reservoirs of China National Petroleum Corporation. *Energy Explor. Exploit.* **2021**, *39*, 1713–1726. [CrossRef]

11. Wang, B.; Zhao, Y.; Tian, Y.; Kong, C.; Ye, Q.; Zhao, S.; Li, Z.; Suo, Y. Numerical Simulation Study of Pressure-Driven Water Injection and Optimization Development Schemes for Low-Permeability Reservoirs in the G Block of Daqing Oilfield. *Processes* **2013**, *12*, 1. [CrossRef]
12. Gao, P.; Qu, L.; Sun, W. Analysis of Reinjection Water to Improve Water-Flooding Effect for Low-Permeability Sandstone Reservoirs. In *IOP Conference Series: Earth and Environmental Science*; IOP Publishing: Bristol, UK, 2019.
13. Shahryar, J. *Petroleum Waste Treatment and Pollution Control*; Butterworth-Heinemann: Oxford, UK, 2016.
14. Nordvik, A.B.; Simmons, J.L.; Bitting, K.R.; Lewis, A.; Strøm-Kristiansen, T. Oil and Water Separation in Marine Oil Spill Clean-up Operations. *Spill Sci. Technol. Bull.* **1996**, *3*, 107–122. [CrossRef]
15. Pintor, A.M.A.; Vilar, V.J.P.; Botelho, C.M.S.; Boaventura, R.A.R. Optimization of a Primary Gravity Separation Treatment for Vegetable Oil Refinery Wastewaters. *Clean Technol. Environ. Policy* **2014**, *16*, 1725–1734. [CrossRef]
16. Liao, X.-H.; Yang, S.-J.; Qin, M.-W.; Dou, M.-Y.; Feng, Q.; Li, H.-M.; Zou, S. Overview of Oil-Water Separation Equipment Technology of Refined Oil. In *IOP Conference Series: Earth and Environmental Science*; IOP Publishing: Bristol, UK, 2020.
17. Clark, S.E.; Roenning, C.D.; Elligson, J.C.; Mikula, J.B. Inclined Plate Settlers to Treat Storm-Water Solids. *J. Environ. Eng.* **2009**, *135*, 621–626. [CrossRef]
18. Liu, Y.; Lu, H.; Li, Y.; Xu, H.; Pan, Z.; Dai, P.; Wang, H.; Yang, Q. A Review of Treatment Technologies for Produced Water in Offshore Oil and Gas Fields. *Sci. Total Environ.* **2021**, *775*, 145485. [CrossRef] [PubMed]
19. Hafskjold, B.; Morrow, T.B.; Celius, H.K.B.; Johnson, D.R. Drop-Drop Coalescence in Oil/Water Separation. In Proceedings of the SPE Annual Technical Conference and Exhibition, New Orleans, LA, USA, 25–28 September 1994.
20. Frising, T.; Noïk, C.; Dalmazzone, C. The Liquid/Liquid Sedimentation Process: From Droplet Coalescence to Technologically Enhanced Water/Oil Emulsion Gravity Separators: A Review. *J. Dispers. Sci. Technol.* **2006**, *27*, 1035–1057. [CrossRef]
21. Lu, H.; Liu, Y.; Cai, J.; Xu, X.; Xie, L.; Yang, Q.; Li, Y.; Zhu, K. Treatment of Offshore Oily Produced Water: Research and Application of a Novel Fibrous Coalescence Technique. *J. Pet. Sci. Eng.* **2019**, *178*, 602–608. [CrossRef]
22. Al-Shamrani, A.A.; James, A.; Xiao, H. Separation of Oil from Water by Dissolved Air Flotation. *Colloids Surf. A Physicochem. Eng. Asp.* **2002**, *209*, 15–26. [CrossRef]
23. Al-Shamrani, A.A.; James, A.; Xiao, H. Destabilisation of Oil–Water Emulsions and Separation by Dissolved Air Flotation. *Water research* **2002**, *36*, 1503–1512. [CrossRef] [PubMed]
24. Etchepare, R.; Oliveira, H.; Azevedo, A.; Rubio, J. Separation of Emulsified Crude Oil in Saline Water by Dissolved Air Flotation with Micro and Nanobubbles. *Sep. Purif. Technol.* **2017**, *186*, 326–332. [CrossRef]
25. Liu, M.; Chen, J.; Cai, X.; Han, Y.; Xiong, S. Oil–Water Pre-Separation with a Novel Axial Hydrocyclone. *Chin. J. Chem. Eng.* **2018**, *26*, 60–66. [CrossRef]
26. de Araújo, C.A.O.; Scheid, C.M.; Loureiro, J.B.; Klein, T.S.; Medronho, R.A. Hydrocylone for Oil-Water Separations with High Oil Content: Comparison between Cfd Simulations and Experimental Data. *J. Pet. Sci. Eng.* **2020**, *187*, 106788. [CrossRef]
27. Zhang, J.; He, Y.-T.; Liu, S.; Xu, J.-Y. Oil-Water Separation in a Cylindrical Cyclone with Vortex Finder. *Phys. Fluids* **2022**, *34*, 033314. [CrossRef]
28. Hu, W.; Wei, Y.; Bao, J. Development of the Theory and Technology for Low Permeability Reservoirs in China. *Pet. Explor. Dev.* **2018**, *45*, 685–697. [CrossRef]
29. Al Moajil, A.; Almubarak, T.; Alotaibi, F.; Alhamad, L. Re-Injection of Produced Water: The Effect of Oil Micro-Emulsion Suspensions. In Proceedings of the Middle East Oil, Gas and Geosciences Show 2023, Manama, Bahrain, 19–21 February 2023.
30. *SY/T 5329-2022;* Water Quality Specification and Practice for Analysis of Oilfield Injecting Waters in Clastic Reservoirs. National Energy Administration: Beijing, China, 2022.
31. Qiang, Y.; Wang, X.; Wang, Y.; Zhang, X. Study on Evaluation Method of Water Injection Efficiency in Low-Permeability Reservoir. *Geofluids* **2021**, *2021*, 1–6.
32. Zhao, C.; Zhou, J.; Yan, Y.; Yang, L.; Xing, G.; Li, H.; Wu, P.; Wang, M.; Zheng, H. Application of Coagulation/Flocculation in Oily Wastewater Treatment: A Review. *Sci. Total Environ.* **2021**, *765*, 142795. [CrossRef] [PubMed]
33. You, Z.; Xu, H.; Sun, Y.; Zhang, S.; Zhang, L. Effective Treatment of Emulsified Oil Wastewater by the Coagulation–Flotation Process. *RSC Adv.* **2018**, *8*, 40639–40646. [CrossRef] [PubMed]
34. Rawlins, C.H.; Erickson, A.E.; Ly, C. Characterization of Deep Bed Filter Media for Oil Removal from Produced Water. In Proceedings of the SME Annual Meeting, Phoenix, AZ, USA, 28 February–3 March 2010.
35. Amakiri, K.T.; Canon, A.R.; Molinari, M.; Angelis-Dimakis, A. Review of Oilfield Produced Water Treatment Technologies. *Chemosphere* **2022**, *298*, 134064. [CrossRef] [PubMed]
36. Liu, A.; Liu, S. Study on Performance of Three Backwashing Modes of Filtration Media for Oilfield Wastewater Filter. *Desalination Water Treat.* **2016**, *57*, 10498–10505. [CrossRef]
37. Kim, S.-H.; Lim, H.-K.; Jeong, W.-C.; Park, N.-S. Optimum Backwash Method for Granular Media Filtration of Seawater. *Desalination Water Treat.* **2011**, *32*, 431–436. [CrossRef]
38. *SY/T 0523-2023;* Oilfield Water Treatment Filters. National Energy Administration: Beijing, China, 2023.
39. Beshr, S.; Moustafa, M.; Fayed, M.; Aly, S. Evaluation of Water Consumption in Rapid Sand Filters Backwashed under Varied Physical Conditions. *Alex. Eng. J.* **2023**, *64*, 601–613. [CrossRef]
40. Samuel, O.; Othman, M.H.D.; Kamaludin, R.; Sinsamphanh, O.; Abdullah, H.; Puteh, M.H.; Kurniawan, T.A.; Li, T.; Ismail, A.F.; Rahman, M.A. Oilfield-Produced Water Treatment Using Conventional and Membrane-Based Technologies for Beneficial Reuse: A Critical Review. *J. Environ. Manag.* **2022**, *308*, 114556. [CrossRef] [PubMed]

41. Munirasu, S.; Haija, M.A.; Banat, F. Use of Membrane Technology for Oil Field and Refinery Produced Water Treatment—A Review. *Process Saf. Environ. Prot.* **2016**, *100*, 183–202. [CrossRef]
42. Ebrahimi, M.; Willershausen, D.; Ashaghi, K.S.; Engel, L.; Placido, L.; Mund, P.; Bolduan, P.; Czermak, P. Investigations on the Use of Different Ceramic Membranes for Efficient Oil-Field Produced Water Treatment. *Desalination* **2010**, *250*, 991–996. [CrossRef]
43. Sun, Y.; Zhou, S.; Chiang, P.-C.; Shah, K.J. Evaluation and Optimization of Enhanced Coagulation Process: Water and Energy Nexus. *Water-Energy Nexus* **2019**, *2*, 25–36. [CrossRef]
44. Behin, J.; Aghajari, M. Influence of Water Level on Oil–Water Separation by Residence Time Distribution Curves Investigations. *Sep. Purif. Technol.* **2008**, *64*, 48–55. [CrossRef]
45. Li, J.; Gu, Y. Coalescence of Oil-in-Water Emulsions in Fibrous and Granular Beds. *Sep. Purif. Technol.* **2005**, *42*, 1–13. [CrossRef]
46. Yang, L.; Yu, J.; Ju, M.; Luo, B.; Wang, Z.; Hu, G. Intelligent Monitoring and Controlling Technology of Water Injection in Ultra-Low Permeability Reservoir of Ordos Basin. In *Proceedings of the International Field Exploration and Development Conference 2017*; Springer: Berlin/Heidelberg, Germany, 2018.
47. Silva, G.; Correia, B.; Cunha, A.; Santos, B.; Lima, A. Water Injection for Oil Recovery by Using Reservoir Simulation Via Cfd. *Int. J. Multiphysics* **2017**, *11*, 83–96.
48. El-Sayed, M.; Ramzi, M.; Hosny, R.; Fathy, M.; Moghny, T.A. Breakthrough Curves of Oil Adsorption on Novel Amorphous Carbon Thin Film. *Water Sci. Technol.* **2016**, *73*, 2361–2369. [CrossRef]
49. El-Maghrabi, H.H.; Hosny, R.; Ramzi, M.; Fathy, M. Novel Mesoporous Silica (Mcm-41) and Its Characterization for Oil Adsorption from Produced Water Injected in Water Injection Projects Using Fixed Bed Column Processes. *Desalin Water Treat* **2017**, *60*, 70–77. [CrossRef]
50. Hollanda, L.R.; Santos, S.B.F.; Faustino, J.G.A.A.; Dotto, G.L.; Foletto, E.L.; Chiavone-Filho, O. Oil Field–Produced Water Treatment: Characterization, Photochemical Systems, and Combined Processes. *Environ. Sci. Pollut. Res.* **2021**, *28*, 52744–52763. [CrossRef]
51. Das, N.; Rajput, H.; Hassan, A.A.; Kumar, S. Application of Different Coagulants and Cost Evaluation for the Treatment of Oil and Gas Produced Water. *Water* **2023**, *15*, 464. [CrossRef]
52. Khor, C.M.; Wang, J.; Li, M.; Oettel, B.A.; Kaner, R.B.; Jassby, D.; Hoek, E.M.V. Performance, Energy and Cost of Produced Water Treatment by Chemical and Electrochemical Coagulation. *Water* **2020**, *12*, 3426. [CrossRef]
53. Al Hawli, B.; Benamor, A.; Hawari, A.A. A Hybrid Electro-Coagulation/Forward Osmosis System for Treatment of Produced Water. *Chem. Eng. Process.-Process Intensif.* **2019**, *143*, 107621. [CrossRef]
54. Shah, M.T.; Parmar, H.B.; Rhyne, L.D.; Kalli, C.; Utikar, R.P.; Pareek, V.K. A Novel Settling Tank for Produced Water Treatment: Cfd Simulations and Piv Experiments. *J. Pet. Sci. Eng.* **2019**, *182*, 106352. [CrossRef]
55. Zeliff, T.; Gallegos, C.; Alghamdi, K.; Goettsche, D.; Bernards, M. Hydrocyclone Separation in Combination with Mature Separation Techniques to Treat Produced Water. In *Waste-Management Education Research Conference (WERC) 2021*; IEEE: Piscataway, NJ, USA, 2021.
56. Abdelrazeq, H.; Khraisheh, M.; Ashraf, H.M.; Ebrahimi, P.; Kunju, A. Sustainable Innovation in Membrane Technologies for Produced Water Treatment: Challenges and Limitations. *Sustainability* **2021**, *13*, 6759. [CrossRef]
57. Wang, Z.; Wu, M.; Liang, X.; Huang, N.; Li, X. A Novel Vortex Flocculation Reactor for Efficient Water Treatment: Kinetic Modeling and Experimental Verification. *Chem. Eng. Process.-Process Intensif.* **2023**, *183*, 109245. [CrossRef]
58. Kolmogorov, A.N. A Refinement of Previous Hypotheses Concerning the Local Structure of Turbulence in a Viscous Incompressible Fluid at High Reynolds Number. *J. Fluid Mech.* **1962**, *13*, 82–85. [CrossRef]
59. Wang, L.-P.; Chen, S.; Brasseur, J.G.; Wyngaard, J.C. Examination of Hypotheses in the Kolmogorov Refined Turbulence Theory through High-Resolution Simulations. Part 1. Velocity Field. *J. Fluid Mech.* **1996**, *309*, 113–156. [CrossRef]
60. Ban, Y.; Liu, L.; Du, J.; Ma, C. Investigation of the Treatment Efficiency and Mechanism of Microporous Flocculation Magnetic Fluidized Bed (Mfmfb) Reactor for Pb (Ii)-Containing Wastewater. *Sep. Purif. Technol.* **2024**, *334*, 125963. [CrossRef]
61. Zhengong, T. Study on the Technology of Vortex Clarification and Its Application in Wastewater Treatment. *Nat. Environ. Pollut. Technol.* **2013**, *12*, 537.
62. Dai, H.; Qiu, Z.; Hu, F.; Gao, C.; Chen, Y.F.; Zhou, Z. Floc Performance Parameters during Water Treatment in a Micro-Vortex Flocculation Process Determined by Machine Vision. *Environ. Technol.* **2019**, *40*, 3062–3071. [CrossRef] [PubMed]
63. Meng, Q. The Impacts of Fracking on the Environment: A Total Environmental Study Paradigm. *Sci. Total Environ.* **2017**, *580*, 953–957. [CrossRef] [PubMed]
64. *GB 50428-2015*; Code for Design of Oil Field Produced Water Treatment. Ministry of Housing and Urban-Rural Development of the People's Republic of China (MOHURD): Beijing, China, 2015.
65. Dai, H.; Cao, Z.; Hu, F.; Huang, B.; Wang, Y.; Luo, W. Research on Water Treatment of Kongmu-Lake Based on Micro Vortex Flocculation Technology with the Combination of Vortex Reactor. In *IOP Conference Series: Earth and Environmental Science*; IOP Publishing: Bristol, UK, 2019.
66. Vendramel, S.; Bassin, J.P.; Dezotti, M.; Sant'Anna, G.L., Jr. Treatment of Petroleum Refinery Wastewater Containing Heavily Polluting Substances in an Aerobic Submerged Fixed-Bed Reactor. *Environ. Technol.* **2015**, *36*, 2052–2059. [CrossRef] [PubMed]
67. Xie, W.; Li, B.; Xu, Y.; Hui, X.; Tang, L.; Zhang, W.; Wang, Z.; Zhu, J. Produced Water Treatment System and Process. Available online: https://patents.google.com/patent/CN109455831B/en (accessed on 31 August 2021).

68. Xie, W.; Li, B.; Zhuang, D.; Hui, X.; Yang, Q.; Liu, H.; Lv, C.; Tang, L.; Li, F.; Zhang, W. A Kind of Degreasing Unit. Available online: https://patents.google.com/patent/CN207566940U/en (accessed on 3 July 2018).
69. Xie, W.; Li, B.; Xu, Y.; Hui, X.; Tang, L.; Zhang, W.; Wang, Z.; Zhu, J. A Kind of Degreasing Unit. Available online: https://patents.google.com/patent/CN207175538U/en (accessed on 3 April 2018).

processes

MDPI

Article

Molecular Insight into the Occurrence Characteristics of Deep Oil with Associated Gas Methane and the Displacement Resistance in Water Flooding in Nano-Pore Throat

Lixia Zhou [1], Weiming Wang [2], Youguo Yan [3,*], Chengen Zhao [1], Jiahao Zhong [1] and Yuqi Liu [1]

1 College of Science, China University of Petroleum, Qingdao 266580, China; zhoulx@upc.edu.cn (L.Z.); 2009020228@s.upc.edu.cn (C.Z.); 2009020229@s.upc.edu.cn (J.Z.); YuqiLiu0621@gmail.com (Y.L.)
2 School of Geosciences, China University of Petroleum, Qingdao 266580, China; wangwm@upc.edu.cn
3 School of Materials Science and Engineering, China University of Petroleum, Qingdao 266580, China
* Correspondence: yyg@upc.edu.cn

Abstract: In deep oil reservoirs, the existence of associated gas generally has a crucial impact on crude oil properties and flow performance. In this work, adopting molecular dynamic simulation, we studied the occurrence characteristics of oil with associate gas methane (the molar ratio of methane to oil $r_{m/o}$ were 1/4, 2/3, 3/2, and 4/1) in nano-pore throat and the displacement behavior of oil and methane in the water flooding process. Simulation results indicated: (1) an increasing replacement of the adsorption-status oil by methane as the methane content increased; (2) the oil and methane displacement efficiency was enhanced as the methane content increased in the water displacement oil and gas process; (3) the threshold displacement pressure gradually decreases as the methane content increases. The microscopic characteristics of the occurrence features and displacement performance of crude oil with associated methane in nano-pore throat were discussed in detail, and the underlying mechanism was discussed at the length concerning the interaction between different components. Our work provides an in-depth understanding of the occurrence characteristics and flow resistance of oil with associated gas in deep oil reservoirs.

Keywords: the occurrence characteristics; oil with associated gas; water displacement oil and gas process; molecular dynamic simulation

Citation: Zhou, L.; Wang, W.; Yan, Y.; Zhao, C.; Zhong, J.; Liu, Y. Molecular Insight into the Occurrence Characteristics of Deep Oil with Associated Gas Methane and the Displacement Resistance in Water Flooding in Nano-Pore Throat. *Processes* **2023**, *11*, 2529. https://doi.org/10.3390/pr11092529

Academic Editors: Zheng Sun, Tao Zhang, Dong Feng, Wen Zhao and Hung Vo Thanh

Received: 14 July 2023
Revised: 17 August 2023
Accepted: 21 August 2023
Published: 23 August 2023

1. Introduction

In recent years, with the increasing scarcity of conventional oil and gas resources, unconventional oil and gas have received extensive attention [1–3]. The exploration practice in recent years has proved that there are rich oil and gas resources in the deep layer of the basin under the current main series of exploration [4–6]. Deep oil and gas reservoirs have been found to exist all over the world. More than 1000 oil and gas fields have been discovered within deep reservoirs globally, which have huge storage content [7]. In deep oil and gas reservoirs, subsurface natural gas is mainly composed of hydrocarbon gases and several kinds of nonhydrocarbon gases [8,9]. Hydrocarbon gases, which are the most common compounds in natural gases, can be generated at all stages of organic matter thermal maturation processes.

The occurrence state characteristics of oil and gas in deep reservoirs are an important basis for enhanced oil recovery, which can provide basic data and theoretical reference for the evaluation of reservoir reserves, recoverable capacity, and the optimization of development plans [10,11]. Some studies have found that the occurrence states of deep oil and gas in the reservoir are mainly found in two forms: the free state of bulk phase in natural pores and fractures and the adsorbed oil and gas on the surface of minerals [12,13]. The proportion of oil and gas, the porosity and permeability characteristics of the reservoir, etc., affect the migration of oil and gas and the development of deep oil and gas. In terms of the occurrence characteristics of oil and gas in nano-pores, previous researchers have carried out a lot of experimental and theoretical research and achieved many valuable results. Zhao et al. measured the adsorption isotherms of methane, ethane, propane, n-butane, iso-butane, and carbon dioxide for two kinds of shales and the kerogens isolated from the two shales. They found that the adsorption of n-butane was one order of magnitude higher than that of methane (0.2 MPa and 35–65 °C) [14]. Rezlerova et al. employed Monte Carlo and molecular dynamic simulation to simulate the adsorption, self-, collective, and transport diffusivities of methane, ethane, propane, and carbon dioxide in shale kerogen micropores and mesopores [15]. The simulation results showed that the total adsorption increases with increasing pressure and decreases with increasing temperature as the pressure drives the fluid molecules into the micropores and mesopores. In the micropores, methane displays the highest adsorption, followed by ethane and propane. Wu et al. studied the wettability impact on nanoconfined methane adsorption behavior in organic-rich and inorganic-rich nano-pores [16]. They obtained the relationship between solid–methane interactions and the surface contact angle and found that the adsorption capacity in organic-rich shale is 1.72 times that in inorganic-rich shale.

The occurrence characteristics of oil and gas in nano-pores significantly affect the fluidity of oil and gas reservoirs. The intermolecular interactions and the interactions between the molecules and the pore surface could alter the fluid properties such as critical properties, phase behavior, solubility, viscosity, and interfacial tension among them because the pore surface available per unit volume increases [17–21]. Zhang Jun et al. employed molecular dynamics simulation to study the interfacial tension of oil/methane/water systems [22]. They found that the interface tension (IFT) decreased as the methane mole fraction increased. When liquid or gas flows through nano-pores, the adsorption phenomenon is very important and has a significant influence on flow enhancement [23–25]. Wang et al. studied the methane flow character as fast methane transport through different types of nano-pore [26]. Their research results showed that nanoscale flow patterns strongly depend on the adsorbed layers–solid interactions and the nature of the substrate surface. Xiong et al. investigated the displacement behavior of nanoslit oil by injected gases (CO_2, CH_4, N_2, and C_3H_8) [27]. They found that the dissolubility of gas, the adsorption strength of oil, and the detachment capability of injected gas play crucial roles in displacement performance. Zhang et al. investigated the transport of an oil/gas mixture droplet through the aqueous nano-pore [28]. They concluded that the dissolution of methane in the oil droplet reduces the oil–water interfacial tension and increases the diffusivity of oil, the merged methane inside the oil droplet reduces the oil conformation energy and promotes oil deformation, and methane reduces the oil–pore interaction and decreases the resistance as the oil droplet passes through the pore throat.

Although previous researchers have conducted lots of theoretical and experimental studies on the occurrence and fluidity of oil and gas in nano-pores, the microscopic occurrence and flow characteristics of deep oil and gas reservoirs and the complex interaction among oil, gas, and rock are still not clear, especially in the oil and gas displacement process. In this paper, using molecular dynamics simulation, the occurrence characteristics and fluidity characteristics in water injection were used to explore the process of deep oil and gas in nano-pores. The mixture fluid of octane and methane (the molar ratio of gas to oil were 1/4, 2/3, 3/2, and 4/1) was selected to simulate oil and gas in deep oil and gas reservoirs. The adsorption characteristics of oil and methane on the rock surface have been intensively studied. Further, the flow characteristics and the flow resistance in the water displacement oil and gas process were investigated. The density distribution of oil and methane, microscopic dynamic behavior, interaction between molecules, migration behavior, oil and gas displacement efficiency, and the threshold displacement pressure of oil and gas were analyzed. Our work provides an in-depth understanding of the occurrence characteristics and flow resistance of oil with associated gas in deep oil and gas reservoirs, and provides a valuable reference for the storage and the exploration and development of deep oil and gas.

2. Simulation Model and Method

2.1. Simulation Models

According to previous research [21,27–30], quartz is often used to represent the inorganic surface of oil reservoirs. Here, we construct a quartz nano-pore throat with a diameter of 6 nm. The silica surface is generated by cleaving the α-quartz along the (0 0 1) crystallographic orientation. Then, the surface is methylated with methyl and forms a hydrophobic surface (Figure 1b) [31]. The dimensions of the silica slab are $2.46 \times 1.4 \times 11.8$ nm^3 (XYZ). Octane is used to represent oil since deep oil is mainly light oil. The mixture fluid of octane and methane molecules (the molar ratio of methane to oil $r_{m/o}$ are 1/4, 2/3, 3/2, and 4/1) are confined within the quartz nano-pore. A mixture fluid of octane and methane with a thickness of 2 nm is placed to the left and side of the nano-pore throat, respectively. The molecule number of oil and methane can be obtained based on the density of the mixture fluid of octane and methane during construct models. The density and molecule numbers of octane and methane in different mixture fluids of octane and methane systems can be found in Table 1 (the density of the mixture fluid of octane and methane are from the NIST-REFPROP database [32]). Then, the water phase is placed to the left side of the nano-pore throat with a length of 10 nm along the z-axis. Additionally, two Argon sheets are placed on the left side of the water phase and the right side of the oil phase, respectively, defined as rigid bodies in the simulation. Vacuum areas with sizes of 3 nm and 27.2 nm are placed on the left and right sides of the model to eliminate the effects of periodic boundary conditions on the interactions between fluids. To compare, the octane/water/silica system is also built. The final dimension of the simulated boxes for different models is $2.46 \times 8.8 \times 53.0$ nm^3 (XYZ). The initial model of water/oil is shown in Figure 1a, and the initial model of the water/oil/methane nano-pore throat system at the molar ratio of methane to oil $r_{m/o} = 3/2$ is also shown in Figure 1a as an example. Other models can be found in Supplementary Materials (Figure S1). A temperature of 353 K and a pressure of 20 MPa are selected in our simulation.

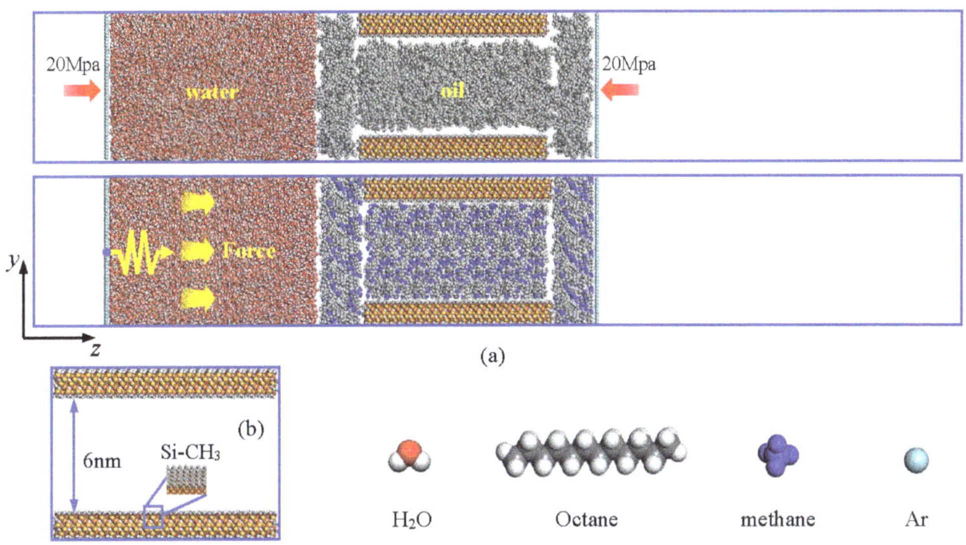

Figure 1. (**a**) The initial model of water/oil and water/oil/methane (the molar ratio of methane to oil $r_{m/o}$ is 3/2) nano-pore throat system; (**b**) hydrophobic quartz surface and nano-pore throat.

Table 1. The density and molecule numbers of the mixture fluid of octane and methane used in our models (353 K, 20 MPa).

Molar Ratio of Methane to Oil	Density [a] (g/cm^3)	Octane Number	Methane Number
0	0.676	788	0
1/4	0.624	768	192
2/3	0.581	690	460
3/2	0.520	600	900
4/1	0.519	461	1844

[a] The density of the mixture fluid of octane and methane is from the NIST-REFPROP database [32].

2.2. Force Fields

The simple point-charge SPC/E model [33] describes the water molecules, which is the most extensive force field for water molecules [22,28]. The consistent valence force field (CVFF) [34] is used for silica, and previous researchers have proved that the CVFF force field could successfully describe hydrophobic quartz surfaces [35]. The optimized potentials for liquid simulations all-atom (OPLS-AA) force field [36] is applied for octane and methane, which has been successfully applied to simulate alkane [22,26–28]. For the bonded potential, it includes bond stretching, angular bending, and dihedral angle torsion. The non-bonding interactions between molecules include short-range van der Waals (vdW) and long-range electrostatic interaction. The vdW and electrostatic interactions are represented by Lennard-Jones 12-6 and Coulombic potentials, respectively, as shown in Equation (1).

$$V_{non-bond} = \varepsilon_{ij}\left[\left(\frac{\sigma_{ij}}{r_{ij}}\right)^{12} - \left(\frac{\sigma_{ij}}{r_{ij}}\right)^{6}\right] + \frac{q_i q_j}{4\pi\varepsilon_0 r_{ij}}, \quad (1)$$

where ε_{ij} and σ_{ij} are the potential well depth and collision diameter, and r_{ij} is the distance between atoms i and j. q_i represents the charge of atom i, and $\varepsilon_0 = 8.8542 \times 10^{-12}\ C^2\ N^{-1}\ m^{-2}$ is the permittivity of the vacuum. The detailed interaction parameters of oil, methane, water, and quartz are listed in Table S1 in the Supplementary Materials.

2.3. Equilibrium Molecular Dynamics (EMD)

All MD simulations are performed with the LAMMPS package [37]. The three-dimensional periodic boundary condition is applied with a cutoff radius of 10 Å and a time step of 1 fs. The Nosé–Hoover thermostat [38] controls the temperature during all the above MD simulations. The long-range electrostatic interactions are calculated by the particle–particle–particle-mesh (PPPM) method [39]. First, the initial configuration is optimized using the steepest descent method. Then, a 5 ns equilibrium molecular dynamics (EMD) simulation under the NVT ensemble at 353 K and 20 MPa is conducted on each nano-pore throat system with the quartz surfaces and water are fixed until the system reaches equilibrium. The constant pressure 20 MPa is applied in opposite directions of the corresponding forces to the two Argon sheets to control the fluid pressure. All the dynamic trajectories and snapshots of the interfacial configurations are rendered using VMD (Visual molecular dynamics) software [40]. After the EMD simulation of oil and methane, a 2 ns EMD simulation is conducted to each nano-pore throat system with the quartz surfaces, oil and methane fixed until the water phase reaches equilibrium.

2.4. Steered Molecular Dynamics (SMD)

The steered molecular dynamics (SMD) simulation is performed by exerting a spring-pulling force onto the COM (center of mass) of the left Argon sheet along the positive direction of the z-axis to mimic the process of water displacement oil and gas in the nano-pore throat [41,42]. The spring force is expressed as:

$$F = K(Z_0 + vt - Z_{COM}), \quad (2)$$

where F is the spring pulling force; K is the spring constant, which is 0.001 Kcal mol^{-1} nm^{-2} in the simulation; Z_0 is the initial COM position of the left Argon sheet in the Z-axis direction; v is a constant velocity along the Z-axis direction, which is 0.001 nm·ps^{-1} in the simulation; t is the simulation time; Z_{com} is the real-time position of the left Argon sheet COM along the Z-axis direction. The position at $(Z_0 + vt)$ is a reference point, which moves along the Z-axis at a constant speed. Each system experiences a 10 ns SMD simulation, and all other simulation details are identical to those in the EMD simulation. The applied spring force is shown schematically in Figure 1a.

3. Results and Discussion

3.1. Microstates and Distribution of the Oil and Methane in the Quartz Slits

The final EMD configurations of water/oil/methane nano-pore throat systems are shown in Figure 2a–e. The molar ratio of methane to oil $r_{m/o}$ is (a) 0, (b) 1/4, (c) 2/3, (d) 3/2, and (e) 4/1, respectively. It can be seen from the EMD configurations that as the methane content increases, methane molecules tend to cluster together, which is not obvious at $r_{m/o}$ = 1/4 and $r_{m/o}$ = 2/3. However, when $r_{m/o}$ increases to 3/2 and 4/1, methane molecules cluster together significantly. At $r_{m/o}$ = 4/1, the aggregation of the methane molecules even forms a gas channel within the nano-pore throat, which indicates that compared to methane, oil molecules are more likely to adsorb on the rock surface, while methane molecules tend to aggregate in the center of the nano-pore throat. The adsorption behavior of methane and octane is ascribed to the adsorption propensity of n-alkanes with more carbon atoms per molecule; therefore, the vicinity of the solid walls is preferentially adsorbed by longer alkanes [43,44]. In addition, as the methane content increases, the overall volume of the oil and gas mixing area increases, which is due to the swelling effect of the gas.

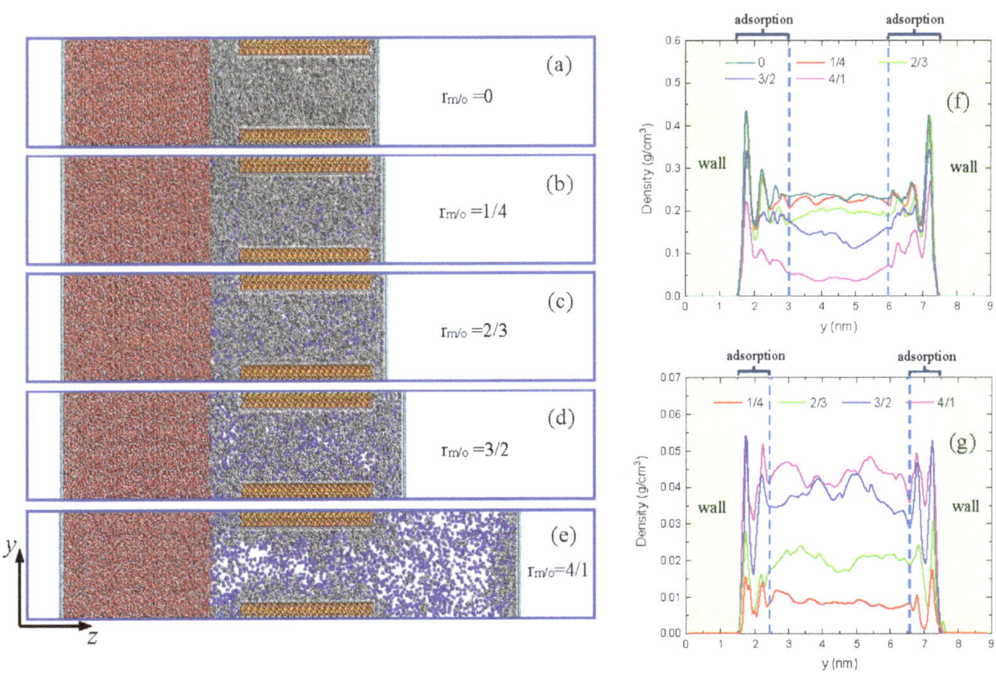

Figure 2. (**a–e**) The oil and methane EMD configurations of water/oil/methane nano-pore throat systems. The molar ratio of methane to oil $r_{m/o}$ is (**a**) 0, (**b**) 1/4, (**c**) 2/3, (**d**) 3/2, and (**e**) 4/1. (**f**) The density distribution of oil and (**g**) the density distribution of methane along y-axis for different $r_{m/o}$ systems.

To quantitatively analyze the occurrence status of formation oil and methane under different formation methane saturations, the density distribution profiles of oil and methane along the nano-pore throat width were calculated, as shown in Figure 2f,g. It can be seen from the figures that both for oil and methane, the density profiles show adsorption peaks near the nano-pore throat surface. The interval of the adsorption peak of oil molecules is larger than that of methane molecules due to the difference in the molecular size of oil and methane. The adsorption peaks of oil molecules appear in the region of 1.5–3 nm and 6–7.5 nm, and the free-state zone of oil molecules is in the region of 3–6 nm. The adsorption peaks of methane molecules appear in the region of 1.5–2.4 nm and 6.6–7.5 nm, and the free-state zone of methane molecules is in the region of 2.4–6.6 nm. The adsorption zone of oil and methane is consistent with previous research [20,45–47]. It can be seen from Figure 2f that with the increase in methane content, the density of oil molecules shows a downward trend. When the methane content was low ($r_{m/o}$ = 1/4 and 2/3), we found that the amplitude of the first adsorption peak (near the rock surface) was almost the same with the pure oil system, which indicates that oil molecules tend to adsorb near the nano-pore throat surface, and have reached saturation adsorption at $r_{m/o}$ = 2/3. From Figure 2g, it can be seen that as the content of methane molecules increases, the density of methane molecules gradually increases. The first adsorption peaks near the nano-pore throat surface increase when the methane content increases from $r_{m/o}$ = 1/4 to 2/3. When the methane content increases to $r_{m/o}$ = 3/2 and 4/1, the amplitude of the first adsorption peaks shows no more changes. The amplitudes of the first methane adsorption peaks of $r_{m/o}$ = 3/2 and 4/1 are almost equal, i.e., the adsorption of methane on the nano-pore throat surface reaches saturation at methane content $r_{m/o}$ = 3/2.

From the above analysis, oil or methane occurrence status could be divided into two types. One is free-status molecules, which locate the internal nano-pore throat. The other is adsorption-status molecules, which are adsorbed on the nano-pore throat surface and located near the surface. Among them, the free-status molecules could be easily replaced by a displacement agent, while adsorption-status molecules interact more with the rock surface, making it difficult to be displaced out of the reservoir. To provide a clearer description of the adsorption-status molecules of oil and methane, we calculated the atom numbers of adsorbed oil (1.5–3 nm and 6–7.5 nm) and methane (1.5–2.4 nm and 6.6–7.5 nm), as shown in Figure 3. It can be seen that as the methane content increases and the oil content decreases, the number of adsorbed oil atoms gradually decreases, and the number of adsorbed methane atoms gradually increases. These phenomena indicate that although oil molecules tend to adsorb on the rock surface, some adsorbed oil molecules are replaced by methane when the methane content increases, i.e., when the methane content increases, in addition to forming clusters or gas channels in the center of the pores, the adsorbed methane molecules on the rock surface also increase. Previous studies by Moh et al. in their research of decane extraction from nano-pores by CH_4 also showed that when methane molecules are mixed with decane molecules in nano-pores, a large amount of adsorbed oil molecules would be replaced by methane molecules [20].

To obtain a deep understanding of the competitive adsorption of oil and methane mentioned above, we calculate the interaction energy between the oil and rock surface, methane and rock surface, and the interaction energy between oil and methane along with simulation time, as shown in Figure 4. It can be seen from the figure that the methane and rock surface has weak interaction, the oil and rock surface has strong interaction, and methane and oil have a very strong interaction. Therefore, less adsorbed oil and more adsorbed methane are beneficial for displacement by the injected liquid. From the above analysis, it can be inferred that in the next water flooding process, the displacement rate D of oil and methane would follow $D_{rm/o=4/1} > D_{rm/o=3/2} > D_{rm/o=2/3} > D_{rm/o=1/4} > D_{rm/o=0}$.

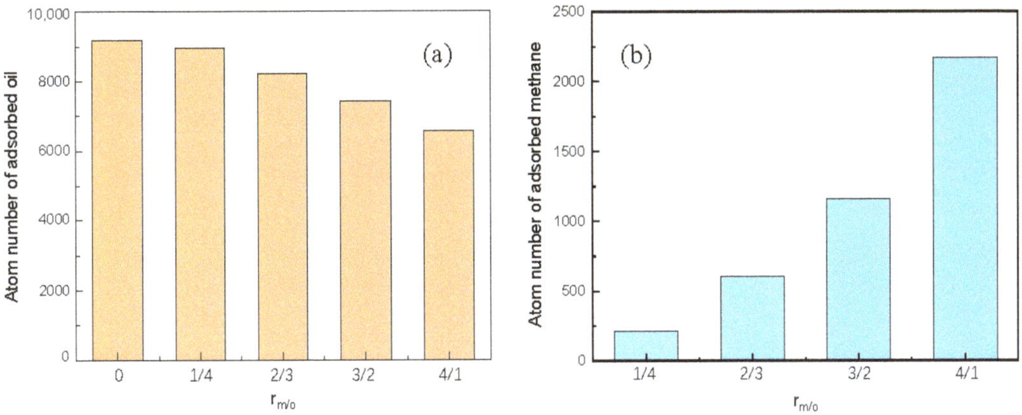

Figure 3. The atom number of adsorbed (**a**) oil and (**b**) methane for different $r_{m/o}$ systems.

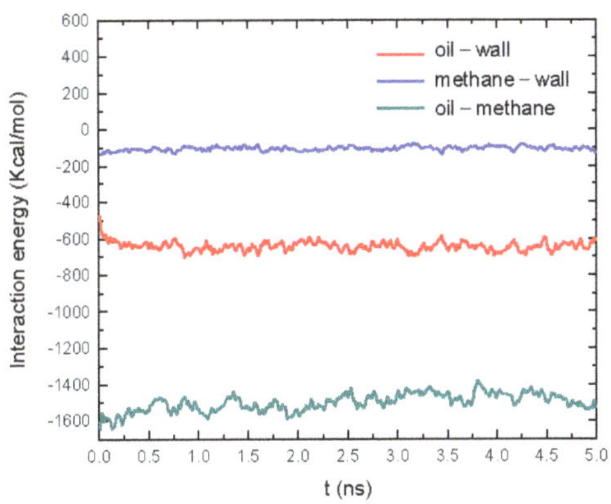

Figure 4. The interaction energy between rock surface, methane, and oil in the equilibrium molecular dynamics process.

3.2. Water Displacement Oil and Gas Process in Nano-Pore Throat

In Section 3.1, we analyzed the occurrence state of the pure oil, mixture fluid of oil and methane ($r_{m/o}$ = 1/4, 2/3, 3/2, and 4/1) in the quartz nano-pore throat. When oil and methane coexist in the nano-pore throat, the content of methane and oil affects the occurrence state of free-status and adsorption-status of oil or methane, which results in different displacement rates of oil and methane for different $r_{m/o}$ systems in the water injection process. Therefore, in this section, we further investigate the flow behavior of oil and methane in the water flooding process of these systems to determine the influence of the content of oil and methane on the water displacement oil and methane process in the nano-pore throat.

3.2.1. Water Displacement Oil and Gas Process

Figure 5 shows the dynamic process of the water displacement oil and gas process at 2 ns, 4 ns, 6 ns, and 8 ns for pure oil and $r_{m/o}$ = 1/4, 2/3, 3/2, and 4/1 systems. It can be seen that with the simulation processes, under spring pulling force on the left Argon sheet, the water moved forward along the z-axis and gradually entered the nano-pore throat. As a result, oil or methane were gradually pulled out of the nano-pore throat. Comparing the water displacement oil and methane process of different methane content systems, it can be seen that at the same simulation time, the front of the water tends towards larger z-coordinates as the methane content increases (oil content decreases), which indicates a gradual decrease in resistance and gradual increase in displacement rate with the increase of $r_{m/o}$. This is consistent with our previous analysis in Section 3.1. The rapid self-diffusion of methane and oil molecules as the methane content increase have also been reported by Moh et al. in their research of decane molecules extraction from nano-pores by CH_4 [20]. In addition, the clusters of methane molecules and the gas channel at the center of the pore throat ($r_{m/o}$ = 3/2 and 4/1) are beneficial for faster expulsion of methane from the nano-pore throat. Therefore, it can be seen via the simulation process that numerous methane molecules quickly pass through the nano-pore throat and form larger gas clusters on the right side at $r_{m/o}$ = 3/2 and 4/1.

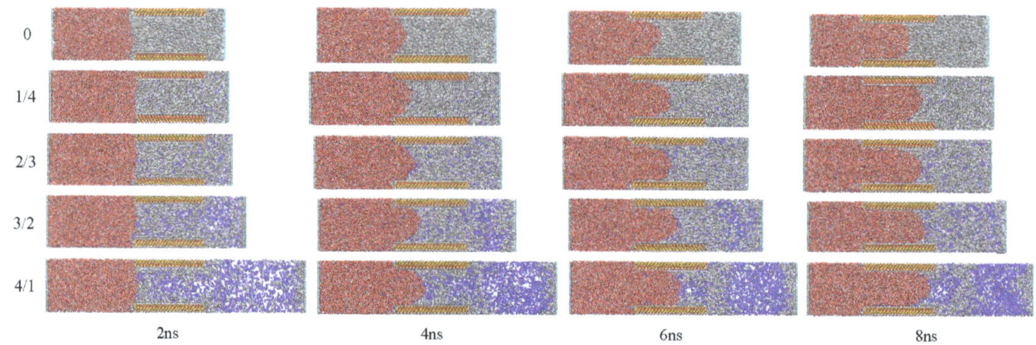

Figure 5. The dynamic process of water displacement oil and methane process at 2 ns, 4 ns, 6 ns, and 8 ns for different $r_{m/o}$ systems.

To further reveal the displacement process of free-status and adsorption-status oil and methane during water displacement, we calculated the density distribution of oil and methane in the z-direction and y-direction. The density profiles of oil and methane at $r_{m/o}$ = 3/2 are shown in Figure 6 as an example. Figure 6a,b shows the density distribution of oil and methane along the z-direction at 2 ns, 4 ns, 6 ns, and 8 ns, respectively. The position of the nano-pore throat is at approximately z = 18–29.8 nm. It can be seen that with the simulation process, both oil and methane move forward along the z-axis, indicating that oil and methane are gradually being displacement out of the nano-pore throat by water. In Figure 6a, at 2 ns, oil has an appreciable density distribution within the nano-pore throat, and there is a deep valley appears between z = 32–35 nm on the right side of the nano-pore throat, indicating a small density distribution of oil in this region. In Figure 6b, it can be seen that the methane density distribution on the right side of the nano-pore throat is greater than that in the nano-pore throat, and there is a large peak in the region z = 32–35 nm at 2 ns, i.e., a large amount of methane molecules aggregates in this region. These phenomena show that methane could be more easily displaced out of the nano-pore throat than oil. Along with the simulation time, as the oil is gradually pulled out of the nano-pore throat, the density of oil gradually increases in the region z = 32–35 nm, indicating that the expelled oil has filled this area.

Figure 6c shows the density distribution of oil along y-axis at 2 ns, 4 ns, 6 ns, and 8 ns at $r_{m/o}$ = 3/2. The density profiles of oil and methane of other systems along the y-axis can be found in Supplementary Materials (Figure S2). It can be seen that the amplitude of the first adsorption oil peaks near the rock surface (1.7 nm and 7.2 nm) is almost the same at 2 ns, 4 ns, and 6 ns, and they are decreased at 8 ns, yet the adsorption peaks in the region 2–3 nm and 6–7 nm gradually decrease with time. These phenomena indicate that in the water displacement process, the oil molecules adsorbed on the rock surface are not easy to be stripped, and the oil molecules away from the rock surface are easier to be pulled out, which could be ascribed to the strong interaction between oil and rock surface and very strong interaction between oil and free-status methane in the center of the nano-pore throat (Figure 7). Therefore, the oil molecules away from the rock surface could be driven out of the nano-pore throat along with the free-status oil and methane molecules. Figure 6d shows the density distribution of methane molecules along the y-axis at 2 ns, 4 ns, 6 ns, and 8 ns, respectively. The peaks near 1.7 nm and 7.2 nm represent the methane molecules adsorbed on the rock surface. It can be seen that their amplitude gradually decreases with time. The second adsorption peaks near 2.2 nm and 6.7 nm represent the methane molecules adsorbed on the oil and methane near the rock surface, which reach the highest at 4 ns and then decrease gradually with the simulation process. These phenomena indicate that in the water displacement process, methane molecules adsorbed on the rock surface could be more easily stripped off the rock surface than the second adsorbed layer of methane molecules. The above behavior of methane molecules could be ascribed to the weak interaction between methane and the rock surface and the strong interaction between oil and methane (Figure 7). Therefore, for these adsorption methane molecules, methane molecules depart from the rock surface first, then they are adsorbed on the near rock surface oil and methane, and these methane molecules could be pulled out of the nano-pore throat along with the free-status oil and methane.

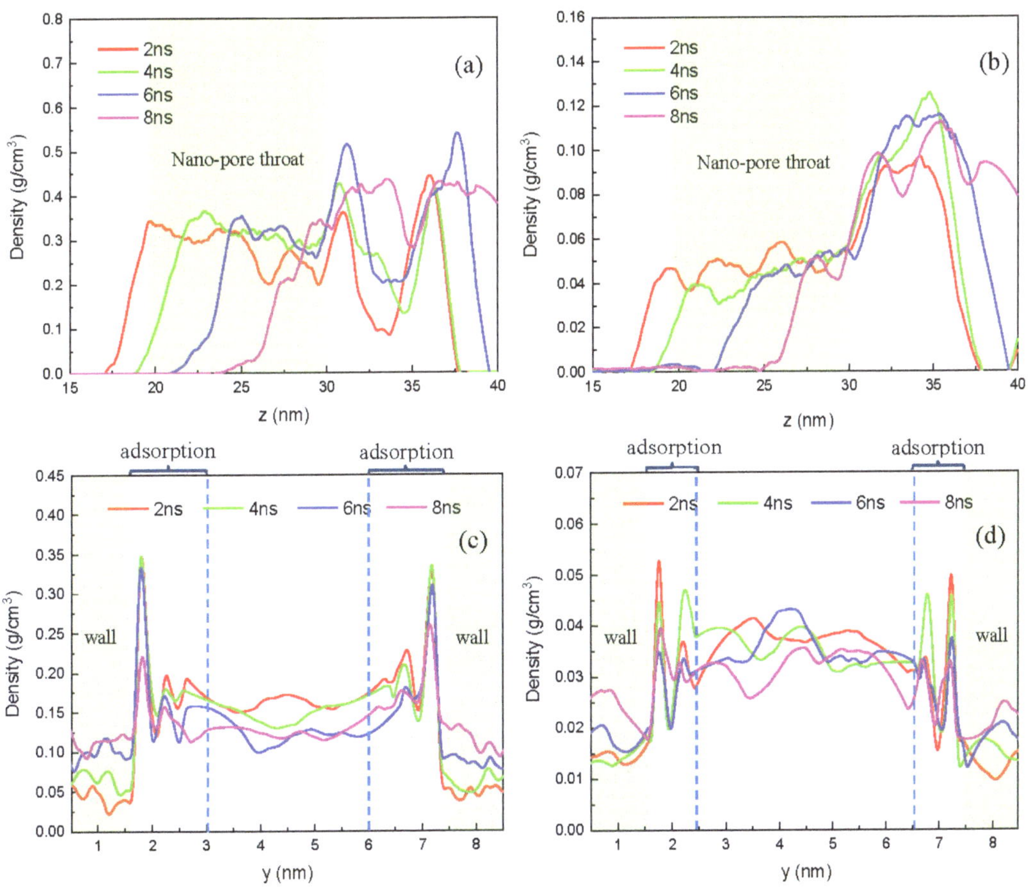

Figure 6. The density distribution of oil and methane along the z-axis and y-axis for $r_{m/o} = 3/2$ system. (**a**,**c**) Oil density; (**b**,**d**) methane density.

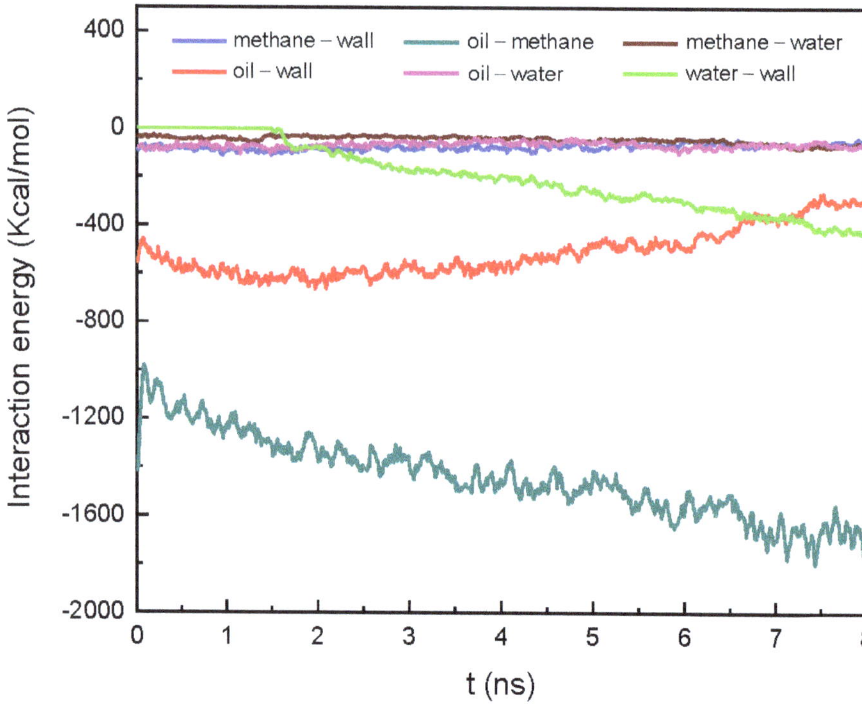

Figure 7. The interaction energy between different components in water displacement oil and methane process.

3.2.2. The Threshold Displacement Pressure

In the water injection process, an important parameter is the threshold displacement pressure. The threshold displacement pressure is crucial to quantify the pressure required for fluids being transported into the capillary [48]. In order to investigate the effect of the content ratio of methane and oil on the threshold displacement pressure during the water displacement oil process, we calculated the pulling force (exert on the left Argon sheet) as a function of simulation time, as displayed in Figure 8a,b. Additionally, the evaluable threshold displacement pressure for different $r_{m/o}$ systems is shown in Figure 8c. The pressure conversion equation and conversion parameters can be found in Supplementary Materials (Table S2).

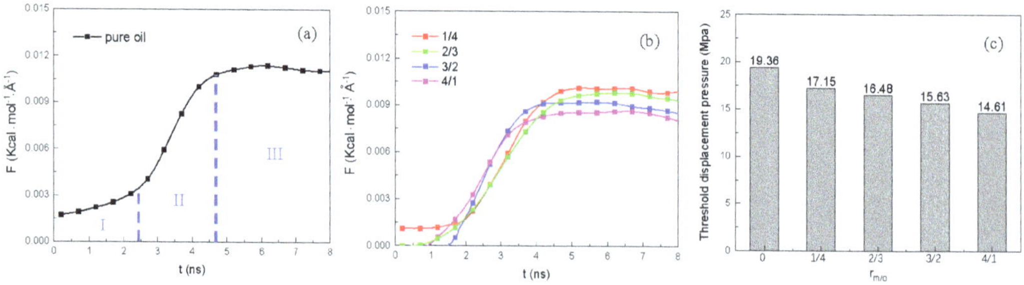

Figure 8. Pulling force as a function of simulation time for (**a**) pure oil system and (**b**) different $r_{m/o}$ systems; (**c**) the threshold displacement pressure for different $r_{m/o}$ systems.

In 1923, Bosanquet [49] thought that in the capillary action of fluids flowing through a nano-pore, the contributions of inertial drag and viscous resistance counteracted the capillary pressure. Oyarzua et al. [50] used the Bosanquet model of capillarity to describe the nanoscale imbibition of water through a nano-pore. In their study, the capillary filling kinetics underwent three stages: the beginning of the imbibition (inviscid stage), the capillary filling stage (inertial drag and viscous resistance both exit stage), and the purely viscous regime (no inertial drag stage). The three stages concluded by Oyarzua et al. [50] can also be seen in Figure 8a,b. In Figure 8a,b, it can be seen that the water displacement pure oil or oil and methane process has gone through three stages. From Figure 8a, in stage I, the pulling force increases slowly. In this stage, water molecules gradually approach the nano-pore throat under the pressure of the left Argon sheet. Since the pulling force is small during this stage, we could conclude that, in this stage, the resistance is small. In stage II, water molecules begin to charge into the nano-pore throat. The pulling force increases quickly and finally reaches a turning point at 4.7 ns, which indicates a difficult entrance of water into the nano-pore throat. In stage III, the pulling force almost maintains a constant value, indicating a steady propel of water molecules. In this stage, resistance is counteracted by the pulling force, i.e., when the pulling force in the water displacement oil and gas process is greater than this pulling force value, oil and gas could be driven out from the reservoirs. So this pulling force value in stage III could be considered the threshold displacement force and be used to evaluate the threshold displacement pressure in oil and gas recovery. Comparing Figure 8a,b, it can be seen that as the methane content increases and the oil content decreases, the experience time for stage I and II decrease gradually, indicating that the displacement rate gradually accelerates as the methane content increases. In addition, the threshold displacement pressure gradually decreases with the increase in methane content. The threshold displacement pressure is 19.36, 17.15, 16.48, 15.63, and 14.61 MPa when $r_{m/o}$ = 0, 1/4, 2/3, 3/2, and 4/1, respectively (Figure 8c). Similar research results have been reported by Zhang et al. in their research of the migration of octane/methane mixture droplets from a wide area into a nano-pore throat [28]. They also found the threshold pulling force was decreased with the increasing methane molar fraction.

As mentioned above, when water enters into the nano-pore throat from the wide area (stage II), there is big resistance which causes the pulling force to continue to increase until it reaches the threshold displacement force. According to Oyarzua et al. [50], in this stage, the inertial drag resistance and viscous resistance both exist, and the Jamin effect plays an important role [51,52]. The shrunk droplet diameter causes an additional pressure difference, hindering the droplet transport. The Jamin effect could be represented by the following equation:

$$\Delta P = 2\gamma(\frac{1}{R_1} - \frac{1}{R_2}), \tag{3}$$

where ΔP is the pressure difference of water charging into the pore throat, and γ is the IFT of the oil–water. R_1 and R_2 refer to the radii of water on the outer side and the inner side of the pore throat, respectively. The schematic diagram of the Jamin effect can be seen in Figure S3. In our simulation, $R_1 \gg R_2$. Therefore, $\Delta P < 0$, which means the pressure caused by the Jamin effect hinders the water phase charging into the nano-pore throat. Since ΔP is proportional to the oil–water interfacial tension γ, we calculated the oil–water interfacial tension of pure oil and different $r_{m/o}$ oil and methane systems, as shown in Figure 9. The IFT of pure oil–water is 48.1 mN/m, consistent with previous experiments and calculations [28,52]. It can be seen that as the methane content increases and the oil content decreases, the IFT gradually decreases, which is consistent with the research results of Zhang et al. [28]. Therefore, from Equation (3), as the methane content increases and the oil content decreases, the corresponding Jamin effect gradually weakens, which is the reason why the threshold displacement pressure follows $P_{rm/o=0} > P_{rm/o=1/4} > P_{rm/o=2/3} > P_{rm/o=3/2} > P_{rm/o=4/1}$. The calculation method for interfacial tension is shown in Figure S4.

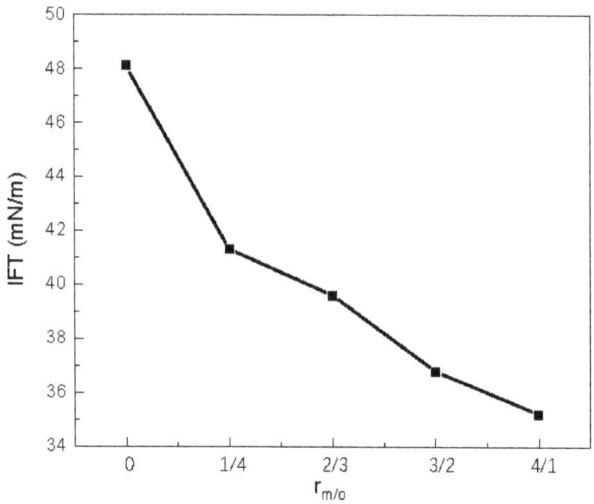

Figure 9. The oil–water interfacial tension (IFT) of different $r_{m/o}$ systems.

In addition, the fluidity of oil and gas will change with the methane content. In stag II and III, as Oyarzua et al. [50] reported, viscous resistance plays an important role. Zhan et al. found that the viscosity of oil gradually decreases with the increase of methar content [28]; that is, the fluidity of oil and methane increases with increasing methar content. Therefore, oil and methane in a high methane-content system are more easil displaced by water.

3.2.3. The Displacement Rate of Oil and Gas

In order to intuitively demonstrate the effect of the content ratio of oil and methar on the oil and gas displacement efficiency, we calculated the atom number of water i the nano-pore throat, as shown in Figure 10. In Figure 10, the atom numbers of water i the nano-pore throat first experience a period of time equal to zero, which indicates th water has not yet entered the nano-pore throat. Then, the atom numbers of water in th nano-pore throat gradually increase, indicating that water molecules begin to enter th nano-pore throat and drive oil and methane out of the nano-pore throat. It can be see from Figure 10 that it takes a shorter time for water to enter into the nano-pore throat whe the methane content increases. And the atom number N of water in the nano-pore thro follows $N_{rm/o=4/1} > N_{rm/o=3/2} > N_{rm/o=2/3} > N_{rm/o=4/1} > N_{rm/o=0}$ at the same simulatio time, indicating the displacement efficiency of oil and gas increases with the increase methane content.

To explain the mechanisms of accelerating the oil and gas displacement process i water flooding as the methane content increases, we calculated the atom number of th adsorbed oil and methane along with the simulation time for different $r_{m/o}$ systems, a shown in Figure 11a,b. The adsorption region of oil and methane could be delimited base on the adsorption peak interval in Figures 2 and 6. It can be seen in Figure 11a,b that th atom number of adsorbed oil decreases with the increasing content of methane, and th atom number of adsorbed methane increases with the increasing content of methane. A mentioned in Section 3.2.1 and Figure 7, since the interaction between oil and rock surfa is strong and the interaction between the methane and rock surface is weak, a decrea in adsorbed oil and an increase in adsorbed methane is beneficial for the detachment adsorbed oil and methane. Therefore, the larger the $r_{m/o}$, the greater the displacement ra of oil and methane.

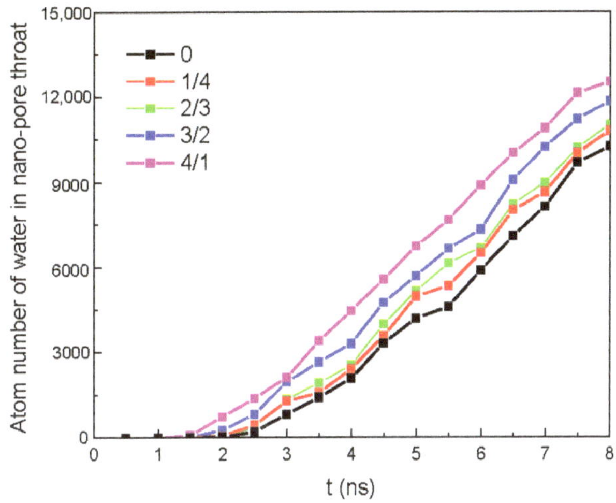

Figure 10. Atom number of water in the nano-pore throat for different $r_{m/o}$ systems in the water displacement oil and methane process.

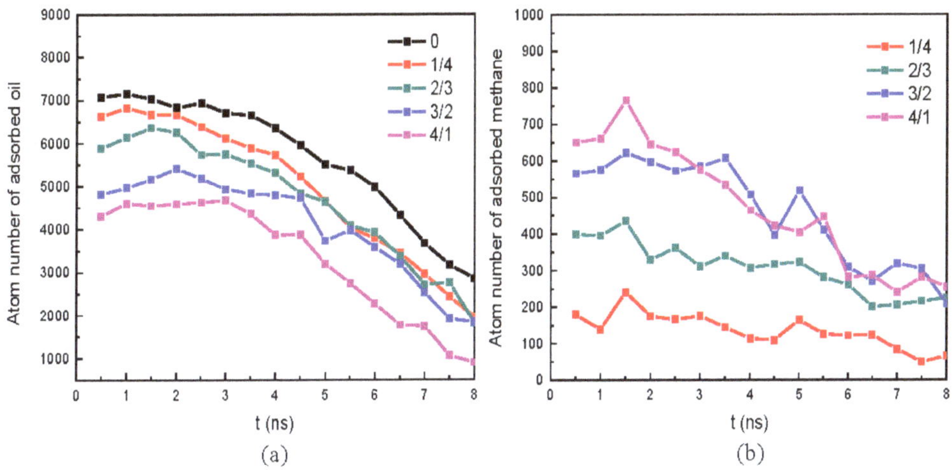

Figure 11. Atom number of adsorbed (**a**) oil and (**b**) methane in the nano-pore throat for different $r_{m/o}$ systems in water displacement oil and methane process.

4. Conclusions

In this work, adopting molecular dynamic simulation, we studied the occurrence characteristics of oil with associated gas methane in the nano-pore throat and the displacement behavior of oil and methane in the water flooding process. Simulation results showed that:

(1). The occurrence characteristics of oil with associated gas methane in the nano-pore throat indicated that there were two occurrence statuses of oil and methane: free status located within the nano-pore throat and adsorption status on the rock surface. We found methane molecules tend to cluster together and form a free status. Compared to methane, oil molecules were more likely to adsorb on the rock surface. We also found that the adsorption-status oil molecules gradually decrease and the adsorption-status methane molecules gradually increase as the methane content increases, which would be beneficial to improving oil and methane displacement efficiency in the water flooding process.

(2). In the water displacement oil and gas process, the trajectory of the displacement process and the atom number of water in the nano-pore throat showed that the oil and methane displacement efficiency was enhanced as the methane content increased, which could be ascribed to a decreasing amount of adsorbed oil and an increasing amount of adsorbed methane as the methane content increased. In addition, we found that the threshold displacement pressure in the displacement process gradually decreased as the methane content increased as follows: $P_{rm/o=0} > P_{rm/o=1/4} > P_{rm/o=2/3} > P_{rm/o=3/2} > P_{rm/o=4/1}$, which could be ascribed to the decrease of the oil/water IFT and the oil viscosity as the methane content increases.

Our work provides an in-depth understanding of the occurrence characteristics and flow resistance of oil with associated gas in deep oil reservoirs, which may help to promote the storage and development of deep oil and gas and provide valuable references to optimize oil and gas extraction processes. Due to limitations in computational power, our research can only be conducted on the size of nano-pores. However, there are many micron-scale or larger-size pores in the oil and gas reservoir. In addition to this, the impact of heterogeneity in oil and gas reservoirs on the displacement process was not considered in this work. Relevant in-depth research is expected to be conducted from the perspective of seepage mechanics.

Supplementary Materials: The following supporting information can be downloaded at: https://www.mdpi.com/article/10.3390/pr11092529/s1, Table S1. Force field parameters for octane, methane, water and quartz. Table S2. Conversion parameters of threshold displacement pressure. Figure S1. The initial model of water/oil/methane nano-pore throat systems. Figure S2. The density profiles of oil and methane along y axis in water displacement oil and methane process. Figure S3. Schematic diagram of Jamin effect (Reference [52]). Figure S4. Snapshot of MD simulation of oil-water interfacial tension. (References [53–55] are cited in the Supplementary Materials).

Author Contributions: Conceptualization, L.Z. and Y.Y.; methodology, L.Z., W.W. and Y.Y.; validation, C.Z., J.Z. and Y.L.; formal analysis, L.Z. and Y.Y.; investigation, C.Z., L.Z. and Y.Y.; data curation C.Z. and L.Z.; writing—original draft preparation, L.Z.; writing—review and editing, L.Z. and Y.Y. visualization, L.Z., C.Z. and J.Z.; funding acquisition, L.Z., W.W. and Y.Y. All authors have read and agreed to the published version of the manuscript.

Funding: This work was financially supported by the Natural Science Foundation of Shandong Province (ZR2022ME138) and the National Natural Science Foundations of China (52073305, 42272145)

Conflicts of Interest: The authors declare no conflict of interest.

References

1. Jiao, F.Z. Re-recognition of "unconventional" in unconventional oil and gas. *Petrol. Explor. Dev.* **2019**, *46*, 847–855. [CrossRef]
2. Tian, Y.X.; Qi, Y.M.; Chen, S.J.; Han, H.J.; Wang, H.Y.; Gong, X.Z.; Zhang, M.; Jiang, X.D.; Chen, Y.G. Recombination of hydrogen bonds clipping interfacial film effectively for dehydrated tight oil. *Sep. Purif. Technol.* **2023**, *319*, 124093. [CrossRef]
3. Lior, N. Exergy, Energy, and Gas Flow Analysis of Hydrofractured Shale Gas Extraction. *J. Energy Resour. Technol.* **2016**, *138*, 061601. [CrossRef]
4. Liu, C.Y.; Huang, L.; Zhang, D.D.; Zhao, J.F.; Deng, Y.; Guo, P.; Huang, Y.J.; Wang, J.Q. Genetic causes of oil-rich and oil-poor reservoirs: Implications from two Cenozoic basins in the eastern North China Craton. *Sci. China Earth Sci.* **2018**, *61*, 1910–1931. [CrossRef]
5. You, X.L.; Wu, S.N.; Xu, F. The characteristics and main control factors of hydrocarbon accumulation of ultra-deep marine carbonates in the Tarim Basin, Nw China—A review. *Carpath. J. Earth Environ.* **2018**, *13*, 135–146. [CrossRef]
6. Nurbekova, R.; Smirnova, N.; Goncharev, I.; Sachsenhofer, R.F.; Hazlett, R.; Smirnov, G.; Yensepbayev, T.; Mametov, S.; Fustic, M. High-quality source rocks in an underexplored basin: The upper Carboniferous-Permian succession in the Zaysan Basin (Kazakhstan). *Int. J. Coal. Geol.* **2023**, *272*, 104254. [CrossRef]
7. Zhu, G.Y.; Zou, C.N.; Yang, H.J.; Wang, K.; Zheng, D.M.; Zhu, Y.F.; Wang, Y. Hydrocarbon accumulation mechanisms and industrial exploration depth of large-area fracture-cavity carbonates in the Tarim Basin, western China. *J. Petrol. Sci. Eng.* **2015**, *133*, 889–907. [CrossRef]
8. Sutton, R.P. Fundamental PVT Calculations for Associated and Gas/Condensate Natural-Gas Systems. *SPE Res. Eval. Eng.* **2007**, *10*, 270–284. [CrossRef]

. Tao, K.; Cao, J.; Chen, X.; Nueraili, Z.; Hu, W. Deep hydrocarbons in the northwestern Junggar Basin (NW China): Geochemistry, origin, and implications for the oil vs gas generation potential of post-mature saline lacustrine source rocks. *Mar. Petrol. Geol.* **2019**, *109*, 623–640. [CrossRef]

0. Su, Y.L.; Zhang, X.; Li, L.; Hao, Y.M.; Zhan, S.Y.; Wang, W.D.; Wu, Z.Y.; Zhang, W.J. Experimental study on microscopic mechanisms and displacement efficiency of N-2 flooding in deep-buried clastic reservoirs. *J. Petrol. Sci. Eng.* **2022**, *208*, 109789. [CrossRef]

1. Zhu, G.Y.; Zhang, Z.Y.; Zhou, X.X.; Yan, L.; Sun, C.H.; Zhao, B. Preservation of Ultradeep Liquid Oil and Its Exploration Limit. *Energy Fuels* **2018**, *32*, 11165–11176. [CrossRef]

2. Zhang, P.F.; Lu, S.F.; Li, J.Q.; Chang, X.C.; Lin, Z.Z.; Chen, G.; Li, J.J.; Liu, J.Z.; Tian, S.S. Evaluating microdistribution of adsorbed and free oil in a lacustrine shale using nuclear magnetic resonance: A theoretical and experimental study. *J. Petrol. Sci. Eng.* **2022**, *212*, 110208. [CrossRef]

3. Li, J.; Wu, Q.Z.; Jin, W.J.; Lu, J.; Nan, Z.Y. Logging evaluation of free-gas saturation and volume content in Wufeng-Longmaxi organic-rich shales in the Upper Yangtze Platform, China. *Mar. Petrol. Geol.* **2019**, *100*, 530–539. [CrossRef]

4. Zhao, H.J.; Lai, Z.P.; Firoozabadi, A. Sorption Hysteresis of Light Hydrocarbons and Carbon Dioxide in Shale and Kerogen. *Sci. Rep.* **2017**, *7*, 16209. [CrossRef] [PubMed]

5. Rezlerová, E.; Jain, S.K.; Lísal, M. Adsorption, Diffusion, and Transport of C1 to C3 Alkanes and Carbon Dioxide in Dual-Porosity Kerogens: Insights from Molecular Simulations. *Energy Fuels* **2023**, *37*, 492–508. [CrossRef]

6. Wu, S.; Luo, W.N.; Fang, S.D.; Sun, Z.; Yan, S.H.; Li, Y.H. Wettability Impact on Nanoconfined Methane Adsorption Behavior: A Simplified Local Density Investigation. *Energy Fuels* **2022**, *36*, 14204–14219. [CrossRef]

7. Devegowda, D.; Sapmanee, K.; Civan, F.; Sigal, R. Phase Behavior of Gas Condensates in Shales Due to Pore Proximity Effects: Implications for Transport, Reserves and Well Productivity. In Proceedings of the SPE Annual Technical Conference and Exhibition, San Antonio, TX, USA, 8–10 October 2012. [CrossRef]

8. Beal, C. The viscosity of air, water, natural gas, crude oil and its associated gases at oil field temperatures and pressures. *Trans. AIME* **1946**, *165*, 94–115. [CrossRef]

9. Jennings, H.Y., Jr.; Newman, G.H. The effect of temperature and pressure on the interfacial tension of water against methanenormal decane mixtures. *SPEJ Soc. Pet. Eng. J.* **1971**, *11*, 171–175. [CrossRef]

0. Moh, D.Y.; Zhang, H.W.; Sun, S.Y.; Qiao, R. Molecular anatomy and macroscopic behavior of oil extraction from nanopores by CO_2 and CH_4. *Fuel* **2022**, *324*, 124662. [CrossRef]

1. Fang, T.M.; Li, S.J.; Zhang, Y.N.; Su, Y.L.; Yan, Y.G.; Zhang, J. How the oil recovery in deep oil reservoirs is affected by injected gas types: A molecular dynamics simulation study. *Chem. Eng. Sci.* **2021**, *231*, 116286. [CrossRef]

2. Zhang, J.; Dong, Z.H.; Wang, M.H.; Yan, Y.G. Effects of the Methane Content on the Water−Oil Interface: Insights from the Molecular Level. *Energy Fuels* **2017**, *31*, 7026–7032. [CrossRef]

3. Wu, K.; Chen, Z.; Li, J.; Li, X.; Xu, J.; Dong, X. Wettability effect on nanoconfined water flow. *Proc. Natl. Acad. Sci. USA* **2017**, *114*, 3358–3363. [CrossRef]

4. Akkutlu, I.Y.; Fathi, E. Multiscale gas transport in shales with local kerogen heterogeneities. *SPE J.* **2012**, *17*, 1002–1011. [CrossRef]

5. Yang, T.; Li, X.; Zhang, D. Quantitative dynamic analysis of gas desorption contribution to production in shale gas reservoirs. *J. Unconv. Oil Gas Resour.* **2015**, *9*, 18–30. [CrossRef]

6. Wang, S.; Feng, Q.H.; Javadpour, F.; Yang, Y.B. Breakdown of Fast Mass Transport of Methane through Calcite Nanopores. *J. Phys. Chem. C* **2016**, *120*, 14260–14269. [CrossRef]

7. Xiong, C.M.; Li, S.J.; Ding, B.; Geng, X.F.; Zhang, J.; Yan, Y.G. Molecular insight into the oil displacement mechanism of gas flooding in deep oil reservoir. *Chem. Phys. Lett.* **2021**, *783*, 139044. [CrossRef]

8. Zhang, Y.N.; Fang, T.M.; Ding, B.; Wang, W.D.; Yan, Y.G.; Li, Z.; Guo, W.Y.; Zhang, J. Migration of oil/methane mixture in shale inorganic nano-pore throat: A molecular dynamics simulation study. *J. Petrol. Sci. Eng.* **2020**, *187*, 106784. [CrossRef]

9. Wang, S.; Javadpour, F.; Feng, Q.H. Molecular dynamics simulations of oil transport through inorganic nanopores in shale. *Fuel* **2016**, *171*, 74–86. [CrossRef]

0. Liang, S.X.; Fang, T.M.; Xiong, W.; Ding, B.; Yan, Y.G.; Zhang, J. Oil detachment by modified nanoparticles: A molecular dynamics simulation study. *Comput. Mater. Sci.* **2019**, *170*, 109177. [CrossRef]

1. Qin, Y.; Yang, X.N.; Zhu, Y.F.; Ping, J.L. Molecular Dynamics Simulation of Interaction between Supercritical CO_2 Fluid and Modified Silica Surfaces. *J. Phys. Chem. C* **2008**, *112*, 12815–12824. [CrossRef]

2. Lavoie, R.G.; Ouellet, M.; Hamelin, J.; Benard, P. Numerical Implementation of the Multicomponent Potential Theory of Adsorption in Python Using the NIST Refprop Database. *Commun. Comput. Phys.* **2018**, *23*, 1602–1625. [CrossRef]

3. Berendsen, H.J.C.; Grigera, J.R.; Straatsma, T.P. The missing term in effective pair potentials. *J. Phys. Chem.* **1987**, *91*, 6269–6271. [CrossRef]

4. Dauber-Osguthorpe, P.; Roberts, V.A.; Osguthorpe, D.J.; Wolff, J.; Genest, M.; Hagler, A.T. Structure and energetics of ligand binding to proteins: Escherichia coli dihydrofolate reductase-trimethoprim, a drug-receptor system. *Proteins Struct. Funct. Genet.* **1988**, *4*, 31–47. [CrossRef] [PubMed]

5. Forte, G.; Grassi, A.; Marletta, G. Molecular Modeling of Oligopeptide Adsorption onto Functionalized Quartz Surfaces. *J. Phys. Chem. B* **2007**, *111*, 11237–11243. [CrossRef]

36. Jorgensen, W.L.; Maxwell, D.S.; Tirado-Rives, J. Development and testing of the OPLS all-atom force field on conformationa energetics and properties of organic liquids. *J. Am. Chem. Soc.* **1996**, *118*, 11225–11236. [CrossRef]
37. Plimpton, S. Fast parallel algorithms for short-range molecular dynamics. *J. Comput. Phys.* **1995**, *117*, 1–19. [CrossRef]
38. Nosé, S. A molecular dynamics method for simulations in the canonical ensemble. *Mol. Phys.* **1984**, *52*, 255–268. [CrossRef]
39. Hockney, R.W.; Eastwood, J.W. *Computer Simulation Using Particles*; CRC Press: Boca Raton, FL, USA, 1988. [CrossRef]
40. Humphrey, W.; Dalke, A.; Schulten, K. VMD: Visual molecular dynamics. *J. Mol. Graph. Model.* **1996**, *14*, 33–38. [CrossRef]
41. Izrailev, S.; Stepaniants, S.; Isralewitz, B.; Kosztin, D.; Lu, H.; Molnar, F.; Wriggers, W.; Schulten, K. *Steered Molecular Dynamic Computational Molecular Dynamics: Challenges, Methods, Ideas*; Springer: Berlin/Heidelberg, Germany, 1999; pp. 39–65.
42. Chen, W.; Foster, A.S.; Alava, M.J.; Laurson, L. Stick-slip control in nanoscale boundary lubrication by surface wettability. *Phy Rev. Lett.* **2015**, *114*, 095502. [CrossRef]
43. Castro, M.A.; Clarke, S.M.; Inaba, A.; Arnold, T.; Thomas, R.K. Competitive adsorption of simple linear alkane mixtures ont graphite. *J. Phys. Chem. B* **1998**, *102*, 10528–10534. [CrossRef]
44. Harrison, A.; Cracknell, R.F.; Krueger-Venus, J.; Sarkisov, L. Branched versus linear alkane adsorption in carbonaceous slit pore *Adsorption* **2014**, *20*, 427–437. [CrossRef]
45. Dong, H.; Zhu, Q.Y.; Wang, L.; Yue, X.K.; Fang, H.X.; Wang, Z.J.; Liu, S.Y.; Wei, S.X.; Lu, X.Q. Effects of Shale Pore Size an Connectivity on $scCO_2$ Enhanced Oil Recovery: A Molecular Dynamics Simulation Investigation. *Langmuir* **2023**, *39*, 6287–629 [CrossRef] [PubMed]
46. Sui, H.G.; Zhang, F.Y.; Wang, Z.Q.; Wang, D.S.; Wang, Y.D. Molecular simulations of oil adsorption and transport behavior i inorganic shale. *J. Mol. Liq.* **2020**, *305*, 112745. [CrossRef]
47. Hao, X.S.; Jia, Z.H.; Zhang, L.L.; Wang, X.; Zhang, J.; Zhou, L.X.; Yan, Y.G. Molecular Insight into the Methane Occurrence insid a Shale Nanochannel with Formation Water. *Energy Fuel* **2023**, *37*, 1207–1215. [CrossRef]
48. Wang, X.; Zhang, Z.; Zhang, J.; He, J. Insight into the pressure-induced displacement mechanism for selecting efficient nanofluid in various capillaries. *Environ. Sci. Nano* **2020**, *7*, 2785. [CrossRef]
49. Bosanquet, C.H. On the flow of liquids into capillary tubes. *Philos. Mag. Ser. 6* **1923**, *45*, 525–531. [CrossRef]
50. Oyarzua, E.; Walther, J.H.; Mejia, A.; Zambrano, H.A. Early regimes of water capillary flow in slit silica nanochannels. *Phys. Chem Chem. Phys.* **2015**, *17*, 14731–14739. [CrossRef]
51. Smith, W.O.; Crane, M.D. The Jamin effect in cylindrical tubes. *J. Am. Chem. Soc.* **1930**, *52*, 1345–1349. [CrossRef]
52. Liang, M.; Yang, S.; Miao, T.; Yu, B. Minimum applied pressure for a drop through an abruptly constricted capillary. *Microflui Nanofluidics* **2015**, *19*, 1–8. [CrossRef]
53. Jang, S.S.; Lin, S.T.; Maiti, P.K.; Blanco, M.; Goddard, W.A.; Shuler, P.; Tang, Y.C. Molecular Dynamics Study of a Surfactan Mediated Decane-Water Interface: Effect of Molecular Architecture of Alkyl Benzene Sulfonate. *J. Phys. Chem. B* **2004**, *10* 12130–12140. [CrossRef]
54. Ndao, M.; Goujon, F.; Ghoufi, A.; Patrice, M. Coarse-grained modeling of the oil-water-surfactant interface through the loca definition of the pressure tensor and interfacial tension. *Theor. Chem. Acc.* **2017**, *136*, 21. [CrossRef]
55. Kunieda, M.; Nakaoka, K.; Liang, Y.; Miranda, C.; Ueda, A.; Takahashi, S.; Okabe, H.; Matsuoka, T. Self-Accumulation c Aromatics at the Oil-Water Interface through Weak Hydrogen Bonding. *J. Am. Chem. Soc.* **2010**, *132*, 18281–18286. [CrossRef] [PubMed]

processes MDPI

Article

Transport Behavior of Methane Confined in Nanoscale Porous Media: Impact of Pore Evolution Characteristics

Shan Wu [1,2], Sidong Fang [1], Liang Ji [3], Feng Wen [4], Zheng Sun [1,5,*], Shuhui Yan [5] and Yaohui Li [5]

1 State Key Laboratory of Shale Oil and Gas Enrichment Mechanism and Effective Development, Beijing 100101, China
2 School of Chemical Engineering and Technology, China University of Mining and Technology, Xuzhou 221116, China
3 PetroChina Coalbed Methane Company Limited, Beijing 100028, China
4 China Oil & Gas Pipeline Network Corporation, Beijing 100007, China
5 School of Mining, China University of Mining and Technology, Xuzhou 221116, China
* Correspondence: sunzheng@cumt.edu.cn; Tel.: +86-185-0061-0758

Abstract: As a key technical aspect contributing to shale gas development, nanoconfined methane flow behavior has received tremendous research interest, which remains challenging to understand clearly. The majority of previous contributions put emphasis on the mechanism model for methane confined in a single nanopore; at the same time, the other part focusing on an upscaling approach fails to capture the spatial pore-network characteristics as well as the way to assign pressure conditions to methane flow behavior. In light of the current knowledge gap, pore-network modeling is performed, in which a pore coordination number, indicating the maximum pores a specified pore can connect, gas flow regimes classified by Knudsen numbers, as well as different assigned pressure conditions, are incorporated. Notably, the pore-network modeling is completely self-coded, which is more flexible in adjusting the spatial features of a constructed pore network than a traditional one. In this paper, the nanoconfined methane flow behavior is elaborated first, then the pore network modeling method based on the mass conservation principle is introduced for upscaling, and in-depth analysis is implemented after that. Results show that (a) as for porous media with pore sizes ranging from 5~80 nm, dramatic advancement on apparent gas permeability takes place while pressure is less than 1 MPa; (b) apparent gas permeability evaluated at a specified pressure shall be underestimated by as much as 31.1% on average compared with that under the pressure-difference condition; (c) both a large pore size and a high coordination number are beneficial for strong gas flow capacity through nanoscale porous media, and the rising ratio can reach about 6 times by altering the coordination number from 3 to 7, which is quantified and presented for the first time.

Keywords: nanoconfined methane; slip; pore network modeling; coordination number; coupling

Citation: Wu, S.; Fang, S.; Ji, L.; Wen, F.; Sun, Z.; Yan, S.; Li, Y. Transport Behavior of Methane Confined in Nanoscale Porous Media: Impact of Pore Evolution Characteristics. *Processes* **2022**, *10*, 2746. https://doi.org/10.3390/pr10122746

Academic Editor: Albert Ratner

Received: 16 November 2022
Accepted: 16 December 2022
Published: 19 December 2022

1. Introduction

The development of unconventional gas reservoirs, such as shale gas, coal seam gas, and tight gas, is a feasible approach to reduce coal combustion as well as CO_2 emissions [1,2]. As reported by the EIA at the end of 2020, the geological reserve of shale gas reaches 317.76 trillion cubic feet, accounting for 70.68% of total proved natural gas reserves in the United States. As a result, both the petroleum industry and the ecological environment will benefit from the successful exploitation of shale gas [3,4]. With the combination of hydraulic fracturing technology and multi-lateral horizontal wells [5,6], gas production performance from ultra-tight shale formations with rich nanoscale pores has been surprisingly enhanced in the past few decades. At the same time, theoretical breakthroughs in shale gas production, particularly for methane confined in nanoscale pore space, are still challenging. Although profound knowledge of multi-scale gas flow physics in shale is the best guarantee for shale gas development, upscaling the method from nanoconfined flow to that in porous media remains undeveloped [7,8], and flow behavior at the shale core scale is still vague.

Basically, existing research focusing on shale gas production can be classified into three types according to scale, suggesting the nanoscale, core scale, and reservoir scale. Massive research efforts are devoted to revealing the nanoconfined gas flow mechanism [9,10], which is fairly mature to date. In accordance with the Knudsen number, a parameter evaluating the gas sparsity under nanoconfinement, there are four different flow regimes, including continuum flow, slip flow, transition flow, and free molecular flow, in the order of the rising gas rarefaction effect. Naturally, a great deal of researchers proposed unified models, attempting to cover all the mentioned flow regimes. Insightfully, Song et al. (2016) proposed an apparent gas permeability model coupling bulk-gas flow and surface diffusion contributed by adsorption molecules [11]. The key correlation reproducing the above four flow regimes is referred to in Beskok and Karniadakis (1999) [12]. After that, Wu et al (2016) noticed the intermolecular collisions would be responsible for alteration of flow regimes [13], they developed a model for bulk gas transport, and its reliability has been well clarified by MD (molecular simulation) and experimental data. In addition, the model for bulk gas transport is further developed by considering inherent water distribution features [14,15], surface roughness [16], various gas types [17,18], and other influential factors [19,20]. After decades of development, a solid theoretical background for bulk gas transport at the nanoscale can be demonstrated. As for research on the reservoir scale, numerical simulation is the primary tool; however, the majority of which fail to couple dynamic shale matrix permeability due to the alteration of the four mentioned flow regimes Therefore, the prediction output from numerical simulation is hard to match with the actual production performance of realistic shale gas wells. Thus, the link between research on the nanoscale and that on the reservoir scale is significantly necessary, which is believed to be research on the core scale. The apparent permeability evaluated at the shale core scale can be utilized as the macroscopic value for the reservoir scale, which also couples microscopic flow mechanisms in each nanopore constructing the core porous media. In this work, a self-coded pore network modeling framework is presented for the first time sharing the advantage of controlling every key aspect of the constructed pore network and formulas representing nanoconfined flow behavior, expecting to shed light on gas flow performance and capacity through ultra-tight porous media.

Actually, to reach characterization of gas flow at the core scale other than through pore network modeling, there are additional approaches such as the classic capillary bundle model [21] and the lattice Boltzmann model (LBM) [22]. Pore network modeling is a direct technological advancement against the classic capillary bundle. LBM is able to capture gas flow mechanisms from a microscopic perspective and can also represent gas flow behavior in a complex pore structure. However, the calculation of LBM upon gas flow in porous media is time-consuming, as it requires sufficient computational resources to handle gas flow complexity in each nanopore [23,24]. In contrast, pore network modeling incorporates the theoretical model for gas transport in nanopores, greatly accelerating the calculation efficiency. Song et al. (2017) utilized pore network modeling to study gas transport behavior and identify the contributions of each flow mechanism; its reliability was verified in combination with LBM results [25]. Yang et al. (2019) extended the applicability of pore network modeling from the gas phase to the oil phase, considering oil positive slip in shale organic nanopores and negative slip in inorganic nanopores [26]. Wang and Sheng (2019) simulated gas-water two-phase flow at the shale core scale [27], in which the organic and inorganic parts are distinguished and generated by a newly proposed stochastic algorithm. Cui et al. (2019) investigated oil-water flow in nanoscale porous media [28], caused by fracturing water invasion, and a computational fluid dynamics (CFD) method was applied to quantify pore cross-section on fluid flow flux. Yi et al. (2021 performed an in-depth analysis on the way to extract the pore network from the shale sample [29], presenting the feasibility of the apparent pore network (APN) over the intrinsic pore network (IPN) after simulating gas transport in nanoscale porous media. A brief summary of the main current research related to gas-phase PNM is provided in Table 1.

Table 1. Literature review of recent main research about gas phase PNM.

Researchers	Purposes	Key Influential Factors	Limitations
Song et al. (2017) [25]	Gas transport behavior in porous media constructed by shale organic nanopores.	microscopic gas flow regimes.	Impact of spatial structure of pore network is not elaborated.
Yang et al. (2019) [26]	Oil transport behavior in shale porous media.	microscopic oil flow mechanism and volumetric toc.	The way pressure condition is assigned on flow has not been revealed.
Wang and Sheng (2019) [27]	Gas-water flow in shale porous media.	nanopore tortuosity, pore size distribution of organic matter and inorganic matter.	Impact of spatial structure of pore network is not elaborated.
Cui et al. (2019) [28]	Oil-water flow in shale porous media.	nanopore shape, spatial structure of pore network.	The way pressure condition is assigned on flow has not been revealed.
Yi et al. (2021) [29]	Gas transport in nanoscale porous media.	microscopic gas flow mechanism, spatial structure of pore network.	The way pressure condition is assigned on flow has not been revealed.

Based on the above analysis, although current research on PNM has achieved fairly accurate prediction of fluid flow in nanoscale porous media, several key shortcomings remain to be overcome. First, the effect of the spatial structure of the pore network, such as the pore size distribution (PSD) and coordination number, on the gas flow behavior has not been investigated completely. Furthermore, the most common method for determining fluid flow capacity is to assign a typical pressure with a negligible pressure difference, while the presence of a pressure difference during the depressurization development process is obvious. As a result, the frequently utilized way to assign pressure condition is different from that in field condition, while the associated impact resulting from the difference is unclear. In light of current research updates, research is performed to try to fill the above-mentioned knowledge gap. It should be underlined that the pore network modeling in this work is totally self-coded, with great flexibility to alter the spatial structure of the pore network, which facilitates identifying the influences of each parameter. Both the PSD and the coordination number of gas transport in nanoscale porous media are investigated. Moreover, the way of assigning pressure conditions is discussed, and a detailed analysis is provided. Therefore, the novel technical part of this research basically resides in the incorporation of spatial pore-network structure and sensible pressure conditions for realistic field exploitation.

The whole paper context is arranged in the following way. The physical model making clear the essential advantage of pore network modeling over the traditional mechanism model or capillary bundle model is presented first. Then, reliable formulas reproducing the nanoconfined gas flow behavior are introduced, and the primary concepts establishing the pore network modeling are briefly provided, including the part to generate concrete spatial structure and that for flow capacity calculation. After that, gas transport through nanoscale porous media is seriously discussed. Finally, several conclusions are drawn.

2. Physical Model

In order to reach a good understanding of gas flow in the shale matrix, which is ultra tight and rich in nanopores with pore sizes ranging from several nanometers to hundreds of nanometers, typical procedures to achieve the upscaling objective are presented in Figure 1. Generally, research efforts start with the primary mechanism model, representing the gas flow behavior in a single nanopore. As for this technical part, molecular dynamics (MD), theoretical modeling, as well as advanced experimental measurement, are used to investigate the nanoconfined gas transport behavior. With the intention to bridge nanoconfined flow and reservoir simulation, it entails an effective tool to transition the mechanism model to that suitable for shale core scale, which can be applied as reservoi permeability. At the very beginning, the classic capillary bundle model is utilized, consisting of a certain number of nanopores with various pore radii, pore lengths, and tortuosities. Although the capillary bundle model is a fairly direct upscaling method, there is an evident difference between the capillary bundle model and the realistic shale matrix, limiting it accuracy on gas flow in porous media. In the capillary bundle model, gas flux in a specifi nanopore does not affect that in neighboring nanopores, which seriously conflicts with the actual case. In light of this, pore network modeling is developed, which generally include two parts. The first part is the construction of the pore network, which is a spatial structure composed of massive nanopores with assigned pore network information, including pore size distribution, coordination number, and so on. Then, the second part is the simulation of gas flow in porous media, which is inherently dependent on the mechanism mode and the mass conservation principle. From the perspective of flow physics, pore networ modeling is evidently superior to the capillary bundle model, which is able to capture ga flux exchange between nanopores and has a more complicated spatial structure.

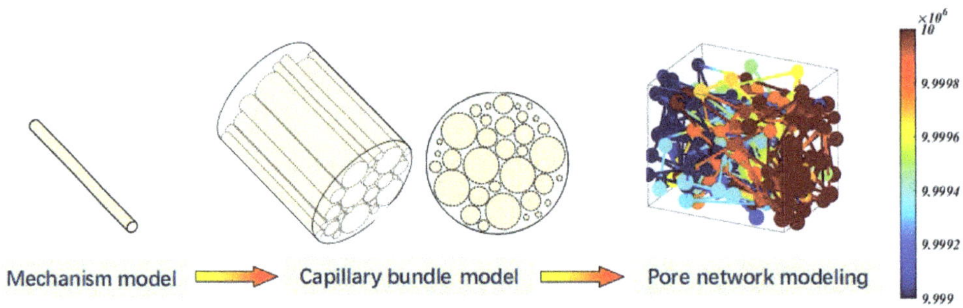

Figure 1. Mimicking methane flow behavior with different approaches.

In this work, self-coded pore network modeling is utilized so that pore size distribution coordination number, and skeleton porosity can be accurately and conveniently controlled Moreover, microscopic flow physics is fully coupled in this pore network modeling, and gas mass flux between nanopores is considered to ensure the mass conservation of the whole system by an efficient algorithm. Notably, the gas flow capacity through porous media is evaluated in a steady state, which is suitable to describe gas permeability in shal matrix during the long-term depressurization development process. In the modeling c this research, the calculation node, representing pores, is over 200, and the number c connecting throats is over 300, which mitigate the uncertainty caused by a discontinuou medium and ensure the reliability of the obtained results.

3. Model Establishment

3.1. Nanoconfined Flow Behavior

The mathematic part of this research starts with the characterization of nanoconfine gas flow behavior, which is apparently different from regular fluid flow described b the classic Navier–Stokes equation [30]. The origin of the difference is caused by th nanoconfinement effect induced by the greatly reduced pore size. From the microscopic viewpoint, the molecular flow velocity near the wall is close to 0 as in the case of bulk scale, and its movability is considered negligible. In contrast, when collisions between gas molecules and the wall surface become dominant, molecular slip takes place at the boundary, manifesting that the movability of molecules near the wall cannot be neglected. It is experimentally reported that nanoconfined flow capacity can reach as great as several orders of magnitude as evaluated by the no-slip equation [31,32].

A number of theoretical models have been proposed to elucidate the discrepancy compared with regular equations for gas flow, and four flow regimes are summarized, consisting of continuum flow, slip flow, transition flow, and free molecular diffusion. The flow regime is heavily dependent on the surrounding pressure and temperature conditions, which are identified by the Knudsen number, with the following formula:

$$Kn = \frac{\lambda}{r} \tag{1}$$

$$\lambda = \frac{\mu}{p}\sqrt{\frac{\pi ZRT}{2M}} \tag{2}$$

where Kn is the Knudsen number, dimensionless; λ is the mean free path, representing the distance a gas molecule can move between two adjacent collision events, its calculation formula is Equation (2), m; r is the pore radius, m; μ is gas viscosity, cp; p is the pressure, MPa; R is the gas universal constant, which is equal to 8.314 J/mol/K; Z is the gas compressibility factor, dimensionless; T is the temperature, K; M is the gas molar mass, 16 g/mol.

After carefully checking existing models developed for nanoconfined gas flow, the primary constitutive equation for the majority of current models describing bulk-gas flow in nanopores is Beskok and Karniadakis (1999), covering the entire Knudsen number range. Thus, the reliable model is adopted in this work, notably the stress dependence and surface diffusion of the adsorption phase, which are not considered currently but can be incorporated and discussed in the future. Concrete formulas for gas transport in nanopores considering gas slip are provided below.

$$K = \frac{r^2}{8}(1+\alpha Kn)(1+\frac{4Kn}{1+Kn}) \tag{3}$$

$$\alpha = \alpha_o \frac{2}{\pi}\tan^{-1}(\alpha_1 Kn^{\beta}) \tag{4}$$

where α is the rarefaction coefficient, dimensionless; α_o is the rarefaction coefficient with infinite Kn, dimensionless; α_1 and β are constants, which are 1.19 and 0.4, respectively.

On the basis of Equations (1)–(4), the apparent permeability for gas transport through a nanopore can be calculated. Furthermore, in order to implement the conservation principle of gas mass rate, it is crucial to adopt a suitable approach to obtaining gas physical properties, such as viscosity and density. The classic PVT equation [33,34] is provided below as a beginning point.

$$P = \frac{RT}{v-b} - \frac{a}{v(v+b)+b(v-b)} \tag{5}$$

where v is the molar volume, m^3/mol.

Notably, a and b are basic PVT parameters, which are described as a function of critical pressure and critical temperature in the bulk state previously. The authors have noticed variations in gas critical properties under the nanoconfinement effect, also its effect on gas density can result in errors as high as 23% [35]. Thus, the shift of critical properties shall be incorporated in the calculation of gas density. Formulas describing a and b are below.

$$a = 0.45724\frac{R^2T_c^2}{P_c}[1+\kappa(1-\sqrt{T_r})] \tag{6}$$

$$b = 0.0778\frac{RT_c}{P_c} \tag{7}$$

$$\kappa = 0.37464 + 1.54226\omega - 0.26992\omega^2 \tag{8}$$

where P_c, T_c are the critical nanoconfined gas pressure and temperature, concrete formulas suggesting the shift compared with the original critical properties may refer to our previous research. T_r is the reduced temperature, dimensionless, and ω is the methane acentric factor, dimensionless. Then, the gas density can be sensibly obtained with the given pressure and temperature conditions.

After calculating sensible gas density under the nanoconfinement effect, the formulas below are available for gas viscosity [36,37], which is described as a function of gas density.

$$\mu = (1 \times 10^{-4})\xi \exp(X\rho^Y) \tag{9}$$

$$\xi = \frac{(22.65 + 0.0388)T^{1.5}}{209.2 + 19.26M + 1.8T} \tag{10}$$

$$X = 3.448 + 548 + 0.01M \tag{11}$$

$$Y = 2.447 - 0.224X \tag{12}$$

where ρ is gas density, kg/m^3; μ is gas viscosity, cp.

Then, for a single nanopore with gas transport, the mass flow rate has the following expression.

$$q = \frac{\rho\pi r^4(1 + \alpha Kn)(1 + \frac{4Kn}{1+Kn})\Delta p}{8\mu l} \tag{13}$$

where Δp is the pressure difference between the two ends of the nanopore, Pa; l is the length of nanopore, m; q is the mass flow rate, kg/s.

3.2. Pore Network Modeling

First, it is necessary to elaborate pore network construction, in which numerous nanopores with various pore sizes and lengths are connected; the number of nanopores exceeds 200 to mitigate the uncertainty caused by discontinuous medium. Notably, pore size distribution, porosity, and coordination of the ultimate pore network are controlled by a powerful algorithm, ensuring that the generated spatial parameters can have excellent matches against the assigned targets. At the beginning of construction, an assigned number of nodes, suggesting the pores in a tight matrix, are placed randomly in a specific space Then, the interpore connections take place, suggesting that the nodes find their neighboring nodes and try to connect them at a certain distance. At the same time, the formation of throats is subject to a coordination number, which indicates the average number of throats a node should connect. After that, the basic pore network is established, and a fine algorithm shall complete the assignments of pore size to all throats and pores.

The inherent physics driving proper modeling is the mass conservation of gas flow rates. As for the case of i-th node in the pore network, it may have several connections, like n, indicating that the inflow and outflow of the i-th node must happen in the n connections Furthermore, for a stable state, the mass flow rates from inflow and outflow should be equal after the whole balance is reached. On the basis of Equation (13), which describes the mass flow rate in a single connection, then the summation of the mass flow rates from n connections has the following expression.

$$\sum_{i=1}^{i=n} q_i = 0 \tag{14}$$

Once the pressure of two connected nodes is given, the mass flow rate between them can be calculated, in which gas density and viscosity are evaluated at the arithmetic mean pressure of the two connected nodes. Notably, for the whole pore network, where the total number of nodes is m, as the pressure at each node is unknown from the beginning of the calculation, there exist m unknown pressure values. Moreover, for each node, there exists Equation (14) as a function of its unknown pressure; therefore, m equations exist for m unknown values, indicating the solutions of all nodes can be obtained properly.

After all the unknown pressures are calculated, the nodes connecting to the boundary nodes at the exit side of the whole pore network are identified, and the mass flow rates from these nodes are summarized and regarded as the total gas mass rate. On the basis of the total rate, the apparent permeability of the entire porous medium is obtained using the equation below [38–40], in which gas density and viscosity are evaluated at the mean pressures of inlet pressure and outlet pressure.

$$K_{app} = \frac{q_t \mu l}{\rho \pi r^2 \Delta p} \tag{15}$$

where q_t is the total gas mass rate through the pore network, kg/s; K_{app} is apparent gas permeability in the constructed pore network, nD.

4. Results and Discussion

In this research, the slip feature and spatial characteristics of nanoscale porous media on apparent gas permeability at the core scale are focused. According to a typical tight gas reservoir located in Xinjiang, China, the pore size in this work falls in the range from 5~80 nm, and the pressure ranges of 0.1~50 MPa [41], covering the entire pressure variation during actual production.

4.1. Rarefaction Phenomenon

The unique feature of nanoconfined flow is the molecular slip at the boundary; other than that, gas flow in nanoscale porous media is affected by network structure. In this section, we put attention on the slip at the boundary, also called the rarefaction phenomenon, the main mechanism contributing to the enhancement of gas flow capacity. The first step in performing pore network modeling is constructing a fancy network with desirable key parameters, like pore size distribution and coordination number. At this point, in order to remove the influence induced by network characteristics, relevant parameters are assigned in advance.

As presented in Figure 2a, from the comparison between the assigned pore size distribution and the actual one generated by the self-coded algorithm provided, it can be demonstrated the algorithm is able to reach a good agreement with the assigned goal. From Figure 2b, it can be observed that the pore size in the network ranges from 5 to 80 nm, also different pores connect in such a way that the coordination number is 3.

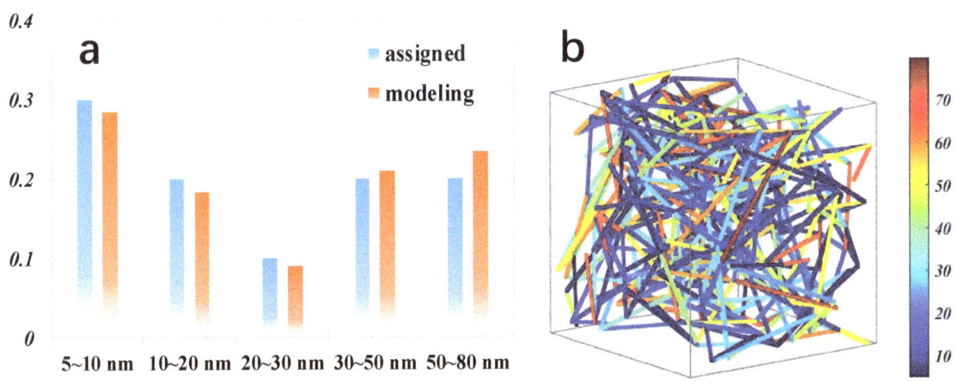

Figure 2. Construction of porous media with assigned spatial pore network attributes (**a**): Pore size distribution; (**b**): Spatial pore structure.

The gas flow through nanoscale porous media is simulated under the condition that the pressure difference between the inlet and outlet is set at 0.001 MPa. In this case, the gas flow capacity at a specified pressure condition can be mimicked, which makes it fairly clea to explore the rarefaction phenomenon with regards to pressure variation. In Figure 3a the red dotted line represents the apparent permeability for gas flow in porous medi without considering the slip boundary, which remains a constant value and is insensitive t pressure change as indicated. The phenomenon that the permeability remains unchanged i in line with the classic Darcy's law, suggesting that the permeability is an inherent propert of reservoirs and shall not change with pressure variation. After considering boundary sli taking place under the nanoconfinement effect, the traditional principle about unchange permeability breaks down. As illustrated in Figure 3a, the apparent permeability (blu line) shows an evident negative trend with the pressure rise; the apparent permeabilit can reach 8103 nD at 0.1 MPa while declining to 626 nD at 10 MPa, close to the valu without considering boundary slip. Firstly, at a low-pressure range, the rarefaction effec plays as a greatly positive role in enhancing the apparent permeability. The gas velocity i nanopores is greatly advanced due to gas slippage at boundary, in which the molecule close to the boundary gain a certain degree of mobility. Meanwhile, the sharp decrease i apparent permeability with increasing pressure indicates that high pressure will inhib the rarefaction effect greatly. As can be observed in Figure 3a, the apparent permeability a 50 MPa is exactly equal to the value from the classic theory, suggesting the gas slip vanishe at high pressure.

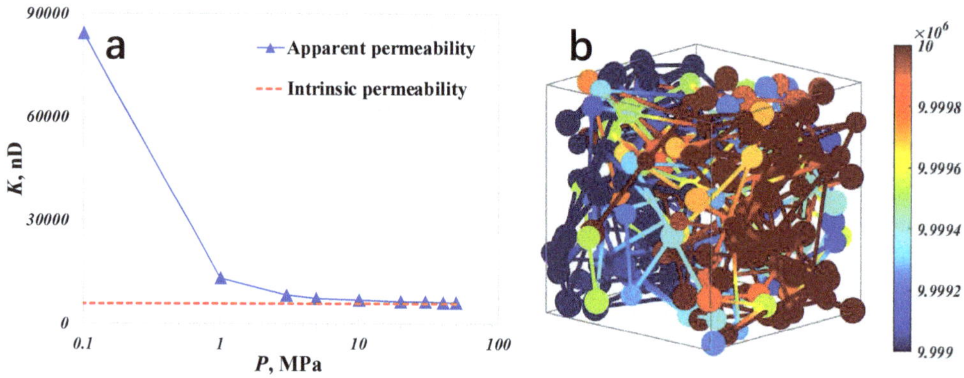

Figure 3. Apparent gas permeability under the specified pressure condition (**a**): Gas flow capacit (**b**): Spatial pressure distribution feature.

Notably, the assigned pressure condition of negligible difference between the inle and outlet pressure failed to fit the realistic pressure condition during the depressurizatio development process. The pressure difference may make a difference on gas flow capacit as nanoconfined gas flow is sensitive to pressure variation as indicated in Figure 3a. Th spatial pressure distribution with the given inlet pressure (10 MPa) and outlet pressu (9.999 MPa) is illustrated in Figure 3b. In order to address this issue, in the followin investigation, outlet pressure is assigned as a constant of 0.001 MPa, at the same tim the inlet pressure ranges from 0.1~50 MPa. As presented in Figure 4a, similar to tha under the specified pressure condition, apparent permeability rises with the pressur reduction, and the rise amplitude increases in the low-pressure condition. Compared wi the specified pressure condition that the gas flow regime in the whole network is almo the same, the large pressure difference in this case will result in several flow regimes in th network. For example, the continuum flow dominates the region close to the inlet side c the network, while slip flow or transition flow is the main flow mechanism for the regio close to the outlet side. As a result, the apparent gas permeability is controlled by multip flow regimes in the network in this pressure condition rather than a simple single flow regime with a negligible pressure difference.

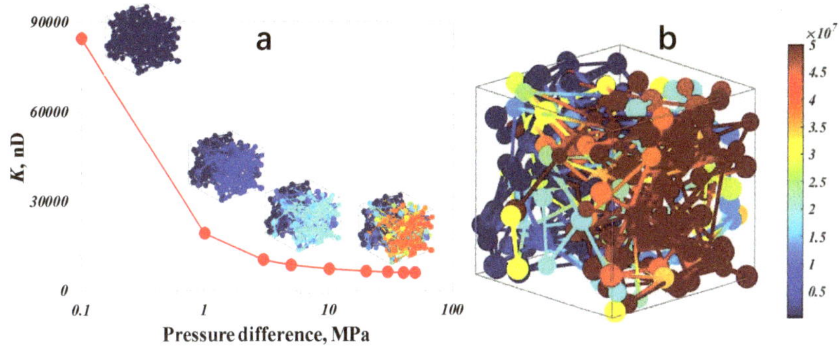

Figure 4. Apparent gas permeability under the displacement pressure-difference condition (**a**): Gas transport capacity; (**b**): Spatial pressure distribution feature.

To highlight the difference in gas flow capacity under both pressure conditions, a comparison is performed. The discrepancy, the vertical axis value in Figure 5, is calculated by the following equation, quantitatively displaying the difference in apparent permeability.

$$Dis = \frac{K_{app\_l} - K_{app\_n}}{K_{app\_n}} \tag{16}$$

where K_{app_l} is the apparent permeability through the pore network with a large pressure difference, nd; K_{app_n} is the apparent permeability with a negligible pressure difference, nD; *Dis* is the discrepancy.

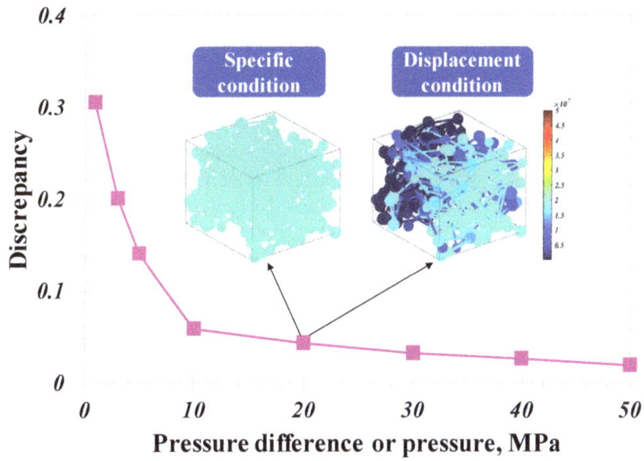

Figure 5. Discrepancy caused by different ways to assign pressure condition.

As indicated in Figure 5, there exists a negative relationship between pressure augment and discrepancy; the discrepancy can be as great as 31.1% at 1 MPa, which ranges from 2.8% to 31.1%. It means that the apparent permeability evaluated with the neglected pressure difference will underestimate the essential gas flow capacity through nanoscale porous media. Moreover, it can be observed from Figure 5 that the decline rate is greatly mitigated after pressure exceeds 10 MPa. Furthermore, the discrepancy can be overlooked when the pressure is over 30 MPa, in which case, the slip phenomenon is drastically weakened.

4.2. Pore-Network Characteristics

Various approaches have been applied to characterize microscopic pore structures, particularly for shale gas reservoirs or coalbed methane reservoirs, which is the inherent factor resulting in the nanoconfined gas flow. In this research, pore-network characteristics, including pore size distribution and coordination number, are studied. Specifically, as displayed in Figure 6, three sets of pore size distribution are provided, the case_1 symbol represents porous media dominated by small nanopores, the case_2 symbols represent the medium nanopores, and case_3 symbols represent the majority of large nanopores.

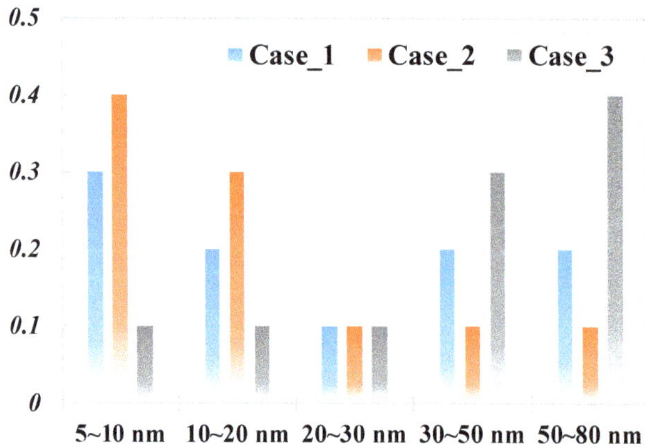

Figure 6. Different pore size distributions of the nanoscale porous media.

Then, the impact of pore size distribution on apparent gas permeability is illustrated in Figure 7. Similarly, the general variation trend remains that the apparent gas permeability will decline with increasing pressure, regardless of the alteration of pore size distribution In addition, at any pressure, apparent gas permeability with large pore size is always stronger than that with small pore size. The discrepancy caused by different pore size distributions has a negative correlation with increasing pressure. The containment effect of boundary slip mechanisms caused by increasing pressure may take responsibility for the above phenomenon. At low pressure, the slip phenomenon intensifies gas permeability more in small nanopores than in large nanopores. As a result, the difference between case_1 and case_2 increases due to the low-pressure conditions.

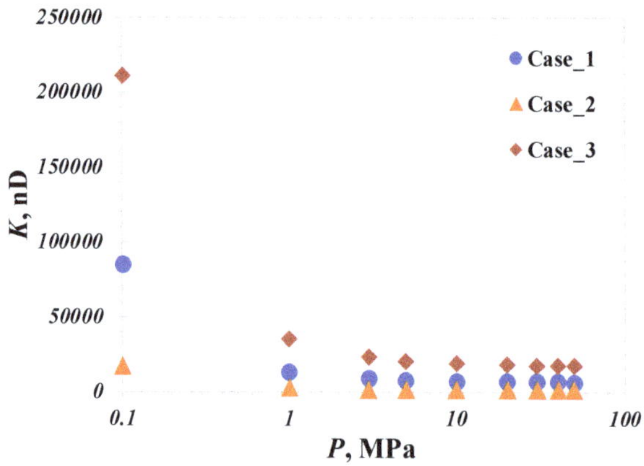

Figure 7. Impact of PSD on apparent gas permeability versus pressure.

The coordination number denotes the maximum number of pores a single pore can connect to in the pore network, which is generally believed to be in the range of 3 to 7. In this research, three sets of coordination numbers are assigned, and the pore size distribution remains unchanged on the basis of case_2, as presented in Figure 7. It can be observed from Figure 8 that the apparent permeability in the case of a coordination number of 7 can reach as high as 6164 nD. There is an evident tendency that apparent permeability rises dramatically with increasing coordination number. The increasing coordination number undoubtedly enhances the connectivity of the whole pore network; as a result, the possibility for the presence of isolated pores is greatly reduced, which can be attributed as a positive aspect for permeability enhancement. In addition, for a molecule flowing in the network with a large coordination number, there will be more passageways available, and the optimal one would be selected based on the minimum resistance principle.

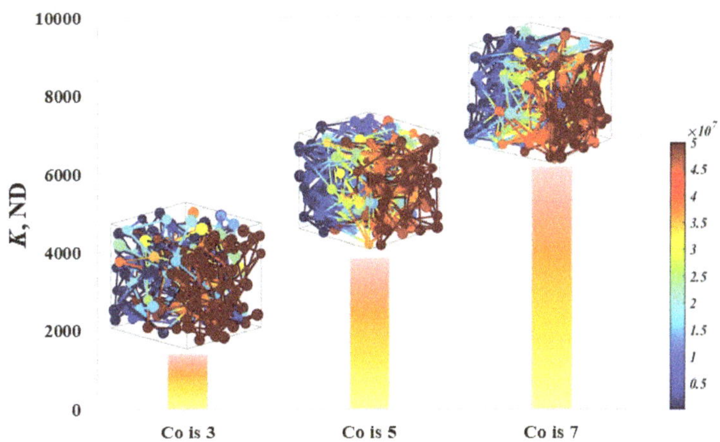

Figure 8. Impact of pore coordination number on apparent gas permeability.

5. Conclusions

(1) A self-coded pore network modeling platform for gas flow in nanoscale porous media is established, in which nanoconfined flow mechanisms covering the entire Knudsen number range and pore evolution features, such as pore size distribution and coordination number, are well incorporated. The platform is more flexible on adjusting key parameters compared with the existing tools for PNM.

(2) The rarefaction phenomenon acts as a positive factor in advancing gas flow capacity through porous media, particularly in the case of low pressure, where the apparent permeability shall sharply decline from 8103 nD to 598 nD after taking no account of the boundary slip. Moreover, the pressure difference across the inlet and outlet imposes a great influence on gas flow; the apparent permeability would be underestimated by as much as 31.1% with a neglected pressure difference.

(3) The variation trend for apparent gas permeability versus pressure remains regardless of the assigned different pore size distribution. Moreover, apparent gas permeability with a large pore size is always stronger than that with a small pore size, discrepancy caused by alteration of pore size distribution has a negative correlation with increasing pressure. Apparent gas permeability rises dramatically with the increasing coordination number, which can be attributed to the reduction of isolated pores as well as the optimized gas flow passageways triggered by the minimum resistance principle.

Author Contributions: Supervision: Z.S.; Methodology: S.W.; Investigation: S.F. and L.J.; Resources: F.W.; Formal analysis: S.Y. and Y.L. All authors have read and agreed to the published version of the manuscript.

Funding: The research was supported by the Open Fund of State Key Laboratory of Shale Oil and Gas Enrichment Mechanisms and Effective Development (No. 33550000-21-ZC0613-0337), National Natural Science Foundation Projects of China (No. 52104099), and Natural Science Foundation Projects of Jiangsu Province (No. BK20210508). The first author also acknowledges the Fundamental Research Funds for Central Universities (No. 2021QN1059) to support part of this work. We also acknowledge China University of Mining & Technology for the permission to publish this work.

Conflicts of Interest: The authors declare no conflict of interest.

References

1. Yu, P.; Dempsey, D.; Archer, R. A three-dimensional coupled thermo-hydro-mechanical numerical model with partially bridging multi-stage contact fractures in horizontal-well enhanced geothermal system. *Int. J. Rock Mech. Min. Sci.* **2021**, *143*, 10478. [CrossRef]
2. Wu, Y.; Zhu, Q.; Zhu, B. Decoupling analysis of world economic growth and CO_2 emissions: A study comparing developed and developing countries. *J. Clean. Prod.* **2018**, *190*, 94–103. [CrossRef]
3. Cao, M.; Sharma, M.M. The impact of changes in natural fracture fluid pressure on the creation of fracture networks. *J. Pet. Sci. Eng.* **2022**, *216*, 110783. [CrossRef]
4. Ma, D.; Duan, H.Y.; Zhang, Q.; Zhang, J.X.; Li, W.X.; Zhou, Z.L.; Liu, W.T. A numerical gas fracturing model of coupled thermal flowing and mechanical effects. *Comput. Mater. Contin.* **2020**, *65*, 2123–2141. [CrossRef]
5. Deng, H.; Sheng, G.; Zhao, H.; Meng, F.; Zhang, H.; Ma, J.; Gong, J.; Ruan, J. Integrated optimization of fracture parameters for subdivision cutting fractured horizontal wells in shale oil reservoirs. *J. Pet. Sci. Eng.* **2022**, *212*, 110205. [CrossRef]
6. Wang, Y.; Liu, H.; Guo, M.; Shen, X.; Han, B.; Zhou, Y. Image recognition model based on deep learning for remaining oil recognition from visualization experiment. *Fuel* **2021**, *291*, 120216. [CrossRef]
7. Sun, Z.; Wang, S.; Xiong, H.; Wu, K.; Shi, J. Optimal nanocone geometry for water flow. *AIChE J.* **2022**, *68*, e17543. [CrossRef]
8. Pang, Y.; Hu, X.; Wang, S.; Chen, S.; Soliman, M.Y.; Deng, H. Characterization of adsorption isotherm and density profile in cylindrical nanopores: Modeling and measurement. *Chem. Eng. J.* **2020**, *396*, 125212. [CrossRef]
9. Sun, Z.; Huang, B.; Yan, S.; Wang, S.; Wu, K.; Yu, W.; Li, Y.; Wang, S. Nanoconfined Methane Thermodynamic Behavior below Critical Temperature: Liquid–Vapor Coexistence Curve under Wettability Effect. *Ind. Eng. Chem. Res.* **2022**, *61*, 4971–497 [CrossRef]
10. Sheng, G.; Su, Y.; Wang, W. A new fractal approach for describing induced-fracture porosity/permeability/ compressibility in stimulated unconventional reservoirs. *J. Pet. Sci. Eng.* **2019**, *179*, 855–866. [CrossRef]
11. Song, W.; Yao, J.; Li, Y.; Sun, H.; Zhang, L.; Yang, Y.; Zhao, J.; Sui, H. Apparent gas permeability in an organic-rich shale reservoir. *Fuel* **2016**, *181*, 973–984. [CrossRef]
12. Beskok, A.; Karniadakis, G.E. Report: A model for flows in channels, pipes, and ducts at micro and nano scales. *Microscale Thermophys. Eng.* **1999**, *3*, 43–77.
13. Wu, K.; Chen, Z.; Li, X.; Guo, C.; Wei, M. A model for multiple transport mechanisms through nanopores of shale gas reservoirs with real gas effect–adsorption-mechanic coupling. *Int. J. Heat Mass Transf.* **2016**, *93*, 408–426. [CrossRef]
14. Li, J.; Li, X.; Wu, K.; Feng, D.; Zhang, T.; Zhang, Y. Thickness and stability of water film confined inside nanoslits and nanocapillaries of shale and clay. *Int. J. Coal Geol.* **2017**, *179*, 253–268. [CrossRef]
15. Li, J.; Li, X.; Wu, K.; Wang, X.; Shi, J.; Yang, L.; Zhang, H.; Sun, Z.; Wang, R.; Feng, D. Water sorption and distribution characteristics in clay and shale: Effect of surface force. *Energy Fuels* **2016**, *30*, 8863–8874. [CrossRef]
16. Zhang, T.; Javadpour, F.; Li, J.; Zhao, Y.; Zhang, L.; Li, X. Pore-scale perspective of gas/water two-phase flow in shale. *SPE J.* **2021**, *26*, 828–846. [CrossRef]
17. Feng, D.; Bakhshian, S.; Wu, K.; Song, Z.; Ren, B.; Li, J.; Hosseini, S.A.; Li, X. Wettability effects on phase behavior and interfacial tension in shale nanopores. *Fuel* **2021**, *290*, 119983. [CrossRef]
18. Feng, D.; Li, X.; Wang, X.; Li, J.; Zhang, T.; Sun, Z.; He, M.; Liu, Q.; Qin, J.; Han, S. Capillary filling of confined water in nanopores: Coupling the increased viscosity and slippage. *Chem. Eng. Sci.* **2018**, *186*, 228–239. [CrossRef]
19. Yu, P.; Dempsey, D.; Archer, R. Techno-Economic feasibility of enhanced geothermal systems (EGS) with partially bridging Multi-Stage fractures for district heating applications. *Energy Convers. Manag.* **2022**, *257*, 115405. [CrossRef]
20. Pang, Y.; Fan, D.; Chen, S. A Novel Approach To Predict Gas Flow in Entire Knudsen Number Regime through Nanochannels with Various Geometries. *SPE J.* **2021**, *26*, 3265–3284. [CrossRef]
21. Sun, Z.; Shi, J.; Wu, K.; Xu, B.; Zhang, T.; Chang, Y.; Li, X. Transport capacity of gas confined in nanoporous ultra-tight gas reservoirs with real gas effect and water storage mechanisms coupling. *Int. J. Heat Mass Transf.* **2018**, *126*, 1007–1018. [CrossRef]
22. Huang, J.; Yin, X.; Barrufet, M.; Killough, J. Lattice Boltzmann simulation of phase equilibrium of methane in nanopores under effects of adsorption. *Chem. Eng. J.* **2021**, *419*, 129625. [CrossRef]
23. Grunau, D.; Chen, S.; Eggert, K. A lattice Boltzmann model for multiphase fluid flows. *Phys. Fluids A: Fluid Dyn.* **1993**, 2557–2562. [CrossRef]
24. Saadat, M.H.; Hosseini, S.A.; Dorschner, B.; Karlin, I.V. Extended lattice Boltzmann model for gas dynamics. *Phys. Fluids* **2021**, 3 046104. [CrossRef]

25. Song, W.; Yao, J.; Ma, J.; Couples, G.; Li, Y. Assessing relative contributions of transport mechanisms and real gas properties to gas flow in nanoscale organic pores in shales by pore network modelling. *Int. J. Heat Mass Transf.* **2017**, *113*, 524–537. [CrossRef]
26. Yang, Y.; Wang, K.; Zhang, L.; Sun, H.; Zhang, K.; Ma, J. Pore-scale simulation of shale oil flow based on pore network model. *Fuel* **2019**, *251*, 683–692. [CrossRef]
27. Wang, X.; Sheng, J.J. Multi-scaled pore network modeling of gas-water flow in shale formations. *J. Pet. Sci. Eng.* **2019**, *177*, 899–908. [CrossRef]
28. Cui, R.; Feng, Q.; Chen, H.; Zhang, W.; Wang, S. Multiscale random pore network modeling of oil-water two-phase slip flow in shale matrix. *J. Pet. Sci. Eng.* **2019**, *175*, 46–59. [CrossRef]
29. Yi, Z.; Hu, S.; Wu, S.; Ma, J.; Gao, J.; Yuan, Y. Pore network extraction for shale gas flow in nanoporous media. *Mar. Pet. Geol.* **2021**, *126*, 104896. [CrossRef]
30. Sun, Z.; Huang, B.; Liu, Y.; Jiang, Y.; Zhang, Z.; Hou, M.; Li, Y. Gas-phase production equation for CBM reservoirs: Interaction between hydraulic fracturing and coal orthotropic feature. *J. Pet. Sci. Eng.* **2022**, *213*, 110428. [CrossRef]
31. Xu, Y.; Sheng, G.; Zhao, H.; Hui, Y.; Zhou, Y.; Ma, J.; Rao, X.; Zhong, X.; Gong, J. A new approach for gas-water flow simulation in multi-fractured horizontal wells of shale gas reservoirs. *J. Pet. Sci. Eng.* **2021**, *199*, 108292. [CrossRef]
32. Wang, Y.; Dai, Z.; Liu, H. Pore-scale mechanisms and simulations for gas–water two-phase transport processes in natural gas reservoirs. *J. Nat. Gas Sci. Eng.* **2021**, *96*, 104314. [CrossRef]
33. Wu, Y.; Cheng, L.; Ma, L.; Huang, S.; Fang, S.; Killough, J.; Jia, P.; Wang, S. Transient two-phase flow model for production prediction of tight gas wells with fracturing fluid-induced formation damage. *J. Pet. Sci. Eng.* **2021**, *199*, 108351. [CrossRef]
34. Ma, D.; Zhang, J.X.; Duan, H.Y.; Huang, Y.L.; Li, M.; Sun, Q.; Zhou, N. Reutilization of gangue wastes in underground backfilling mining: Overburden aquifer protection. *Chemosphere* **2021**, *264*, 128400. [CrossRef] [PubMed]
35. Sun, Z.; Huang, B.; Wu, K.; Shi, S.; Wu, Z.; Hou, M.; Wang, H. Nanoconfined methane density over pressure and temperature: Wettability effect. *J. Nat. Gas Sci. Eng.* **2022**, *99*, 104426. [CrossRef]
36. Cao, M.; Hirose, S.; Sharma, M.M. Factors controlling the formation of complex fracture networks in naturally fractured geothermal reservoirs. *J. Pet. Sci. Eng.* **2022**, *208*, 109642. [CrossRef]
37. Shen, W.J.; Li, X.Z.; Ma, T.R.; Cai, J.C.; Lu, X.B.; Zhou, S.W. High-pressure methane adsorption behavior on deep shales: Experiments and modeling. *Phys. Fluids* **2021**, *33*, 063103. [CrossRef]
38. Zhao, W.; Jia, C.; Zhang, T.; Jiang, L.; Li, X.; Jiang, Z.; Zhang, F. Effects of nanopore geometry on confined water flow: A view of lattice Boltzmann simulation. *Chem. Eng. Sci.* **2021**, *230*, 116183. [CrossRef]
39. Wang, S.; Liang, Y.; Feng, Q.; Javadpour, F. Sticky layers affect oil transport through the nanopores of realistic shale kerogen. *Fuel* **2022**, *310*, 122480. [CrossRef]
40. Wang, S.; Yao, X.; Feng, Q.; Javadpour, F.; Yang, Y.; Xue, Q.; Li, X. Molecular insights into carbon dioxide enhanced multi-component shale gas recovery and its sequestration in realistic kerogen. *Chem. Eng. J.* **2021**, *425*, 130292. [CrossRef]
41. Wang, S.; Feng, Q.; Javadpour, F.; Zha, M.; Cui, R. Multiscale modeling of gas transport in shale matrix: An integrated study of molecular dynamics and rigid-pore-network model. *SPE J.* **2020**, *25*, 1416–1442. [CrossRef]

processes

MDPI

Article

Unusual Water Flow in Ultra-Tight Porous Media: Integration of Profession and Innovation

Yanglu Wan [1], Na Niu [2,*], Wei Lu [1], Yushuang Zhou [3], Bin Wang [3] and Shan Lu [1]

[1] School of Innovation and Entrepreneurship, Wuchang University of Technology, Wuhan 430200, China
[2] Southwest Oil and Gas Company, SINOPEC, Chengdu 610041, China
[3] Wuxi Research Institute of Petroleum Geology, SINOPEC, Wuxi 214126, China
* Correspondence: wutcxcy@163.com

Abstract: Hydraulic fracturing is an effective method for stimulating reservoirs, making the economic development of ultra-tight shale gas and coalbed methane reservoirs possible. These formations are rich in nanopores, in which the fracturing fluid, such as fresh water, the flow, and the behavior of this flow differ significantly from those described in the classic Navier-Stokes formula. In bulk space, the interaction force exerted by the solid phase can be ignored, but the solid–fluid interaction plays a dominant role in nanoconfinement spaces in which the pore size is comparable to the molecular diameter. Nanoconfined water molecules tend to approach the water-wet pore surface, enhancing the water viscosity, which is a key parameter affecting the water flow capacity. Conversely, water molecules tend to stay in the middle of nanopores when subjected to a hydrophobic surface, leading to a decrease in viscosity. Thus, nanoconfined water viscosity is a function of the strength of the surface–fluid interaction, rather than a constant parameter, in classic theory. However, the influence of varying the viscosity on the nanoscale water flow behavior is still not fully understood. In this research, we incorporate wettability-dependent viscosity into a pore network modeling framework for stable flow for the first time. Our results show that: (a) the increase in viscosity under hydrophilic nanoconfinement could reduce the water flow capacity by as much as 11.3%; (b) the boundary slip is the primary mechanism for boosting the water flow in hydrophobic nanopores, as opposed to the slight enhancement contributed by a viscosity decline; and (c) water flow characterization in nanoscale porous media must consider both the pore size and surface wettability. Revealing the varying viscosity of water flow confined in nanopores can advance our microscopic understanding of water behavior and lay a solid theoretical foundation for fracturing-water invasion or flowback simulation.

Keywords: viscosity; nanoconfinement; wettability; pore network; simulation

Citation: Wan, Y.; Niu, N.; Lu, W.; Zhou, Y.; Wang, B.; Lu, S. Unusual Water Flow in Ultra-Tight Porous Media: Integration of Profession and Innovation. *Processes* **2023**, *11*, 1245. https://doi.org/10.3390/pr11041245

Academic Editors: Tao Zhang, Zheng Sun, Dong Feng, Hung Vo Thanh and Wen Zhao

Received: 28 March 2023
Revised: 13 April 2023
Accepted: 14 April 2023
Published: 18 April 2023

1. Introduction

In recent decades, advances in hydraulic fracturing technology have enabled the economic development of oil and gas resources from ultra-tight shale and coal [1–3]. The pores in shale can be broadly classified as organic and inorganic types, with pore sizes ranging between two and several hundred nanometers. Coal can be considered a composite medium consisting of a coal matrix and cleats [4,5], with cleat widths in the range of several microns and the pores in the matrix in the nanoscale range. During hydraulic fracturing, a large amount of fracturing fluid is injected into the shale or coal at high pressure, which can reach up to 50 MPa or more. As a result, water invasion unavoidably encounters the nanopores, and the water flow in the nanopores affects the overall invasion efficiency [6,7]. Nanoconfined fluids behave differently from conventional fluids in terms of their physical properties, such as the decline of the bubble point for hydrocarbon phase behavior in pores smaller than 20 nm [8–10]. The gas flow capacity through nanopores is also surprisingly higher than that evaluated by the classic Navier-Stokes (N-S) formula [11,12], with the magnitude exceeding several orders. Consequently, conventional methods for describing water flow in nanopores can lead to inaccurate characterization, which requires further research.

The physical properties of nanoconfined water, including the viscosity and no-slip boundary conditions, are mainly governed by solid–fluid interactions, rather than intermolecular interactions in the bulk state. When water molecules are confined to a nanopore with a size comparable to their molecular diameter, the attraction or repulsive force induced by the surrounding pore wall cannot be ignored, which is fundamentally different from the precondition of the classic theoretical framework that is purely based on molecule–molecule interactions [13–15]. Therefore, focusing on the physical properties under solid–fluid interactions is crucial for understanding nanoconfined water flow behavior. Advanced experimental techniques have been used to directly examine water flow conductivity in nanotubes [16,17], indicating the inapplicability of the classic Navier-Stokes formula for describing nanoconfined water flow. Feibelman (2013) investigated the viscosity of water film on a kaolinite surface, a weak hydrophilic solid phase [18], using molecular simulation, and found that the water film viscosity was several times that of bulk water. Secchi (2016) experimentally examined the water flow capacity through a single nanotube [19] and found a significant enhancement in the flow efficiency compared to the conventional knowledge, which was attributed to a radius-dependent boundary slip as well as the viscosity variation. Using molecular dynamics, Thomas and McGaughey (2008) [20] simulated pressure-driven nanoconfined water flow and found that both the water viscosity in a nanopore and the slip length can be described as a function of the pore size. To understand the microscopic water molecular structure confined between two kaolinite surfaces, molecular dynamics simulations were performed, suggesting that the nanoconfined water viscosity could be one order larger than that of bulk water [21–23]. While the unusual alteration in the water viscosity confined inside nanopores has been recognized by both advanced experiments and molecular simulations, its impact on pressure-driven flow is still unclear. In addition to water viscosity variation, the boundary slip phenomenon, in which water molecules near the surface have some mobility, instead of always being stationary, in bulk water flow, must be incorporated to achieve a relatively accurate characterization of water flow at the nanoscale.

A great deal of effort has been devoted to investigating the behavior of water at the nanoscale [24–26]. However, little attention has been given to the water flow dynamics in porous media consisting of nanopores, which is crucial for understanding fracturing-water invasions inside a shale or coal matrix. A theoretical model for water flow in a nanopore was proposed, taking into account the surface wettability on the water viscosity variation and slip length. It suggests that the water flow capacity with a super-hydrophobic surface could be up to seven orders of magnitude stronger than that evaluated by the Hagen-Poiseuille equation [27]. Sun (2019) presented that the viscosity of the water close to the pore wall is extremely sensitive to surface–molecule interaction, while the water viscosity in the middle of the nanopore space is the same as bulk water [28]. In line with the continuum flow assumption, the water velocity profile at the nanoscale was analyzed, demonstrating the considerable effect of viscosity variation on the water flow capacity [29–31]. Molecular dynamics (MD) and the Lattice Boltzmann method (LBM) are widely utilized to explore nanoconfined water behavior [32–34]. However, the current fatal deficiency in MD research is the use of specific solid phase molecules to simulate the shale or coal nanopore surface [35–37], which fails to capture the surface–molecule interaction on the flow behavior of water confined in nanopores. Similarly, it is challenging to mimic the nanopore surface over a wide range of wettability using the LDM. Moreover, its time-consuming calculation and complex iteration hinder its applicability [38–40]. In regard to a shale or coal matrix, where the spatial pore structure is multi-factor-dependent and challenging to characterize, there is still no finite solution for pore network construction, although advanced approaches [41–43], including SEM image technology, CT scanning, as well as X-ray diffraction, are applied. Furthermore, the majority of pore network modeling (PNM) studies have focused on gas flow, while few studies on water flow have considered the viscosity variation and boundary slip induced by the surface–molecule interaction. Therefore, a PNM investigation on water flow at the nanoscale is urgently needed, aiming to reveal the fracturing water invasion or flow-back behavior in a shale or coal matrix.

To clarify the research, the content is arranged as follows: First, a physical model illustrating the difference between nanoconfined water viscosity and bulk water is presented. Then, a theoretical model is developed based on the physical model to describe the water flow in a nanopore, taking into account the water viscosity variation and water slip at the boundary. Subsequently, a pore network model (PNM) is established to couple the spatial pore structure and the proposed pore-scale water flow model. The PNM is applied to identify the influence of the water viscosity, slip mechanism, and pore–structure evolution on the water flow capacity. Finally, several conclusions are drawn.

2. Model Description

The focus of this research is to investigate the effect of water viscosity variation on the water flow behavior in nanopores and nanoscale porous media. As shown in Figure 1, water viscosity is affected by surface wettability when water molecules are confined in a nano-space. In the bulk region, where the interaction force exerted by the pore surface can be neglected, the water viscosity is a function of the surrounding pressure and temperature [44,45]. However, for nanoconfined molecules, the surface–water interaction forces intensify, and water molecules tend to gather close to a hydrophilic surface, resulting in an increase in the water viscosity. Conversely, the repulsive force induced by a hydrophobic surface prevents water molecules from moving towards the surface, leading to a decrease in the water viscosity [46–48]. From a theoretical perspective, the interaction force by a mid-wet surface on water molecules is zero; therefore, the water viscosity remains the same as bulk water. Experimental findings in recent decades have shown that nanoconfined water flows surprisingly faster than bulk water, and this is attributed to the boundary slip phenomenon. This suggests that nanoconfined water molecules near the solid surface have fairly strong mobility, in contrast to the predictions made by the traditional N-S equation. Thus, the difference between nanoconfined water behavior and bulk water is evident [49,50]. Both the slip phenomenon and viscosity variation play prominent roles in affecting the water flow in nanopores, yet they have received little research attention. Additionally, an upscaling investigation on water flow in nanoscale porous media is rare, which solidifies the fundamental basis for revealing the fracturing fluid loss mechanisms and water production performance from shale reservoirs.

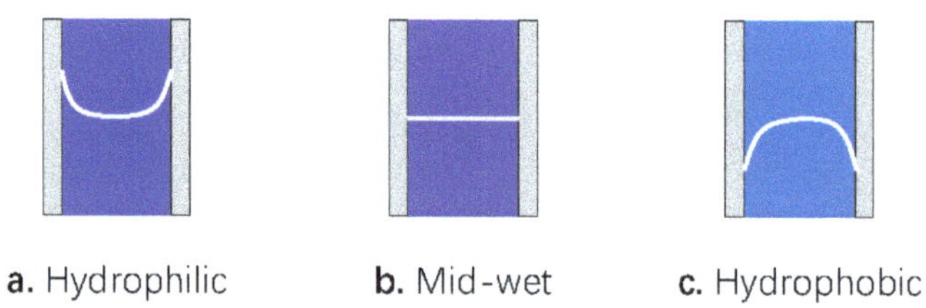

Figure 1. Water viscosity behavior inside nanopores with different wettability: (**a**) Hydrophilic surface; (**b**) Mid-wet surface; (**c**) Hydrophobic surface.

3. Model Establishment

To accurately simulate water flow behavior in porous media containing numerous nanopores, it is imperative to have a clear understanding of the water flow in a single nanopore. At the nanoscale, the slip phenomenon and viscosity variation are two prominent factors that affect water behavior and must be taken into account. The slip phenomenon describes the movement of water molecules close to the pore surface, while a viscosity change in nanoconfined water alters the flow resistance. Both the slip phenomenon and viscosity are known to be sensitive to temperature variation, but in order to focus on the impact of unusual viscosity in this research, a constant temperature of 20 °C is used in this paper.

3.1. Nanoconfined Water Flow

In this section, the water flow capacity in a single nanopore is investigated, taking into account the effects of water viscosity variation and the boundary slip. The apparent permeability, which represents the water flow flux per displacement pressure difference, is derived, providing a basis for subsequent pore network modeling.

In regard to nanoconfined water viscosity, due to the limited distance where the fluid–surface interaction takes place, the viscosity of the water molecules close to the surface is drastically different compared to bulk water. In contrast, the water molecules away from the surface behave the same as bulk water. Therefore, the concept of the interfacial region, where unusual water viscosity occurs under nanoconfinement impact, is introduced in this work. The thickness of the interfacial region remains in debate, ranging between 0.5~1.5 nm, which is mainly related to the surface attributes and fluid type. In terms of the water molecules, the thickness is assigned as 0.7 nm, indicating that water molecules 0.7 nm away from the surface are beyond the control of the surface-induced force. Additionally, Wu et al. (2017) proposed a fancy correlation linking the surface contact angle to the water viscosity inside the interfacial region [27], and its reliability was verified by experimental data and molecular simulation results. It should be highlighted that the water viscosity in the whole interfacial region is regarded as a single value, instead of a varying parameter, in Wu et al. (2017). Actually, due to the close relationship between the fluid–surface interaction and water viscosity and the rapid rise in the interaction strength while water molecules approach the surface, the water viscosity will vary over the decreasing distance towards the surface. Utmost caution should be paid when using the correlation [27].

$$\mu_i = \mu_b(-0.018\theta + 3.25) \tag{1}$$

where μ_i denotes the water viscosity in the interfacial region, cp; μ_b denotes the bulk water viscosity, cp; θ denotes the surface contact angle.

The geometry of nanopores has been found to affect the behavior of confined water, but for simplicity, this study focuses on cylindrical nanopores and does not consider the effects of geometry changes. Using Equation (1), the viscosity of the water molecules near the surface can be calculated, while the viscosity of the rest of the water molecules is assumed to be the same as that of bulk water. To determine the overall water viscosity within the entire nanopore, a volume-average method is proposed, which sums up the viscosity contributions from both the interfacial region and the rest of the nanopore based on their respective volumes.

$$\mu = \mu_i \frac{A_i(d)}{A_t(d)} + \mu_b \left[1 - \frac{A_i(d)}{A_t(d)}\right] \tag{2}$$

where μ denotes the nanoconfined water viscosity, cp; d denotes the nanopore diameter, nm; A_i denotes the volume of the interfacial region, m^3; A_t denotes the entire volume of the nanopore, m^3.

The behavior of water in nanopores is affected by both the pore size and surface wettability, as demonstrated by the combination of Equations (1) and (2). In particular, the water viscosity in the interfacial region is highly sensitive to surface wettability, especially for nanopores with a size smaller than 10 nm, where the effect of wettability on the viscosity dominates. For larger nanopores, the impact of wettability on the nanoconfined water viscosity decreases significantly due to the smaller volume proportion of the interfacial region.

To quantify the slip phenomenon, the slip length is introduced, which represents the ratio of water molecular velocity to the shear rate at the boundary position. The strength of the slip phenomenon is determined by the slip length, with a longer slip length indicating a stronger boundary slip. Molecular simulations and microscopic displacement experiments have confirmed the correlation between the slip length and nanopore size. In this work, the correlation describing the water slip length inside carbon nanotubes [20] is used, as the shale or coal matrix contains organic matter that can transform into graphene, a form of carbon nanotubes, under high geological thermal maturity.

$$L_s = L_\infty + \frac{C}{d^3} \tag{3}$$

where L_s denotes the slip length, nm; L_∞ denotes the water slip length on a perfect flat carbon sheet, nm; C is a fitted parameter, related to the surface properties and pore size. In this work, for the physical model of water flowing through carbon nanopores, L_∞ is 30 nm and C is equal to 352 nm^4.

It has been reported that the continuum hydrodynamics theory remains valid for water flow in nanopores with a pore size of less than 2 nm [51,52]. Based on this, and taking into account the characterization of the nanoconfined water viscosity and slip length, the water flow equation at the nanoscale can be derived. The conventional water flow flux per pressure gradient in a pore with a pore diameter of d is provided first.

$$Q_c = \frac{\pi}{8\mu_b}(d/2)^4\frac{dp}{dz} \tag{4}$$

where Q_c denotes the water flow flux using the conventional N-S equation, m^3/s.

Considering the viscosity variation and boundary slip length, Equation (4) becomes the following equation.

$$Q_n = \frac{\pi}{8\mu}\left[(d/2)^4 + 4(d/2)^3 L_s\right]\frac{dp}{dz} \tag{5}$$

where Q_n denotes the nanoconfined water flow flux, m^3. In terms of a large pore size, the second term in the bracket of Equation (5) becomes negligible compared to the first term and the nanoconfined water viscosity approaches bulk water viscosity; thus, Equation (5) can degenerate to Equation (4) successfully. In terms of nanopores, both the water viscosity and slip length affected by nanoconfinement are coupled in Equation (5), which could shed light on its discrepancy compared to the bulk water flow.

3.2. Pore Network Modeling

Water flow at the nanoscale can have important practical applications in various industries, such as the characterization of fracturing fluid loss or multiphase flow in the shale/coal matrix. However, upscaling the behavior of water flow from the nanoscale to the core scale is necessary to make it relevant to industry applications. Pore network modeling is a commonly used upscaling tool, where the core sample is considered to be a large number of nodes interconnected to each other, and the connections between any two nodes represent the nanopores. In this work, the nanopores were assigned specific pore sizes and surface contact angles. To establish the spatial pore network, the pore size distribution, which denotes the range of nanopore sizes, as well as the proportion of nanopores at each pore size, was provided. Furthermore, the co-ordination number which denotes the number of nanopores that a single nanopore could connect to at the core scale, was also given. A higher co-ordination number implies a stronger connectivity of the constructed pore network. The spatial pore network was established using a self-coded program, based on the provided pore size distribution and co-ordination number.

This work focused on investigating stable water flow, rather than time-dependent flow issues. For a stable flow, the fluid pressure at each node in the developed pore network remains unchanged, as does the water flow flux in the nanopores. According to the mass conservation principle, the total water inflow flux towards a specified node should be equal to the water mass variation in the node. However, considering the stable flow characteristics, there is no water mass variation in each node. Therefore, the control equation for the mass conservation of a node connecting n surrounding nodes is as follows:

$$\sum_{i=1}^{i=n} q_i = 0 \quad (6)$$

where q_i denotes the water mass flow flux from the i node to the calculation node, kg/s.

In Section 3.1, we investigated the capacity of nanoconfined water flow by considering the effects of the viscosity and slip length. By coupling the pressure difference between two connected nodes in the pore network, the mass flow flux can be calculated using the following expression.

$$q_i = \frac{\pi}{8\mu}\left[(d/2)^4 + 4(d/2)^3 L_s\right]\frac{(P_i - P)}{L} \quad (7)$$

where P_i and P denote the node i pressure and calculation node pressure, respectively, MPa.

To obtain the water flow flux through the entire pore network, it is necessary to calculate the pressure of each node, which satisfies the correlation Equation (6). As the solutions for all nodes in the pore network are unique, the water flow flux can be obtained after solving all of the node pressures. The first step is to assign the inflow pressure and outflow pressure to the nodes located at the inlet and outlet of the pore network, respectively. Then, the pressure values ranging between the inlet pressure and outlet pressure are randomly generated, and the number of values is equal to the total number of nodes, excluding the inlet and outlet nodes. These generated pressure values are then assigned to the rest of the nodes, and each node has an initial pressure value. Next, the pressure of each node, except for the inlet and outlet nodes, is updated based on the mass conservation principle. During each iteration, the overall water flow flux is calculated after updating the pressure of all the nodes. The iteration continues until the difference in the water flow flux between successive pressure updates is within the desirable accuracy. The calculation procedures are illustrated in Figure 2. It should be noted that the calculation method is aimed at achieving a stable flow, and time-dependent flow issues are not considered in this work.

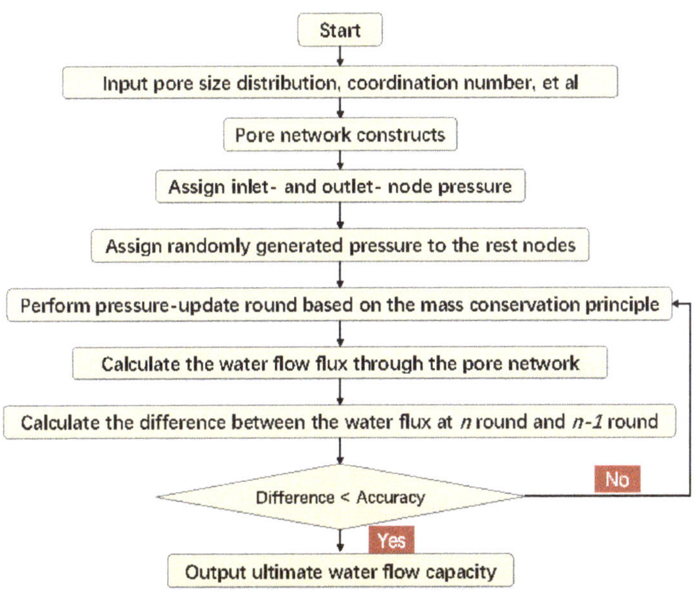

Figure 2. Calculation procedures for pore network modeling on stable water flow.

4. Results and Discussion

In this section, we aim to investigate the water flow behavior in nanoscale porou media while considering the effects of nanoconfinement, such as unusual water viscosit and the slip phenomenon. However, the spatial pore structure evolution characteristic in a shale or coal matrix add complexity to the accurate simulation of the water flow. T address this challenge, we use a pore network modeling approach that can reproduc the key parameters of porous media, such as the pore size distribution and co-ordinatio number. We also perform a sensitivity analysis to identify the influence of each influentia factor on the water flow performance in the constructed pore network.

4.1. Nanoconfinement Effect

In this section, the focus is on quantifying the influential extent of the viscosity var ation or boundary slip on the water flow capacity. Understanding this effect is essentia to developing approaches that can enhance or contain water flow efficiency, dependin on the dominant factor. For instance, if viscosity plays a prominent role in controlling th water flow under a specific condition, manipulating the water temperature to enhance o decline the water viscosity is a desirable method. On the other hand, surface modificatio becomes the effective approach when the boundary slip governs the water flow behavio Thus, to gain insight into the nanoconfinement effect, the self-coded program develope for a stable water flow in the pore network in Section 3.2 is utilized.

The implemented pore network modeling couples the nanoscale water flow mech nisms discussed in Section 3.1, making it a unique approach. The pore size distributio is assigned to each nanopore in the constructed pore network, and the pore size range between 5~80 nm, which is a typical distribution for a shale or coal matrix. As depicted i Figure 3, the generated pore size distribution by the self-coded program has a favorabl match with the assigned pore size distribution. This agreement verifies the reliability c the modeling results to some extent. The pore size distribution and co-ordination numbe remain unchanged to identify the impact of the water viscosity change subjected to th nanopores on the spatial pore characteristics. In this regard, the water flow performance i analyzed by varying the surface contact angle, which affects the boundary slip phenomeno The quantification of the water flow capacity through porous media is conducted by utili ing the apparent permeability, which is the ratio of the cross-sectional water flow velocit to the pressure difference.

$$K = \frac{Q_{final}\mu_b L_t}{A(P_{inlet} - P_{outlet})} \quad (8$$

where K denotes the apparent permeability, nD; Q_{final} denotes the final stable water flo flux, illustrated in Figure 2, m^3; L_t denotes the total pore network length at the water flo direction, m; A denotes the flow cross-sectional area, m^2; P_{inlet} and P_{outlet} denote the inl pressure and outlet pressure, respectively.

The investigation in Figure 4 provides an insight into the impact of the surface conta angle on the water flow behavior. In this case, the surface contact angle of all the nanopore is the same, ranging between 0 and 180 degrees. The inherent permeability, as shown i Figure 4a, is constant at 5231 nD due to the assumption of the bulk water viscosity and n slip hydrodynamic boundary condition, which overlooks the influence of the unusual wat viscosity in the nanopores. However, the effect of the surface wettability on the appare permeability is remarkable, particularly when the contact angle exceeds 100 degrees. Th relationship between the apparent permeability and surface wettability can be divided int two parts based on the critical surface contact angle, which is 100 degrees.

When the contact angle is less than the critical value, indicating that the interactio force exerted by the surface on the water molecules is in attraction form, the appare permeability gently enhances with the increasing contact angle. The apparent permeabili may be less than the inherent permeability. On the other hand, the apparent permeabili boosts when the contact angle exceeds the critical value, and the flow capacity could rea up to four orders of magnitude higher than the inherent permeability, indicating the supe fast flow phenomenon as a result of the hydrophobic surface property. In this case, the surface imposes repulsive force on the water molecules to prevent their molecular approach to the surface, resulting in the decline of the whole water viscosity, and subsequently promoting water flow.

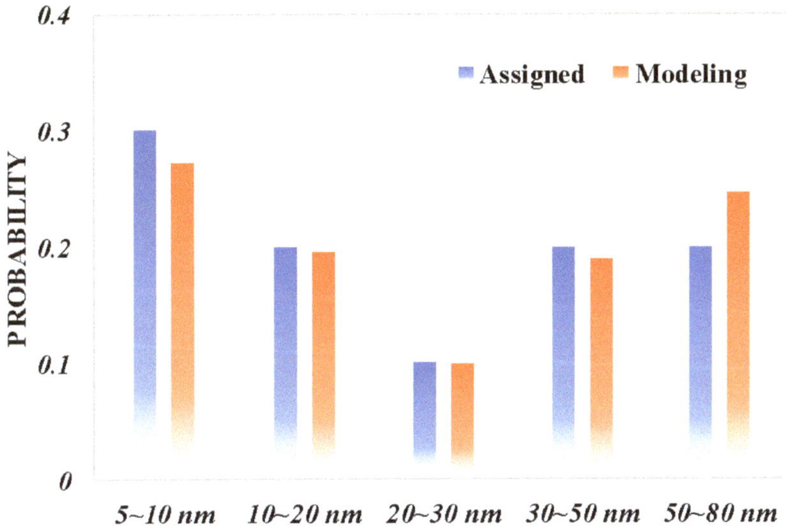

Figure 3. Comparison of the target pore size distribution and that generated by the program.

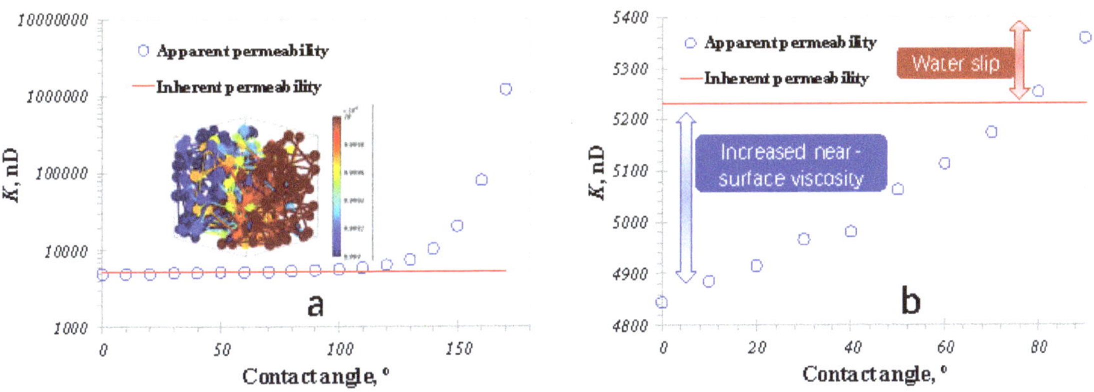

Figure 4. Water flow capacity in porous media consisting of nanopores: (**a**) Apparent permeability versus surface contact angle; (**b**) Underlying mechanism governing water flow capacity.

The permeability variation in the contact angle range, between 0 and 100 degrees, as presented in Figure 4b, shows that the apparent permeability is less than the inherent permeability when the contact angle is less than 80 degrees, which can be attributed to the increased viscosity in the hydrophilic nanopores. Moreover, the water slip contributes to the enhancement of the water flow capacity. The effect of the water slip length at the boundary is a function of the nanopore size, indicating that the slip length remains unchanged in this calculation case as the assigned pore size is unchanged. However, the idea that the impact of the boundary slip remains unchanged is misleading. According to the continuum flow theory, the viscosity variation affects the influence of the water slip. The constant slip length indicates the constant flow velocity of the water molecules at the boundary. The impact of the water slip enlarges with the decreasing water viscosity and declines with the enhancing water viscosity on the water flow capacity. Therefore, the dramatic enhancement of the apparent permeability observed in Figure 4a is a result of both the water viscosity variation and water slip. It is insightful to recognize the changing role of the water slip with unusual water viscosity in nanopores.

To investigate the impact of the spatial porous characteristics on the water flow, a pore size of 20 nm was assigned to all nanopores in the constructed pore network. Three cases of surface wettability were then considered, as shown in Figure 5, and the associated apparent permeability for the water flow through the porous media was calculated. Case_1 represented hydrophilic porous media, with 40% of the total nanopores having contact angles of less than 30 degrees. In contrast, Case_3 represented hydrophobic porous media, where the water viscosity was significantly reduced. Case_2 had the same surface wettability as the base case, which had a wide pore size distribution, shown in Figure 3, but with an unchanged pore size.

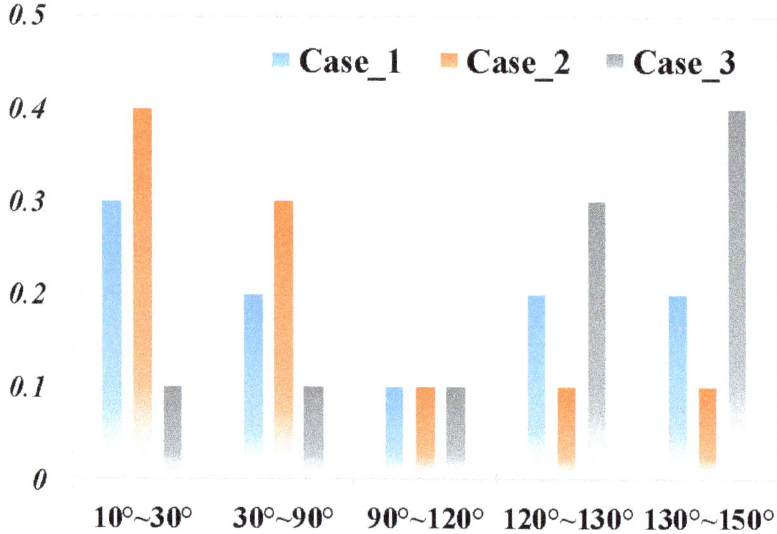

Figure 5. Surface contact angle variation range of nanopores in porous media.

As shown in Figure 6, the difference in the flow capacity between Case_2 and the base case is relatively small, indicating that the pore size distribution has a limited effect on the flow capacity when the majority of the pore sizes are in the nanoscale range. The apparent permeability in Case_2 is slightly stronger than the base case, which may be due to the fact that the average pore size in the base case is less than 20 nm. Additionally larger pore sizes contribute to a stronger water flow capacity, assuming the same surface wettability feature. The apparent permeability for the water flow through the pore network increases in the order of Case_2, Case_1, and Case_3. Similar to the water flow performance in nanopores, the water flow in nanopores with more hydrophobic surfaces can achieve a stronger flow efficiency. According to the surface contact angle distribution shown in Figure 5, 70% of the nanopores in Case_3 have surface contact angles over 120°, which is much higher than in Case_1 and Case_2. As a result, the water flow resistance in the hydrophobic nanopores decreases dramatically, leading to the lowest overall resistance to water flow in Case_3. Conversely, the porous media of Case_2, consisting of many hydrophilic nanopores, imposes a lower water flow efficiency due to the higher water flow resistance.

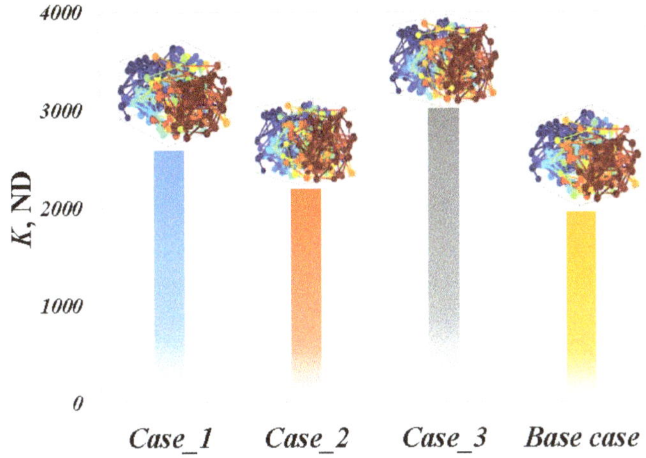

Figure 6. Wettability effect on apparent permeability for water flow at core scale.

4.2. Pore Network Characteristics

Advanced experimental devices have made it possible to study the spatial evolution of pores in a coal or shale matrix as they vary with the geological depth and thermal maturity of the organic content. The ability to explore the complex spatial pore network characteristic and its related impact on fluid flow behavior is of utmost importance. To this end, numerous studies have been conducted to reproduce and present the spatial evolution characteristics of coal or shale matrix systems. In this article, we use an established pore network modeling tool coupled with nanoconfined water flow mechanisms to study the pore network characteristics, particularly the pore size distribution and co-ordination number, on water flow at the core scale. One of the primary objectives of this study is to understand the impact of the pore size distribution and co-ordination number on the fluid flow in coal or shale matrix systems. In this section, we focus on the impact of the surface contact angle of the nanopores in the pore network on the fluid flow. To exclude the effect of wettability, we set the surface contact angle of all the nanopores in the pore network to be 80°. This enabled us to study the effect of other pore network characteristics on the fluid flow without the confounding factor of varying wettability. The use of pore network modeling tools in conjunction with experimental techniques has enabled researchers to study the pore-scale mechanisms governing fluid flow in coal or shale matrix systems. The ability to control the wettability of the nanopores in the pore network provides a unique opportunity to study the impact of other pore network characteristics on the fluid flow, which can ultimately help to optimize the extraction of hydrocarbons from these systems.

In the exploration of the impact of pore size distribution on the fluid flow behavior, Figure 7 depicts three sets of pore size distributions. In these cases, other parameters describing the pore network, such as the positions of the nodes and the way the nodes connect to each other, remain unchanged. Specifically, Case_1 presents a pore network where 70% of the nanopores have a pore size less than 20 nm, while 70% of the nanopores in the pore network of Case_3 possess relatively large pore sizes, exceeding 30 nm. The simulation of water flow inside the pore network is then performed to identify the differences induced by varying the pore size distribution. As shown in Figure 8, it can be observed that the apparent permeability declines in the order of Case_3, Case_2, and Case_1. In other words, the water flow capacity for Case_3 is stronger than in Case_2 and Case_1 by as much as 18.3% and 51.7%, respectively, demonstrating the great positive effect of enlarging the pore size. Furthermore, the difference between Case_2 and Case_3 is far less than that between Case_1 and Case_2, indicating that enlarging the pore size could contribute more to the water flow efficiency when the average pore size is close to 10~20 nm than when it approaches 50 nm. This finding suggests that a formation stimulation creating fractures or tremendously large pores could effectively advance the fluid flow capacity. The enhancement magnitude is more desirable for pore sizes less than 20 nm. The results of this study provide insights into the optimization of stimulation strategies for tight reservoirs. The formation of fractures or the creation of large pores can help to increase the average pore size and thus improve the water flow efficiency, particularly when the majority of the pore sizes are in the nanoscale range.

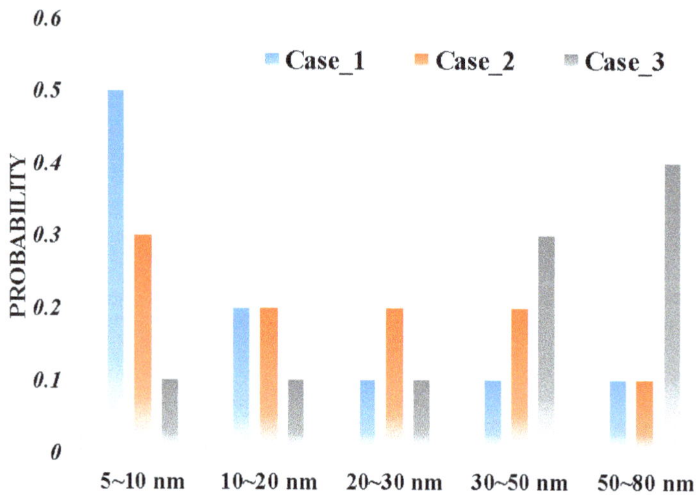

Figure 7. Different pore size distributions assigned to pore network.

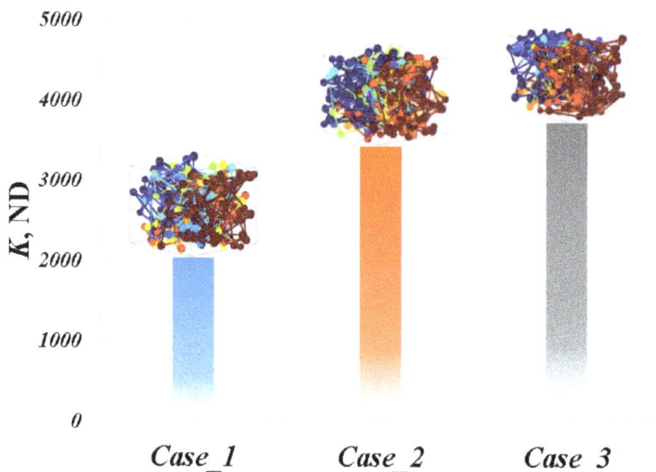

Figure 8. Impact of pore size distribution on apparent permeability.

In the construction process of the pore network, the co-ordination number plays crucial role in determining the potential for water flow. The co-ordination number refer to the number of nodes that each node can connect to, and this parameter determine the availability of nanopores for water transport. The higher the co-ordination numbe the more interconnected the pore network becomes, leading to enhanced water transpo through the network. Therefore, increasing the co-ordination number can effectivel improve the connectivity of the pore network, thereby promoting the water flow capaci in nanoscale porous media. In this study, the researchers kept all parameters of th pore network constant, except for the co-ordination number. Figure 9 shows that the apparent permeability increases with an increase in the co-ordination number, confirming the importance of connectivity in promoting the water flow capacity. The findings of this study highlight the need to focus on improving the connectivity of nanoscale porous media to enhance their fluid flow behavior. By increasing the co-ordination number or finding other means to improve the connectivity, the potential for water transport through the pore network can be increased, leading to a more efficient and effective use of these porous media. Overall, these findings provide insights into the design and optimization of porous media for various applications, including energy and environmental engineering.

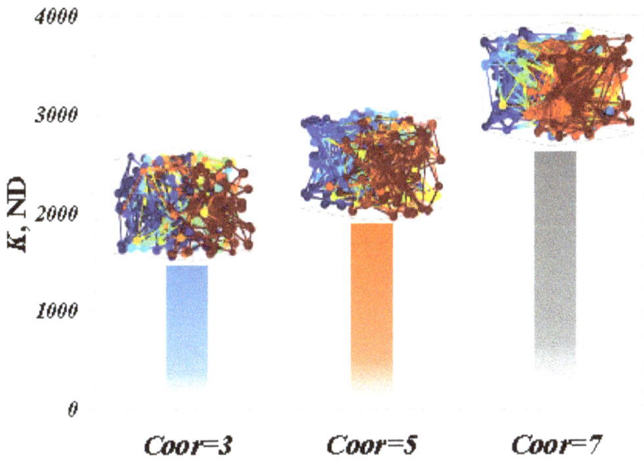

Figure 9. Impact of co-ordination number on apparent permeability.

5. Conclusions

(1) We establish a model for the viscosity of water confined in nanoscale pores using the volume-average concept. This model considers variations in the viscosity occurring in the interfacial region and assumes bulk-like behavior in the rest of the region. Additionally, we develop a model for pore-scale water flow capacity and a pore network model that ensures a stable water flow. Both models are developed based on the continuum hydrodynamic principles.

(2) The impact of hydrophilic and super-hydrophobic nanopore surfaces on water viscosity and permeability is significant. A hydrophilic surface leads to an increase in the viscosity and a decrease in the apparent permeability, while a super-hydrophobic surface results in a significant increase in the permeability due to a decline in the viscosity, coupled with the boundary slip phenomenon.

(3) There are several methods for improving the water flow efficiency in nanoscale porous media, including enlarging the pore size, enhancing the co-ordination number, and modifying the surface properties to intensify the wettability. Enlarging the pore size contributes more significantly to the flow efficiency when the average pore size is close to 10–20 nm than when it approaches 50 nm.

Author Contributions: Investigation, Y.W.; supervision, N.N.; data curation, W.L.; draft preparation, Y.Z.; formal analysis, B.W.; validation, S.L. All authors have read and agreed to the published version of the manuscript.

Funding: This research was funded by National Natural Science Foundation of China grant number 52204031 and Natural Science Foundation of Sichuan Province grant number 23NSFSC4527. The APC was funded by 52204031.

Institutional Review Board Statement: Not applicable.

Informed Consent Statement: Not applicable.

Data Availability Statement: Data will be available on request.

Acknowledgments: This work was supported by the National Natural Science Foundation of China (Grant No. 52204031) and the Natural Science Foundation of Sichuan Province (Grant No. 23NS-FSC4527).

Conflicts of Interest: The authors declare no conflict of interest.

References

1. Hou, B.; Chang, Z.; Fu, W.; Muhadasi, Y.; Chen, M. Fracture initiation and propagation in a deep shale gas reservoir subject to an alternating-fluid-injection hydraulic-fracturing treatment. *SPE J.* **2019**, *24*, 1839–1855. [CrossRef]
2. Huang, T.; Li, Z.; Mayer, B.; Nightingale, M.; Li, X.; Li, G.; Long, Y.; Pang, Z. Identification of geochemical processes during hydraulic fracturing of a shale gas reservoir: A controlled field and laboratory water-rock interaction experiment. *Geophys. Res. Lett.* **2020**, *47*, e2020GL090420. [CrossRef]
3. Sun, Z.; Huang, B.; Liu, Y.; Jiang, Y.; Zhang, Z.; Hou, M.; Li, Y. Gas-phase production equation for CBM reservoirs: Interaction between hydraulic fracturing and coal orthotropic feature. *J. Pet. Sci. Eng.* **2022**, *213*, 110428. [CrossRef]
4. Akhondzadeh, H.; Keshavarz, A.; Al-Yaseri, A.Z.; Ali, M.; Awan, F.U.; Wang, X.; Yang, Y.; Iglauer, S.; Lebedev, M. Pore-scale analysis of coal cleat network evolution through liquid nitrogen treatment: A Micro-Computed Tomography investigation. *Int. J. Coal Geol.* **2020**, *219*, 103370. [CrossRef]
5. Roslin, A.; Pokrajac, D.; Zhou, Y. Cleat structure analysis and permeability simulation of coal samples based on micro-computed tomography (micro-CT) and scan electron microscopy (SEM) technology. *Fuel* **2019**, *254*, 115579. [CrossRef]
6. Wijaya, N.; Sheng, J.J. Mitigating near-fracture blockage and enhancing oil recovery in tight reservoirs by adding surfactants in hydraulic fracturing fluid. *J. Pet. Sci. Eng.* **2020**, *185*, 106611. [CrossRef]
7. Fernandez, C.A.; Gupta, V.; Dai, G.L.; Kuprat, A.P.; Bonneville, A.; Appriou, D.; Horner, J.A.; Martin, P.F.; Burghardt, J.A. Insights into a greener stimuli-responsive fracturing fluid for geothermal energy recovery. *ACS Sustain. Chem. Eng.* **2019**, *7*, 19660–19668. [CrossRef]
8. Chen, S.; Jiang, J.; Guo, B. A pore-network-based upscaling framework for the nanoconfined phase behavior in shale rocks. *Chem. Eng. J.* **2021**, *417*, 129210. [CrossRef]
9. Sun, Z.; Wang, S.; Xiong, H.; Wu, K.; Shi, J. Optimal nanocone geometry for water flow. *AIChE J.* **2022**, *68*, e17543. [CrossRef]
10. Sun, Z.; Huang, B.; Yan, S.; Wang, S.; Wu, K.; Yu, W.; Li, Y.; Wang, S. Nanoconfined methane thermodynamic behavior below critical temperature: Liquid–vapor coexistence curve under wettability effect. *Ind. Eng. Chem. Res.* **2022**, *61*, 4971–4979. [CrossRef]
11. Yang, X. Numerical approximations of the Navier–Stokes equation coupled with volume-conserved multi-phase-field vesicles system: Fully-decoupled, linear, unconditionally energy stable and second-order time-accurate numerical scheme. *Comput. Methods Appl. Mech. Eng.* **2021**, *375*, 113600. [CrossRef]
12. Coron, J.M.; Marbach, F.; Sueur, F.; Zhang, P. Controllability of the Navier–Stokes equation in a rectangle with a little help of a distributed phantom force. *Ann. PDE* **2019**, *5*, 17. [CrossRef]
13. Gong, C.; Sun, S.; Zhang, Y.; Sun, L.; Su, Z.; Wu, A.; Wei, G. Hierarchical nanomaterials via biomolecular self-assembly and bioinspiration for energy and environmental applications. *Nanoscale* **2019**, *11*, 4147–4182. [CrossRef]
14. Geagea, E.; Jeannoutot, J.; Morgenthaler, L.; Lamare, S.; Palmino, F.; Chérioux, F. On-Surface Synthesis of Ligands to Elaborate Coordination Polymers on an Au (111) Surface. *Nanomaterials* **2021**, *11*, 2102. [CrossRef]
15. Wang, L.; Gong, C.; Yuan, X.; Wei, G. Controlling the self-assembly of biomolecules into functional nanomaterials through internal interactions and external stimulations: A review. *Nanomaterials* **2019**, *9*, 285. [CrossRef]
16. Shrestha, R.G.; Kubota, Y.; Sakamoto, Y.; Kawakita, J. Quick and sensitive detection of water using galvanic-coupled arrays with a submicron gap for the advanced prediction of dew condensation. *Sensors* **2020**, *20*, 3314. [CrossRef]
17. Mekawy, M.; Noguchi, H.; Kawakita, J. Quantitative and qualitative studies for real monitoring of interfacial molecular water. *J. Colloid Interface Sci.* **2022**, *613*, 311–319. [CrossRef]
18. Feibelman, P.J. Viscosity of ultrathin water films confined between aluminol surfaces of kaolinite: Ab initio simulations. *J. Phys. Chem. C* **2013**, *117*, 6088–6095. [CrossRef]
19. Secchi, E.; Marbach, S.; Niguès, A.; Stein, D.; Siria, A.; Bocquet, L. Massive radius-dependent flow slippage in carbon nanotubes. *Nature* **2016**, *537*, 210–213. [CrossRef]
20. Thomas, J.A.; McGaughey, A.J. Reassessing fast water transport through carbon nanotubes. *Nano Lett.* **2008**, *8*, 2788–2793. [CrossRef]
21. Haria, N.R.; Grest, G.S.; Lorenz, C.D. Viscosity of nanoconfined water between hydroxyl basal surfaces of kaolinite: Classical simulation results. *J. Phys. Chem. C* **2013**, *117*, 6096–6104. [CrossRef]
22. Bakli, C.; Chakraborty, S. Anomalous interplay of slip, shear and wettability in nanoconfined water. *Nanoscale* **2019**, *11*, 11254–11261. [CrossRef] [PubMed]
23. Deißenbeck, F.; Freysoldt, C.; Todorova, M.; Neugebauer, J.; Wippermann, S. Dielectric properties of nanoconfined water: A canonical thermopotentiostat approach. *Phys. Rev. Lett.* **2021**, *126*, 136803. [CrossRef] [PubMed]

4. Pan, J.; Xiao, S.; Zhang, Z.; Wei, N.; He, J.; Zhao, J. Nanoconfined water dynamics in multilayer graphene nanopores. *J. Phys. Chem. C* **2020**, *124*, 17819–17828. [CrossRef]
5. Zhao, Z.; Xue, M.; Qiu, H.; Guo, W.; Zhang, Z. Giant mechanocaloric effect of nanoconfined water near room temperature. *Cell Rep. Phys. Sci.* **2022**, *3*, 100822. [CrossRef]
6. Moid, M.; Finkelstein, Y.; Moreh, R.; Maiti, P.K. Microscopic study of proton kinetic energy anomaly for nanoconfined water. *J. Phys. Chem. B* **2019**, *124*, 190–198. [CrossRef]
7. Wu, K.; Chen, Z.; Li, J.; Li, X.; Xu, J.; Dong, X. Wettability effect on nanoconfined water flow. *Proc. Natl. Acad. Sci. USA* **2017**, *114*, 3358–3363. [CrossRef]
8. Sun, Z.; Wu, K.; Shi, J.; Zhang, T.; Feng, D.; Huang, L.; Li, X. An analytical model for transport capacity of water confined in nanopores. *Int. J. Heat Mass Transf.* **2019**, *138*, 620–630. [CrossRef]
9. Corti, H.R.; Appignanesi, G.A.; Barbosa, M.C.; Bordin, J.R.; Calero, C.; Camisasca, G.; Elola, M.D.; Franzese, G.; Gallo, P.; Hassanali, A.; et al. Structure and dynamics of nanoconfined water and aqueous solutions. *Eur. Phys. J. E* **2021**, *44*, 136. [CrossRef]
0. Kapil, V.; Schran, C.; Zen, A.; Chen, J.; Pickard, C.J.; Michaelides, A. The first-principles phase diagram of monolayer nanoconfined water. *Nature* **2022**, *609*, 512–516. [CrossRef]
1. Paniagua-Guerra, L.E.; Gonzalez-Valle, C.U.; Ramos-Alvarado, B. Effects of the interfacial modeling approach on equilibrium calculations of slip length for nanoconfined water in carbon slits. *Langmuir* **2020**, *36*, 14772–14781. [CrossRef]
2. Cheng, Z.; Ning, Z.; Kang, D.H. Lattice Boltzmann simulation of water flow through rough nanopores. *Chem. Eng. Sci.* **2021**, *236*, 116329. [CrossRef]
3. Rao, Q.; Xia, Y.; Li, J.; Deo, M.; Li, Z. Flow reduction of hydrocarbon liquid in silica nanochannel: Insight from many-body dissipative particle dynamics simulations. *J. Mol. Liq.* **2021**, *344*, 117673. [CrossRef]
4. Lubbers, N.; Agarwal, A.; Chen, Y.; Son, S.; Mehana, M.; Kang, Q.; Karra, S.; Junghans, C.; Germann, T.C.; Viswanathan, H.S. Modeling and scale-bridging using machine learning: Nanoconfinement effects in porous media. *Sci. Rep.* **2020**, *10*, 13312. [CrossRef]
5. Jin, L.; Zhou, L.; Du, X. A Molecular Analysis of Critical Factors for Interface and Size Effects on Heat Conduction in Nanoconfined Water Film. *J. Therm. Sci.* **2022**, *31*, 1155–1166. [CrossRef]
6. Lynch, C.I.; Klesse, G.; Rao, S.; Tucker, S.J.; Sansom, M.S. Water nanoconfined in a hydrophobic pore: Molecular dynamics simulations of transmembrane protein 175 and the influence of water models. *ACS Nano* **2021**, *15*, 19098–19108. [CrossRef]
7. Sun, Z.; Li, X.; Liu, W.; Zhang, T.; He, M.; Nasrabadi, H. Molecular dynamics of methane flow behavior through realistic organic nanopores under geologic shale condition: Pore size and kerogen types. *Chem. Eng. J.* **2020**, *398*, 124341. [CrossRef]
8. Santos, J.E.; Mehana, M.; Wu, H.; Prodanovic, M.; Kang, Q.; Lubbers, N.; Viswanathan, H.; Pyrcz, M.J. Modeling nanoconfinement effects using active learning. *J. Phys. Chem. C* **2020**, *124*, 22200–22211. [CrossRef]
9. Yi, J.; Liu, L.; Xia, Z.; Wang, J.; Jing, Y.; Duan, L. Effects of wettability on relative permeability of rough-walled fracture at pore-scale: A lattice Boltzmann analysis. *Appl. Therm. Eng.* **2021**, *194*, 117100. [CrossRef]
0. Sun, Z.; Li, X.; Shi, J.; Zhang, T.; Sun, F. Apparent permeability model for real gas transport through shale gas reservoirs considering water distribution characteristic. *Int. J. Heat Mass Transf.* **2017**, *115*, 1008–1019. [CrossRef]
1. Blunt, M.J.; Jackson, M.D.; Piri, M.; Valvatne, P.H. Detailed physics, predictive capabilities and macroscopic consequences for pore-network models of multiphase flow. *Adv. Water Resour.* **2002**, *25*, 1069–1089. [CrossRef]
2. Joekar-Niasar, V.; Hassanizadeh, S.M. Analysis of fundamentals of two-phase flow in porous media using dynamic pore-network models: A review. *Crit. Rev. Environ. Sci. Technol.* **2012**, *42*, 1895–1976. [CrossRef]
3. Medici, E.F.; Zenyuk, I.V.; Parkinson, D.Y.; Weber, A.Z.; Allen, J.S. Understanding Water Transport in Polymer Electrolyte Fuel Cells Using Coupled Continuum and Pore-Network Models. *Fuel Cells* **2016**, *16*, 725–733. [CrossRef]
4. Shaat, M. Viscosity of water interfaces with hydrophobic nanopores: Application to water flow in carbon nanotubes. *Langmuir* **2017**, *33*, 12814–12819. [CrossRef]
5. Nalaparaju, A.; Wang, J.; Jiang, J. Enhancing water permeation through alumina membranes by changing from cylindrical to conical nanopores. *Nanoscale* **2019**, *11*, 9869–9878. [CrossRef]
6. Martyushev, D.A.; Ponomareva, I.N.; Chukhlov, A.S.; Davoodi, S.; Osovetsky, B.M.; Kazymov, K.P.; Yang, Y. Study of void space structure and its influence on carbonate reservoir properties: X-ray microtomography, electron microscopy, and well testing. *Mar. Pet. Geol.* **2023**, *151*, 106192. [CrossRef]
7. Rashid, M.; Luo, M.; Ashraf, U.; Hussain, W.; Ali, N.; Rahman, N.; Hussain, S.; Martyushev, D.A.; Vo Thanh, H.; Anees, A. Reservoir Quality Prediction of Gas-Bearing Carbonate Sediments in the Qadirpur Field: Insights from Advanced Machine Learning Approaches of SOM and Cluster Analysis. *Minerals* **2023**, *13*, 29. [CrossRef]
8. Galkin, S.V.; Martyushev, D.A.; Osovetsky, B.M.; Kazymov, K.P. Huaisen Song Evaluation of void space of complicated potentially oil-bearing carbonate formation using X-ray tomography and electron microscopy methods. *Energy Rep.* **2022**, *8*, 6245–6257. [CrossRef]
9. Cong, Z.; Li, Y.; Tang, J.; Martyushev, D.A.; Yang, F. Numerical simulation of hydraulic fracture height layer-through propagation based on three-dimensional lattice method. *Eng. Fract. Mech.* **2022**, *264*, 108331. [CrossRef]
0. Makarian, E.; Abad, A.B.M.N.; Manaman, N.S.; Mansourian, D.; Elyasi, A.; Namazifard, P.; Martyushev, D. An efficient and comprehensive poroelastic analysis of hydrocarbon systems using multiple data sets through laboratory tests and geophysical logs: A case study in an iranian hydrocarbon reservoir. *Carbonates Evaporites* **2023**, *38*, 37. [CrossRef]

51. Zhang, T.; Zhang, L.; Zhao, Y.; Zhang, R.; Zhang, D.; He, X.; Ge, F.; Wu, J.; Javadpour, F. Ganglia dynamics during imbibition and drainage processes in nanoporous systems. *Phys. Fluids* **2022**, *34*, 042016. [CrossRef]
52. Zhang, T.; Zhang, B.; Zhao, Y.; Javadpour, F.; He, X.; Ge, F.; Wu, J.; Zhang, D. Simulation of water flow in a nanochannel with a sudden contraction or expansion. *Langmuir* **2022**, *38*, 6720–6730. [CrossRef]

processes

MDPI

Article

Study on the Flow Behavior of Gas and Water in Fractured Tight Gas Reservoirs Considering Matrix Imbibition Using the Digital Core Method

Feifei Chen [1,*], Yonggang Duan [1] and Kun Wang [1,2]

[1] State Key Laboratory of Oil and Gas Reservoir Geology and Exploitation, Southwest Petroleum University, Chengdu 610500, China; nanchongdyg@163.com (Y.D.); wangkun25@cnooc.com.cn (K.W.)
[2] Nanhai East Petroleum Research Institute, Shenzhen Branch of CNOOC Limited, Shenzhen 518000, China
* Correspondence: cff1004@163.com

Abstract: Tight gas reservoirs possess unique pore structures and fluid flow mechanisms. Delving into the flow and imbibition mechanisms of water in fractured tight gas reservoirs is crucial for understanding and enhancing the development efficiency of such reservoirs. The flow of water in fractured tight gas reservoirs encompasses the flow within fractures and the imbibition flow within the matrix. However, conventional methods typically separate these two types of flow for study, failing to accurately reflect the true flow characteristics of water. In this study, micro-CT imaging techniques were utilized to evaluate the impact of matrix absorption and to examine water movement in fractured tight gas deposits. Water flooding experiments were conducted on tight sandstone cores with different fracture morphologies. Micro-CT scanning was performed on the cores after water injection and subsequent static conditions, simulating the process of water displacement gas in fractures and the displacement of gas in matrix pores by water through imbibition under reservoir conditions. Changes in gas–water distribution within fractures were observed, and the impact of fracture morphology on water displacement recovery was analyzed. Additionally, the recovery rates of fractures and matrix imbibition at different displacement stages were studied, along with the depth of water infiltration into the matrix along fracture walls. The insights gained from this investigation enhance our comprehension of the dynamics of fluid movement within tight gas deposits, laying a scientific foundation for crafting targeted development plans and boosting operational efficiency in such environments.

Keywords: tight gas reservoirs; fractures; digital cores; imbibition; gas–water flow

Citation: Chen, F.; Duan, Y.; Wang, K. Study on the Flow Behavior of Gas and Water in Fractured Tight Gas Reservoirs Considering Matrix Imbibition Using the Digital Core Method. *Processes* **2024**, *12*, 709. https://doi.org/10.3390/pr12040709

Academic Editors: Adam Smoliński and Youguo Yan

Received: 23 February 2024
Revised: 21 March 2024
Accepted: 28 March 2024
Published: 30 March 2024

1. Introduction

Traditional natural gas reserves are progressively being exhausted; the large-scale development of tight gas resources has become an important approach to supplementing traditional gas sources [1–3]. Their notable low permeability and intricate pore formations characterize these reservoirs; tight gas reservoirs exhibit poor gas mobility, resulting in the low efficiency of traditional oil and gas extraction techniques in tight reservoir development [4–7]. Hydraulic fracturing technology, which fractures reservoir rocks with high-pressure fluids to increase rock permeability, enabling gas to flow and be effectively extracted, is a key technology for developing tight gas reservoirs [8–11]. Inherent water saturation exists within tight sandstone deposits, and the process of fracturing introduces substantial water volumes into these deposits. This alters the typical single-phase flow in the matrix to a more complex two-phase flow of gas and water within both fractures and the matrix [12,13]. The dynamics between gas and water in such tight environments significantly surpass the complexity found in standard reservoirs, underlining the importance of a thorough grasp of gas and water movement patterns for strategic development and efficiency enhancement.

A considerable amount of research has been conducted on the flow patterns of gas and water in sandstone reservoirs. For the flow in the matrix of sandstone reservoirs, predecessors have studied the effects of factors such as pressure gradient, stress, and water saturation on the flow patterns of gas and water through numerical simulation methods [14–20]. For fluid flow in the fractures of tight reservoirs, scholars have conducted numerous studies using numerical simulation and experimental methods. Saboorian-Jooybari, H. and Wu et al. [17] proposed an innovative analytical approach to accurately determine gas–liquid relative permeability within fractures. Liu et al. [21] crafted a mathematical formulation for modeling water and gas flow within the natural fracture–shale matrix setup, triggered by hydraulic fracturing activities, which encompass aspects like hydraulic, capillary, and permeation convection, along with gas adsorption and the sealing of natural fractures. Utilizing the volume of the fluid model, Huang et al. [22] applied CFD (computational fluid dynamics) software for the delineation of the gas–liquid two-phase flow interface within micro-fracture networks, exploring gas flow traits at the micro-level and the influence of fluidic parameters on flow regimes. Wang [23] and his team's visual experiments revealed diverse gas–water two-phase flow patterns in fractures, identifying several flow typologies like bubble, stratified, slug, wave, annular-mist, and mist flows, influenced by various gas and liquid rate combinations. Chen et al. [24] adopted a method that combines large-scale physical simulation with discrete fracture numerical simulation to study the patterns of water invasion and customize water control strategies based on the water invasion patterns.

Additionally, water imbibition in tight gas reservoirs is also a critically important issue of concern. Water imbibition can affect the flow behavior of gas in reservoirs, thus impacting the productivity of gas reservoirs [25,26]. For water-bearing tight sandstone gas reservoirs, water imbibition may serve as a potential mechanism for enhanced production [27]. By thoroughly studying the laws of water imbibition, water flooding techniques can be explored to increase the recovery rate of gas reservoirs, facilitating the efficient development of oil and gas resources. Scholars both domestically and abroad have conducted extensive and in-depth experimental research as well as numerical simulation studies gradually revealing the mechanisms and influencing factors of water imbibition in tight gas reservoirs [28–31].

In the context of fractured tight sandstone gas deposits, water traverses through fractures and simultaneously seeps into the matrix via imbibition. However, most scholars tend to separate these two phenomena for research, which fails to accurately reflect water movement patterns within these reservoirs. In this study, the digital core method was employed to conduct water displacement gas experiments on tight sandstone cores with different fracture morphologies. By performing CT scans on the cores after displacement at different displacement nodes and after static conditions, the water displacement gas process in fractures and the water imbibition process in the matrix under reservoir conditions were simulated separately. Monitoring the spatial distribution of gas and water within fracture systems offers valuable insights. The impact of fracture morphology on water displacement recovery rate was analyzed. The water displacement gas recovery rate in fractures and the imbibition recovery rate in the matrix at different displacement stages were studied, and the depth of water infiltration into the matrix along the fracture walls was determined. This research approach provides a new perspective for studying the flow patterns of gas and water in fractures and the matrix in fractured gas reservoirs, and the results of the study provide data support for the evaluation of reservoir productivity and reserve estimation.

2. Methodology

2.1. Core Preparation

The research specimens were sourced from the tight sandstone formations within the Keshen sector of the Tarim Oilfield, situated in China. The reservoir burial depth is greater than 6000 m, with a pressure coefficient of about 1.8. The porosity is usually less than 8%, and the permeability is generally lower than 0.1×10^{-3} μm. Using six cylindrical core samples of tight sandstone with a diameter of 25 mm as the research objects artificial fractures were created to conduct water flooding gas displacement experiments. To simulate different types of fractures found in real reservoirs, the artificially created fractures included transverse fractures (TF), conjugate fractures (CF), and longitudinal fractures (LF). CT scans were performed on the fractured cores at a resolution of 17 μm, as shown in Figure 1. The basic information of the cores is presented in Table 1, where it is noteworthy that the fracture porosity in Table 1 represents the dry sample porosity observed under 15 MPa confining pressure in subsequent displacement experiments.

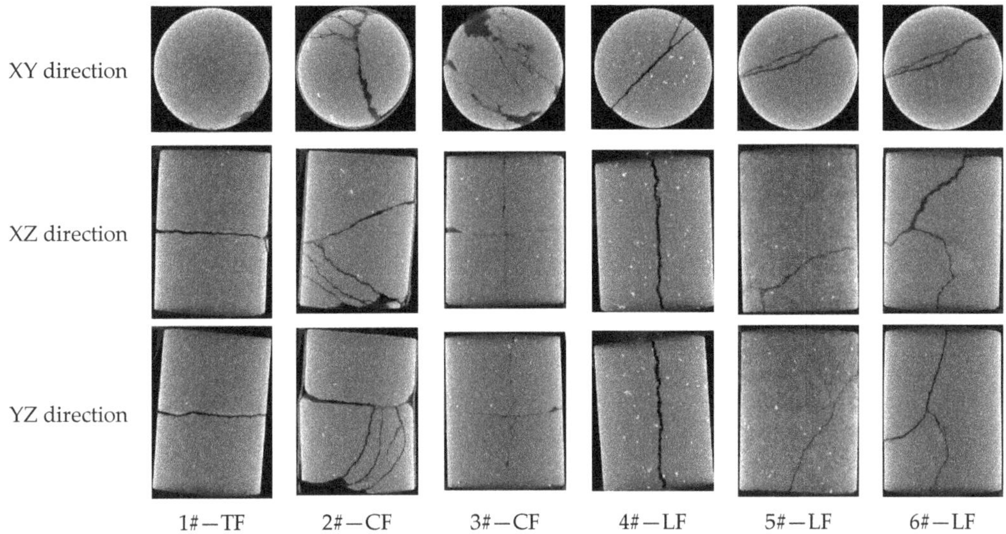

Figure 1. CT scan images of artificially fractured core samples.

Table 1. Experimental core sample information.

Number	Length, mm	Matrix Permeability of Dry Sample, 10^{-3} μm^3	Matrix Porosity of Dry Sample, %	Fracture Porosity of Dry Samples, %	Type of Fracture
No. 1	38.42	0.089	7.08	0.21	TF
No. 2	35.22	0.079	6.30	1.88	CF
No. 3	33.12	0.038	5.08	1.57	CF
No. 4	36.68	0.068	4.79	1.88	LF
No. 5	31.23	0.083	8.06	0.75	LF
No. 6	38.49	0.097	5.44	1.14	LF

2.2. Experimental Plan

Conduct water flooding displacement experiments on the above six artificially fractured core samples. To enhance the contrast of gas–water phases during displacement, the displacement solution is a water solution containing an added contrast agent. The primary steps involved in the experiment are as follows:

(1) Secure the core samples within the clamp, adjust the confining pressure to 15 MPa, the outlet pressure to 12 MPa, and set the temperature to 90 °C. Once equilibrium is achieved, execute micro-CT scans on both the top and bottom sections of the dry core samples, achieving a resolution of 17 μm.

(2) Set the injection rate to 0.005 mL/min, injecting from bottom to top. Measure the injected volume at the inlet of the core samples. Inject 0.15 pore volume (PV), 0.3 PV, 0.5 PV, 1 PV cumulative pore volume of water. After each injection, perform CT scans on the upper and lower sections of the core samples, then allow them to stand for 12 h before performing another set of CT scans. Repeat the process for each volume multiple. The

experimental process involves injecting the predetermined volume of water, performing CT scans, allowing for a 12 h static period, and conducting another CT scan. It should be noted that there are differences in the water injection nodes for each sample. Sample 6# does not undergo a 12 h static period after each water injection before injecting the next volume multiple directly.

2.3. CT Scan Image Processing

The images from CT scans at different water injection nodes were reconstructed, filtered, and denoised [32,33], followed by segmentation of the three phases: gas, water, and rock matrix. In this study, WEKA software was used for image threshold segmentation. The image threshold segmentation function in WEKA software [34,35] is mainly based on machine learning algorithms. Image segmentation progresses through stages including the extraction of grayscale features, preparation of data, creation of labeled datasets, choice of classifiers, training of the segmentation model, evaluation of the model's performance, segmentation of additional images, and their subsequent refinement, all facilitated by advanced algorithms and machine learning techniques; manual intervention is greatly reduced, thereby enhancing the accuracy and consistency of image segmentation.

2.4. Analysis of Water Flooding Flow Patterns

Based on the gas–water–rock images at different displacement stages, the process of water infiltration can be intuitively observed. Due to the limitation of scanning resolution (17 μm) and the small pore size of the tight rock matrix, CT imaging primarily reveals alterations in gas and water distribution within fractures, with changes within the matrix pores being less discernible. The post-displacement static period is the process where water infiltrates into the matrix through imbibition. The difference between the water volume in the fractures after a certain displacement node injection and the water volume in the fractures after 12 h of a static period represents the amount of water infiltrated into the matrix. At any given stage, displacement efficiency is determined by the ratio of the cumulative water volume within the fractures post-static period plus the water volume absorbed by the matrix against the core's total pore volume. It is important to note that prior to fracturing, the core's bound water saturation was established between 30% and 40%. The core pore volume used in calculations excludes the volume of bound water. By statistically analyzing the gas and water content during the water flooding process, changes in water saturation can be analyzed, thus obtaining the water flooding recovery curve.

By separately counting the water volume entering the fractures and the water volume infiltrating into the matrix during the displacement process, we can determine the contributions of fractures and matrix to the water flooding recovery. The recovery contribution from fractures is calculated as the ratio of the total water volume in the fractures after each displacement node's static period to the total volume of core pores. The matrix's contribution to recovery is quantified by the proportion of the total water volume that permeates the matrix during each displacement node's static period to the total volume of core pores.

2.5. Calculation of Water Imbibition Depth into the Matrix

During the water injection (WI) stabilization process (i.e., static period), a reduction in water volume within the fractures is indicative of water seeping into the matrix pores via imbibition. Assuming that water ingresses into the matrix uniformly along the fracture pore walls observed by micro-CT scanning, the infiltration depth of water into the matrix is calculated as the ratio of the water volume entering the matrix against the combined product of the fracture surface area and the matrix's porosity(see Equation (1)), derived by aggregating the fracture surface area, water volume entering the matrix, and matrix porosity. The depth of water infiltration into the matrix along the fracture surface can be calculated as follows:

$$h = \frac{V_w}{S_f \times \varnothing_m} \quad (1)$$

In the above equation: h represents the depth of water infiltration into the matrix along the fracture surface, m; V_w is the metric for the water volume entering the matrix, m^3; S_f represents the fracture surface area, m^2; $\varnothing m$ indicates the core matrix's porosity.

3. Results and Discussion

3.1. Distribution of Fluids at Different Displacement Stages

3.1.1. Observation of Gas and Water Distribution Images

By conducting threshold segmentation of gas, water, and rock skeleton phases on the CT scan images at different displacement nodes, observations reveal variations in gas and water distributions within fractures throughout the displacement activity, with samples 1#, 3#, and 5# illustrating these changes, as shown in Figures 2–4. In these illustrations, water is depicted in blue, gas in red, and the rock matrix in gray.

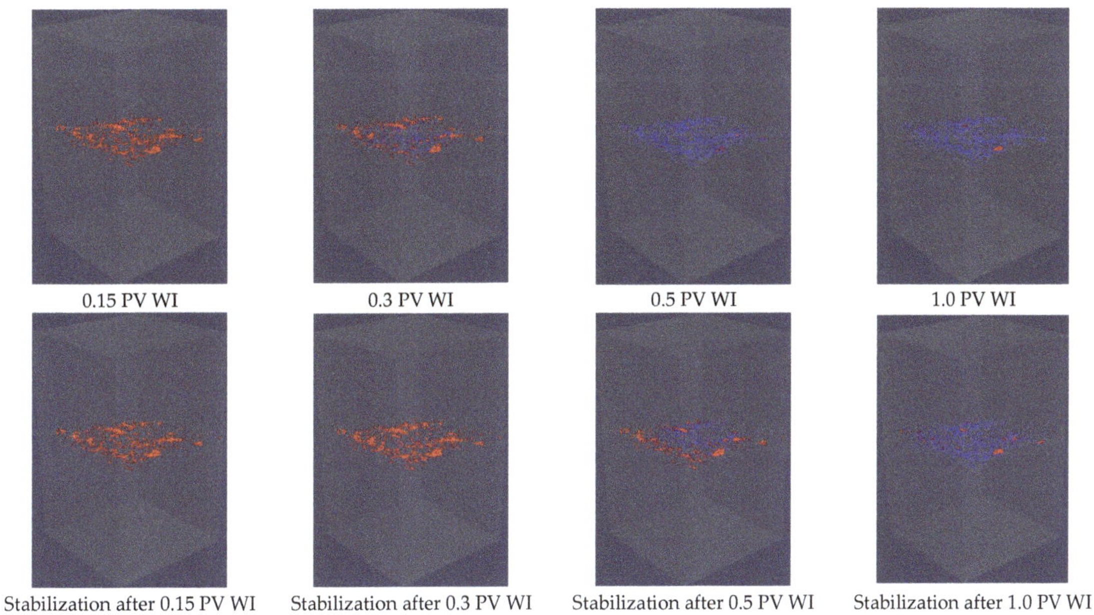

Figure 2. Gas and water distribution at different displacement nodes of sample 1#.

For the transverse fracture (Figure 2), even after injecting 0.15 PV of water and allowing for a static period, there is still no water phase observed in the fracture; this indicates the water front has yet to penetrate the fracture, necessitating additional water injections to detect the presence of water within the fracture, particularly in scenarios involving longitudinal fractures (Figure 3) and conjugate fractures (Figure 4). Water enters the fractures from the beginning of injection (0.15 PV) and gradually fills the fractures. During the static period after water injection, the water in the fractures decreases, infiltrating into the matrix of the core.

3.1.2. Water Saturation Changes

Data on the water content within fractures are gathered post-displacement across various stages and the water content in fractures after the static period is shown in Figure 5.

Compared to the artificially fractured cores where fractures do not completely penetrate longitudinally (Figure 5a,b), for cores where fractures fully penetrate longitudinally (Figure 5c–f), the water flooding front arrives at the fractures earlier. With increased volumes of the water injected, the imbibition effect transfers water from fractures into the surrounding matrix, although the extent of the water absorbed by the matrix through imbibition diminishes progressively. For the 6 tight sandstone cores in the experiment, at the moment of injecting 0.5 PV of water, over 90% of the fracture volume is filled with water, and, thereafter, there is no significant change in fracture water saturation with continued water injection.

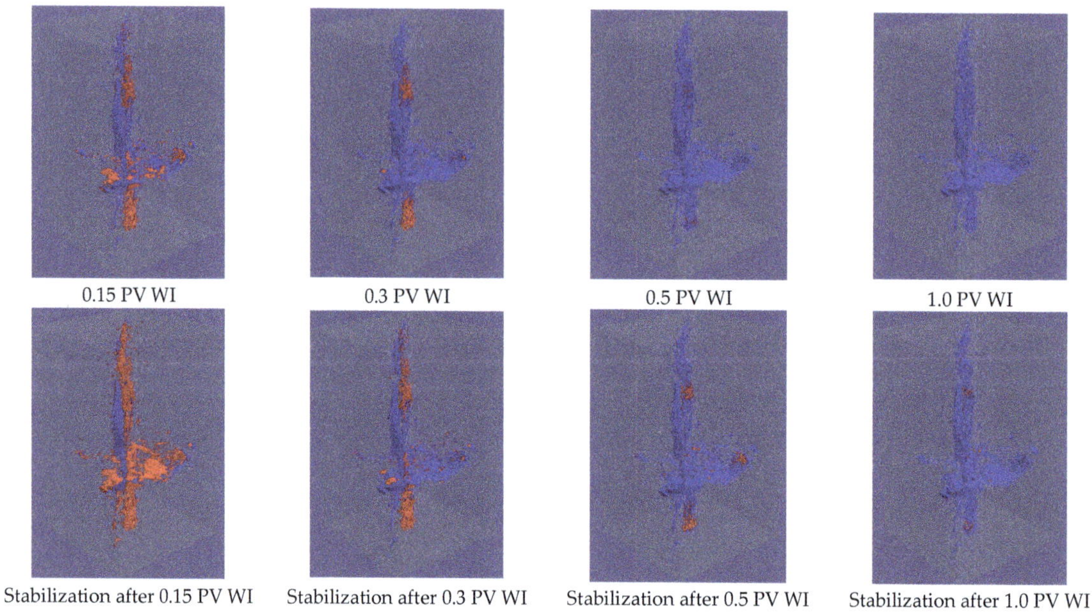

Figure 3. Gas and water distribution at different displacement nodes of sample #3.

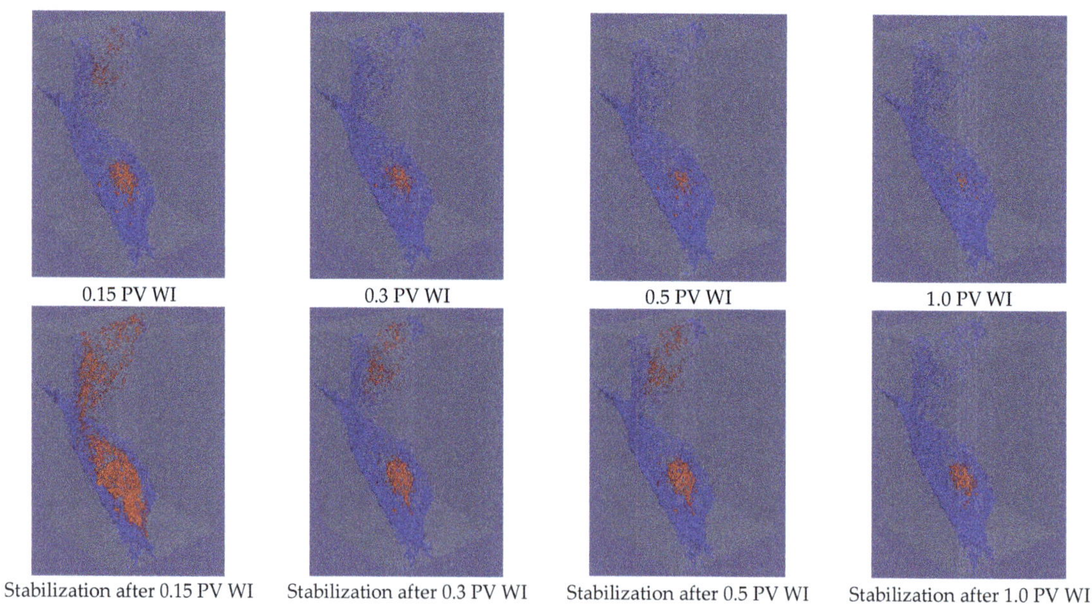

Figure 4. Gas and water distribution at different displacement nodes of sample 5#.

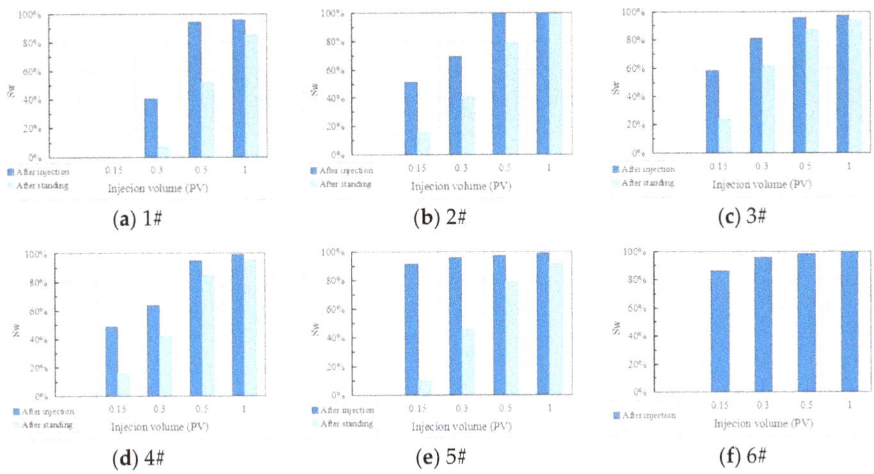

Figure 5. Water saturation in fractures at different displacement nodes.

3.2. Displacement Efficiency

The water flooding efficiency for each sample is shown in Figure 6. Before injecting 0.3 PV of water, the water flooding recovery rate increases linearly with the increasing water injection volume. At 0.5 PV of water injection, a turning point appears in the water flooding recovery rate, where the increase slows down, and the continued water injection has little effect on enhanced recovery. This indicates that, before injecting 0.5 PV of water, it is the main stage of water flooding for gas recovery. The water flooding efficiency is higher for samples 2#, 3# (CF), and 4# (LF), exceeding 50%; samples 5# and 6# (LF) have intermediate water flooding efficiency, ranging from 20% to 30%; sample 1# (TF) has the lowest water flooding efficiency, less than 10%. Therefore, conjugate fractures are most conducive to water flooding gas recovery, followed by longitudinal fractures, while transverse fractures exhibit the poorest water flooding effect. Sample 6# did not undergo a static period after water injection, thus failing to utilize the capillary action of matrix pores through imbibition. Therefore, water saturation within the fractures rapidly attains its maximum capacity, leading to negligible increases in the rate of recovery from water flooding and, ultimately, a significantly reduced efficiency in water flooding recovery.

Weighing the core samples before and after the flooding experiment allows for the calculation of water absorption by comparing their mass difference, which represents the amount of water that penetrated the core. Then, based on the core volume and porosity, the water flooding recovery rate could be calculated. The water flooding recovery rate obtained through weighing was compared with the water flooding recovery rate obtained through CT scanning using digital core technology, as shown in Figure 7.

The similarity in recovery rates determined through digital core analysis and traditional weighing methods underscores the efficacy of employing CT scanning to monitor and quantify changes in water saturation within fractures across different stages of water flooding. For instance, in sample 1# with transverse fractures, the initial water phase enters the matrix at the onset of injection, reaching the fractures only after achieving a certain level of saturation in the matrix. Given the limitations in resolution, this process is less observable in the matrix, leading to a discrepancy in recovery rates measured by CT scanning and those determined through experimental weighing.

The recovery rates through imbibition in the matrix and water flooding in the fractures are examined separately, as shown in Figure 8. It should be noted that since sample 6# did not undergo a static period after water injection, the recovery rate through imbibition is not considered.

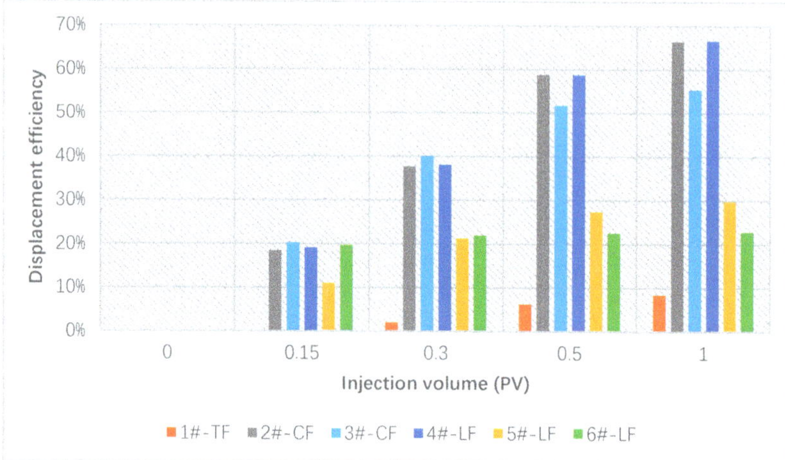

Figure 6. Water flooding recovery rate curves.

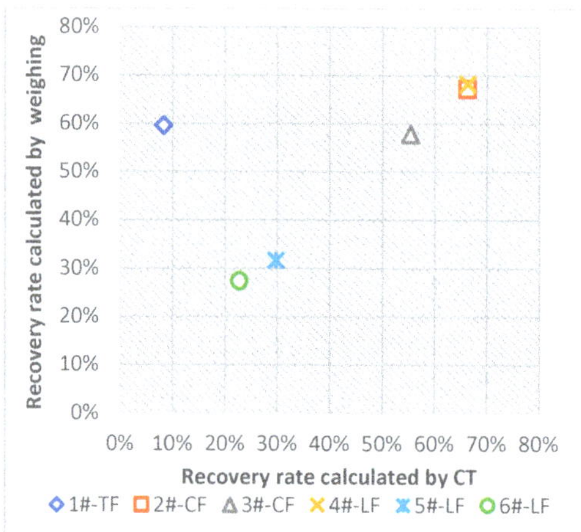

Figure 7. Recovery rate comparison between CT-simulated and laboratory weighing.

Displacement efficiency within the matrix can be categorized into three phases i Figure 8a. Initially, a rapid linear increase in imbibition recovery rate is observed unt 0.3 PV. This is followed by a deceleration in recovery rate growth between 0.3 PV an 0.5 PV. The third stage is the late stage of imbibition, where the imbibition recovery rat basically remains unchanged over time, indicating that the imbibition process is essentiall completed. The segmentation of the imbibition process into distinct stages arises from th interplay of multiple forces, including capillary, viscous, and gravitational forces, whic dominate at different phases of the imbibition, especially noticeable during the initia stage; although there are viscous and gravitational forces, the capillary force predominate and the resistance can be ignored, resulting in a strong driving force for imbibition an thus, a rapid imbibition rate. After a period of imbibition, as gravity gradually increase and cannot be ignored, the combined force of the three forces decreases, resulting in slower imbibition rate. In the late stage of imbibition, the resistance becomes increasingl significant, and the capillary force cannot overcome the resistance to drive the imbibitio process effectively, resulting in an insignificant imbibition effect.

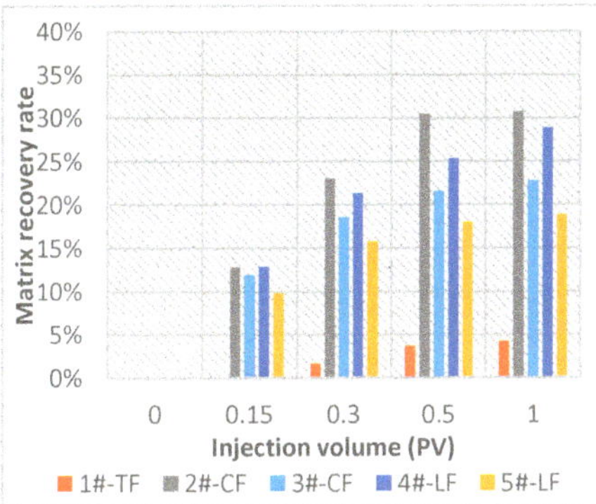

(**a**) Matrix recovery rate curves

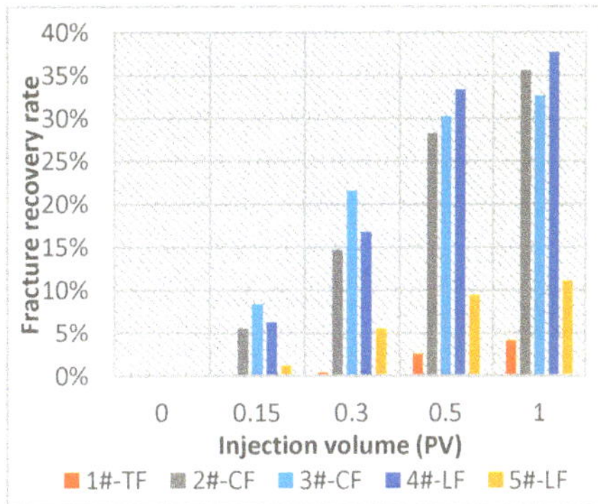

(**b**) Fracture recovery rate curves

Figure 8. Matrix and fracture recovery rate curves.

Figure 9 shows the correlation between matrix and fracture recovery rates and fracture porosity. It can be observed that both matrix and fracture recovery rates are positively correlated with fracture porosity. Except for sample 5#, the overall matrix recovery rate of the other samples is either equal to or slightly lower than the fracture recovery rate. Notably, sample 5# demonstrates a matrix recovery rate that markedly surpasses the recovery rate from fractures, while the porosity of fractures in sample 5# is relatively low. This indicates that, in reservoirs with small fracture volumes, it may be beneficial to extend the soaking time appropriately to fully utilize the imbibition effect of matrix porosity, thereby increasing reservoir recovery.

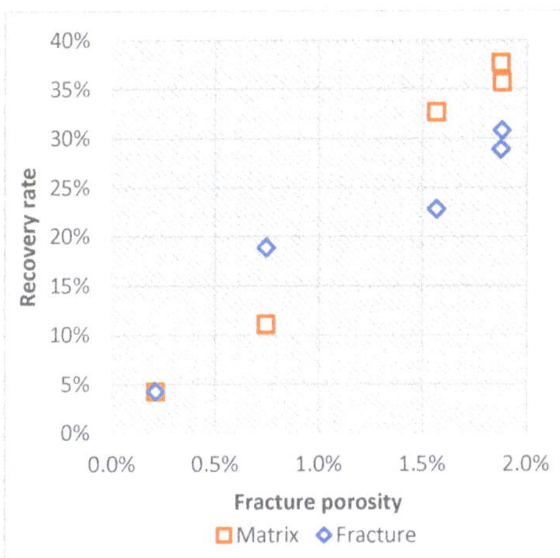

Figure 9. Relationship between matrix/fracture and fracture porosity.

3.3. The Depth of Water Invasion into the Matrix Due to Water Imbibition

The depth of water infiltration into the matrix for the rock core samples is presented in Table 2. During the static phase after water injection, the average depth of water infiltration into the matrix is 0.674 cm. Sample 1#, characterized by a very small fracture area, exhibits the greatest infiltration depth, while sample 5#, owing to its higher matrix porosity, shows a relatively smaller infiltration depth.

Table 2. The depth of water invasion into the matrix.

Core Number	Depth of Invasion (10^{-2} m)
1#	1.569
2#	0.408
3#	0.343
4#	0.691
5#	0.193

4. Conclusions

This study conducted water flooding experiments on tight sandstone cores with different fracture morphologies under reservoir conditions. Micro-CT scans were performed to observe the gas displacement process in the fractures and the process of gas displacement by imbibition. Changes in gas–water distribution in the fractures during displacement were observed, and the study closely examined how the shape and structure of fractures impact the efficacy of water flooding recovery. Moreover, the recovery rates of matrix imbibition and fracture flooding at different displacement stages were investigated, and the depth of water infiltration into the matrix along the fracture walls was calculated. Through these comparisons, the following understandings were obtained:

(1) A comparison between the water flooding recovery rates obtained using digital rock CT scanning technology and those obtained traditionally through core weighing showed good consistency, verifying the reliability of studying the water flooding process using digital rock technology.

(2) For the artificial fractured tight sandstone cores used in the experiments, the period before injecting 0.5 PV of water was the main stage of gas displacement by water flooding. Conjugate fractures were most favorable for gas displacement by water flooding, followed by longitudinal fractures, while transverse fractures showed the poorest water flooding effectiveness due to the lack of effective flow channels.

(3) Water flooding recovery rates were significantly positively correlated with fracture volume. For reservoirs with small fracture volumes, the subsequent static period after water injection (soaking period), which allowed the capillary action of the matrix pores to come into play, can significantly increase the overall water flooding recovery rates.

(4) The process of water being absorbed into the matrix is systematically divided into three distinct phases. In the initial stage, the dominate capillary forces will result in linear growth in imbibition recovery rates. During the mid-stage of imbibition (0.3 PV~0.5 PV), as the imbibition advances, the acceleration in recovery rates observed initially tends to decrease. In the late stage of imbibition (after 0.3 PV), the capillary forces are unable to overcome the resistance to drive the imbibition process, resulting in imbibition recovery rates essentially no longer changing over time.

(5) The average depth of water infiltration into the matrix along the fracture surface in this study was 0.674 cm, providing data support for guiding the evaluation of reservoir productivity and reserves and formulating production strategies for this type of gas reservoir.

Author Contributions: The research outlined herein benefitted from the combined efforts of F.C. and Y.D. in conceptualization; F.C. in methodology, software usage, validation (alongside Y.D. and K.W.), formal analysis, investigation, resource gathering, data management; and F.C.'s contributions to the original draft, review, visualization, under the supervision and project management of Y.D. All involved authors have reviewed and consented to the manuscript's published form. All authors have read and agreed to the published version of the manuscript.

Funding: This research received no external funding.

Data Availability Statement: Data will be made available on request.

Conflicts of Interest: Author Kun Wang was employed by the company Nanhai East Petroleum Research Institute, Shenzhen Branch of CNOOC Limited. The remaining authors declare that the research was conducted in the absence of any commercial or financial relationships that could be construed as a potential conflict of interest.

eferences

McGlade, C.; Speirs, J.; Sorrell, S. Unconventional Gas—A Review of Regional and Global Resource Estimates. *Energy* **2013**, *55*, 571–584. [CrossRef]

Szabó, N.P.; Remeczki, F.; Jobbik, A.; Kiss, K.; Dobróka, M. Interval Inversion Based Well Log Analysis Assisted by Petrophysical Laboratory Measurements for Evaluating Tight Gas Formations in Derecske through, Pannonian Basin, East Hungary. *J. Pet. Sci. Eng.* **2022**, *208*, 109607. [CrossRef]

Zou, C.; Tao, S.; Han, W.; Zhao, Z.; Ma, W.; Li, C.; Bai, B.; Gao, X. Geological and Geochemical Characteristics and Exploration Prospect of Coal-Derived Tight Sandstone Gas in China: Case Study of the Ordos, Sichuan, and Tarim Basins. *Acta Geol. Sin.-Engl. Ed.* **2018**, *92*, 1609–1626. [CrossRef]

Afagwu, C.; Alafnan, S.; Mahmoud, M.A.; Patil, S. Permeability Model for Shale and Ultra-Tight Gas Formations: Critical Insights into the Impact of Dynamic Adsorption. *Energy Rep.* **2021**, *7*, 3302–3316. [CrossRef]

Akilu, S.; Padmanabhan, E.; Sun, Z. A Review of Transport Mechanisms and Models for Unconventional Tight Shale Gas Reservoir Systems. *Int. J. Heat Mass Transf.* **2021**, *175*, 121125. [CrossRef]

Liu, G.; Xie, S.; Tian, W.; Wang, J.; Li, S.; Wang, Y.; Yang, D. Effect of Pore-Throat Structure on Gas-Water Seepage Behaviour in a Tight Sandstone Gas Reservoir. *Fuel* **2022**, *310*, 121901. [CrossRef]

Shilov, E.; Dorhjie, D.B.; Mukhina, E.; Zvada, M.; Kasyanenko, A.; Cheremisin, A. Experimental and Numerical Studies of Rich Gas Huff-n-Puff Injection in Tight Formation. *J. Pet. Sci. Eng.* **2022**, *208*, 109420. [CrossRef]

Verdugo, M.; Doster, F. Impact of Capillary Pressure and Flowback Design on the Clean up and Productivity of Hydraulically Fractured Tight Gas Wells. *J. Pet. Sci. Eng.* **2022**, *208*, 109465. [CrossRef]

Yekeen, N.; Padmanabhan, E.; Idris, A.K. A Review of Recent Advances in Foam-Based Fracturing Fluid Application in Unconventional Reservoirs. *J. Ind. Eng. Chem.* **2018**, *66*, 45–71. [CrossRef]

Hui, G.; Chen, S.; Chen, Z.; Jing, G.; Hu, D.; Gu, F. Role of Fluid Diffusivity in the Spatiotemporal Migration of Induced Earthquakes during Hydraulic Fracturing in Unconventional Reservoirs. *Energy Fuels* **2021**, *35*, 17685–17697. [CrossRef]

Wang, Q.; Zhou, F.; Su, H.; Zhang, S.; Dong, R.; Yang, D.; Wang, Y.; Chen, Z.; Li, J. Experimental Evaluations of Nano High-Viscosity Friction Reducers to Improve Acid Fracturing Efficiency in Low-Permeability Carbonate Reservoirs. *Chem. Eng. J.* **2024**, *483*, 149358. [CrossRef]

Gao, Y.; Chen, S.; Huang, F.; Gao, Y.; Song, T.; Wang, S.; Jia, P.; Liu, C. Micro-Occurrence of Formation Water in Tight Sandstone Gas Reservoirs of Low Hydrocarbon Generating Intensity: Case Study of Northern Tianhuan Depression in the Ordos Basin, NW China. *J. Nat. Gas Geosci.* **2021**, *6*, 215–229. [CrossRef]

Shen, W.; Xu, Y.; Li, X.; Huang, W.; Gu, J. Numerical Simulation of Gas and Water Flow Mechanism in Hydraulically Fractured Shale Gas Reservoirs. *J. Nat. Gas Sci. Eng.* **2016**, *35*, 726–735. [CrossRef]

Cao, R.; Ye, L.; Lei, Q.; Chen, X.; Ma, Y.Z.; Huang, X. Gas-Water Flow Behavior in Water-Bearing Tight Gas Reservoirs. *Geofluids* **2017**, *2017*, 9745795. [CrossRef]

Li, J.; Chen, Z.; Wu, K.; Li, R.; Xu, J.; Liu, Q.; Qu, S.; Li, X. Effect of Water Saturation on Gas Slippage in Tight Rocks. *Fuel* **2018**, *225*, 519–532. [CrossRef]

Sun, Z.; Shi, J.; Wang, K.; Miao, Y.; Zhang, T.; Feng, D.; Sun, F.; Wang, S.; Han, S.; Li, X. The Gas-Water Two Phase Flow Behavior in Low-Permeability CBM Reservoirs with Multiple Mechanisms Coupling. *J. Nat. Gas Sci. Eng.* **2018**, *52*, 82–93. [CrossRef]

Wu, Y.; Cheng, L.; Huang, S.; Fang, S.; Killough, J.E.; Jia, P.; Wang, S. A Transient Two-Phase Flow Model for Production Prediction of Tight Gas Wells with Fracturing Fluid-Induced Formation Damage. In Proceedings of the SPE Western Regional Meeting, San Jose, CA, USA, 24 April 2019; p. D021S006R006.

Li, L.; Sheng, G.; Su, Y. Water-Gas Two-Phase Flow Behavior of Multi-Fractured Horizontal Wells in Shale Gas Reservoirs. *Processes* **2019**, *7*, 664. [CrossRef]

Yao, J.; Ding, Y.; Sun, H.; Fan, D.; Wang, M.; Jia, C. Productivity Analysis of Fractured Horizontal Wells in Tight Gas Reservoirs Using a Gas–Water Two-Phase Flow Model with Consideration of a Threshold Pressure Gradient. *Energy Fuels* **2023**, *37*, 8190–8198. [CrossRef]

Zhao, X.; Song, H.; Wang, M.; Shi, X. Identification of Gas-Water Two-Phase Flow Patterns in Horizontal Wells of Shale Gas Reservoirs Based on Production Logging Data. *Geofluids* **2023**, *2023*, 6178829. [CrossRef]

Liu, H.; Hu, X.; Guo, Y.; Ma, X.; Wang, F.; Chen, Q. Fracture Characterization Using Flowback Water Transients from Hydraulically Fractured Shale Gas Wells. *ACS Omega* **2019**, *4*, 14688–14698. [CrossRef]

Huang, X.; Zhang, L.; Zhang, R.; Chen, X.; Zhao, Y.; Yuan, S. Numerical Simulation of Gas-Liquid Two-Phase Flow in the Micro-Fracture Networks in Fractured Reservoirs. *J. Nat. Gas Sci. Eng.* **2021**, *94*, 104101. [CrossRef]

Wang, Q.; Su, X.; Feng, Y.; Wang, H.; Song, J. Experimental Study of Gas-Water Two-Phase Flow Patterns in Fracture: Implication for Enhancing Coalbed Methane Production. *J. Pet. Sci. Eng.* **2021**, *207*, 109168. [CrossRef]

Chen, D.; Zhang, C.; Yang, M.; Li, H.; Wang, C.; Diwu, P.; Jiang, H.; Wang, Y. Research on Water Invasion Law and Control Measures for Ultradeep, Fractured, and Low-Porosity Sandstone Gas Reservoirs: A Case Study of Kelasu Gas Reservoirs in Tarim Basin. *Processes* **2024**, *12*, 310. [CrossRef]

25. Guo, J.; Li, M.; Chen, C.; Tao, L.; Liu, Z.; Zhou, D. Experimental Investigation of Spontaneous Imbibition in Tight Sandston Reservoirs. *J. Pet. Sci. Eng.* **2020**, *193*, 107395. [CrossRef]
26. Zhang, L.; Yu, X.; Chen, Z.; Li, J.; Hui, G.; Yang, M.; Yu, R. Capillary Dynamics of Confined Water in Nanopores: The Impact c Precursor Films. *Chem. Eng. J.* **2021**, *409*, 128113. [CrossRef]
27. Yang, L.; Wang, S.; Tao, Z.; Leng, R.; Yang, J. The Characteristics of Oil Migration Due to Water Imbibition in Tight Oil Reservoir *Energies* **2019**, *12*, 4199. [CrossRef]
28. Li, H.; Wang, C.; Li, B.; Wen, X.; Li, J.; Tian, L. Experimental Investigation on the Imbibition Behavior of Nanofluids in the Tigh Oil and Gas Reservoir through the Application of Nuclear Magnetic Resonance Method. *Energies* **2022**, *16*, 454. [CrossRef]
29. Sergeev, V.; Abe, M. Laboratory Evaluation of Reversibility Feature of Physico-Chemical Water Shut-Off Agent for Enhanced Ga Recovery. In Proceedings of the SPE Annual Technical Conference and Exhibition, San Antonio, TX, USA, 9 October 2023; D031S041R002.
30. Yang, X.; Fei, H.; Shi, J.; Sheng, L.; Guo, D.; Yao, E. Enhancing Gas Recovery in Tight Dolomite Reservoirs: A Study of Water-Loc Damage and Chemical Drainage. *Processes* **2023**, *11*, 3228. [CrossRef]
31. Zhang, X.; Xiao, K.; Wang, C.; Li, X.; Mmbuji, A.O. CO2-Assisted Technologies for the Development of Tight Gas Reservoirs: Th Implication on CCUS. *Front. Energy Res.* **2023**, *10*, 1102774. [CrossRef]
32. Buades, A.; Coll, B.; Morel, J.M. A Review of Image Denoising Algorithms, with a New One. *Multiscale Model. Simul.* **2005**, 490–530. [CrossRef]
33. Wang, Z.; Bovik, A.C.; Sheikh, H.R.; Simoncelli, E.P. Image Quality Assessment: From Error Visibility to Structural Similarit *IEEE Trans. Image Process.* **2004**, *13*, 600–612. [CrossRef] [PubMed]
34. Bultreys, T.; De Boever, W.; Cnudde, V. Imaging and Image-Based Fluid Transport Modeling at the Pore Scale in Geologica Materials: A Practical Introduction to the Current State-of-the-Art. *Earth-Sci. Rev.* **2016**, *155*, 93–128. [CrossRef]
35. Walczak, M.; Binkowski, M.; Sulikowska-Drozd, A.; Wróbel, Z. Maximum Sphere Method for Shell Patency Measurements i Viviparous Land Snails Based on X-Ray Microcomputed Tomography Imaging. *Comput. Biol. Med.* **2015**, *64*, 187–196. [CrossRe [PubMed]

processes

MDPI

Article

Digital-Rock Construction of Shale Oil Reservoir and Microscopic Flow Behavior Characterization

Jianguang Wei [1], Jiangtao Li [1,*], Ying Yang [1], Ao Zhang [1], Anlun Wang [1], Xiaofeng Zhou [1], Quanshu Zeng [2] and Demiao Shang [1]

1 Key Laboratory of Continental Shale Hydrocarbon Accumulation and Efficient Development (Northeast Petroleum University), Ministry of Education, Northeast Petroleum University, Daqing 163318, China
2 College of Petroleum Engineering, China University of Petroleum (Beijing), Beijing 102249, China
* Correspondence: lijiangtao112@163.com

Abstract: In shale oil reservoirs, nano-scale pores and micro-scale fractures serve as the primary fluid storage and migration space, while the associated flow mechanism remains vague and is hard to understand. In this research, a three-dimensional (3D) reconstruction of the shale core and micro-pore structure description technique is established; digital core technology for shale reservoirs was developed using X-ray computed tomography (X-CT), scanning electron microscope (SEM) and a focused ion beam scanning electron microscope (FIB-SEM). Microscopic oil–water two-phase flow is mimicked using the lattice Boltzmann method (LBM), a well-acknowledged approach to exploring nanoconfined fluid dynamics. In addition, coupled with digital cores, the flow characteristics of shale reservoirs are characterized. The total porosities of bedding fractures in shale and lamellar shale are 2.042% and 1.085%, respectively. The single-phase oil flow inside bedding fractures follows Darcy's linear flow principle. This work can deepen the understanding of the microscopic flow characteristics of continental shale reservoirs and provide a reference for similar problems that may be encountered.

Keywords: continental shale reservoir; digital core; lattice Boltzmann method; micro-nano scale; flow characteristics

Citation: Wei, J.; Li, J.; Yang, Y.; Zhang, A.; Wang, A.; Zhou, X.; Zeng, Q.; Shang, D. Digital-Rock Construction of Shale Oil Reservoir and Microscopic Flow Behavior Characterization. *Processes* **2023**, *11*, 697. https://doi.org/10.3390/pr11030697

Academic Editor: Albert Ratner

Received: 27 January 2023
Revised: 17 February 2023
Accepted: 23 February 2023
Published: 25 February 2023

1. Introduction

An apparent difference between shale and conventional reservoirs is that the pores in the shale matrix are mainly nano-pores with very low permeability. There are many types of shale pores with complex structures and obvious multi-scale [1]. Because of the micro-nano scale characteristics of shale pores, using traditional methods [2] to study the rock properties generally lacks quantitative description and cannot shed light on the fluid flow behavior and mechanism in the core microstructure. In the early 1990s, Dunsmuir [3] advanced X-ray computed tomography (X-CT) scanning technology and introduced it to the petroleum industry for the first time. Scholars have investigated the flow mechanism in the microstructure porous media in the petroleum engineering industry using digital core technology [4–6]. Based on the computed tomography (CT) scanning images of the realistic core sample collected from the field [7], digital core technology and a 3D model of the physical structure are established using reverse reconstruction software. Rosenberg et al. [8] introduced X-CT scanning imaging technology into the field of the digital core and established the 3D digital core of Fontainebleau sandstone. Gao [9] verified the accuracy and reliability of digital cores constructed by CT and suggested that a stricter constraint shall be applied to the alteration principle of average throat radius. Based on CT images, Liu [10] used the pore-scale reconstruction model to simulate water and CO_2 displacement, revealing the influence of pore characteristics on water displacement efficiency and the two-phase flow mechanisms.

Compared to traditional numerical methods, the lattice Boltzmann method (LBM) [11,12] is an efficient method widely used to simulate fluid percolation characteristics in a digital core. The LBM contains different discrete models for mathematical and physical problems one may face in fluid flow modeling [13,14], and its complete parallelism dramatically improves calculation efficiency. In the 1990s, Martys [15] used the LBM model of two-phase immiscible fluid to estimate the relative permeability of 3D digital cores, and its small model size shall take responsibility for the undesirable reliability of calculation results. Walsh [16] established a new partial boundary regression algorithm, which has mass conservation in heterogeneous media and derived the analytical expression for calculating the permeability of different models. Leclaire [17] performed 3D experiments on water absorption and drainage in digital Berea sandstone, and three primary states of fluid invasion are reproduced [18]. Based on the LBM and the multi-component Shan–Chen method, Lautenschlaeger [19] proposed homogenization method to simulate the fluid–fluid and solid–fluid interactions in the porous media. To solve the problem of inaccurate throat simulation with the help of machine learning, Rabbani [20] combined the advantages of pore network modeling (PNM) and LBM for the first time, and LBM is utilized to calculate throat permeability. Zhao [21] established an improved PNM to simulate single-phase flow in porous media, using LBM to simulate the fluid flow in realistic throats. Suo et al. [22] combined LBM and PNM, PNM modeled the coarse scale multi-phase transport in porous media; meanwhile, LBM solved the multi-component fluid dynamics.

To sum up, the digital core technology based on CT can genuinely reproduce the internal microstructure inside the core, and its visual accuracy advantage is irreplaceable by other methods. LBM has developed into a widely used and efficient numerical simulation method, which will play an increasingly important role in stimulating the flow characteristics of rocks. In this paper, the digital core constructed by the CT scanning method realizes the realistic microscopic pore structure of the shale core, presented in Section 1. After that, in Section 2, scanning electron microscope (SEM) images are processed to obtain the characteristic parameters of the pore structure in the shale matrix. In Section 3, the PNM of different matrix cores was obtained. Subsequently, LBM is used to quantitatively describe fluid flow characteristics in the microscale pore network in Section 4. At last, conclusions are drawn in Section 5.

2. 3D Reconstruction of Digital Shale Core by CT Scanning

The digital core technology has been widely used to advance oilfield development [23,24] including micro-flow mechanism research, oil displacement mechanism research, and reservoir production performance simulation and prediction. Meanwhile, it is realized that CT scanning is a favorable method for reconstructing the digital core, as it can reproduce the micro-pore structure of the actual core with high accuracy.

2.1. Digital Core Reconstruction Method

The core target selected in this paper is the typical shale reservoir. There are six cores divided into pure shale and Lamellar shale, and the basic parameters of the above six cores are summarized in Table 1. CT detection and digital core reconstruction using the above core mainly include the following six steps: a. Drill 25 mm diameter core on the full-diameter core with wire cutters. b. Use a CT scanner to conduct 3D scanning of a 25 mm diameter core with a scanning accuracy of 13.6 μm. c. Processing the measurement results using Avizo software, identifying identifiable bedding cracks by threshold segmentation, and calculating the width, spacing, and porosity of bedding cracks. d. Drill a core with a diameter of 2 mm at the center of the core with a diameter of 25 mm. e. Use a CT scanner to scan the core with a diameter of 2 mm in three dimensions and a scanning accuracy of 1 μm. f. Processing and calculating the measurement results by using Avizo software.

Table 1. Basic parameter of collected shale cores.

Shale-Type	Well Name	Core No.	Depth/m	Porosity/%	Permeability/mD
Pure shale	Y-47	CZ-Y47-1	2380.25	7.35	0.012
	G-851	CZ-G851-1	2479.91	6.85	0.009
	G-33	CZ-G33-1	2466	7.66	0.019
Lamellar shale	Y-47	ST-Y47-2	2379.95	7.52	0.011
	G-851	ST-G851-2	2478.21	4.72	0.008
	G-33	ST-G33-4	2398.4	7.29	0.005

2.2. Digital Core Reconstruction Method

Figure 1 shows the 3D reconstruction results of 25 mm and 2 mm diameter shale cores; the dark flake and strip areas are shale bedding fractures, and the dark spots are matrix pores; and Table 2 shows the statistical table of bedding fracture characteristic parameters of shale. The porosity of fractures in shale cores with a width greater than 13.6μm is about 1.153%, and the average fracture spacing is about 5.15 mm. The porosity of fractures in lamellar shale cores with a width greater than 13.6 μm is about 0.571%.

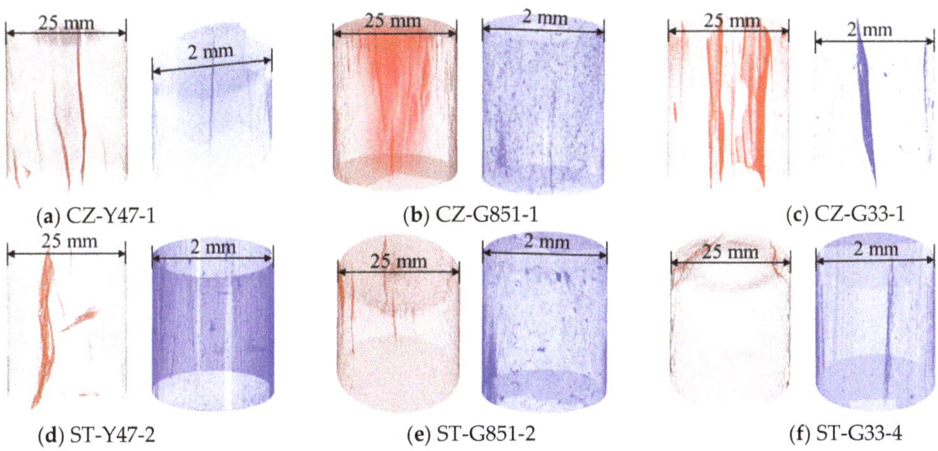

Figure 1. 3D CT reconstruction results of shale cores.

Table 2. Characteristic parameters of shale bedding fractures.

Sample No.		CZ-Y47-1	CZ-G851-1	CZ-G33-1	ST-Y47-2	ST-G851-2	ST-G33-4
Core diameter: 25 mm	Average fracture width/μm	39.4	45.5	54.2	54.6	36.5	31.2
	Average fracture spacing/mm	4.5	3.3	7.6	10.71	5.8	15.17
	Porosity/%	1.115	1.756	0.908	0.649	0.802	0.262
Core diameter: 2 mm	Average fracture width/μm	3.9	2.4	4.5	2.8	2.4	4.2
	Average fracture spacing/mm	0.6	0.3	0.8	0.5	0.8	1.2
	Porosity/%	0.828	1.019	0.717	0.713	0.382	0.446
Total porosity/%		1.943	2.776	1.625	1.363	1.184	0.708

3. Microporous Structure and Digital Core Construction

3.1. Scanning Images of SEM

Taking core fragments, the shale samples polished by argon ion were scanned b scanning electron microscope in the direction parallel to the bedding plane and perpendi ular to the bedding plane of the core, and five viewing zones were taken in each directio for scanning. Figure 2 shows the approximate position of the viewing areas in differen directions. The characteristic parameters of matrix pore structure can be obtained throug image processing and statistics. In this chapter, six cores in the first Section are scanne by SEM. Figures 3–5 shows the scanning results of CZ-Y 47-1. Although the resolution c Figure 3d is 20 μm, its HFW value increases accordingly, so these images can be reasonabl used in the process of digital core reconstruction.

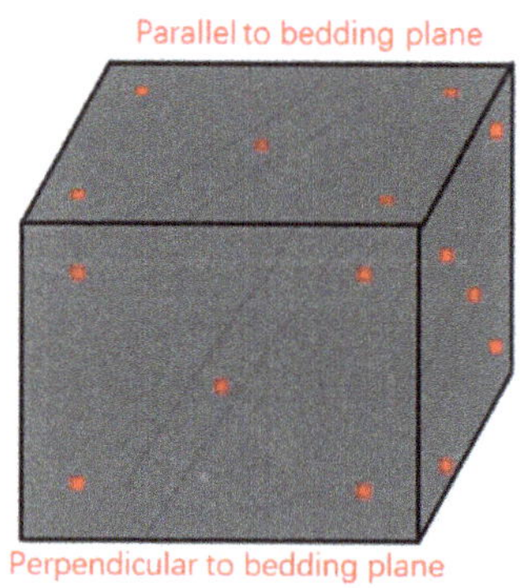

Figure 2. Schematic diagram of SEM observation position of the rock sample.

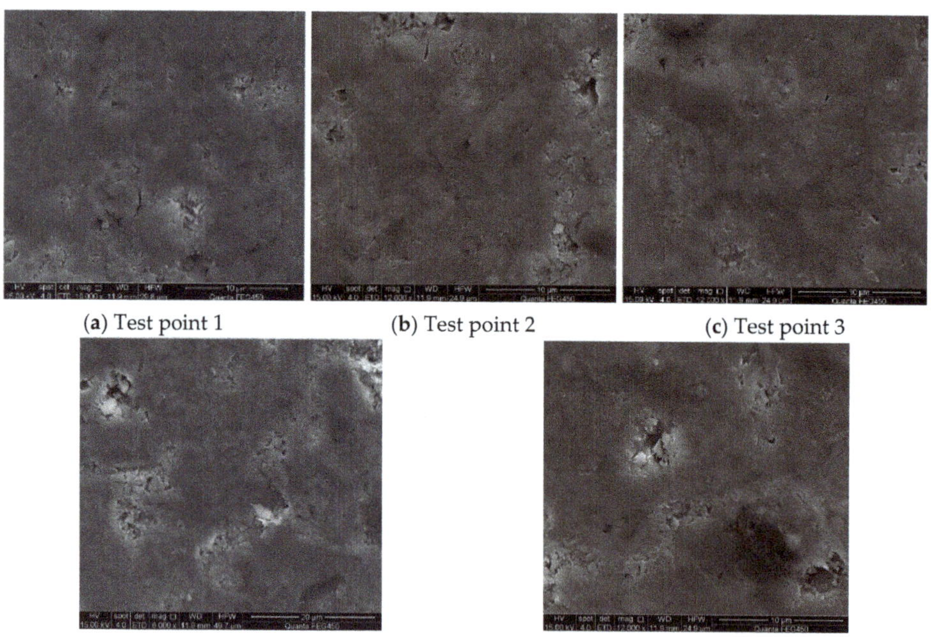

(**a**) Test point 1 (**b**) Test point 2 (**c**) Test point 3

(**d**) Test point 4 (**e**) Test point 5

Figure 3. SEM image of CZ-Y47-1 in the direction parallel to the bedding plane.

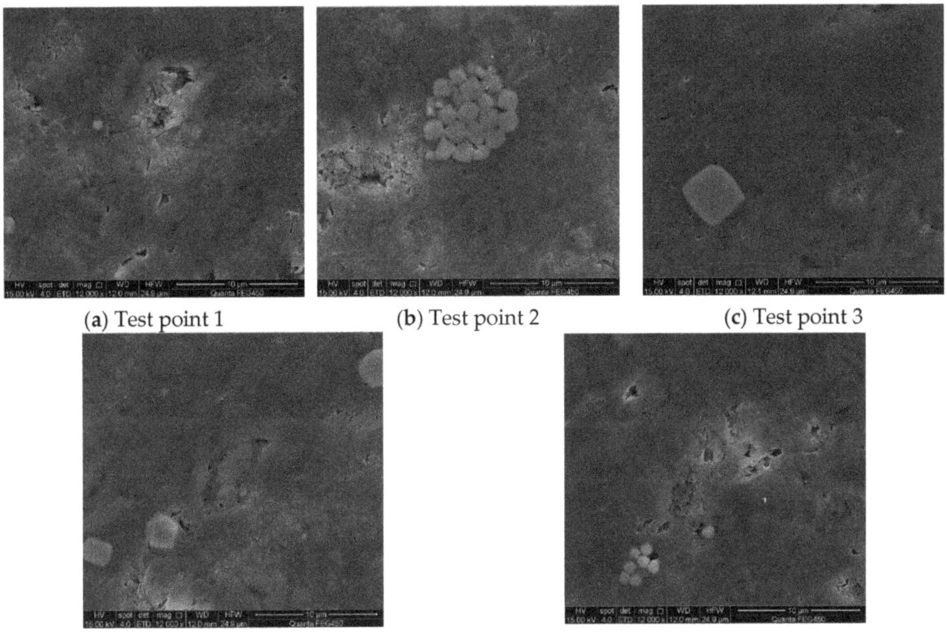

(**a**) Test point 1 (**b**) Test point 2 (**c**) Test point 3 (**d**) Test point 4 (**e**) Test point 5

Figure 4. SEM image of CZ-Y47-1 in the direction perpendicular #1 to the bedding plane.

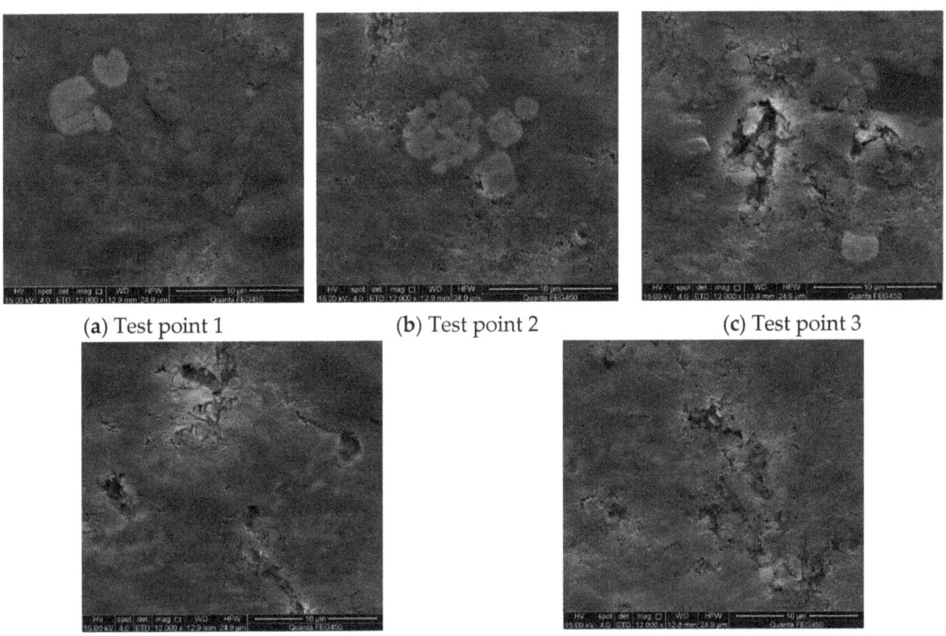

(**a**) Test point 1 (**b**) Test point 2 (**c**) Test point 3 (**d**) Test point 4 (**e**) Test point 5

Figure 5. SEM image of CZ-Y47-1 in the direction perpendicular #2 to the bedding plane.

3.2. Pore Structure Analysis

After processing the SEM images, the parameters such as specific surface area, pore–throat radius, tortuosity, shape factor and pore—throat coordination number were obtained. In this Section, CZ-Y47-1, one of the six shale samples is taken as an example, after binarizing the SEM images; as shown in Figure 6, the black pixels are matrix pores, and the matrix surface porosity results in different directions are shown in the title. The plane porosity is the ratio of the pore area to the total image area. Because of the anisotropy of shale, the core porosity is about the average porosity of each plane porosity. Firstly, the processed images are statistically analyzed. According to Equation (1), the plane porosity of three different observation directions can be calculated: 1.8%, 3.6%, and 3.4%, as shown in Figure 6:

$$\Phi_f = \frac{A_p}{A_{all}} \quad \Phi = (\Phi_{fv1} + \Phi_{fv2} + \Phi_{fp})/3 \tag{1}$$

where Φ_f is the plane porosity; A_p is the total pore area, μm^2; A_{all} is the total area of view, μm^2; Φ *is* core porosity; Φ_{fv1} is the plane porosity in #1 vertical bedding direction; Φ_{fv2} is the plane porosity in #2 vertical bedding direction; Φ_{fp} is the plane porosity in parallel bedding direction.

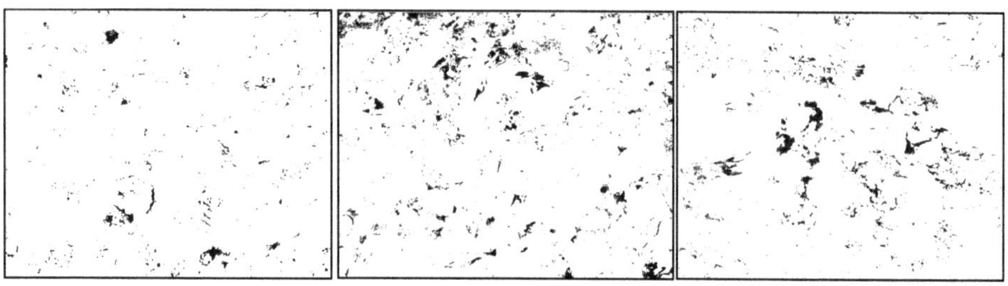

(**a**) Parallel bedding direction (**b**) #1 vertical bedding direction (**c**) #2 vertical bedding direction

Figure 6. Matrix surface porosity measured in different directions of CZ-Y47-1. (**a**) Parallel bedding direction: 1.8%, (**b**) #1 perpendicular bedding direction: 3.6%, (**c**) #2 perpendicular bedding direction: 3.4%

Then, the average core porosity of the CZ-Y47-1 core is calculated as 2.9% according to Equation (1). According to the above statistical results and Equation (2), the pore diameter distribution is the ratio of the number of pores with different diameters to the total number of pores; it can be estimated that the average pore diameter of the shale matrix is close to 0.11 μm:

$$D_{pi} = 2\sqrt{A_{pi}/\pi} \quad \overline{D}_p = \sum_{i=1}^{N} D_{pi}/N \tag{2}$$

where D_{pi} is the diameter of a single pore, μm; A_{pi} is the area of a single pore, μm^2; $\overline{D}_p$ is the average pore diameter, μm; N is the total number of pores.

The throat diameter distribution refers to the ratio of the number of throats with different diameters to the total number:

$$D_{ti} = 2\sqrt{A_{ti}/\pi} \quad \overline{D}_t = \sum_{i=1}^{M} D_{ti}/M \tag{3}$$

where D_{ti} is the diameter of a single throat, μm; A_{ti} is the area of a single throat, μm^2; $\overline{D}_t$ is the average throat diameter, μm; M is the total throat number. The tortuosity is the ratio of the actual length of the flow channel to the apparent length through the flow medium, which indicates the complexity of the fluid flow channel:

$$\tau_{ti} = \frac{S_i}{L_i} \quad \overline{\tau}_{t} = \sum_{i=1}^{M} \tau_{ti}/M \tag{4}$$

where τ_{ti} is the tortuosity of a single throat; S_i is the actual length of a single throat between two pores, μm; L_i is the linear length of a single throat between two pores, μm; $\overline{\tau}_t$ is the average tortuosity. The coordination number refers to the number of connected throats of a single pore, indicating the pore network's complexity:

$$\overline{CN}_{t} = \sum_{i=1}^{N} CN_{pi}/N \tag{5}$$

where CN_{pi} is the coordination number of a single pore. The shape factor is the ratio of the cross-sectional area of the pore and throats to the square of the perimeter, which can quantitatively show the shape characteristics of the pore-throat unit:

$$F_{pi} = \frac{A_{pi}}{S^2_{perim,i}} \quad \overline{F}_{p} = \sum_{i=1}^{N} F_{pi}/N \tag{6}$$

where $\overline{F}_p$ is the average shape factor.

By calculation according to Equations (3)–(6), the average throat diameter of the shale matrix approaches 0.034 μm; the above results are shown in Figures 7 and 8. The average tortuosity of the shale matrix is 6.53, and the average coordination number of the shale matrix is 1.77, and using the above formulas and image statistics, the parameters are summarized in Table 3.

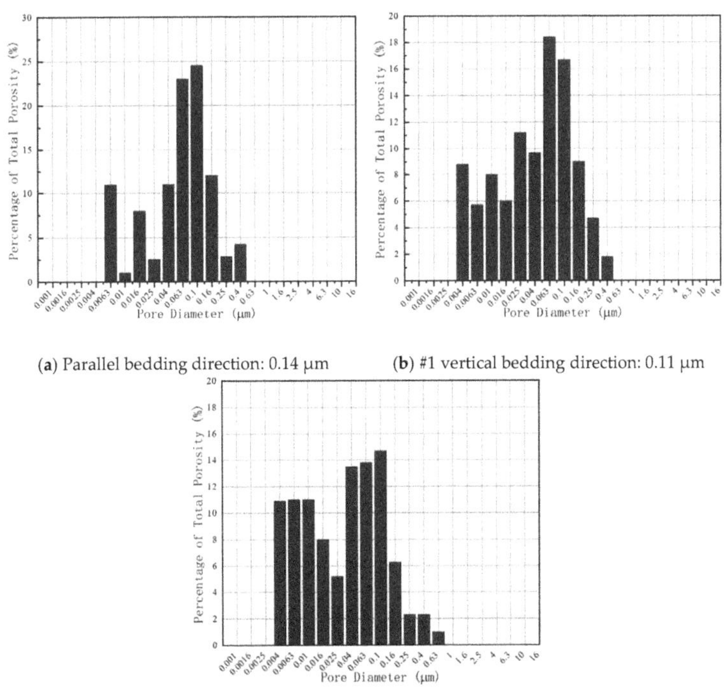

(**a**) Parallel bedding direction: 0.14 μm

(**b**) #1 vertical bedding direction: 0.11 μm

(**c**) #2 vertical bedding direction: 0.10 μm

Figure 7. Average pore diameter in different directions of CZ-Y47-1.

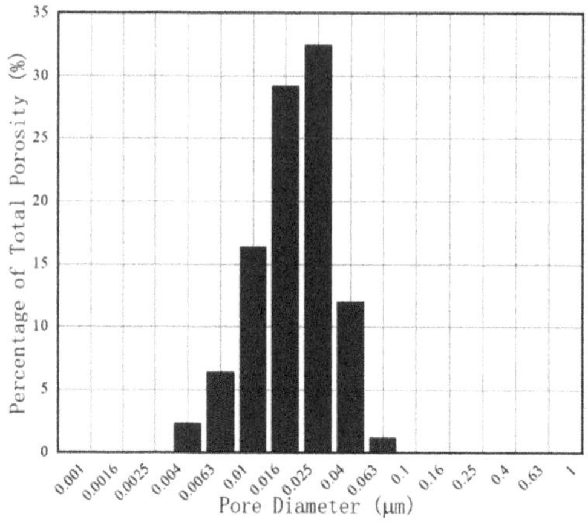

(**a**) Parallel bedding direction: 0.033 μm

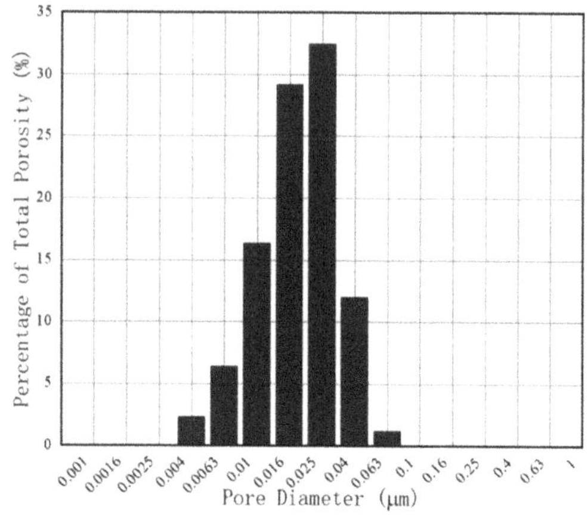

(**b**) #1 perpendicular bedding direction: 0.034 μm

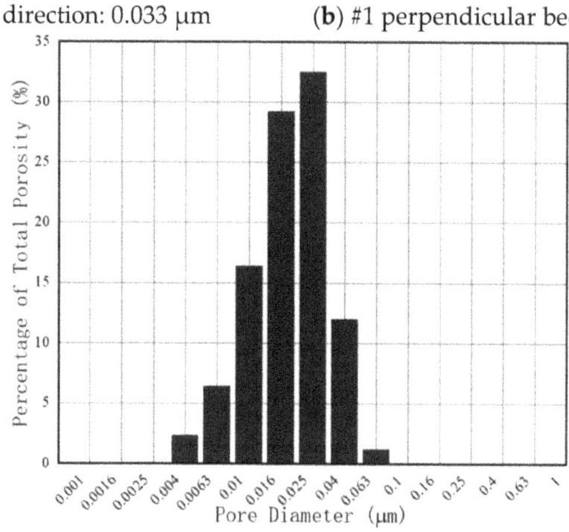

(**c**) #2 perpendicular bedding direction: 0.034 μm

Figure 8. Average throat diameter in different directions of CZ-Y47-1.

Table 3. Summary of pore and throat structure parameters.

Shale Type	No.	Sample No.	Porosity/%	Average Pore Diameter/nm	Average Throat Diameter/nm	Tortuosity	Coordination Number	Shape Factor
Pure Shale	1	CZ-Y47-1	2.9	113	34	6.53	1.77	0.037
	2	CZ-G851-1	2.8	128	32	4.78	1.85	0.034
	3	CZ-G33-1	3.3	99	32	5.65	1.89	0.033
	average value		3	113	32.6	5.65	1.837	0.035
Lamellar shale	4	ST-Y47-2	2.6	150	35	5.1	1.89	0.049
	5	ST-G851-2	1.4	163	34	4.98	1.93	0.054
	6	ST-G33-4	2.7	149	31	5.11	1.88	0.058
	average value		2.23	154	33.3	5.06	1.9	0.054

3.3. Digital Core Construction Based on SEM Images

Figures 9 and 10 present digital cores of shale matrix with average pore–throat diameters of 130 and 30 nm, respectively. They are reconstructed by the random growth method, which provides a model for subsequent micro-flow simulation with less than 5% error.

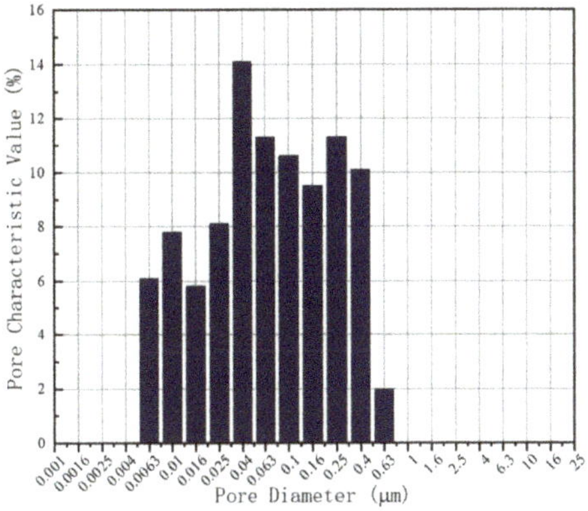

(a) The average pore diameter: 130 nm

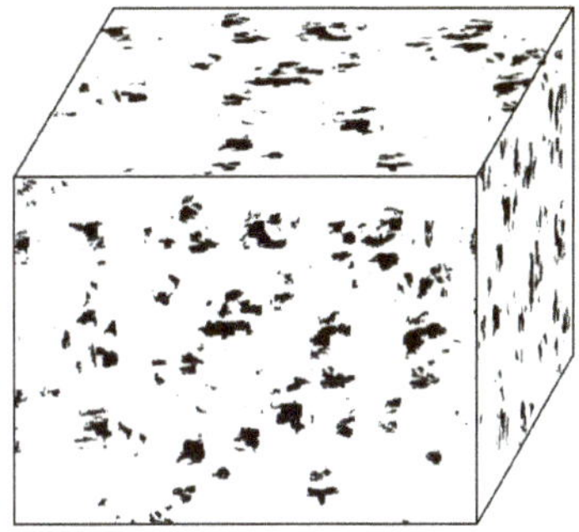

(b) Digital core

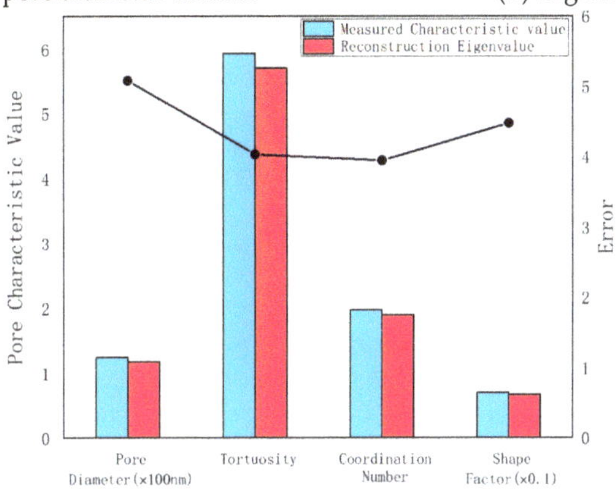

(c) The average reconstruction error: 4.3%

Figure 9. Reconstructed digital core of shale with an average pore–throat diameter of 130 nm.

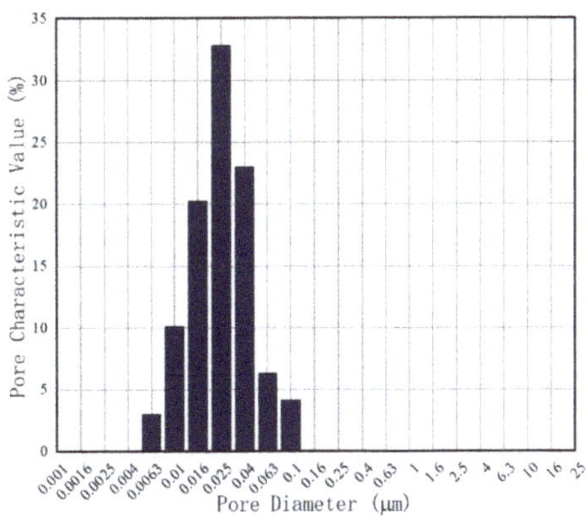

(a) The average pore diameter: 30 nm

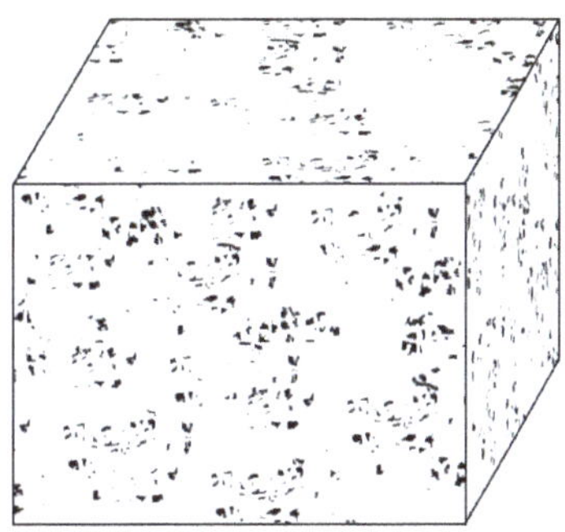

(b) Digital core

Figure 10. *Cont.*

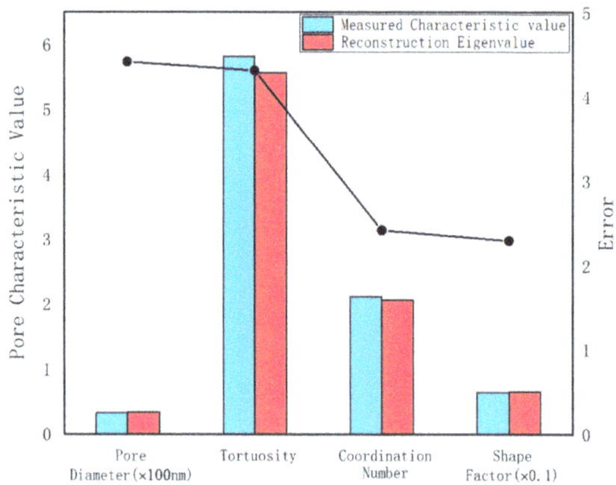

(**c**) The average reconstruction error: 3.4%

Figure 10. Reconstructed digital core of shale with an average pore–throat diameter of 30 nm.

4. Construction of Pore Network Model Based on FIB-SEM Tests

PNM, pore network modeling, can reflect the topological structure of the core pore space and has geometric parameters equivalent to the real core pore space [25,26]. Shale cores were scanned by focused ion beam scanning electron microscope (FIB-SEM) technology, and the slice spacing was close to 10 nm. Then, the PNM of different matrix cores is obtained through image processing and reconstruction.

4.1. Construction of PNM

The pixel size is 5.42 nm, and the effective measurement range is 9100 × 6476 × 3990 nm. The digital core of CZ-Y47-1 was obtained by cutting the image using Avizo software. Image segmentation of organic matter, pore, and matrix is performed to extract the structure of different components, as shown in Figure 11.

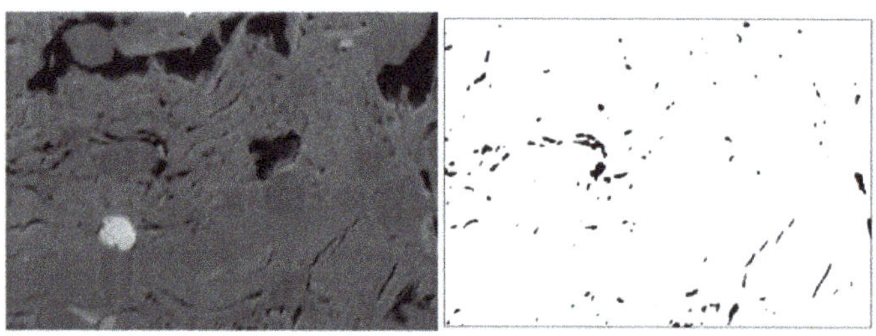

Original scanned image (**a**) Binary image of pores

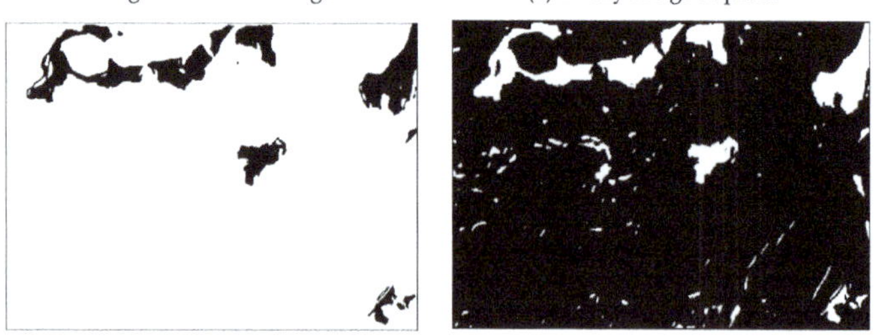

(**b**) Binary image of organic matter (**c**) Binary image of matrix minerals

Figure 11. Image segmentation and structure extraction with different compositions.

Reconstructing the binarized images to obtain the segmented 3D photos, as shown in Figure 12, and pore network modeling is generated, as presented in Figure 13.

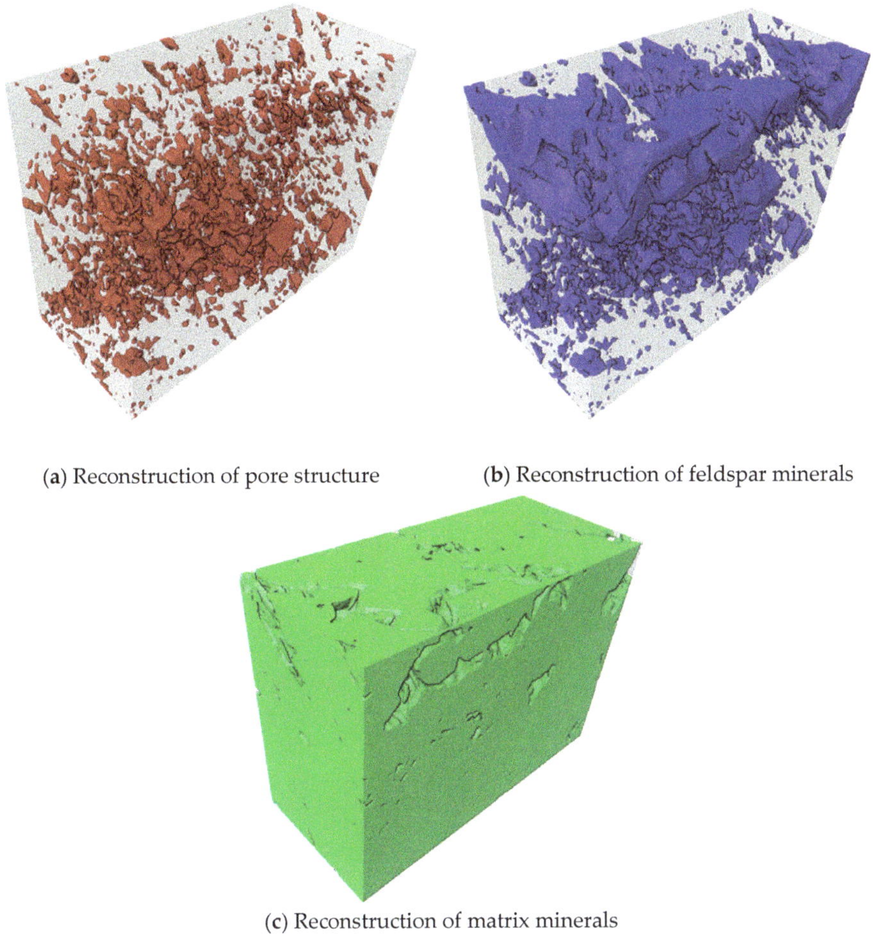

Figure 12. Reconstruction of different pores and minerals.

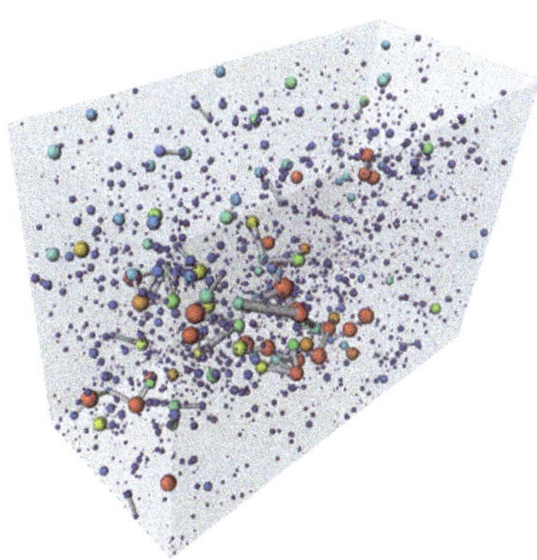

Figure 13. PNM of CZ-Y47-1.

4.2. Structural Characteristics Analysis of PNM

The classification of reservoir space type depends on the structural characteristics o the core pore network, which can be displayed intuitively and clearly by extracting th digital core pore network model. The microstructure of pores can be evaluated in detail b statistical analysis of structural parameters [27].

According to the 3D digital core of CZ-Y47-1, volume percentage statistics of th pore, organic matter, and core matrix are 1.85%, 4.12%, and 95.03%, respectively. Th fractal dimension of the generated PNM is 1.95, and the tortuosity is 8.72. The statistica distribution of core structure characteristics is shown in Figure 14.

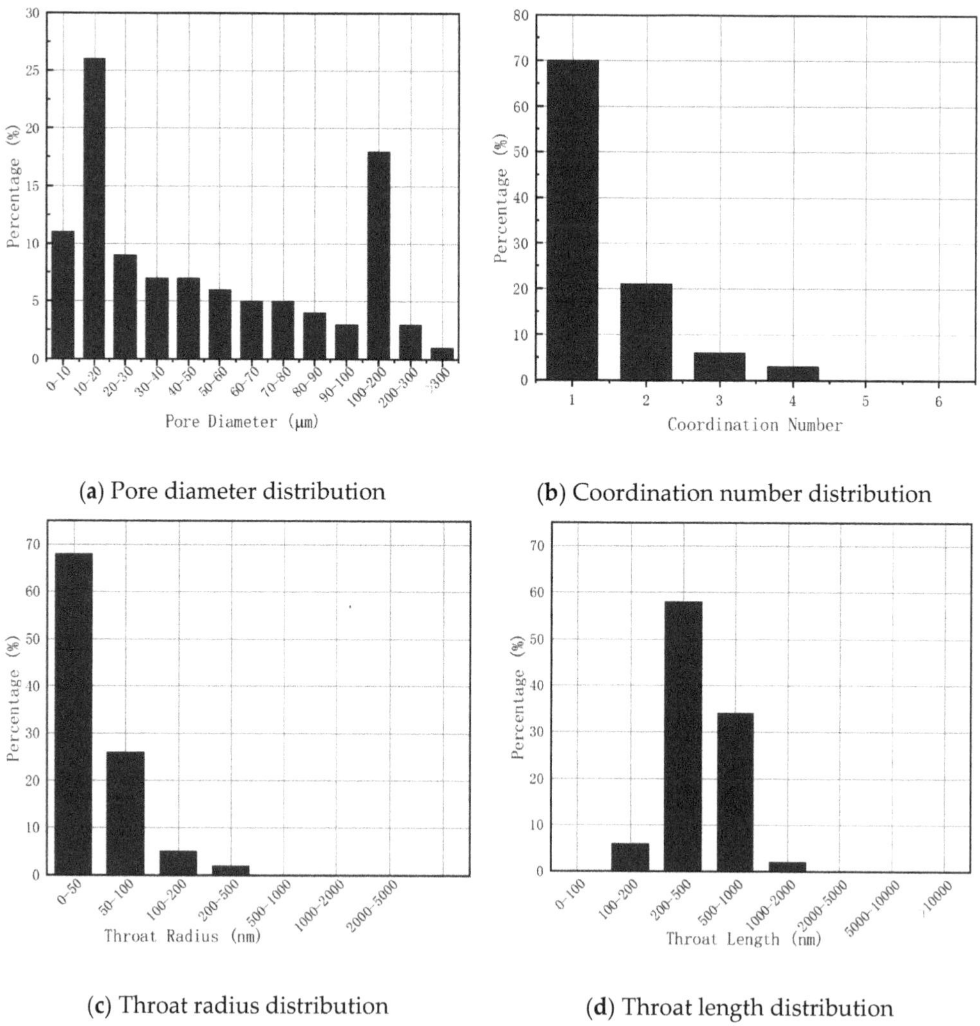

(**a**) Pore diameter distribution (**b**) Coordination number distribution

(**c**) Throat radius distribution (**d**) Throat length distribution

Figure 14. Distribution of structural characteristics of CZ-Y47-1.

The PNM parameters are shown in Table 4, and it can be observed that the porosity micropores is relatively large; the increase of micropores enhances the connectivity betwee some isolated macropores.

Table 4. Pore network parameters (unit, volume: ml; radius: nm).

Throat Number	88	Total Number of Pores	3866
Maximum throat radius	220.26	Maximum pore diameter	848.43
Average throat radius	44.58	Average pore diameter	60.07
Maximum throat volume	8.37×10^7	Maximum pore volume	3.20×10^8
Average throat volume	1.60×10^7	Average pore volume	1.13×10^6
Maximum throat length	1160.05	Maximum coordination number	4
Average throat length	454.48	Average coordination number	1.41

5. Microscopic Flow Characteristics

Based on the reconstructed digital core, LBM can be used to simulate the displacement process and quantitatively characterize the flow characteristics of the fluid in the microporous network [28]. The underlying influence of wettability, interfacial tension, and fluid viscosity on the single-phase oil flow characteristics are investigated, expecting to guide shale oil production prediction and development scheme optimization.

5.1. Establishment of the Flow Model

In 1992, Qian [29] developed the D_dQ_m model that laid a foundation for the extensive application of the LBM. The LBM can also be deduced from the continuous Boltzmann Equation:

$$\frac{\partial f}{\partial t} + e \cdot \nabla_x f + \boldsymbol{a} \cdot \nabla_\xi f = \Omega(f) \tag{7}$$

where f is the particle number density distribution function, e is the particle velocity, a is the particle motion acceleration, and $\Omega(f)$ is the collision operator.

The collision operator is a complex nonlinear integral form. A linear collision operator is proposed based on Maxwell equilibrium distribution, and the Boltzmann–BGK Equation is obtained:

$$\frac{\partial f}{\partial t} + e \cdot \nabla_x f + \boldsymbol{a} \cdot \nabla_\xi f = v(f^{eq} - f) \tag{8}$$

BGK is a continuous equation, and the discretization of velocity and space obtains the discretized lattice Boltzmann Equation. From the macroscopic statistical theory, it can be considered that all molecules are moving in specified directions, so the rate of particle movement is simplified to several rates along the specified direction $\{e_0, e_1, \ldots, e_N\}$, and N is the direction number of the rate. The distribution function f of particle number density is also discretized into $\{f_0, f_1, \ldots, f_N\}$, which defines $f_\alpha = f_\alpha\,(r, e, \alpha, t)$, $\alpha = 0, 1, \ldots, N$. The Boltzmann Equation for velocity dispersion can be obtained as follows:

$$\frac{\partial f_\alpha}{\partial t} + e_\alpha \cdot \nabla f_\alpha = \frac{1}{\tau_0}(f_\alpha^{eq} - f_\alpha) + (\boldsymbol{a} \cdot \nabla_\xi f)_\alpha \tag{9}$$

The equation is discretized in time and space, and the fully discretized LB Equation is obtained:

$$f_\alpha(\boldsymbol{r} + \boldsymbol{e}_\alpha \delta_t, t + \delta_t) - f_\alpha(\boldsymbol{r}_t, t) = -\frac{1}{\tau}\left[f_\alpha(\boldsymbol{r}_t, t) - f_\alpha^{eq}(\boldsymbol{r}_t, t)\right] + \delta_t F_\alpha(\boldsymbol{r}_t, t) \tag{10}$$

The last term on the right side of the equal sign is the external force term, and the above formula is the Lattice–BGK Equation considering external force. The LBM comprises lattice type, evolution equation of particle number density distribution, equilibrium distribution function, and boundary condition. Using the D_3Q_{19} lattice, the evolution equation is:

$$f_\alpha(\boldsymbol{r} + \boldsymbol{e}_\alpha \delta_t, t + \delta_t) = f_\alpha(r, t) - \frac{1}{\tau}(f_\alpha(r, t) - f_\alpha{}^{eq}(r, t)) \tag{11}$$

The equilibrium distribution function is:

$$f_\alpha{}^{eq} = \omega_\alpha \rho \left[1 + \frac{e_\alpha \cdot u}{c_s{}^2} + \frac{(e_\alpha \cdot u)^2}{2c_s{}^4} - \frac{u^2}{2c_s{}^2}\right] \tag{12}$$

where, c_s is the lattice sound speed, $c_s = \frac{c}{\sqrt{3}}$, and ω_α is the weight coefficient:

$$\omega_\alpha = \begin{cases} 4/9, & c_\alpha{}^2 = 0 \\ 1/9, & c_\alpha{}^2 = c^2 \\ 1/36, & c_\alpha{}^2 = 2c^2 \end{cases} \tag{13}$$

Consider the boundary conditions of flow:

$$\begin{cases} \rho = \int f de = \sum f_\alpha & \alpha = 1, 2, \cdots, 19 \\ u = \frac{\int f e de}{\rho} = \frac{\sum f_\alpha e_\alpha}{\rho} & \alpha = 1, 2, \cdots, 19 \\ P = c_s^2 \rho \end{cases} \tag{14}$$

Based on the above equations, a single-phase flow LBM model is established considering the interfacial tension and liquid–solid wettability. The viscosity of the oil phase can influence the relaxation time:

$$\tau = \frac{v M_w}{RT\delta_t} + 0.5 \tag{15}$$

The oil phase adsorption layer on the pore wall affects the oil phase flow in the pore The oil–solid interfacial tension (γ) and contact angle (θ) jointly determine the adhesive force. The contact angle is only used to describe the strength of fluid–solid force and does not indicate the existence of multiphase fluid in porous media; additionally, with the increase of the contact angle, the wettability of the wall gradually tends to water.

$$F_s = r\gamma_{so}(1 + \cos\theta) \tag{16}$$

$$F_\alpha = \left(1 - \frac{1}{2\tau}\right) \frac{(c_\alpha - u) \cdot F_s}{\rho c_s^2} f_\alpha^{eq} \tag{17}$$

5.2. Percolation Characteristics of Crude Oil in Matrix Pores

Based on the established shale matrix digital cores with average pore–throat diameters of 130 nm and 30 nm, the permeability of the digital cores is calculated, and the matrix permeability of the shale core is measured by a matrix permeability meter. The simulation parameters are shown in Table 5. Compared with the two models, the error between the model verification measurement results and the simulation results is only 6.2%, clarifying the accuracy and reliability of the proposed model. In this paper, the length scale L_0 equals 5×10^{-8} m, the mass scale M_0 equals 1.51×10^{-19} kg, and the time scale T_0 equals 1.51×10^{-10} s. Then, the physical pore length ($L_{physics}$) can be calculated by $L_{physics} = L_0 \times L_{lattice}$, where $L_{lattice}$ is the lattice length in LB simulation. The physical viscosity ($v_{physics}$) can be calculated by $v_{physics} = v_{lattice}\frac{L_0^2}{T_0}$, where $v_{lattice}$ is lattice viscosity. The physical density ($\rho_{physics}$) can be calculated by $\rho_{physics} = \rho_{lattice}\frac{M_0}{L_0^3}$, where $\rho_{lattice}$ is the lattice density.

Table 5. Flow simulation parameters.

Physical Parameters	Value
Pressure gradient (MPa/m)	1, 5, 10, 20, 50, 100, 150, 200, 250, 300
Oil viscosity (mPa·s)	0.7, 0.9, 1.1, 1.3
Oil density (g/cm^3)	0.83
Contact angle (degree)	10, 30, 50, 70
Oil solid interfacial tension (mN/m)	0.5, 0.7, 0.9, 1.1
The proportion of water-wetted minerals (%)	0, 30, 70, 100

With increased hydrophilic mineral proportion, the starting pressure gradient of single-phase oil flow in the shale matrix decreases, as seen in Figure 15.

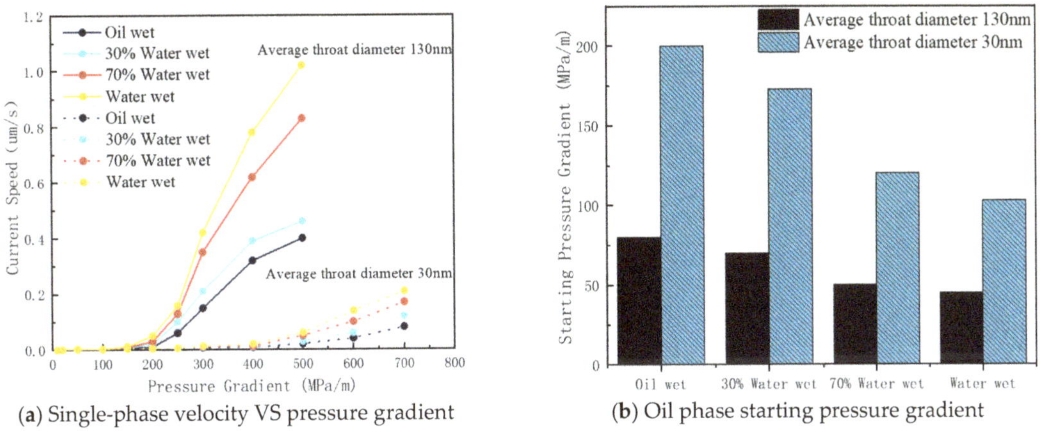

Figure 15. Oil pressure diagram of hydrophilic minerals with different proportions.

As illustrated in Figure 16, with the increase in the proportion of hydrophilic minerals, the adequate area oil flow increases, and the pressure decreases in large-size pores. The adsorption force the water-wet surface imposes on oil molecules is less than that with the oil-wet surface; as a result, oil-flowing resistance in water-wet pores is reduced compared to oil-wet pores, leading to relatively high flow efficiency.

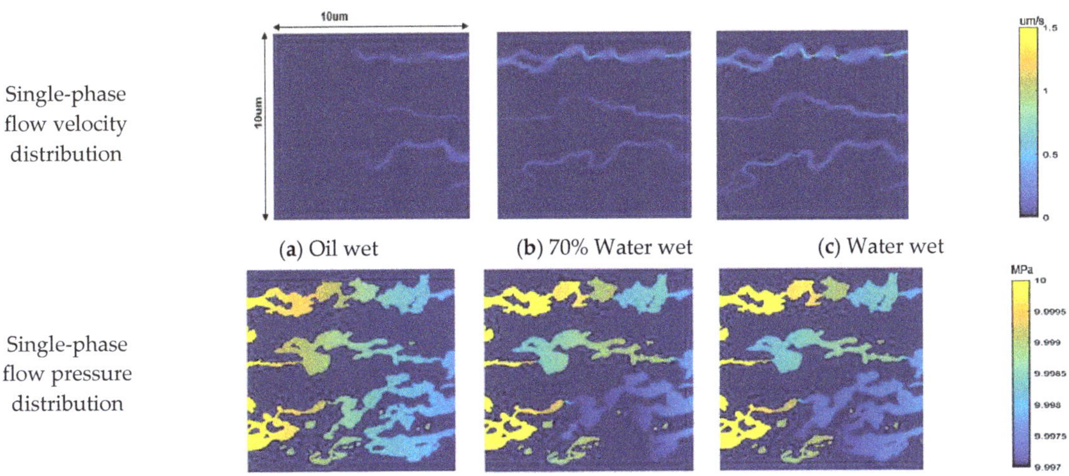

Figure 16. Simulation results of hydrophilic minerals with different proportions.

It can be seen from Figure 17 that the starting pressure gradient of single-phase oil flow in the shale matrix decreases with the increase of wetting angle.

As can be seen from Figure 18, with the increase of wetting angle, the area where oil can flow increases, and the pressure in large-size pores decreases.

Figure 19 shows that the starting pressure gradient of single-phase oil flow in the shale matrix increases with the increase of interfacial tension.

It can be seen from Figure 20 that with the increase of interfacial tension, the oil-flow area decreases, and the pressure in large-size pores increases. The adsorption force of oil on the pore wall increases, and the flow characteristics become more complicated.

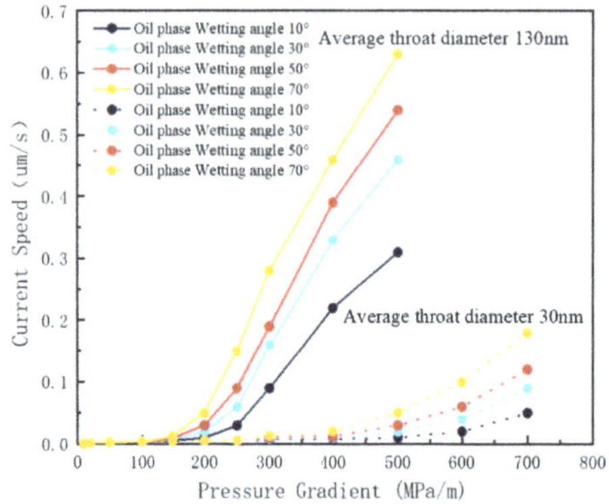

(**a**) Single-phase velocity VS pressure gradient

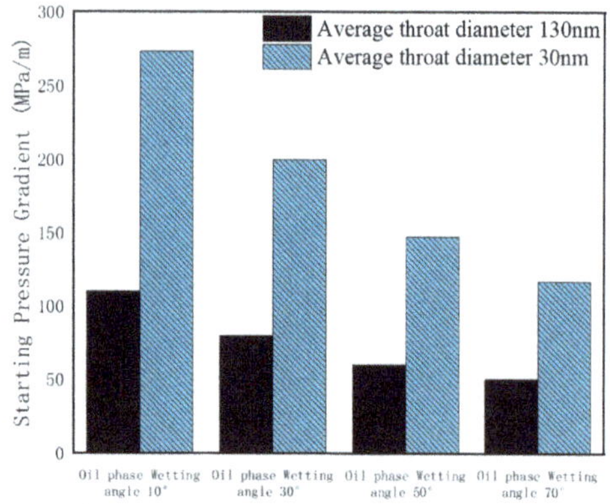

(**b**) Oil phase starting pressure gradient

Figure 17. Fluid pressure diagram under different wetting angles of oil.

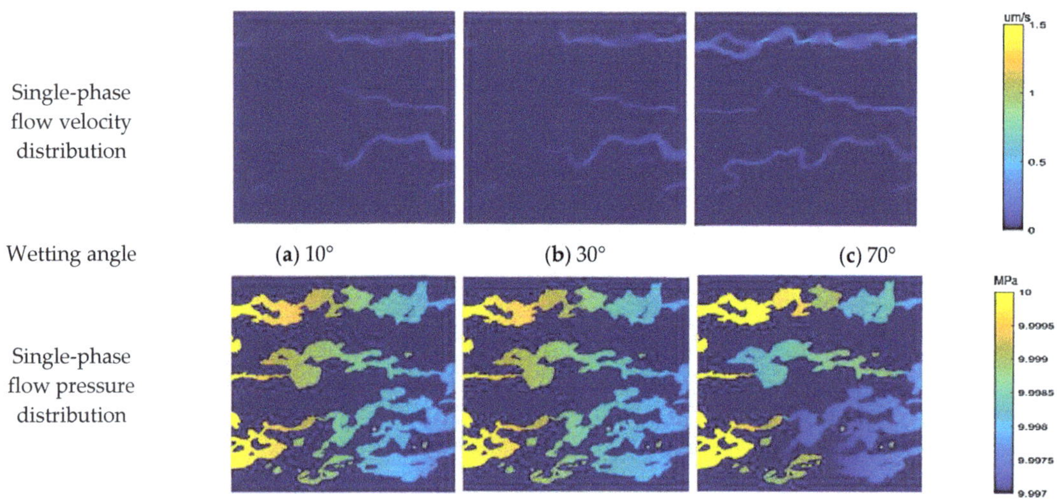

Figure 18. Simulation results under different wetting angles of the oil phase.

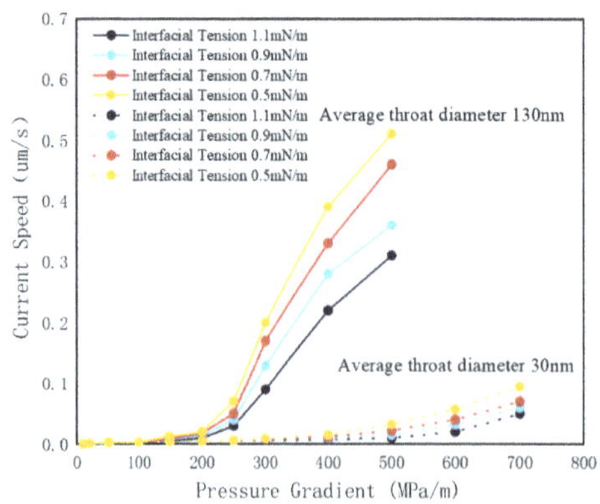

(**a**) Single-phase velocity VS pressure gradient

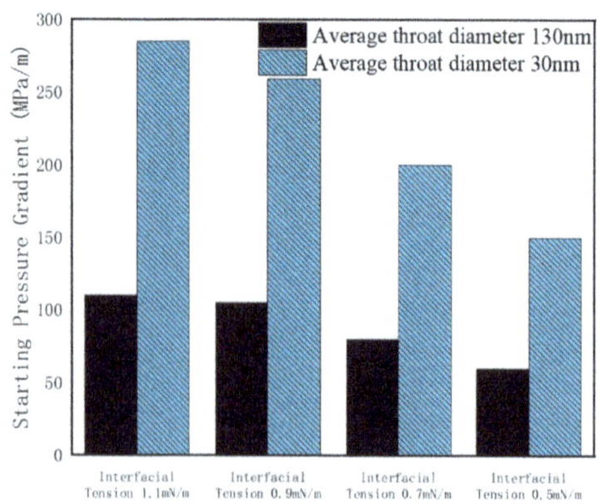

(**b**) Oil phase starting pressure gradient

Figure 19. Fluid pressure diagram under different interfacial tension.

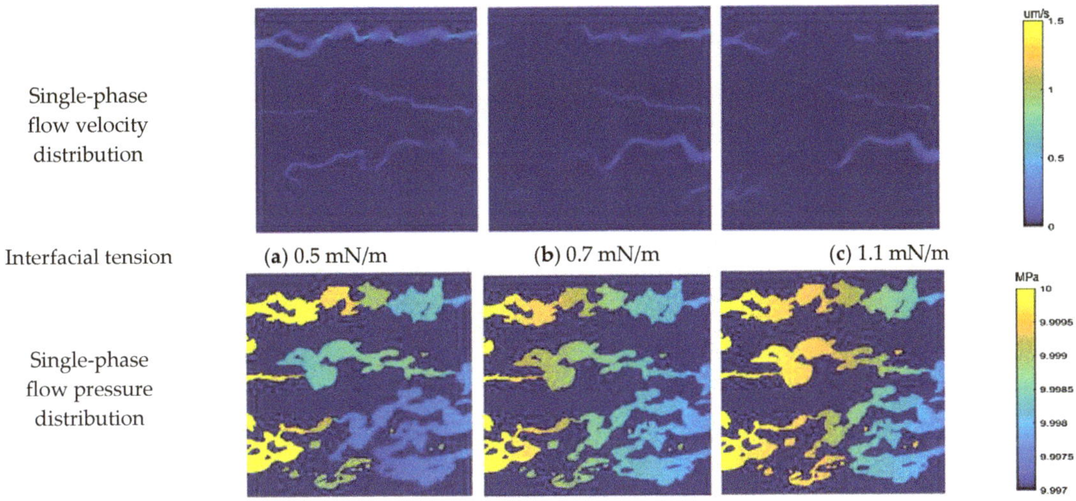

Figure 20. Simulation results under different interfacial tension.

Figure 21 shows that as the viscosity of the oil phase increases, the starting pressure gradient of single-phase oil flow in the shale matrix increases.

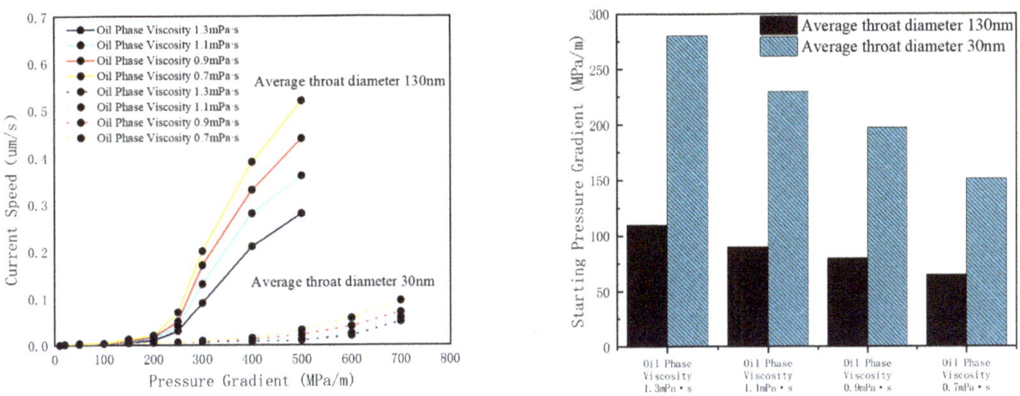

(**a**) Single-phase velocity vs. pressure gradient (**b**) Oil phase starting pressure gradient

Figure 21. Fluid pressure diagram under different oil phase viscosities.

It can be seen from Figure 22 that when the viscosity of the oil phase increases, the oil-flow area decreases, and the pressure in large-size pores decreases. The increasing oil viscosity makes it more difficult to flow in the pores.

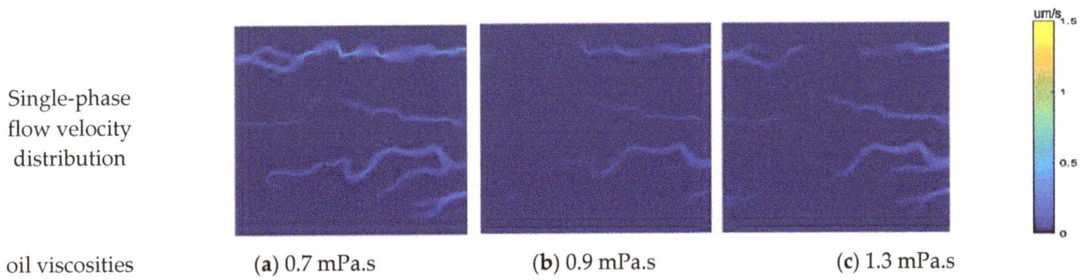

Figure 22. *Cont.*

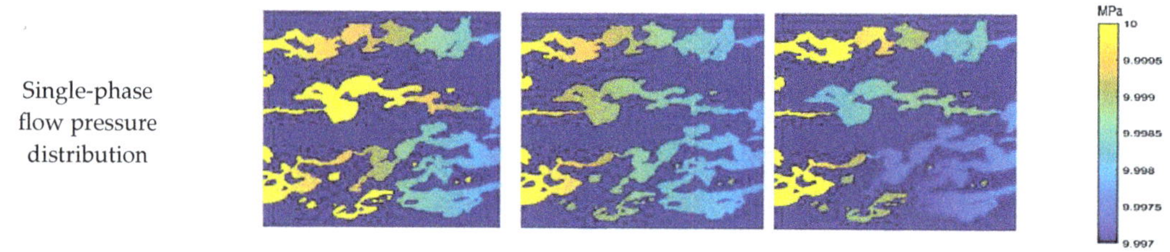

Figure 22. Simulation results under different oil phase viscosities.

According to the single-phase starting pressure gradient curve, the average pore-throat diameters are 30 nm and 130 nm, and the starting pressure gradient of flow in the matrix is 0.103–0.285 and 0.045–0.110 MPa/mm, respectively.

5.3. Flow Characteristics of the Oil Phase in Bedding Fractures

Based on this digital core, the percolation characteristics of single-phase oil in the bedding fractures are simulated. As seen from Figure 23, there is no starting pressure gradient in the bedding fracture flow, while the effect of stress is not considered, and the velocity has a linear relationship with the pressure gradient.

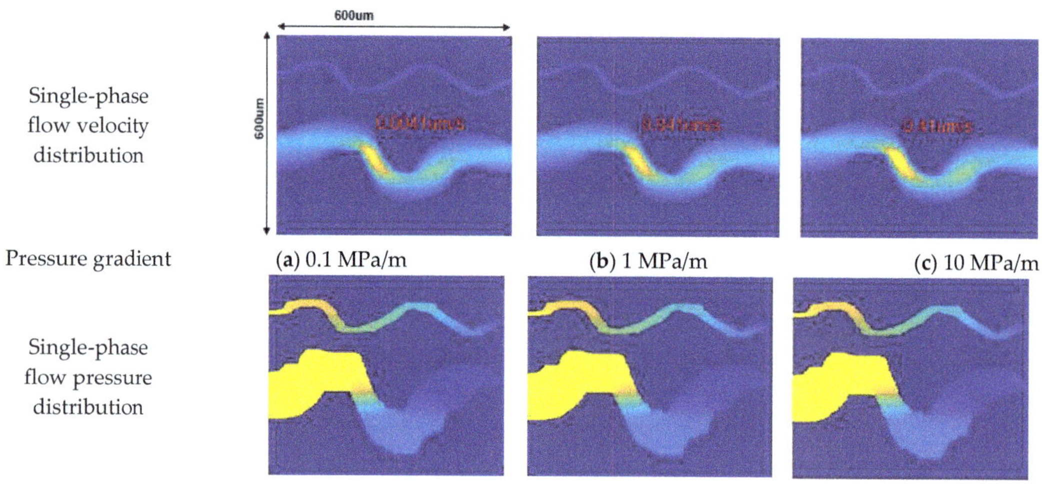

Figure 23. Simulation results of oil in bedding fractures under different pressure gradient conditions

Figure 24 shows the relationship between the single-phase flow velocity of crude oil and pressure gradient under different influential factors. The results show that the single-phase flow follows Darcy's principle, and there is no threshold pressure gradient. The more significant the proportion of hydrophilic minerals is, the greater the wetting angle of the oil phase is, the smaller the interfacial tension is, and the higher the flow velocity of the oil phase is. In addition, the higher the viscosity of the oil phase, the lower the flow rate of the oil phase.

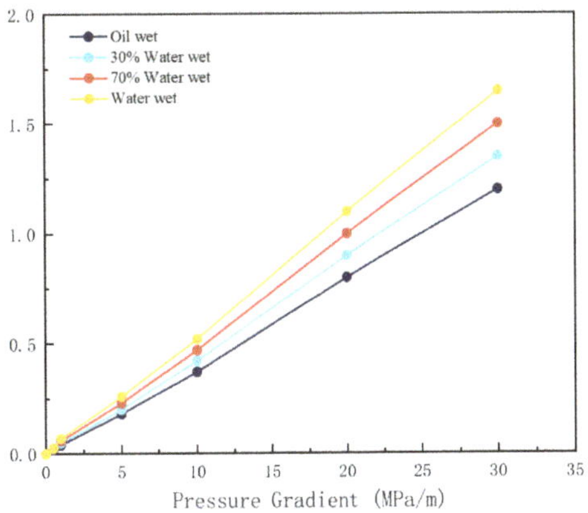

(**a**) Proportion of hydrophilic minerals

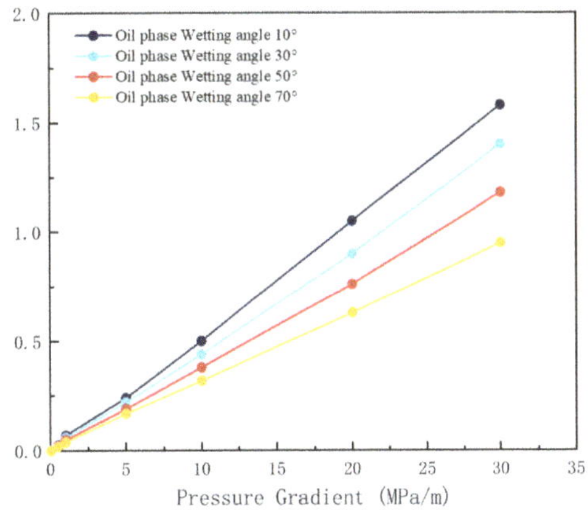

(**b**) Oil phase contact angle

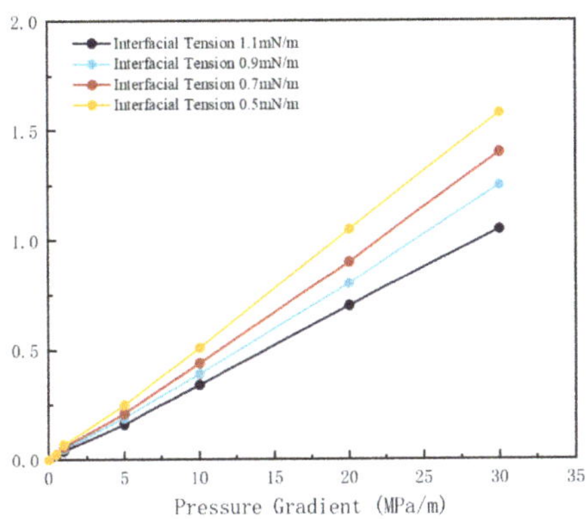

(**c**) Interfacial tension

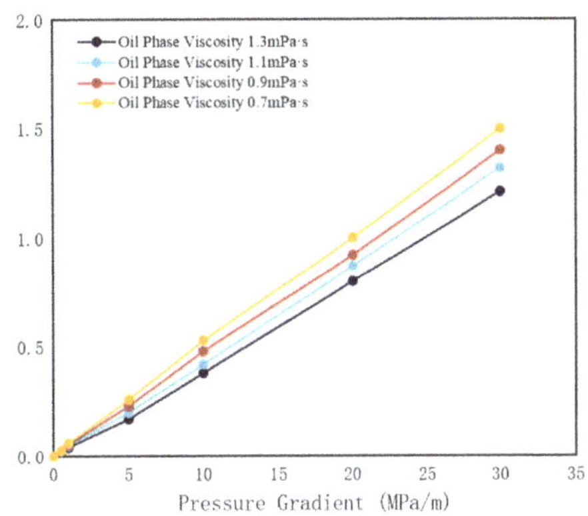

(**d**) Oil phase viscosity

Figure 24. Relationship curve between oil phase flow rate and pressure gradient under different conditions.

6. Conclusions

In this paper, the CT scans of cores with diameters of 25 mm and 2 mm were carried out, six 3D digital cores were reconstructed, and the characteristics of shales and lamellar shales were obtained. SEM and FIB-SEM were used to rebuild the digital core, obtain the characteristic parameters of the pore structure of the shale matrix, and reconstruct the 3D pore network model of the shale matrix. Considering the viscosity, interfacial tension, and wettability of fluid, the LBM model of shale matrix/bedding fracture flow is established, and the microscopic flow characteristics of shale oil under different conditions are calculated. The following conclusions were drawn:

(1) Total porosities of bedding fractures in shale and lamellar shale are 2.042% and 1.085%, respectively; the average fracture spacing is 0.5 and 0.9 mm, and the average fracture widths are 6.8 and 5.5 μm.

(2) The average porosities of the shale and lamellar shale matrix cores obtained by SEM are 3% and 2.23%, the average pore diameters are 113 and 154 nm, the average throat diameters are 32.6 and 33.3 nm, the average tortuosities are 5.65 and 5.06, and the average shape factors are 0.035 and 0.054, respectively.

(3) The average pore throat diameter is 30 nm, and the starting pressure gradients in the matrix are 0.103~0.285 and 0.045~0.110 MPa/mm, respectively. The mechanism of single-phase flow in shale bedding fractures is the same as Darcy's linear flow.

(4) The accuracy of the digital core model constructed by the three modeling method needs further improvement. Although it can reflect the structural characteristics of th reservoir, the single structure method and mixed numerical reconstruction metho still need to be improved to simulate the pore structure and improve the applicatio results accurately.

Author Contributions: Methodology, J.W. and Q.Z.; Investigation, Y.Y.; Resources, X.Z.; Data cura tion, A.W.; Writing—original draft, D.S.; Visualization, A.Z.; Supervision, J.L. All authors have rea and agreed to the published version of the manuscript.

Funding: This work was supported by the School Youth Foundation of Northeast Petroleum Universit (Grant No. 2019QNL-40), the National Natural Science Foundation of China (Grant No. 51474070).

Data Availability Statement: Data will be made available on request.

Acknowledgments: This work was supported by the School Youth Foundation of Northea Petroleum University (Grant No.2019QNL-40), the National Natural Science Foundation of Chin (Grant No. 51474070).

Conflicts of Interest: The author(s) declare(s) that there is no conflict of interest regarding th publication of this paper.

References

1. Hou, P.; Liang, X.; Zhang, Y.; He, J.; Gao, F.; Liu, J. 3D multi-scale reconstruction of fractured shale and influence of fractu morphology on shale gas flow. *Nat. Resour. Res.* **2021**, *30*, 2463–2481. [CrossRef]
2. Soulaine, C.; Creux, P.; Tchelepi, H. Micro-continuum framework for pore-scale multi-phase fluid transport in shale formatio *Transp. Porous Media* **2019**, *127*, 85–112. [CrossRef]
3. Dunsmuir, J.H.; Ferguson, S.R.; D'Amico, K.L.; Stokes, J.P. X-ray microtomography: A new tool for the characterization porous media. In Proceedings of the SPE annual technical conference and exhibition, Dallas, TX, USA, 6 October 1991; OnePetr Dallas, TX, USA, 1991.
4. Bera, A.; Shah, S. A review on modern imaging techniques for characterization of nanoporous unconventional reservoi Challenges and prospects. *Mar. Pet. Geol.* **2021**, *133*, 105287. [CrossRef]
5. Walls, J.; Diaz, E.; Cavanaugh, T. Shale reservoir properties from digital rock physics. In Proceedings of the SPE/EAGE Europea Unconventional Resources Conference and Exhibition, Vienna, Austria, 20 March 2012; OnePetro, 2012.
6. Li, G.; Diaz, E.; Nur, A. Rock physical properties computed from digital core and cuttings with applications to deep gas exploratic and development. In Proceedings of the SPE Deep Gas Conference and Exhibition, Manama, Bahrain, 24–26 January 20 OnePetro, 2010.
7. Albert, K.; Franz, C.; Koenigsdorff, R.; Zosseder, K. Inverse estimation of rock thermal conductivity based on numerical microsca modeling from sandstone thin sections. *Eng. Geol.* **2017**, *231*, 1–8. [CrossRef]
8. Rosenberg, E.; Lynch, J.; Gueroult, P.; Bisiaux, M.; De Paiva, R.F. High resolution 3D reconstructions of rocks and composites. C *Gas Sci. Technol.* **1999**, *54*, 497–511. [CrossRef]
9. Gao, X.; Qi, Y.; Song, X.; Zhao, T. Parallel comparison of digital core analysis and real core test. *Spec. Oil Gas Reserv.* **2015**, 2 93–96+145.
10. Liu, J.; Song, R. Investigation of water and CO2 flooding using pore-scale reconstructed model based on micro-CT images Berea sandstone core. Progress in Computational Fluid Dynamics. *Int. J.* **2015**, *15*, 317–326.
11. Han, Y.; Liu, K. Numerical simulation workflow using carbonate digital rock built from integrated lattice Boltzmann and krigi method. In Proceedings of the SPE Asia Pacific Oil & Gas Conference and Exhibition, Virtual, 17–19 November 2020; OnePetro, 20
12. Yue, W.; Tian, B.; Lu, Y. Investigation on pore-scale transport properties of sonic wave using numerical simulation with digit rock. In Proceedings of the 2017 SEG International Exposition and Annual Meeting, Houston, TX, USA, 24–27 September 20 OnePetro, 2017.
13. Lautenschlaeger, M.P.; Prifling, B.; Kellers, B.; Weinmiller, J.; Danner, T.; Schmidt, V.; Latz, A. Understanding Electrolyte Filli of Lithium-Ion Battery Electrodes on the Pore Scale Using the Lattice Boltzmann Method. *Batter. Supercaps* **2022**, *5*, e202200 [CrossRef]
14. Shodiev, A.; Primo, E.; Arcelus, O.; Chouchane, M.; Osenberg, M.; Hilger, A.; Franco, A.A. Insight on electrolyte infiltration lithium ion battery electrodes by means of a new three-dimensional-resolved lattice Boltzmann model. *Energy Storage Mater.* **20** *38*, 80–92. [CrossRef]
15. Martys, N.; Douglas, J. Critical properties and phase separation in Lattice Boltzmann fluid mixtures. *Phys. Rev. E* **2001**, *63*, 312 [CrossRef]
16. Walsh, S.; Burwinkle, H.; Saar, M. A new partial-bounceback lattice-Boltzmann method for fluid flow through heterogeneo media. *Comput. Geosci.* **2009**, *35*, 1186–1193. [CrossRef]

17. Leclaire, S.; Parmigiani, A.; Chopard, B.; Latt, J. Three-dimensional lattice Boltzmann method benchmarks between color-gradient and pseudo-potential immiscible multi-component models. *Int. J. Mod. Phys. C* **2017**, *28*, 1750085. [CrossRef]
18. Afsharpoor, A.; Javadpour, F. Pore connectivity between organic and inorganic matter in shale: Network modeling of mercury capillary pressure. *Transp. Porous Media* **2018**, *125*, 503–519. [CrossRef]
19. Lautenschlaeger, M.P.; Weinmiller, J.; Kellers, B.; Danner, T.; Latz, A. Homogenized lattice Boltzmann model for simulating multi-phase flows in heterogeneous porous media. *Adv. Water Resour.* **2022**, *170*, 104320. [CrossRef]
20. Rabbani, A.; Babaei, M. Hybrid pore-network and lattice-Boltzmann permeability modelling accelerated by machine learning. *Adv. Water Resour.* **2019**, *126*, 116–128. [CrossRef]
21. Zhao, J.; Qin, F.; Derome, D.; Kang, Q.; Carmeliet, J. Improved pore network models to simulate single-phase flow in porous media by coupling with lattice Boltzmann method. *Adv. Water Resour.* **2020**, *145*, 103738. [CrossRef]
22. Suo, S.; Liu, M.; Gan, Y. An LBM-PNM framework for immiscible flow: With applications to droplet spreading on porous surfaces. *Chem. Eng. Sci.* **2020**, *218*, 115577. [CrossRef]
23. Ly, C.V.; Mintz, J.; Bian, A.; Andrew, M.; Bhattiprolu, S. Inter-Scaled Digital Core Analysis: From Core to Pore and Back. In Proceedings of the SPE/AAPG/SEG Unconventional Resources Technology Conference, Austin, TX, USA, 24–26 July 2017; OnePetro, 2017.
24. Sun, Z.; Li, X.; Liu, W.; Zhang, T.; He, M.; Nasrabadi, H. Molecular dynamics of methane flow behavior through realistic organic nanopores under geologic shale condition: Pore size and kerogen types. *Chem. Eng. J.* **2020**, *398*, 124341. [CrossRef]
25. Xu, M. *Method of Digital Core Construction and Pore Network Extraction*; Southwest Petroleum University: Chengdu, China, 2017.
26. Sun, Z.; Huang, B.; Liu, Y.; Jiang, Y.; Zhang, Z.; Hou, M.; Li, Y. Gas-phase production equation for CBM reservoirs: Interaction between hydraulic fracturing and coal orthotropic feature. *J Petrol. Sci. Eng.* **2022**, *213*, 110428. [CrossRef]
27. Jianpeng, Z.; Likai, C.; Hui, C.; Ning, L.; Ziliang, W.; Yao, M.; Guichao, D. Quantitative characterization of rock microstructure of digital core based on CT scanning. *Geoscience* **2020**, *34*, 1205–1213.
28. Sun, Z.; Wang, S.; Xiong, H.; Wu, K.; Shi, J. Optimal nanocone geometry for water flow. *AIChE J.* **2022**, *68*, e17543. [CrossRef]
29. Qian, Y.; d'Humières, D.; Lallemand, P. Lattice BGK models for Navier-Stokes equation. *EPL Europhys. Lett.* **1992**, *17*, 479. [CrossRef]

processes

MDPI

Article

Study on Gas Invasion Behavior of Gas–Liquid Displacement in Fractured Reservoirs

Cheng Ye [1], Jiaqin Gong [1], Kecheng Liu [1], Jingjing Pei [1], Shengjiang Xu [1] and Peng Xu [2,*]

1 PetroChina Xinjiang Oilfield Engineering Technology Research Institute, Karamay 834000, China
2 School of Petroleum Engineering, Yangtze University, Wuhan 430100, China
* Correspondence: cdxupeng@yangtzeu.edu.cn

Abstract: When drilling or exploiting fractured formations, gas fluid displacement and invasion often occur, and gas invasion is very subtle and difficult to find. The gas in the fracture enters the wellbore and arrives near the wellhead with the drilling fluid. Improper treatment may lead to serious accidents such as lost circulation and blowout. In this study, using computational fluid dynamics (CFD) simulation software for modeling and grid generation, based on the volume of fluid (VOF) method, the gas invasion behavior under different conditions was simulated to explore the flow process and characteristics of gas invasion, and the effects of different drilling fluid properties and fracture morphology on gas invasion were analyzed. The experimental results show that the drilling fluid enters the fracture to compress the gas, making the pressure in the fracture greater than that in the wellbore, thus leading to the occurrence of gas invasion. The viscosity and density of the drilling fluid have different effects on the gas invasion process. The higher the viscosity, the smaller the possibility of gas invasion. However, when the viscosity of the drilling fluid gradually increases from 10–50 MPa·s, the change of gas invasion rate is small, all within 1.0–1.2 m/s. The higher the density, the more conducive to the occurrence of gas invasion. The inlet pressure has no obvious effect on the occurrence of gas invasion, and the occurrence time of the gas invasion fluctuates in 0.35 s at 0.5–2.5 MPa. With the increase in the fracture width and length, the possibility of gas invasion decreases, but there is an extreme value for the fracture height. The time of gas invasion does not change beyond this extreme value. When the fracture height is 100–700 mm, the time of gas invasion increases with the increase in the height; when the height is 700–900 mm, the gas invasion time does not change. These results provide a practical and effective method for enhancing oil recovery, preventing and treating gas invasion in gas–liquid flooding.

Keywords: gas invasion; gas–liquid displacement; gas–liquid two-phase flow; fracture; CFD simulation

Citation: Ye, C.; Gong, J.; Liu, K.; Pei, J.; Xu, S.; Xu, P. Study on Gas Invasion Behavior of Gas–Liquid Displacement in Fractured Reservoirs. *Processes* **2022**, *10*, 2533. https://doi.org/10.3390/pr10122533

Academic Editors: Tao Zhang, Zheng Sun, Dong Feng, Hung Vo Thanh and Wen Zhao

Received: 2 November 2022
Accepted: 25 November 2022
Published: 29 November 2022

1. Introduction

In the process of oil and gas exploration drilling, when drilling multi-fracture and multi-cave formations, the drilling fluid will leak, and the gas in the formation will invade the wellbore [1]. When the drilling fluid is oil-based, under a certain pressure and temperature, the gas will dissolve in the oil-based drilling fluid to form dissolved gas, which is not easy to find in the early overflow detection [2,3]. When it rises to the wellhead, it can easily cause blowout and other accidents, reducing the oil and gas recovery rate. When the drilling fluid is water-based, the ground overflow monitoring is obvious, and the gas will slide and expand when entering the wellbore, and the upper overshoes may be damaged if the well is shut in for a long time [4–7]. The invasion of drilling fluid gas not only brings difficulties to drilling engineering, but also causes casualties and economic losses [8–15]. In recent years, scholars at home and abroad have also conducted a lot of research on the problem of gas invasion in fractured gas reservoirs. In order to reduce the harm caused by gas invasion, it is integral to systematically study the occurrence conditions mechanism, influencing factors, and prevention measures of gas invasion in gas–liquid displacement [16–18].

It is generally believed that gas invasion will occur when the local formation pressure is higher than the wellbore pressure. However, in the process of drilling, when the bottom hole is in a positive differential pressure state, gas invasion will also occur [19]. Beenion [20] analyzed and studied the drilling and completion process of a fractured gas reservoir. The research results show that in fractured formations, drilling fluid leakage and gas invasion can easily occur under negative pressure due to the difference in the density of the gas and the liquid, that is, blowout and leakage occur simultaneously. Nickens [21] considered the effects of pump speed, formation gas invasion, blowout preventer, throttle valve closing, and throttle valve adjustment. During gas invasion, the model can predict the change of flow rate and pressure with time and wellbore location. Zhang [22] assumed that gas invasion in gas–liquid displacement occurred at the bottom of the well, determining the gas invasion rate. The process of gas invasion from gas–liquid displacement to underbalanced gas invasion was studied. Zhang [23] assumed that the gas invasion occurred at the bottom hole during gas–liquid displacement and measured the gas invasion rate. The process of gas invasion from gas–liquid displacement to under equilibrium gas invasion is studied. Hou [24] simulated a real fracture by establishing a fracture space, and carried out visual experimental research on the gas invasion of gas–liquid displacement according to the gas–liquid displacement mechanism of fracture formation. On the basis of analyzing the reservoir geological characteristics of Shunnan Block in the northern slope of Tazhong, Fang [25] established a gas invasion model for gas–liquid displacement, simulated the gas invasion process of gas–liquid displacement with fluid computing software, and analyzed the main factors affecting the gas invasion of gas–liquid displacement [26].

The gas–liquid displacement invasion is a typical gas–liquid two-phase flow, which is very complicated in calculation, and it is difficult to observe the state of gas invasion intuitively in the laboratory. In this paper, Ansys Fluent is used for numerical simulation, the VOF multiphase flow model is used based on volume fraction equation and momentum equation (referring to the official instruction manual: https://ansyshelp.ansys.com/account/secured?returnurl==/Views/Secured/prod_page.html?pn=Fluent&pid=Fluent&lang=en&prodver=22, accessed on 10 September 2022), the gas invasion behavior of gas–liquid displacement is studied, and the effects of drilling fluid viscosity, density, fracture inlet pressure, fracture width, fracture height, and fracture length on gas invasion are discussed. In order to improve the safety of exploiting fractured gas reservoirs and increase the oil and gas recovery rate, it is necessary to clarify the influencing factors of gas invasion in gas–liquid displacement and the state characteristics of gas invasion.

2. Mathematical Model of Gas–Liquid Two-Phase Flow

There are many calculation models that can be used to calculate multi-phase flow problems in fluid, and selecting an appropriate calculation model is crucial to the final simulation results. The paper uses the volume of fluid (VOF) model, which is mainly used to track the interface position of two or more incompatible fluids. In the VOF model, the interface tracking is completed by solving the phase continuity equation, and the location of the interface is determined by finding the sharp change points in the volume component. The momentum equation of the mixed fluid is solved by the material properties of the mixed material, so the material properties of the mixed fluid will have sudden changes at the interface. The VOF model is mainly applied to the simulation of stratified flow, free surface flow, sloshing, flow with large bubbles in the liquid, dam break, and other phenomena. It can calculate the space–time distribution of the interface in the flow process. In the VOF model, the phases are completely independent, and there is no mutual penetration between them. The sum of the volume fractions of the phases is 1 in each governing equation. Therefore, if the volume fraction of the q phase is expressed by Vq, then Vq0 indicates that there is no q phase in the control volume; Vq1 indicates that the unit is full of q; $0 < Vq < 1$ indicates that the q phase and one or more other phases are included in the control volume.

2.1. Volume Fraction Equation

In the VOF model, the interphase interface can be located by solving the mass equa-tion for the volume fraction of multi-phase flow. For the qth phase, the volume fractio expression is shown in the formula:

$$\frac{\partial x}{\partial t} + v_q \times \alpha_q = \frac{S_{\pm q}}{\rho_q} + \frac{1}{\rho_q}\sum_{p=1}^{n}(m_{pq} - m_{qp})$$

where v_q is the velocity of the qth phase, α_q is the volume fraction of the qth phase, S_a is the source phase of the qth phase, m_{pq} is the mass transfer from the p phase to the phase, and m_{pq} is the mass transfer from the q phase to the p phase. It should be noted tha the above equation is not the solution of the main phase, and the solution of the volum fraction of the main phase is as follows:

$$\sum_{p=1}^{n} \alpha_q = 1$$

The attribute is determined by the phase separation of different control bodies in th flow field. In the two-phase system, if the subscripts c and d are used to express, and th volume fraction of the second phase is located and tracked, the density expression in eac unit is as follows:

$$\rho = \alpha_d \rho_d + (1 - \alpha_d)\rho_c$$

In general, for the n-phase system, the average density of the volume fraction as follows:

$$\rho = \sum \alpha_q \rho_q$$

Other properties, such as viscosity, are calculated in the same way [27,28].

2.2. Liquid-Phase Control Equation

In the process of gas invasion, the fluid flow at the junction of the fracture and wellbo changes from single-phase flow to gas–liquid two-phase flow, and follows the gas–liqui continuous phase equation and momentum equation. The continuity equation of the liqui phase is shown in the formula:

$$\frac{\partial}{\partial t}\left[\rho_m(1-\gamma)\right] + \frac{\partial}{\partial z}\left[\rho_m v_m(1-\gamma)\right] = 0$$

where ρ_m is the density of the drilling fluid, kg/m^3; v_m is the drilling fluid velocity, m/s; is the gas content, dimensionless [29].

2.3. Gas–Liquid Two-Phase Momentum Equation

In the interface between the gas phase and the liquid phase, the mass exchange pha will appear in the mass conservation equation of the gas phase and liquid phase. Th following is the equation set describing the gas–liquid two-phase flow in the wellbore:

$$\frac{\partial}{\partial t}\left[\rho_m(1-\gamma) + \rho_g v_g \gamma\right] + \frac{\partial}{\partial z}\left[\rho_m v_m^2(1-\gamma) + \rho_g \rho_g^2 \gamma\right] + \frac{\partial P}{\partial z} + \frac{\tau_0 P}{A} + \left[\rho_m(1-\gamma) + \rho_g \gamma\right] g = 0$$

where ρ_g is the gas density, kg/m^3; v_g is the real gas velocity, m/s; P is the node pressu MPa; τ_0 is the shear force between the fluid and the pipe wall, N/m^2.

2.4. Physical Model

There are many fractures and caves in the underground reservoir, providing hu storage space for natural gas and oil. When this formation is encountered during drillir under the conditions of near balance and over balance, the annulus drilling fluid ente the fracture under the action of differential pressure, compressing the gas in the fracture, causing the gas pressure in the fracture to increase. When the gas pressure in the fracture is greater than the hydrostatic fluid column pressure in the wellbore, local negative pressure will be generated between the two, causing gas invasion into the wellbore. The physical model is shown in Figure 1, and the parameters are shown in Table 1.

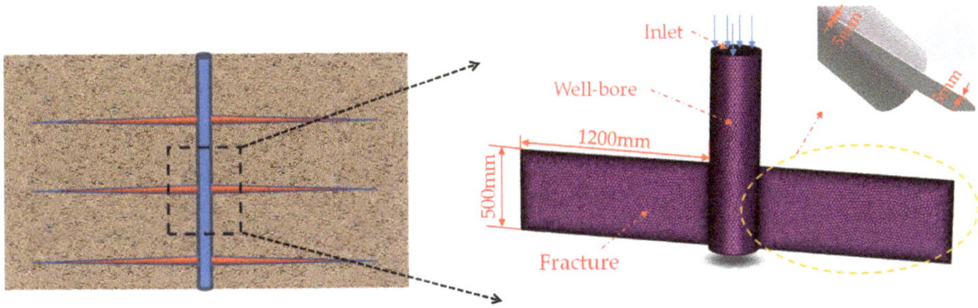

Figure 1. Wellbore fracture model.

In this model, the wellbore radius is 150 mm and the height is 1200 mm. On both sides of the wellbore are symmetrical ladder-shaped fractures with a length of 1200 mm and a height of 500 mm. The width of the fractures near the wellbore is 5 mm and that far from the wellbore is 3 mm. The minimum grid size is 0.5 mm, the maximum grid size is 20 mm, and the volume grid-filling method is poly-hexcore. The grid quality of all cells is above 0.7. The upper part of the wellbore is set as the pressure inlet, and the rest are set as the wall. Compared with the actual working condition, the model is small, and the fluid velocity changes greatly and is almost a straight line. Therefore, the laminar flow mode is adopted, and the time step is set as 0.01 s. Before simulation, the wellbore is fully filled with drilling fluid, and the fractures are fully filled with gas methane. The gas density is 0.2669 g/cm^3, and the viscosity is 0.0361 mPa·s. The density, viscosity, inlet pressure, fracture width, fracture height, and fracture length of the drilling fluid are changed to explore the effects of drilling fluid performance and fracture morphology changes. According to the gas invasion process, the dynamic image of the gas phase diagram over time is intercepted, which directly shows that the gas invasion process in the wellbore and the leakage process of the drilling fluid simulate the gas invasion law under different conditions. The specific parameters are shown in the table below.

Table 1. Analog input parameters.

Parameter	Gradient				
Viscosity of drilling fluid (mPa·s)	10	20	30	40	50
Drilling fluid density (g/cm^3)	1	1.2	1.8	2	2.4
Inlet pressure (MPa)	0.5	1	1.5	2	2.5
Fracture width (mm)	5–1	5–2	5–3	5–4	5–5
Fracture length (mm)		600	1200	1800	
Fracture height (mm)	100	300	500	700	900

3. Results and Discussion

Gas invasion in gas–liquid displacement generally occurs in vertically fractured reservoirs, and the safety density window of the reservoirs is generally narrow, even 0, which belongs to typical blowout leakage reservoirs of the same layer type. During the exploration and development of the reservoirs, a slightly higher drilling fluid density will lead to fracture opening and drilling fluid leakage, and a slightly lower drilling fluid concentration will lead to gas influx, which will lead to overflow, kick, and blowout. It is difficult to implement drilling using conventional drilling technology. In addition, this phenomenon is difficult to identify in the initial detection, which can easily cause the phenomenon c "initial neglect, later difficult to deal with", and even lead to serious downhole accidents

The fundamental reason for gas invasion in gas–liquid displacement is that unde near-equilibrium and overbalance conditions, annular drilling fluid leaks into the fractur under the effect of differential pressure, compressing the gas in the fracture, leading to th increase in gas pressure. When the gas pressure in the fracture increases to more than th annulus liquid column pressure, local negative pressure will be generated between then forming gas intrusion.

3.1. Working Condition Simulation

The simulated wellbore has a diameter of 300 mm, a height of 1200 mm, a fractur height of 500 mm, a length of 1200 mm, and a fracture width of 5–3 mm. The volum fraction distribution of the gas invasion process is shown in Figure 2 (in the figure, the colc represents the percentage content of gas in the fracture, the blue represents the volum fraction of 0, the red represents the volume fraction of 1, and the total volume fraction c the drilling fluid and gas is 1, so the deepest blue drilling fluid has the highest liquid-phas content). The wellbore pressure of the drilling fluid is 1 MPa, and the drilling fluid is base on the actual data on site, with a density of 1.2 g/cm^3 and a viscosity of 30 mPa·s. Th wellbore is full of drilling fluid, and the fractures are full of methane gas.

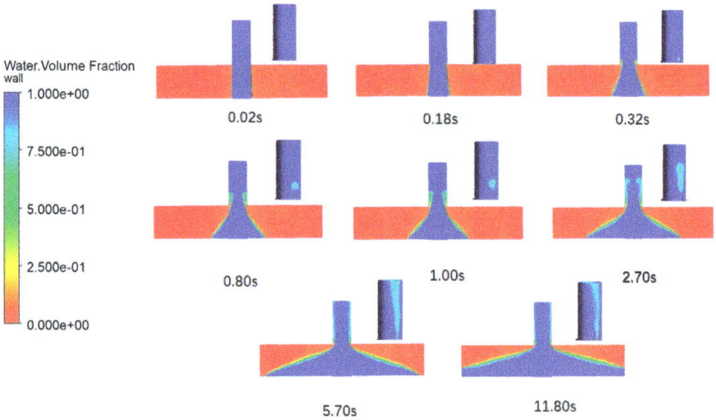

Figure 2. Gas invasion simulation volume fraction distribution.

Figure 2 shows the volume fraction distribution of the drilling fluid rapidly invadin and occupying the fractures. The speed is fast in the first 1 s, and then the leakage o drilling fluid is slow; the pressure difference between the drilling fluid and gas decrease and gradually reaches equilibrium. The drilling fluid reaches half of the fracture capacit in about 5.7 s, the gas pressure in the fracture tends to be gentle, and the gas-phase velocit tends to be gentle at about 10 s. It can be observed from Figure 2 that the drilling fluid in th wellbore compresses the gas in the fracture with a pressure of 1 MPa. About 0.18 s later, th gas is extruded from the fracture and gradually rises along the inner wall of the wellbore t the wellhead. The migration rule of gas in the wellbore can be observed through the profil

The pressure is recorded at the wellbore midpoint (0, 0, 450) and fracture midpoint (160, 450) and the gas-phase velocity at the fracture midpoint. The velocity change is show in Figure 3 and the pressure curve is shown in Figure 4. At the beginning, the pressure i the fracture and wellbore increases rapidly, and the pressure in the fracture exceeds th pressure in the wellbore. As the drilling fluid compresses the gas, the differential pressur between the fracture and the wellbore decreases. Until around 7 s, the differential pressur between the wellbore and the fracture is the smallest, and the pressure in the fracture i higher and lower than the pressure in the wellbore. This indicates that when the pressur in the wellbore and the fracture reaches a relative balance, gas invasion into the wellbor occurs intermittently, and this process repeats endlessly. This can also be proven from the gas-phase fluid velocity diagram in Figure 3. At the beginning, the pressure difference between the two sides is large, and the gas velocity increases rapidly. The time taken is very short, about 0.12 s. With the differential pressure between the two sides decreasing, the gas-phase velocity gradually decreases, until about 7 s when the differential pressure reaches equilibrium, and the gas-phase velocity finally becomes stable.

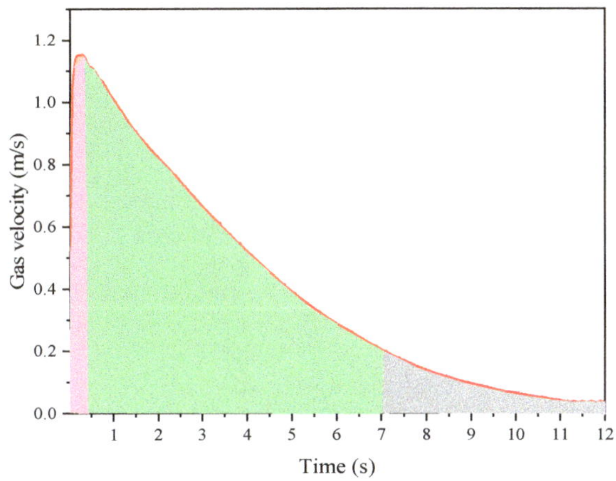

Figure 3. Velocity of gas phase.

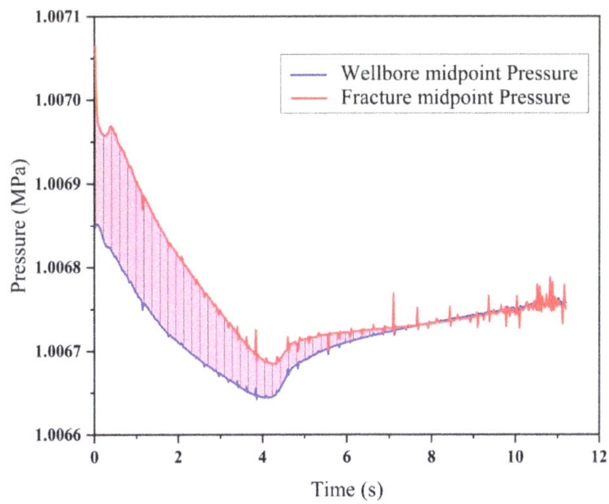

Figure 4. Pressure change of fracture and wellbore midpoint.

3.2. Factors Affecting Gas Invasion

According to the data analysis in the previous section, the constant-pressure drilling fluid flows into the fracture and compresses the gas in the fracture, causing the gas pressure to be higher than that of the drilling fluid in the wellbore, thus allowing the gas to enter the wellbore. The gas invasion velocity is affected by the drilling fluid characteristics, fracture morphology, wellbore pressure, and other aspects. Therefore, it is necessary to explore the influence of these factors on gas invasion.

3.2.1. Effect of Drilling Fluid Viscosity

In this simulation, the wellbore is 300 mm in diameter, 1200 mm in height, 500 mr in fracture height, 1200 mm in length, 5–3 mm in fracture width, 1.2 g/cm^3 in drillin fluid density, and 1 MPa in inlet pressure to explore the impact of different drilling flui viscosities on gas invasion.

It can be seen from Figures 5 and 6 that the viscosity of the drilling fluid is proportiona to the time required for gas–liquid displacement and gas invasion after the drilling flui enters the fracture. The higher the viscosity of the drilling fluid, the longer the time i takes for the pressure in the fracture to rise to the effective fluid column pressure of th drilling fluid at the upper end of the fracture. With the increase in viscosity, the invasio rate of the gas decreases. This is because the greater the viscosity of the drilling flui the greater the flow resistance in the fracture, and the less drilling fluid will enter th fracture in a certain period of time, thus reducing the gas invasion velocity and delayin the time of gas invasion. During on-site construction, the viscosity of the drilling fluid ca be properly increased to prevent gas invasion, but the viscosity should not be too hig as this will significantly reduce the drilling speed and reduce the service life of the mu pump. Therefore, the viscosity of the drilling fluid should be reasonably selected.

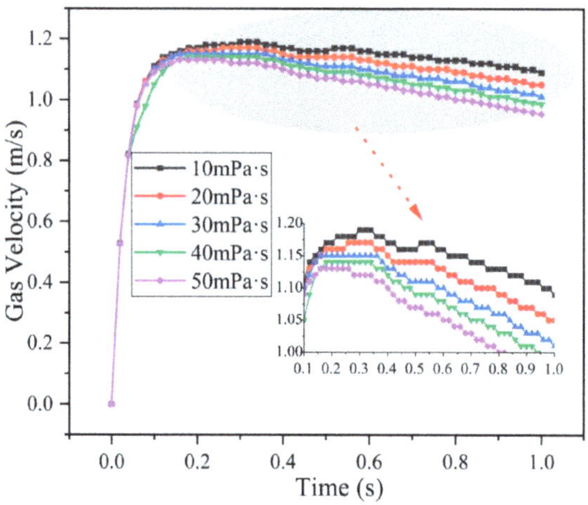

Figure 5. Gas-phase velocity under different drilling fluid viscosities.

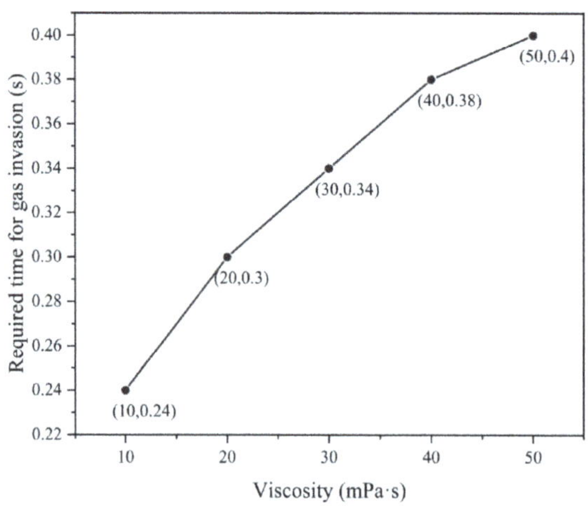

Figure 6. Time required for gas invasion under different drilling fluid viscosities.

3.2.2. Effect of Drilling Fluid Density

The simulated wellbore is 300 mm in diameter, 1200 mm in height, 500 mm in fracture height, 1200 mm in length, 5–3 mm in fracture width, 30 mPa·s in drilling fluid viscosity, and 1 MPa in initial wellbore pressure to explore the impact of different drilling fluid densities on gas invasion.

Figure 7 shows the pressure profile of the wellbore and fracture model when the gas influx occurs under different drilling fluid densities. It can be seen intuitively from the gas invasion pressure distribution that with the increasing density of the drilling fluid, the pressure distribution at each point when the gas influx occurs and the part with the largest pressure are distributed at the bottom of the model. The maximum pressure at 1 g/cm^3 is 1.009 MPa; the maximum pressure is 1.011 MPa at 1.2 g/cm^3; the maximum pressure is 1.018 MPa at 1.8 g/cm^3; the maximum pressure is 1.020 MPa at 2 g/cm^3; and the maximum pressure is 1.025 MPa at 2.4 g/cm^3.

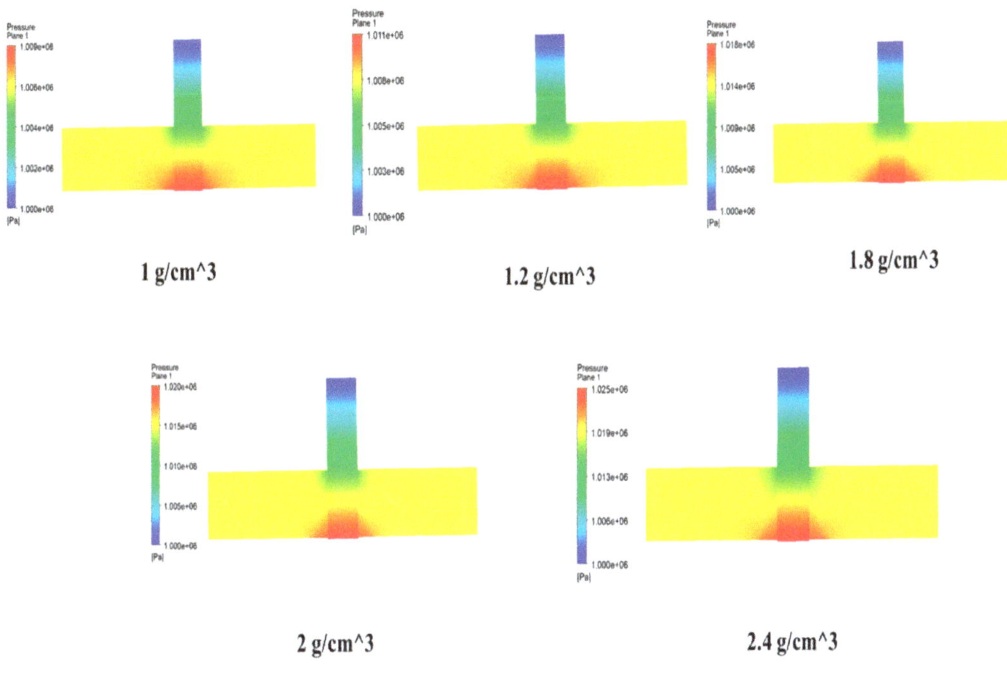

Figure 7. Gas invasion pressure distribution at different drilling fluid densities.

This makes the gas velocity at a point (0,160,450) in the fracture in Figure 8 increase with the increase in drilling fluid density. When the drilling fluid density is below 1.2 g/cm^3, the maximum gas invasion velocity is less than 1.2 m/s; when the drilling fluid density is more than 1.2 g/cm^3, the maximum gas invasion velocity is more than 1.6 m/s. It can be seen intuitively from Figure 9 that with the increasing density of the drilling fluid, the fluid continuously enters into the fracture, squeezing the gas in the fracture, and the time required for gas invasion gradually decreases. This is because when density of the drilling fluid increases, the corresponding hydrostatic column pressure of the fluid in the wellbore increases. The drilling fluid has a stronger impact on the gas in the fracture, and the drilling fluid enters the fracture at a faster speed. In the same fracture model, it takes less time to squeeze the gas in the fracture to raise it to the same pressure range. The change in the drilling fluid density is positively correlated with the change in the gas invasion velocity. When gas influx is encountered during drilling or production, the density of the drilling fluid can be appropriately reduced to delay the occurrence of this phenomenon, and win time for well control.

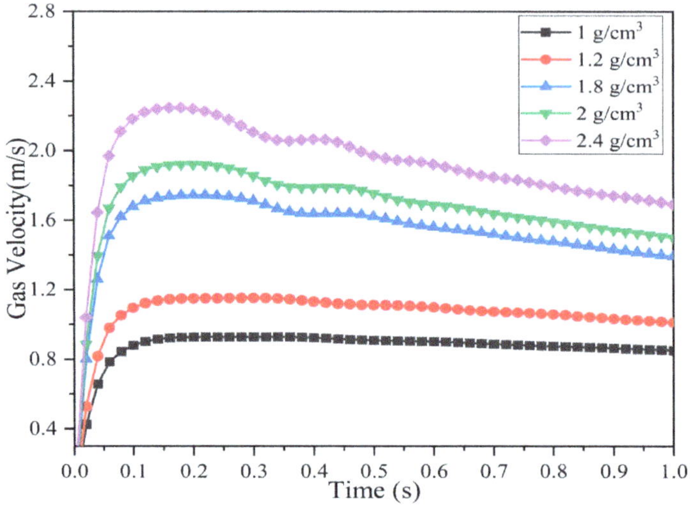

Figure 8. Velocity of a point in the fracture under different drilling fluid densities.

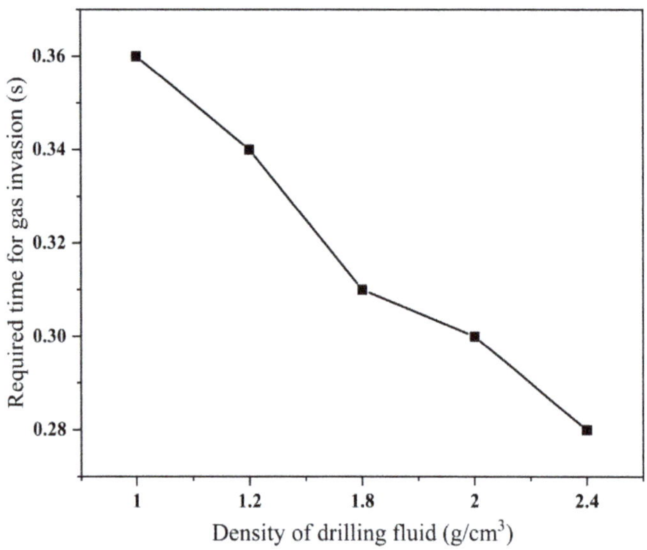

Figure 9. Time required for gas invasion under different drilling fluid densities.

3.2.3. Effect of Pressure Difference

In this simulation, the wellbore is 300 mm in diameter, 1200 mm in height, 500 mm in fracture height, 1200 mm in length, 5–3 mm in fracture width, 30 mPa·s in drilling fluid viscosity, and 1.2 g/cm^3 in drilling fluid density to explore the impact of different inlet pressures on gas invasion.

It can be seen from Figure 10 that the influence of the wellbore inlet pressure on gas–liquid displacement in fractures is weak. The inlet pressure is within 0.5–2.5 MPa and the time required for gas invasion is mostly kept between 0.33–0.34 s. When the drilling fluid enters the wellbore at a lower pressure through the inlet, the drilling fluid compresses the gas in the fracture. When gas invasion occurs, the pressure of the gas in the fracture must be greater than the "barrier pressure" formed by the static fluid column in the wellbore. When the drilling fluid enters the wellbore at a higher pressure through the inlet, the drilling fluid has a strong impact on the fracture. However, if gas invasion occurs, the gas pressure must be greater than the "pressure barrier" of the wellbore's static liquid column at this time.

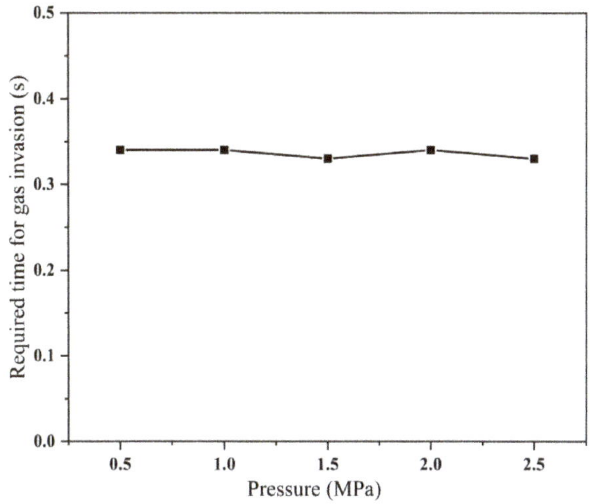

Figure 10. Time required for gas invasion under different pressures.

Figure 11 reflects the curve of gas velocity change with time at a point (0,160,450) in the fracture. It can be seen from the figure that the gas velocity change is basically independent of the inlet pressure, that is, the greater the inlet pressure, the greater the hydrostatic column "barrier pressure", the smaller the inlet pressure, and the smaller the hydrostatic column "barrier pressure".

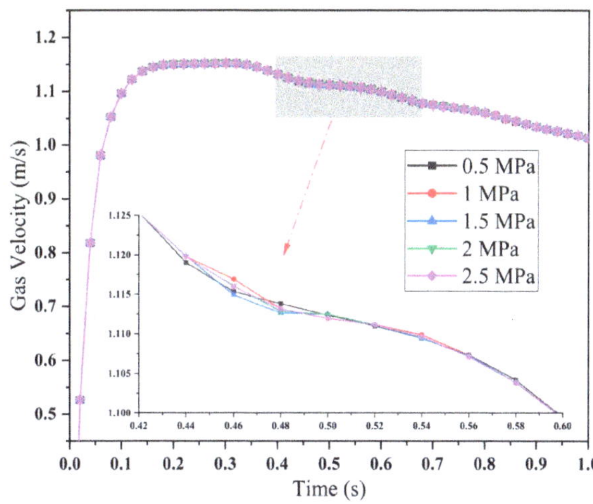

Figure 11. Velocity of a point in the fracture under different pressures.

3.2.4. Effect of Fracture Width

The simulated wellbore adopts a diameter of 300 mm, a height of 1200 mm, a fracture height of 500 mm, a length of 1200 mm, an inlet pressure of 1 MPa, a drilling fluid viscosity of 30 mPa·s, and a drilling fluid density of 1.2 g/cm^3 to explore the impact of different fracture widths on gas invasion.

It can be seen from Figure 12 that the time required for gas invasion in gas–liquic displacement is negatively correlated with the fracture width. The time for gas invasior in the 5–1 mm wide fracture is 0.39 s, while the time for gas invasion in the 5–5 mm wid fracture is 0.26 s, which indicates that the fracture width has a great influence on ga invasion in gas–liquid displacement.

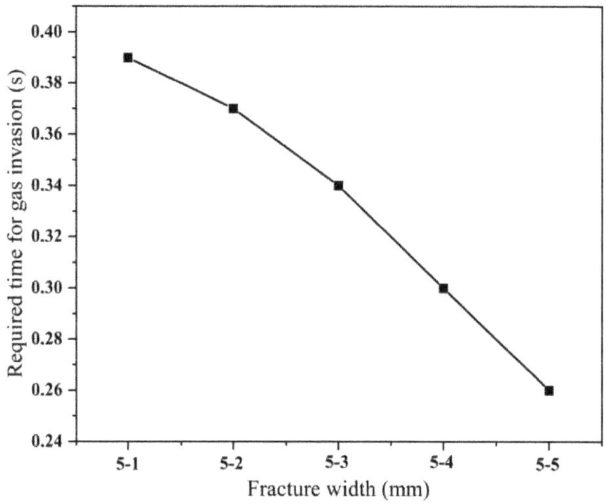

Figure 12. Time required for gas invasion under different fracture widths.

Because the time for gas invasion of the above-mentioned fractures is less than 0.4 Figure 13 shows the morphological distribution of the gas and liquid phases in the wellbor and fractures at 0.4 s of five fracture widths. With the increase in the fracture width, th volume occupied by the liquid phase in the model becomes larger and larger, and th volume of the gas phase entering the wellbore also increases [29]. This is because a fractur with larger width has less flow resistance. The drilling fluid can easily enter the fractur and squeeze the gas. The "barrier pressure" of the wellbore is the same. The fluid wit smaller flow resistance has a stronger impact on the gas in the fracture, and it is easier fc the gas to enter the wellbore. The velocity at a point (0,160,450) in the fracture in Figure 1 also proves the previous viewpoints. The wider the fracture, the smaller the fluid resistanc and the greater the gas flow velocity. This also provides a solution to prevent gas invasio The key to effectively reducing gas invasion is to use appropriate plugging particles, chang fractures into holes, change holes into gaps, and reduce the width of the fractures.

3.2.5. Effect of Fracture Height

The simulated wellbore has a diameter of 300 mm, a height of 1200 mm, a fractur width of 5–3 mm, a length of 1200 mm, an inlet pressure of 1 MPa, a drilling fluid viscosi of 30 mPa·s, and a drilling fluid density of 1.2 g/cm^3. The impact of different fractu heights on gas invasion is explored. The results are shown in the following figure.

Figure 15 shows the time required for gas invasion at different fracture heights. Th lower the fracture height, the easier it is for the gas invasion to occur, and the impact wi be more obvious. With the increase in the height, the time for gas invasion will gradual flatten, and the time for gas invasion to occur in the 700 mm high fracture and 900 m high fracture is 0.36 s. With the increase in the fracture height, the time for gas invasio will not be extended. However, under the same conditions, the higher the fracture heigh the larger the fracture volume. Once the gas is squeezed, its flow resistance is smaller ar the flow speed is larger, as shown in Figure 16.

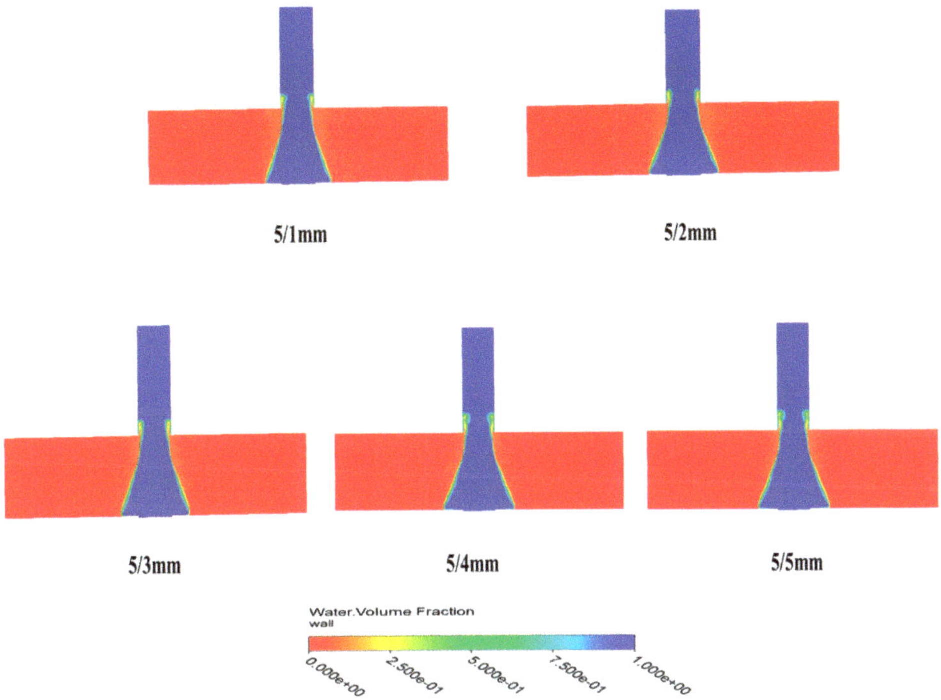

Figure 13. Volume fraction distribution of gas–liquid two-phase flow with different fracture widths at 0.4 s.

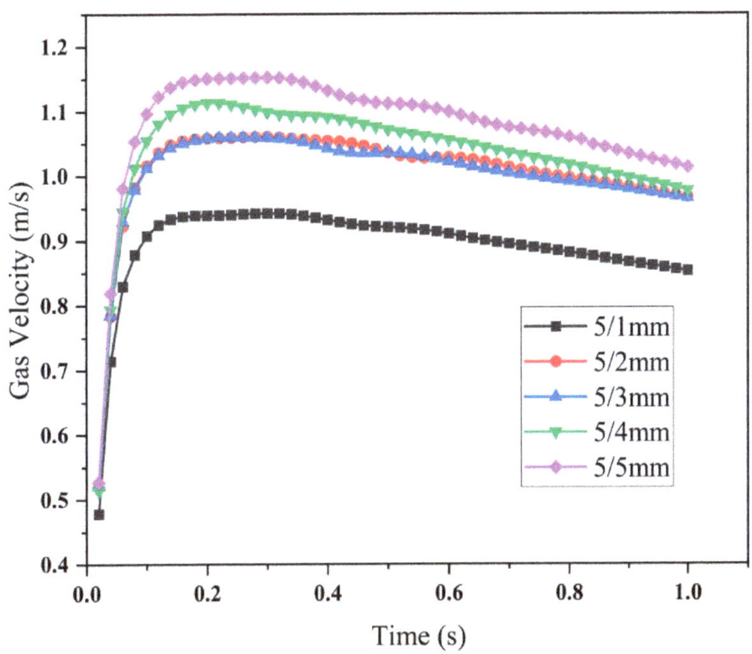

Figure 14. Velocity of a point in a fracture with different fracture widths.

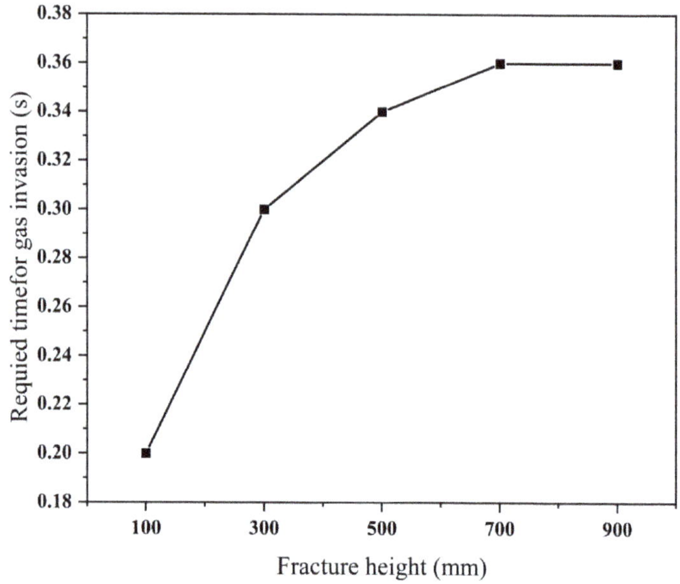

Figure 15. Time required for gas invasion at different fracture heights.

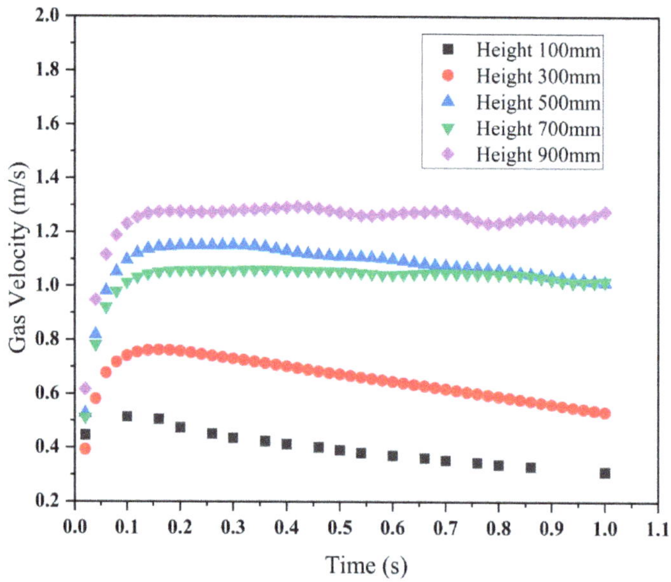

Figure 16. Velocity of a point in a fracture at different fracture heights.

3.2.6. Effect of Fracture Length

The diameter of the simulated wellbore is 300 mm, the fracture width is 5–3 mm, the length is 1200 mm, the height is 500 mm, the inlet pressure is 1 MPa, the viscosity of the drilling fluid is 30 mP·s, and the density of the drilling fluid is 1.2 g/cm^3. The impact o different fracture lengths on gas invasion is explored, and the results are shown in the following figures.

It can be seen from Figure 17 that there is a negative correlation between the time required for gas invasion and the length of the fractures. The length of the fracture ha obvious influence on gas invasion in gas–liquid displacement. The longer the fracture, the more difficult it is for gas invasion to occur. This is because under the same conditions, the longer the fracture, the greater the resistance along the fracture, the longer the drilling fluid enters the fracture, and the longer the time taken to reach the "barrier pressure", that is, the longer the time taken to start the gas–liquid displacement of the gas invasion. Figure 18 shows the velocity distribution at 0.5 s under each fracture. The longer the fracture, the harder it is for the drilling fluid to enter the fracture to squeeze the gas. Under the same conditions, the shorter the fracture, the more the drilling fluid enters the fracture, and the easier it is for the gas to be squeezed out.

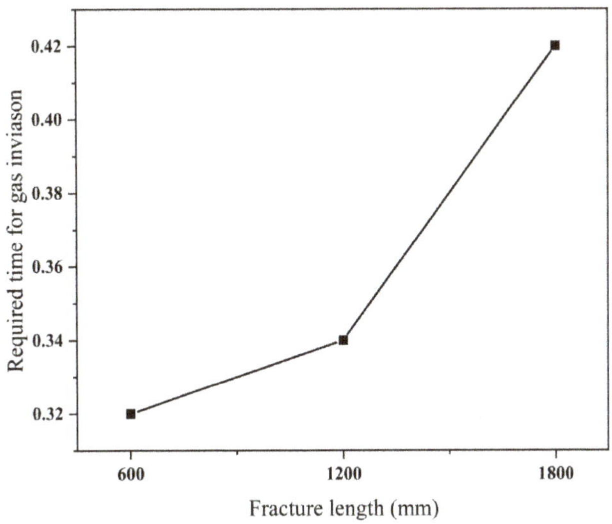

Figure 17. Time required for gas invasion under different fracture lengths.

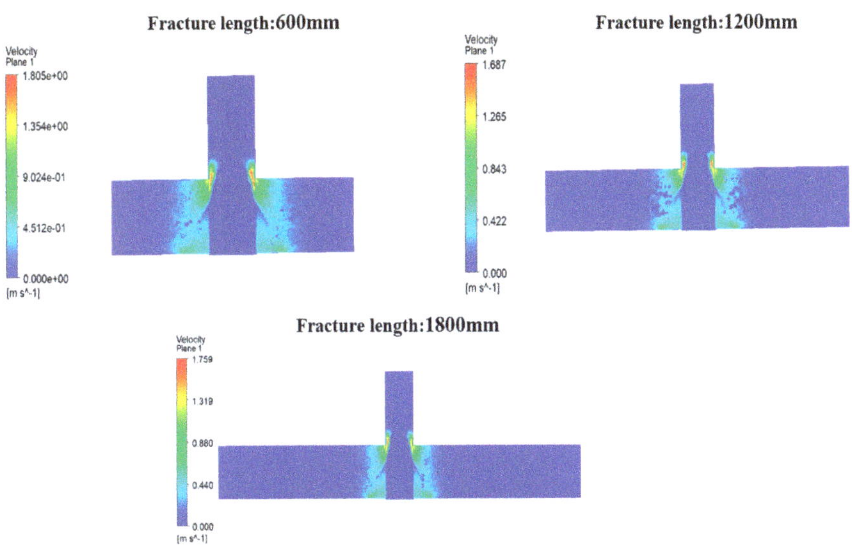

Figure 18. Velocity distribution at different fracture lengths in 0.5 s.

4. Conclusions

1. Gas–liquid displacement gas invasion occurs intermittently. When the gas pressure in the fracture is close to the wellbore pressure, the gas pressure in the fracture rises and falls around the wellbore pressure. The amount of gas being squeezed out is small and not easy to see.

2. The viscosity of the drilling fluid mainly affects the occurrence of gas invasion through flow resistance. The higher the viscosity, the less likely it is that the gas invasion will occur, but the viscosity has no obvious effect on the gas invasion. The change in the drilling viscosity on the gas invasion velocity is small when it gradually increases from 10–50 mPa·s, all within 1.0–1.2 m/s. The density of the drilling fluid mainly affects the gas invasion phenomenon by changing the pressure distribution in the wellbore, and the change in the density is positively correlated with the occurrence time of the gas invasion. The inlet pressure has the least significant effect on the occurrence of gas invasion: the pressure gradually increases from 0.5–2.5 MPa, and the occurrence time of the gas invasion is always about 0.35 s.

3. The influence of fracture morphology is very obvious for gas invasion. The fracture morphology mainly affects the occurrence of gas invasion through the resistance of the fluid along the fracture. The longer the fracture, the less likely it is to for gas invasion to occur. Similarly, the wider the fracture, the less likely it is to experience gas invasion. The influence of fracture height on gas invasion is different from the former. The height is 300–700 mm. With the increase in the fracture height, the time required for gas invasion is longer. When the height is 700–900 mm, the time required for gas invasion is basicall[y] unchanged with the increase in the height.

Author Contributions: Conceptualization, C.Y. and J.G.; methodology, C.Y. and K.L.; software, C.Y.; validation, C.Y., J.P. and S.X.; formal analysis, P.X.; investigation, C.Y. and K.L.; resources, S.X.; data curation, P.X.; writing—original draft preparation, C.Y.; writing—review and editing, C.Y.; visualization, J.G.; supervision, J.P.; project administration, C.Y.; funding acquisition, C.Y. All authors have read and agreed to the published version of the manuscript.

Funding: The authors gratefully acknowledge the financial support from the major engineering technology field test of China National Petroleum Corporation's "Integration and Test of Excellent and Fast Drilling and Completion Technology in Key Areas such as the Southern Margin of Junggar and Mahu" (No. 2019F-33)and the National Natural Science Foundation of China (No. 51804044).

Institutional Review Board Statement: Not applicable.

Informed Consent Statement: Written informed consent has been obtained from the patient(s) to publish this paper.

Data Availability Statement: Data available on request due to restrictions eg privacy or ethicalTh[e] data presented in this study are available on request from the corresponding author.

Conflicts of Interest: No conflict of interest exists in the submission of this manuscript, and a[ll] authors approve the manuscript for publication. I want to declare on behalf of my co-authors th[at] the work described is original research that has not been published previously and is not und[er] consideration for publication elsewhere, in whole or in part. All of the authors listed have approve[d] the manuscript.

References

1. Lin, A.C.; Han, L.X. Interpretation of runaway blowout in well Luojia 16—Well control problem in overbalanced drilling. *Dr[ill.] Prod. Technol.* **2006**, *29*, 3.
2. Li, Y.; He, M.; Xia, W.; Li, X. Study on swelling and migration law of dissolved gas during gas kick of oil-based drilling fluid horizontal well. *China Saf. Prod. Sci. Technol.* **2016**, *12*, 6.
3. Gravdal, J.E.; Nikolaou, M.; Breyholtz, Ø.; Carlsen, L.A. Improved kick management during MPD by real-time pore-pressu[re] estimation. *SPE Drill. Complet.* **2010**, *25*, 577–584. [CrossRef]
4. Rommetveit, R.; Vefring, E.H. CoMParison of Results from an Advanced Gas Kick Simulator with Surface and Downhole Da[ta] from Full Scale Gas Kick Experiments in an Inclined Well. In Proceedings of the SPE Annual Technical Conference and Exhibitic[on,] Dallas, TX, USA, 6–9 October 1991.

5. Baoping, L.U.; Hou, X.; Xing, S. Asphalt Displacement Mechanism and Pore Pressure Fluctuation in Yadavaran Oilfield, Iran. *J. China Univ. Pet.* **2017**, *41*, 88–93.
6. Louis, C.; Maini, Y.N. Determination of in-situ hydraulic parameters in jointed rock. *Int. Soc. Rock Mech. Proc.* **1970**, *1*, 1–19.
7. Majidi, R.; Miska, S.Z.; Yu, M.; Thompson, L.G.; Zhang, J. Quantitative Analysis of Mud Losses in Naturally Fractured Reservoirs: The Effect of Rheology. *SPE Drill. Complet.* **2010**, *25*, 509–517. [CrossRef]
8. Zhang, Y.; Zhao, W.; Zhang, X.; Li, S. Movement features of gas in different types of drilling fluid during kicking in deepwater wells. *Oil Drill. Prod. Technol.* **2016**, *38*, 5.
9. Gugl, R.; Kharrat, R.; Shariat, A.; Ott, H. Evaluation of Gas-Based EOR Methods in Gas-Invaded Zones of Fractured Carbonate Reservoir. *Energies* **2022**, *15*, 4921. [CrossRef]
10. Gandomkar, A.; Nasriani, H.R.; Enick, R.M.; Torabi, F. The effect of CO_2-philic thickeners on gravity drainage mechanism in gas invaded zone. *Fuel* **2023**, *331*, 125760. [CrossRef]
11. Cheraghian, G.; Rostami, S.; Afrand, M. Nanotechnology in enhanced oil recovery. *Processes* **2020**, *8*, 1073. [CrossRef]
12. Shen, J.; Du, Z.; Li, H.; Li, L.; Wang, F.; Ni, J.; Teng, G. Fracture gas invasion mechanism and plugging method of shale gas layer. *J. Chengdu Univ. Technol.* **2022**, *5*, 579–585.
13. Biao, Q.; Junwei, F.; Lixin, Z.; Shengming, H.; Jun, Z.; Wei, H. Simulation of the lost circulation in fractured formation under the gas invasion conditions. *Sci. Technol. Eng.* **2021**, *17*, 7059–7066.
14. Jiang, R.; Cui, Y.; Zhang, C.; Chen, X. Numerical simulation of cuttings transport during gas invasion in eccentric horizontal Wells of shale gas. *Technol. Ind.* **2022**, *6*, 283–289.
15. Liu, F.; Cheng, C.; Qiao, S.; Xing, X.; Wu, Y.; Yang, H. Transient changes of annulus pressure at the initial stage of gas influx. *Drill. Prod. Technol.* **2022**, *2*, 1–7.
16. Yue, H.; Wu, P.; Liang, J.; Zhong, C.; Zhang, Z.; Li, Z.; Li, H. Gas-liquid gravity displacement laws of horizontal during well drilling in shale formations with developed fractures. *Nat. Gas Ind.* **2021**, *12*, 90–98.
17. Wang, J.; Li, J.; Liu, G.; Luo, X. Study on the change law of annular outlet flow rate in new-type of dual-gradient drilling under gas cut condition. *Pet. Drill. Tech.* **2020**, *4*, 43–49.
18. Huang, G.; He, S.; Tang, M.; Liu, Y.L. A study on the effect of displacement gas cut on fractured reservoirs in Shunnan Block. *Pet. Drill. Tech.* **2018**, *5*, 21–25.
19. Zhao, Z.; Pu, X.; Wang, G.; Li, Z.; Chen, L.; Cao, C. Numerical study of the gas-fluid displacement invasion behavior in a fractured carbonate reservoir. *J. Nat. Gas Sci. Eng.* **2015**, *27*, 686–691. [CrossRef]
20. Bennion, D.B.; Lunan, B.; Saponja, J. Underbalanced drilling and completion operations to minimize formation damage-reservoir screening criteria for optimum application. *J. Can. Pet. Technol.* **1998**, *37*, 9. [CrossRef]
21. Nickens, H.V. A dynamic computer model of a kicking well. *SPE Drill. Eng.* **1987**, *2*, 159–173. [CrossRef]
22. Zhang, X.; Zhou, Y.; Liu, W.; Guo, Q.; Cui, T. Characters of gravity replacement gas kick in carbonate formation. *Acta Pet. Sin.* **2014**, *35*, 958–962.
23. Zhang, X.; Zhou, Y.; Liu, W.; Guo, Q. A method for characterization and identification of gas kicks caused by underbalanced pressure and gravity displacement. *J. China Univ. Pet.* **2015**, *39*, 95–102.
24. Hou, X.; Zhao, X.; Meng, Y.; Yang, S.; Li, G.; Liu, W. Liquid-Liquid Gravity Displacement Test Based on Experimental Apparatus for Real Fractures. *Pet. Drill. Tech.* **2018**, *1*, 30–36.
25. Fang, J.; Zhu, L.; Luo, F.; Zhang, J.; Wang, Y.; Huang, W.; Niu, X. Simulation study on the effects of drilling fluid on gas invasion from fractured formations. *Drill. Complet. Fluid* **2019**, *3*, 287–292.
26. Shu, G.; Meng, Y.; Li, G.; Wei, N.; Zhao, X.; Yang, M. Mechanism of mud loss and well kick due to gravity displacement. *Pet. Drill. Tech.* **2011**, *1*, 6–11.
27. Zhao, X.; Hou, X.; Yang, S.; Bao, H. Partern and control of gravity displacement between asphaltic heavy oil and drilling fluid. *Oil Drill. Prod. Technol.* **2016**, *38*, 622–627.
28. Song, R.; Sun, B.; Liu, X.; Wang, Z. Calculation and analysis of bottom hole pressure in wellbore after gas invasion. *Duankuai Youqitian Fault Block Oil Gas Field* **2011**, *4*, 486–488.
29. Li, Z. The Vertical Cracks Formation Gas Liquid Displacement and Prevent Gas Invasion Drilling Fluid Plugging Technology Research. Master's Thesis, Southwest Petroleum University, Chengdu, China, 2014. Available online: https://kns.cnki.net/KCMS/detail/detail.aspx?dbname=CDFDLAST2015&filename=1014419528.nh (accessed on 10 September 2022).

processes

Article

Effects of Pore Water Content on Stress Sensitivity of Tight Sandstone Oil Reservoirs: A Study of the Mahu Block (Xinjiang Province, China)

Xiaoshan Li [1], Kaifang Gu [2], Wenxiu Xu [2], Junqiang Song [2], Hong Pan [2], Yan Dong [2], Xu Yang [2], Haoyu You [2], Li Wang [2], Zheng Fu [3], Lingqi Liu [4] and Ke Wang [5,*]

1 Development Department, Xinjiang Oilfield Company, Petro China, Karamay 834000, China; lxiaoshan@petrochina.com.cn
2 Exploration and Development Institute, Xinjiang Oilfield Company, Petro China, Karamay 834000, China; ghgukf@petrochina.com.cn (K.G.); fcxwx@petrochina.com.cn (W.X.); sjq_xj@petrochina.com.cn (J.S.); fcph1@petrochina.com.cn (H.P.); dongyan66@petrochina.com.cn (Y.D.); xjyx@petrochina.com.cn (X.Y.); you-hy@petrochina.com.cn (H.Y.); fcwl2@petrochina.com.cn (L.W.)
3 School of Geoscience, China University of Petroleum (Huadong), Qingdao 266580, China; b20010019@s.upc.edu.cn
4 Sanya Offshore Oil & Gas Research Institute, Northeast Petroleum University, Sanya 572024, China; liulingqi2001@163.com
5 College of Resources and Environmental Engineering, Guizhou University, Guiyang 550025, China
* Correspondence: kwang5@gzu.edu.cn

Abstract: Traditional stress sensitivity experiments are typically conducted under dry conditions, without considering the reservoir's water content. In reality, the presence of water within pores significantly influences the extent of stress sensitivity damage in tight sandstone oil formations, subsequently affecting the determination of stress sensitivity coefficients during experimentation. By investigating sandstone samples from wells in the Mahu Block of China's Xinjiang province we observed that increasing water saturation reduces the stress sensitivity of tight sandstone. By conducting stress sensitivity experiments under varying water content conditions, we found that the stress sensitivity coefficient is not a constant value but decreases as water saturation increases. Based on experimental comparisons, an optimized power-law model for stress-sensitive damage assessment was refined. By conducting stress-sensitive damage assessment experiments under different water content conditions and integrating the concept of comprehensive compression coefficient an improved stress-sensitive power-law model was established allowing for the influence of water content. The accuracy of this improved model was increased by 46.98% compared to the original power-law model through experimental validation. The research outcomes can enhance the accuracy of permeability and productivity evaluation, providing valuable guidance for unconventional oil and gas development.

Keywords: Mahu Block; tight sandstone; pore water content; stress sensitivity; improved power-law model

Citation: Li, X.; Gu, K.; Xu, W.; Song, J.; Pan, H.; Dong, Y.; Yang, X.; You, H.; Wang, L.; Fu, Z.; et al. Effects of Pore Water Content on Stress Sensitivity of Tight Sandstone Oil Reservoirs: A Study of the Mahu Block (Xinjiang Province, China). *Processes* **2023**, *11*, 3153. https://doi.org/10.3390/pr11113153

Academic Editors: Qingbang Meng, Zheng Sun, Tao Zhang, Dong Feng, Wen Zhao and Hung Vo Thanh

Received: 23 August 2023
Revised: 12 October 2023
Accepted: 16 October 2023
Published: 4 November 2023

1. Introduction

Tight oil will be a highly significant unconventional hydrocarbon resource for increasing oil reserves and production in the future. The development of tight oil has been a major driving force behind the recent rise in crude oil production in the United States. Data based on the information provided by the U.S. Energy Information Administration shows that about 90% of the growth in oil production in the U.S. originates from six tight oil areas [1]. In 2018, tight oil production reached 310 million metric tons, accounting for 60% of the total production [2]. In accordance with the fourth oil and gas resource assessment conducted by PetroChina (Beijing, China), the national geological reserves of tight oil in China are about 12.5 billion metric tons, with recoverable resources of 1.3 billion metric tons and confirmed reserves of roughly 300 million metric tons [3]. Tight oil has become one of the key project areas when exploring China's unconventional petroleum resources and is a major contributor to China's increased oil recoverable reserves and production.

In the case of oil and gas reservoirs, the original equilibrium involves a balancing of pressures between the overlying rock formation pressure, pore pressure, and rock framework support pressure. Nevertheless, as the fluids are extracted from the rock pores, this equilibrium is disrupted [4]. Consequently, the rock framework undergoes compression, leading to a gradual reduction in pore size and a subsequent decrease in permeability [5]. The greater the degree of extraction, the greater the compression of the rock pore size, resulting in a greater reduction in permeability [6]. In simpler terms, the rock experiences a more signification level of stress-induced damage. Based on prior studies, if a rock is softer, indicating a weaker rock framework supporting capacity, it will experience more pronounced pore compression, resulting in a more substantial decrease in permeability when subjected to the same overlying rock formation pressure [7]. Typically, soft rocks tend to display a more pronounced stress sensitivity, whereas dense rocks exhibit a less significant stress sensitivity effect. The extent of stress-induced damage is directly proportional to the rock's compressibility coefficient [8,9]. Hence, rocks that are more easily compressible will experience more severe stress-induced damage.

Permeability is one of the important parameters describing reservoir properties, serving as fundamental data for reservoir evaluation, development scheme design, numerical simulation, and productivity assessment. Permeability development and evolution can both be a product of original sedimentary depositional processes and variability, as well as of post-depositional processes and modifications such as compaction, fracturing, and faulting and even of human activity [10–14]. As fluid is extracted from reservoirs, the phenomenon of decreasing reservoir permeability due to increasing effective stress is known as stress sensitivity [15]. Prior research assessed stress sensitivity by examining changes in rock fracture permeability or porosity [16]. Dong et al. [9] conducted experimental studies on the stress sensitivity of tight sandstone and shale based on the changes in porosity and permeability, suggesting that shale permeability stress sensitivity is 2–3 times that of sandstone. Kassis et al. [17] derived experimental results of permeability which indicated that shale stress sensitivity is jointly influenced by factors such as effective stress loading mode, fracture characteristics, and support materials. Kang et al. [18] designed and tested experiments to investigate the influence of variable effective stress conditions and loading/unloading sustained stable time on rock sample permeability. The study determined that shale composition and multi-scale pore-fracture structures are the main causes of the differential deformation characteristics of shale pores. Geng et al. [19] conducted multiple studies which indicated that tight sandstone reservoirs generally demonstrate strong stress sensitivity during production, posing challenges for restoring permeability. Fernandes et al. [20] established a flow model considered the stress-sensitive effect and monitored its role in the permeability change to enhance oil recovery. Chang et al. [21] examined the impacts of the stress loading methods, cyclic loading, and loading rates on reservoir permeability using Ordos Basin tight sandstones; they observed a significant decrease in permeability for tight reservoirs with increasing loading time.

Mathematical models for assessing stress-sensitive damage to porosity or permeability in tight rocks are scarce [22,23], with the majority being empirical in nature. Cui et al. [24] observed that shale porosity exhibited a linear decrease with increasing effective stress, while permeability exhibited a power-law decay trend with increasing effective stress based on pressure pulse decay tests. Zhang et al. [25] investigated the applicability and physical significance of power-law formulas and their constants based on the physical meaning of the Walsh model and the intrinsic mathematical relationship between the Walsh model and power-law formulas. The researchers verified their findings through experimentation and classified stress-sensitive formulas based on five different pore types. Duan et al. [7] experimentally examined the sensitivity of matrix, micro-fracture, and artificial fracture permeability to stress. They summarized the fitting and correction results of pore-permeability power-law index model, Gangi model, and Walsh model for fracture permeability. The research findings showed that micro-fractures and unfilled artificial fractures have the strongest stress sensitivity in permeability, exhibiting exponential decay with increasing effective stress. The fitting accuracy of both the Gangi and Walsh models is above 97%.

Nevertheless, all these experiments were carried out using dry rock samples, which do not align with actual water-bearing reservoir conditions [26]. Firstly, sandstone reservoirs contain a certain amount of original water. Gao et al. [27] analyzed and discovered the presence of free water in large pores, abundant capillary water in medium and small pores, and full water saturation in tiny pores within the He [8]—Shan [1] reservoir interval in the northern Tianhuan region of the Ordos Basin. Shi et al. [28] and Ma [29] identified four types of water, including bounded water, interlayer water, edge–bottom water, and isolated water, in the tight sandstone of the Sulige area. Secondly, effective production from tight sandstone oil reservoirs requires extensive hydraulic fracturing under current technological conditions. During the fracturing process, a significant amount of fracturing fluid infiltrates the sandstone reservoir, leading to a rapid increase in reservoir water saturation in the near-fracture zone [30–35], further influencing the sensitivity of sandstone permeability to stress. Prior research has shown, through numerous experiments, that the comprehensive compression coefficient of sandstone is indeed influenced by water saturation [26]. Therefore, the degree of stress-sensitive damage in sandstone will also be affected by water saturation.

Based on the previous analysis, it is evident that actual reservoir rocks exist within a water-containing environment. Despite these previous analyses, and despite the major implications of better characterizing the effects of water on stress sensitivity for both oil and gas systems on land and offshore as well as emerging resource systems such as marine hydrates and geothermal reservoirs [36–40], there are still numerous shortcomings in the current understanding of the impact of water on the sensitivity of rocks to stress. This study aims to explore the effect of water content on stress sensitivity in tight sandstone. First, by conducting experiments to assess the extent of stress sensitivity damage under different water saturation conditions, we examine the impact of water content on stress sensitivity damage. Second, we evaluate the practicality of these models for assessing stress sensitivity damage in tight sandstone reservoirs by comparing the fitting performance of commonly used exponential models and the original power law model. Finally, we improve the original power-law model by addressing its limitations and incorporating the comprehensive compression coefficient. Moreover, it has been determined that the stress sensitivity coefficient is not a constant value; instead, it decreases as water content increases. Building upon these experimental findings and the concept of comprehensive compression coefficient, a correction formula for the stress sensitivity coefficient under the influence of water content was established. The accuracy of this formula was validated through experimental and referenced results. The research outcomes enhance the accuracy of permeability and productivity evaluation, providing effective guidance for unconventional oil and gas development.

2. Geological Background

Tight sandstone samples of groups A and B were collected, each originating from wells A and B in the Mahu Block, Xinjiang Province, China, as shown in Figure 1. The Mahu Block is located on the northern slope of the western ring belt within the Mahu Depression, featuring a southwest-dipping monocline [41,42]. The fault blocks exhibit dip angles of less than 10 degrees. Reservoir depths vary from 2440 m in the northernmost to 3305 m in the southernmost. The faults are predominantly reverse faults with separation distances ranging from 10 to 50 m and lengths spanning from 4 to 28.5 km [43]. Baikouquan Formation 2 (T_1b_2), from northeast to southwest, gradually transitions from a plain facies to a foreland facies. Baikouquan Formation 3 (T_1b_3), from northeast to southwest, h…

the main reservoir in the foreland facies, with the southern part entering the foredeep delta. As the transition occurs from plain facies to foreland facies, the grain size of the reservoir becomes finer, evolving from medium conglomerates to fine conglomerates and conglomeratic sandstones. While grain flows in the plain facies do not yield oil, sheet-like turbidites and distributary channels in the foreland facies exhibit the most promising oil-bearing characteristics [44]. The geological strata in the Madong Slope area exhibit a well-developed sequence, progressing from the bottom upwards and encompassing the Carboniferous, Permian, Triassic, Jurassic, and Cretaceous systems. Notably, there are regional unconformities present between the Permian and Triassic systems, the Triassic and Jurassic systems, as well as between the Jurassic and Cretaceous systems.

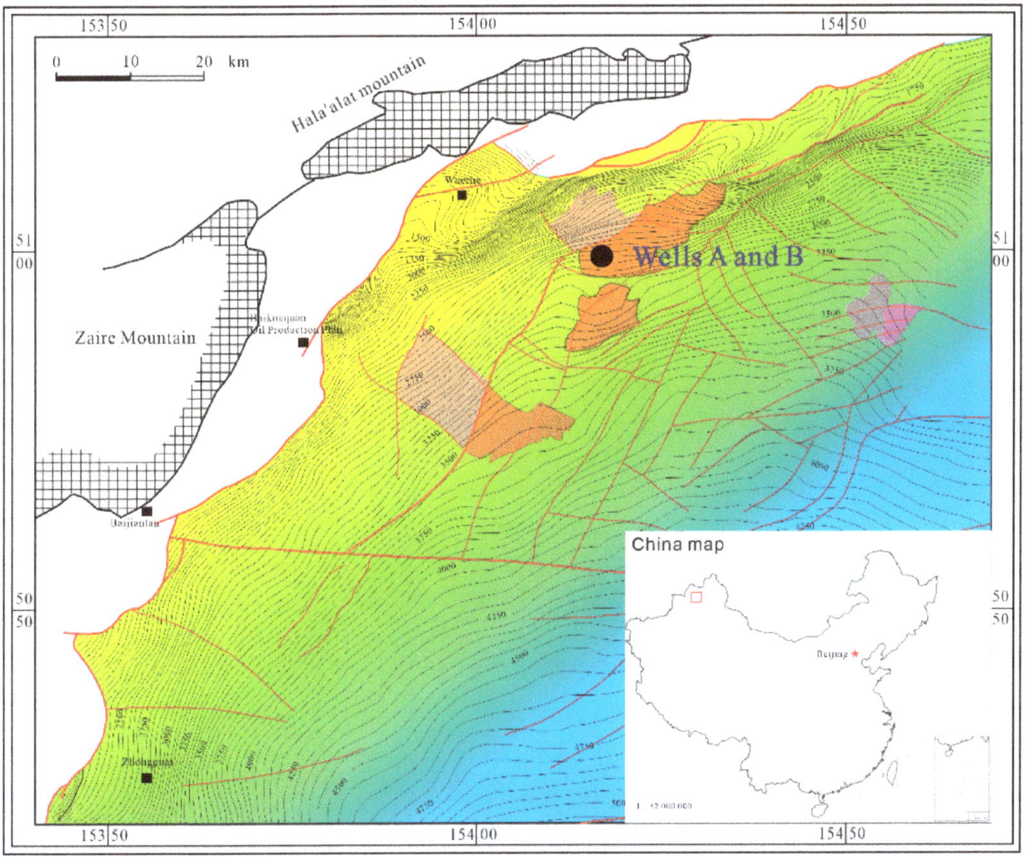

Figure 1. The sampling location in Mahu Block, Xinjiang, China (The area of red rectangle is the study area).

The target reservoir layers in wells A and B originate from the Triassic Baikouquan Formation, which can be divided into three intervals from bottom to top: Baikouquan Formation 1 (T_1b_1), Baikouquan Formation 2 (T_1b_2), and Baikouquan Formation 3 (T_1b_3), as shown in Figure 2 [42,44]. The primary oil-bearing strata are situated within Baikouquan Formation 2.

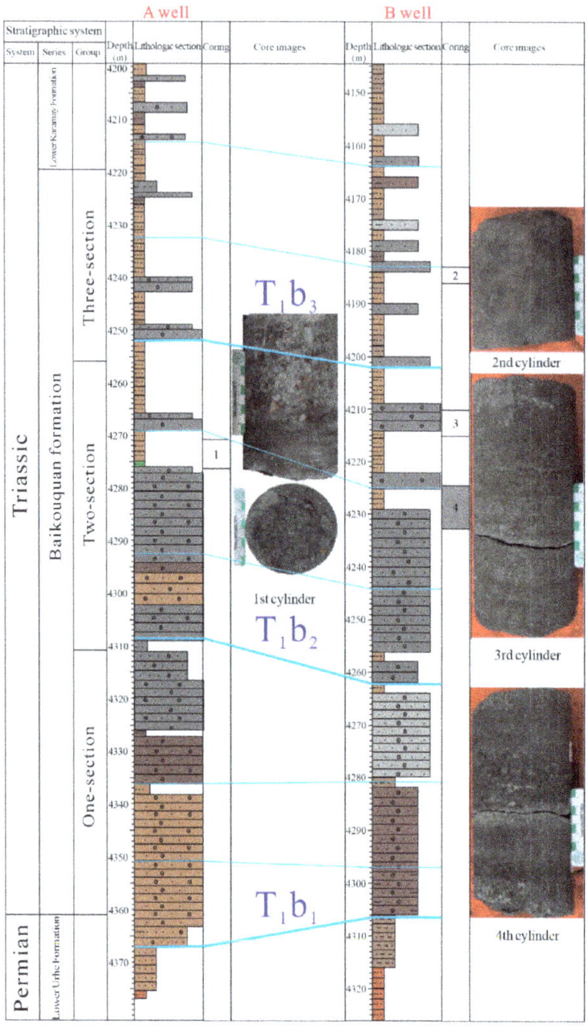

Figure 2. Sedimentary facies diagram of the target strata in well A and well B.

Baikouquan Formation 1: This interval is predominantly composed of blocky gray and brown pebble conglomerates, occasionally interbedded with brown mudstone, conglomeratic mudstone, and muddy sandstone. T_1b_1 has a thickness ranging from 16.7 to 60.6 m, with an average of 45.2 m.

Baikouquan Formation 2: The lower portion is characterized by gray blocky mudstone with pebbles, and is relatively compact. The upper part consists of gray blocky sandy conglomerates, sandy pebble conglomerates, and conglomeratic medium to coarse sandstones with intermittent layers of brown blocky mudstone. T_1b_2 has a thickness ranging from 51.7 to 67.2 m, with an average of 59.3 m.

Baikouquan Formation 3: This interval is primarily composed of brown mudstone with thin layers of gray mudstone and silty sandstone. T_1b_3 has a thickness ranging from 55.4 to 63.3 m, with an average of 56.8 m.

The core sampling depths for well A range from 4271 to 4277 m, while for well B, the core sampling depths are 4183 to 4186 m, 4210 to 4215 m, and 4224 to 4232 m, as shown in Figure 2.

3. Methods

3.1. Comparison of Stress Sensitivity Experiments in Water-Bearing and Dry Conditions for Tight Sandstone

Firstly, a section of downhole cores from wells A and B were precision-cut into standard core samples measuring 5 cm × 2.54 cm. Four samples meeting the criteria of being "intact, free of cracks, and possessing similar permeability" were chosen from each well. Samples from well A were designated as group A, and those from well B as group B. Ultimately, the initial permeability under dry conditions measured approximately 0.8 mD for group A cores and approximately 0.0024 mD for group B cores.

Secondly, the two sets of core samples were saturated with different amounts of water using the self-imbibition method [34]. For group A, the samples were subjected to different conditions: dry, water saturation (S_W) of 30%, S_W of 60%, and S_W of 80%. Group B samples were subjected to dry conditions, S_W of 25%, S_W of 52%, and S_W of 80%.

Thirdly, we utilized the MHY-TY3 instrument to gauge the permeability of each set of core samples under diverse confining pressures and at room temperature, while also assessing how fluctuations in water content within the cores impact stress sensitivity. The minimum confining pressure was 2.75 MPa, and it was incrementally increased to a maximum confining pressure of 15 MPa [9]. The gas permeability results for both sandstone groups under different confining pressures and water saturation conditions are shown in Figure 3.

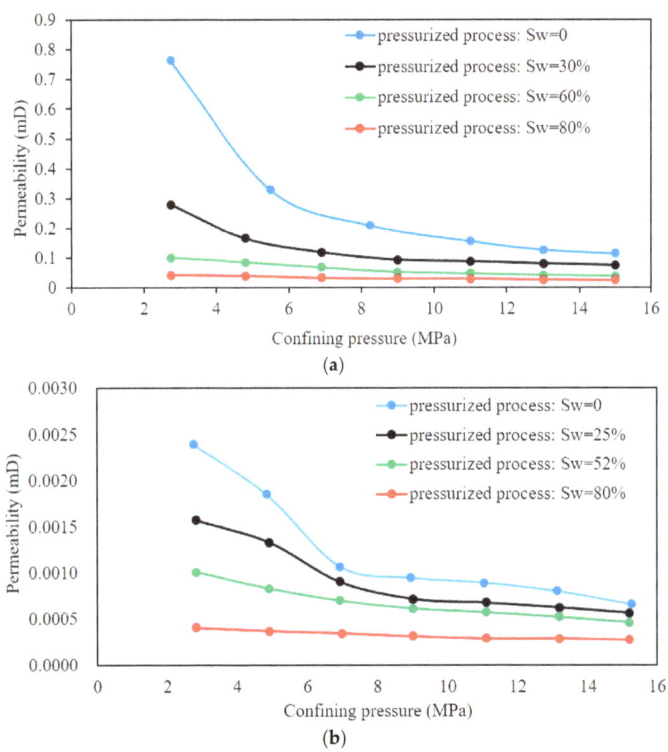

Figure 3. The experimental results of stress sensitivity for tight sandstone with different water content. (**a**) The experimental results of group A and (**b**) the experimental results of group B.

From Figure 3a,b, we observed a significant decrease in gas permeability as the confining pressure gradually increases. Beyond 8 MPa, the rate of permeability reduction levels off. Moreover, as the water content in the samples increases, the gas permeability gradually decreases. Furthermore, as the water saturation within the rock increases, the decline rate of permeability also decreases. This phenomenon can be attributed to the rise in water content, which leads to a decrease in the overall compressibility coefficient of the rock. Consequently, the rock's resistance to compression increases. These findings align with conventional expectations and represent typical experimental outcomes.

3.2. Limitations of Traditional Stress Sensitivity Models in Fitting the Experimental Results under Water-Bearing Conditions

Currently, the type of models used for assessing stress sensitivity mainly include the exponential model [8,45] and the power-law model [9,46], as showed in Table 1. Most of the models are the improvements of the two types of assessing stress sensitivity models. These two models are the most classic models and are frequently applied in the assessment of stress-sensitive damage levels.

Table 1. The models mostly used for stress sensitivity assessment.

Type	Detailed Models	Equations	References
Exponential model	Traditional model	$k = k_0 e^{-C_k(P_c - P)}$	[45,47]
	Seidle's model	$k = k_0 e^{-3C_t \sigma_{eff}}$	[8]
	S and D model	$k = k_0 e^{\frac{3C_t v}{1-v}(P-P_0)}$	[48]
Power-law model	Traditional model	$k = k_0 (\frac{P_c - P}{P_c - P_0})^{-n}$	[9,25]
	Walsh's model	$k = k_0 \left[1 - \frac{\sqrt{2}h}{a_0} \mathrm{Ln}\left(\frac{P}{P_0}\right)\right]^3$	[49]
	Gangi's model	$k = k_0 \left[1 - \left(\frac{\sigma_{eff}}{\lambda}\right)^m\right]^3$	[50]
	Han's model	$k = k_0 \left[1 - S_s \lg\left(\frac{\sigma_{eff}}{\sigma_{eff0}}\right)\right]^3$	[26]

where k and k_0 are the permeability and the initial permeability, respectively, mD; P_c and P are the overburden pressure and pore pressure, respectively, MPa; C_k is the stress sensitivity coefficient of permeability for the exponential model, MPa^{-1}; n is the fitting coefficient for the power-law model, dimensionless; C_t is the compressibility factor, MPa^{-1}; σ_{eff} is the effective stress, MPa; v is the Poisson's ratio; h and a_0 are the intersecting diameters of the pore, respectively; m is the parameter related to the pore roughness, dimensionless; λ is the model parameter, MPa; P_e is the effective pressure, MPa; S_s refers to stress sensitivity coefficient, dimensionless; σ_{eff} means the effective stress, MPa; and σ_{eff0} is the minimum effective stress, MPa.

Using the aforementioned two equations to fit stress-sensitive experimental data of tight sandstone under dry conditions, we obtained the stress sensitivity coefficients for samples A and B under dry conditions. Specifically, the stress sensitivity coefficients obtained through the traditional exponential model (M1) were 0.162 MPa^{-1} and 0.1022 MPa^{-1} for samples A and B, respectively. Additionally, the fitting coefficients "n" obtained through the traditional power-law model (M2) were 1.1695 and 0.7195 for samples A and B, respectively. The corresponding fitting results are shown in Figure 4. Obtaining the stress sensitivity coefficients for samples under dry conditions through data fitting is a widely utilized method in both scientific research and practical production applications.

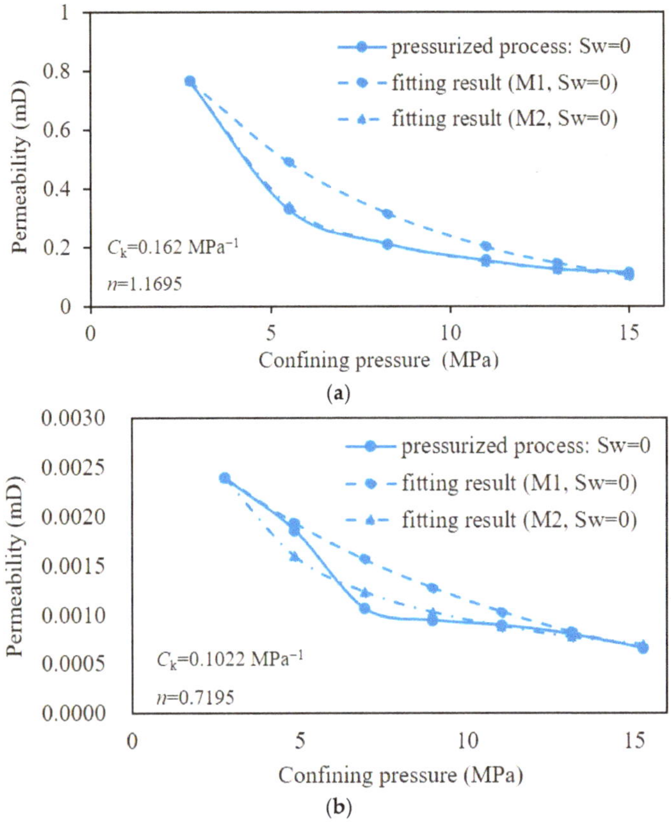

Figure 4. The experimental results in dry conditions and the fitting results of stress sensitivity using the power-law model and the exponential model. (**a**) The experimental and fitting results of group A in dry conditions and (**b**) the experimental and fitting results of group B in dry conditions.

It is evident that discrepancies exist between the stress sensitivity coefficients obtained using the two methods and the actual reservoir conditions. This discrepancy arises because oil and gas reservoirs typically exist in water-bearing conditions, whereas the experimental results obtained from samples under dry conditions naturally deviate from the real scenario. This observation is supported by Figure 5, where we substituted the stress sensitivity coefficient C_k obtained from the exponential model under dry conditions and the fitting coefficient n obtained from the power-law model under dry conditions for fitting the experimental data under various water-containing conditions, respectively. Notably, the results calculated using the two models are difficult to fit with the experimental results under water-containing conditions, consistently yielding values smaller than the actual experimental observations, as demonstrated in Figure 5. The observation concurrently implies that the overall compressibility coefficient under dry conditions exceeds that under water-saturated conditions. In simpler terms, rocks experience greater stress-induced damage under dry conditions compared to water-saturated conditions.

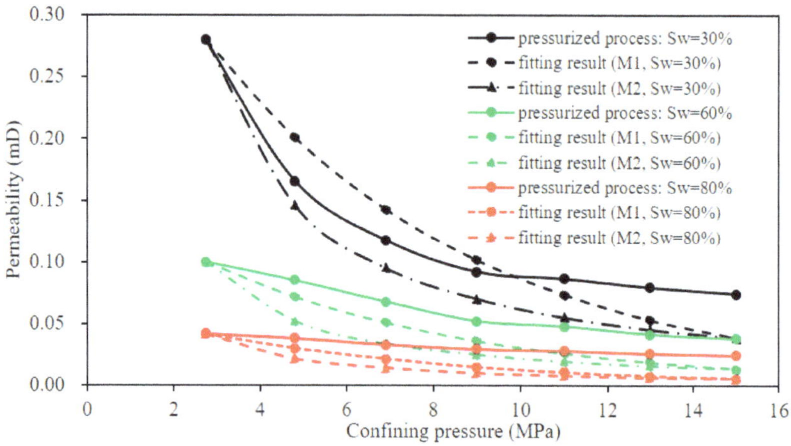

(a)

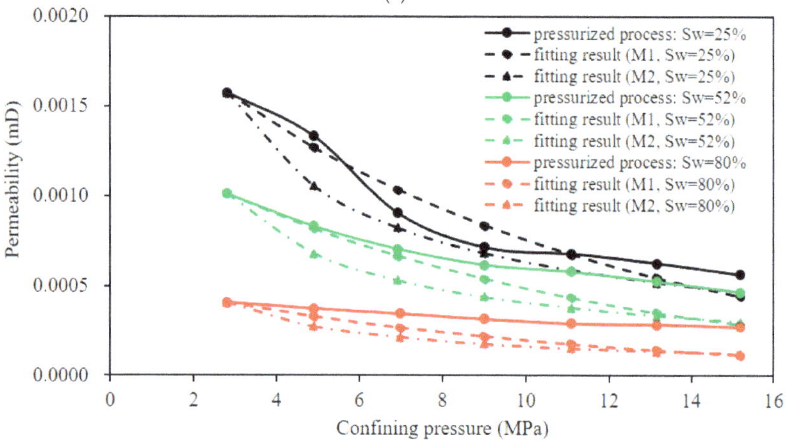

(b)

Figure 5. The experimental results with different water content and the fitting results of stress sensitivity using the power-law model and the exponential model. (**a**) The experimental and fitting results of group A with different water content and (**b**) the experimental and fitting results of group B with different water content.

In the traditional numerical simulation process, the influence of water-containing conditions on the magnitude of stress sensitivity damage has often been overlooked despite the ability to derive the initial permeability under such conditions from the relative permeability curves of water–gas two phases, denoted as k_0 in Table 1. Furthermore, most studies have focused on addressing stress sensitivity damage, but these data are typically derived from experimental procedures conducted under dry conditions. Based on the experimental and fitting results obtained using the power-law and exponential models mentioned above, it is evident that the stress sensitivity function obtained under dry conditions is not applicable to water-containing conditions. This highlights the urgent need for corrections to existing models.

3.3. Improvement to Power-Law Function Model

By fitting the stress sensitivity data under dry conditions and different water content conditions as shown in Figures 4 and 5, we can observe that the power-law model (M2) provides significantly better fitting results when compared to the exponential function (M1). This indicates that the power-law model is more suitable for evaluating the degree of stress-sensitive damage in tight sandstone. This application pattern is further substantiated by [25], who noted that rock cores following the power-law model exhibit a rapid decrease in permeability during the initial increase in effective stress, followed by a more gradual change in permeability during the middle and later stages of effective stress increase, commonly referred to as the 'steep-then-gentle pattern.' This distinctive pattern aligns with the characteristics of low-permeability reservoirs governed by fractures, where the reservoir permeability is primarily influenced by fractures. The initial closure of fractures due to the increase in effective stress results in a sharp decrease in permeability. As the fractures are completely closed, the extremely low matrix permeability causes a gradual reduction in permeability as the effective stress continues to increase. Consequently, this paper opts to optimize the model based on the power-law formulation, given its compatibility with the observed behavior.

However, the power-law model is an empirical model, and the fitting coefficient "n" does not have a specific physical interpretation, making it difficult to modify based on concrete meanings. Nevertheless, the fitting coefficient in the power-law model is directly proportional to the sample's compressibility factor. The relationship between the fitting coefficient under dry and water content conditions and the sample's compressibility factor can be expressed using the following equations:

$$n \propto C_{\text{t-drying}} = C_p + C_g \tag{1}$$

$$n_w \propto C_{\text{t-w}} = C_p + S_w C_w + S_g C_g \tag{2}$$

where n and n_w are the dimensionless fitting coefficients of the power-law stress-sensitive model under dry and water content conditions, respectively; $C_{\text{t-drying}}$ and $C_{\text{t-w}}$ are the total compressibility coefficients of the sandstone under dry and water content conditions, respectively, MPa^{-1}; C_p, C_g, and C_w are the compressibility coefficients of the rock, gas, and water, respectively, MPa^{-1}; and S_w and S_g are the dimensionless water saturation and gas saturation, respectively.

Therefore, a relationship is

$$\frac{n_w}{n} = \frac{C_p + S_w C_w + S_g C_g}{C_p + C_g} = 1 + \frac{S_w (C_w - C_g)}{C_{\text{t-drying}}} \tag{3}$$

Therefore, the power-law model can be transformed into

$$k = k_0 \left(\frac{P_c - P}{P_c - P_0}\right)^{-n\left[1 + \frac{S_w (C_w - C_g)}{C_{\text{t-drying}}}\right]} \tag{4}$$

The compressibility coefficient of water is very small and can be neglected. Therefore, the above equation can be simplified to

$$k = k_0 \left(\frac{P_c - P}{P_c - P_0}\right)^{-\kappa n} \tag{5}$$

$$\kappa = 1 - \frac{C_g}{C_{\text{t-drying}}} S_w \tag{6}$$

where κ is the correction coefficient of water saturation.

4. Results

4.1. Verification of the Improved Stress Sensitivity Assessment Model Using Experimental Data

Under atmospheric pressure conditions, the compressibility coefficient of gas (C_g) is much greater than the rock compressibility coefficient (C_p). Thus, the correction factor $C_g/C_{t\text{-drying}}$ in Equation (6) is approximately one. Consequently, the improved fitted coefficient of the power-law model becomes $(1 - S_w)\, n$.

The comparison between the corrected fitting results, original data, and uncorrected fitting results is shown in Figures 6–8. The fitting results in Figure 6 were based on Figure 3. In Figure 7, the fitting index 'n' under drying conditions is 1.4867, while in Figure 8, the fitting index under drying conditions is 1.7867. It can be observed that:

(1) As the water content in the sample increases, the disparity between most of the computed results of the fitted coefficients obtained under dry conditions according to the original model and the experimental results becomes more pronounced. This result corroborates the speculation in the preceding section that an increase in water content reduces the total compressibility factor of the sample. Moreover, the larger the water content, the smaller the total compressibility factor of the sample. Certainly, this disparity does not continue to increase, primarily during the high water content stages, as shown in Figures 7 and 8. Nevertheless, on the whole, it remains evident that utilizing the stress-sensitive damage coefficient fitted under dry conditions to evaluate the extent of stress-sensitive damage under wet conditions will result in an overestimation of the degree of stress sensitivity damage.

(2) The improved model, which takes into account the correction of the fitting coefficients of the power-law model for water content, exhibits improved fitting accuracy compared to the actual data, even though some disparity still remains. Comparing the fitting results of the original M2 for the sample group A, the accuracy of the improved model's fitting results has increased by 35.46%, 60.56%, and 70.94% for the samples with S_w = 30%, S_w = 60%, and S_w = 80%, respectively. In the case of the sample group B, the accuracy of the improved model's fitting results has increased by 21.06%, 39.63%, and 54.20% for the samples with S_w = 25%, S_w = 52%, and S_w = 80%, respectively. The average increase in accuracy for these experiments is 46.98%. As for Figure 7, the accuracy of the improved model's fitting results has increased by 6.01%, 4.18%, and −48.96% for the samples with S_w = 20%, S_w = 40%, and S_w = 70%, respectively. As for Figure 8, the accuracy of the improved model's fitting results has increased by 9.23% and −42.90% for the samples with S_w = 30% and S_w = 70%, respectively. The fitting results for the improved model have decreased the errors for the samples with low water saturation, but increased the errors for the samples with S_w = 70%. These comparisons highlight that the improved model performs well in assessing the degree of stress-sensitive damage in low water content samples. However, when evaluating the extent of stress-sensitive damage in high water content samples, there is some degree of error present.

(3) As the water content in the sample increases, the rate of permeability decrease during the initial pressurization stage gradually slows down. This means that the type of permeability decline curve changes from a "rapid–slow" similar to an "L-shaped" decline to a more linear decline due to the fact that the increased water content in the sample counteracts the rapid compression effect of confining pressure on the pores. This phenomenon further validates the earlier speculation, indicating that as the water saturation within the rock decreases, the gas saturation increases. Therefore, under the same overlying rock formation pressure, the framework bears a greater stress load, making the rock framework more susceptible to compression.

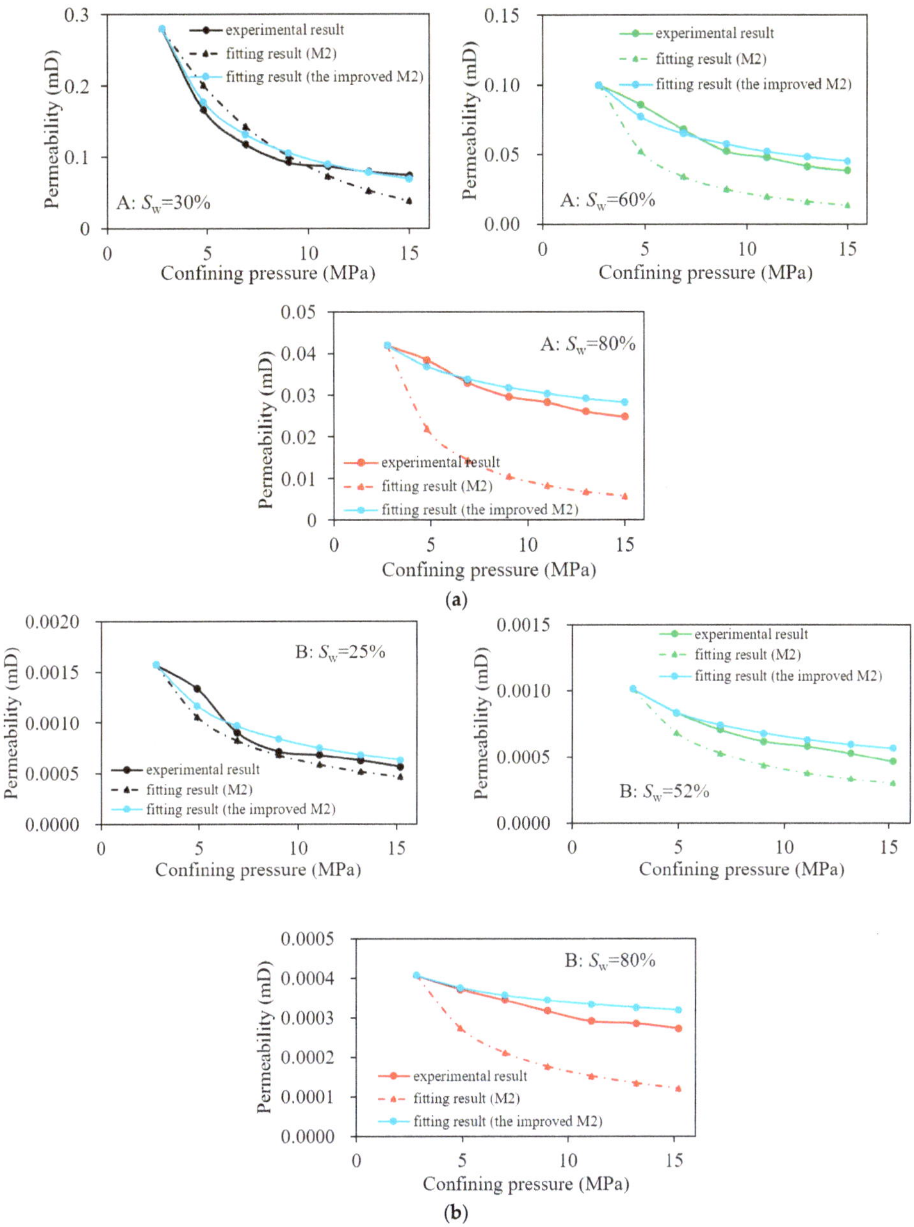

Figure 6. The comparison of the fitting results using the original M2 and the improved M2. (**a**) The comparison of the original M2 and the improved M2 with the experimental results of the sample group A and (**b**) the comparison of the original M2 and the improved M2 with the experimental results of the sample group A.

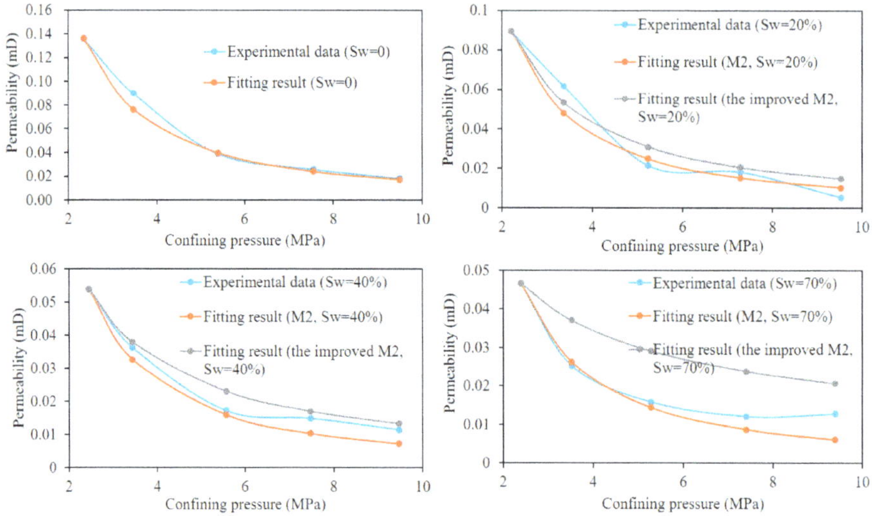

Figure 7. The comparison of the fitting results using the original M2 and the improved M2 for example 1.

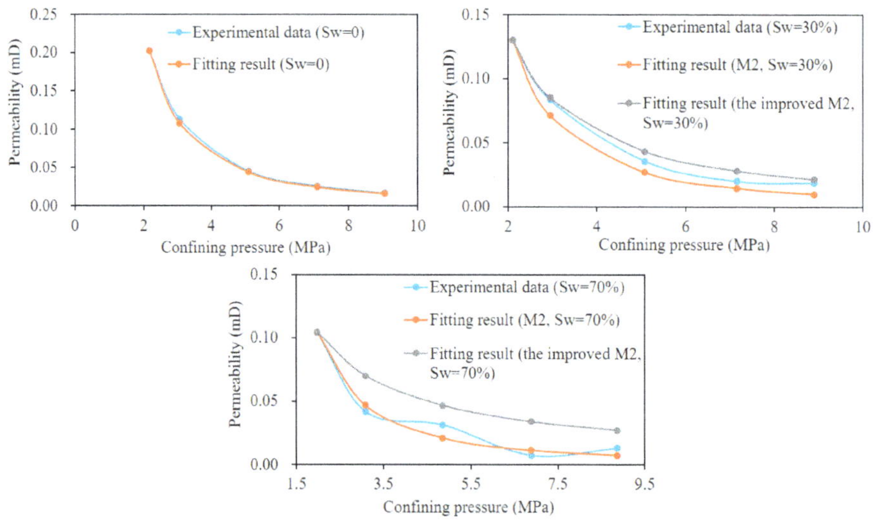

Figure 8. The comparison of the fitting results using the original M2 and the improved M2 fo example 2.

4.2. Verification of the Improved Stress Sensitivity Assessment Model Using Referenced Data

Drawing on some experimental data from Han et al. [26], we assessed the degree o rock stress-sensitive damage under various water content conditions and further validatec the accuracy of the model proposed in this paper. Although Han et al. [26] conducted stres sensitivity experiments on coal rock, which may disrupt the original structure of the coa rock during the repeated pressurization experiments, the observed phenomenon align broadly with this paper's approach. Therefore, their experimental results can still be usec to validate the model presented in this paper.

Firstly, we utilized a power-law model to fit the stress sensitivity experimental dat under dry conditions, yielding a stress-sensitive damage coefficient 'n' of 2.1167. Secondl we separately employed the original power-law model and the improved power-law model that considers the influence of water saturation to fit the stress sensitivity experimental data under different water saturation conditions. The comparative results are depicted in Figure 9.

(1) It can be observed that using the stress-sensitive damage coefficient obtained from fitting under dry conditions to fit stress sensitivity experimental data under different water content conditions often leads to an overestimation of the degree of stress-sensitive damage. In other words, the assessment results tend to be higher than the experimental results. Overall, as the water content increases, the original model exhibits a larger error in overestimating the degree of stress-sensitive damage.

(2) For low water-content samples, the improved power-law model that considers the influence of water saturation performs better than the original power-law model. However, for high water-content samples, the fitting performance of the improved power-law model considering water saturation effects is worse than that of the original power-law model. The reason for this phenomenon is that, compared to dense sandstone samples, coal rock samples exhibit plastic behavior, resulting in greater damage to permeability during the pressurization process, thus, causing errors in the experimental results for high water-content samples.

However, overall, utilizing previous experimental data to compare the original power-law model and the improved power-law model still demonstrates the following: (1) fitting stress sensitivity experimental data under different water content conditions with the stress-sensitive damage coefficient obtained from fitting under dry conditions often results in an overestimation of the degree of stress-sensitive damage and (2) the improved model exhibits superior fitting performance for low water content samples.

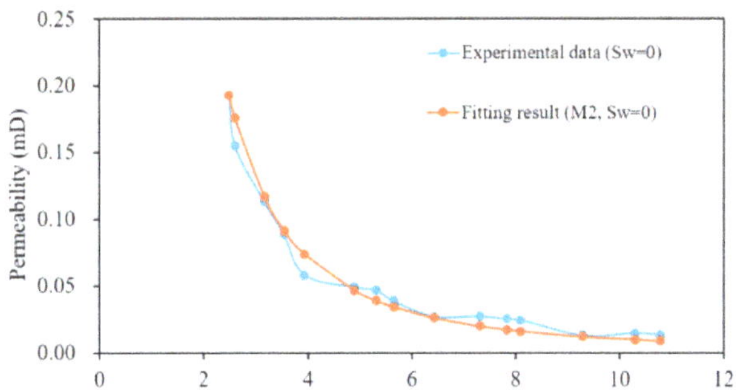

(a)

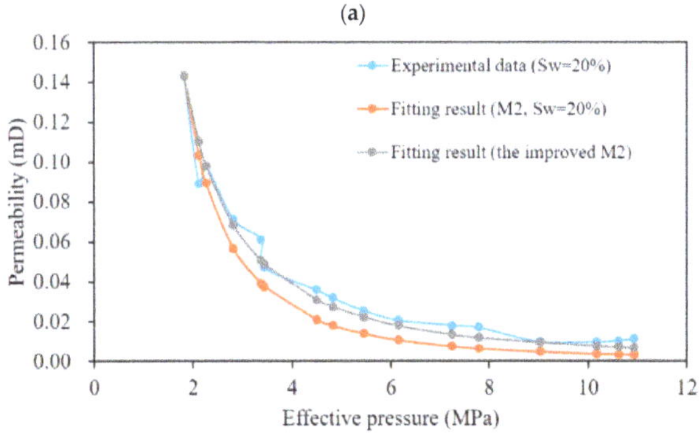

(b)

Figure 9. *Cont.*

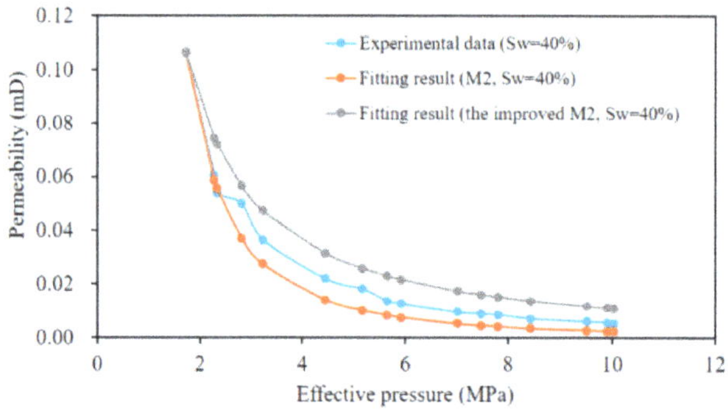

(**c**)

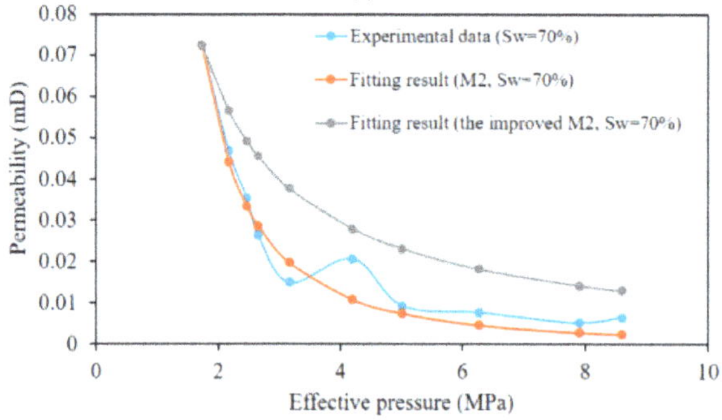

(**d**)

Figure 9. The comparison of the fitting results with the original M2 and the improved M2. (**a**) Th results for $S_w = 0$ and (**b**) The results for $S_w = 20\%$ and (**c**) The results for $S_w = 40\%$ and (**d**) The result for $S_w = 70\%$.

5. Discussion and Analysis

5.1. Model Application and Analysis

Taking well A in the Mahu Block, Xinjiang province, China, as an example, this sectio presents the relevant parameters of the model, as detailed in Table 2. Additionally, th physical model is visually depicted in Figure 10. The simulation calculations were carrie out using the analytical Hz oil multifrac-enhanced frac region model. The power-law mode was employed in the stress sensitivity damage assessment, with the fitting exponent ' under different water saturation conditions using group A experimental data. Specificall for well A, the stress sensitivity fitting exponent '*n*' was 1.1695 under dry conditions, 0.818 at 30% water saturation, 0.4678 at 60% water saturation, and 0.2339 at 80% water saturatio. Apart from the differences in the water content and the stress sensitivity fitting exponen under varying water saturation conditions, we maintained consistency in other paramete values, including initial permeability. Subsequently, we evaluated the impact of stres sensitivity fitting exponent variations on permeability and productivity. The comparativ results are visually presented in Figures 11 and 12.

The primary objective of this study is to explore the repercussions of overlooking th impact of water saturation on the stress sensitivity coefficient during actual application an to gauge the magnitude of this oversight on production capacity. The simulation conditior involve relatively idealized assumptions, such as assuming the same reservoir permeabili under different water saturation conditions. However, there are differences in permeabili and crude oil reserves under different water saturation conditions. By employing thes assumed simulation conditions, it becomes evident how much the stress-sensitive err caused by water saturation can affect production capacity, thereby capturing the attentic and emphasis of reservoir engineers on the research problem addressed in this paper.

Table 2. Parameters for the production simulation using RTA-analytical Hz oil multifrac-enhanced frac region model.

Parameters	Value	Parameters	Value
Initial reservoir pressure, kPa	40,000	Well depth, m	4800
Reservoir temperature, °C	80	Vertical depth, m	3000
Net pay thickness, m	46.5	Fracture half length, m	100
Wellbore radius, m	0.06	Horizontal well length, m	2200
Total porosity	0.1	Number of fractures	20
Initial formation compressibility, 1/kPa	7.05×10^{-7}	Distance from fracture to permeability boundary, m	15
Initial total compressibility (S_W = 0), 1/kPa	1.14×10^{-6}	Reservoir width, m	200
Water compressibility, 1/kPa	4.02×10^{-7}	Oil compressibility, 1/kPa	4.34×10^{-7}
Initial permeability, mD	10	Dimensionless fracture conductivity	0.03

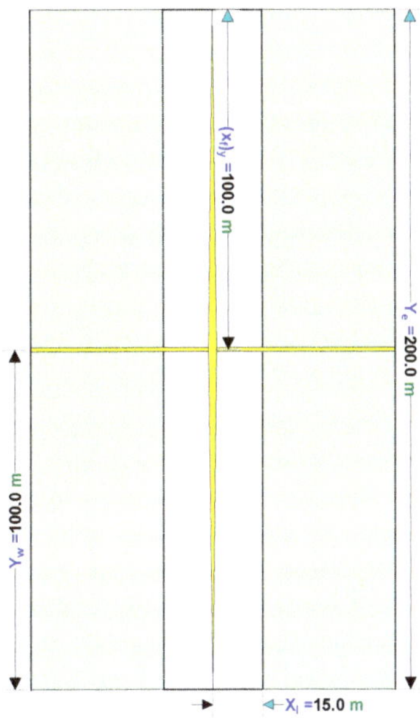

Figure 10. The model of analytical Hz oil multifrac-enhanced frac region in harmony-RTA.

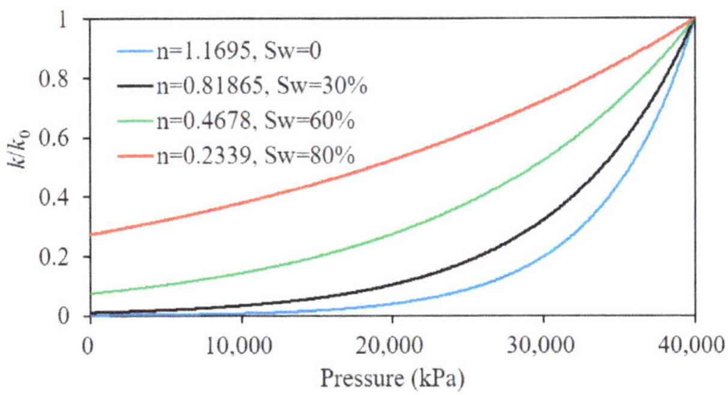

Figure 11. The relation between k/k_0 and pore pressure.

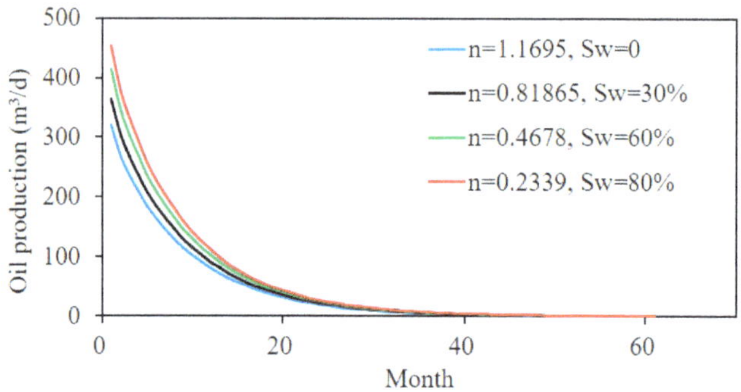

Figure 12. The effect of water content and fitting coefficient of stress sensitivity on tight oil production.

From Figure 11, we observed that the degree of stress-sensitive damage under different water content conditions is much smaller when compared to that under dry conditions. The greater the water content, the smaller the degree of stress-sensitive damage. Therefore, utilizing stress-sensitive data from dry conditions in numerical simulation of oil production can lead to an exaggerated perception of stress sensitivity impact on production. Meanwhile, as shown in Figure 12, the usage of inaccurate stress-sensitive data severely underestimates the well's production capacity. This underestimation becomes more pronounced as the actual water content increases, particularly during the early production stages. Such inaccuracies can significantly affect the assessment of production capability during the rapid decline phase. Furthermore, when fitting historical production data using production capacity simulation software, the usage of stress-sensitive data from dry conditions can affect the simulation and inversion accuracy of other engineering parameters.

To illustrate this point further, consider the application results shown in Figure 12. When the stress-sensitive fitting coefficients from dry conditions are applied to tight oil reservoirs with water saturations of 30%, 60%, and 80%, respectively, the underestimation of the ultimate recoverable reserves reaches 12.45%, 27.33%, and 38.85%, respectively. These simulation results clearly demonstrate that using stress-sensitive data from dry conditions to assess oil well production significantly underestimates the well's output. Such inaccuracies can lead to increased project investment when incorrect production values are utilized for oil well redevelopment assessments.

5.2. Signification and Limitations

Based on the experimental findings presented in Section 3.2 of this study, as well as the analysis of fitting outcomes using conventional methods, it becomes evident that traditional models for assessing stress sensitivity indices, including stress sensitivity power-law models, tend to disregard the influence of water content on the extent of stress-induced damage in tight sandstone [51–55]. This oversight results in an overestimation of the stress sensitivity coefficient obtained through fitting processes. In actual reservoir conditions, rock compositions exhibit varying water-content levels [56]. Consequently, evaluating the degree of stress sensitivity damage in moist tight sandstone based on an inflated stress sensitivity coefficient significantly amplifies the impact of stress sensitivity on reservoir rocks. In simpler terms, under similar pressure differential conditions, the assessments made using traditional models suggest a more pronounced pore compression and a larger decline in permeability. Moreover, as illustrated in Section 5.1 of this paper, when comparing the outcomes of production capacity simulations, an overestimated stress sensitivity coefficient leads to an underestimation of initial production rates, ultimately resulting in a lower estimation of the recoverable reserves [57].

Hence, when performing stress sensitivity assessments for reservoirs through laboratory experiments, it is crucial to preset the water content of samples according to the actual water saturation in the reservoir rocks [26]. This ensures that the experimental result

closely replicate reservoir conditions. Furthermore, in the production capacity simulation process, the stress sensitivity damage coefficient should be adjusted using correction factors, as illustrated in Equation (6). Only by conducting such simulations can we achieve a more precise assessment of stress sensitivity's influence on reservoir production capacity.

Based on experimental findings from downhole core samples in the Mahu Block of Xinjiang and an improved stress sensitivity power-law model, this paper first emphasizes the necessity of conducting stress sensitivity experiments under realistic reservoir water saturation conditions. Secondly, it rectifies the shortcomings of traditional models, providing guidance for production development in the Mahu Block of Xinjiang and even offering insights for the development of tight oil and gas fields.

Nevertheless, the two frequently referenced models discussed in this paper, the stress sensitivity index model and the power-law model, are both empirical in nature. While the correction coefficient is derived from the overall compressibility of the rock, the models themselves remain fundamentally empirical. For regions outside of the Mahu Block in Xinjiang, and even for other unconventional oil and gas reservoirs, future researchers may need to conduct additional refinements and precise data quantifications based on the improved model presented in this paper. Furthermore, through practical application it has been observed that the improved model still exhibits significant errors in assessing high water-content samples and highly plastic samples. Therefore, in later-stage applications, it is essential to consider the inherent characteristics of the samples and apply the improved model appropriately.

6. Conclusions

This paper, utilizing core samples from the Mahu Block, conducted stress sensitivity experiments on tight sandstones with differing water content levels. It refined the stress sensitivity power-law model by incorporating the comprehensive compressibility coefficient. Experimental validation and comparison confirmed the reliability of this enhanced model. These studies closely replicate actual reservoir conditions and address limitations in prior research. They have the potential to improve the precision of permeability, stress sensitivity, and production capacity assessments, thereby facilitating the efficient development of tight oil resources.

(1) The water content in sandstone samples has a discernible impact on the extent of stress sensitivity damage. Conducting stress sensitivity experiments on the sandstone samples under varying water content conditions revealed a consistent trend: higher water content leads to a reduction in the magnitude of stress sensitivity damage experienced by the samples.
(2) Through a comprehensive analysis involving experimental results, fitted outcomes, and numerical simulations, a clear pattern emerged. Employing stress-sensitive fitting coefficients from dry conditions to fit stress-sensitive experimental data under wet conditions resulted in calculated data significantly lower than the experimental data. Likewise, when using these coefficients to assess the actual oil well production, the estimated production output was notably lower than the actual production of the oil well.
(3) This study has improved the power-law model used for stress sensitivity damage assessment. The model takes into account the influence of water saturation on the degree of stress sensitivity damage, enhancing the fitting accuracy of the power-law model.

Author Contributions: Conceptualization, K.W.; Methodology, Y.D. and Z.F.; Software, H.Y. and L.W.; Validation, L.L.; Investigation, J.S.; Resources, X.L. and H.P.; Data curation, K.G., W.X. and X.Y. All authors have read and agreed to the published version of the manuscript.

Funding: This work was supported by the Major Science and Technology Project of China National Petroleum Corporation (2017E-0405) and the National Natural Science Foundation of China (42272156).

Data Availability Statement: The data underlying this article will be shared on reasonable reques to the corresponding author.

Conflicts of Interest: The authors declare no conflict of interest.

References

1. Aguilera, R.; Ramirez, J.F. Factors controlling fluid migration and distribution in the Eagle Ford shale. *SPE Reserv. Eval. Eng.* **2016** *19*, 403–414.
2. EIA (U.S. Energy Information Administration). *Annual Energy Outlook 2018 with Projections to 2040*; EIA: Washington, DC USA, 2018.
3. Sun, L.D.; Zhou, C.N.; Jia, A.L.; Wei, Y.S.; Zhu, R.K.; Wu, S.T.; Guo, Z. Characteristics and direction of tight oil and ga development in China. *Oil Explor. Dev.* **2019**, *46*, 1015–1026.
4. Zhang, T.; Zhang, L.H.; Zhao, Y.L.; Zhang, R.H.; Zhang, D.X.; He, X.; Ge, F.; Wu, J.F. Javadpour Farzam. Ganglia dynamics during imbibition and drainage processes in nanoporous systems. *Phys. Fluids* **2022**, *34*, 042016. [CrossRef]
5. Berumen, S.; Tiab, D. Effect of Pore Pressure on Conductivity and Permeability of Fractured Rocks. In Proceedings of the SP Annual Western Regional Meeting, Anchorage, AK, USA, 22–24 May 1996; SPE 35694.
6. Zhang, Z.; He, S.L.; Liu, G.F.; Guo, X.J.; Mo, S.Y. Pressure buildup behavior of vertically fractured wells with stress-sensitiv conductivity. *J. Pet. Sci. Eng.* **2014**, *122*, 48–55. [CrossRef]
7. Duan, X.G.; An, W.G.; Hu, Z.M.; Gao, S.S.; Ye, L.Y.; Chang, J. Experimental study on fracture stress sensitivity of Siluria Longmaxi shale formation, Sichuan Basin. *Nat. Gas Geosci.* **2017**, *28*, 1416–1424.
8. Seidle, J.P.; Jeansonne, M.W.; Erickson, D.J. Application of matchstick geometry to stress dependent permeability in coal In Proceedings of the SPE Rocky Mountain Regional Meeting, Casper, WY, USA, 18–21 May 1992; Paper Number: SPE-24361-M
9. Dong, J.J.; Hsu, J.Y.; Wu, W.J.; Shimamoto, T.; Hung, J.H.; Yeh, E.C.; Wu, Y.H.; Sone, H. Stress-dependence of the permeability an porosity of sandstone and shale from TCDP Hole-A. *Int. J. Rock Mech. Min. Sci.* **2010**, *47*, 1141–1157. [CrossRef]
10. Athy, L.F. Density, porosity, and compaction of sedimentary rocks. *AAPG Bull.* **1930**, *14*, 1–24.
11. David, C.; Menendez, B.; Zhu, W.; Wong, T.F. Mechanical compaction, microstructures and permeability evolution in sandstone *Phys. Chem. Earth Part A Solid Earth Geod.* **2001**, *26*, 45–51. [CrossRef]
12. Burton, Z.F.; Moldowan, J.M.; Magoon, L.B.; Sykes, R.; Graham, S.A. Interpretation of source rock depositional environment an age from seep oil, east coast of New Zealand. *Int. J. Earth Sci.* **2019**, *108*, 1079–1091. [CrossRef]
13. Burton, Z.F.M.; McHargue, T.; Kremer, C.H.; Bloch, R.B.; Gooley, J.T.; Jaikla, C.; Harrington, J.; Graham, S.A. Peak Cenozo warmth enabled deep-sea sand deposition. *Sci. Rep.* **2023**, *13*, 1276. [CrossRef]
14. Al-Khdheeawi, E.A.; Mahdi, D.S.; Ali, M.; Iglauer, S.; Barifcani, A. Reservoir scale porosity-permeability evolution in sandsto due to CO_2 geological storage. In Proceedings of the 15th Greenhouse Gas Control Technologies Conference, Abu Dhabi, Unite Arab Emirates, 5–8 October 2020; pp. 15–18.
15. Ju, W.; Huang, P.M.; Zhong, Y.; Hu, H.H.; Liang, Y.; Liu, B.; Zhang, X.L. Experimental study of sandstone stress sensitivity und different fluids: Characteristics and mechanisms. *Geoenergy Sci. Eng.* **2023**, *223*, 211537. [CrossRef]
16. Wang, H.L.; Tian, L.; Jie, Q.; Fei, S.X.; Wan, M.; Gao, X.X.; Zhang, K.Q. Stress-sensitivity analysis of geological confined pores wi ultrasonics. *Int. J. Rock Mech. Min. Sci.* **2023**, *170*, 105426. [CrossRef]
17. Kassis, S.; Sondergeld, C.H. Fracture permeability of gas shale: Effects of roughness, fracture offset, proppant, and effecti stress. In Proceedings of the International Oil and Gas Conference and Exhibition in China, Beijing, China, 8–10 June 201 SPE-131376-MS.
18. Kang, Y.L.; Lai, Z.H.; Chen, M.J.; Hou, T.F.; You, L.J.; Bai, J.J.; Yu, Z.H. Stress sensitivity experiments of shale gas diffusic coefficients based on the pressure decay method. *Nat. Gas Ind.* **2022**, *42*, 59–70.
19. Geng, S.Y.; Li, C.Y.; Li, Y.; Zhai, S.; Xu, T.H.; Gong, Y.F.; Jing, M. Pressure transient analysis for multi-stage fractured horizont wells considering threshold pressure gradient and stress sensitivity in tight sandstone gas reservoirs. *Gas Sci. Eng.* **2023**, *11* 205030. [CrossRef]
20. Fernandes, F.B.; Braga, A.M.B.; de Souza, A.L.S.; Soares, A.S. Mechanical formation damage control in permeability Biot's effecti stress-sensitive oil reservoirs with source/sink term. *J. Pet. Sci. Eng.* **2023**, *220*, 111180. [CrossRef]
21. Chang, Y.L.; Yang, Z.M.; Zhang, Y.P.; Niu, Z.K.; Chen, X.L. Permeability characterization and its correlation with pore microstr ture of stress-sensitive tight sandstone: Take Chang 6 in Ordos Basin for example. *Geofluids* **2022**, *2022*, 3334658. [CrossRef]
22. Warpinski, N.R.; Teufel, L.W. Determination of the effective-stress law for permeability and deformation in low-permeabili rocks. *SPE Form. Eval.* **1992**, *7*, 123–131. [CrossRef]
23. Dassanayake, A.B.N.; Fujii, Y.; Fukuda, D.; Kodama, J.-I. A new approach to evaluate effective stress coefficient for strength Kimachi sandstone. *J. Petrol. Sci. Eng.* **2015**, *131*, 70–79. [CrossRef]
24. Cui, X.A.; Bustin, R.M.; Brezovski, R.; Nassichuk, B.; Glover, K.; Pathi, V. A new method to simultaneously measure situ permeability and porosity under reservoir conditions: Implications for characterization of unconventional gas reservoi In Proceedings of the Canadian Unconventional Resources and International Petroleum Conference, Calgary, AB, Canada, 19– October 2010; SPE-138148-MS.

25. Zhang, R.; Ning, Z.F.; Zhang, H.S.; Xie, Q. New insights and discussions on stress sensitivity of fractured tight reservoir. *Nat. Gas Geosci.* **2016**, *27*, 918–923.
26. Han, J.; Wu, C.F.; Jiang, X.M.; Fang, X.J.; Zahng, S.S. Investigation on effective stress coefficients and stress sensitivity of different water-saturated coals using the response surface method. *Fuel* **2022**, *316*, 123238. [CrossRef]
27. Gao, Y.; Chen, S.S.; Tian, J.; She, Y.Q.; Huang, F.X.; Song, T.; Wang, S.Y.; Lv, W.N.; Jia, P.; Liu, C. Micro-occurrence of formation water in tight sandstone gas reservoir of north Tianhuan in Ordos Basin. *Nat. Gas Geosci.* **2020**, *31*, 1717–1732.
28. Shi, Y.J.; Yang, X.M.; Zhang, H.T.; Liu, T.D. Water-cut characteristic analysis and well logging identification methods for low-permeability lithologic gas reservoirs: A case study of the Sulige Gas Field. *Nat. Gas Ind.* **2011**, *31*, 25–28.
29. Ma, X.H. Discussion on characteristics and reservoiring mechanism of deep basin gas in Upper Paleozoic in Ordos Basin. *Oil Gas Geol.* **2005**, *26*, 230–236.
30. Cai, J.C.; Hu, X.Y.; Standnes, D.C.; You, L.J. An analytical model for spontaneous imbibition in fractal porous media including gravity. *Colloids Surf. A* **2012**, *414*, 228–233. [CrossRef]
31. Li, J.; Li, X.F.; Wang, X.Z.; Li, Y.Y.; Wu, K.L.; Shi, J.T.; Yang, L.; Feng, D.; Zhang, T.; Yu, P.L. Water distribution characteristic and effect on methane adsorption capacity in shale clay. *Int. J. Coal Geol.* **2016**, *159*, 135–154. [CrossRef]
32. Tian, J.; Kang, Y.L.; Jia, N.; Luo, P.Y.; You, L.J. Investigation of the controlling rock petrophysical factors on water phase trapping damage in tight gas reservoirs. *Energy Sci. Eng.* **2020**, *8*, 647–660. [CrossRef]
33. Peng, X.L.; Wang, X.Z.; Zhou, X.; Lin, Z.Y.; Zeng, F.H.; Huang, X.L. Lab-on-a-chip systems in imbibition processes: A review and applications/issues for studying tight formations. *Fuel* **2021**, *306*, 121603. [CrossRef]
34. Wang, K.; Jiang, B.B.; Li, H.T.; Ye, K.R.; Tan, Y.S. Spontaneous imbibition model for micro–nano-scale pores in coalbed methane reservoirs considering gas–water interaction. *J. Pet. Sci. Eng.* **2022**, *209*, 109893. [CrossRef]
35. Zhang, T.; Luo, S.; Zhou, H.; Hu, H.R.; Zhang, L.H.; Zhao, Y.L.; Li, J.; Javadpour, F. Pore-scale modelling of water sorption in nanopore systems of shale. *Int. J. Coal Geol.* **2023**, *27*, 104266. [CrossRef]
36. Lorenz, J.C. Stress-sensitive reservoirs. *J. Pet. Technol.* **1999**, *51*, 61–63. [CrossRef]
37. Osorio, J.G.; Chen, H.Y.; Teufel, L.W. Numerical simulation of the impact of flow-induced geomechanical response on the productivity of stress-sensitive reservoirs. In Proceedings of the SPE Reservoir Simulation Symposium, Houston, TX, USA, 14–17 February 1999; Paper Number SPE-51929.
38. Burton, Z.F.M.; Kroeger, K.F.; Hosford Scheirer, A.; Seol, Y.; Burgreen-Chan, B.; Graham, S.A. Tectonic uplift destabilizes subsea gas hydrate: A model example from Hikurangi Margin, New Zealand. *Geophys. Res. Lett.* **2020**, *47*, e2020GL087150. [CrossRef]
39. Bohnsack, D.; Potten, M.; Freitag, S.; Einsiedl, F.; Zosseder, K. Stress sensitivity of porosity and permeability under varying hydrostatic stress conditions for different carbonate rock types of the geothermal Malm reservoir in Southern Germany. *Geotherm. Energy* **2021**, *9*, 15. [CrossRef]
40. Burton, Z.; Dafov, L.N. Salt diapir-driven recycling of gas hydrate. *Geochem. Geophys. Geosyst.* **2023**, *24*, e2022GC010704. [CrossRef]
41. Liu, G.D.; Gao, G.; Huang, Z.L.; Yang, H.F. Oil source and accumulation in the overthrust belt in the Ke-Bai region, Junggar Basin, west China. *Pet. Sci.* **2009**, *7*, 31–33. [CrossRef]
42. Liang, Y.Y.; Zhang, Y.Y.; Chen, S.; Guo, Z.J.; Tang, W.B. Controls of a strike-slip fault system on the tectonic inversion of the Mahu depression at the northwestern margin of the Junggar Basin, NW China. *J. Asian Earth Sci.* **2020**, *198*, 104229. [CrossRef]
43. Chen, Y.B.; Cheng, X.G.; Zhang, H.; Li, C.Y.; Ma, Y.P.; Wang, G.D. Fault characteristics and control on hydrocarbon accumulation of middle-shallow layers in the slope zone of Mahu sag, Junggar Basin, NW China. *Pet. Explor. Dev.* **2018**, *45*, 1050–1060. [CrossRef]
44. Dang, W.L.; Gao, G.; You, X.C.; Wu, J.; Liu, S.J.; Yan, Q.; He, W.J.; Guo, L.L.B. Genesis and distribution of oils in Mahu Sag, Junggar Basin, NW China. *Pet. Explor. Dev.* **2023**, *50*, 840–850. [CrossRef]
45. Yves, G.; Amin, G.; Cuss, R.; Zoback, M. Gas storage capacity and transport in shale gas reservoirs—A review. Part A: Transport processes. *J. Unconv. Oil Gas Resour.* **2015**, *12*, 87–122.
46. Wu, K.L.; Li, X.F.; Chen, Z.X. Micro-scale effects of gas transport in organic nanopores of shale gas reservoirs. *Nat. Gas Ind.* **2016**, *36*, 51–64.
47. Jones, S.C. A technique for faster pulse-decay permeability measurements in tight rocks. *SPE Form. Eval.* **1997**, *12*, 19–26. [CrossRef]
48. Durucan, S.; Edwards, J.S. The effects of stress and fracturing on permeability of coal. *Min. Sci. Technol.* **1986**, *3*, 205–216. [CrossRef]
49. Walsh, J.B. Effect of pore pressure and confining pressure on fracture permeability. *Int. J. Rock Mech. Min. Sci. Geomech. Abstr.* **1981**, *18*, 429–435. [CrossRef]
50. Gangi, A.F. Variation of whole and fractured porous rock permeability with confining pressure. *Int. J. Rock Mech. Min. Sci. Geomech. Abstr.* **1978**, *15*, 249–257. [CrossRef]
51. Kazak, E.S.; Kazak, A.V. A novel laboratory method for reliable water content determination of shale reservoir rocks. *J. Pet. Sci. Eng.* **2019**, *183*, 106301. [CrossRef]
52. Gurevich, B. A simple derivation of the effective stress coefficient for seismic velocities in porous rocks. *Geophysics* **2004**, *69*, 393–397. [CrossRef]

53. Glubokovskikh, S.; Gurevich, B. Effect of micro-inhomogeneity on the effective stress coefficients and undrained bulk modulus of a poroelastic medium: A double spherical shell model: Micro-inhomogeneity on stress coefficient. *Geophys. Prospect.* **2015**, *63*, 656–668. [CrossRef]
54. Moghadam, J.N.; Mondol, N.H.; Aagaard, P.; Hellevang, H. Effective stress law for the permeability of clay-bearing sandstones by the Modified Clay Shell model. *Greenhouse Gas. Sci. Technol.* **2016**, *6*, 752–774. [CrossRef]
55. Fink, R.; Krooss, B.M.; Gensterblum, Y.; Amann-Hildenbrand, A. Apparent permeability of gas shales—Superposition of fluid-dynamic and poro-elastic effects. *Fuel* **2017**, *199*, 532–550. [CrossRef]
56. Kranzz, R.L.; Frankel, A.D.; Engelder, T.; Scholz, C.H. The permeability of whole and jointed Barre Granite. *Int. J. Rock Mech. Min. Sci. Geomech. Abstr.* **1979**, *16*, 225–234. [CrossRef]
57. Chen, Y.; Liu, D.; Yao, Y.; Cai, Y.; Chen, L. Dynamic permeability change during coalbed methane production and its controlling factors. *J. Nat. Gas Sci. Eng.* **2015**, *25*, 335–346. [CrossRef]

processes MDPI

ırticle

\ew Method for Capacity Evaluation of Offshore ̄.ow-Permeability Reservoirs with Natural Fractures

Kun Wang [1,2,*], Mingying Xie [2], Weixin Liu [2], Li Li [2], Siyu Liu [2], Ruijie Huang [2], Shasha Feng [2], Guotao Liu [2] nd Min Li [1,*]

1 State Key Laboratory of Oil and Gas Reservoir Geology and Exploitation, Southwest Petroleum University, Chengdu 610500, China
2 Nanhai East Petroleum Research Institute, Shenzhen Branch of CNOOC Limited, Shenzhen 518000, China; xiemy2@cnooc.com.cn (M.X.); liuwx@cnooc.com.cn (W.L.); lili8@cnooc.com.cn (L.L.); liusy20@cnooc.com.cn (S.L.); huangrj10@cnooc.com.cn (R.H.); fengshsh@cnooc.com.cn (S.F.); liugt@cnooc.com.cn (G.L.)
* Correspondence: wangkun25@cnooc.com.cn (K.W.); hytlxf@126.com (M.L.)

Abstract: In recent years, the development of two offshore low-permeability oil fields has revealed unexpected challenges. The actual productivity of these fields significantly deviates from the designed capacity. Some wells even outperform the expectations for low-permeability limestone fields. This discrepancy primarily stems from a lack of accurate understanding of natural fractures before and after drilling, resulting in substantial errors in capacity assessment. This paper addresses these challenges by proposing a new production capacity model and evaluation method for both vertical and horizontal wells in low-permeability limestone reservoirs. The method leverages logging curve data, incorporating vertical gradation and fractal analysis to effectively represent the fracture's complexity and connectivity. It uniquely considers factors such as fracture fractal dimensions, threshold pressure, and stress sensitivity, significantly enhancing prediction accuracy. Furthermore, by analyzing the longitudinal gradient in logging curves, the method effectively identifies strong heterogeneity, leading to more accurate capacity evaluations in actual fields. The results demonstrate that our model reduces the average prediction error to less than 15%, markedly outperforming traditional methods. Calculation results of the newly developed capacity formula align closely with actual production data and tracer test results, showcasing its practical applicability and potential for widespread use. This study notably advances the evaluation of reasonable production capacity in similar offshore reservoirs.

Keywords: offshore low-permeability reservoirs; natural fractures; fractal parameters; capacity evaluation; threshold pressure gradient

tation: Wang, K.; Xie, M.; Liu, W.; L.; Liu, S.; Huang, R.; Feng, S.; Liu, · Li, M. New Method for Capacity aluation of Offshore w-Permeability Reservoirs with ıtural Fractures. *Processes* **2024**, *12*, 7. https://doi.org/10.3390/ .2020347

ademic Editor: Qingbang Meng

ceived: 9 January 2024
vised: 4 February 2024
cepted: 5 February 2024
blished: 6 February 2024

1. Introduction

Oil and gas resources play an important role in global energy. As conventional oil and gas fields gradually deplete, offshore low-permeability reservoirs are gaining industry attention for their significant development potential [1]. Despite years of development, these reservoirs remain under-exploited. Capacity evaluation of offshore low-permeability reservoirs faces challenges due to high drilling costs, limited production wells, and a complex subsurface environment. The notable discrepancy between predicted and actual capacities in these oil fields presents development challenges. Effective capacity evaluation of directional and horizontal wells in reservoirs with natural fractures is vital for enhancing reserves and achieving rational, efficient development.

Taking two typical low-permeability oil fields in the offshore FA region as examples, the production capacity, according to the conventional Darcy formula, is proportional to the flow coefficient and differential pressure. However, this formula fails to explain certain anomalies. First, the low-permeability limestone Zone A in the FA15 field has a lower flow coefficient than the FA14 field, yet its production capacity exceeds FA14 by more than six times. Second, although Zone A has lower flow coefficient and production differentia pressure than Zone B, their production capacities are nearly the same. Analyzing th reasons, it is concluded that the low-permeability limestone Zone A in the FA15 fielc developed natural fractures making a significant contribution to production capacity which is confirmed by imaging logging and tracer tests. Therefore, conventional capacit models are unsuited for the capacity prediction of this type of reservoir. This highlight the importance of accurately characterizing permeability and capacity in low-permeabilit reservoirs with natural fractures, which forms the basis of this study.

Many scholars [2–5] have conducted a lot of research on capacity forecasting anc evaluation. There are four main types of methods for capacity forecasting. The first i the method based on statistical analysis [6–9], which uses historical data and technique such as trend analysis to predict future capacity through data analysis and modelling. I typically requires a substantial amount of high-quality data, making it unsuitable for earl capacity assessments, especially in scenarios with limited data. The second is the empirica equation-based approach [10–12], which uses correlation equations to extrapolate futur capacity based on past experience and observations. The currently widely used empirica analysis method is the modified Arps decreasing curve, which has a large error. The thir is the analytical or semi-analytical capacity model [13–15] based on physical assumption which is derived with rigorous ideas, but the current model cannot reasonably characteriz the actual seepage characteristics of heavy oil reservoirs, and it is more difficult to apply a the mine site. The fourth is the simulation-based method [16,17], usually using numerica simulation software to predict the production capacity. This method can consider th effects of multiple factors on the production capacity, but it depends on the computationa power, model accuracy, and input data accuracy, which is relatively cumbersome. Thes methods have also achieved some results in practical applications, but they also hav obvious limitations. In addition, for offshore low-permeability reservoirs with stron heterogeneity and natural fractures, there are relatively few studies on capacity evaluatio and a large number of scholars refer to the Jubilee formula [18] for capacity calculatio Under the influence of sedimentation, diagenesis, and tectonic movement, natural fractur in offshore low-permeability limestone reservoirs are more developed, and the presence c fractures improves the physical properties of the reservoir and increases the productio capacity of oil wells. The production of oil wells is related to the fracture developmer degree, and there is no good solution internationally on how to quantitatively descrit the relationship between natural fractures and oil well production. The development c low-permeability reservoirs generally requires fracturing construction operations, and the are many production formulas [19–22] corresponding to them. Although these productic capacity formulas take into account the threshold pressure gradient or fracture conductivit none account for natural fracture. Due to the lack of comprehensive consideration of th seepage characteristics (fractal characteristics of natural fractures, threshold pressure, ar stress sensitivity) and the refined consideration of the strong longitudinal heterogeneit existing capacity models cannot well reflect the interference caused by the longitudin permeability gradient. It leads to a capacity model that cannot better reflect the actu seepage characteristics of the reservoir, and the limitations of the model lead to the inabili to accurately predict the capacity of vertical and horizontal wells in this type of reservoi

Targeting the unique offshore geological structures and complex fluid dynamic traditional capacity evaluation methods show clear limitations in such reservoirs. Th study systematically examines how natural fractures affect the capacity of low-permeabili reservoirs. Utilizing actual drilling data, it highlights the stark differences between pr drilling models and actual post-drilling fracture distributions. These differences challen the efficacy of traditional models and underscore the need for new evaluative approach

Based on the above considerations, a novel capacity evaluation model is proposed this study, which incorporates the theory of fractal geometry for rational characterizati of natural fractures based on the traditional theory, and introduces the fractal dimensic

D and the connectivity coefficient θ to represent the permeability behavior of the fracture network more realistically. Additionally, in order to more finely characterize the strong longitudinal heterogeneity, the longitudinal heterogeneity based on the logging curves is considered in this paper. In the end, in order to improve the accuracy of production capacity prediction of low-permeability reservoirs with natural fractures, a new production capacity model that comprehensively considers the influence of fracture fractal, longitudinal heterogeneity, threshold pressure, and stress sensitivity is established in this paper, which forms a new production capacity evaluation method to provide a reference for the development of the same type of reservoirs. The new production capacity evaluation method proposed in this study which combines seepage and geological characteristics of offshore low-permeability reservoirs with developed natural fractures can provide theoretical guidance for the formulation of reasonable development technology policies of early project development.

2. Materials and Methods

2.1. Degree of Fracture Development and Fractal Dimension

A fractal is a form in nature in which the local and the whole are similar in some way [23]. The complexity of a fractal can be described quantitatively by the size of the fractal dimension, which is denoted by the letter D.

Before applying fractal theory, it should be judged first whether the distribution of the research object has fractal characteristics. For example, when studying the reservoir fractures, the grid coverage method is used on the core to judge whether the reservoir fractures conform to fractal characteristics. The specific method is as follows. First of all, cover the core profile with a square grid of side length r, then count the number of grids containing fractures $N(r)$. Secondly, change the scale of r, count the corresponding $N(r)$ values, and analyze the statistical data by least squares regression analysis in the double logarithmic coordinate system.

Research shows that there is a linear relationship between $\lg N(r)$ and $\lg r$. Therefore, the distribution of reservoir fractures has fractal characteristics, so the fractal dimension D can be used to quantitatively characterize the degree of fracture development. Some scholars have studied the relationship between fractal dimension and natural fracture density and found that the two satisfy certain quantitative relationships [24,25], which can be expressed as Equation (1):

$$s = f(D), \tag{1}$$

where s is the natural fracture density, 1/cm; D is the fractal dimension.

In general, the higher the value of D, the greater the natural fracture density and the more developed the natural fractures are in relative terms. For cores, the grid coverage method can be used to determine the value of D. There are many methods to calculate the fractal dimension, of which the most commonly applied is the box-counting dimension method [26,27], obtaining the box-counting dimension by processing the image and then fitting the function using the least squares method. Therefore, the fractal dimension of the fracture network can be obtained by the box-counting dimension method after obtaining the shape of the fracture network based on the inversion of microseismic data [28].

2.2. Formations Consideration of Natural Fracture Evolution in Low-Permeability Reservoir Capacity Modeling

2.2.1. Permeability Characterization of Natural Fracture Networks

In this paper, fractal theory is introduced to characterize the permeability of the natural fracture network, considering the stress-sensitive properties of the fracture network. Considering the wellbore and surrounding natural fractures in a reservoir can connect to form a dominant seepage channel, this indicates that the closer from the wellbore it is, the greater the total permeability of the reservoir. Based on earlier studies [29,30], the fractal permeability of natural fractures considering the stress sensitivity can be obtained as shown in Equation (2).

$$k = k_{iT}\left(\frac{r}{r_w}\right)^{n'} e^{-\beta_k(p-p_w)}, \quad (2)$$

where $n' = -(D-2-\theta)$, k_{iT} is the initial total permeability of natural fractures, darcy; β_k is reservoir pressure sensitivity coefficient, MPa^{-1}; θ is anomaly diffusion coefficient, which indicates the connectivity of the fractal network (the smaller θ is, the better the connectivity of the network); p is the formation pressure, MPa; and p_w is the pressure at the reference point, MPa; r is the radial distance between the wellbore and the volume node of the fluid storage space unit in a medium containing natural fractures, m; r_w is the radial distance between the wellbore and the selected reference point position, m.

It is generally accepted that a dual-media reservoir consists of two types of media, matrix blocks and fractures, and that these two media are equally distributed. If the fractures are sufficiently developed and the connectivity between the fractures is good, fluid flow will follow the seepage law of fluids in dual-media reservoirs, but in low-permeability reservoirs, the connectivity between natural fractures is poor, so low-permeability reservoirs with natural fractures cannot be considered dual-media reservoirs. Total reservoir permeability is not a simple superposition of matrix permeability and fracture permeability but is related to the degree of development of natural fractures. Based on the above understanding, combined with the fractal dimension, the formula for calculating the total permeability of a low-permeability reservoir considering natural fractures is established. It can be represented as Equation (3):

$$k_t = k_m + (D-2-\theta)k_f, \quad (3)$$

where k_m is matrix permeability, darcy; k_f is natural fracture permeability, darcy; k_t is total permeability of the reservoir, darcy. k_t is related to the size of D value. Usually, the D value of reservoir fracture is 2 to 3; when the D value is 2, the natural fracture of reservoir is not developed, k_t is equal to k_m, which is a single-medium reservoir. When the D value is 3 k_t is the sum of k_m and k_f, which is a double-medium reservoir. When the D value is between 2 and 3, the reservoir is in the transition zone from single-medium to double-medium. The value of D is given between 2 and 3 in this paper while considering the development of natural fractures in low-permeability reservoirs.

The development of natural fractures in low-permeability reservoirs increases the permeability of the reservoir, assuming that fluid flow in the fractures is between parallel flat plates [31], thus the permeability of natural fractures can be expressed as Equation (4):

$$k_f = 8.33 \times 10^{-6} s b^3, \quad (4)$$

where b is the average width of natural fractures, μm.

Combining Equations (2)–(4), the expression for permeability characterization of the reservoir considering natural fractures can be obtained as the Equation (5):

$$k = \left[k_m + 8.33 \times 10^{-6} s b^3 n\right] e^{-\beta_k(p-p_w)} \left(\frac{r}{r_w}\right)^{n'}, \quad (5)$$

where $n = D - 2 - \theta$.

2.2.2. Modeling of Production Capacity of Vertical Wells Containing Natural Fractures

Figure 1 shows a schematic diagram of a vertical well in an infinite formation. In this paper, the vertical well seepage process is considered as the radial flow from the reservoir supply boundary to the vertical wellbore.

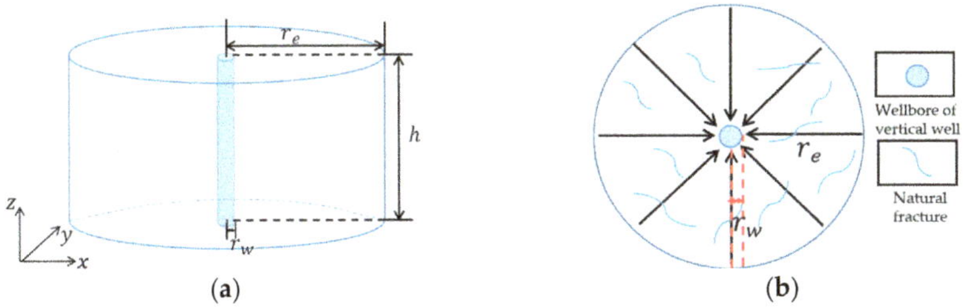

Figure 1. Physical model of the vertical well in low-permeability reservoir: (**a**) Schematic of a three-dimensional model of a vertical well; (**b**) Plane seepage pattern diagram of a vertical well.

The equation of motion for an offshore low-permeability reservoir considering the threshold pressure gradient is given as the Equation (6):

$$v = \frac{k}{\mu_{eff}}\left(\frac{\partial p}{\partial r} - G\right), \tag{6}$$

where μ_{eff} is the effective fluid viscosity, mP as; v is the seepage velocity, m/s; and G is the starting pressure gradient, MPa/m.

The above equation of motion comprehensively reflects the seepage characteristics of offshore low-permeability reservoirs, and the planar radial flow equation for vertical wells can be obtained as shown in Equation (7).

$$\frac{qB_o}{2\pi rh} = \frac{86.4}{\mu_{eff}}\left[k_m + 8.33 \times 10^{-6} sb^3 n\right] e^{-\beta_k (p - p_w)} \left(\frac{r}{r_w}\right)^{n'} \left(\frac{\partial p}{\partial r} - G\right), \tag{7}$$

where q is the production rate, m^3/d; B_o is the crude oil volume factor; h is the effective thickness of the reservoir, m.

Introduce $\zeta = e^{\beta_k p}$, $\zeta_i = e^{\beta_k p_e}$, and let $n = D - 2 - \theta$, $\Omega = \frac{qB_o}{2\pi h}\frac{\mu_{eff}\beta_k r_w^{n'}\zeta_i}{86.4(k_m + 8.33\times10^{-6}sb^3 n)}$, from Equation (7), we can obtain the Equation (8) by simplification:

$$\frac{d\zeta}{dr} - \beta_k G\zeta - \frac{\Omega}{r^{n'+1}} = 0, \tag{8}$$

where ζ is the introduced pressure transformation coefficient; ζ_i is the corresponding pressure transformation coefficient at the supply boundary $r = r_e$; and Ω is the integrated reservoir parameter used for simplification.

Equation (8) is a typical first-order nonlinear differential equation, ζ is expressed as Equation (9):

$$\zeta = e^{\beta_k G r}\left(\int e^{-\beta_k G r} \cdot \frac{\Omega}{r^{n'+1}} dr + C\right). \tag{9}$$

Since $\beta_k G << 1$, $(\beta_k G)^2 << 1$, substituting into Equation (9), we can obtain Equation (10):

$$\zeta = e^{\beta_k G r}\left(-\frac{\Omega}{n'}\frac{e^{-\beta_k G r}}{r^{n'}} + C\right). \tag{10}$$

When $r = r_e$, $\zeta_i = e^{\beta_k p_e}$, based on Equation (10), we can obtain Equation (11):

$$C = \left(\zeta_i + \frac{\Omega}{n'}\frac{1}{r_e^{n'}}\right) / e^{\beta_k G r_e}. \tag{11}$$

Combining Equations (10) and (11), the constant term can be eliminated, and Equatior (12) can be derived:

$$\xi = -\frac{\Omega}{n'}\frac{1}{r^{n'}} + \left(\frac{\Omega}{n'}\frac{1}{r_e^{n'}} + \xi_i\right)e^{\beta_k G(r-r_e)}. \quad (12$$

To obtain an expression for the capacity, isolating Ω, and we can acquire Equation (13

$$\Omega = \frac{\left[\zeta - \zeta_i e^{\beta_k G(r-r_e)}\right]n'}{\left[\frac{e^{\beta_k G(r-r_e)}}{r_e^{n'}} - \frac{1}{r^{n'}}\right]}. \quad (13$$

The express Ω in Equation (13) is related to the production capacity, thus furthe deriving the production capacity equation for vertical wells in low-permeability reservoir with natural fracture, as shown in Equation (14):

$$q = 86.4\frac{2\pi h}{B_o}\left\{\frac{\left(k_m + 8.33\times10^{-6}sb^3n\right)\left[e^{-\beta_k[(p_e-p_w)-G(r_e-r_w)]}-1\right]}{\mu_{eff}\beta_k\left[\frac{1}{n'}\left(\frac{r_w}{r_e}\right)^{n'} - \frac{e^{\beta_k G(r_e-r_w)}}{n'}\right]}\right\}, \quad (14$$

where p_w is the bottomhole pressure of the vertical well, MPa; p_e is the reservoir pressur MPa.

In the above equation, $n = D - 2 - \theta$, when $D = 2$ and $\theta = 0$, then $n = 0$, indicates single pore medium. As a result, Equation (14) becomes the capacity equation of a singl pore medium considering only the threshold pressure and stress sensitivity. When G and β converge to 0, Equation (14) can be transformed into the commonly used Joshi equation [32 When $D = 3$, $\theta = 1$, then $n = 0$, similar to the single medium, indicates that despite th development of fractures, the natural fractures are mainly developed as isolated seam without good connectivity, and the contribution to the capacity remains negligible. Whe $D = 3$, $\theta = 0$, then $n = 1$, indicating that the fractures are more developed, the connectivit between the fractures is good, which can improve the physical properties of the reservo and have a greater contribution to the production capacity. Equation (14) demonstrates th contribution of the fracture development degree and fracture connectivity to the productio capacity. The traditional method is difficult to well characterize the permeability of th fracture, and the matrix permeability is usually used to calculate the production capacit which leads to a pessimistic assessment of the production capacity of low-permeabili reservoirs containing natural fractures. The method in this paper can better characteriz the production capacity of low-permeability reservoirs with developed natural fractures

At present, offshore oil fields are still developed with directional wells due to the larg number of longitudinal layers as well as economic considerations. But low-permeabili reservoirs tend to exhibit strong heterogeneity in the longitudinal direction. In order accurately characterize this heterogeneity, a production capacity method combined wi logging curve analysis based on Equation (14) is proposed. In this method, a logging poi is set every 0.125 m on the logging curve, and the production capacity calculated fro each logging point is superimposed. Due to the certain similarity between the differenti equation of seepage of porous media in reservoirs and the differential equation of electr charge flow through conductor materials, the reservoir production capacity can be solve in parallel using the hydroelectric similarity principle [33], which can effectively show th effect caused by the longitudinal physical property grade difference.

The indoor experimental evaluation method has been used by previous researchers [3 who conducted an experimental analysis of drilling-fluid-contaminated core to quanti the effect of drilling fluid on production capacity, and concluded that the effect of drillir fluid contamination on production capacity is large and non-negligible. In this study, th epidermal coefficient was introduced to account for drilling fluid contamination, and we can obtain the Equation (15):

$$Q=\sum_{\mathrm{j}=1}^{N} q_{\mathrm{j}}=\sum_{\mathrm{j}=1}^{N} 86.4\frac{2\pi h_j}{B_{\mathrm{o}}}\left\{\frac{\left(k_{mj}+8.33\times 10^{-6} s b^3 n\right)\left[e^{-\beta_{kj}[(p_{\mathrm{e}}-p_{\mathrm{w}})-G_j(r_{\mathrm{e}}-r_w)]}-1\right]}{\mu_{eff}\beta_{kj}\left[\frac{1}{n'}\left(\frac{r_w}{r_e}\right)^{n'}-\frac{e^{\beta_{kj}G_j(r_{\mathrm{e}}-r_w)}}{n'}+S\right]}\right\}, \quad (15)$$

where Q is the final calculated production rate, m^3/d; q_j is the calculated production rate at logging point j, m^3/d; h_j is the effective thickness at logging point j (0.125 m in this paper), m; G_j is the threshold pressure gradient at logging point j, MPa/m; β_{kj} is the stress sensitivity factor at logging point j, 1/MPa; K_{mj} is the initial permeability at logging point j (generally based on the interpreted permeability on the logging curve), darcy; r_{w} is the wellbore radius, m; S is the skin factor, dimensionless.

Equation (15) is the production capacity equation for offshore low-permeability reservoirs considering the effects of fractal natural fracture, longitudinal heterogeneity, threshold pressure gradient, stress sensitivity, and skin factor.

The laboratory results show that there is an exponential relationship between the threshold pressure gradient and the measured fluid permeability of the low-permeability reservoirs in the study area, which can be characterized by Equation (16):

$$G=0.0755{K_{mj}}^{-1.006}. \quad (16)$$

The stress sensitivity coefficients were obtained from the core experiments in the study area, and the experimental results showed that the correlation between stress sensitivity coefficients and the permeability was good, which can be expressed by Equation (17).

$$\beta_k=0.007{K_{mj}}^{-0.25}. \quad (17)$$

When the average permeability interpreted from logging is used to directly evaluate the production capacity, it cannot well reflect the inhibiting effect of longitudinal physical property differences on the production capacity, which often leads to large errors in the production capacity prediction results. Applying the new capacity formula derived in this paper to calculate the capacity of each measurement point based on the relevant parameters can effectively characterize the contribution of the natural fracture development degree, fracture connectivity, and strong reservoir longitudinal heterogeneity to the capacity of low-permeability reservoirs, which is more in line with the seepage laws and characteristics of this type of reservoir, and is of better value for popularization and application in the evaluation of capacity.

In the calculation process of this method, it is necessary to obtain the data of a single well logging curve; at the same time, the selection of other parameters also has some influence on the results of capacity evaluation. In order to improve the operability of the method, it is necessary to explain how some parameters are taken: ① permeability is mainly based on the results of the well logging interpretation of each measurement point and the results of the well logging interpretation of core calibration; ② the threshold pressure gradient and stress sensitivity coefficient are mainly obtained by combining the regional characteristics of the offshore oil field with the results of core experiments; ③ the supply radius and skin coefficient are determined by the results of well test interpretation; ④ fluid viscosity and density are obtained from the results of oil sampling experiments.

2.2.3. Capacity Modeling of Horizontal Wells with Natural Fractures

Figure 2 is a schematic diagram of a horizontal well in an infinite formation in three-dimensional space. In this paper, the horizontal well seepage process is regarded as the proposed radial flow from the reservoir supply boundary to the horizontal wellbore. Based on the research results in the literature [35,36], the seepage field during the stable production of a horizontal well are divided into an external seepage zone in the *xy* plane and an internal seepage zone in the *xz* plane.

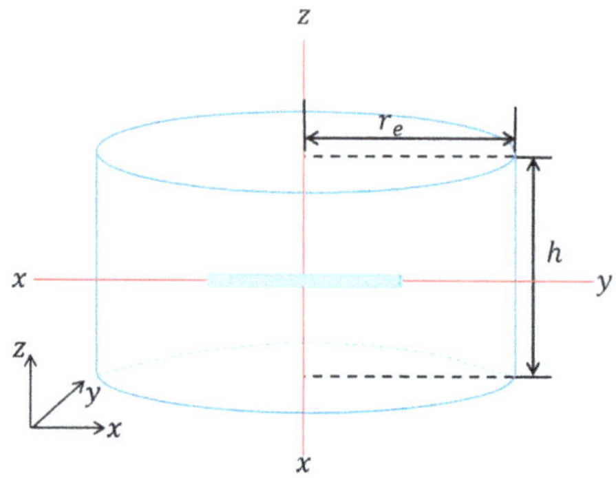

Figure 2. Three-dimensional view of horizontal wells in an infinity formation.

- *xy*-plane external seepage area

Figure 3 shows the 2D seepage pattern diagram of a horizontal well in *xy* plane. The fluid flows from the reservoir supply boundary to the reservoir surrounding horizontal well in a simulated planar radial flow. According to the principle of production equivalence [37], the horizontal section of the horizontal well can be regarded as a proposed circular production pit, then the planar radial seepage process of the fluid from the supply boundary to the proposed circular production pit in the horizontal section constitutes the external seepage field in the *xy*-plane.

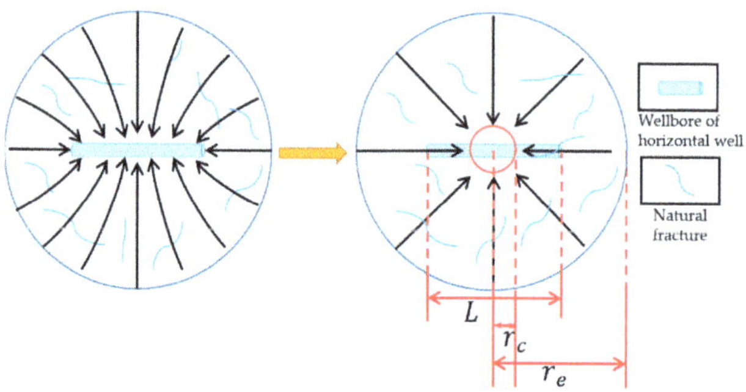

Figure 3. 2D xy-plane seepage pattern diagram for horizontal wells.

Considering the circular radius $h/2$ of the production pit space itself, and then according to the research results [32,38], the radius of the proposed circular production pit in the horizontal section is considered as Equation (18):

$$r_c = L/4 + h/2, \tag{18}$$

where L is the length of horizontal well section, m; h is the reservoir effective thickness, m

- *xz*-plane internal seepage zone

Figure 4 shows the two-dimensional seepage pattern of the horizontal well in the *xz* plane. The fluid flows from the outer boundary of the proposed circular production pit of the horizontal section to the bottom of the horizontal well in a planar radial flow vertically and around the horizontal well section, and the flow process in this section constitutes the internal seepage field in the *xz*-plane.

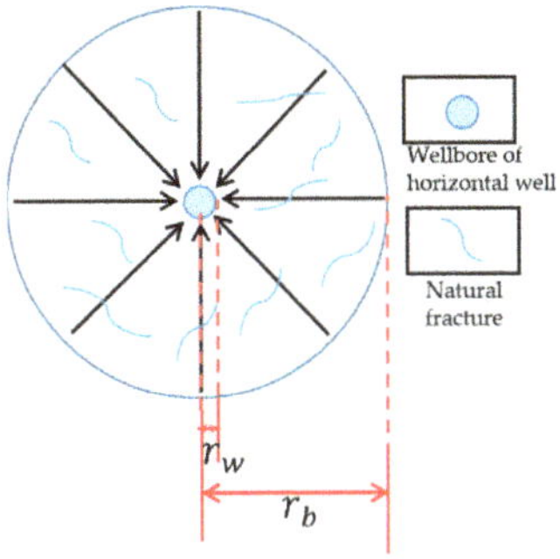

Figure 4. 2D xz-plane seepage pattern diagram for horizontal wells.

Since the internal seepage process occurs at a spatial height of reservoir thickness h, the seepage radius of the internal seepage zone can be expressed as Equation (19)

$$r_b = h/2. \quad (19)$$

- Horizontal well steady-state capacity equation

Considering the influence of fractal characteristics of natural fracture, threshold pressure gradient, stress sensitivity, and skin factor on the production capacity of low-permeability reservoirs, the derivation of the steady-state production capacity formula for horizontal wells is the same as that for vertical wells, except that the seepage process of horizontal wells is divided into two parts, the horizontal external seepage and the vertical internal seepage. Ignoring the repetition of the derivation process, the volume flows Q_1 and Q_2 are given for the horizontal external seepage region and the vertical internal seepage region of the horizontal well respectively, which can be expressed as Equations (20) and (21):

$$Q_1 = 86.4\frac{2\pi h}{B_o}\left\{\frac{\left(k_{mj} + 8.33 \times 10^{-6} s b^3 n\right)\left[e^{-\beta_{kj}[(p_e - p_w) - G_j(r_e - r_c)]} - 1\right]}{\mu_{eff}\beta_{kj}\left[\frac{1}{n'}\left(\frac{r_c}{r_e}\right)^{n'} - \frac{e^{\beta_{kj}G_j(r_e - r_c)}}{n'} + S\right]}\right\}, \quad (20)$$

$$Q_2 = 86.4\frac{2\pi h}{B_o}\left\{\frac{\left(k_{mj} + 8.33 \times 10^{-6} s b^3 n\right)\left[e^{-\beta_{kj}[(p_e - p_w) - G_j(r_b - r_w)]} - 1\right]}{\mu_{eff}\beta_{kj}\left[\frac{1}{n'}\left(\frac{r_w}{r_b}\right)^{n'} - \frac{e^{\beta_{kj}G_j(r_b - r_w)}}{n'} + S\right]}\right\}. \quad (21)$$

The capacity of the horizontal well is the sum of the volume flow rates in the horizontal external seepage zone and the vertical internal seepage zone, so the production capacity Equation (22) is obtained as follows.

$$Q' = Q_1 + Q_2. \quad (22)$$

3. Results

3.1. Example Applications

Section 3.1.1 mainly discusses the natural fractures in offshore low-permeability oil fields and highlights the limitations of conventional evaluation methods. It aligns with the main objectives of the study, emphasizing the need to improve capacity assessment methods in these oil fields.

Section 3.1.2 focuses on comparing and validating the proposed method against traditional evaluation methods in actual capacity assessment. The main discussion focuses on the practical application of the two well-type capacity formulas derived earlier. This part aligns with the study's main objectives, demonstrating the new method's superiority in enhancing accuracy in capacity evaluation, thereby highlighting how each section contributes towards achieving the overall research goals.

3.1.1. Overview of Foundations

Take the low-permeability limestone reservoir of the FA15 oil field in the eastern part of the South China Sea as an example. The reservoir of this oil field is vertically divided into two reservoir calculation units, the upper Zone A and the lower Zone B. Zone A is a localized point-reef deposition; the reservoir is extremely heterogeneous, and the lateral distribution of the reservoir is unstable compared to Zone B. The pre-drilling exploration well is considered to have undeveloped fractures, while the post-drilling well is considered to have more developed natural fractures (e.g., Figure 5). Figure 5 illustrates the more developed natural fractures around this well. The real drilling imaging logging shows that several development wells encountered high-angle fractures, and the fracture density is becoming lower in Zone A from top to bottom. The fracture in Zone B is slightly developed with a fracture dip over 80°, and the fracture in Zone A is more developed with a fracture dip over 80°. The reservoir parameters are shown in Table 1 as follows.

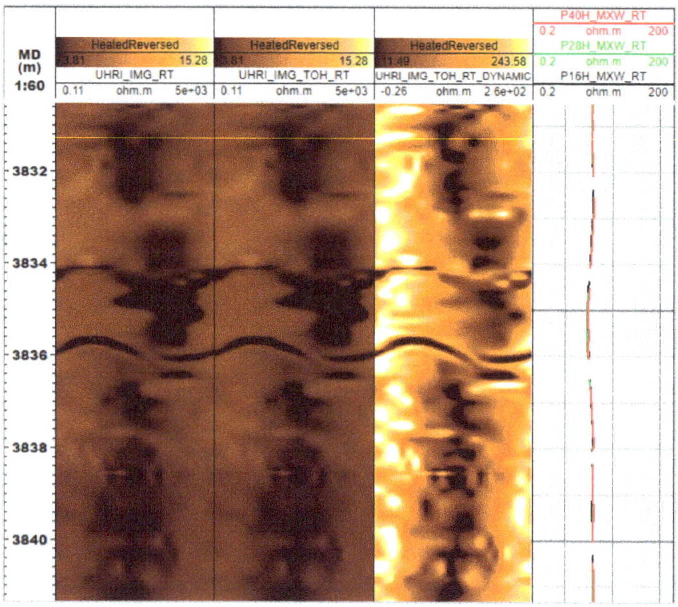

Figure 5. Horizontal section resistivity imaging logging with drilling identifies fractures.

Table 1. Basic reservoir parameters.

Parameters	Reservoir		Unit (of Measure)
	Zone A	Zone B	
Crude oil volume factor	1.041	1.041	-
Formation Crude Oil Viscosity	5.6	5.6	mPa·s
Effective reservoir thickness	4.9	5.7	m
Permeability of matrix block	0.0179	0.0412	darcy
Natural fracture density	0.006~0.0187	0.0005~0.001	1/cm
Average width of natural fracture	186~778	189	μm
Stress sensitivity factor	0.0016~0.0034	0.0012~0.0028	MPa^{-1}
Threshold pressure gradient	0.004145499	0.001792092	MPa/m
Initial formation pressure	17.970~18.511	17.970~18.511	MPa

The reservoir has a total of 11 development wells, including 9 production wells and 2 water injection wells, and the production capacity of each production well varies widely. In addition, the thickness of the reservoir drilled in Zone A is much thinner, the physical properties of the reservoir are not as expected, and the field is the first low-permeability limestone reservoir in the eastern part of the South China Sea with few analogue data. These differences in understanding before and after drilling and the lack of analogous data increase the difficulty of capacity evaluation. And the use of conventional capacity evaluation methods will lead to a large deviation in the capacity evaluation, which is not conducive to decision-making on the development plan.

3.1.2. Comparison and Validation of Capacity Forecasting Methods

A tracer test means injecting tracers into the reservoir, and the tracers are detectable substances. These substances then flow with the reservoir fluids, allowing their movement to be tracked. By monitoring the arrival and concentration of tracers in production wells, we gain insights into the fluid flow paths and the reservoir's heterogeneity, which is crucial for accurate capacity assessment.

The tracer test plays a crucial role in understanding fluid production profiles and contributes significantly to capacity evaluation. They offer unique insights into the movement and distribution of fluids within the reservoir, providing a more accurate assessment of the reservoir's production capacity. By analyzing tracer test data, we can better understand the dynamics of fluid flow and incorporate the understanding into our capacity evaluation models, thus enhancing the accuracy and reliability of the proposed capacity evaluation method.

Taking the X well as an example, the length of the horizontal section drilled is 1362 m, of which Zone B is drilled for 1120 m and Zone A is drilled for 242 m, and the actual initial production is 142 m^3/d. In order to better understand the well capacity, the well was tracer tested. In order to effectively test the different sections of the fluid production profile, taking into account the properties of the reservoir and the pressure, the well was divided into four sections, and a packer was used between the sections to seal the well. The positions of the tracer in the horizontal section are shown in Table 2.

Table 2. Position of short sections of tracer.

Serial Number	Serial Number	Water-Soluble Tracer	Actual Lowering Position (m)		Permeability (mD)	Stratum (Geology)
1	ZFFA-SRT01	DWT-6	4050.8	4058.9	0.1–1	Zone A
2	ZFFA-SRT02	DWT-10	3570.5	3578.6	350–400	Zone B
3	ZFFA-SRT03	DWT-11	3248.5	3256.6	40–60	Zone B
4	ZFFA-SRT04	DWT-12	2786.3	2794.4	0.5–1	Zone A

Among them, section 4 was drilled in the stratigraphic section and encountered Zone A. Sections 2 and 3 were mainly drilled in Zone B, which had better physical characteristics. And section 1 was drilled in the end of the horizontal section encountering Zone A, which had poorer physical characteristics, but a large amount of leakage occurred in this section, confirming that the natural fracture was more developed. According to the conventional understanding and evaluation result of the capacity formula, the capacity should mainly come from sections 2 and 3, but the actual situation of the tracer shows that although the permeability of section 1 is 0.1–1 mD, section 1 has the largest contribution to the production of well X, which is 59.3%. However, Zone B with the permeability of 40–400 mD has a small contribution to the production of the X well of only 16.5%, as shown in Figure 6.

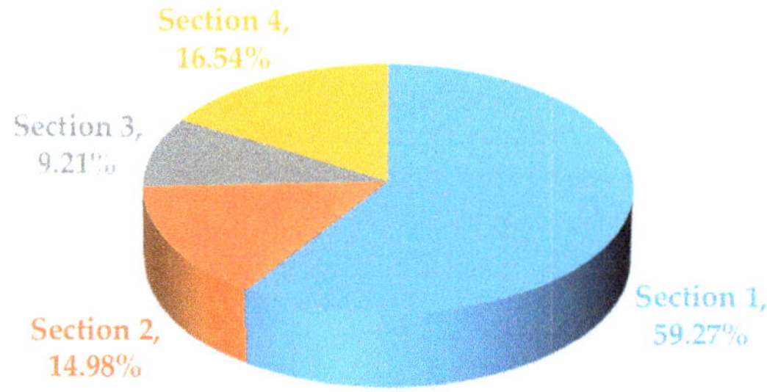

Figure 6. Cumulative percentage of oil production from each section of the tracer.

The tracer test results further confirm that the contribution of natural fractures to the well capacity should not be ignored. The existence of natural fractures leads to a larger overall permeability of the reservoir, and a higher contribution to the well capacity. The conventional production capacity formula with considering the permeability of the reservoir as a constant value of the permeability of the matrix does not meet the needs of capacity evaluation, and the new production capacity formula to consider the existence of natural fractures is more fit in the actual situation.

- Example application and verification of horizontal wells

This study introduces a groundbreaking capacity prediction model for offshore low-permeability reservoirs with natural fractures. Unlike traditional models, our approach innovatively considers the natural fracture development and connectivity, offering a more nuanced characterization of the total permeability of the reservoir. The model integrates key factors such as fractal dimensions, threshold pressure gradient, and stress sensitivity, providing a holistic view of reservoir dynamics.

Comparing the difference in capacity calculation between the method of this paper and the conventional method, taking well X as an example, the basic parameters of which are shown in Table 3. Substituting the drilling data of the X well into the capacity correction formula of Joshi [32], the calculated initial capacity is 72.79 m^3/d, of which the error is 48.7% compared with the well's actual capacity of 142 m^3/d. Substituting the actual drilling data of the X well into the new capacity formula of this paper, the calculated initial capacity is 148.9 m^3/d, and the error is only 5% compared with the actual capacity of 142 m^3/d of the well, which is highly accurate. It is analyzed and believed that the capacity contribution of the X well mainly comes from Zone A, and the Joshi formula is difficult to properly characterize the natural fracture permeability of Zone A. The new capacity formula has made great improvement in this aspect, which effectively reduces the capacity calculation error and confirms the reliability of the method promoted in this paper.

To further verify the accuracy and applicability of the method promoted in this paper, all other wells in the field were also calculated. There are a total of 9 production wells in this reservoir, and the capacity of the 9 wells is calculated respectively using the new method in this study and the conventional method, and the results are as shown in Table 4. In Table 4, error 1, 2, and 3 respectively represents the calculation error between Joshi correction formula [32], Chen formula [35], new capacity formula, and the actual production of the wells. It is known from the table, for horizontal wells, the calculation error of the Joshi correction formula is within 50%, and for multi-branch wells, the calculation error of the Joshi correction formula is 50~80%. For horizontal wells, the calculation error of the Chen formula is within 70%, and for multi-branch wells, the calculation error of the Joshi correction formula is 70~90%. For horizontal wells, the calculation error of the new capacity formula is within 10%, and for multi-branch wells, the calculation error is within 15%. It should be noted here that the capacity of multi-branch wells is calculated as roughly 1.5 times that of horizontal wells, according to the empirical relationship of production between these two types of wells. The capacity calculated by the conventional method seriously underestimates the actual reservoir capacity, and the capacity formula following the matrix permeability leads to a large error. The capacity formula considering the natural fracture in this paper can greatly reduce the error, and the prediction result is much closer to the actual capacity, which confirms the accuracy and effectiveness of this method.

Table 3. Basic parameters of well X.

Notation	Physical Meaning	Parameter Value (Zone A)	Parameter Value (Zone B)	Unit (of Measure)
B_o	Crude oil volume factor	1.041	1.041	dimensionless
h	Effective reservoir thickness	4.9	5.7	m
k_m	Permeability of matrix block	0.001	0.0412	darcy
s	Natural fracture density	0.0095	0.0016	1/cm
b	Average width of natural fracture	380	210	µm
D	Fractal dimension (math.)	2.1	2.1	dimensionless
θ	Abnormal diffusion coefficient (physics)	0.05	0.05	dimensionless
β_k	Stress sensitivity factor	0.0083	0.0027	1/MPa
μ_{eff}	Formation Crude Oil Viscosity	5.6	5.6	mPa·s
p_e	Initial formation pressure	17.97	17.97	MPa
p_w	Bottom hole pressure	12.95	12.95	MPa
G	Threshold pressure gradient	0.0074	0.0018	MPa/m
r_e	Supply radius	720	2240	m
L	Horizontal section length	242	1120	m
r_w	Wellbore radius	0.2159	0.2159	m

Table 4. Comparison of the results of capacity calculation by different methods.

Name of Well	Stratum (Geology)	Actual Capacity (m^3/d)	Joshi Correction Formula (m^3/d)	Chen Capacity Formula (m^3/d)	New Capacity Formula (m^3/d)	Error 1 (%)	Error 2 (%)	Error 3 (%)
3H	Zone A/B	90.20	48.24	32.26	82.56	−46.52	−64.23	−8.47
4M	Zone A/B	416.50	197.75	111.14	469.20	−52.52	−73.32	12.65
5M	Zone A/B	459.60	218.61	135.37	393.68	−52.43	−70.55	−14.34
6H	Zone A/B	142.00	72.79	46.16	148.92	−48.74	−67.49	4.87
7M	Zone A	239.00	53.37	37.43	207.52	−77.67	−84.34	−13.17
8M	Zone A	398.00	88.70	67.03	454.59	−77.71	−83.16	14.22
9H	Zone A/B	288.00	192.45	104.83	307.15	−33.18	−63.60	6.65
10M	Zone A	447.00	82.06	54.94	508.36	−81.64	−87.71	13.73
11H	Zone A/B	307.30	154.20	101.70	324.51	−49.82	−66.90	5.60

The novel capacity formula presented here marks a significant advancement over existing methods. It accurately captures the complex interplay between natural fractures and reservoir properties, leading to a more precise prediction of production capacity. This enhancement is evident in our results, where the model demonstrates an average error reduction to less than 15%, which is a substantial improvement compared to conventional methods.

- Example application and verification of vertical wells

  The conventional generalized capacity formula is depicted as Equation (23).

$$Q = 86.4\frac{2\pi h k_m}{B_o \mu_{eff} \beta_k} \frac{p_e - p_w}{\ln r_e - \ln r_w}. \tag{23}$$

Based on Equation (15), when the natural fracture is undeveloped in the low-permeability reservoir, the capacity formula of the vertical well is expressed as Equation (24).

$$Q = \sum_{j=1}^{N} q_j = \sum_{j=1}^{N} 86.4\frac{2\pi h_j k_{mj}}{B_o \mu_{eff} \beta_{kj}} \left[\frac{e^{-\beta_{kj}[(p_e - p_w) - G_j(r_e - r_w)]} - 1}{e^{\beta_{kj} G_j (r_e - r_w)} \ln r_w - \ln r_e} \right]. \tag{24}$$

The development case data of vertical wells in offshore low-permeability reservoir with natural fractures are few; two wells are applied to verify the correctness of the new capacity formula in this paper. The FA15-Y well were shot holes in Zone A and Zone B, respectively; the basic parameters are as shown in Table 5. This well shows strong longitudinally heterogeneity, the production of DST test is 127.3 m^3/d, and the capacity calculated using the conventional generalized capacity Equation (23) is 78.5 m^3/d, which indicates a hug error. Applying the method in the paper, based on the post-drilling physical propertie of the limestone reservoir, and substituting the basic parameters of the FA15-Y well int the production capacity Equation (15) for vertical wells of a low-permeability reservoi with development of natural fractures, a production capacity is calculated based on eacl measurement point in the logging curve. The final steady-state production capacity of single well of 115.47 m^3/d is obtained by integration and summation, of which the error i only about 9.30% compared with the DST test, which is much closer to the test data.

Table 5. Basic Parameters of well FA15-Y.

Notation	Physical Meaning	Parameter Value (Zone A)	Parameter Value (Zone B)	Unit (of Measure)
B_o	Crude oil volume factor	1.041	1.041	dimensionless
h	Effective reservoir thickness	16	7	m
k_m	Permeability of matrix block	0.0231	0.0279	darcy
s	Natural fracture density	0.0172	0.0016	1/cm
b	Average width of natural fractures	385	210	μm
D	Fractal dimension (math.)	2.2	2.2	dimensionless
θ	Abnormal diffusion coefficient (physics)	0.05	0.1	dimensionless
β_k	Stress sensitivity factor	0.0032	0.003	1/MPa
μ_{eff}	Formation Crude Oil Viscosity	5.6	5.6	mPa·s
p_e	Initial formation pressure	17.97	17.97	MPa
p_w	Bottom hole pressure	12.636	12.636	MPa
G	Threshold pressure gradient	0.0032	0.0027	MPa/m
r_e	Supply radius	1000	1000	m
r_w	Wellbore radius	0.155	0.155	m

Substituting the actual drilling data of the FA14-Z well into the capacity Equation (2 for vertical wells of a low-permeability reservoir with undeveloped natural fractures, th calculated result of the initial capacity is 32.2 m^3/d, of which the error is 9.98% compare with the actual capacity of 29.3 m^3/d. The calculated results of the capacity predictio of vertical wells by the method promoted in this paper are relatively consistent with th actual production, and the errors are all less than 10%, which confirms the correctness an effectiveness of this method.

Based on the validation examples of vertical wells, it is concluded that the derive vertical well capacity formula agrees well with the actual initial production of the wells wit an error of about 10%, which means it can be applied to the initial production prediction vertical wells in low-permeability reservoirs with developed natural fractures.

These findings validate the efficacy of our approach and provide a more optimize capacity assessment method for this type of reservoir. Furthermore, it opens avenues for i application in other reservoir types.

4. Discussion

The analysis of the factors influencing the production of vertical and horizontal wel in low-permeability limestone reservoirs based on actual data is crucial for developme strategies and field management, as it can reveal the reasons for changes in productio under different conditions and provide a scientific basis for future reservoir manageme

Figure 7 shows the relationship between natural fracture density and capacity at diffe ent reservoir thicknesses. This mainly represents the effect of the degree of natural fractu development on capacity. At the same thickness, the higher the density of natural fractur the higher the production capacity. It indicates that the development of natural fractur has a positive effect on the well capacity, and only when the distribution of natural fractur is well understood can the production of wells in low-permeability reservoirs be reasonab assessed, and wells can be considered to be designed exploiting the region developi natural fractures. Natural fracture density reflects the degree of fracture developme when the fracture density is greater, vertical and horizontal wells are more likely to interse with these fractures. In low-permeability reservoirs, the development of natural fractur can significantly improve the permeability of the rock, thus increasing the fluid mobility. Understanding and evaluating the development of natural fractures in the development of low-permeability reservoirs is critical to predicting and increasing well capacity. Choosing areas of higher fracture density to locate wells can improve economic benefits.

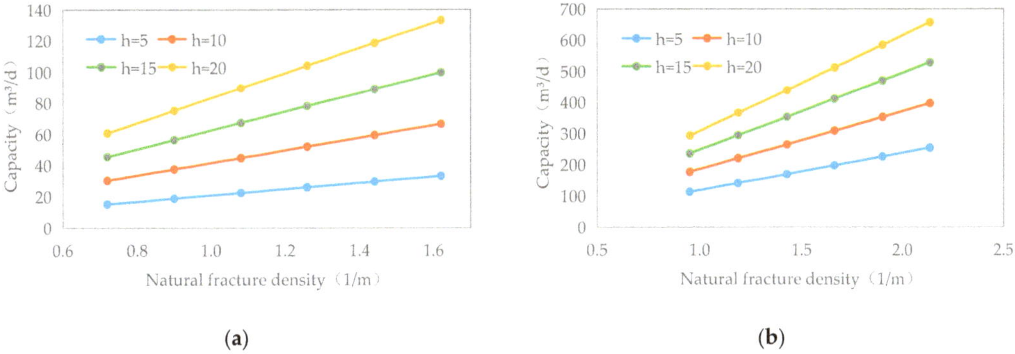

Figure 7. Relationship between natural fracture density and capacity at different thicknesses: (**a**) Vertical wells; (**b**) Horizontal wells.

Figure 8 shows the relationship between the fractal dimension and the capacity at different reservoir thicknesses, which mainly represents the effect of fractal dimension on capacity. It can be seen that the higher the fractal dimension, the higher the capacity for the same thickness. This suggests that fractal dimension, as a measure of fracture complexity, has a positive effect on well capacity. In the early evaluation of reservoirs, the study of reservoir soil stress and fracture network complexity should be emphasized. The fractal dimension D reflects the complexity and spatial distribution characteristics of the fracture network. When the fractal dimension is high, it indicates that the fracture network is more developed and the connection between fractures is more complicated, which increases the probability of intersection between effective fractures. In low-permeability reservoirs, a complex fracture network helps to increase the permeability of the rock, thereby promoting the flow of crude oil.

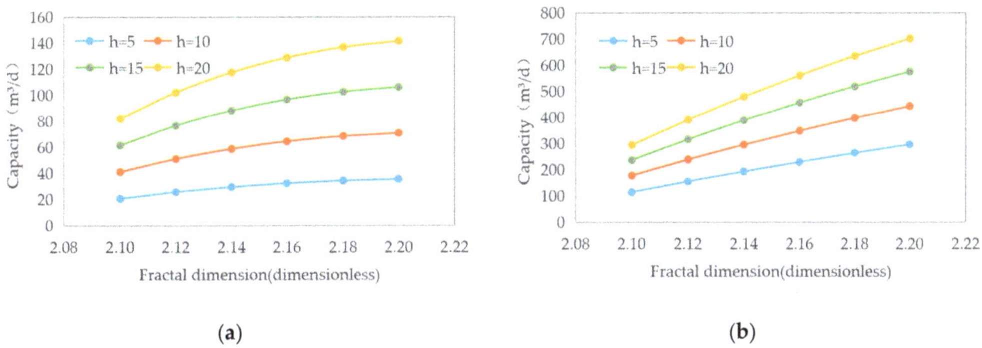

Figure 8. Relationship between fractal dimension D and capacity for different thicknesses: (**a**) Vertical wells; (**b**) Horizontal wells.

Figure 9 shows the relationship between connectivity coefficient and capacity at different reservoir thicknesses, which mainly represents the effect of natural fracture connectivity on capacity. The smaller the connectivity coefficient θ, the higher is the capacity at the same reservoir thickness. As mentioned earlier, the smaller the θ, the better the network connectivity, which may lead to a higher fluid storage capacity in the local area and means that the distribution of these fractures in the reservoir may be more favorable for the accumulation and seepage of fluid. A higher connectivity coefficient indicates that the connectivity between the matrix and fractures is poor, and the fractures formed in this case tend to be isolated, which does not improve fluid mobility in the reservoir.

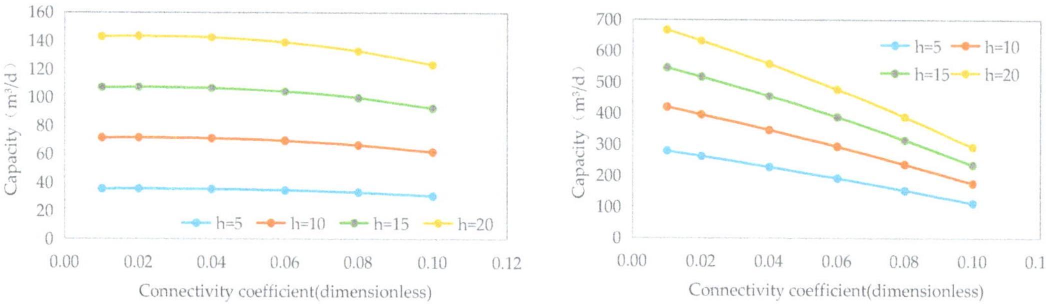

(a) (b)

Figure 9. Curve of connectivity coefficient θ versus capacity at different thicknesses: (**a**) Vertical wells; (**b**) Horizontal wells.

Figure 10 depicts the relationship between stress sensitivity and horizontal well capacity at different reservoir thicknesses, which mainly demonstrates the effect of the stress sensitivity of the seepage characteristic parameters on capacity. As the stress sensitivity coefficient increases, the capacity of horizontal wells at different reservoir thicknesses tends to decrease. It indicates that reservoirs with a lower stress sensitivity coefficient are more favorable for oil production of horizontal wells because the permeability of the reservoir changes less under pressure changes, thus maintaining higher fluid mobility.

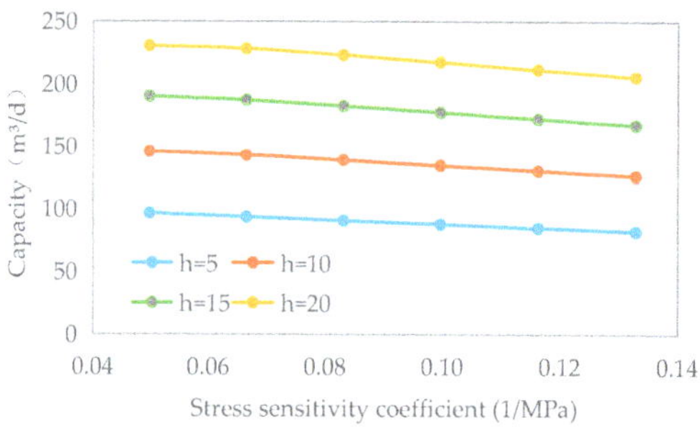

Figure 10. Relationship between stress sensitivity and horizontal well capacity at different thicknesses.

Figure 11 shows the relationship between the producing pressure difference and productivity index under different threshold pressure gradients. This figure primarily reflects the effect of the initiating pressure gradient in the seepage parameters on capacity. From the figure, as the drawdown increases, the productivity index shows an increasing trend under all conditions of threshold pressure gradient. For the same producing pressure difference, the smaller the threshold pressure gradient, the higher the productivity index, indicating that a higher oil recovery efficiency can be realized with a relatively small drawdown at a lower threshold pressure gradient. Horizontal wells with a drawdown of less than 5 MPa have a greater increase in the productivity index, and a drawdown of more than 5 MPa has a relatively small increase in the productivity index. Therefore, it is not recommended that horizontal wells in low-permeability reservoirs containing natural fractures produce at an excessive producing pressure difference. The threshold pressure gradient is the minimum pressure difference required for the reservoir fluid to start flowing. A lower initial pressure gradient means that the oil and gas in the reservoir can start to flow at a lower producing pressure difference, which is very favorable for improving oil recovery efficiency and reducing production costs. Therefore, when formulating oil field development plans, the threshold pressure gradient of the reservoir should be considered, and the appropriate producing pressure difference should be selected to optimize oil and gas recovery and increase economic benefits.

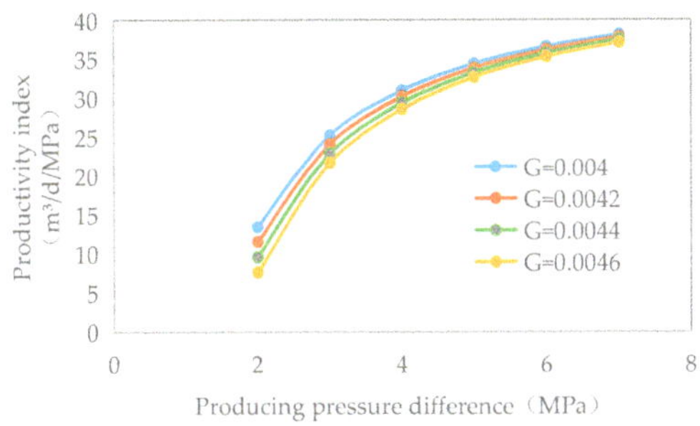

Figure 11. Relationship between producing pressure difference and productivity index of horizontal wells at different threshold pressure gradients.

The analysis and discussion above contribute significantly to understanding the complex dynamics within low-permeability reservoirs, especially those with natural fractures. The results demonstrate the importance of accurately characterizing the effects of natural fractures, stress sensitivity, and threshold pressure gradients on reservoir performance. The enhanced understanding allows for more precise capacity predictions, which is crucial for optimal reservoir management.

Applying this knowledge in real-world scenarios could lead to more effective development strategies for low-permeability reservoirs. It enables better planning and optimization of drilling and production operations, potentially improving recovery rates. The findings can guide industry professionals in making more informed decisions, especially in challenging offshore environments.

5. Conclusions

Taking into account the non-Darcy flow characteristics of fluids, stress sensitivity characteristics of porous media, random distribution characteristics of natural fractures, and strong longitudinal heterogeneity of reservoirs, this study proposes new capacity evaluation methods for both vertical and horizontal wells in low-permeability oil reservoirs containing natural fractures. It is verified that the established capacity evaluation models can be applied to accurately predict initial productivity capacity of oil wells by fitting actual production data. Based on the study of capacity evaluation models and the sensitivity analysis, several conclusions are summarized as follows.

- Aiming at the problems of traditional methods in predicting the production of offshore low-permeability reservoirs with natural fractures and the inappropriateness of traditional methods in capacity evaluation, the fractal theory is introduced to reasonably characterize the permeability of low-permeability reservoirs with natural fractures, and the steady-state capacity prediction model of vertical and horizontal wells are established, which comprehensively considers the effects of fractal fractures, threshold pressure gradient, and stress sensitivity of low-permeability limestone reservoirs. To further reveal the strong longitudinal heterogeneity during the drilling of vertical wells, it is suggested that considering the longitudinal gradient through logging curves in the evaluation of vertical well capacity can more accurately evaluate the capacity of

this type of reservoir and characterize the strong longitudinal heterogeneity of this type of reservoir.
- Application examples demonstrate that our method significantly outperforms conventional models, with an average error reduction to less than 15%.
- A high-density fracture network increases the permeability and fluid mobility of the rock, which significantly improves well capacity. In addition, the increase in fractal dimension reflects the increased complexity of the fracture network, which further increases the probability of intersection between horizontal wells and effective fractures, thus facilitating the oil recovery. A lower connectivity coefficient θ is associated with higher production capacity, indicating that optimizing the connectivity of the fracture network is critical for the aggregation and seepage of crude oil. A lower threshold pressure gradient is favorable to achieve higher oil recovery efficiency under smaller producing pressure difference. The relationship between the production pressure difference and productivity index is not a simple linear relationship; it is not recommended that horizontal wells in low-permeability reservoirs containing natural fractures produce under an excessive producing pressure difference.

Therefore, a comprehensive consideration of the non-Darcy flow, stress sensitivity of porous media, random distribution of natural fractures, and strong longitudinal heterogeneity is essential for the capacity evaluation of low-permeability reservoirs containing natural fractures, which can establish a theoretical foundation for the efficient development of low-permeability reservoirs and assist in upgrading and transforming low-permeability reserves. Additionally, the derived production capacity formulas can be applied to different types of oil reservoirs including low-permeability reservoirs with or without natural fractures while changing some prerequisites. Except for vertical and horizontal wells, the capacity evaluation model of multi-lateral wells of low-permeability reservoirs containing natural fractures will be further studied in our future work.

Author Contributions: Conceptualization, K.W.; methodology, K.W.; validation, S.L., K.W. and R.H.; formal analysis, K.W.; investigation, K.W. and G.L.; resources, M.X., W.L., L.L., S.F. and M.L.; data curation, S.L.; writing—original draft preparation, K.W.; writing—review and editing, S.L.; visualization, S.L.; supervision, M.X., W.L., L.L. and S.F.; project administration, M.X. and L.L.; funding acquisition, L.L. All authors have read and agreed to the published version of the manuscript.

Funding: This research was funded by Shenzhen Branch of CNOOC Limited Production Research Project (No. SCKY-2024-SZ-11). The APC was funded by Nanhai East Petroleum Research Institute, Shenzhen Branch of CNOOC Limited, China.

Data Availability Statement: Restrictions apply to the availability of these data. Data were partly obtained from Shenzhen Branch of CNOOC Limited and partly generated during the process of calculations. Oilfield data are available from the lead author with the permission of Shenzhen Branch of CNOOC Limited and calculation data can be acquired from the paper or the lead author directly.

Conflicts of Interest: Authors Kun Wang, Mingying Xie, Weixin Liu, Li Li, Siyu Liu, Ruijie Huang, Shasha Feng and Guotao Liu were employed by the company Nanhai East Petroleum Research Institute, Shenzhen Branch of CNOOC Limited. The remaining authors declare that the research was conducted in the absence of any commercial or financial relationships that could be construed as a potential conflict of interest. The authors declare that this study received funding from Nanhai East Petroleum Research Institute, Shenzhen Branch of CNOOC Limited. The funder was not involved in the study design, collection, analysis, interpretation of data, the writing of this article or the decision to submit it for publication.

References

1. Zhu, W. Advancements of Oil and Gas Exploration in China Offshore Area: Overview and Perspective. *Strateg. Study CAE* **201**[illegible] *12*, 18–24.
2. Ashraf, U.; Zhang, H.; Anees, A.; Ali, M.; Zhang, X.; Abbasi, S.S.; Mangi, H.N. Controls on Reservoir Heterogeneity of a Shallow Marine Reservoir in Sawan Gas Field, SE Pakistan: Implications for Reservoir Quality Prediction Using Acoustic Impedance Inversion. *Water* **2020**, *12*, 2972. [CrossRef]

3. Liu, J.; Liu, J. Machine Learning Technique for the Quantitative Evaluation of Tight Sandstone Reservoirs Using High-Pressure Mercury-Injection Data; Oral, 2020. Available online: https://meetingorganizer.copernicus.org/EGU2020/EGU2020-17007.html (accessed on 8 January 2024).
4. Nzekwu, A.I.; Abraham-A, R.M. Reservoir Sands Characterisation Involving Capacity Prediction in NZ Oil and Gas Field, Offshore Niger Delta, Nigeria. *AIMS Geosci.* **2022**, *8*, 159–174. [CrossRef]
5. Wang, K.; Li, L.; Chen, X.; Liang, W.; Yang, Y.; Yan, Z.; Dai, J.; Li, W.; Wang, Y.; Wang, W.; et al. A New Method for Research on Unsteady Pressure Dynamics and Productivity of Ultralow-Permeability Reservoirs. *Geofluids* **2021**, *2021*, 9926853. [CrossRef]
6. Alali, Z.H.; Horne, R.N. A Comparative Study of Deep Learning Models and Traditional Methods in Forecasting Oil Production in the Volve Field. In Proceedings of the SPE Annual Technical Conference and Exhibition, San Antonio, TX, USA, 21–23 November 2023; SPE: San Antonio, Texas, USA, 2023; p. D031S032R003.
7. Chen, Y.; Lan, W.; Wang, C. Application of MES System in Offshore Oil and Gas Field Production Management. *J. Big Data* **2019**, *1*, 47–54. [CrossRef]
8. Babanezhad, M.; Behroyan, I.; Nakhjiri, A.T.; Marjani, A.; Shirazian, S. Performance and application analysis of ANFIS artificial intelligence for pressure prediction of nanofluid convective flow in a heated pipe. *Sci. Rep.* **2021**, *11*, 902. [CrossRef]
9. He, X.; Zhu, W.; Santoso, R.; Alsinan, M.; Kwak, H.; Hoteit, H. *Fracture Permeability Estimation Under Complex Physics: A Data-Driven Model Using Machine Learning*; OnePetro: Richardson, TX, USA, 2021.
10. Gao, L.; Tang, Y.T.; Wu, S.; Deev, A.; Olson, R.; Mosca, F.; Zhang, S.; Ni, Y.; Qu, Q.; LaFollette, R.; et al. The gas isotope interpretation tool: A novel method to better predict production decline. *AAPG Bull.* **2017**, *101*, 1263–1275. [CrossRef]
11. Arps, J. Analysis of Decline Curves. *Trans. AIME* **1945**, *160*, 228–247. [CrossRef]
12. Shin, H.-J.; Lim, J.-S.; Shin, S.-H. Estimated ultimate recovery prediction using oil and gas production decline curve analysis and cash flow analysis for resource play. *Geosystem Eng.* **2014**, *17*, 78–87. [CrossRef]
13. Harsh, A. Semi-Analytical Modeling of Water Injection under Fracturing Conditions by Capturing Effects of Formation Damage Mechanisms to Predict Fracture Propagation. Master's Thesis, Delft University of Technology, Delft, The Netherlands, 2017.
14. Li, M.; Bi, G.; Shi, Y.; Zhao, K. A semi-analytical two-phase flow model of fractured horizontal well with complex fracture networks in natural fractured reservoirs. *Energy Explor. Exploit.* **2022**, *40*, 174–192. [CrossRef]
15. Salih, I.S.; Baker, H.A. Development of a Semi-Analytical Type Curve of Transient Pressure Response in Complex Well-Reservoir Architectures. *Iraqi J. Chem. Pet. Eng.* **2019**, *20*, 61–69. [CrossRef]
16. Jing, W.; Xiao, L.; Haixia, H.; Wei, L. Productivity Prediction and Simulation Verification of Fishbone Multilateral Wells. *Geofluids* **2022**, *2022*, 3415409. [CrossRef]
17. Liu, Z.; Li, S.; Li, L. Research on oil and gas production prediction process based on machine learning. *Int. J. Energy* **2023**, *2*, 76–79. [CrossRef]
18. Zhang, H.; Peng, W.; Hao, P.; Du, M. A heterogeneous model for simulating fluid flow in naturally fractured-vuggy carbonate reservoirs. *Int. J. Oil. Gas Coal Technol.* **2019**, *20*, 397–419. [CrossRef]
19. Gong, J.; Cao, L.; Guan, Y.; Tian, J.; Wang, T. Productivity Prediction Methods of Multi-Fractured Horizontal Wells in Low-Permeability Reservoirs: Progress and Challenges. *Front. Energy Res.* **2022**, *10*, 861333. [CrossRef]
20. Li, J.; Xu, H.; Xiong, Q.; Yi, Y.; Zhou, W.; Zhang, L.; Chen, S.; Lian, L.; Zeng, M.; Yang, X.; et al. Study on the Influence of Natural Fracture Modeling on Hydraulic Fracture Propagation for Horizontal Wells in Unconventional Resource—A Case Study from China. In Proceedings of the Abu Dhabi International Petroleum Exhibition and Conference, Dubai, United Arab Emirates, 31 October–2 November 2022; SPE: Dubai, United Arab Emirates, 2022; p. D032S180R007.
21. Xia, D.; Yang, Z.; Li, D.; Zhang, Y.; He, Y.; Luo, Y.; Zhang, A.; Wang, W.; Zhao, X. Evaluation Method of the Vertical Well Hydraulic Fracturing Effect Based on Production Data. *Geofluids* **2021**, *2021*, 9952719. [CrossRef]
22. Yang, Y.; Xiong, W.; Liao, G.; Gao, S.; Shen, R.; Zhang, J.; Li, Q. Flow mechanism of production decline during natural depletion after hydraulic fracturing of horizontal wells in tight oil reservoirs. *Pet. Sci. Technol.* **2022**, *40*, 383–400. [CrossRef]
23. Wang, F. Hydrogeological and Engineering Geological Significance of Fault Debris Fractal Dimension. In Proceedings of the GeoCongress 2012, American Society of Civil Engineers, Oakland, CA, USA, 29 March 2012; pp. 3543–3552.
24. Babadagli, T.; Develi, K. Fractal Analysis of Natural and Synthetic Fracture Surfaces of Geothermal Reservoir Rocks. In Proceedings of the World Geothermal Congress, Kyushu-Tohoku, Japan, 28 May 2000.
25. Yang, Y.; Liu, Z.; Yao, J.; Zhang, L.; Ma, J.; Hejazi, S.H.; Luquot, L.; Ngarta, T.D. Flow Simulation of Artificially Induced Microfractures Using Digital Rock and Lattice Boltzmann Methods. *Energies* **2018**, *11*, 2145. [CrossRef]
26. Sheng, G.; Su, Y.; Wang, W.; Javadpour, F.; Tang, M. Application of Fractal Geometry in Evaluation of Effective Stimulated Reservoir Volume in Shale Gas Reservoirs. *Fractals* **2017**, *25*, 1740007. [CrossRef]
27. Zhu, W.; Lei, G.; He, X.; Patzek, T.W.; Wang, M. Fractal and multifractal characterization of stochastic fracture networks and real outcrops. *J. Struct. Geol.* **2022**, *155*, 104508. [CrossRef]
28. Cao, J.; Song, C. Calculation and improvement of the fractal dimension. *Inf. Technol. Informatiz.* **2017**, *10*, 19–23. [CrossRef]
29. Xie, B.; Jiang, J.; Jia, J.; Ren, L.; Huang, B.; Huang, X. Partition seepage model and productivity analysis of fractured horizontal wells in tight reservoirs. *Fault-Block Oil Gas Field* **2019**, *26*, 324–328. [CrossRef]
30. Li, X.; Liu, S.; Chen, Q.; Su, Y.; Sheng, G. An Evaluation of the Stimulation Effect of Horizontal Well Volumetric Fracturing in Tight Reservoirs with Complex Fracture Networks. *Pet. Drill. Tech.* **2019**, *47*, 73–82. [CrossRef]
31. He, G.; Tang, H. *Petrophysics*; Petroleum Industry Press: Beijing, China, 2011.

32. Joshi, S. Augmentation of Well Productivity With Slant and Horizontal Wells (includes associated papers 24547 and 25308). *J. Pet. Technol.* **1988**, *40*, 729–739. [CrossRef]
33. Li, L. A boundary modification for water and electricity analogy experiment of fractured horizontal wells. *Pet. Reserv. Eval. Dev.* **2019**, *9*, 61–63+71. [CrossRef]
34. Tian, W.; Deng, R.; Liu, Y.; Wang, M.; Gao, Z.; Hu, C.; Qiu, Y.; Zhao, Z. Effect of Drilling Fluid on Reservoir Damage and Productivity of Marine Reservoir. *Sci. Technol. Eng.* **2022**, *22*, 129–135. [CrossRef]
35. Chen, Y. Derivation and Correlation of Production Rate Formula for Horizontal Well. *Xinjiang Pet. Geol.* **2008**, *29*, 68–71.
36. Shan, B.; Fu, Y. Discussion and Analysis about Horizontal Well Productivity Formulas. *Xinjiang Pet. Geol.* **2015**, *36*, 624–628. [CrossRef]
37. Chen, Y. *Calculation Methods for Oil and Gas Reservoir Engineering*; Petroleum Industry Press: Beijing, China, 1990.
38. Borisov, J.P. *Oil Production Using Horizontal and Multiple Deviation Wells*; Nedra: Moscow, Russia, 1964; Volume 364.

processes

MDPI

Article

A Semi-Analytical Model for Gas–Water Two-Phase Productivity Prediction of Carbonate Gas Reservoirs

Dayong Chen [1,2] and Zheng Sun [1,2,*]

[1] State Key Laboratory of Coal Resources and Safe Mining, China University of Mining and Technology, Xuzhou 221116, China

[2] School of Mining, China University of Mining and Technology, Xuzhou 221116, China

* Correspondence: sunzheng@cumt.edu.cn; Tel.: +86-185-0061-0758

Abstract: The productivity prediction of gas wells in carbonate gas reservoirs is greatly affected by the characteristics of gas–water two-phase flow and fracture seepage parameters. Compared with numerical simulation, the productivity prediction based on the analytical model is fast and widely used, but the traditional analytical model is fairly simplified while dealing with the nonlinear problem of the two-phase seepage equation, leading to a large discrepancy in the results of dynamic analysis. To solve this problem, this paper considers the characteristics of gas–water two-phase flow in the reservoir and fracture, uses the dual-medium model to characterize the stress sensitivity of the fracture and reservoir, and establishes a gas–water two-phase productivity prediction model for carbonate gas reservoirs. Combining the flowing material balance equation with the Newton iteration method, the nonlinear parameters of the percolation model are updated step by step with the use of average formation pressure, and the gas–water two-phase model is linearized through successive iterations to obtain the semi-analytical solution of the model. The accuracy of the model was verified using a comparison with the results of commercial numerical simulation software and field application, the gas–water two-phase productivity prediction curve was obtained, and the influence of sensitive parameters on productivity was analyzed. The results show that: (1) the semi-analytical solution method can efficiently deal with the gas–water two-phase nonlinear seepage problem and obtain the productivity prediction curve of carbonate gas wells rapidly and (2) the water production of the carbonate gas reservoir seriously affects the productivity of gas wells. During the development process, the production pressure difference should be reasonably controlled to reduce the negative impact of stress sensitivity on productivity performance.

Keywords: carbonate gas reservoir; gas–water two-phase; stress sensitivity; semi-analytical model

Citation: Chen, D.; Sun, Z. A Semi-Analytical Model for Gas–Water Two-Phase Productivity Prediction of Carbonate Gas Reservoirs. *Processes* **2023**, *11*, 591. https://doi.org/10.3390/pr11020591

Academic Editor: Qingbang Meng

Received: 8 January 2023
Revised: 2 February 2023
Accepted: 14 February 2023
Published: 15 February 2023

1. Introduction

A carbonate gas reservoir is rich in natural fractures and shares the feature of strong stress sensitivity. The reservoirs essentially have high water saturation; therefore, water production has a prominent role in limiting gas-well production [1–3]. The productivity prediction is greatly affected by the characteristics of gas–water two-phase flow as well as the fracture seepage parameters. At present, the gas–water two-phase productivity prediction methods of carbonate gas wells mainly include analytical methods and numerical simulation methods [4–6]. The analytical method is usually based on the steady-state seepage theory. The equation is linearized by introducing the gas–water two-phase pseudo pressure, and the productivity equation of water-producing gas wells is derived using the conformal transformation and potential superposition principle to convert the water production into gas production for evaluation [7–9]. The numerical simulation method can explicitly characterize the artificial fracture parameters and deal with multiphase fluid flow problems, but the process of model establishment and solution is complex [10–12]. For carbonate gas reservoirs, the development process falls in the unsteady flow stage for a long time, and the productivity equation established on the basis of steady-state seepage theory cannot reflect the production process of the gas reservoir [13–15]. In addition, the analytical method usually only introduces two-phase pseudo pressure to simplify the equation when dealing with the nonlinear problem of the gas–water two-phase seepage equation [16–18] and ignores the influence of nonlinear seepage parameters, and, in addition, the calculation results have large errors. The pre-treatment process of the numerical simulation method is complex. In order to obtain high simulation accuracy, it is necessary to densify the grid of cracks, resulting in a large number of grids. When dealing with thousands of case studies the computational cost for numerical simulation is unfavorable.

In this paper, a gas–water two-phase productivity prediction model for carbonate gas reservoirs is established based on the dual medium unsteady crossflow model, and an efficient solution method is proposed to handle the nonlinear seepage problem caused by gas–water two-phase flow. First, the seepage equation is normalized by introducing two-phase pseudo pressure and pseudo time, and the analytical solution is obtained using Laplace transform and other methods. Then, combined with the flow material balance and Newton iteration method, the nonlinear seepage parameters of the model are updated using the average formation pressure and saturation at each timestep, and the linearization of the seepage model is gradually realized, so as to obtain the semi-analytical solution of the model. The accuracy of the model is verified using a comparison with commercial numerical simulation software, and the influence of key seepage parameters of fractures and reservoirs on productivity prediction is analyzed. Then, the productivity prediction and analysis of field wells are carried out.

2. Establishment of Model

2.1. Physical Model

The dual-medium unsteady crossflow model [19,20] is used to characterize the fractured porous carbonate reservoir. The fluid flow in the carbonate reservoir is in line with the dual-porosity and single permeability flow characteristics, as shown in Figure 1. The model assumes that the natural fractures are directly connected with the wellbore, and the fluid only flows into the production wellbore through the fractures, and the fluid in the matrix continuously flows to the fractures to provide an energy supply. Considering that gas and water are produced at the same time, gas–water two-phase flow takes place at both the matrix and fractures, where isothermal Darcy flow is applied. The assumptions of other physical models are: (1) the thickness of the production layer is fully opened, and the fluid flows into the wellbore radially; (2) considering the permeability loss of natural fractures caused by stress sensitivity; (3) compared with gas phase, the compressibility of formation water is small and can be ignored; and (4) the effects of gravity and capillary force are not considered.

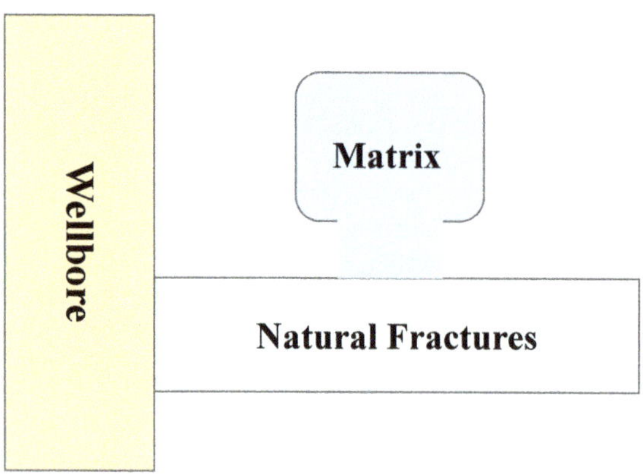

Figure 1. Schematic diagram of the well model in a fractured carbonate gas reservoir.

2.2. Mathematical Model

Based on the assumption of the physical model, seepage mathematical models are established for natural fractures and matrix systems, respectively. For the convenience of derivation, pseudo pressure and pseudo time are introduced as:

$$\psi = 2\int_0^p \frac{p}{\mu_g Z} dp \tag{1}$$

$$t_a = \int_0^t \frac{\mu_{gi} c_{ti}}{\mu_g(\hat{p}) c_t(\hat{p})} dt \tag{2}$$

where p is the formation pressure, MPa; ψ is the formation pseudo pressure, $\mathrm{MPa}^2/(\mathrm{mPa}\cdot\mathrm{s})$; C_t is the comprehensive compression coefficient, 1/MPa; Z is the gas deviation factor, $\mathrm{m}^3/\mathrm{m}^3$; μ_g is the gas viscosity, mPa · s; T is the time, h; t_a is the pseudo time; and $\hat{p}$ is the average formation pressure.

The gas phase flow equation adopts pseudo pressure and pseudo time treatments, and the gas phase seepage control equation of the fracture system under the radial cylindrical coordinates is:

$$\frac{1}{r}\frac{\partial}{\partial r}\left(r\frac{\partial \psi_f}{\partial r}\right) - \frac{3k_m k_{rgm}}{k_f k_{rgf} r_1}\left(\frac{\partial \psi_m}{\partial r_m}\right)\Big|_{r_m=r_1} = \frac{1}{\eta_{gf}}\frac{\partial \psi_f}{\partial t} \tag{3}$$

$$\frac{1}{\eta_{gf}} = \frac{\phi_f}{0.0864 k_f k_{rgf}} \frac{1}{\mu_g(\hat{p}) B_g(\hat{p})}\left(-\frac{S_g}{B_g{}^2}\frac{dB_g}{d\psi_f}\right) \tag{4}$$

where η_{gf} is the gas transmission factor in the fracture, m^2/s.

The initial conditions are:

$$\psi_f|_{t=0} = \psi_m|_{t=0} = \psi_i \tag{5}$$

The internal boundary conditions are:

$$\lim_{r\to r_w} \psi_f = \psi_{wf} \tag{6}$$

The outer boundary conditions are:

$$\left.\frac{\partial \psi_f}{\partial r}\right|_{r=r_e} = 0 \tag{7}$$

The gas phase seepage control equation of the matrix system in radial spherical coordinates is:

$$\frac{1}{r_m{}^2}\frac{\partial}{\partial r_m}\left[r_m{}^2\frac{\partial \psi_m}{\partial r_m}\right] = \frac{1}{\eta_{gm}}\frac{\partial \psi_m}{\partial t} \tag{8}$$

$$\frac{1}{\eta_{gm}} = \frac{\phi_m}{0.0864 k_m k_{rgm}} \frac{1}{\mu_g(\hat{p}) B_g(\hat{p})}\left(-\frac{S_g}{B_g{}^2}\frac{dB_g}{d\psi_m}\right) \tag{9}$$

where η_{gm} is the gas transmission factor in the matrix, m^2/s.

The initial conditions are:

$$\psi_f|_{t=0} = \psi_m|_{t=0} = \psi_i \tag{10}$$

The internal boundary conditions are:

$$\frac{\partial \psi_m}{\partial r_m}\Big|_{r_m=0} = 0 \tag{11}$$

The outer boundary connection conditions are:

$$\psi_{\mathrm{m}}(r_{\mathrm{m}}, t)|_{r_{\mathrm{m}}=r_1} = \psi_{\mathrm{f}} \tag{12}$$

where, ψ_f is the pseudo pressure of the natural fracture system, $MPa^2/(mPa \cdot s)$; ψ_m is the pseudo pressure of the matrix system, $MPa^2/(mPa \cdot s)$; ψ is the original formation pseudo pressure, $MPa^2/(mPa \cdot s)$; ψ_{wf} is the pseudo bottom hole flow pressure, $MPa^2/(mPa \cdot s)$; K_f is the natural fracture permeability, mD; K_m is matrix permeability, mD; K_{rgf} is the relative permeability of the gas phase in the fracture system; K_{rgm} is the relative permeability of the gas phase in the matrix system; B_g is the gas volume coefficient, m^3/m^3; S_g is the gas saturation; r is the radial distance, m; r_m is the distance from the inner diameter of the matrix block, m; r_1 is the width of the matrix block, m; and r_e is the outer boundary distance, m.

The water phase flow equation is treated in real-time, and the control equation of water phase seepage of the fracture system under radial cylindrical coordinates is:

$$\frac{1}{r}\frac{\partial}{\partial r}\left(r\frac{\partial p_{\mathrm{f}}}{\partial r}\right) - \frac{3k_{\mathrm{m}}k_{\mathrm{rwm}}}{k_{\mathrm{f}}k_{\mathrm{rwf}}r_1}\left(\frac{\partial p_{\mathrm{m}}}{\partial r_{\mathrm{m}}}\right)|_{r_{\mathrm{m}}=r_1} = \frac{1}{\eta_{\mathrm{wf}}}\frac{\partial p_{\mathrm{f}}}{\partial t} \tag{13}$$

$$\frac{1}{\eta_{\mathrm{wf}}} = \frac{\phi_{\mathrm{f}}}{0.0864k_{\mathrm{f}}k_{\mathrm{rwf}}}\frac{1}{\mu_{\mathrm{w}}(\hat{p})B_{\mathrm{w}}(\hat{p})}\left(-\frac{S_{\mathrm{w}}}{B_{\mathrm{w}}{}^2}\frac{\mathrm{d}B_{\mathrm{w}}}{\mathrm{d}p_{\mathrm{f}}}\right) \tag{14}$$

where η_{wf} is the water transmission factor in fracture, m^2/s.

The initial conditions are:

$$p_{\mathrm{f}}|_{t=0} = p_{\mathrm{m}}|_{t=0} = p_{\mathrm{i}} \tag{15}$$

The internal boundary conditions are:

$$\lim_{r \to r_{\mathrm{w}}} p_{\mathrm{f}} = p_{\mathrm{wf}} \tag{16}$$

The outer boundary conditions are:

$$\left.\frac{\partial p_{\mathrm{f}}}{\partial r}\right|_{r=r_{\mathrm{e}}} = 0 \tag{17}$$

The gas phase seepage control equation of the matrix system in radial spherical coordinates is:

$$\frac{1}{r_{\mathrm{m}}{}^2}\frac{\partial}{\partial r_{\mathrm{m}}}\left[r_{\mathrm{m}}{}^2\frac{\partial p_{\mathrm{m}}}{\partial r_{\mathrm{m}}}\right] = \frac{1}{\eta_{\mathrm{wm}}}\frac{\partial p_{\mathrm{m}}}{\partial t} \tag{18}$$

$$\frac{1}{\eta_{\mathrm{wm}}} = \frac{\phi_{\mathrm{m}}}{0.0864k_{\mathrm{m}}k_{\mathrm{rwm}}}\frac{1}{\mu_{\mathrm{w}}(\hat{p})B_{\mathrm{w}}(\hat{p})}\left(-\frac{S_{\mathrm{w}}}{B_{\mathrm{w}}{}^2}\frac{\mathrm{d}B_{\mathrm{w}}}{\mathrm{d}p_{\mathrm{m}}}\right) \tag{19}$$

where η_{wm} is the water transmission factor in the matrix, m^2/s.

The initial conditions are:

$$p_{\mathrm{f}}|_{t=0} = p_{\mathrm{m}}|_{t=0} = p_{\mathrm{i}} \tag{20}$$

The internal boundary conditions are:

$$\frac{\partial p_{\mathrm{m}}}{\partial r_{\mathrm{m}}}|_{r_{\mathrm{m}}=0} = 0 \tag{21}$$

The outer boundary connection conditions are:

$$p_{\mathrm{m}}(r_{\mathrm{m}}, t)|_{r_{\mathrm{m}}=r_1} = p_{\mathrm{f}} \tag{22}$$

where p_f is the natural fracture system pressure, MPa; p_m is the matrix system pressure, MPa; p_i is the original formation pressure, MPa; p_{wf} is bottom hole flow pressure, MPa; K_{rwf} is the relative permeability of the water phase in the fracture system; K_{rwm} is the relative permeability of the water phase in the matrix system; μ_w is the viscosity of formation water, mPa · s; B_w is the volume coefficient of formation water, m^3/m^3; and S_w is water the saturation.

Comparing Formulas (3) and (13), it can be found that the relative permeability in the seepage equation is a function of water saturation, and the gas physical property parameter is a function of pressure, which leads to the strong nonlinearity of the seepage equation, so it cannot be directly solved. In order to obtain the solution of the model, the prediction years are divided into multiple periods in this paper. The parameters related to saturation are treated explicitly, the saturation is considered as a fixed value at a specific period, and the parameters related to pressure are treated implicitly with pseudo pressure. The nonlinear parameters of the model are updated step by step using the average pressure and average saturation of the gas reservoir, and a gas–water two-phase production solution is obtained using iterative calculation. In the next section, the specific solution process of the gas–water two-phase flow model for carbonate gas wells is elaborated.

3. Semi-Analytical Solution of the Gas–Water Two-Phase Productivity Prediction

The entire production time is divided into multiple time steps. In each time step, parameters related to pressure (μ_g, B_g) and the saturation-related parameters (k_{rgm}, k_{rwm}, k_{rgf}, k_{rwf}) are updated and replaced with the average pressure and average saturation within the production range, respectively. Therefore, the nonlinear parameters in the above formula can be treated as fixed values in each time step. After dealing with the nonlinear seepage problem, the gas and water phase production in each time step can be obtained by solving the equation directly.

3.1. Solution of the Gas Phase Flow Equation

Similar to the solution process of the single-phase model, the gas phase seepage equation of the fracture system is first transformed using the Laplace transform.

The gas phase seepage control equation of the fracture system in radial cylindrical coordinates is:

$$\frac{\partial^2 \Delta\overline{\psi}_f}{\partial r^2} + \frac{1}{r}\frac{\partial \Delta\overline{\psi}_f}{\partial r} - \frac{3k_m k_{rgm}}{k_f k_{rgf} r_1}\left(\frac{\partial \Delta\overline{\psi}_m}{\partial r_m}\right)\Big|_{r_m=r_1} = \frac{1}{\eta_{gf}} u \Delta\overline{\psi}_f \tag{23}$$

where u is the Laplace operator.

The initial conditions are:

$$\Delta\overline{\psi}_f\big|_{t=0} = \Delta\overline{\psi}_m\big|_{t=0} = 0 \tag{24}$$

The internal boundary conditions are:

$$\lim_{r\to r_w} \Delta\overline{\psi}_f = \frac{1}{u}\Delta\psi_{wf} \tag{25}$$

The outer boundary conditions are:

$$\left.\frac{\partial \Delta\overline{\psi}_f}{\partial r}\right|_{r=r_e} = 0 \tag{26}$$

The gas phase seepage control equation of the matrix system in radial spherical coordinates is:

$$\frac{\partial^2 \Delta\overline{\psi}_m}{\partial {r_m}^2} + \frac{2}{r_m}\frac{\partial \Delta\overline{\psi}_m}{\partial r_m} = \frac{1}{\eta_{gm}} u \Delta\overline{\psi}_m \tag{27}$$

The initial conditions are:

$$\Delta\overline{\psi}_{\mathrm{f}}\big|_{t=0} = \Delta\overline{\psi}_{\mathrm{m}}\big|_{t=0} = 0 \tag{28}$$

The internal boundary conditions are:

$$\frac{\partial\Delta\overline{\psi}_{\mathrm{m}}}{\partial r_{\mathrm{m}}}\Big|_{r_{\mathrm{m}}=0} = 0 \tag{29}$$

The outer boundary conditions are:

$$\Delta\overline{\psi}_{\mathrm{m}}(r_{\mathrm{m}},t)\big|_{r_{\mathrm{m}}=r_1} = \Delta\overline{\psi}_{\mathrm{f}} \tag{30}$$

Next, solve the partial differential equation of the matrix system under its definite solution condition. The general solution of Equation (27) is:

$$\Delta\overline{\psi}_{\mathrm{m}} = \frac{A_{\mathrm{m}}\mathrm{e}^{\sigma_{\mathrm{m}}r_{\mathrm{m}}} + B_{\mathrm{m}}\mathrm{e}^{-\sigma_{\mathrm{m}}r_{\mathrm{m}}}}{r_{\mathrm{m}}} \tag{31}$$

$$\sigma_{\mathrm{m}} = \sqrt{\frac{u}{\eta_{\mathrm{gm}}}} \tag{32}$$

Then, substitute the general solution Equation (31) into the definite solution conditions Equations (29) and (30) to obtain:

$$A_{\mathrm{m}} = \frac{1}{\mathrm{e}^{\sigma_{\mathrm{m}}} - \mathrm{e}^{-\sigma_{\mathrm{m}}}}\Delta\overline{\psi}_{\mathrm{f}} \tag{33}$$

$$B_{\mathrm{m}} = -\frac{1}{\mathrm{e}^{\sigma_{\mathrm{m}}} - \mathrm{e}^{-\sigma_{\mathrm{m}}}}\Delta\overline{\psi}_{\mathrm{f}} \tag{34}$$

Therefore, the solution of Equation (31) is:

$$\Delta\overline{\psi}_{\mathrm{m}} = \frac{\Delta\overline{\psi}_{\mathrm{f}}}{r_{\mathrm{m}}}\frac{\mathrm{e}^{\sigma_{\mathrm{m}}r_{\mathrm{m}}} - \mathrm{e}^{-\sigma_{\mathrm{m}}r_{\mathrm{m}}}}{\mathrm{e}^{\sigma_{\mathrm{m}}} - \mathrm{e}^{-\sigma_{\mathrm{m}}}} \tag{35}$$

Take the derivative of rm in Equation (35) to obtain:

$$\frac{\partial\Delta\overline{\psi}_{\mathrm{m}}}{\partial r_{\mathrm{m}}}\Big|_{r_{\mathrm{m}}=r_1} = \frac{\Delta\overline{\psi}_{\mathrm{f}}}{{r_1}^2}\frac{(\sigma_{\mathrm{m}}r_1 - 1)\mathrm{e}^{\sigma_{\mathrm{m}}r_1} + (\sigma_{\mathrm{m}}r_1 + 1)\mathrm{e}^{-\sigma_{\mathrm{m}}r_1}}{\mathrm{e}^{\sigma_{\mathrm{m}}} - \mathrm{e}^{-\sigma_{\mathrm{m}}}} \tag{36}$$

Substituting Equation (36) into the seepage Equation (23) of the fracture system obtains:

$$\frac{\partial^2\Delta\overline{\psi}_{\mathrm{f}}}{\partial r^2} + \frac{1}{r}\frac{\partial\Delta\overline{\psi}_{\mathrm{f}}}{\partial r} - f(\xi)\Delta\overline{\psi}_{\mathrm{f}} = 0 \tag{37}$$

$$f(\xi) = \frac{3k_{\mathrm{m}}k_{\mathrm{rgm}}}{k_{\mathrm{f}}k_{\mathrm{rgf}}r_1}\frac{1}{{r_1}^2}\frac{(\sigma_{\mathrm{m}}r_1 - 1)\mathrm{e}^{\sigma_{\mathrm{m}}r_1} + (\sigma_{\mathrm{m}}r_1 + 1)\mathrm{e}^{-\sigma_{\mathrm{m}}r_1}}{\mathrm{e}^{\sigma_{\mathrm{m}}} - \mathrm{e}^{-\sigma_{\mathrm{m}}}} + \frac{1}{\eta_{\mathrm{gf}}}u \tag{38}$$

Equation (37) is the virtual argument Bessel equation, whose general solution is:

$$\Delta\overline{\psi}_{\mathrm{f}} = AI_0\left(r\sqrt{f(\xi)}\right) + BK_0\left(r\sqrt{f(\xi)}\right) \tag{39}$$

$$\frac{\mathrm{d}\Delta\overline{\psi}_{\mathrm{f}}}{\mathrm{d}r} = A\sqrt{f(\xi)}I_1\left(r\sqrt{f(\xi)}\right) - B\sqrt{f(\xi)}K_1\left(r\sqrt{f(\xi)}\right) \tag{40}$$

Then, substitute Equations (39) and (40) into Equation (25) for the inner boundary conditions and Equation (26) for the outer boundary conditions, respectively:

$$\left.\frac{\mathrm{d}\Delta\overline{\psi}_{\mathrm{f}}}{\mathrm{d}r}\right|_{r=r_{\mathrm{e}}} = A\sqrt{f(\xi)}I_1\left(r_{\mathrm{e}}\sqrt{f(\xi)}\right) - B\sqrt{f(\xi)}K_1\left(r_{\mathrm{e}}\sqrt{f(\xi)}\right) = 0 \tag{41}$$

$$\lim_{r\to r_{\mathrm{w}}}\Delta\overline{\psi}_{\mathrm{f}} = AI_0\left(r_{\mathrm{w}}\sqrt{f(\xi)}\right) + BK_0\left(r_{\mathrm{w}}\sqrt{f(\xi)}\right) = \frac{1}{u}\Delta\psi_{\mathrm{wf}} \tag{42}$$

Simultaneous type (39) and type (40):

$$A = \frac{\Delta\psi_{\mathrm{wf}}}{u}\frac{K_1(r_{\mathrm{e}}\sqrt{f(\xi)})}{I_0(r_{\mathrm{w}}\sqrt{f(\xi)})K_1(r_{\mathrm{e}}\sqrt{f(\xi)}) + I_1(r_{\mathrm{e}}\sqrt{f(\xi)})K_0(r_{\mathrm{w}}\sqrt{f(\xi)})} \tag{43}$$

$$B = \frac{\Delta\psi_{\mathrm{wf}}}{u}\frac{I_1(r_{\mathrm{e}}\sqrt{f(\xi)})}{I_0(r_{\mathrm{w}}\sqrt{f(\xi)})K_1(r_{\mathrm{e}}\sqrt{f(\xi)}) + I_1(r_{\mathrm{e}}\sqrt{f(\xi)})K_0(r_{\mathrm{w}}\sqrt{f(\xi)})} \tag{44}$$

Substituting Equation (41) and Equation (42) into Equation (39), we obtain:

$$\Delta\overline{\psi}_{\mathrm{f}} = \frac{\Delta\psi_{\mathrm{wf}}}{u}\frac{K_1(r_{\mathrm{e}}\sqrt{f(\xi)})}{\alpha}I_0\left(r\sqrt{f(\xi)}\right) + \frac{\Delta\psi_{\mathrm{wf}}}{u}\frac{I_1(r_{\mathrm{e}}\sqrt{f(\xi)})}{\alpha}K_0\left(r\sqrt{f(\xi)}\right) \tag{45}$$

$$\frac{\partial\Delta\overline{\psi}_{\mathrm{f}}}{\partial r} = \frac{\Delta\psi_{\mathrm{wf}}}{u}\frac{K_1(r_{\mathrm{e}}\sqrt{f(\xi)})}{\alpha}\sqrt{f(\xi)}I_1\left(r\sqrt{f(\xi)}\right) - \frac{\Delta\psi_{\mathrm{wf}}}{u}\frac{I_1(r_{\mathrm{e}}\sqrt{f(\xi)})}{\alpha}\sqrt{f(\xi)}K_1\left(r\sqrt{f(\xi)}\right) \tag{46}$$

The expression of gas production by a single well is:

$$\overline{q}_{\mathrm{gf}} = -2\pi k_{\mathrm{f}}k_{\mathrm{rgf}}hr_{\mathrm{w}}\left.\frac{\partial\Delta\overline{\psi}_{\mathrm{f}}}{\partial r}\right|_{r\to r_{\mathrm{w}}} \tag{47}$$

Combined with Formula (44), the solution of the gas phase production is:

$$\overline{q}_{\mathrm{gf}} = \frac{-2\pi k_{\mathrm{f}}k_{\mathrm{rgf}}hr_{\mathrm{w}}\Delta\psi_{\mathrm{wf}}}{u}\sqrt{f(\xi)}\left[\frac{\frac{K_1(r_{\mathrm{e}}\sqrt{f(\xi)})I_1\left(r_{\mathrm{w}}\sqrt{f(\xi)}\right)}{I_0(r_{\mathrm{w}}\sqrt{f(\xi)})I_1(r_{\mathrm{e}}\sqrt{f(\xi)})} - \frac{K_1\left(r_{\mathrm{w}}\sqrt{f(\xi)}\right)}{I_0(r_{\mathrm{w}}\sqrt{f(\xi)})}}{\frac{K_1(r_{\mathrm{e}}\sqrt{f(\xi)})}{I_1(r_{\mathrm{e}}\sqrt{f(\xi)})} + \frac{K_0(r_{\mathrm{w}}\sqrt{f(\xi)})}{I_0(r_{\mathrm{w}}\sqrt{f(\xi)})}}\right] \tag{48}$$

The gas phase production solution in Equation (46) is obtained in Laplace space, and the solution in real space can be obtained using the numerical inversion of Stehfest [21].

3.2. Solution of the Aqueous Phase Flow Equation

The Laplace transformation about dimensionless time is adopted for the water phase seepage equation, which is similar to the solution process for the gas phase seepage model.

The control equation of the water phase seepage of the fracture system in radial cylindrical coordinates is:

$$\frac{\partial^2\Delta\overline{p}_{\mathrm{f}}}{\partial r^2} + \frac{1}{r}\frac{\partial\Delta\overline{p}_{\mathrm{f}}}{\partial r} - \frac{3k_{\mathrm{m}}k_{\mathrm{rwm}}}{k_{\mathrm{f}}k_{\mathrm{rwf}}r_1}\left(\frac{\partial\Delta\overline{p}_{\mathrm{m}}}{\partial r_{\mathrm{m}}}\right)|_{r_{\mathrm{m}}=r_1} = \frac{1}{\eta_{\mathrm{wf}}}u\Delta\overline{p}_{\mathrm{f}} \tag{49}$$

The initial conditions are:

$$\Delta\,\overline{p}_{\mathrm{f}}|_{t=0} = \Delta\,\overline{p}_{\mathrm{m}}|_{t=0} = 0 \tag{50}$$

The internal boundary conditions are:

$$\lim_{r\to r_{\mathrm{w}}}\Delta\overline{p}_{\mathrm{f}} = \frac{1}{u}\Delta p_{\mathrm{wf}} \tag{51}$$

The outer boundary conditions are:

$$\left.\frac{\partial\Delta\overline{p}_{\mathrm{f}}}{\partial r}\right|_{r=r_{\mathrm{e}}} = 0 \tag{52}$$

The seepage control equation of the matrix system in radial spherical coordinates is:

$$\frac{\partial^2 \Delta \overline{p}_{\mathrm{m}}}{\partial r_{\mathrm{m}}{}^2} + \frac{2}{r_{\mathrm{m}}} \frac{\partial \Delta \overline{p}_{\mathrm{m}}}{\partial r_{\mathrm{m}}} = \frac{1}{\eta_{\mathrm{wm}}} u \Delta \overline{p}_{\mathrm{m}} \tag{53}$$

The initial conditions are:

$$\Delta \overline{p}_{\mathrm{f}}|_{t=0} = \Delta \overline{p}_{\mathrm{m}}|_{t=0} = 0 \tag{54}$$

The internal boundary conditions are:

$$\frac{\partial \Delta \overline{p}_{\mathrm{m}}}{\partial r_{\mathrm{m}}}|_{r_{\mathrm{m}}=0} = 0 \tag{55}$$

The outer boundary conditions are:

$$\Delta \overline{p}_{\mathrm{m}}(r_{\mathrm{m}}, t)|_{r_{\mathrm{m}}=r_1} = \Delta \overline{p}_{\mathrm{f}} \tag{56}$$

Next, solve the partial differential equation of the matrix system under its definite solution condition. The general solution of Equation (51) is:

$$\Delta \overline{p}_{\mathrm{m}} = \frac{C_{\mathrm{m}} e^{\beta_{\mathrm{m}} r_{\mathrm{m}}} + D_{\mathrm{m}} e^{-\beta_{\mathrm{m}} r_{\mathrm{m}}}}{r_{\mathrm{m}}} \tag{57}$$

$$\beta_{\mathrm{m}} = \sqrt{\frac{u}{\eta_{\mathrm{wm}}}} \tag{58}$$

Substitute the general solution Equation (55) into the definite solution condition Equations (53) and (54) to obtain:

$$C_{\mathrm{m}} = \frac{1}{e^{\beta_{\mathrm{m}}} - e^{-\beta_{\mathrm{m}}}} \Delta \overline{p}_{\mathrm{f}} \tag{59}$$

$$D_{\mathrm{m}} = -\frac{1}{e^{\beta_{\mathrm{m}}} - e^{-\beta_{\mathrm{m}}}} \Delta \overline{p}_{\mathrm{f}} \tag{60}$$

Therefore, the solution of Equation (55) is:

$$\Delta \overline{p}_{\mathrm{m}} = \frac{\Delta \overline{p}_{\mathrm{f}}}{r_{\mathrm{m}}} \frac{e^{\beta_{\mathrm{m}} r_{\mathrm{m}}} - e^{-\beta_{\mathrm{m}} r_{\mathrm{m}}}}{e^{\beta_{\mathrm{m}}} - e^{-\beta_{\mathrm{m}}}} \tag{61}$$

Take the derivative of rm in Equation (59) to obtain:

$$\frac{\partial \Delta \overline{p}_{\mathrm{m}}}{\partial r_{\mathrm{m}}}|_{r_{\mathrm{m}}=r_1} = \frac{\Delta \overline{p}_{\mathrm{f}}}{r_1{}^2} \frac{(\beta_{\mathrm{m}} r_1 - 1) e^{\beta_{\mathrm{m}} r_1} + (\beta_{\mathrm{m}} r_1 + 1) e^{-\beta_{\mathrm{m}} r_1}}{e^{\beta_{\mathrm{m}}} - e^{-\beta_{\mathrm{m}}}} \tag{62}$$

Substituting Equation (60) into the seepage Equation (47) of fracture system gives:

$$\frac{\partial^2 \Delta \overline{p}_{\mathrm{f}}}{\partial r^2} + \frac{1}{r} \frac{\partial \Delta \overline{p}_{\mathrm{f}}}{\partial r} - f(\zeta) \Delta \overline{p}_{\mathrm{f}} = 0 \tag{63}$$

$$f(\zeta) = \frac{3 k_{\mathrm{m}} k_{\mathrm{rwm}}}{k_{\mathrm{f}} k_{\mathrm{rwf}} r_1} \frac{1}{r_1{}^2} \frac{(\beta_{\mathrm{m}} r_1 - 1) e^{\beta_{\mathrm{m}} r_1} + (\beta_{\mathrm{m}} r_1 + 1) e^{-\beta_{\mathrm{m}} r_1}}{e^{\beta_{\mathrm{m}}} - e^{-\beta_{\mathrm{m}}}} + \frac{1}{\eta_{\mathrm{wf}}} u \tag{64}$$

Equation (61) is the virtual argument Bessel equation, whose general solution is:

$$\Delta \overline{p}_{\mathrm{f}} = C I_0\left(r\sqrt{f(\zeta)}\right) + D K_0\left(r\sqrt{f(\zeta)}\right) \tag{65}$$

$$\frac{\mathrm{d}\Delta \overline{p}_{\mathrm{f}}}{\mathrm{d}r} = C\sqrt{f(\zeta)} I_1\left(r\sqrt{f(\zeta)}\right) - D\sqrt{f(\zeta)} K_1\left(r\sqrt{f(\zeta)}\right) \tag{66}$$

Substituting Equations (63) and (64) into Equation (49) for the inner boundary conditions and Equation (50) for the outer boundary conditions, respectively, gives:

$$\left.\frac{\mathrm{d}\Delta \overline{p}_{\mathrm{f}}}{\mathrm{d}r}\right|_{r=r_{\mathrm{e}}} = C\sqrt{f(\zeta)} I_1\left(r_{\mathrm{e}}\sqrt{f(\zeta)}\right) - D\sqrt{f(\zeta)} K_1\left(r_{\mathrm{e}}\sqrt{f(\zeta)}\right) = 0 \tag{67}$$

$$\lim_{r \to r_{\mathrm{w}}} \Delta \overline{p}_{\mathrm{f}} = C I_0\left(r_{\mathrm{w}}\sqrt{f(\zeta)}\right) + D K_0\left(r_{\mathrm{w}}\sqrt{f(\zeta)}\right) = \frac{1}{u} \Delta p_{\mathrm{wf}} \tag{68}$$

Simultaneous (63) and (64):

$$C = \frac{\Delta p_{\mathrm{wf}}}{u}\frac{K_1(r_{\mathrm{e}}\sqrt{f(\zeta)})}{I_0(r_{\mathrm{w}}\sqrt{f(\zeta)})K_1(r_{\mathrm{e}}\sqrt{f(\zeta)}) + I_1(r_{\mathrm{e}}\sqrt{f(\zeta)})K_0(r_{\mathrm{w}}\sqrt{f(\zeta)})} \tag{69}$$

$$D = \frac{\Delta p_{\mathrm{wf}}}{u}\frac{I_1(r_{\mathrm{e}}\sqrt{f(\zeta)})}{I_0(r_{\mathrm{w}}\sqrt{f(\zeta)})K_1(r_{\mathrm{e}}\sqrt{f(\zeta)}) + I_1(r_{\mathrm{e}}\sqrt{f(\zeta)})K_0(r_{\mathrm{w}}\sqrt{f(\zeta)})} \tag{70}$$

Equation (65) and Equation (66) are then substituted into Equation (63) to obtain:

$$\Delta\overline{p}_{\mathrm{f}} = \frac{\Delta p_{\mathrm{wf}}}{u}\frac{K_1(r_{\mathrm{e}}\sqrt{f(\zeta)})}{\chi}I_0\left(r\sqrt{f(\zeta)}\right) + \frac{\Delta\psi_{\mathrm{wf}}}{u}\frac{I_1(r_{\mathrm{e}}\sqrt{f(\zeta)})}{\chi}K_0\left(r\sqrt{f(\zeta)}\right) \tag{71}$$

$$\frac{\partial\Delta\overline{p}_{\mathrm{f}}}{\partial r} = \frac{\Delta p_{\mathrm{wf}}}{u}\frac{K_1(r_{\mathrm{e}}\sqrt{f(\zeta)})}{\chi}\sqrt{f(\zeta)}I_1\left(r\sqrt{f(\zeta)}\right) - \frac{\Delta p_{\mathrm{wf}}}{u}\frac{I_1(r_{\mathrm{e}}\sqrt{f(\zeta)})}{\chi}\sqrt{f(\zeta)}K_1\left(r\sqrt{f(\zeta)}\right) \tag{72}$$

The water yield expression of a single well is:

$$\overline{q}_{\mathrm{wf}} = -2\pi k_{\mathrm{f}}k_{\mathrm{rwf}}hr_{\mathrm{w}}\left.\frac{\partial\Delta\overline{p}_{\mathrm{f}}}{\partial r}\right|_{r\to r_{\mathrm{w}}} \tag{73}$$

Combined with Formula (70), the solution of the aqueous phase yield is:

$$\overline{q}_{\mathrm{wf}} = \frac{-2\pi k_{\mathrm{f}}k_{\mathrm{rwf}}hr_{\mathrm{w}}\Delta p_{\mathrm{wf}}}{u}\sqrt{f(\zeta)}\left[\frac{\frac{K_1(r_{\mathrm{e}}\sqrt{f(\zeta)})I_1\left(r_{\mathrm{w}}\sqrt{f(\zeta)}\right)}{I_0(r_{\mathrm{w}}\sqrt{f(\zeta)})I_1(r_{\mathrm{e}}\sqrt{f(\zeta)})} - \frac{K_1\left(r_{\mathrm{w}}\sqrt{f(\zeta)}\right)}{I_0(r_{\mathrm{w}}\sqrt{f(\zeta)})}}{\frac{K_1(r_{\mathrm{e}}\sqrt{f(\zeta)})}{I_1(r_{\mathrm{e}}\sqrt{f(\zeta)})} + \frac{K_0(r_{\mathrm{w}}\sqrt{f(\zeta)})}{I_0(r_{\mathrm{w}}\sqrt{f(\zeta)})}}\right] \tag{74}$$

The gas–water two-phase production solution in Equations (46) and (72) is obtained in Laplace space, and the solution in real space can be obtained using the numerical inversion of Stehfest [21].

There are still parameters related to pressure and saturation in Equations (46) and (72). In addition, Equation (73) gives the permeability expression considering stress sensitivity. In this paper, all stress sensitivity terms are integrated into the transmissivity coefficient as a function of average formation pressure [22–24]. In the process of solving the model, the average formation pressure and saturation within the production range are used to update the nonlinear parameters in each time step, and the solution of the model is obtained with step-by-step iteration, while the average formation pressure and average saturation are calculated using the flow material balance method.

$$k_{\mathrm{f}} = k_{\mathrm{fi}}e^{-\gamma(p_{\mathrm{i}}-\hat{p})} \tag{75}$$

where k_{fi} is the initial fracture permeability, mD and γ is permeability modulus, MPa^{-1}.

3.3. Establishment of the Fluid Mass Balance Equation

The gas phase mass balance equation is:

$$\frac{\hat{S}_{\mathrm{g}}}{\hat{B}_{\mathrm{g}}} = \frac{S_{\mathrm{gi}}}{B_{\mathrm{gi}}} - \frac{\int_0^t q_{\mathrm{g}}\mathrm{d}t}{\pi r_{\mathrm{inv}}{}^2 h\phi_{\mathrm{m}}} \tag{76}$$

where, S_{gi} represents the initial gas saturation; $\hat{S}_{\mathrm{g}}$ is the average gas saturation; B_{gi} represents the initial gas volume coefficient; $\hat{B}_{\mathrm{g}}$ is the average gas volume coefficient; r_{inv} is the utilization range, m; ϕ_{m} is the matrix porosity, h is the effective reservoir thickness, m; and Q_g is the daily gas production, $10^4\ \mathrm{m}^3/\mathrm{d}$.

The formula for calculating the detection radius of a dual-medium reservoir is [25–27]:

$$r_{\mathrm{inv}} = 0.03248\sqrt{\frac{k_{\mathrm{m}}}{\phi_{\mathrm{m}}\mu_{\mathrm{gi}}c_{\mathrm{ti}}}t} \tag{77}$$

The water phase material balance equation is:

$$\frac{\hat{S}_{\mathrm{w}}}{\hat{B}_{\mathrm{w}}} = \frac{S_{\mathrm{wi}}}{B_{\mathrm{wi}}} - \frac{\int_{0}^{t} q_{\mathrm{w}} \mathrm{d}t}{\pi r_{\mathrm{inv}}{}^{2} h \phi_{\mathrm{m}}} \tag{78}$$

where S_{wi} represents the initial water saturation, decimal; $\hat{S}_{\mathrm{w}}$ represents average water saturation, decimal; and Q_w is the daily water yield, m^3/d.

Gas–water two-phase saturation satisfies the following relationship:

$$\hat{S}_{\mathrm{w}} + \hat{S}_{\mathrm{g}} = 1 \tag{79}$$

Formulas (74), (76) and (77) are used to construct the average pressure function, including:

$$f(\hat{p}) = \frac{1}{\hat{B}_{\mathrm{g}}} + \left(-\frac{\hat{B}_{\mathrm{w}}}{\hat{B}_{\mathrm{g}}}\right) E - F \tag{80}$$

$$E = \frac{S_{\mathrm{wi}}}{B_{\mathrm{wi}}} - \frac{\int_{0}^{t} q_{\mathrm{w}} \mathrm{d}t}{\pi r_{\mathrm{inv}}{}^{2} h \phi_{\mathrm{m}}} \tag{81}$$

$$F = \frac{S_{\mathrm{gi}}}{B_{\mathrm{gi}}} - \frac{\int_{0}^{t} q_{\mathrm{g}} \mathrm{d}t}{\pi r_{\mathrm{inv}}{}^{2} h \phi_{\mathrm{m}}} \tag{82}$$

The derivation of Equation (78) yields:

$$f'(\hat{p}) = -\frac{1}{\hat{B}_{\mathrm{g}}^{2}} \frac{\mathrm{d}B_{\mathrm{g}}}{\mathrm{d}p} + A_1 \left(-\frac{1}{\hat{B}_{\mathrm{g}}} \frac{\mathrm{d}B_{\mathrm{w}}}{\mathrm{d}p} + \frac{B_{\mathrm{w}}}{\hat{B}_{\mathrm{g}}^{2}} \frac{\mathrm{d}B_{\mathrm{g}}}{\mathrm{d}p}\right) \tag{83}$$

The average pressure Newton iteration scheme is constructed from Equation (81), including:

$$\hat{p}_{\mathrm{k}+1} = \hat{p}_{\mathrm{k}} - \omega \frac{f(\hat{p}_{\mathrm{k}})}{f'(\hat{p}_{\mathrm{k}})} \tag{84}$$

where k represents the last time step; k + 1 represents the current time step; and ω is th iteration factor.

The average pressure is obtained with Newton iteration, and then the average saturation can b calculated by substituting Equation (74) or Equation (76). The average pressure and saturation withi the production range are used to update the nonlinear parameters in each time step, and the solutio of the gas–water two-phase seepage model of carbonate gas wells can be obtained using step-by-ste iterative calculation, and then the gas–water two-phase productivity curve can be programmed t predict the gas water production performance.

4. Dynamic Analysis of Gas–Water Two-Phase Production

4.1. Model Validation

In order to verify the accuracy of the proposed semi-analytical model, this paper compares th calculation results with the commercial numerical simulation software Eclipse. First, a numeric model corresponding to the physical model in this paper is established: a straight well in the cent of the carbonate dual-medium reservoir is used to produce at a fixed bottom flowing pressure. In th example, both the matrix and fracture system consider gas–water two-phase flow, that is, gas–wat two-phase flow occurs at the beginning of production. The relationship curve of gas high-pressu physical parameters used in the model is shown in Figure 2. Table 1 shows the basic parameters us by the two methods, and Figure 3 shows the gas–water two-phase relative permeability curve natural fractures [14].

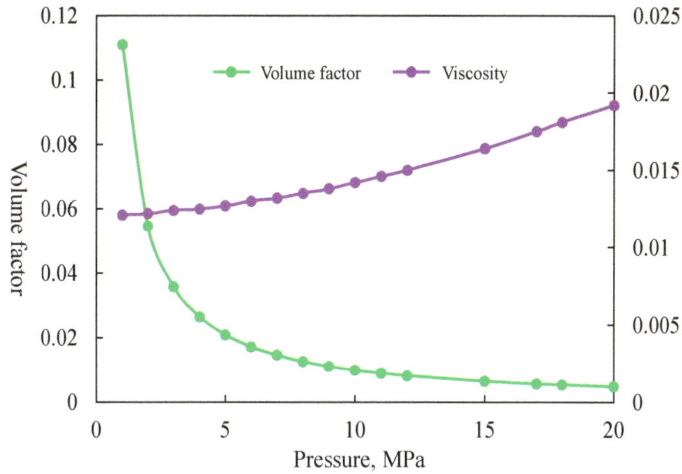

Figure 2. Schematic of the gas PVT properties.

Table 1. Input parameters for model validation.

Parameter	Value	Parameter	Value
Reservoir temperature, K	373 K	Original formation pressure, MPa	22
Fracture permeability, mD	50	Reservoir radius, m	500
Bottom hole flow pressure, MPa	6	Initial gas saturation, %	65
Compressibility coefficient of rock, MPa^{-1}	1×10^{-4}	Permeability modulus, MPa^{-1}	0.02
Matrix porosity, %	9	Matrix permeability, mD	0.025
Effective thickness of reservoir, m	29	Wellbore radius, m	0.07

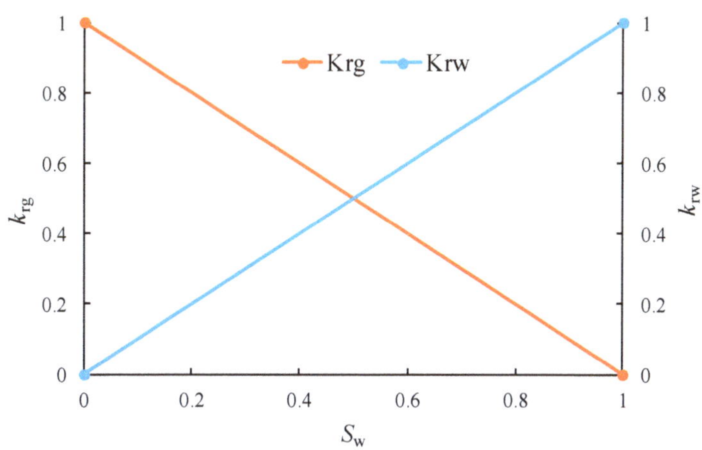

Figure 3. Relative permeability of the natural fracture system.

The comparison results of the semi-analytical model and Eclipse in this paper are shown in Figure 4. It can be seen that the output curves obtained using the two methods are somewhat different in the early production period, and the results are basically the same in the later period. Under the same reservoir and fracture parameters, the productivity of the gas–water two-phase flow is significantly lower than that of the single-phase flow. This is mainly because in the early stage of production, the pressure and saturation near the well vary greatly, and production is very sensitive to the parameters related to pressure and saturation. In the semi-analytical model, pseudo pressure is used to implicitly process some parameters related to pressure, but the parameters related to saturation are treated explicitly, so the early error of the single-phase seepage model is not obvious, while the early error of the two-phase flow model is obvious. However, the average relative error calculated is less than 10%, which is within the allowable error range of the project, indicating that the semi-analytical model proposed in this paper can be used for production data analysis and prediction. Compared with the numerical simulation method, the semi-analytical method proposed in this paper has faster calculation speed and is more conducive to the application of large-scale case analysis in mines.

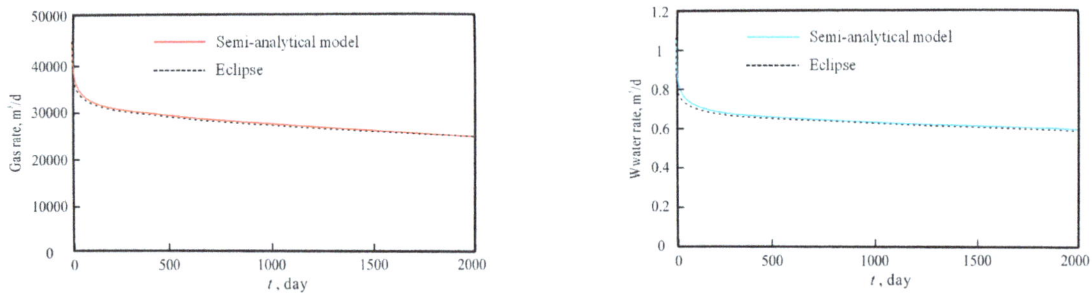

(a) Comparison results of daily gas production **(b)** Comparison results of daily water production

Figure 4. Comparison of production performance from the semi-analytical method and numerical simulation method.

4.2. Analysis of Factors Affecting Production Dynamics

Based on the semi-analytical model, this paper focuses on the influence of the stress sensitivity coefficient, fracture porosity, and outer boundary distance on the productivity of carbonate gas wells. The basic input parameters of the model are collected in Table 2.

Table 2. Range of values for sensitive parameters.

Parameter	Value
Permeability modulus, MPa^{-1}	0.01, 0.05, 0.1
Fracture porosity, %	0.1, 1, 5
Outer boundary distance, m	400, 500, 600

Figure 5 shows the influence of fracture porosity on production performance. It can be seen from Figure 5 that the greater the fracture porosity, the slower the decline rate in the gas–water two-phase production. In addition, the fracture porosity has a greater impact on the early and middle production stages, but less impact on the late stage. This is mainly because the fracture porosity reflects the fluid storage capacity in the fracture system. The larger the fracture porosity, the stronger the reservoir capacity. When the fracture porosity is large, production can be maintained with a small production decline under the production condition of constant bottom hole flow pressure. In addition, in the late production period, the supply capacity of the matrix to fractures plays a leading role, so the fracture porosity has little impact on the late production period.

Figure 6 shows the influence of the permeability modulus on production performance. The permeability modulus reflects the strength of stress sensitivity. It can be found that the permeability stress sensitivity of natural fractures affects the whole production process, especially in the early and middle production stages, and has little impact on the later production stages. In addition, with the increase in the permeability modulus, the position of the gas–water production curve moves downward. This is mainly because the strong stress sensitivity leads to the serious loss of fracture permeability and the increase in fluid seepage resistance, and more gas cannot be recovered, which ultimately leads to the reduction in gas production and the slowing down of the production decline trend.

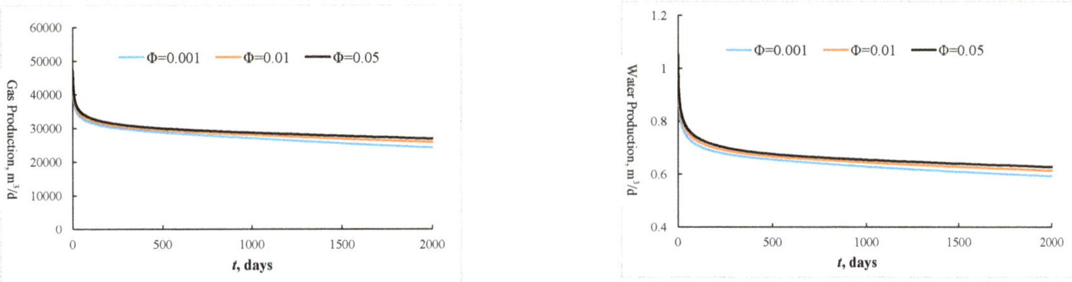

(**a**) Daily gas yield curve (**b**) Daily water yield curve

Figure 5. Effects of fracture porosity on the gas and water production rates.

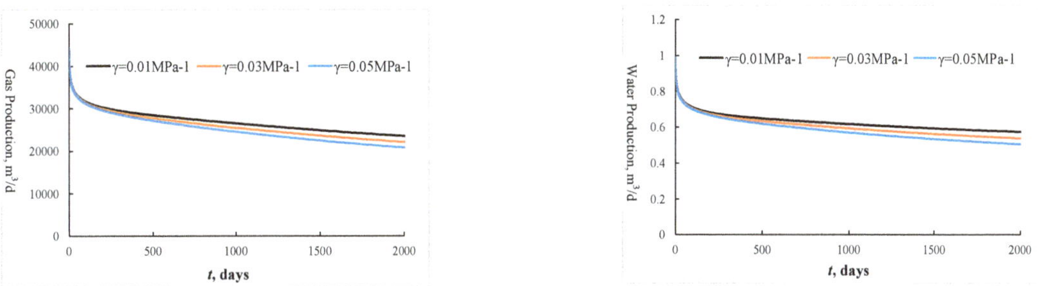

(**a**) Daily gas yield curve (**b**) Daily water yield curve

Figure 6. Effects of the stress sensitivity of the fracture on the gas and water production rates.

Figure 7 shows the influence of the outer boundary distance on the production dynamics. It can be seen that the smaller the distance from the outer boundary, the lower the position of the gas and water production curve, and the faster the production decline. In addition, the distance from the outer boundary has a greater impact on the early and middle production stages but less impact on the late stage. The size of the outer boundary distance reflects the time of the boundary reaction. The greater the outer boundary distance, the later the boundary reaction time. When the outer boundary distance is larger, the production can be maintained with a small production decline under the production condition of constant bottom hole flowing pressure.

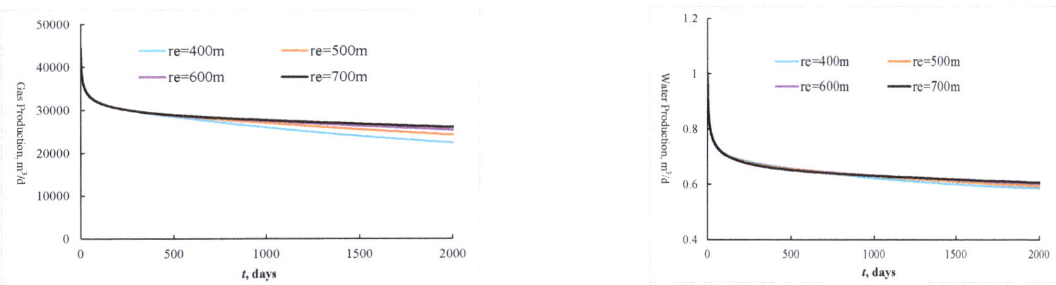

(**a**) Daily gas yield curve (**b**) Daily water yield curve

Figure 7. Effects of the radial distance of the external boundary on the gas and water production rates.

4.3. Case Analysis

This paper takes a water-producing carbonate gas well in a basin as an example to illustrate the application effect of the model. The well was put into production in May 2018, and the initial water saturation of the reservoir was about 65%. As of April 2022, the cumulative gas production was 3543×10^4 m^3, the cumulative water production was 3866 m^3, the daily gas production was 9557 m^3/d, and the daily water production was 1.25 m^3/d. The basic data of the well are shown in Table 2. The PVT curve of gas and the gas–water relative permeability curve are shown in Figures 2 and 3, respectively. The semi-analytical model proposed in this paper is used to fit and interpret the gas–water two-phase production data of the well, as shown in Figure 8. It can be seen that the fitting effect of the theoretical curve and the measured curve is good. Although there is a certain deviation in the fitting result, it is within the allowable range of engineering error. The fitting interpretation results of this well are summarized in Table 3, where the interpreted fracture porosity is 0.02, the outer boundary distance is 495 m, and the reservoir stress sensitivity coefficient is 0.035 MPa^{-1}, which are consistent with the actual gas reservoir. The production of the well is predicted, and the gas production in 30 years is 7870×10^4 m^3, and the produced water is 8.21×10^4 m^3.

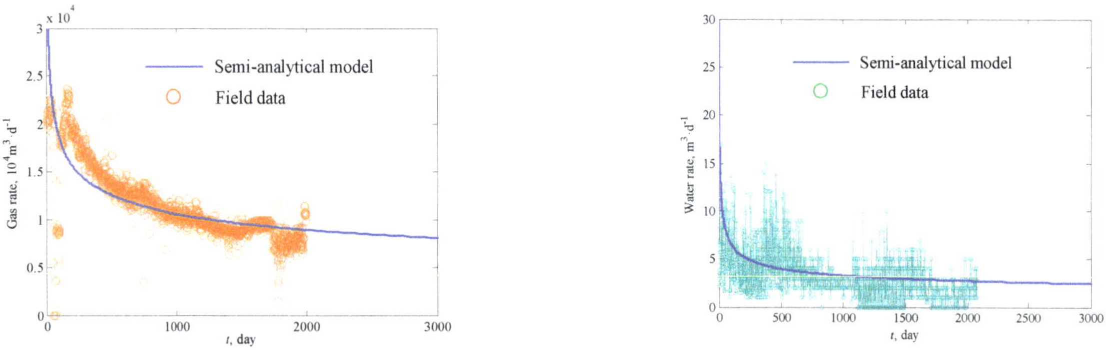

(a) Fitting results of gas production **(b)** Fitting results of water production

Figure 8. Interpretation and prediction of the production decline in the example wells.

Table 3. Matching Interpretation Results of Example Wells.

Parameter	Value
Fracture porosity	0.02
Permeability modulus, MPa^{-1}	0.035
Outer boundary distance, m	495

5. Conclusions

In this paper, a productivity model considering complex fractures, the stress sensitivity effec and the gas–water two-phase flow characteristics of the reservoir are established, and a semi-analytic solution method for the model is established by incorporating the flow material balance method. Th productivity prediction of gas–water two-phase flow in carbonate gas wells can be realized. Mai conclusions are now summarized:

(1) The average pressure and average saturation of the reservoir are calculated using the flo material balance method, and the nonlinear parameters in the seepage model are update step by step, which can deal with the gas–water two-phase nonlinear seepage problem wi higher accuracy.

(2) Validation work using the numerical model and field application showed that the sem analytical method proposed in this paper has high prediction accuracy and can be used predict the gas–water two-phase production of carbonate gas wells.

(3) Fracture and key percolation parameters of the reservoir play important roles in gas–wat two-phase production performance. Since the water production of carbonate gas reservoi dramatically affects the productivity of a gas well, under the same reservoir and fractu parameters, the productivity with gas–water two-phase flow is significantly lower than that single-phase flow.

(4) The stress sensitivity of fractures affects the productivity of carbonate gas wells. In the production process, it is necessary to control the production pressure difference carefully to inhibit the negative impact of the stress sensitivity effect.

Author Contributions: Software and investigation, D.C.; supervision, Z.S. All authors have read and agreed to the published version of the manuscript.

Funding: The authors acknowledge financial support from the National Natural Science Foundation Projects of China (No. 51604261 and No. 52104099) and the Natural Science Foundation Projects of Jiangsu Province (No. BK20210508).

Institutional Review Board Statement: On behalf of all the co-authors, the corresponding author states that there are no ethical statements contained in the manuscripts.

Data Availability Statement: Data will be available on request.

Acknowledgments: The authors acknowledge financial support from the National Natural Science Foundation Projects of China (No. 51604261 and No. 52104099) and the Natural Science Foundation Projects of Jiangsu Province (No. BK20210508). We also acknowledge China University of Mining & Technology for the permission to publish this work.

Conflicts of Interest: The authors declare no conflict of interest.

eferences

Wu, Y.; Cheng, L.; Huang, S.; Fang, S.; Jia, P.; Wang, S. A semianalytical model for simulating fluid flow in naturally fractured reservoirs with non-homogeneous vugs and fractures. *SPE J.* **2019**, *24*, 334–348. [CrossRef]

Jones, F.O. A Laboratory Study of the Effects of Confining Pressure on Fracture Flow and Storage Capacity in Carbonate Rocks. *Soc. Pet. Eng.* **1975**, *27*, 21–27.

Duan, Y.; Meng, Y.; Luo, P.; Su, W. Stress Sensitivity of Naturally Fractured-porous Reservoir with Dual-porosity. In *SPE International Oil and Gas Conference and Exhibition in China*; OnePetro: Beijing, China, 1998.

Ayala, L.F.; Kouassi, J.P. The similarity theory applied to the analysis of two-phase flow in gas-condensate reservoirs. *Energy Fuels* **2007**, *21*, 1592–1600. [CrossRef]

Sheng, G.; Zhao, H.; Su, Y.; Javadpour, F.; Wang, C.; Zhou, Y.; Liu, J.; Wang, H. An analytical model to couple gas storage and transport capacity in organic matter with noncircular pores. *Fuel* **2020**, *268*, 117288. [CrossRef]

Sheng, G.; Su, Y.; Wang, W. A new fractal approach for describing induced-fracture porosity/permeability/ compressibility in stimulated unconventional reservoirs. *J. Pet. Sci. Eng.* **2019**, *179*, 855–866. [CrossRef]

Wang, S.; Cheng, L.; Huang, S.; Xue, Y.; Bai, M.; Wu, Y.; Wang, J. A semi-analytical method for modeling two-phase flow behavior in fractured carbonate oil reservoirs. *J. Energy Resour. Technol.* **2019**, *141*, 072902. [CrossRef]

O'Sullivan, M.J. A similarity method for geothermal well test analysis. *Water Resour. Res.* **1981**, *17*, 390–398. [CrossRef]

Pruess, K.; Narasimhan, T.N. Practical method for modeling fluid and heat flow in fractured porous media. *Soc. Pet. Eng. J.* **1985**, *25*, 14–26. [CrossRef]

Qu, M.; Hou, J.; Zhao, F.; Song, Z.; Ma, S.; Wang, Q.; Li, M.; Yang, M. 3-D visual experiments on fluid flow behavior of water flooding and gas flooding in fractured-vuggy carbonate reservoir. In Proceedings of the Paper SPE-187273-MS Presented at SPE Annual Technical Conference and Exhibition, San Antonio, TX, USA, 9–11 October 2017.

Jia, P.; Cheng, L.; Huang, S.; Xue, Y.; Clarkson, C.R.; Williams-Kovacs, J.D.; Wang, S.; Wang, D. Dynamic coupling of analytical linear flow solution and numerical fracture model for simulating early-time flowback of fractured tight oil wells (planar fracture and complex fracture network). *J. Pet. Sci. Eng.* **2019**, *177*, 1–23. [CrossRef]

Matthai, S.K.; Mezentsev, A.A.; Belayneh, M. Finite element-node-centered finite-volume two-phase-flow experiments with fractured rock represented by unstructured hybrid-element meshes. *SPE Reserv. Eval. Eng.* **2007**, *10*, 740–756. [CrossRef]

Wu, Y.S.; Li, J.; Ding, D.Y.; Wang, C.; Di, Y. A Generalized Framework Model for Simulation of Gas Production in Unconventional Gas Reservoirs. *SPE J.* **2014**, *19*, 845–857. [CrossRef]

Wu, Y.; Cheng, L.; Huang, S.; Jia, P. A green element method-based discrete fracture model for simulation of the transient flow in heterogeneous fractured porous media. *Adv. Water Resour.* **2020**, *136*, 103489. [CrossRef]

Xu, Y.; Sheng, G.; Zhao, H.; Hui, Y.; Zhou, Y.; Ma, J.; Rao, X.; Zhong, X.; Gong, J. A new approach for gas-water flow simulation in multi-fractured horizontal wells of shale gas reservoirs. *J. Pet. Sci. Eng.* **2021**, *199*, 108292. [CrossRef]

Wu, Y.; Cheng, L.; Ma, L.; Huang, S.; Fang, S.; Killough, J.E.; Jia, P.; Wang, S. Transient two-phase flow model for production prediction of tight gas wells with fracturing fluid-induced formation damage. *J. Pet. Sci. Eng.* **2021**, *199*, 108351. [CrossRef]

De Swaan, A.O. Analytical solution for determining naturally fractured reservoir properties by well testing. *SPE J.* **1976**, *16*, 117–122.

Sun, Z.; Li, X.; Liu, W.; Zhang, T.; He, M.; Nasrabadi, H. Molecular Dynamics of Methane Flow Behavior through Realistic Organic Nanopores under Geologic Shale Condition: Pore size and Kerogen Types. *Chem. Eng. J.* **2020**, *398*, 124341. [CrossRef]

19. Zhou, W.; Banerjee, R.; Poe, B.; Spath, J.; Thambynayagam, M. Semi-analytical production simulation of complex hydrauli fracture networks. *SPE J.* **2014**, *19*, 6–18. [CrossRef]
20. Jia, P.; Cheng, L.; Huang, S.; Liu, H. Transient behavior of complex fracture networks. *J. Pet. Sci. Eng.* **2015**, *132*, 1–17. [CrossRef
21. Stehfest, H. Numerical Inversion of Laplace Transforms. *ACM Commun.* **1970**, *13*, 47–49. [CrossRef]
22. Clarkson, C.R.; Qanbari, F. A semi-analytical forecasting method for unconventional gas and light oil wells: A hybrid approac for addressing the limitations of existing empirical and analytical methods. *SPE Reserv. Eval. Eng.* **2014**, *18*, 260–263.
23. Sun, Z.; Huang, B.; Liu, Y.; Jiang, Y.; Zhang, Z.; Hou, M.; Li, Y. Gas-phase production equation for CBM reservoirs: Interactio between hydraulic fracturing and coal orthotropic feature. *J. Pet. Sci. Eng.* **2022**, *213*, 110428. [CrossRef]
24. Sun, Z.; Wang, S.; Xiong, H.; Wu, K.; Shi, J. Optimal nanocone geometry for water flow. *AIChE J.* **2022**, *68*, e17543. [CrossRef]
25. Wattenbarger, R.A.; El-Banbi, A.H.; Villegas, M.E.; Maggard, J.B. Production analysis of linear flow into fractured tight gas well *Gas Field* **1998**.
26. Clarkson, C.R.; Qanbari, F. History-matching and forecasting tight gas condensate and oil wells using the dynamic drainage are concept. In Proceedings of the SPE/CSUR Unconventional Resources Conference, Calgary, AB, Canada, 20–22 October 2015 Society of Petroleum Engineers: Calgary, AB, Canada, 2015.
27. Wang, S.; Cheng, L.; Xue, Y.; Huang, S.; Wu, Y.; Jia, P.; Sun, Z. A semi-analytical method for simulating two-phase flov performance of horizontal volatile oil wells in fractured carbonate reservoirs. *Energies* **2018**, *11*, 2700. [CrossRef]

processes

Article

A Semi-Analytical Model for Production Prediction of Deep CBM Wells Considering Gas-Water Two-Phase Flow

Suran Wang *, Dongjun Li and Wenlan Li

CNOOC Research Institute Co., Ltd., Beijing 100028, China; lidj26@cnooc.com.cn (D.L.); liwl37@cnooc.com.cn (W.L.)
* Correspondence: wangsr@cnooc.com.cn; Tel.: +86-13551249096

Abstract: The productivity prediction of deep coalbed methane (CBM) wells is significantly influenced by gas-water two-phase flow characteristics and seepage parameters of the fracture network. While numerical simulations offer a comprehensive approach, analytical models are favored for their faster and broader applicability. However, conventional analytical models often oversimplify the complex problem of two-phase seepage equations, leading to substantial errors in dynamic analysis outcomes. Addressing this shortcoming, we establish a gas-water two-phase productivity prediction model for deep CBM reservoirs. This model takes into account the two-phase flow characteristics within the reservoir and fracture network, as well as the stress sensitivity of the reservoir and fractures. Additionally, a modified trilinear flow model characterizes the fractured modification body. By integrating the flowing material balance equation with the Newton Iteration method, we gradually update the seepage model's nonlinear parameters using the average formation pressure. We also linearize the gas-water two-phase model through successive iterations to derive a semi-analytical solution. The accuracy of the model was verified through comparison with commercial numerical simulation software results and field application. The model also enabled us to scrutinize the influence of reservoir and fracture network parameters on productivity. Our research findings suggest that the semi-analytical solution approach can efficiently address the nonlinear seepage problem of gas-water two-phase flow, enabling quick and accurate prediction of deep CBM well productivity. Moreover, appropriate fracture network parameters are paramount for enhancing the productivity of deep CBM wells. Lastly, during the development of deep CBM reservoirs, it is crucial to control the production pressure difference appropriately to minimize the stress sensitivity impact on production capacity.

Keywords: deep CBM reservoir; fractured horizontal wells; two-phase; semi-analytical model; production prediction

Citation: Wang, S.; Li, D.; Li, W. A Semi-Analytical Model for Production Prediction of Deep CBM Wells Considering Gas-Water Two-Phase Flow. *Processes* **2023**, *11*, 3022. https://doi.org/10.3390/pr11103022

Academic Editors: Zheng Sun, Udo Fritsching, Tao Zhang, Dong Feng, Wen Zhao and Hung Vo Thanh

Received: 17 August 2023
Revised: 7 October 2023
Accepted: 17 October 2023
Published: 20 October 2023

1. Introduction

Deep coalbed methane (CBM) reservoirs, in contrast to shallow counterparts, are typified by adverse physical properties and high in-situ stress. Consequently, conventional fracturing technologies have limited efficacy, necessitating the deployment of long horizontal wells and large-scale hydraulic fracturing technologies for their development [1–6]. CBM reservoirs, inherently comprising a dual pore media system (matrix and cleat), harbor extensive natural fractures that compound into complex networks post-fracturing [7–9]. The degree of fracturing is pivotal in defining the productivity of deep CBM wells, making the comprehension of fracturing modification imperative for the design and efficient development of deep CBM reservoirs [10–12]. Moreover, accurate prediction of deep CBM well productivity necessitates an understanding of the adsorption and desorption characteristics of deep CBM and the nonlinear flow mechanism in micro-nano pores [13–15]. CBM development invariably involves the two-phase flow stage of gas and water, the characteristics of which are notably intricate. The high in-situ stress and intense stress sensitivity of deep CBM reservoirs negatively impact gas-water seepage capacity, directly influencing the steady productivity of gas wells [16–20]. Hence, to predict deep CBM well capacity accurately, it is crucial to consider the impact of stress sensitivity and gas-water two-phase flow characteristics. Therefore, establishing a dynamic prediction method for gas-water two-phase production in deep CBM reservoirs is paramount for deep CBM development planning, fracturing evaluation, production dynamic analysis, and prediction.

Currently, deep CBM well productivity prediction methods include analytical, semi-analytical, and numerical simulation techniques. The analytical method is typically based on steady-state seepage theory, often simplifying the equation through the introduction of the gas-water two-phase pseudo-pressure [21–25]. However, this method struggles to characterize complex seam meshes, leading to considerable errors in productivity prediction. The semi-analytical method is based on the linear flow assumption, effective in depicting the fracturing mesh modification body and expedient for calculations [26–29]. However yet, it only applies to single-phase flow capacity prediction, and its utility diminishes fo the highly nonlinear gas-water two-phase flow that occurs during CBM development. The numerical simulation method can handle multiphase fluid flow problems and accurately characterize artificial fractures' parameter characteristics, but it is computationally inten sive [30–34]. To achieve high simulation precision, the fractures require grid encryption which can lead to an extensive mesh network, slowing the calculation speed, especially when dealing with thousands of case studies. Hence, it is of vital importance to develop dynamic prediction method for gas-water two-phase production in deep CBM reservoirs Such an approach would not only accurately represent the fracturing network's charac teristics in deep CBM reservoirs but also elucidate the seepage flow patterns of gas-wate two-phase in these networks. Moreover, it would enable fast and accurate predictions o the production dynamics of the gas-water two-phase.

This study strives to develop a semi-analytical model for predicting the productio capacity of deep CBM wells, taking into account the behavior of gas-water two-phas flow. We first construct a physical model for the dynamic analysis of gas-water two-phas production in deep CBM reservoirs, guided by the fundamental characteristics of th fracturing network formed through hydraulic fracturing technology and the gas-water tw phase seepage mechanism within this network. Next, we employ a modified three-linea flow model to represent the fracturing network, wherein complex-induced fractures, coa cleats, and the matrix are approximated as equivalent continuous media. When solving th proposed mathematical model, we explore an efficient method to handle the non-linea seepage flow issue brought about by the two-phase flow of gas and water. We verify th accuracy of our model by contrasting its results with those obtained from commerci numerical simulation software. Furthermore, based on the actual measurement data repo of the production wells in the study area, and the semi-analytical solution devised i this study, we analyze the influence of several critical parameters related to fractures an reservoirs on production prediction. There are many factors that affect the productivi of CBM wells, we will continue to study the influence of the sensitivity parameters o rate response in detail in the following study [35–37]. Finally, to demonstrate the practic utility of our proposed method, we conduct production capacity prediction and analys on a sample well.

2. Methodology

2.1. Physical Model Assumption

In this paper, we adopt a three-linear flow model to reflect the fracturing modificatio in deep CBM wells. As depicted in Figure 1, the proposed model entails three distin regions: the inner zone where fractures exhibit linear flow; the formation fluid's linear flo perpendicular to the fractures; and the outer zone where fluids flow linearly parallel to t fractures [38]. The model presumes that the artificial fracture intersects perpendicular with the reservoir, intersecting only with the perforation of the horizontal well, whi the remaining sections of the horizontal well remain closed. The artificial fractur is directly linked to the wellbore, and fluids enter the production wellbore solely v the fracture. Consequently, the fluid of the reservoir continuously flows towards the fracture, replenishing its energy. We also assume that gas and water's two-phase flow takes place in both the fracture network and cleat systems, with the flow behavior adhering to Darcy's law [28,29]. The adsorption of CBM in the coal matrix obeys the Langmuir equation [31], while CBM's diffusion within the coal matrix conforms to Fick's first law, with the pseudo-steady-state diffusion model taken into account [34]. The following simplifying assumptions are adopted for the derivation of the governing equation: (1) The top, bottom, and lateral boundaries of the reservoir are all closed, the distance of external boundary is denoted by y_e; (2) The fracturing fracture is symmetrical with the wellbore, where the fracture's half-length is denoted by y_F, and the fracture width is w_F; (3) The permeability of the reservoir is sensitive to stress; (4) The influences of gravity and capillary pressure are not considered.

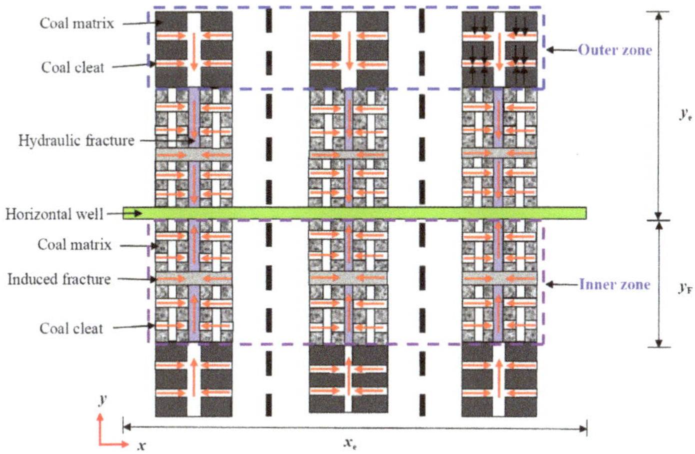

Figure 1. Schematic diagram of fractured horizontal well model in the deep CBM reservoir.

2.2. Mathematical Model

Informed by the presuppositions of the physical model, we devise a mathematical model within this section to elucidate the fluid flow in both the coal reservoir and associated fracture networks. The Langmuir equation is employed to capture the dynamics of CBM adsorption and desorption within the inner and outer zones. To describe CBM diffusion within these regions, we incorporate a pseudo-steady-state diffusion model. Darcy's seepage flow equation provides the framework for portraying the transient flow of gas and water within the cleats and fractures. Following the mechanistic model, mathematical models are correspondingly established for the fluid flow within each of the three distinct flow regions.

2.2.1. Fluid Flow in the Outer Zone

For fluid flow in the outer zone, using pseudo-pressure and pseudo-time, the governing equation for coal gas flow in the cleats is approximately written as:

$$\frac{\partial^2 \psi_{f2}}{\partial x^2} = \frac{1}{0.0864}\left[\frac{\phi_{f2} c_{tf2} \mu}{k_{f2} k_{frg}} \frac{\partial \psi_{f2}}{\partial t} - \frac{\alpha k_m}{k_{f2}}(\psi_m - \psi_{f2})\right] \tag{1}$$

The flow equation in the coal matrix is written as:

$$-\frac{\alpha k_m}{k_{f2}}(\psi_m - \psi_{f2}) = \frac{\phi_m \mu c_{tmi}}{k_{f2}} \frac{\partial \psi_m}{\partial t} + \frac{2RT}{k_{f2}} \frac{6D\pi^2}{R_m^2}[V_E(\psi_m) - V_m] \tag{2}$$

Coal gas adsorption follows the Langmuir isotherm principle:

$$V_{\mathrm{E}} = \frac{V_{\mathrm{L}}\psi(p)}{\psi_{\mathrm{L}}(p_{\mathrm{L}}) + \psi(p)} \tag{3}$$

Coal gas diffuses in the matrix system obeying Fick's diffusion law. The diffusion equation of shale gas is given by:

$$\frac{\partial V_{\mathrm{m}}}{\partial t} = -3.6D\alpha(V_{\mathrm{m}} - V_{\mathrm{E}}) \tag{4}$$

The initial adsorption amount is:

$$V_{\mathrm{m}}|_{t=0} = V_{\mathrm{Ei}}(\psi_{\mathrm{i}}) \tag{5}$$

where, α is the shape factor of the matrix, $1/\mathrm{m}^2$; D is the diffusion coefficient, m^2/h; V_{E} is the equilibrium adsorption concentration, $\mathrm{m}^3/\mathrm{m}^3$; V_{Ei} is the equilibrium adsorption concentration at initial condition, $\mathrm{m}^3/\mathrm{m}^3$; V_{L} is the Langmuir volume, $\mathrm{m}^3/\mathrm{m}^3$; p_{L} is the Langmuir volume, MPa; Ψ_{L} is the Langmuir pseudo-pressure, MPa/(mPa·s).

To simplify the form of the mathematical model, the dimensionless parameters and definitions are introduced in Table 1.

Table 1. Definition of dimensionless parameters.

Parameters	Symbol	Definition
Dimensionless pseudo pressure of the hydraulic fracture	ψ_{FD}	$\psi_{\mathrm{FD}} = \frac{\psi_{\mathrm{i}}-\psi_{\mathrm{F}}}{\psi_{\mathrm{i}}-\psi_{\mathrm{w}}}$
Dimensionless pseudo pressure of the cleat in the outer zone	ψ_{f2D}	$\psi_{\mathrm{f2D}} = \frac{\psi_{\mathrm{i}}-\psi_{\mathrm{f2}}}{\psi_{\mathrm{i}}-\psi_{\mathrm{w}}}$
Dimensionless pseudo pressure of the cleat in the inner zone	ψ_{f1D}	$\psi_{\mathrm{f1D}} = \frac{\psi_{\mathrm{i}}-\psi_{\mathrm{f1}}}{\psi_{\mathrm{i}}-\psi_{\mathrm{w}}}$
Dimensionless pseudo pressure of the coal matrix in the outer zone	ψ_{m2D}	$\psi_{\mathrm{f2D}} = \frac{\psi_{\mathrm{i}}-\psi_{\mathrm{m2}}}{\psi_{\mathrm{i}}-\psi_{\mathrm{w}}}$
Dimensionless pseudo pressure of the coal matrix in the inner zone	ψ_{m1D}	$\psi_{\mathrm{m1D}} = \frac{\psi_{\mathrm{i}}-\psi_{\mathrm{m1}}}{\psi_{\mathrm{i}}-\psi_{\mathrm{w}}}$
Dimensionless pseudo time	t_{aD}	$t_{\mathrm{aD}} = \beta_2 \frac{k_{\mathrm{m}}}{\phi_{\mathrm{m}}\mu_{\mathrm{g}}c_{\mathrm{tm}}{l_R}^2} t_{\mathrm{a}}$
Dimensionless pressure of the hydraulic fracture	p_{FD}	$p_{\mathrm{FD}} = \frac{p_{\mathrm{i}}-p_{\mathrm{F}}}{p_{\mathrm{i}}-p_{\mathrm{w}}}$
Dimensionless pressure of the cleat in the outer zone	p_{f2D}	$p_{\mathrm{f2D}} = \frac{p_{\mathrm{i}}-p_{\mathrm{f2}}}{p_{\mathrm{i}}-p_{\mathrm{w}}}$
Dimensionless pressure of the cleat in the inner zone	p_{f1D}	$p_{\mathrm{f1D}} = \frac{p_{\mathrm{i}}-p_{\mathrm{f1}}}{p_{\mathrm{i}}-p_{\mathrm{w}}}$
Dimensionless time	t_{D}	$t_{\mathrm{D}} = \beta_2 \frac{k_{\mathrm{m}}}{\phi_{\mathrm{m}}\mu_{\mathrm{g}}c_{\mathrm{tm}}{l_R}^2} t$
Dimensionless hydraulic fracture conductivity	C_{FD}	$C_{\mathrm{FD}} = \frac{k_{\mathrm{F}}w_{\mathrm{F}}}{k_{\mathrm{m}}l_{\mathrm{R}}}$
Dimensionless hydraulic fracture pressure transmitting coefficient	η_{FD}	$\eta_{\mathrm{FD}} = \frac{k_{\mathrm{F}}}{k_{\mathrm{m}}} \frac{\phi_{\mathrm{m}}c_{\mathrm{tm}}}{\phi_{\mathrm{F}}c_{\mathrm{tF}}}$
Dimensionless fracture aperture	w_{FD}	$w_{\mathrm{FD}} = \frac{w_{\mathrm{F}}}{l_{\mathrm{R}}}$
Dimensionless distance in the x direction	x_{D}	$x_{\mathrm{D}} = \frac{x}{l_{\mathrm{R}}}$
Dimensionless distance in the y direction	y_{D}	$y_{\mathrm{D}} = \frac{y}{l_{\mathrm{R}}}$
Dimensionless distance in the z direction	z_{D}	$z_{\mathrm{D}} = \frac{z}{l_{\mathrm{R}}}$
Dimensionless outer boundary in the x direction	x_{eD}	$x_{\mathrm{eD}} = \frac{x_e}{l_{\mathrm{R}}}$
Dimensionless outer boundary in the y direction	y_{eD}	$y_{\mathrm{eD}} = \frac{y_e}{l_{\mathrm{R}}}$
Dimensionless half-length of the hydraulic fracture	y_{FD}	$y_{\mathrm{FD}} = \frac{y_{\mathrm{F}}}{l_{\mathrm{R}}}$
Dimensionless adsorption concentration of the coal matrix	V_{mD}	$V_{\mathrm{mD}} = \frac{V_{\mathrm{Ei}}(\psi_{\mathrm{i}})-V_{\mathrm{m}}(\psi)}{V_{\mathrm{Ei}}(\psi_{\mathrm{i}})}$
Dimensionless equilibrium adsorption concentration	V_{ED}	$V_{\mathrm{ED}} = \frac{V_{\mathrm{Ei}}(\psi_{\mathrm{i}})-V_{\mathrm{E}}(\psi)}{V_{\mathrm{Ei}}(\psi_{\mathrm{i}})}$
Capacitance coefficient of the coal cleat in the outer zone	ω_2	$\omega_2 = \frac{\phi_{\mathrm{f2}}\mu_{\mathrm{gi}}c_{\mathrm{tfi}}}{\sigma}$
Diffusion flow coefficient	λ	$\lambda = \frac{2D\alpha h p_{\mathrm{sc}} T r_{\mathrm{w}}^2 V_{\mathrm{Ei}}}{\beta_2 q_{\mathrm{sc}} \mu_{\mathrm{g}} T_{\mathrm{sc}}}$
Capacitance coefficient of coal cleat in the inner zone	ω_1	$\omega_1 = \frac{\phi_{\mathrm{f1}}\mu_{\mathrm{gi}}c_{\mathrm{tfi}}}{\sigma}$
Inter-porosity flow factor of coal matrix system into coal cleat system	λ_{mf}	$\lambda_{\mathrm{mf}} = \alpha \frac{k_{\mathrm{m}}}{k_{\mathrm{f}}} l_{\mathrm{R}}^2$

Considering the PVT-dependent properties of the gas phase, the flow equation of the gas phase in the outer zone is treated by pseudo-pressure and pseudo-time. The dimensionless governing equation of the gas phase in the cleat system can be written by:

$$\frac{\partial^2 \psi_{\mathrm{f2D}}}{\partial x_{\mathrm{D}}^2} = \frac{1}{\eta_{\mathrm{fD}} k_{\mathrm{frg}}} \left[\omega_2 \frac{\partial \psi_{\mathrm{f2D}}}{\partial t_{\mathrm{aD}}} + (1-\omega_2) \frac{\partial V_{\mathrm{m2D}}}{\partial t_{\mathrm{aD}}} \right] \tag{6}$$

The pseudo-steady-state diffusion equation of CBM in the outer zone is expressed by:

$$(1-\omega_2) \frac{\partial V_{\mathrm{m2D}}}{\partial t_{\mathrm{aD}}} = -\lambda_2 (V_{\mathrm{m2D}} - V_{\mathrm{ED}}) \tag{7}$$

With the Langmuir isothermal theory, the equation for describing the adsorption of CBM in the outer zone is given below:

$$V_{\mathrm{ED}} = \beta \psi_{\mathrm{f2D}} \tag{8}$$

The boundary conditions, including the initial, the inner and the external boundaries are given by:

$$\begin{cases} V_{\mathrm{m2D}}\big|_{t_{\mathrm{aD}}=0} = 0 \\ \psi_{\mathrm{f2D}}(x_{\mathrm{D}}, t_{\mathrm{aD}})\big|_{t_{\mathrm{aD}}=0} = 0 \\ \psi_{\mathrm{f2D}}(x_{\mathrm{D}}, t_{\mathrm{aD}})\big|_{x_{\mathrm{D}}=1} = \psi_{\mathrm{f1D}}(x_{\mathrm{D}}, t_{\mathrm{aD}})\big|_{x_{\mathrm{D}}=1} \\ \frac{\partial \psi_{\mathrm{f2D}}(x_{\mathrm{D}}, t_{\mathrm{aD}})}{\partial x_{\mathrm{D}}}\big|_{x_{\mathrm{D}}=x_{\mathrm{eD}}} = 0 \end{cases} \tag{9}$$

Because the physical properties of formation water do slightly vary with temperature and pressure, the flow equation of the water phase in the outer zone can be treated by real pressure and time, which is given by:

$$\frac{\partial^2 p_{\mathrm{f2}}}{\partial x^2} = \frac{1}{0.0864} \frac{\phi_{\mathrm{f2}} c_{\mathrm{tf2}} \mu}{k_{\mathrm{f2}} k_{\mathrm{frw}}} \frac{\partial p_{\mathrm{f2}}}{\partial t} \tag{10}$$

Thus, the dimensionless governing equation of the water phase in the cleat system can be written as:

$$\frac{\partial^2 p_{\mathrm{fD}}}{\partial x_{\mathrm{D}}^2} = \frac{1}{\eta_{\mathrm{fwD}} k_{\mathrm{frw}}} \frac{\partial p_{\mathrm{fD}}}{\partial t_{\mathrm{D}}} \tag{11}$$

The boundary conditions are written by:

$$\begin{cases} p_{\mathrm{f2D}}(x_{\mathrm{D}}, t_{\mathrm{D}})\big|_{t_{\mathrm{D}}=0} = 0 \\ p_{\mathrm{f2D}}(x_{\mathrm{D}}, t_{\mathrm{D}})\big|_{x_{\mathrm{D}}=1} = p_{\mathrm{f1D}}(x_{\mathrm{D}}, t_{\mathrm{D}})\big|_{x_{\mathrm{D}}=1} \\ \frac{\partial p_{\mathrm{f2D}}(x_{\mathrm{D}}, t_{\mathrm{D}})}{\partial x_{\mathrm{D}}}\big|_{x_{\mathrm{D}}=x_{\mathrm{eD}}} = 0 \end{cases} \tag{12}$$

where ψ_{f1D} is the dimensionless pseudo-pressure of the cleat system in the inner zone; ψ_{f2D} is the dimensionless pseudo-pressure of the cleat system in the outer zone; p_{f1D} is the pressure of the cleat system in the inner zone; p_{f2D} is the pressure of the cleat system in the outer zone; V_{m2D} is the dimensionless adsorption concentration of the coal matrix in the outer zone; V_{ED} is the dimensionless equilibrium adsorption concentration of the coal matrix in the outer zone; k_{frg} is the relative permeability of the gas phase in the cleat system; k_{frw} is the relative permeability of water phase in the cleat system; x_{D} is the dimensionless distance in the x direction; x_{eD} is the dimensionless outer boundary distance in the x direction; η_{f2D} is the dimensionless conductivity coefficient of the gas phase in the outer zone cleat system; η_{f2wD} is the dimensionless conductivity coefficient of the water phase in the outer zone cleat system; t_{aD} is the dimensionless pseudo-time; t_{D} is the dimensionless time; ω_2 is the storage capacity coefficient of the cleat system in the outer zone; λ_2 is the interporosity flow coefficient of the cleat system in the outer zone; β is the adsorption coefficient of CBM.

2.2.2. Fluid Flow in the Inner Zone

The complex fracture network in the inner zone is composed of micro-fractur secondary fractures and hydraulic fractures. In this section, a dual-porosity model modified to capture the transient flow of the micro-fractures and secondary fractur Using pseudo-pressure and pseudo-time, the dimensionless governing equation of the g phase in the fracture systems of the inner zone is given by:

$$\frac{\partial}{\partial y}\left(\frac{p_{f1}}{Z}\cdot\frac{k_{f1}k_{frg}}{\mu}\frac{\partial p_{f1}}{\partial y}\right)=\frac{1}{0.0864}\left[\frac{\partial\left(\frac{p_f}{Z}S_{f1g}\phi_{f1}\right)}{\partial t}-\frac{\alpha k_m}{k_{f1}}(\psi_m-\psi_{f1})\right] \quad (1$$

The dimensionless governing equation of the gas phase in the inner zone is written

$$\frac{\partial^2\psi_{f1D}}{\partial y_D^2}+\frac{1}{k_{frg}}\frac{\partial\psi_{f2D}}{\partial x_D}\bigg|_{x_D=1}=\frac{1}{\eta_{fD}k_{frg}}\left[\omega_1\frac{\partial\psi_{f1D}}{\partial t_{aD}}+(1-\omega_1)\frac{\partial V_{m1D}}{\partial t_{aD}}\right] \quad (1$$

The pseudo-steady-state diffusion equation of CBM in the inner zone is given by:

$$(1-\omega_1)\frac{\partial V_{m1D}}{\partial t_{aD}}=-\lambda_1(V_{m1D}-V_{ED}) \quad (1$$

According to the Langmuir isothermal adsorption law, the equation for describing t adsorption of CBM in the inner zone is:

$$V_{ED}=\beta\psi_{f1D} \quad (1$$

The boundary conditions of the inner zone are given by:

$$\begin{cases} V_{m1D}\big|_{t_{aD}=0}=0 \\ \psi_{f1D}(y_D,t_{aD})\big|_{t_{aD}=0}=0 \\ \psi_{f1D}(y_D,t_{aD})\big|_{y_D=w_{FD}/2}=\psi_{FD} \\ \frac{\partial\psi_{f1D}(y_D,t_{aD})}{\partial y_D}\Big|_{y_D=y_{eD}}=0 \end{cases} \quad (1$$

The flow equation of the water phase in the inner zone is treated with real pressu and real-time, and the dimensionless governing equation is:

$$\frac{\partial^2 p_{f1D}}{\partial y_D^2}+\frac{1}{k_{frw}}\frac{\partial p_{f2D}}{\partial x_D}\bigg|_{x_D=1}=\frac{1}{\eta_{fwD}k_{frw}}\frac{\partial p_{f1D}}{\partial t_D} \quad (1$$

The boundary conditions are given by:

$$\begin{cases} p_{f1D}(y_D,t_D)\big|_{t_D=0}=0 \\ p_{f1D}(y_D,t_D)\big|_{y_D=w_{FD}/2}=p_{FD} \\ \frac{\partial p_{f1D}(y_D,t_D)}{\partial y_D}\Big|_{y_D=y_{eD}}=0 \end{cases} \quad (1$$

2.2.3. Fluid Flow in the Hydraulic Fractures

In this paper, a two-phase flow of gas and water is considered in hydraulic fractur Using the pseudo-pressure and pseudo-time, the gas flow equation in the hydraulic fractu is written by:

$$\frac{\partial^2\psi_F}{\partial x^2}=\frac{1}{0.0864}\frac{\phi_F c_{tF}\mu}{k_F k_{Frg}}\frac{\partial\psi_F}{\partial t}-\frac{2k_{f1}k_{frg}}{w_F k_F k_{Frg}}\frac{\partial\psi_{f1}}{\partial y}\bigg|_{y_D=w_F/2} \quad (2$$

The dimensionless governing equation is written as:

$$\frac{\partial^2 \psi_{FD}}{\partial x_D^2} + \frac{2k_{frg}}{C_{FD}k_{Frg}} \frac{\partial \psi_{f1D}}{\partial y_D}\Big|_{y_D = w_{FD}/2} = \frac{1}{\eta_{FD}k_{Frg}} \frac{\partial \psi_{FD}}{\partial t_{aD}} \tag{21}$$

The boundary conditions are expressed as:

$$\begin{cases} \psi_{FD}(x_D, t_{aD})\big|_{t_{aD}=0} = 0 \\ \psi_{FD}(x_D, t_{aD})\big|_{x_D=0} = 1 \\ \frac{\partial \psi_{FD}}{\partial x_D}\big|_{x_D=1} = 0 \end{cases} \tag{22}$$

The water phase flow equation is given by:

$$\frac{\partial^2 p_F}{\partial x^2} = \frac{1}{0.0864} \frac{\phi_F c_{tF} \mu}{k_F k_{Frw}} \frac{\partial p_F}{\partial t} - \frac{2k_{f1}k_{frw}}{w_F k_F k_{Frw}} \frac{\partial p_{f1}}{\partial y}\bigg|_{y_D = w_F/2} \tag{23}$$

The dimensionless governing equation can be written by:

$$\frac{\partial^2 p_{FD}}{\partial x_D^2} + \frac{2k_{frw}}{C_{FD}k_{Frw}} \frac{\partial p_{f1D}}{\partial y_D}\Big|_{y_D = w_{FD}/2} = \frac{1}{\eta_{FwD}k_{Frw}} \frac{\partial p_{FD}}{\partial t_D} \tag{24}$$

The boundary conditions are:

$$\begin{cases} p_{FD}(x_D, t_D)\big|_{t_D=0} = 0 \\ p_{FD}(x_D, t_D)\big|_{x_D=0} = 1 \\ \frac{\partial p_{FD}}{\partial x_D}\big|_{x_D=1} = 0 \end{cases} \tag{25}$$

where C_{FD} is the dimensionless conductivity of the hydraulic fracture; η_{FD} is the dimensionless conductivity coefficient of the gas phase in the hydraulic fracture system; η_{FwD} is the dimensionless conductivity coefficient of the water phase in the hydraulic fracture system; w_{FD} is the dimensionless width of the hydraulic fracture; k_{Frg} is the relative permeability of the gas phase in the hydraulic fracture system; k_{Frw} is the relative permeability of the water phase in the hydraulic fracture system.

2.3. Solution to Mathematical Model

Because the parameters including μ_g B_g, k_{Frg}, k_{frg}, k_{mrg}, k_{Frw}, k_{frw} and k_{mrw} are functions of the pressure and saturation, it is difficult to derive the analytical solution of the proposed mathematical model directly. In this paper, an effective and accurate semi-analytical method is presented to solve the problem. In our approach, the production time is discretized into several time steps, and at each time step, the pressure-dependent parameters (μ_g, B_g) and the saturation-dependent parameters (k_{mrg}, k_{mrw}, k_{frg}, k_{frw}, k_{Frg}, k_{Frw}) are updated by the average pressure and average saturation, respectively. The average pressure and saturation are calculated by using the material balance method. Therefore, the nonlinear parameters can be regarded as fixed approximations at each time step, and the nonlinear problems can be converted to linear problems through this method. Thus, the solution of the proposed model can be obtained with the Laplace transformation method directly. In the following sections, the analytical solutions for gas and water production are presented, respectively.

2.3.1. Solution for Gas and Water Flow in the Outer Zone

Using the Laplace transformation, the solution of the outer zone flow model can be obtained and the general solution for Equation (6) is given by:

$$\overline{\psi}_{f2D} = A\sinh\left(\sqrt{\frac{f_2(u)}{\eta_{fD}k_{frg}}} \cdot x_D\right) + B\cosh\left(\sqrt{\frac{f_2(u)}{\eta_{fD}k_{frg}}} \cdot x_D\right) \tag{26}$$

$$f_2(u) = u\omega_2 + u(1-\omega_2)\frac{\lambda_2\beta}{u(1-\omega_2)+\lambda_2} \tag{27}$$

Substituting Equation (9) into Equation (26), one can obtain

$$A = \frac{\sinh\left(\sqrt{\frac{f_2(u)}{\eta_{\mathrm{fD}}k_{\mathrm{frg}}}}\cdot x_{\mathrm{eD}}\right)\cdot\overline{\psi}_{\mathrm{f1D}}}{\sinh\left(\sqrt{\frac{f_2(u)}{\eta_{\mathrm{fD}}k_{\mathrm{frg}}}}\right)\sinh\left(\sqrt{\frac{f_2(u)}{\eta_{\mathrm{fD}}k_{\mathrm{frg}}}}\cdot x_{\mathrm{eD}}\right)-\cosh\left(\sqrt{\frac{f_2(u)}{\eta_{\mathrm{fD}}k_{\mathrm{frg}}}}\right)\cosh\left(\sqrt{\frac{f_2(u)}{\eta_{\mathrm{fD}}k_{\mathrm{frg}}}}\cdot x_{\mathrm{eD}}\right)} \tag{28}$$

$$B = \frac{-\cosh\left(\sqrt{\frac{f_2(u)}{\eta_{\mathrm{fD}}k_{\mathrm{frg}}}}\cdot x_{\mathrm{eD}}\right)\cdot\overline{\psi}_{\mathrm{f1D}}}{\sinh\left(\sqrt{\frac{f_2(u)}{\eta_{\mathrm{fD}}k_{\mathrm{frg}}}}\right)\sinh\left(\sqrt{\frac{f_2(u)}{\eta_{\mathrm{fD}}k_{\mathrm{frg}}}}\cdot x_{\mathrm{eD}}\right)-\cosh\left(\sqrt{\frac{f_2(u)}{\eta_{\mathrm{fD}}k_{\mathrm{frg}}}}\right)\cosh\left(\sqrt{\frac{f_2(u)}{\eta_{\mathrm{fD}}k_{\mathrm{frg}}}}\cdot x_{\mathrm{eD}}\right)} \tag{29}$$

The general solution for Equation (11) in the Laplace domain is given as:

$$\overline{p}_{\mathrm{f2D}} = A_1\sinh\left(\sqrt{\frac{u}{\eta_{\mathrm{fwD}}k_{\mathrm{frw}}}}\cdot x_{\mathrm{D}}\right)+B_1\cosh\left(\sqrt{\frac{u}{\eta_{\mathrm{fwD}}k_{\mathrm{frw}}}}\cdot x_{\mathrm{D}}\right) \tag{30}$$

Substituting Equation (12) into Equation (30), one can obtain

$$A_1 = \frac{\sinh\left(\sqrt{\frac{u}{\eta_{\mathrm{fwD}}k_{\mathrm{frw}}}}\cdot x_{\mathrm{eD}}\right)\cdot\overline{p}_{\mathrm{f1D}}}{\sinh\left(\sqrt{\frac{u}{\eta_{\mathrm{fwD}}k_{\mathrm{frw}}}}\right)\sinh\left(\sqrt{\frac{u}{\eta_{\mathrm{fwD}}k_{\mathrm{frw}}}}\cdot x_{\mathrm{eD}}\right)-\cosh\left(\sqrt{\frac{u}{\eta_{\mathrm{fwD}}k_{\mathrm{frw}}}}\right)\cosh\left(\sqrt{\frac{u}{\eta_{\mathrm{fwD}}k_{\mathrm{frw}}}}\cdot x_{\mathrm{eD}}\right)} \tag{31}$$

$$B_1 = \frac{-\cosh\left(\sqrt{\frac{u}{\eta_{\mathrm{fwD}}k_{\mathrm{frw}}}}x_{\mathrm{eD}}\right)\cdot\overline{p}_{\mathrm{f1D}}}{\sinh\left(\sqrt{\frac{u}{\eta_{\mathrm{fwD}}k_{\mathrm{frw}}}}\right)\sinh\left(\sqrt{\frac{u}{\eta_{\mathrm{fwD}}k_{\mathrm{frw}}}}\cdot x_{\mathrm{eD}}\right)-\cosh\left(\sqrt{\frac{u}{\eta_{\mathrm{fwD}}k_{\mathrm{frw}}}}\right)\cosh\left(\sqrt{\frac{u}{\eta_{\mathrm{fwD}}k_{\mathrm{frw}}}}\cdot x_{\mathrm{eD}}\right)} \tag{32}$$

where, $\overline{\psi}_{\mathrm{f1D}}$ is the dimensionless pseudo-pressure of the inner zone cleat system in the Laplace domain; $\overline{\psi}_{\mathrm{f2D}}$ is the dimensionless pseudo-pressure of the outer zone cleat system in the Laplace domain; $\overline{p}_{\mathrm{f1D}}$ is the dimensionless pressure of the inner zone cleat system in the Laplace domain; $\overline{p}_{\mathrm{f2D}}$ is the dimensionless pressure of the outer zone cleat system in the Laplace domain; u is the Laplace constant.

2.3.2. Solution for Gas and Water Flow in the Inner Zone

The solving method for gas and water flow in the inner zone is similar to that of the outer zone flow model. We perform the Laplace transformation on the equation of gas and water phase in the inner zone. Combined with Equation (26), the solution of the inner zone flow model can be obtained and the general solution for Equation (14) is expressed by:

$$\overline{\psi}_{\mathrm{f1D}} = C\sinh(\sqrt{\sigma}\cdot y_{\mathrm{D}}) + D\cosh(\sqrt{\sigma}\cdot y_{\mathrm{D}}) \tag{33}$$

$$\sigma = \left[\sqrt{\frac{f_1(u)}{\eta_{\mathrm{fD}}k_{\mathrm{frg}}}}+\frac{1}{k_{\mathrm{frg}}}\sqrt{\frac{f_2(u)}{\eta_{\mathrm{fD}}k_{\mathrm{frg}}}}\tanh\sqrt{\frac{f_2(u)}{\eta_{\mathrm{fD}}k_{\mathrm{frg}}}}(x_{\mathrm{eD}}-1)\right] \tag{34}$$

Substituting the boundary conditions Equation (17) into Equation (33), one can obtain

$$C = \frac{\sinh(\sqrt{\sigma}\cdot y_{\mathrm{eD}})\cdot\overline{\psi}_{\mathrm{FD}}}{\sinh(\sqrt{\sigma})\sinh(\sqrt{\sigma}\cdot y_{\mathrm{eD}})-\cosh(\sqrt{\sigma})\cosh(\sqrt{\sigma}\cdot y_{\mathrm{eD}})} \tag{35}$$

$$D = \frac{-\cosh(\sqrt{\sigma} \cdot y_{eD}) \cdot \overline{\psi}_{FD}}{\sinh(\sqrt{\sigma})\sinh(\sqrt{\sigma} \cdot y_{eD}) - \cosh(\sqrt{\sigma})\cosh(\sqrt{\sigma} \cdot y_{eD})} \tag{36}$$

Combined with the solution for water flow in the outer zone, the general solution for Equation (18) in the Laplace domain is given by:

$$\overline{p}_{f1D} = C_1 \sinh\left(\sqrt{\delta} \cdot y_D\right) + D_1 \cosh\left(\sqrt{\delta} \cdot y_D\right) \tag{37}$$

$$\delta = \left[\sqrt{\frac{u}{\eta_{fwD} k_{frw}}} + \frac{1}{k_{frw}} \sqrt{\frac{u}{\eta_{fwD} k_{frw}}} \tanh \sqrt{\frac{u}{\eta_{fwD} k_{frw}}} (x_{eD} - 1)\right] \tag{38}$$

Substituting Equation (30) and the boundary conditions Equation (19) into Equation (37), one can obtain

$$C_1 = \frac{\sinh\left(\sqrt{\delta} \cdot y_{eD}\right) \cdot \overline{p}_{FD}}{\sinh\left(\sqrt{\delta}\right)\sinh\left(\sqrt{\delta} \cdot y_{eD}\right) - \cosh\left(\sqrt{\delta}\right)\cosh\left(\sqrt{\delta} \cdot y_{eD}\right)} \tag{39}$$

$$D_1 = \frac{-\cosh\left(\sqrt{\delta} \cdot y_{eD}\right) \cdot \overline{p}_{FD}}{\sinh\left(\sqrt{\delta}\right)\sinh\left(\sqrt{\delta} \cdot y_{eD}\right) - \cosh\left(\sqrt{\delta}\right)\cosh\left(\sqrt{\delta} \cdot y_{eD}\right)} \tag{40}$$

where, $\overline{\psi}_{FD}$ is the dimensionless pseudo-pressure of the hydraulic fracture in the Laplace domain; $\overline{p}_{FD}$ is the dimensionless pressure of the hydraulic fracture in the Laplace domain.

2.3.3. Solution for Gas and Water Flow in the Hydraulic Fracture

With Laplace transformation, the solution of the hydraulic fracture flow model can be obtained and the general solution for Equation (21) is expressed as:

$$\overline{\psi}_{FD} = -\frac{1}{u}\left(\tanh\sqrt{\alpha}\sinh\sqrt{\alpha}x_D - \cosh\sqrt{\alpha}x_D\right) \tag{41}$$

$$\alpha = \frac{u}{\eta_{FD} k_{Frg}} + \frac{2k_{frg}}{C_{FD} k_{Frg}} \sqrt{\sigma} \tanh\sqrt{\sigma}\left(y_{eD} - \frac{w_{FD}}{2}\right) \tag{42}$$

Combined with the solution for water flow in the inner zone, the general solution for Equation (24) in the Laplace domain is given by:

$$\overline{p}_{FD} = -\frac{1}{u}\left(\tanh\sqrt{\xi}\sinh\sqrt{\xi}x_D - \cosh\sqrt{\xi}x_D\right) \tag{43}$$

$$\xi = \frac{u}{\eta_{FwD} k_{Frw}} + \frac{2k_{frw}}{C_{FD} k_{Frw}} \sqrt{\delta} \tanh\sqrt{\delta}\left(y_{eD} - \frac{w_{FD}}{2}\right) \tag{44}$$

The gas production rate is:

$$\overline{q}_{gD} = -\frac{k_{FD} k_{Frg} w_{FD}}{\pi} \frac{\partial \overline{\psi}_{FD}}{\partial x_D}\Big|_{x_D=0} \tag{45}$$

According to Equation (45), the solution of the gas production rate becomes:

$$\overline{q}_{gD} = -\frac{k_{FD} k_{Frg} w_{FD}}{\pi} \frac{1}{u} \sqrt{\alpha} \tanh\sqrt{\alpha} \tag{46}$$

The water production rate is:

$$\overline{q}_{wD} = -\frac{k_{FD} k_{Frw} w_{FD}}{\pi} \frac{\partial \overline{p}_{FD}}{\partial x_D}\Big|_{x_D=0} \tag{47}$$

According to Equation (47), the solution of the water production rate becomes:

$$\bar{q}_{\mathrm{wD}} = -\frac{k_{\mathrm{FD}}k_{\mathrm{Frw}}w_{\mathrm{FD}}}{\pi}\frac{1}{u}\sqrt{\xi}\tanh\sqrt{\xi} \tag{48}$$

In this paper, the stress-sensitive terms are all integrated into the conductivity, and as a function of the average formation pressure. The theoretical formula for describing the effect of the stress-sensitive is defined as [38]:

$$k_{\mathrm{f}} = k_{\mathrm{fi}}e^{-\gamma(p_{\mathrm{i}}-\hat{p})} \tag{49}$$

where, k_{f} is the permeability of the cleat system, mD; k_{fi} is the initial permeability of the cleat system, mD; γ is the permeability modulus, MPa^{-1}; p_{i} is the original formation pressure, MPa; $\hat{p}$ is the average formation pressure, MPa.

2.3.4. Flowing Material Balance Method

As mentioned in Section 2.3 of this paper, at each time step, the pressure-related parameters (μ_{g}, B_{g}) and saturation-related parameters (k_{mrg}, k_{mrw}, k_{frg}, k_{frw}, k_{Frg}, k_{Frw}) in the gas-water two-phase production solution and the stress-sensitive terms are updated by the average pressure and average saturation within the investigated range, while the average formation pressure and average saturation are calculated using the flowing material balance method. In this section, the process of establishing the flowing material balance equations of gas-water two-phase is presented.

For a single fracture stage, the mass equilibrium equation for the gas phase can be written as:

$$x_{\mathrm{inv}}Hy_{\mathrm{inv}}\phi_{\mathrm{m}}\left(\frac{S_{\mathrm{gi}}}{B_{\mathrm{gi}}}\right) - x_{\mathrm{inv}}Hy_{\mathrm{inv}}\phi_{\mathrm{m}}\left(\frac{\hat{S}_{\mathrm{g}}}{\hat{B}_{\mathrm{g}}}\right) = \int_0^t q_{\mathrm{g}}\mathrm{d}t \tag{50}$$

The mass equilibrium equation for the water phase can be expressed as:

$$x_{\mathrm{inv}}Hy_{\mathrm{inv}}\phi_{\mathrm{m}}\left(\frac{S_{\mathrm{wi}}}{B_{\mathrm{wi}}}\right) - x_{\mathrm{inv}}Hy_{\mathrm{inv}}\phi_{\mathrm{m}}\left(\frac{\hat{S}_{\mathrm{w}}}{\hat{B}_{\mathrm{w}}}\right) = \int_0^t q_{\mathrm{w}}\mathrm{d}t \tag{51}$$

The investigated range for the transient flow in the inner zone is given by [38]:

$$x_{\mathrm{inv}} = 0.5836\sqrt{\frac{k_{\mathrm{f}}t}{\phi_{\mathrm{f}}\mu_{\mathrm{g}}c_{\mathrm{t}}}} + y_{\mathrm{F}} \tag{52}$$

$$y_{\mathrm{inv}} = 0.5836\sqrt{\frac{k_{\mathrm{f}}t}{\phi_{\mathrm{f}}\mu_{\mathrm{g}}c_{\mathrm{t}}}} \tag{53}$$

The saturation relationship between the gas and water phase is:

$$\hat{S}_{\mathrm{g}} + \hat{S}_{\mathrm{w}} = 1 \tag{54}$$

Combining Equations (50)–(54), the average pressure function can be derived as:

$$f(\hat{p}) = \frac{1}{\hat{B}_{\mathrm{g}}} + \left(-\frac{\hat{B}_{\mathrm{w}}}{\hat{B}_{\mathrm{g}}}\right)A_3 - B_3 \tag{55}$$

$$A_3 = \frac{S_{\mathrm{wi}}}{B_{\mathrm{wi}}} - \frac{\int_0^t q_{\mathrm{w}}\mathrm{d}t}{x_{\mathrm{inv}}Hy_{\mathrm{inv}}\phi_{\mathrm{m}}} \tag{56}$$

$$B_3 = \frac{S_{\text{gi}}}{B_{\text{gi}}} - \frac{\int_0^t q_{\text{g}} \text{d}t}{x_{\text{inv}} H y_{\text{inv}} \phi_{\text{m}}} \tag{57}$$

The Newton iteration is used to update the dynamic parameters by coupling the flowing material balance equations, which is given by:

$$\hat{p}_{\text{k}+1} = \hat{p}_{\text{k}} - \chi \frac{f(\hat{p}_{\text{k}})}{f'(\hat{p}_{\text{k}})} \tag{58}$$

where S_{gi} is the initial gas saturation; $\hat{S}_{\text{g}}$ is the average gas saturation; S_{wi} is the initial water saturation; $\hat{S}_{\text{w}}$ is the average water saturation; B_{gi} is the initial gas volume factor; $\hat{B}_{\text{g}}$ is the average gas volume factor; B_{wi} is the initial formation water volume factor; $\hat{B}_{\text{w}}$ is the average formation water volume factor; x_{inv} is the investigated range along the fracture direction in the inner zone; y_{inv} is the investigated range perpendicular to the fracture direction in the inner zone; ϕ_{m} is the matrix porosity; H is the formation thickness; y_{F} is the fracture half-length; t is the production time; q_{g} is the gas production rate; q_{w} is the water production rate; c_{t} is the total formation compressibility factor.

3. Results and Discussion

3.1. Model Validation

To verify the accuracy of the proposed semi-analytical model, a numerical model was established using the commercial simulation software Eclipse (2019), as shown in Figure 2. Table 2 presents the reservoir and fracture parameters used in both methods, and Figure 3 shows the relative permeability curves of the gas-water two-phase flow in the CBM reservoir and the hydraulic fractures. The selection of parameters in Table 2 is obtained through the actual measurement data report of the production well in the study area. Parameters such as Langmuir volume, Langmuir pressure, and coal gas content, etc. can be obtained through gas content reports; parameters such as reservoir original pressure, pressure gradient, temperature, permeability, etc. can be obtained from injection/fall-off well testing report; parameters such as fracture length and fracture direction can be obtained from micro-fracture monitoring report. In this case study, both the coal cleats and the artificial fractures are considered to have gas-water two-phase flow from the early production stage.

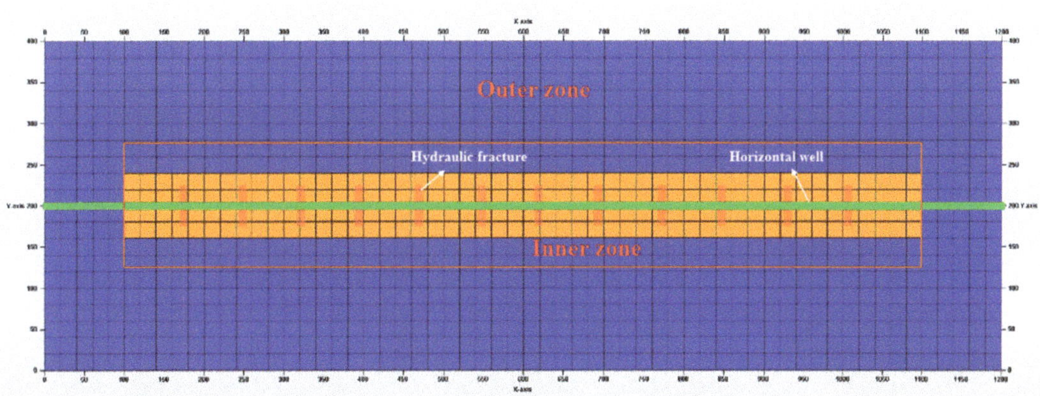

Figure 2. Schematic of the numerical model for a hydraulically fractured horizontal well in the deep CBM reservoir.

Table 2. Input parameters for model validation.

Parameters	Symbol	Units	Value
Initial formation pressure	p_i	MPa	20
Initial formation temperature	T_i	K	323
Bottom-hole pressure	p_{wf}	MPa	4
Formation thickness	h	m	9.1
Permeability of hydraulic fractures	k_F	mD	500
Number of hydraulic fractures	N_F	Dimensionless	12
Half length of hydraulic fractures	y_F	m	100
Width of hydraulic fractures	w_F	m	0.001
Porosity of hydraulic fractures	Φ_F	Fraction	0.4
Length of the horizontal well	L	m	1000
Half-width of the reservoir	y_e	m	200
Porosity of the coal cleat	Φ_f	Fraction	0.05
Permeability of the coal cleat in the inner reservoir	k_f	mD	0.22
Permeability of the outer reservoir	k_o	mD	0.12
Permeability modulus of the coal cleat	γ_f	MPa^{-1}	0.02
Langmuir pressure	p_L	MPa	3.1
Langmuir volume	V_L	m^3/t	12.9
Gas content in coal seams	V	m^3/t	14.8

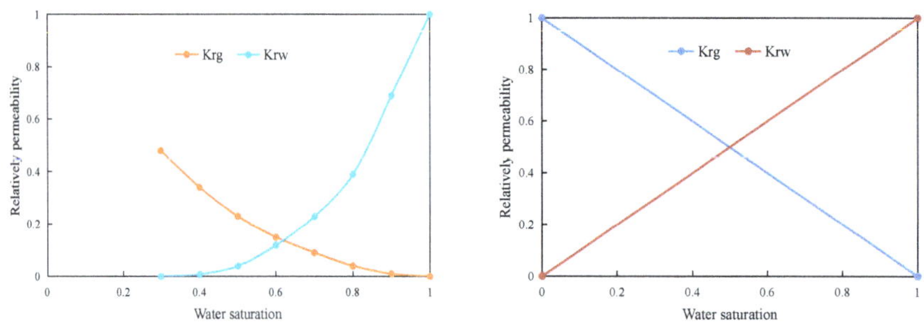

(**a**) (**b**)

Figure 3. Gas-water relative permeability curves used for model validation. (**a**) Relative permeability curve in the reservoir. (**b**) Relative permeability curve in the hydraulic fracture.

The comparison results between the semi-analytical model proposed in this paper and Eclipse are shown in Figure 4. It can be seen that there are some differences in the production curves obtained by the two methods in the early production stage, but the results are basically consistent in the later stage. This is mainly because, in the early production stage, the pressure and saturation in the near wellbore zone change greatly and production is very sensitive to parameters related to pressure and saturation. In the semi-analytical model, some parameters related to pressure are implicitly handled using a pseudo-pressure approach, but parameters related to saturation are explicitly handled. Therefore, the early errors in the single-phase flow model are not obvious, while the errors in the two-phase flow model are more significant. However, the calculated average relative error is less than 10%, which is within the allowable engineering error range, indicating that the semi-analytical model proposed in this paper can be used for production data analysis and prediction. We should admit that numerical simulation is a powerful tool for evaluating the performance of CBM Wells. However, the process of establishing such fracture networks using commercial numerical simulation software is relatively complex. In addition, the use of localized refinement or non-uniform grids can lead to an increase in the number of grids and significant differences in size between grids, thereby substantially reducing calculation efficiency. Conventional analytical models are often used to model the flow behavior of fractured wells. However, the traditional analytical models are limited to single-phase flow problems, and seldom consider the combined effect of hydraulic fractures, induced fractures, coal cleats, and the matrix, leading to considerable errors in productivity prediction. The proposed semi-analytical method in this paper is based on the linear flow assumption, effective in depicting the fracturing mesh modification body and expedient for calculations. In addition, when solving the proposed mathematical model, we explore an efficient method to handle the non-linear seepage flow issue brought about by the two-phase flow of gas and water. Employing the flowing material balance method for computing the average pressure and average saturation of the reservoir, and iteratively updating the non-linear parameters in the seepage model, facilitates the accurate handling of the non-linear seepage problem associated with gas-water dual-phase flow. The proposed semi-analytical model captures the flow behavior in the hydraulic fracture network and coal reservoir, and has much higher computational efficiency than numerical simulation methods. This is because the fracture and matrix systems do not need to be divided into meshes. In addition, the solution is not limited by the time step, so the early simulation accuracy is better and the computational efficiency is higher. To obtain the calculation results shown in Figure 4, the proposed semi-analytical method developed in this study was used to calculate on the same platform, and the time was less than 5 s, while Eclipse's calculation time was more than 200 s. Compared with numerical simulation methods, the semi-analytical method proposed in this paper has a faster calculation speed and is more suitable for large-scale case analysis and application in mining fields.

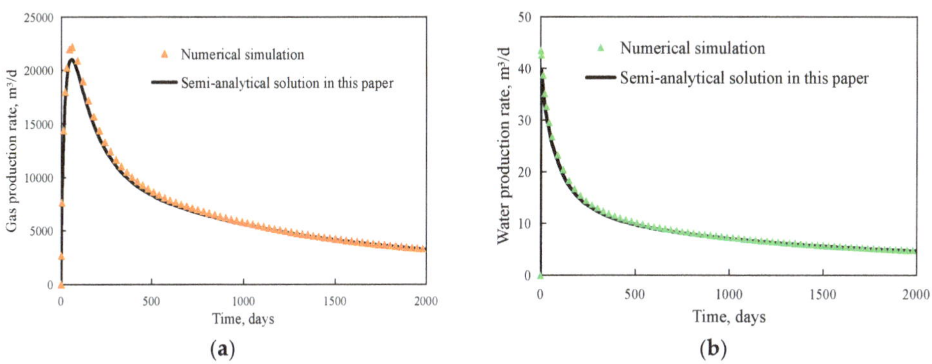

Figure 4. Comparison of gas and water rate between the semi-analytical model in this paper and the Eclipse. (**a**) Gas rate validation with the numerical solution. (**b**) Water rate validation with the numerical solution.

The main difference between the proposed model and the traditional analytical models is that the two-phase flow behaviour is incorporated in the proposed model. In this section, we simultaneously simulated the rate transients of both the proposed and conventional analytical models using the same group of formation and well parameters in Table 2. The initial water saturation in the cleats and the hydraulic fractures are given 1 and 0 in the proposed and conventional models, respectively. The modelling results are shown in Figure 5. It is shown that two-phase flow in the cleat and hydraulic fracture systems significantly influences the gas production rate during the whole production period. In addition, the gas production rate is smaller if the water-phase flow is considered in the two systems. This is possible because the gas production from the coal matrix acts as a displacement effect for the limited water in the cleat and hydraulic fracture systems, so the two-phase flow in the cleat and hydraulic fracture systems affects the gas production during the whole period. Besides, the gas production rate will be overestimated without considering the two-phase flow in the cleats and hydraulic fractures.

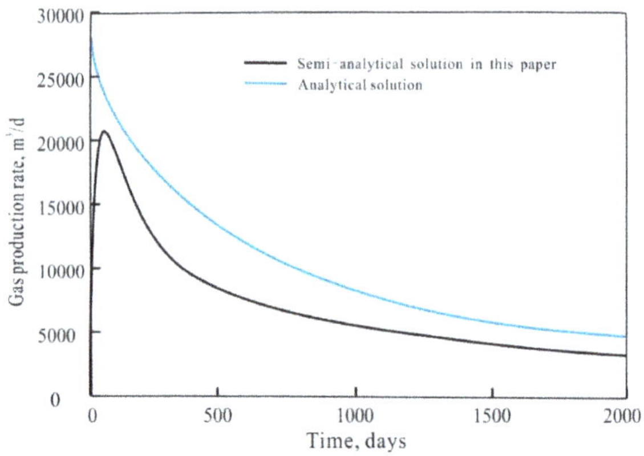

Figure 5. Comparison of gas production rate between the proposed semi-analytical model and th analytical model.

3.2. *Sensitivity Analysis*

Given the inherently low natural productivity of deep coal reservoirs, hydrauli fracturing has emerged as a pivotal strategy to augment production. Drawing from th investigation into the semi-analytical model, this paper scrutinizes the influence of bot the artificial fracturing capacity and the fracture half-length on the production capacity o deep CBM wells. Moreover, grounded in the empirical research findings of the reservoi seepage mechanism, we assess the impact of reservoir properties and stress sensitivit on the production capacity of both gas and water phases. The basic input parameter integral to the model are delineated in Table 2, while Table 3 illustrates the range c sensitivity parameters. The selection of parameters in Table 3 is obtained through th actual measurement data report of the production wells in the study area. Accordin to previous fracturing reports and actual well production performance, the number c hydraulic fractures generally ranges from 8 to 16, and the half-length of hydraulic fracture is generally between 50 to 120 m. In addition, due to the strong heterogeneity of coa reservoirs, it was found through well testing on multiple wells within the same block th the range of permeability of the coal cleat in the inner reservoir is from 0.1 mD to 0.5 m and the range of permeability of the outer reservoir is between 0.1 mD and 0.5 mD, an the range of permeability modulus of the coal cleat is from 0.01 MPa^{-1} to 0.04 MPa^{-} Therefore, parameters of conductivity of hydraulic fractures, half length of hydraul fractures, number of hydraulic fractures, permeability of the coal cleat in the inner reservo permeability of the outer reservoir, and permeability modulus of the coal cleat used f sensitivity analysis are shown in Table 3.

Table 3. Parameters used for sensitivity analysis.

Parameters	Symbol	Units	Value
Conductivity of hydraulic fractures	η_F	mD·m	0.3, 0.5, 1
Half length of hydraulic fractures	y_F	m	50, 80, 100
Number of hydraulic fractures	N_F	Dimensionless	9, 12, 15
Permeability of the coal cleat in the inner reservoir	k_f	mD	0.2, 0.3, 0.4
Permeability of the outer reservoir	k_o	mD	0.1, 0.2, 0.3
Permeability modulus of the coal cleat	γ_f	MPa^{-1}	0.01, 0.02, 0.03

Figure 6 shows the influence of fracture conductivity on the gas productivity of de CBM wells. It becomes apparent that an increased fracture diversion capacity results both a higher initial production rate and a more gradual production decline. The prima cause of this is that greater fracture diversion capacity yields elevated production rat

Under conditions of constant bottom-hole flowing pressure production, a more moderate production decline is adequate to sustain production.

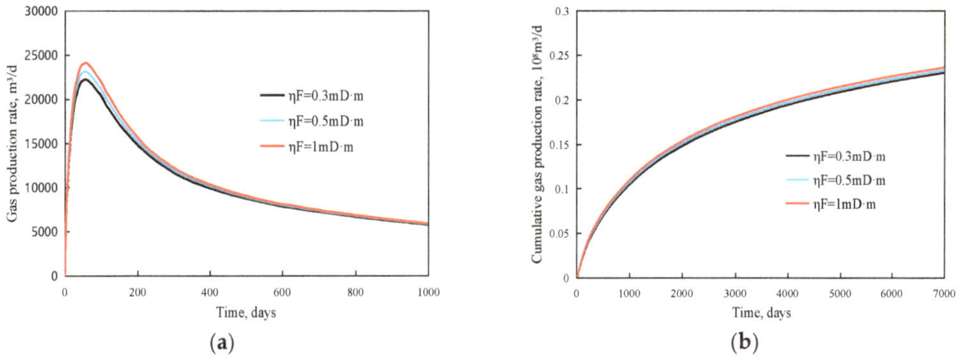

Figure 6. Effect of fracture conductivity on gas production rate. (**a**) Influence of ηF on gas production rate. (**b**) Influence of ηF on cumulative gas production rate.

Figure 7 illustrates the impact of fracture half-length on the gas productivity of deep CBM wells. It conveys that alterations in fracture half-length influence the entire development stage, particularly during the initial and intermediate production phases. Moreover, gas production increases with the expansion of the fracture half-length, albeit at a progressively diminishing rate. This is chiefly attributed to the fact that fracture half-length not only signifies the extent of reservoir stimulation volume but also mirrors the well-regulated reserves and drainage area. A more substantial fracture half-length equates to a larger linear flow area of the fracture and an increased drainage area, which in turn leads to higher production rates and a more gradual production decline.

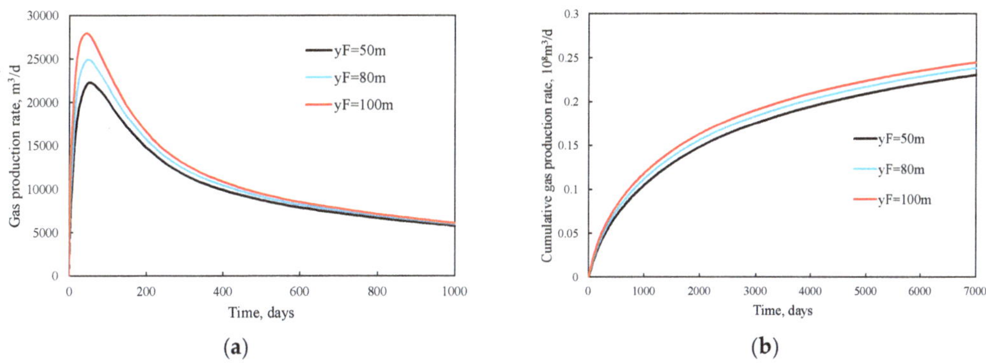

Figure 7. Effect of fracture half-length on gas production rate. (**a**) Influence of yF on gas production rate. (**b**) Influence of yF on cumulative gas production rate.

Figure 8 demonstrates the influence of the number of hydraulic fracturing stages on the gas production capacity of deep CBM wells. The diagram clearly indicates that the quantity of hydraulic fracturing stages substantially impacts production. An increase in the number of hydraulic fracturing stages results in a greater initial production capacity and cumulative production. This phenomenon is primarily due to the fact that an increased number of fracturing stages, accompanied by reduced intervals between them, amplifies the pressure drop in the inner zone, stimulating the desorption of adsorbed gas. Moreover, an increased number of fracturing stages also enhances the contact area between artificial fractures and reservoirs, resulting in an increase in production. However, it is worth noting that the augmentation in production capacity progressively diminishes as the number of fracturing stages increases.

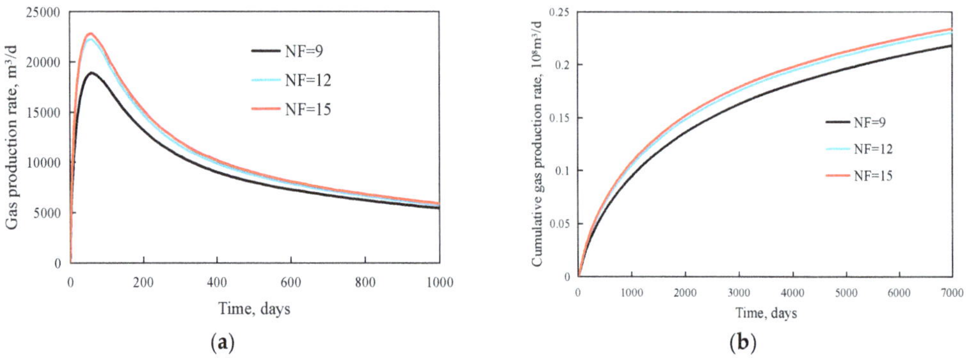

Figure 8. Effect of number of hydraulic fracturing stages on gas production rate. (**a**) Influence of NF on gas production rate. (**b**) Influence of NF on cumulative gas production rate.

Figure 9 illustrates the impact of the permeability of the coal cleat in the inner reservoir on the gas production capacity of deep CBM wells. The figure clearly demonstrates that higher coal cleat permeability in the inner reservoir results in greater initial production capacity and cumulative production. The primary reason for this is that an increase in the permeability of the coal cleat in the inner reservoir simultaneously enhances the flow capacity of free gas and desorption. As a result, the ease with which a pressure drop influences the interior of the inner zone matrix is increased, stimulating the desorption of adsorbed gas in the inner zone. Moreover, as the permeability of the coal cleat in the inner reservoir increases, the gas in the inner zone finds it increasingly conducive to flow towards secondary and artificial fractures.

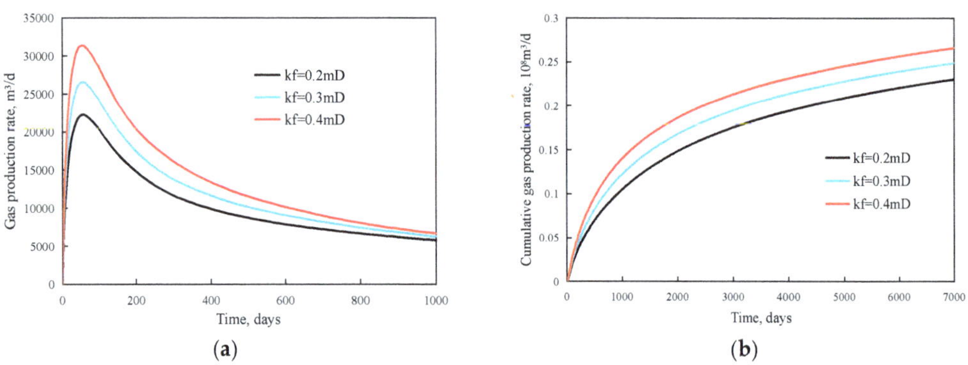

Figure 9. Effect of permeability of the coal cleat in the inner reservoir on gas production rate. (**a**) Influence of kf on gas production rate. (**b**) Influence of kf on cumulative gas production rate.

Figure 10 presents the effect of the permeability of the coal cleat in the outer reservoir on the gas production capacity of deep CBM wells. The data suggests that higher coal cleat permeability in the outer reservoir significantly influences the initial production capacity. More precisely, as the permeability of the coal cleat in the outer reservoir increases, the time required for the pressure drop to impact the interior of the outer zone matrix decreases, leading to a greater initial production capacity. The underlying reason for this is that higher coal cleat permeability in the outer reservoir allows the pressure drop to more easily impact the exterior of the outer zone matrix, stimulating the desorption of adsorbed gas in the outer zone. However, it is important to note that the permeability of the coal cleat in the inner reservoir tends to have a more substantial impact on the desorption of gas than the permeability of the coal cleat in the outer reservoir.

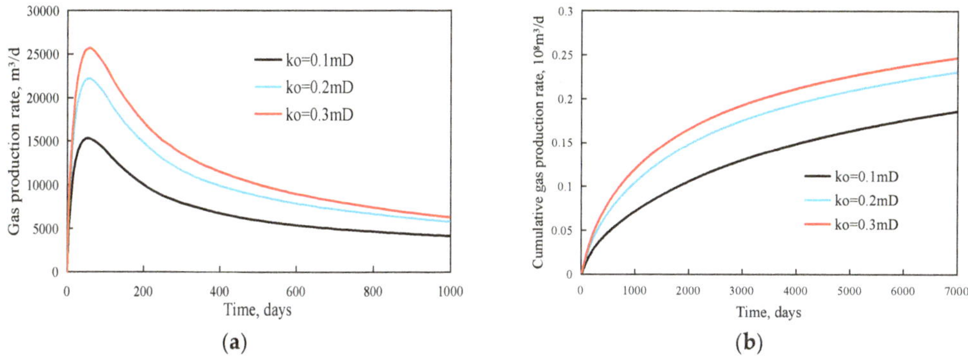

Figure 10. Effect of permeability of the coal cleat in the outer reservoir on gas production rate. (**a**) Influence of ko on gas production rate. (**b**) Influence of ko on cumulative gas production rate.

Figure 11 demonstrates the influence of the coal cleat's permeability stress sensitivity on the gas productivity of deep CBM wells. As shown in Figure 11, stress sensitivity impacts the entire development lifecycle of the deep CBM reservoir, particularly during the initial and middle stages of production. As the production pressure difference escalates, the permeability loss intensifies, leading to an increase in fluid flow resistance and, ultimately, a reduction in production accompanied by a faster rate of decline. Consequently, judicious management of the production pressure difference for deep CBM reservoirs should be maintained throughout the production process to mitigate the effects of stress sensitivity on production capacity.

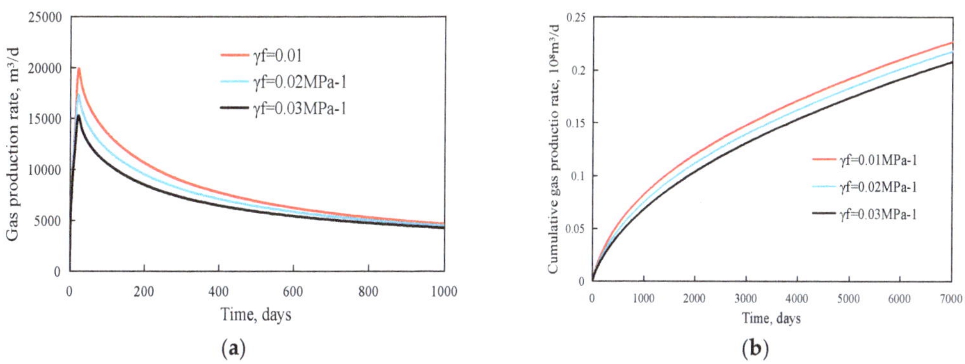

Figure 11. Effect of permeability stress sensitivity of the coal cleat on gas production rate. (**a**) Influence of γf on gas production rate. (**b**) Influence of γf on cumulative gas production rate.

3.3. Field Application

This study employs a deep CBM well situated to demonstrate the effectiveness of the proposed model. The production well is located on the northeast edge of the Ordos Basin, and its tectonic evolution is controlled by the regional tectonic evolution of the Ordos Basin. The burial depth of the main coal seam is controlled by the structure and surface influence, and it is between 1500 m and 2500 m. The burial depth of the coal seam in the southeast direction becomes shallower. The structure in the region is relatively stable, with no development of major faults and more development of inter-layer small faults. The average coal seam thickness is 11 m, the average coal seam Langmuir volume is 15.0 m^3/t, and the average coal seam gas content is 14 m^3/t. Table 4 details the basic data utilized in the production data analysis. The selection of parameters in Table 4 is obtained through the actual measurement data report of the example well. From the well logging report, the burial depth of the example well is 2230 m, the coal seam thickness is 13.2 m, and the porosity of the coal cleat is 3%. The pressure gradient of the production well area is 1 MPa/100 m, thus, the initial formation pressure is 22.3 MPa. From the fracturing design report, the number of hydraulic fractures is 14, the length of the horizontal well is 1200 m, and the width of hydraulic fractures is 0.001 m. The initial formation temperature is 320.2 K can be obtained from the well testing report. Moreover, core taking experiment is conducted on the production well, and the parameters of Langmuir pressure, Langmuir volume, and gas content in coal seams are obtained from the gas content report. Production for the well commenced on 1 January 2023. The production dynamic data, spanning from 3 January 2023, to 21 June 2023, were gathered for fitting and interpretation, followed by a prediction of the gas-water two-phase production capacity based on the fitting results.

Table 4. Data used in production data analysis.

Parameters	Symbol	Units	Value
Initial formation pressure	p_i	MPa	22.3
Initial formation temperature	T_i	K	320.2
Bottom-hole pressure	p_{wf}	MPa	4
Formation thickness	h	m	13.2
Number of hydraulic fractures	N_F	Dimensionless	14
Width of hydraulic fractures	w_F	m	0.001
Porosity of hydraulic fractures	Φ_F	Fraction	0.4
Length of the horizontal well	L	m	1200
Half-width of the reservoir	y_e	m	220
Porosity of the coal cleat	Φ_f	Fraction	0.03
Langmuir pressure	p_L	MPa	2.9
Langmuir volume	V_L	m^3/t	13.3
Gas content in coal seams	V	m^3/t	15.7

The proposed medium gas-water two-phase production dynamic prediction metho tailored for deep CBM reservoirs, was employed to fit the production dynamic data c the sample well. Figure 12 illustrates the fitting curves, evidencing a strong fit with th gas-water two-phase production of the deep CBM well. We also show the period wit data between 0–200 days in Figure 13. Besides, numerical simulation is also conducted c the basis of production history match. The calculated results with the Eclipse model ar compared with the proposed model, which is shown in Figure 13. Because the solution c the proposed model is not limited by the time step, and the fracture and matrix systen do not need to be divided into meshes, the early simulation accuracy is better and th computational efficiency is higher than the Eclipse model. The parameters of the reservo and fracture, acquired through inversion post-fitting, are listed in Table 5. A compariso between the inversion results, the geological attributes of the original gas reservoir, an actual production outcomes suggest that the key permeability parameters, as identified f the reservoir and fracture, offer theoretical guidance in predicting the production capaci of the CBM well, optimizing the design of fracturing parameters, and evaluating the po fracturing impacts. Utilizing this model to predict subsequent production, the cumulati gas and water production are projected to be 0.36 × 10^8 m^3 and 3.7 × 10^4 m^3, respective These findings offer valuable insights and guidance in adjusting the development plans f deep CBM reservoirs.

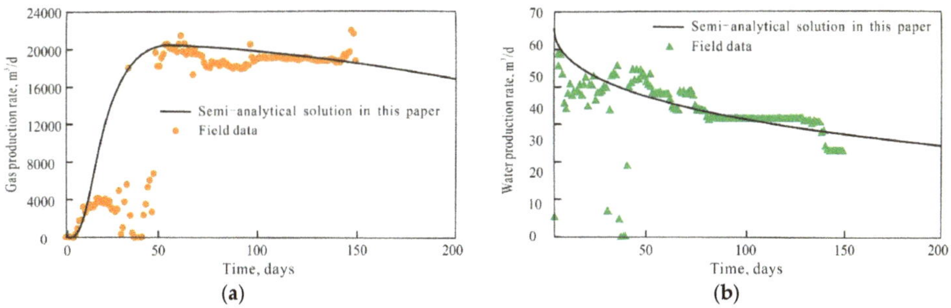

Figure 12. Type curve fitting for the production data with the proposed model. (**a**) Type curve fitting for the gas production data. (**b**) Type curve fitting for the water production data.

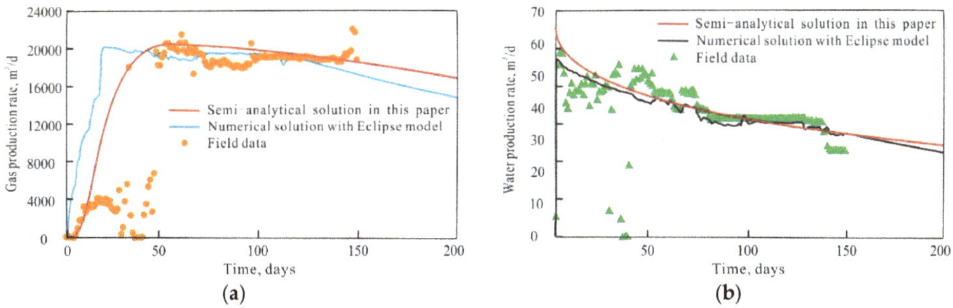

Figure 13. Type curve fitting for the production data with the proposed model and the Eclipse model. (**a**) Type curve fitting for the gas production data. (**b**) Type curve fitting for the water production data.

Table 5. Fitting results of the proposed model.

Parameters	Symbol	Units	Value
Conductivity of hydraulic fractures	η_F	mD·m	0.32
Half length of hydraulic fractures	y_F	m	82
Permeability of the coal cleat in the inner reservoir	k_f	mD	0.25
Permeability of the outer reservoir	k_o	mD	0.13
Permeability modulus of the coal cleat	γ_f	MPa^{-1}	0.05

4. Conclusions

This study introduces a novel methodology for predicting production in deep CBM wells, accounting for intricate fracture characteristics of reservoirs, effects of stress sensitivity, and dual-phase gas-water flow characteristics. The semi-analytical method proposed here, founded on the flowing material balance approach and iterative substitution, effectively addresses gas-water dual-phase non-linear flow issues with appreciable accuracy and computational efficiency. Key conclusions drawn from this study include:

1. Employing the flowing material balance method for computing the average pressure and average saturation of the reservoir, and iteratively updating the non-linear parameters in the seepage model, facilitates the accurate handling of the non-linear seepage problem associated with gas-water dual-phase flow. This results in an efficient and precise resolution of the seepage model;
2. The verification of the numerical model and field applications together suggest that the semi-analytical method investigated in this study offers a high degree of prediction accuracy and can be employed to forecast the production rate of deep CBM wells;
3. Critical seepage parameters of hydraulic fractures significantly influence the production dynamics of gas-water dual-phase. Therefore, hydraulic fracturing should aim

to enhance the degree and extent of modification to boost the production capacity of deep CBM wells;

4. The stress-sensitive effect of the coal cleats has an impact on the production capacity of deep CBM wells. Therefore, it is necessary to judiciously control the production pressure differential during production to effectively mitigate the adverse influence of stress-sensitive effects.

Author Contributions: Methodology, S.W.; Software, S.W.; Validation, D.L.; Investigation, W.L. All authors have read and agreed to the published version of the manuscript.

Funding: This research was partially funded by CNOOC Research Institute Co., Ltd., Science and Technology Projects (No. CNOOC-YXKJ-QZJC-TJ-2020-01).

Conflicts of Interest: The authors declare no conflict of interest.

References

1. Yang, Z.; Wang, H.; Liu, N.; Li, J. Simulation study of characteristics of hydraulic fracturing propagation of low permeability coal seam. *Disaster Adv.* **2012**, *5*, 717–720.
2. Sun, Z.; Shi, J.; Wu, K.; Zhang, T.; Feng, D.; Li, X. Effect of pressure-propagation behavior on production performance: Implication for advancing low-permeability coalbed-methane recovery. *SPE J.* **2019**, *24*, 681–697. [CrossRef]
3. Chen, Y.; Pan, Z. Reservoir properties of Chinese tectonic coal: A review. *Fuel* **2020**, *260*, 116350. [CrossRef]
4. Ren, J.; Zhang, L.; Ren, S.; Lin, J.; Meng, S.; Ren, G.; Gentzis, T. Multi-branched horizontal wells for coalbed methane production: Field performance and well structure analysis. *Int. J. Coal Geol.* **2014**, *131*, 52–64. [CrossRef]
5. Ma, J.; Chen, J.; Guan, J.; Lin, Y.; Chen, W.; Huang, L. Implementation of Johnson-Holmquist-Beissel model in four-dimensional lattice spring model and its application in projectile penetration. *Int. J. Impact Eng.* **2022**, *170*, 104340. [CrossRef]
6. Ma, J.; Zhao, J.; Lin, Y.; Liang, J.; Chen, J.; Chen, W.; Huang, L. Study on Tamped Spherical Detonation-Induced Dynamic Responses of Rock and PMMA Through Mini-Chemical Explosion Tests and a Four-Dimensional Lattice Spring Model. *Rock Mech. Rock Eng.* **2023**, *56*, 7357–7375. [CrossRef]
7. Wu, X.; Xi, C.; Wang, G. The mathematic model research of complicated fractures system in coalbed methane wells. *Nat. Gas Ind.* **2006**, *26*, 124–126.
8. Pant, L.M.; Huang, H.; Secanell, M.; Larter, S.; Mitra, S.K. Multi scale characterization of coal structure for mass transport. *Fuel* **2015**, *159*, 315–323. [CrossRef]
9. Pan, Z.; Wood, D.A. Coalbed methane (CBM) exploration, reservoir characterisation, production, and modelling: A collection of published research (2009–2015). *J. Nat. Gas Sci. Eng.* **2015**, *26*, 1472–1484. [CrossRef]
10. Zhang, J.; Feng, Q.; Zhang, X.; Hu, Q.; Wen, S.; Chen, D.; Yan, X. Multi-fractured horizontal well for improved coalbed methane production in eastern Ordos basin, China: Field observations and numerical simulations. *J. Pet. Sci. Eng.* **2020**, *194*, 107488 [CrossRef]
11. Yang, Y.; Liu, S. Integrated modeling of multi-scale transport in coal and its application for coalbed methane recovery. *Fuel* **2021**, *300*, 120971. [CrossRef]
12. Sun, Z.; Huang, B.; Li, Y.; Yu, W.; Ji, L. Production forecast of fractured vertical wells in coalbed methane reservoirs: Coupling dynamic drainage area. *Arab. J. Geosci.* **2022**, *15*, 7. [CrossRef]
13. Shi, X.; Pan, J.; Hou, Q.; Jin, Y.; Wang, Z.; Niu, Q.; Li, M. Micrometer-scale fractures in coal related to coal rank based on micro-CT scanning and fractal theory. *Fuel* **2018**, *212*, 162–172. [CrossRef]
14. Zhang, T.; Zhang, L.; Zhao, Y.; Zhang, R.; Zhang, D.; He, X.; Ge, F.; Wu, J.; Javadpour, F. Ganglia dynamics during imbibition and drainage processes in nanoporous systems. *Phys Fluids.* **2022**, *34*, 042016. [CrossRef]
15. Zhang, T.; Luo, S.; Zhou, H.; Hu, H.; Zhang, L.; Zhao, Y.; Javadpour, F. Pore-scale modelling of water sorption in nanopore systems of shale. *Int. J. Coal Geol.* **2023**, *27*, 104266. [CrossRef]
16. Clarkson, C.R.; Qanbari, F. Transient flow analysis and partial water relative permeability curve derivation for low permeability undersaturated coalbed methane wells. *Int. J. Coal Geol.* **2015**, *152*, 110–124. [CrossRef]
17. Zhao, J.; Tang, D.; Lin, W.; Xu, H.; Li, Y.; Tao, S.; Lv, Y. Permeability dynamic variation under the action of stress in the medium and high rank coal reservoir. *J. Nat. Gas Sci. Eng.* **2015**, *26*, 1030–1041. [CrossRef]
18. Tang, X.; Wang, Z.; Ripepi, N.; Kang, B.; Yue, G. Adsorption affinity of different types of coal: Mean isosteric heat of adsorption *Energy Fuels* **2015**, *29*, 3609–3615. [CrossRef]
19. Zeng, Q.; Wang, Z. Stress and temperature sensitivity of coal permeability in the Eastern Ordos Basin. *Pet. Sci. Bull.* **2020**, 512–519.
20. Yu, P.; Dempsey, D.; Archer, R. Techno-Economic feasibility of enhanced geothermal systems (EGS) with partially bridging multi-stage fractures for district heating applications. *Energy Convers. Manag.* **2022**, *257*, 115405. [CrossRef]
21. Aminian, K.; Ameri, S. Predicting production performance of CBM reservoirs. *J. Nat. Gas Sci. Eng.* **2009**, *1*, 25–30. [CrossRef]

22. Salmachi, A.; Yarmohammadtooski, Z. Production data analysis of coalbed methane wells to estimate the time required to reach to peak of gas production. *Int. J. Coal Geol.* **2015**, *141*, 33–41. [CrossRef]
23. Yan, X.; Zhang, S.; Tang, S.; Li, Z.; Guan, W.; Zhang, Q.; Wang, J. A prediction model for pressure propagation and production boundary during coalbed methane development. *Energy Fuels* **2021**, *35*, 1219–1233. [CrossRef]
24. Shi, J.; Jia, Y.; Wu, J.; Xu, F.; Sun, Z.; Liu, C.; Liu, C. Dynamic performance prediction of coalbed methane wells under the control of bottom-hole pressure and casing pressure. *J. Pet. Sci. Eng.* **2021**, *196*, 107799. [CrossRef]
25. Yu, P.; Dempsey, D.; Rinaldi, A.P.; Calibugan, A.; Ritz, V.A.; Archer, R. Association between Injection and Microseismicity in Geothermal Fields with Multiple Wells: Data-driven Modelling of Rotokawa, New Zealand, and Húsmúli, Iceland. *J. Geophys. Res. Solid Earth* **2022**, *128*, e2022JB025952. [CrossRef]
26. Ibrahim, A.F.; Nasr-El-Din, H.A. A comprehensive model to history match and predict gas/water production from coal seams. *Int. J. Coal Geol.* **2015**, *146*, 79–90. [CrossRef]
27. Ma, T.; Rutqvist, J.; Oldenburg, C.M.; Liu, W.; Chen, J. Fully coupled two-phase flow and poromechanics modeling of coalbed methane recovery: Impact of geomechanics on production rate. *J. Nat. Gas Sci. Eng.* **2017**, *45*, 474–486. [CrossRef]
28. Sun, Z.; Shi, J.; Zhang, T.; Wu, K.; Feng, D.; Sun, F.; Li, X. A fully-coupled semi-analytical model for effective gas/water phase permeability during coal-bed methane production. *Fuel* **2018**, *223*, 44–52. [CrossRef]
29. Shi, J.; Hou, C.; Wang, S.; Xiong, X.; Wu, S.; Liu, C. The semi-analytical productivity equations for vertically fractured coalbed methane wells considering pressure propagation process, variable mass flow, and fracture conductivity decrease. *J. Pet. Sci. Eng.* **2019**, *178*, 528–543. [CrossRef]
30. Zou, M.; Wei, S.; Huang, Z.; Lv, X.; Guo, B. Simulations on recoverability performances for a coalbed methane field in SE edge of Ordos basin, China. *Fuel* **2018**, *233*, 354–360. [CrossRef]
31. Zhu, S.; Salmachi, A.; Du, Z. Two phase rate-transient analysis of a hydraulically fractured coal seam gas well: A case study from the Ordos Basin, China. *Int. J. Coal Geol.* **2018**, *195*, 47–60. [CrossRef]
32. Peng, X.; Zhu, S.; You, Z.; Du, Z.; Deng, P.; Wang, C.; Wang, M. Numerical simulation study of fines migration impacts on an early water drainage period in undersaturated coal seam gas reservoirs. *Geofluids* **2019**, *2019*, 5723694. [CrossRef]
33. Yu, P.; Dempsey, D.; Archer, R. A three-dimensional coupled thermo-hydro-mechanical numerical model with partially bridging multi-stage contact fractures in horizontal-well enhanced geothermal system. *Int. J. Rock Mech. Min. Sci.* **2021**, *143*, 104787. [CrossRef]
34. Sun, Z.; Huang, B.; Liu, Y.; Jiang, Y.; Zhang, Z.; Hou, M.; Li, Y. Gas-phase production equation for CBM reservoirs: Interaction between hydraulic fracturing and coal orthotropic feature. *J. Pet. Sci. Eng.* **2022**, *213*, 110428. [CrossRef]
35. Song, P.; Li, Y.; Yin, Z.; Yuan, Q. Hydrogen generation from heavy oils via in-situ combustion gasification. In *SPE Western Regional Meeting*; SPE: Kuala Lumpur, Malaysia, 2023; p. D031S007R003.
36. Nossent, J.; Elsen, P.; Bauwens, W. Sobol'sensitivity analysis of a complex environmental model. *Environ. Model. Softw.* **2011**, *26*, 1515–1525. [CrossRef]
37. Yin, Z.; Feng, T.; MacBeth, C. Fast assimilation of frequently acquired 4D seismic data for reservoir history matching. *Comput. Geosci.* **2019**, *128*, 30–40. [CrossRef]
38. Clarkson, C.R.; Qanbari, F. A semi-analytical forecasting method for unconventional gas and light oil wells: A hybrid approach for addressing the limitations of existing empirical and analytical methods. *SPE Reserv. Eval. Eng.* **2014**, *18*, 260–263.

processes

MDPI

Article

Research on Fluid–Solid Coupling Mechanism around Openhole Wellbore under Transient Seepage Conditions

Erhu Liu [1], Desheng Zhou [1,*], Xu Su [1], Haiyang Wang [1,*], Xiong Liu [1] and Jinze Xu [1,2]

1 School of Petroleum Engineering, Xi'an Shiyou University, Xi'an 710065, China; 23111010020@stumail.xsyu.edu.cn (E.L.); 22211010042@stumail.xsyu.edu.cn (X.S.); liuxiong2016@xsyu.edu.cn (X.L.); jinzxu@ucalgary.ca (J.X.)
2 Department of Chemical and Petroleum Engineering, University of Calgary, Alberta, AB T2N 1N4, Canada
* Correspondence: dzhou@xsyu.edu.cn (D.Z.); wanghaiyang@xsyu.edu.cn (H.W.)

Abstract: Hydraulic fracturing is one of the most important enhanced oil recovery technologie currently used to develop unconventional oil and gas reservoirs. During hydraulic fracture initiatior fluid seeps into the reservoir rocks surrounding the wellbore, inducing rock deformation and change in the stress field. Analyzing the fluid–solid coupling mechanism around the wellbore is crucial t the construction design of fracturing technologies such as pulse fracturing and supercritical carbo dioxide fracturing. In this study, a new transient fluid–solid coupling model, capable of simulatin the pore pressure field and effective stress field around the wellbore, was established based on th Biot consolidation theory combined with the finite difference method. The numerical results ar in excellent agreement with the analytical solutions, indicating the reliability of the model and th stability of the computational approach. Using this model, the influence of seepage parameters an reservoir properties on the fluid–solid coupling around the open-hole wellbore was investigated. Th simulation results demonstrate that, during wellbore pressurization, significant changes occur in th pore pressure field and effective stress field near the wellbore. The fluid–solid coupling effect aroun the wellbore returns to its initial state when the distance exceeds four times the radius away from th wellbore. As the fluid viscosity and wellbore pressurization rate decrease, the pore pressure field an effective circumferential stress (ECS) field around the wellbore become stronger. Adjusting the flui viscosity and wellbore pressurization rate can control the effect of seepage forces on the rock skeletc during wellbore fluid injection. For the same injection conditions, rocks with q higher Young modulus and Poisson's ratio exhibit stronger pore pressure fields and ECS fields near the wellbo This model furnishes a dependable numerical framework for examining the fluid–solid couplir mechanism surrounding the open-hole wellbore in the initiation phase of hydraulic fractures.

Keywords: open-hole wellbore; fluid–solid coupling; pore pressure; effective circumferential stres

Citation: Liu, E.; Zhou, D.; Su, X.; Wang, H.; Liu, X.; Xu, J. Research on Fluid–Solid Coupling Mechanism around Openhole Wellbore under Transient Seepage Conditions. *Processes* **2024**, *12*, 412. https://doi.org/10.3390/pr12020412

Academic Editor: Sara Liparoti

Received: 19 January 2024
Revised: 13 February 2024
Accepted: 15 February 2024
Published: 18 February 2024

1. Introduction

Hydraulic fracturing, also known as fracking, is a method used to extract natur gas or oil from underground reservoirs by injecting high-pressure water and chemica into the rock formations to create and widen fractures, thus facilitating the release energy resources. Hydraulic fracturing technology plays a pivotal role in the extraction unconventional hydrocarbon resources, revolutionizing the energy landscape [1,2]. T significance of this technology stems from its ability to enhance well productivity ar unlock vast reservoir potential. In hydraulic fracturing, a thorough understanding fracture initiation and propagation is essential for designing and optimizing fracturi operations effectively. Analyzing the nuanced processes involved in crack initiation a propagation is crucial in tailoring hydraulic fracturing strategies, ultimately influenci the success and efficiency of well stimulation endeavors. Therefore, a comprehensi investigation into the intricacies of fracture nucleation and propagation is essential advancing the state of the art in hydraulic fracturing construction design [3–7].

Drawing upon the linear elastic theory and Terzaghi's effective stress theory, numerous scholars have previously conducted in-depth analyses of the stress field around wellbores during the injection of working fluids. Notable contributions include the work of Haimson and Hubbert [8–11], who systematically analyzed the stress field around open-hole wellbores. They categorized the stress field into three components: one arising from the geostatic stress of the reservoir, another from wellbore pressure, and the third from fluid filtration. Their research suggests that fracturing occurs when the effective total stress around the wellbore reaches the tensile strength of the rock formation, with the wellbore pressure becoming the formation's fracturing pressure. This led to the derivation of widely used formulas for open-hole well formation fracturing pressure, known as the H-W formula and the H-F formula. Subsequent studies have built upon this foundation. For instance, Reisabadi et al. [12] introduced an analytical model that considers the effects of depletion, matrix shrinkage, and wellbore trajectory when estimating the stress distribution around the wellbore. Li et al. [13] developed a heat–fluid–solid coupling numerical model for horizontal wells in NGH reservoirs, taking into account hydrate decomposition. Ren and Cai [14] established a three-dimensional model to calculate temperature and stress distributions in reservoir rock when liquid nitrogen is injected into the wellbore. Raad et al. [15] investigated the dynamics of a CO_2 injection wellbore using a fully coupled non-isothermal transient multiphase wellbore–reservoir flow model. Qiu et al. [16] created a pseudo-3D hydraulic–mechanical coupling finite element model, considering the DM and anisotropy of shale in wellbores. This model allowed for a parametric analysis of the effects of elastic anisotropy, fracture volume fraction, fracture permeability, mass transfer coefficient, and fluid pressure on wellbore stress. Xiao et al. [17] proposed a semi-analytical model for coal wellbore stability during mixture gas injection. Additionally, Rim et al. [18] developed a new seepage model that employs the method of the separation of variables and the Bessel function theory to predict variations in the pore pressure and seepage force around a vertical wellbore during hydraulic fracturing. By integrating fluid dynamics and rock mechanics models, Hou et al. [19] developed a comprehensive data-driven workflow to predict pressure based on fracturing experience from adjacent wells.

Furthermore, by assuming the perforation channel as a small-scale open-hole wellbore, many scholars have analyzed the stress field around the perforation tunnel under the assumption of an impermeable reservoir, deriving analytical solutions for the formation fracturing pressure of deviated perforated wells [20–22]. Wang et al. [6] simplified the perforation channel as an open-hole wellbore and analyzed the impact of fluid seepage on the effective stress field around the perforation tunnel, deriving the initiation pressure for fractures at arbitrary perforation angles. While existing research has provided initial insights into the coupling effect between fluid seepage and rock deformation around open-hole wellbores, most of these studies are based on analytical models established using steady-state Darcy's flow theory. Fluid seepage into reservoir rocks during fluid injection is a transient process that depends on time. Therefore, it is necessary to establish transient fluid–solid coupling models to investigate the influence of factors such as fluid viscosity and wellbore pressurization rate on the fluid–solid coupling around the wellbore.

This study initially established a numerical simulation model for fluid–solid coupling, considering transient fluid seepage effects, based on the Biot consolidation theory and the finite difference method. Subsequently, the reliability and stability of the fluid–solid coupling model were examined using a pore pressure analytical solution model. Finally, this model was employed to comprehensively analyze the impact of seepage-influencing parameters and reservoir properties on the fluid–solid coupling around the open-hole wellbore.

2. Numerical Methodology

2.1. Fluid–Solid Coupling Model

To analyze the fluid–solid coupling effect around the open-hole wellbore under transient seepage conditions, this paper establishes a two-dimensional planar wellbore physical model. As shown in Figure 1, the wellbore has an inner radius of a and an outer radius of c, where r represents the distance from the wellbore axis, θ represents the angle, the water pressure at the inner boundary of the cylinder is P_w, and the initial pore pressure at the outer boundary is P_o. It is assumed that the rocks surrounding the open-hole wellbore are isotropic and homogeneous. Under the influence of the pressure difference between the inner and outer boundaries, fluid seeps radially from the wellbore into the pores of the reservoir rock. By establishing a polar coordinate system, this paper analyzes the effective radial stress σ'_r and ECS σ'_θ formed in the rock around the cylinder due to fluid action. (Note that in this paper, positive values are assigned to compressive stress, while negative values represent tensile stress).

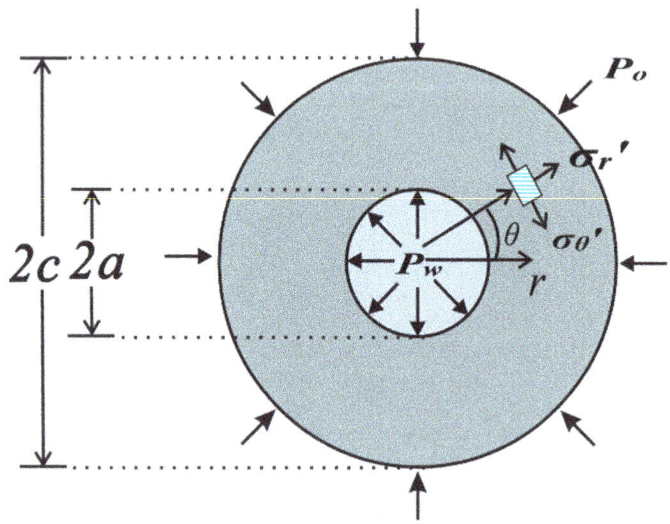

Figure 1. Two-dimensional planar wellbore physical model.

For an open-hole wellbore, the stress equilibrium equation and stress boundary conditions satisfied by the surrounding rock when fluid seepage enters the rock pores are as follows [23]:

$$\begin{array}{ll} \frac{d\sigma_r'}{dr} + \frac{\sigma_r' - \sigma_\theta'}{r} + \alpha \frac{dP}{dr} = 0 & (Stress\ Equilibrium\ Equation) \\ \begin{cases} \sigma_r' = P_w - \alpha P_w & r = a \\ \sigma_r' = P_o - \alpha P_o & r = c \end{cases} & (Stress\ Boundary\ Conditions) \end{array} \quad (1)$$

Here, α is the Biot effective stress coefficient and P represents the pore pressure. As shown in Equation (1), the effective stress field formed in the surrounding reservoir rock due to fluid action around the wellbore can be decomposed into the superposition of stress fields generated by both volume forces and surface forces, including: (1) internal and external boundary surface forces $P_w - \alpha P_w$ and $P_o - \alpha P_o$; and (2) the seepage force term $\alpha dP/dr$.

To establish a finite difference format, the first step is to create a stress function Φ to characterize the radial stress σ'_r and circumferential stress σ'_θ:

$$\begin{cases} \sigma'_{\theta 2} = \frac{d\Phi}{dr} + r\alpha\frac{dP}{dr} \\ \sigma'_{r2} = \frac{\Phi}{r} \end{cases} \quad (2)$$

Due to the axisymmetric nature of the stress field formed by fluid action around the wellbore, the geometric and physical equations satisfied by the rock surrounding the wellbore under fluid influence can be obtained using plane strain conditions:

$$\begin{cases} \varepsilon_\theta = \frac{u_r}{r} \\ \varepsilon_r = \frac{du_r}{dr} \qquad (Geometric\ Equation) \\ \gamma_{r\theta} = \frac{dr\varepsilon_\theta}{dr} \end{cases}$$
$$\begin{cases} \varepsilon_r = \frac{1}{E}\left[\sigma_r(1-\nu^2) - \nu(1+\nu)\sigma_\theta\right] \\ \varepsilon_\theta = \frac{1}{E}\left[\sigma_\theta(1-\nu^2) - \nu(1+\nu)\sigma_r\right] \end{cases} \quad (Physical\ Equation) \qquad (3)$$

Combining Equation (2) with (3) and substituting them into Equation (1), we can derive the compatibility equation and boundary conditions satisfied by the stress function as follows:

$$r^2\frac{d^2\Phi}{dr^2} + r\frac{d\Phi}{dr} - \Phi + r^2\alpha\frac{(2+\nu-\nu^2)}{(\nu^2-1)}\frac{\partial P}{\partial r} + r^3\alpha\frac{\partial^2 P}{\partial r^2} = 0 \quad (Compatibility\ Equation)$$
$$\begin{cases} \Phi = a(P_w - \alpha P_w) & r = a \\ \Phi = c(P_o - \alpha P_o) & r = c \end{cases} \quad (Boundary\ Conditions) \qquad (4)$$

To solve the compatibility equation, it is necessary to calculate the distribution of pore pressure around the wellbore under transient seepage conditions. In the Biot poroelastic consolidation theory [24,25], the governing equation for pore pressure propagation under the combined influence of solid deformation and fluid is as follows:

$$\begin{cases} \kappa\nabla^2 P = \alpha\frac{\partial\varepsilon_{kk}}{\partial t} + \frac{1}{Q}\frac{\partial P}{\partial t} \\ Q = \frac{2G(\nu_u-\nu)}{\alpha^2(1-2\nu)(1-2\nu_u)} \end{cases} \quad (5)$$

In the equation, the parameters are defined as follows: Q is the Biot modulus, when $\alpha = 1$ and only the seepage force is considered, $Q = 1/\phi c_f$, where ϕ is the porosity of the rock, and c_f is the fluid compressibility in Pa^{-1}. G is the shear modulus defined as $G = E/2\ (1 + \nu)$, where E is the Young's modulus in Pa and ν is the drained Poisson's ratio. ν_u represents the undrained Poisson's ratio. t is time measured in seconds. κ is the permeability coefficient defined as $\kappa = k/\mu$, where k is the permeability in m^2 and μ is the fluid viscosity in Pa·s. ε_{kk} represents volumetric expansion.

To facilitate a convenient solution for the pore pressure propagation equation, Geertsma and Yew, among others [26,27], derived a conversion equation relating volumetric expansion (ε_{kk}) to pore pressure (P) based on the Biot consolidation theory:

$$\begin{cases} \varepsilon_{kk} = c_m P \\ c_m = \frac{\alpha(1-2\nu)}{2G(1-\nu)} \end{cases} \quad (6)$$

By utilizing Equation (6) to process Equation (5), we can obtain the pore pressure propagation equation that takes into account the influence of solid deformation:

$$\nabla^2 P = \frac{1}{\kappa}\left(\alpha c_m + \frac{1}{Q}\right)\frac{\partial p}{\partial t} \quad (7)$$

The combination of the Compatibility Equation (4) and Pore Pressure Propagation Equation (7) forms the fundamental control equations for the transient fluid–solid coupling model around a bare wellbore under seepage conditions. By applying the finite difference method to discretize the above equations, we obtain:

$$\begin{cases} r_i{}^2\frac{\Phi_{i+1}+\Phi_{i-1}-2\Phi_i}{h^2}+r_i\frac{\Phi_{i+1}-\Phi_{i-1}}{2h}-\Phi_i-r_i{}^2\alpha\frac{(2+v-v^2)}{(v^2-1)}\frac{P_{i+1}-P_{i-1}}{2h}+r_i{}^3\alpha\frac{P_{i+1}+P_{i-1}-2P_i}{h^2}=0 \\ \frac{P_{i+1}^{j+1}-2P_i^{j+1}+P_{i-1}^{j+1}}{h^2}+\frac{1}{r_i}\frac{P_{i+1}^{j+1}-P_{i-1}^{j+1}}{2h}=\beta\frac{P_i^{j+1}-P_i^j}{\Delta t} \\ \beta=\frac{1}{\kappa}\left(\alpha c_m+\frac{1}{Q}\right) \\ r_i=h\cdot i+a \quad i=0,1,2\ldots n \\ t_j=\Delta t\cdot j \quad j=0,1,2\ldots m \end{cases} \quad (8)$$

where h is the iterative step size, $h = c - a/n$; Δt is the time step size; and (r_i, t_j) represents the grid nodes. The adopted time scheme is backward differencing. Since the solution area is two-dimensional and axially symmetric, there is no need to determine the unit type, just mesh the solution area. To solve Equation (8), we meshed the region within a certain radius around the open-hole wellbore, as illustrated in Figure 2.

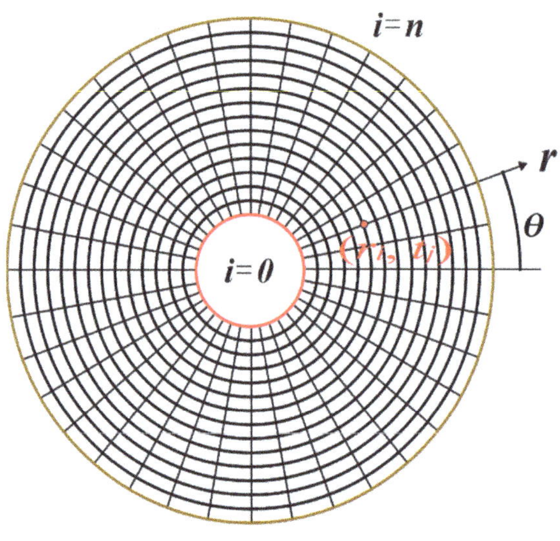

Figure 2. Schematic diagram of grid nodes.

Taking hydraulic fracturing as an example, according to the fracturing constructio curve, the wellbore pressure can be considered to increase at a constant pressurizatio rate [28]. Meanwhile, at the far-field boundary, the pore pressure can be assumed to be th initial pore pressure. Therefore, the boundary conditions for the pore pressure propagatio equation are as follows:

$$\begin{cases} P(0,j)=Ct_j+P_o & r=a \\ P(n,j)=P_o & r=c \end{cases} \quad ($$

where C represents the wellbore's pressurization rate, MPa/s_o.

As indicated by the boundary conditions specified in Equation (9), the wellbore pre sure is directly influenced by both the wellbore pressurization rate and time. Meanwhi at the far-field outer boundary, the pore pressure remains constant and equal to the init pore pressure.

According to the governing Equation (8) and boundary conditions 9, the tridiagonal matrix required to solve the stress function and pore pressure can be established. The tridiagonal matrix **AX** = **B** for pore pressure is:

$$\mathbf{A} = \begin{bmatrix} (1+2A_1) & -\left(A_1+\frac{A_2}{r_1}\right) & 0 & \bullet & \bullet & 0 \\ \left(\frac{A_2}{r_2}-A_1\right) & (1+2A_1) & -\left(A_1+\frac{A_2}{r_2}\right) & \bullet & \bullet & 0 \\ 0 & \bullet & \bullet & \bullet & \bullet & \bullet \\ \bullet & \bullet & \bullet & \bullet & \bullet & \bullet \\ \bullet & \bullet & \bullet & \left(\frac{A_2}{r_{n-2}}-A_1\right) & (1+2A_1) & -\left(A_1+\frac{A_2}{r_{n-2}}\right) \\ 0 & \bullet & \bullet & \bullet & \left(\frac{A_2}{r_{n-1}}-A_1\right) & (1+2A_1) \end{bmatrix} \tag{10}$$

where $A_1 = \Delta t/(h^2\beta)$; $A_2 = \Delta t/(2h\beta)$

$$\mathbf{X} = \begin{bmatrix} P_1^{j+1} \\ P_2^{j+1} \\ \bullet \\ \bullet \\ \bullet \\ P_{n-1}^{j+1} \end{bmatrix} \mathbf{B} = \begin{bmatrix} P_1^j - \left(\frac{A_2}{r_1}-A_1\right)P_0^{j+1} \\ P_2^j \\ \bullet \\ \bullet \\ \bullet \\ P_{n-1}^j + \left(A_1+\frac{A_2}{r_{n-1}}\right)P_n^{j+1} \end{bmatrix} \tag{11}$$

The tridiagonal matrix that the stress function satisfies is **CY** = **D**:

$$\mathbf{C} = \begin{bmatrix} \frac{-2{r_1}^2-h^2}{h^2} & \frac{2{r_1}^2+hr_1}{2h^2} & 0 & \bullet & \bullet & 0 \\ \frac{2{r_2}^2-hr_2}{2h^2} & \frac{-2{r_2}^2-h^2}{h^2} & \frac{2{r_2}^2+hr_2}{2h^2} & \bullet & \bullet & 0 \\ 0 & \bullet & \bullet & \bullet & \bullet & \bullet \\ \bullet & \bullet & \bullet & \bullet & \bullet & \bullet \\ \bullet & \bullet & \bullet & \frac{2{r_{n-2}}^2-hr_{n-2}}{2h^2} & \frac{-2{r_{n-2}}^2-h^2}{h^2} & \frac{2{r_{n-2}}^2+hr_{n-2}}{2h^2} \\ 0 & \bullet & \bullet & \bullet & \frac{2{r_{n-1}}^2-hr_{n-1}}{2h^2} & \frac{-2{r_{n-1}}^2-h^2}{h^2} \end{bmatrix} \tag{12}$$

$$\mathbf{Y} = \begin{bmatrix} \Phi_1 \\ \Phi_2 \\ \bullet \\ \bullet \\ \bullet \\ \Phi_{n-1} \end{bmatrix} \mathbf{D} = \begin{bmatrix} \frac{{r_1}^3}{h^2}\alpha(P_2+P_0-2P_1) - \frac{{r_1}^2}{2h}\alpha\gamma(P_2-P_0) - \frac{2{r_1}^2+hr_1}{2h^2}(aP_o - a\alpha P_o) \\ \frac{{r_2}^3}{h^2}\alpha(P_3+P_1-2P_2) - \frac{{r_2}^2}{2h}\alpha\gamma(P_3-P_1) \\ \bullet \\ \bullet \\ \bullet \\ \frac{{r_{n-1}}^3}{h^2}\alpha(P_n+P_{n-2}-2P_{n-1}) - \frac{{r_{n-1}}^2}{2h}\alpha\gamma(P_n-P_{n-2}) - \frac{2{r_{n-1}}^2+hr_2}{2h^2}(aP_n - a\alpha P_n) \end{bmatrix} \tag{13}$$

Under the boundary conditions specified in Equation (9), we simultaneously solve the aforementioned matrices through programming to obtain the pore pressure field and stress function Subsequently, this stress function can be utilized to solve and acquire the stress field resulting from the fluid interactions around the wellbore under transient fluid–solid coupling conditions. During hydraulic fracturing, the effective circumferential stress directly affects the fracture initiation pressure [29], so subsequent chapters will only focus on the circumferential stress around the wellbore.

2.2. Model Validation

2.2.1. Pore Pressure Field Verification

When considering the absence of rock matrix particle compressibility (Biot coefficient α equals 1) and neglecting the influence of solid deformation on pore pressure, the pore bnpressure propagation Equation (7) can be simplified to:

$$\begin{cases} \frac{\partial^2 P}{\partial r^2} + \frac{1}{r}\frac{\partial P}{\partial r} = \frac{1}{\chi}\frac{\partial P}{\partial t} \\ \chi = \frac{k}{\phi \mu c_f} \end{cases} \quad (14)$$

where χ is the propagation coefficient, m^2/s; this study assumes $c_f = 4.5 \times 10^{-10}$ Pa^{-1}.

Despite not accounting for the influence of solid deformation on pore pressure, Equation (14) remains a transient pore pressure propagation equation dependent on time. Based on Equation (14), Carslaw and Jaeger [30] derived an analytical solution for the radial distribution of pore pressure around a hollow cylindrical body under a constant pressurization rate:

$$P(r,t) = C\int_0^t f(r,s)\,ds \quad (15)$$

where:

$$f(r,t) = 1 + \int_0^\infty \exp\left(-\chi u^2 t\right)\left[\frac{J_0(ur)Y_0(ua) - Y_0(ur)J_0(ua)}{J_0(ua)^2 + Y_0(ua)^2}\right]\frac{du}{u} \quad (16)$$

J_0 and Y_0 are the first and second kinds of zero-order Bessel functions.

By adjusting the parameter β in the finite difference Equation (8) and comparing it with the analytical solution for the transient pore pressure distribution mentioned above, we can verify the accuracy of the difference format established in this work.

Figure 3 compares the analytical solution with the numerical solution of pore pressure distribution around the wellbore under a constant pressurization rate. It is evident from Figure 3 that, starting from the initial moment, the pore pressure distribution near the wellbore undergoes significant variations due to wellbore pressurization. As one moves away from the wellbore, the distribution gradually approaches the initial pore pressure before pressurization.

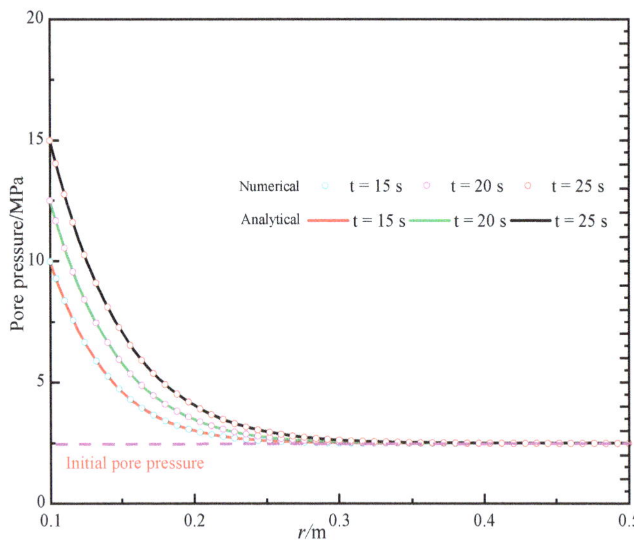

Figure 3. Comparison of radial distribution of pore pressure between analytical solution and numerical solution (χ = 2.2 × 10^{-4} m^2/s, C = 0.5 MPa/s, P_0 = 2.5 MPa).

At different time instances, the analytical solution for transient pore pressure and the numerical solution presented in this study demonstrate a close agreement, with the relative error limited to less than 0.1%. This validation confirms the accuracy of the transient fluid–solid coupling model established in this paper.

2.2.2. Effective Stress Field Verification

There is no analytical solution available for the stress field around the wellbore under transient seepage conditions. However, an analytical solution does exist for the stress field around the wellbore in the case of impermeable reservoirs. When the rock is impermeable, the ECS σ'_θ induced by the fluid pressures (P_w and P_o) at the inner and outer boundaries of the wellbore can be calculated using the Lame formula and Terzaghi's principle of effective stress [31–33].

$$\sigma'_\theta = \left[\frac{a^2c^2(P_o - P_w)}{c^2 - a^2}\frac{1}{r^2} - \frac{a^2P_w - c^2P_o}{c^2 - a^2}\right] - P_o \tag{17}$$

When $c >> a$, Equation (17) can be simplified to $\frac{a^2}{r^2}(P_o - P_w)$.

By varying the reservoir's permeability, we employed the finite difference scheme detailed in Section 2.1 to compute the effective circumferential stress resulting from the fluid flow around the wellbore during transient seepage. We investigated the variations in the transient stress field surrounding the wellbore as the reservoir's permeability gradually decreased, eventually reaching impermeability. As depicted in Figure 4, the results obtained from the numerical solution are compared with those derived from the analytical solution provided by Equation (17).

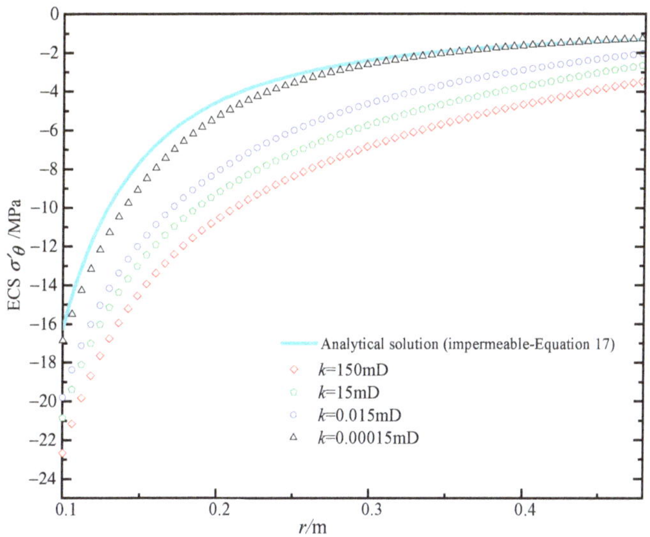

Figure 4. Change in ECS σ'_θ along the radial direction for different permeability (numerical and analytical solutions). $C = 0.15$ MPa/s; $\mu = 100$ mPa·s; $\alpha = 1$; $t = 100$ s; $P_0 = 5.0$ MPa; $\nu = 0.25$; and $E = 5.0 \times 10^4$ MPa.

In the case of an impermeable reservoir, the seepage force term in Equation (1) is negligible, and the fluid primarily exerts a surface force on the rocks surrounding the wellbore. As depicted in Figure 4, for impermeable reservoirs, the effective circumferential stress resulting from fluid interaction around the wellbore is tensile stress. However, when fluid seepage enters the reservoir, the effective circumferential tensile stress induced by the fluid around the wellbore, due to the seepage force, increases significantly. A greater effective circumferential tensile stress increases the susceptibility of the well wall to tensile failure. This demonstrates that the seepage force significantly increases the possibility of tensile failure of the wellbore wall. Under identical injection conditions, an increasing reservoir fluid permeability amplifies the seepage effect, leading to a higher effective circumferential tensile stress around the wellbore.

As the reservoir's permeability gradually diminishes, the numerical solution begins to converge toward the analytical solution. When the reservoir's permeability approaches an impermeable state, the analytical and numerical solution curves essentially overlap. This confirms the stability and accuracy of the finite difference scheme presented in this article for computing the effective circumferential stress field.

3. Numerical Results

3.1. Evolution Characteristics of Pore Pressure Field and Effective Stress Field

In this section, α is set to 1, resulting in the disappearance of the surface force term in Equation (1), with fluid exerting only seepage force on the rock framework. Figure 5 illustrates pseudo-color plots of the pore pressure field and the corresponding ECS field around the wellbore at different time steps during the stable pressurization process of hydraulic fracturing. It shows the variations in pore pressure and ECS along the radial direction (C = 0.25 MPa/s, μ = 10 mPa·s, K = 0.1 mD, ϕ = 0.1, P_0 = 5 MPa, a = 0.1 m, C = 0.5 m, ν = 0.25, and $E = 5.0 \times 10^4$ MPa).

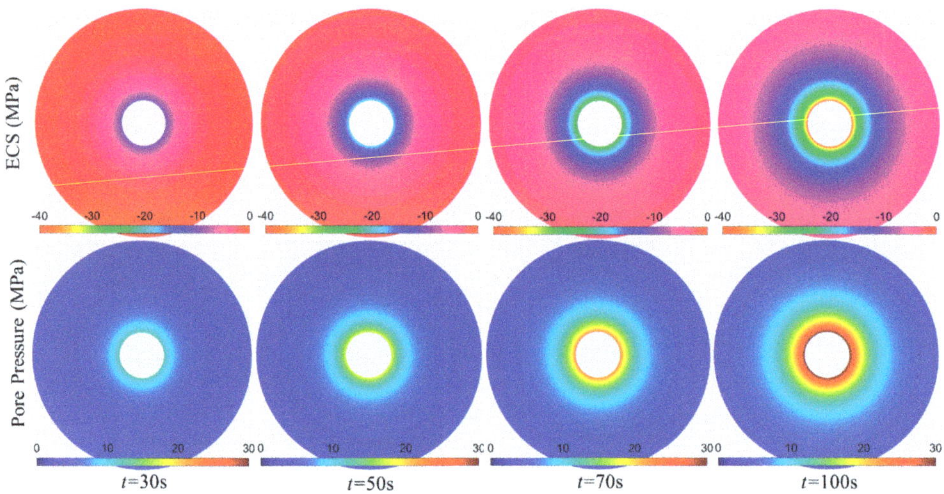

Figure 5. Pseudo-color maps of pore pressure field and ECS field at different time steps.

By combining the observations from Figures 5 and 6, it becomes evident that si nificant changes occur in the pore pressure field around the wellbore and the Effectiv Circumferential Stress (ECS) field formed by the action of seepage forces during the stab pressurization process of the wellbore. The pore pressure and ECS near the wellbore exhib larger variations over time, while moving away from the wellbore results in a rapid d crease in pore pressure towards the initial pore pressure state. Under transient fluid–sol coupling conditions, the ECS formed by the action of seepage forces is maximized ne the wellbore and gradually diminishes towards zero with an increasing distance from th wellbore. The magnitude of changes in the ECS due to seepage forces is more pronounce in the vicinity of the wellbore.

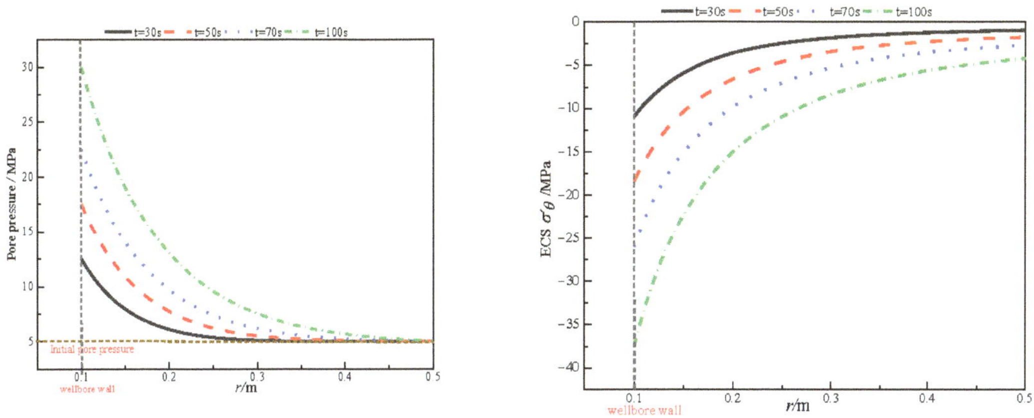

(**a**): Pore pressure (**b**): ECS

Figure 6. Variation in radial pore pressure and ECS with radius at different time steps.

3.2. *Influence of Seepage Parameters*

Using the established transient fluid–solid coupling model, an analysis of two key parameters, fluid viscosity and wellbore pressurization rate, was conducted. Figure 7 presents pseudo-color plots of the pore pressure field around the wellbore and the corresponding ECS field formed by fluid action at the same time step but with varying fluid viscosities. Figure 8 displays pseudo-color plots of the pore pressure field and ECS field when the wellbore pressure is pressurized to the same level using different pressurization rates (α = 1; ν = 0.25, E = 5.0 × 10^4 MPa, K = 0.1 mD, ϕ = 0.1, and P_0 = 5 MPa).

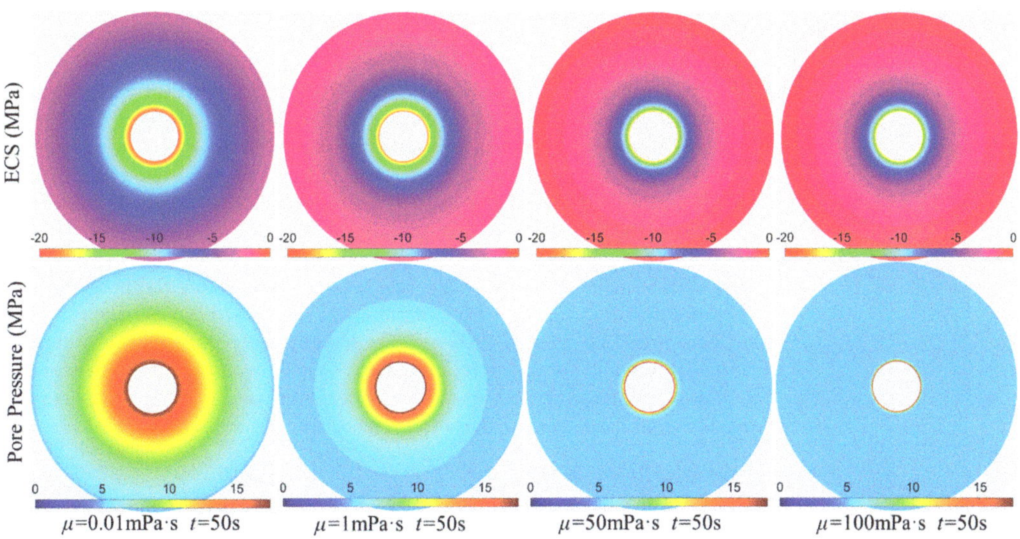

Figure 7. Pseudo-color maps of pore pressure field and stress field under different fluid viscosities (C = 0.25 MPa/s).

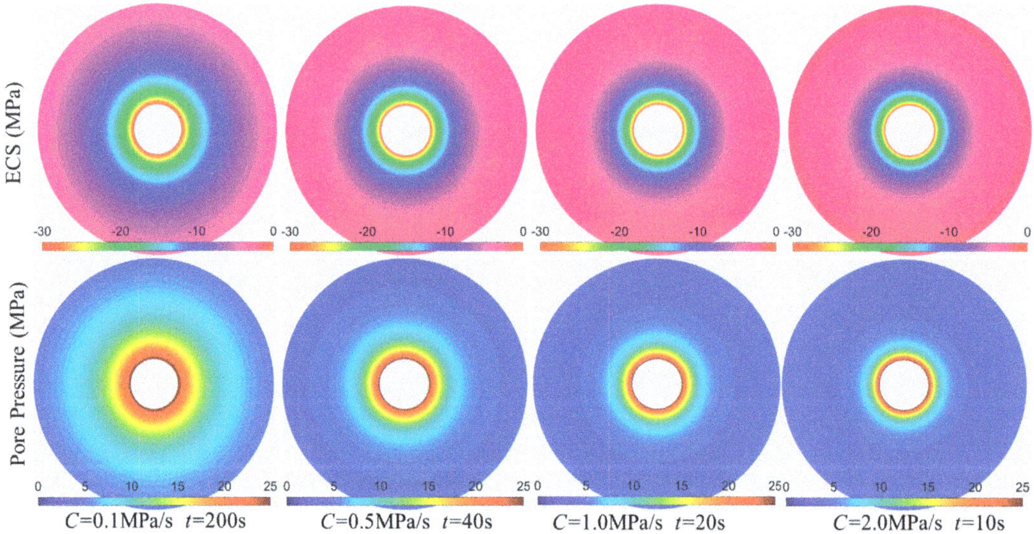

Figure 8. Pseudo-color maps of pore pressure field and ECS field under different pressurization rates (μ = 10 mPa·s).

In Figure 7, it is evident that the fluid viscosity has a significant impact on the pore pressure field around the wellbore and the ECS field formed by seepage forces under transient conditions. At the same time step, when the fluid viscosity is high, the pore pressure rapidly decreases to the initial pore pressure state around the wellbore as fluid seeps into the reservoir. Conversely, as the fluid viscosity decreases, the pore pressure distribution around the wellbore gradually becomes smoother. Additionally, with a decreasing fluid viscosity, the ECS field formed by seepage forces around the wellbore significantly strengthens, resulting in a notable increase in effective circumferential tensile stress values. Therefore, for low-permeability reservoirs, reducing fluid viscosity can significantly enhance the action of seepage forces around the wellbore.

Figure 8 demonstrates that, when pressurized to the same pressure level, the pressurization rate significantly affects the pore pressure field and ECS field. A smaller wellbore pressurization rate results in a longer time required to reach the same pressure level, but leads to higher pore pressure values around the wellbore. Conversely, a higher pressurization rate leads to a faster decline in the pore pressure around the wellbore. Additionally, as the pressurization rate decreases, the ECS values formed by seepage forces around the wellbore gradually increase. Therefore, a lower pressurization rate results in higher ECS values formed by seepage forces around the wellbore when pressurized to the same pressure level. Consequently, for reservoirs with a lower permeability, it may be advisable to consider lower fluid injection rates to enhance the action of seepage forces.

3.3. Influence of Reservoir Parameters

Using the transient fluid–solid coupling model established in this study, we can analyze the effects of key reservoir parameters, such as reservoir permeability and rock mechanics, on the pore pressure field around the wellbore and the ECS field formed by seepage forces (α = 1, μ = 1 mPa·s, ϕ = 0.1, P_0 = 5 MPa, and C = 0.25 MPa/s).

Figure 9 illustrates pseudo-color plots of the pore pressure field and ECS field under different reservoir permeability conditions, while Figure 10 shows the corresponding variations in the pore pressure and ECS along the radial direction. Combining the observations from Figures 8 and 9, it becomes apparent that, under low-viscosity fluid conditions, decreasing the reservoir permeability amplifies the magnitude of pore pressure variation. Near the wellbore, a lower reservoir permeability leads to a faster decay in pore pressure

In the simulation case depicted in Figure 10, with low-viscosity fluid, once the reservoir permeability exceeds 1 mD, the pore pressure distribution tends toward a steady-state linear Darcy seepage distribution, and the influence of permeability on the pore pressure field and ECS field gradually diminishes. Consequently, under low-viscosity fluid conditions, a higher reservoir permeability results in elevated pore pressure values around the wellbore and increased ECS values formed by seepage forces.

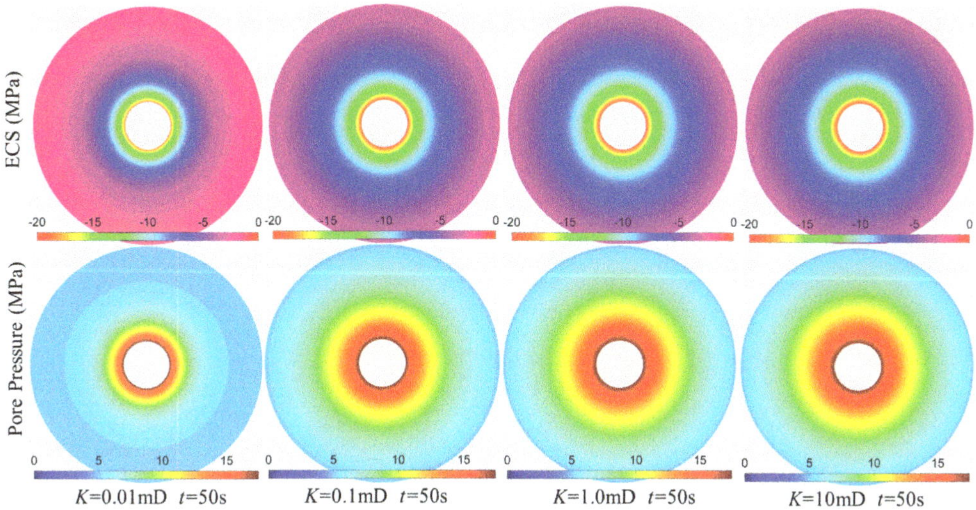

Figure 9. Pseudo-color maps of pore pressure field and stress field under different reservoir permeability ($E = 5.0 \times 10^4$ MPa; $\nu = 0.25$).

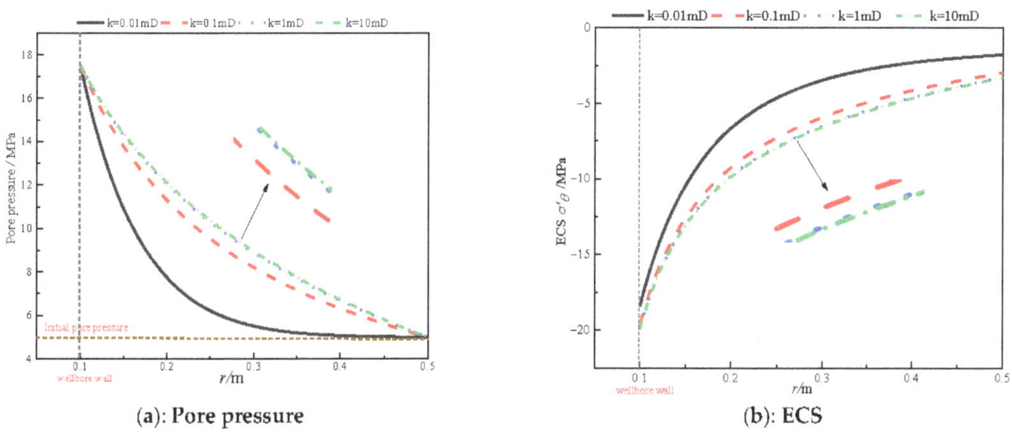

Figure 10. Radial pore pressure and ECS change with radius under different reservoir permeability conditions.

Figures 11 and 12 present pseudo-color plots illustrating the pore pressure field around the wellbore and the ECS field formed by seepage forces under varying conditions of Young's modulus and Poisson's ratio. Correspondingly, Figures 13 and 14 provide radial profiles of the pore pressure and ECS along the radius.

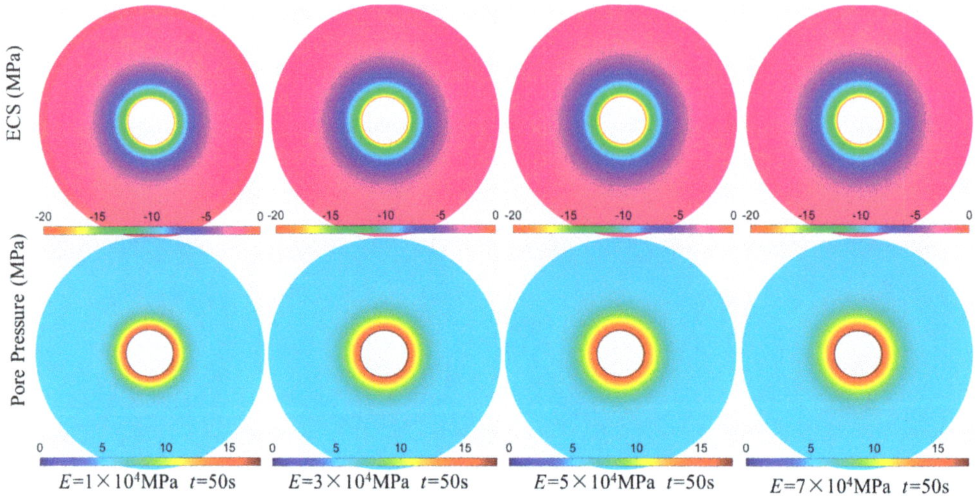

Figure 11. Pseudo-color maps of pore pressure field and stress field under different Young's modul
(K = 0.1 mD; ν = 0.25).

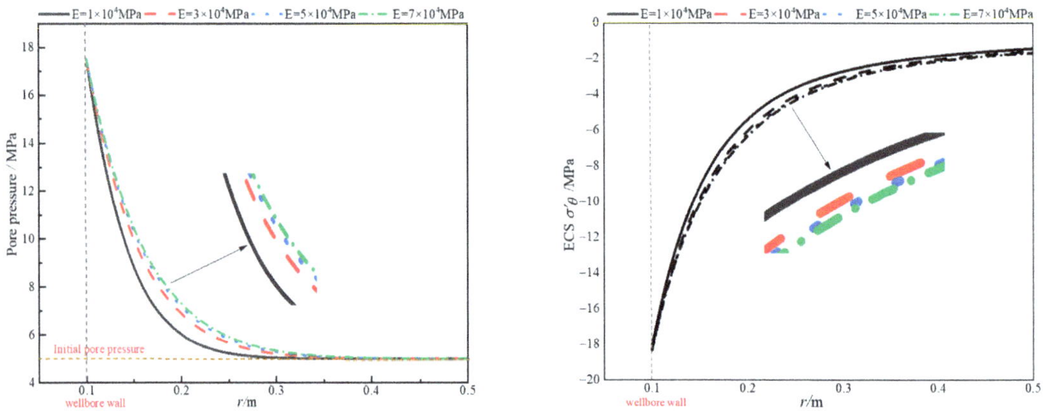

(**a**): Pore pressure (**b**): ECS

Figure 12. Variation in pore pressure and ECS in the radial direction with radius under differe
Young's modulus conditions.

Combining the observations from Figures 11–14 reveals that, under fluid–solid co pling conditions, the elastic deformation of rock elements significantly affects the po pressure field around the wellbore and the ECS field formed by seepage forces. High values of Young's modulus and Poisson's ratio result in elevated pore pressure valu around the wellbore, corresponding to higher ECS values formed by seepage forces. Ov all, Young's modulus (E) has a slightly greater influence on the pore pressure field ar the stress field formed by seepage forces than Poisson's ratio (ν). However, compared seepage parameters, rock mechanics parameters have a much weaker impact on the po pressure field and the ECS field formed by seepage forces around the wellbore. In th paper, there is still room for improvement. Future research will delve into the analysis well wall behavior, specifically in terms of tensile or shear failure, induced by fluid injecti This analysis will be based on the fluid–solid coupling model presented here, combin with rock failure criteria.

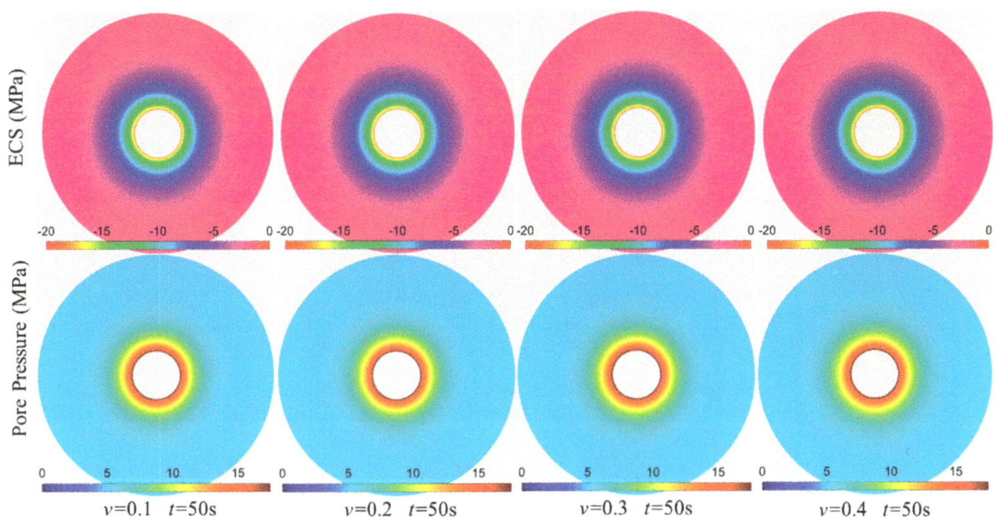

Figure 13. Pseudo-color maps of pore pressure field and ECS field under different Poisson's ratio conditions (K = 0.1 mD; E = 5.0 × 10^4 MPa).

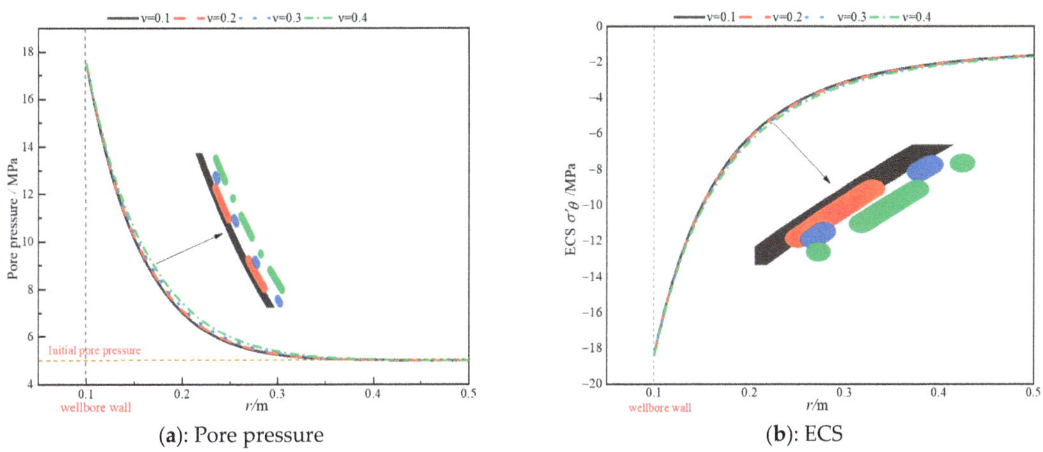

Figure 14. Variation in pore pressure and ECS in the radial direction with radius at different Poisson's ratios.

4. Conclusions

Based on the Biot consolidation theory and finite difference method, this paper established a transient fluid–solid coupling model that can analyze the stress field and pore pressure field around an open-hole wellbore during the initiation of hydraulic fractures. The transient analytical solution of the pore pressure field was used to verify the accuracy of the model and the stability of the calculation method. Using this model, this paper analyzed the effects of reservoir parameters and seepage parameters on open-hole wellbore fluid–solid coupling under different injection conditions. The simulation results show that:

(1) During the stable pressurization of the wellbore, significant changes were observed in the pore pressure field and effective stress field near the wellbore. As the distance from the wellbore increased to more than four times the radius, the fluid–solid coupling effect around the wellbore substantially diminished, returning to its initial state.

(2) With a decrease in fluid viscosity and wellbore pressurization rate, the pore pressure and ECS around the wellbore increased. Adjusting the fluid viscosity and wellbore

pressurization rate allowed control over the action of seepage forces during wellbore fluid injection.

(3) Under conditions of low-viscosity fluid injection, a higher reservoir permeability yielded larger pore pressure values around the wellbore, accompanied by increased ECS values generated by seepage forces.

(4) When the Young's modulus and Poisson's ratio of the rock surrounding the wellbore were large, the pore pressure field and ECS were stronger. Notably, the Young's modulus exhibited a more pronounced impact on the fluid–solid coupling effect around the wellbore compared to Poisson's ratio.

Author Contributions: Conceptualization, E.L., J.X., H.W. and D.Z.; methodology, E.L., X.L. and H.W.; software, X.S.; validation, E.L.; writing—original draft preparation, E.L. and D.Z. All authors have read and agreed to the published version of the manuscript.

Funding: This work was supported by the National Natural Science Foundation of China, grant number 51934005, U23B2089, 52374038; Shaanxi Provincial Natural Science Basic Research Program Project, grant number 2024JC-YBQN-0554, 2023-JC-QN-0427.

Informed Consent Statement: Informed consent was obtained from all subjects involved in the study.

Data Availability Statement: The raw data supporting the conclusions of this article will be made available by the authors on request.

Conflicts of Interest: The authors declare no conflicts of interest.

References

1. Hasan, M.M.F.; First, E.L.; Boukouvala, F.; Floudas, C.A. A multi-scale framework for CO_2 capture, utilization, and sequestration: CCUS and CCU. *Comput. Chem. Eng.* **2015**, *81*, 2–21. [CrossRef]
2. Jiang, K.; Ashworth, P.; Zhang, S.; Liang, X.; Sun, Y.; Angus, D. China's carbon capture, utilization and storage (CCUS) policy: A critical review. *Renew. Sustain. Energy Rev.* **2020**, *119*, 109601. [CrossRef]
3. Pan, L.; Oldenburg, C.; Pruess, K.; Wu, Y.-S. Transient CO_2 leakage and injection in wellbore-reservoir systems for geologic carbon sequestration. *Greenh. Gases Sci. Technol.* **2011**, *1*, 335–350. [CrossRef]
4. Zhu, H.; Deng, J.; Jin, X.; Hu, L.; Luo, B. Hydraulic Fracture Initiation and Propagation from Wellbore with Oriented Perforation. *Rock Mech. Rock Eng.* **2015**, *48*, 585–601. [CrossRef]
5. Zhou, D.; Wang, H.; He, Z.; Liu, Y.; Liu, S.; Ma, X.; Cai, W.; Bao, J. Numerical study of the influence of seepage force on the stress field around a vertical wellbore. *Eng. Appl. Comput. Fluid Mech.* **2020**, *14*, 1489–1500. [CrossRef]
6. Wang, H.; Zhou, D.; Liu, S.; Wang, X.; Ma, X.; Yao, T. Hydraulic fracture initiation for perforated wellbore coupled with the effect of fluid seepage. *Energy Rep.* **2022**, *8*, 10290–10298. [CrossRef]
7. Roy, P.; Morris, J.P.; Walsh, S.D.C.; Iyer, J.; Carroll, S. Effect of thermal stress on wellbore integrity during CO_2 injection. *Int. J. Greenh. Gas Control* **2018**, *77*, 14–26. [CrossRef]
8. Haimson, B.; Fairhurst, C. Initiation and Extension of Hydraulic Fractures in Rocks. *Soc. Pet. Eng. J.* **1967**, *7*, 310–318. [CrossRef]
9. Haimson, B.C. *Hydraulic Fracturing in Porus and Nonporus Rock and Its Potential for Determining In-Situ Stresses at Great Depth* University of Minnesota: Minneapolis, MN, USA, 1968.
10. Haimson, B.; Fairhurst, C. Hydraulic Fracturing in Porous-Permeable Materials. *J. Pet. Technol.* **1969**, *21*, 811–817. [CrossRef]
11. Hubbert, M.K.; Willis, D.G. Mechanics of Hydraulic Fracturing. *Trans. AIME* **1957**, *210*, 153–168. [CrossRef]
12. Reisabadi, M.Z.; Haghighi, M.; Sayyafzadeh, M.; Khaksar, A. Effect of matrix shrinkage on wellbore stresses in coal seam gas: An example from Bowen Basin, east Australia. *J. Nat. Gas Sci. Eng.* **2020**, *77*, 103280. [CrossRef]
13. Li, Y.; Cheng, Y.; Yan, C.; Song, L.; Liu, H.; Tian, W.; Ren, X. Mechanical study on the wellbore stability of horizontal wells in natural gas hydrate reservoirs. *J. Nat. Gas Sci. Eng.* **2020**, *79*, 103359. [CrossRef]
14. Ren, K.; Cai, C. Numerical Investigation into the Distributions of Temperature and Stress around Wellbore during the Injection of Cryogenic Liquid Nitrogen into Hot Dry Rock Reservoir. *Math. Probl. Eng.* **2021**, *2021*, 9913321. [CrossRef]
15. Jafari Raad, S.M.; Lawton, D.; Maidment, G.; Hassanzadeh, H. Transient non-isothermal coupled wellbore-reservoir modeling of CO_2 injection—Application to CO_2 injection tests at the CaMI FRS site, Alberta, Canada. *Int. J. Greenh. Gas Control* **2021**, *111*, 103462. [CrossRef]
16. Qiu, Y.; Ma, T.; Liu, J.; Peng, N.; Ranjith, P.G. Poroelastic response of inclined wellbore geometry in anisotropic dual-medium/media. *Int. J. Rock Mech. Min. Sci.* **2023**, *170*, 105560. [CrossRef]
17. Xiao, H.; Li, W.; Wang, Z.; Yang, S.; Tan, P. Critical Conditions for Wellbore Failure during CO_2-ECBM Considering Sorption Stress. *Sustainability* **2023**, *15*, 3696. [CrossRef]
18. Rim, H.; Chen, Y.; Tokgo, J.; Du, X.; Li, Y.; Wang, S. Time-Dependent Effect of Seepage Force on Initiation of Hydraulic Fracture around a Vertical Wellbore. *Materials* **2023**, *16*, 2012. [CrossRef]

19. Hou, L.; Zhou, L.; Elsworth, D.; Wang, S.; Wang, W. Prediction of Fracturing Pressure and Parameter Evaluations at Field Practical Scales. *Rock Mech. Rock Eng.* **2024**, 1–14. [CrossRef]
20. Hossain, M.M.; Rahman, M.K.; Rahman, S.S. Hydraulic fracture initiation and propagation: Roles of wellbore trajectory, perforation and stress regimes. *J. Pet. Sci. Eng.* **2000**, *27*, 129–149. [CrossRef]
21. Fallahzadeh, S.H.; Shadizadeh, S.R.; Pourafshary, P. Dealing with the Challenges of Hydraulic Fracture Initiation in Deviated-Cased Perforated Boreholes. In Proceedings of the Trinidad and Tobago Energy Resources Conference, Port of Spain, Trinidad and Tobago, 27–30 June 2010.
22. Ding, Y.; Liu, X.; Luo, P. The analytical model of hydraulic fracture initiation for perforated borehole in fractured formation. *J. Pet. Sci. Eng.* **2018**, *162*, 502–512. [CrossRef]
23. Wang, H.Y.; Zhou, D.S.; Gao, Q.; Fan, X.; Xu, J.Z.; Liu, S. Study on the Seepage Force-Induced Stress and Poroelastic Stress by Flow Through Porous Media Around a Vertical Wellbore. *Int. J. Appl. Mech.* **2021**, *13*, 2150065. [CrossRef]
24. Biot, M.A. General theory of three-dimensional consolidation. *J. Appl. Phys.* **1941**, *12*, 155–164. [CrossRef]
25. Rice, J.R.; Cleary, M.P. Some basic stress diffusion solutions for fluid-saturated elastic porous media with compressible constituents. *Rev. Geophys.* **1976**, *14*, 227–241. [CrossRef]
26. Geertsma, J. Problems of Rock Mechanics in Petroleum Production Engineering. In Proceedings of the 1st ISRM Congress, Lisbon, Portugal, 25 September–1 October 1966.
27. Yew, C.H.; Liu, G. Pore Fluid and Wellbore Stabilities. In Proceedings of the International Meeting on Petroleum Engineering, Beijing, China, 24–27 March 1992.
28. Ito, T. Effect of pore pressure gradient on fracture initiation in fluid saturated porous media: Rock. *Eng. Fract. Mech. Eng. Fract. Mech.* **2008**, *75*, 1753–1762. [CrossRef]
29. Wang, H.; Zhou, D. Mechanistic study on the effect of seepage force on hydraulic fracture initiation. *Fatigue Fract. Eng. Mater. Struct.* **2024**. [CrossRef]
30. Carslaw, H.S.; Jaeger, J.C. *Conduction of Heat in Solids*; Clarendon Press: Oxford, UK, 1959; pp. 314–332.
31. Terzaghi, K.; Peck, R.B.; Mesri, G. *Soil Mechanics in Engineering Practice*, 3rd ed.; John Wiley: New York, NY, USA, 1995; 750p.
32. Chen, X.; Li, Y.; Liu, Z.; Tang, Y.; Sui, M. Visualized investigation of the immiscible displacement: Influencing factors, improved method, and EOR effect. *Fuel* **2023**, *331*, 125841. [CrossRef]
33. Hou, L.; Ren, J.; Fang, Y.; Cheng, Y. Data-driven optimization of brittleness index for hydraulic fracturing. *Int. J. Rock Mech. Min. Sci.* **2022**, *159*, 105207. [CrossRef]

processes

MDPI

Article

A Study on the Production Simulation of Coal–Shale Interbedded Coal Measure Superimposed Gas Reservoirs under Different Drainage Methods

Wenkai Wang [1,2,3,4], Shiqi Liu [2,3,*], Shuxun Sang [1,2,3,4], Ruibin Du [1,2,3,4] and Yinghai Liu [1,2,3,4]

1 School of Resources and Geosciences, China University of Mining and Technology, Xuzhou 221008, China; wwk971227@163.com (W.W.); shxsang@cumt.edu.cn (S.S.)
2 Jiangsu Key Laboratory of Coal-Based Greenhouse Gas Control and Utilization, China University of Mining and Technology, Xuzhou 221008, China
3 Carbon Neutrality Institute, China University of Mining and Technology, Xuzhou 221008, China
4 Key Laboratory of Coalbed Methane Resources and Reservoir Formation Process, China University of Mining and Technology, Xuzhou 221008, China
* Correspondence: liushiqi@cumt.edu.cn

Abstract: To study the influence of different drainage methods on the production performance of coa measure gas wells, the interbedded reservoir composed of coal and shale in the Longtan Formation c the Dahebian block was used as the research object. Considering the influence of coal and shale matri shrinkage, effective stress, and interlayer fluid flow on reservoir properties such as fluid migratio behavior and permeability, a fluid–solid coupling mathematical model of coal measure superimpose gas reservoirs was established. Numerical simulations of coal measure gas production under differer drainage and production modes were conducted to analyze the evolution of reservoir pressure, ga content in the matrix, permeability, and other characteristic parameters of the superimposed reservoi as well as differences in interlayer flow. The results showed that, compared to single-layer drainag cumulative gas production increased by 33% under multi-layer drainage. Both drainage method involve interlayer energy and substance transfer. Due to the influence of permeability, porosity, an mechanical properties, significant differences exist in reservoir pressure distribution, preferenti flow direction, gas content in the matrix, and permeability ratio between coal and shale reservoir under different drainage and production modes. Multi-layer drainage effectively alleviates th influence of vertical reservoir pressure differences between reservoir layers, facilitates reservo pressure transmission in shale reservoirs, enhances methane desorption in shale matrices, promote matrix shrinkage, and induces the rebound of shale reservoir permeability, thus improving overa gas production.

Keywords: coal measures gas; drainage method; interlayer fluid flow low; reservoir pressure

Citation: Wang, W.; Liu, S.; Sang, S.; Du, R.; Liu, Y. A Study on the Production Simulation of Coal–Shale Interbedded Coal Measure Superimposed Gas Reservoirs under Different Drainage Methods. *Processes* **2023**, *11*, 3424. https://doi.org/10.3390/pr11123424

Academic Editors: Tao Zhang, Zheng Sun, Dong Feng, Hung Vo Thanh and Wen Zhao

Received: 21 November 2023
Revised: 6 December 2023
Accepted: 9 December 2023
Published: 13 December 2023

1. Introduction

Coal measure gas refers to all kinds of natural gas (i.e., coalbed methane, coal me sure shale gas, and coal measure sandstone gas) in a coal measure reservoir, which an important part of unconventional natural gas [1,2]. Western Guizhou is rich in co measure gas resources and is a key area for coal measure gas development [3–6]. D to the difference in physical properties of superimposed reservoirs, the gas productic law is more complicated than that for single reservoirs, and the development is mo difficult [7,8]. Therefore, studying the gas production effect of coal measure reservoi under different drainage methods is helpful to understand the fluid migration law in cc measure superimposed reservoirs.

With the maturation of computer technology, the research method for coal measu gas development has changed in the direction of numerical simulation [9,10]. Mode such as CMG, Eclipse, COMSOL Multiphysics, and so on are widely used [11–14] for the numerical simulation of coal measure gas development. Some scholars have developed gas and water seepage numerical models for coalbed methane and tight gas reservoirs and have studied the influence of pressure, saturation, and reservoir permeability on combined production [15–18]. Some scholars have also developed mathematical models for coal measure gas seepage in coal–sandstone composite reservoirs that consider interlayer channeling, have analyzed the variation law of reservoir pressure and gas production, and have discussed the gas seepage mechanism in coal–sandstone composite reservoirs [15,19]. In summary, the above models are mainly based on a combination of coal and sandstone reservoirs. There are relatively few studies on the combination of coal and shale reservoirs in coal measure reservoirs and the effect of coal measure gas production under different drainage methods. Research in this area is still in its infancy and needs further exploration.

In this paper, the C409 coal and its shale roof and floor in the Dahebian block of the Liupanshui Coalfield in western Guizhou are used as the research objects. Considering the influence of stress on the adsorption, diffusion, and seepage of CH_4 in coal and shale and the dynamic change characteristics of coal and shale permeability, a fluid–solid coupling mathematical model of coal measure superimposed gas reservoirs was developed. Numerical simulations of coal measure gas output are carried out, and the gas production effect of coal measure gas wells under different drainage modes is discussed, which provides a scientific basis for the development of coal measure gas in the Dahebian block.

2. Coal Measure Superimposed Gas Reservoir Drainage Mathematical Model

2.1. Basic Assumptions

According to the differences in occurrence state and migration mechanism of coal measure gas in different reservoirs and previous studies [20–22], this study puts forward the following assumptions: (1) The reservoir is a 'double pore' medium and homogeneous in all directions. (2) The seepage of CH_4 and water in different reservoirs follows Darcy's law, and water and CH_4 are saturated in the fractures of coal and shale reservoirs. (3) The adsorption and desorption of CH_4 mainly occur in the pores of the coal and shale matrix, and the diffusion process of CH_4 in the matrix follows Fick's first diffusion law. (4) The deformation of rock mass conforms to the assumption of small deformation, while the adsorption, desorption, and effective stress of CH_4 will change the volume of the coal and shale matrix. (5) The reservoir temperature change is not considered in the process of coal measure gas drainage.

2.2. Governing Equations of Mechanical Field

According to assumption (4), considering the effective stress and the strain caused by the shrinkage effect of the coal and shale matrix, the governing equation for the stress field of coal and shale can be expressed as follows [23,24]:

$$G_{(1,2)}u_{i,jj} + \frac{G_{(1,2)}}{1-2v_{(1,2)}}u_{j,ji} + F_i = \alpha_{m(1,2)}P_{m(1,2),i} + \alpha_{f(1,2)}P_{f(1,2),i} + K_{(1,2)}\varepsilon_{a(1,2),i} \quad (1)$$

where

$$\begin{cases} \alpha_m = 1 - \frac{K}{K_s} \\ \alpha_f = 1 - \frac{K}{(aK_n)} \\ \varepsilon_a = a_{sg}V_{sg} \end{cases} \quad (2)$$

where (1, 2) is coal and shale, respectively; G is the shear modulus (Pa); u_i ($i = x,y,z$) is the displacement in the i direction; v is the Poisson's ratio; K is the bulk modulus (Pa); F_i is the force in the i direction; α_m and α_f are Biot effective pressure coefficients; K_n is the stiffness of fracture (Pa); P_f and P_m are the fluid pressure in the fracture and the gas pressure in the matrix (Pa); ε_a is the matrix desorption shrinkage strain; V_{sg} is the adsorbed gas content (m^3/kg); and a_{sg} is the adsorption strain coefficient (kg/m^3).

The fluid pressure (P_f) in the fracture can be expressed as follows [25–27]:

$$P_{f(1,\,2)} = S_{w(1,\,2)} P_{fw(1,\,2)} + S_{g(1,\,2)} P_{fg(1,\,2)} \tag{3}$$

where P_{fw} and P_{fg} are the water pressure and gas pressure in the fracture (Pa), respectively, while S_w and S_g are water saturation and gas saturation, respectively.

The content of adsorbed gas in coal and shale reservoirs can be expressed by the Langmuir volume equation as follows [9,27,28]:

$$V_{sg(1,\,2)} = \frac{V_{L(1,\,2)} P_{m(1,\,2)}}{P_{L(1,\,2)} + P_{m(1,\,2)}} \tag{4}$$

where V_L is the Langmuir volume constant (m^3/kg), and P_L is the Langmuir pressure constant (Pa).

2.3. Governing Equation of Hydraulic Field

Before drainage, CH_4 in the superimposed reservoir is in a dynamic equilibrium state, and the CH_4 pressure in the matrix is equal to the CH_4 pressure in the fracture. After the beginning of drainage, CH_4 in the matrix begins to desorb. According to the Fick's diffusion law and the mass conservation equation of CH_4 in the coal and shale matrix, the CH_4 migration equation in the coal and shale reservoir matrix can be expressed as follows [29]:

$$\frac{\partial m_{m(1,\,2)}}{\partial t} = -\frac{M_g}{\tau_{(1,\,2)} RT}\left(P_{m(1,\,2)} - P_{fg(1,\,2)}\right) \tag{5}$$

where (1, 2) is coal and shale, respectively; m_m is the methane content in the matrix (kg/m^3); τ is the methane desorption time (s); and M_g is the CH_4 molar mass (kg/mol).

$$m_{m(1,\,2)} = \left(\frac{V_{L(1,\,2)} P_{m(1,\,2)}}{P_{L(1,\,2)} + P_{m(1,\,2)}} \rho_{s(1,\,2)} \frac{M_g}{RT_s} P_s + \phi_{m(1,\,2)} \frac{M_g}{RT} P_{m(1,\,2)}\right) \tag{6}$$

where ρ_s is the coal skeleton density (kg/m^3), and ϕ_m is the porosity in the matrix system.

During the exploitation of coal measure gas, the coal and shale matrix continuously provides CH_4 to the fracture. The coal and shale matrix can be considered the mass source of CH_4 in the fracture. The mass conservation equation of CH_4 in the fracture of coal and shale can be expressed as follows [27,28]:

$$\begin{cases} \frac{\partial}{\partial t}\left(S_{g(1,\,2)} \phi_{f(1,\,2)} \frac{M_g P_{fg(1,\,2)}}{RT_{(1,\,2)}}\right) + \nabla\left[-\frac{M_g\left(P_{fg(1,\,2)} + b_1\right)}{RT_{(1,\,2)}} \frac{k_{rg} k_{(1,\,2)}}{\mu_g} \nabla P_{fg(1,\,2)}\right] = Q_g \\ \frac{\partial}{\partial t}\left(S_{w(1,\,2)} \phi_{f(1,\,2)} \rho_w\right) + \nabla\left[-\frac{\rho_w k_{rw} k_{(1,\,2)}}{\mu_w} \nabla P_{fw(1,\,2)}\right] = Q_w \end{cases} \tag{7}$$

where

$$\begin{cases} Q_g = \left(1 - \phi_{f(1,\,2)}\right) \frac{M_g}{\tau RT_{(1,\,2)}}\left(P_{m(1,\,2)} - P_{fg(1,\,2)}\right) \\ Q_w = 0 \end{cases} \tag{8}$$

where ϕ_f is the porosity in the fracture; u_g and u_w are the velocities of gas and water (Pa·s); Q_g and Q_w are gas and water sources or sinks, respectively; and k_{rg} and k_{rw} are the relative permeabilities of gas and water, respectively, which are given as follows [9,30,31]:

$$\begin{cases} k_{rg} = \left(1 - \frac{S_w - S_{wr}}{1 - S_{wr} - S_{gr}}\right)^2 \left[1 - \left(\frac{S_w - S_{wr}}{1 - S_{wr}}\right)^2\right] \\ k_{rw} = \left(\frac{S_w - S_{wr}}{1 - S_{wr}}\right)^4 \end{cases} \tag{9}$$

where S_{wr} is the irreducible water saturation, and S_{gr} is the residual gas saturation.

Under the influence of the vertical pressure difference, the fluid migrates to the reservoir with better permeability through interlayer flow. According to Darcy's law, the interlayer fluid flow equation can be expressed as follows [32]:

$$\frac{\partial}{\partial t}\left(\rho_g \phi_c P_f\right) + \frac{\partial}{\partial z}\left(\rho_g u_z\right) = 0 \tag{10}$$

where u_z is the velocity of interlayer flow, which can be expressed as follows [32]:

$$u_z = \frac{k_c(P_f)}{\mu_g} div(P_z) \tag{11}$$

where k_c is interlayer permeability, and $div(P_z)$ is the pressure gradient.

2.4. Porosity and Permeability Equations in Reservoirs

Considering the influence of effective stress and adsorption strain on porosity, the porosity of coal and shale reservoirs can be expressed as follows [27,33]:

$$\begin{cases} \phi_{(1,2)} = \dfrac{\left[\left(1+s_{0(1,2)}\right)\phi_{0(1,2)} + \alpha_f\left(s_{(1,2)} - s_{0(1,2)}\right)\right]}{\left(1+s_{(1,2)}\right)\phi_{0(1,2)}} \\ s_{(1,2)} = \varepsilon_{v(1,2)} + \dfrac{P_f}{K_{s(1,2)}} + \varepsilon_{a(1,2)} \end{cases} \tag{12}$$

where s_0 is the initial value of the variable, ε_a is the deformation of the coal and shale matrix caused by gas adsorption, ε_v is the volumetric strain, and K_s is the bulk modulus (Pa).

Based on the cubic law, the relationship between permeability and porosity can be derived. Specifically, the fracture permeability of the reservoir can be expressed as follows: [27,34,35]:

$$k_{(1,2)} = k_{0(1,2)}\left(\frac{\phi_{(1,2)}}{\phi_{0(1,2)}}\right)^3 \tag{13}$$

where k_0 is the initial absolute permeability (m^2).

3. Model Validation

3.1. Simulation Case

Liupanshui Coalfield is located in the west of Guizhou Province (Figure 1a). It is an important coal industry base in China and an important CBM exploration and development zone in China [36–39]. The Dahebian block is located in the Liupanshui coalfield (Figure 1b), which is rich in coalbed methane resources and has good development potential for coalbed methane resources [37,38]. The C409 coal seam is the main coal seam with a relatively stable distribution in this area, with a thickness of 1.17~13.8 m and an average thickness of 4.6 m. According to logging data, the gas content of the C409 coal seam in the W1 well is as high as 25.27 m^3/t, and each coal seam in the well field is a methane-rich coal seam (Figure 2). The C409 coal and its roof and floor generally contain gas, and the gas content of the coal seam is much higher than that of other lithologic reservoirs. The peak values of gas logging are generally 5~20 times those of non-coal seams (Figure 1c).

Before the production of coal measure gas wells in the Dahebian block, sand hydraulic fracturing was first carried out. Taking the W1 well as an example, fracture simulation was carried out using Fracpro PT fracturing software. The results of the fracture simulation showed that the half-length of the fracture, the total height of the fracture, and the total height of the supporting fractures were far greater than the thickness of the C409 coal seam and that the fracture extended to the roof and floor of the coal seam (Table 1).

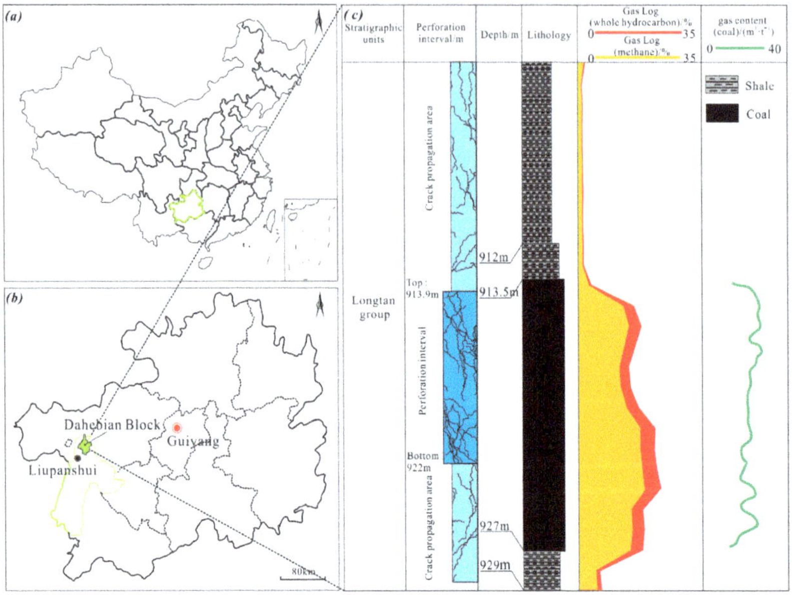

Figure 1. Location and gas content of the study area: (**a**) Guizhou Province in China, (**b**) the Dahebia block, and (**c**) gas content of the W1 well.

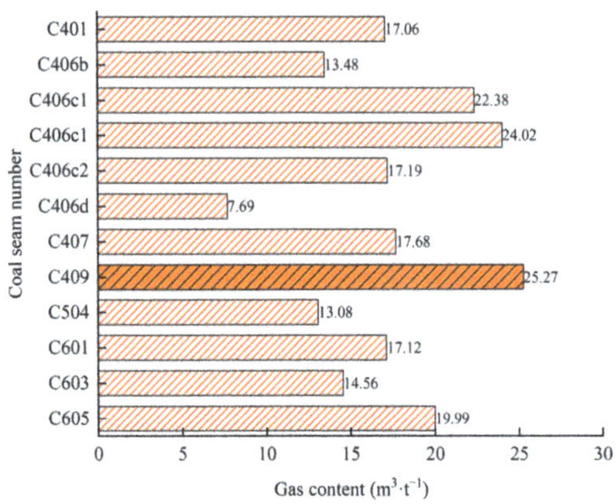

Figure 2. Gas content of each coal seam.

Table 1. Fracture simulation results.

Parameter	Value	Parameter	Value
Fracture half-length (m)	157.9/165.9	Support crack half-length (m)	147.3/160.4
Total crack height (m)	24.2/38.8	Total height of support cracks (m)	22.6/35.1
Vertical depth at the top of the crack (m)	903.2/889.9	Vertical depth of supporting crack top (m)	904.6/893.4
The bottom of the crack is deep (m)	927.4/928.7	The bottom of the support crack is deep (m)	927.2/928.5

In this study, the C409 coal and its roof and the floor coal reservoir of the W1 w in the Dahebian block of the Liupanshui mining area were selected as the objects to car out numerical simulation. The buried depth of the reservoir is 903 m, the thickness the C409 coal seam is 13.5 m, and the thickness of the roof shale is 10.5 m. The botto mud shale is 2 m thick. According to the formation structure of the production well a taking into account the computer memory and running speed, 1/4 of the area was selected according to symmetry to construct the physical model (Figure 3). The length and width were 250 m × 250 m, the height was the actual reservoir thickness of 26 m, and the radius of the gas well was 0.12 m. In order to observe the simulation effect of gas drainage in coal measures, the XZ plane was selected as the observation surface, and A (1,1,21), B (1,1,8), C (1,1,1), A′ (10,10,21), B′ (10,10,8), and C′ (10,10,1) were the spatial position coordinates (m) of the observation points (Figure 3c).

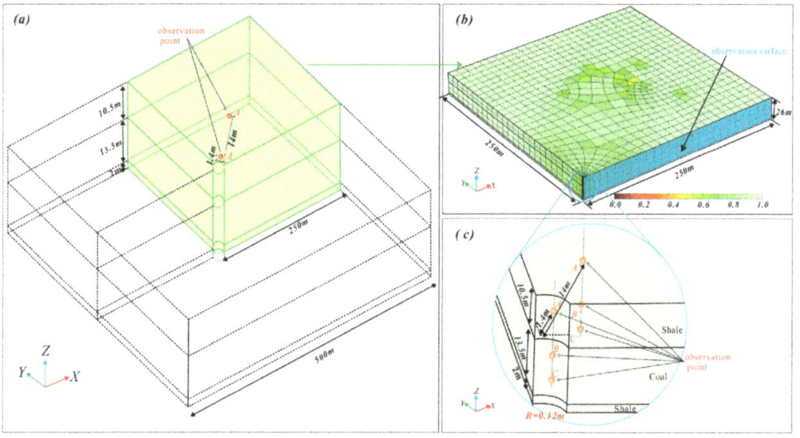

Figure 3. Geological model of coal measure gas drainage: (**a**) geometric model, (**b**) mesh partition, and (**c**) wellbore and observation points.

3.2. Numerical Parameters and Their Schemes

Experimental findings or the relevant literature were the sources for the parameters utilized in the numerical simulation of coal measure gas drainage (Table 2) [27,32,34,40].

Table 2. Key parameters for numerical simulation.

Variable	Parameter	Value	Unit
ϕ_{m10}	Initial porosity for coal matrix	4.50	%
ϕ_{f10}	Initial porosity for coal fracture	2.30	%
ϕ_{m20}	Initial porosity for shale matrix	3.20	%
ϕ_{f20}	Initial porosity for shale fracture	1.12	%
k_{10}	Initial reservoir permeability (Coal)	0.550	10^{-3} μm^2
k_{20}	Initial reservoir permeability (Shale)	0.197	10^{-3} μm^2
K_1	Bulk modulus (Coal)	3.0	GPa
K_2	Bulk modulus (Shale)	6.550	GPa
v_1	Poisson's ratio of coal	0.350	-
v_2	Poisson's ratio of shale	0.280	-
K_s	Skeleton bulk modulus	7.340	GPa
ρ_{s1}	Density of coal skeleton	1470	kg m^{-3}
ρ_{s2}	Density of shale	2660	kg m^{-3}
μ_w	Gas dynamic viscosity	1×10^{-3}	Pa·s
μ_g	Water dynamic viscosity	1.84×10^{-5}	Pa·s
P_{L1}	Langmuir pressure constant (Coal)	2.07	MPa
V_{L1}	Langmuir volume constant (Coal)	0.0256	m^3·kg^{-1}
P_{L2}	Langmuir pressure constant (Shale)	1.01	MPa
V_{L2}	Langmuir volume constant (Shale)	0.02	m^3·kg^{-1}
b_1	Klinkenberg factor	0.76	MPa
R	Gas molar constant	8.314	J·mol^{-1}·K^{-1}
P_s	Standard atmospheric pressure	101	kPa
T_s	Standard temperature	273.5	K

The accuracy of the mathematical model was verified by comparing it to the actual production history of coal measure gas wells. Combined with the actual production characteristics of gas wells, the variation law of reservoir characteristic parameters and differences in interlayer flow during the production process of coal measure gas under different drainage modes were studied. According to the conditions mentioned above, the simulations were divided into two categories: multi-layer drainage (full-stage perforation) and single-layer drainage (coal seam perforation) (Table 3).

Table 3. Schemes for numerical simulation.

Schemes	Drainage Pressure/MPa	Simulation Duration/d
Production history fitting	Actual bottom hole flowing pressure	500
Single-layer drainage (coal seam perforation)	0.16	2000
Multi-layer drainage (full-stage perforation)	0.16	2000

3.3. Boundary Conditions

The initial coal measure reservoir pressure was set as the initial condition (the coal seam and its roof and floor are in the same pressure system under the initial condition), P_{fg0} = 9.203 MPa; the actual bottom hole flow pressure of the gas well was set as the internal boundary condition, and the other boundaries were designated non-flow boundaries. For the physical model, the upper boundary was designated a vertical downward boundary load, the lower boundary was designated a fixed constraint, and the surrounding area was designated a slip boundary.

3.4. History Fitting

By comparing the measured daily gas production of CH_4 in the production well with the numerical simulation results, it can be seen that the simulated daily gas production is in good agreement with the measured daily gas production, with an error of 10.7%. The applicability and accuracy of the mathematical model are verified, which provides a basis for estimating dynamic changes in reservoir parameters in subsequent research and development (Figure 4).

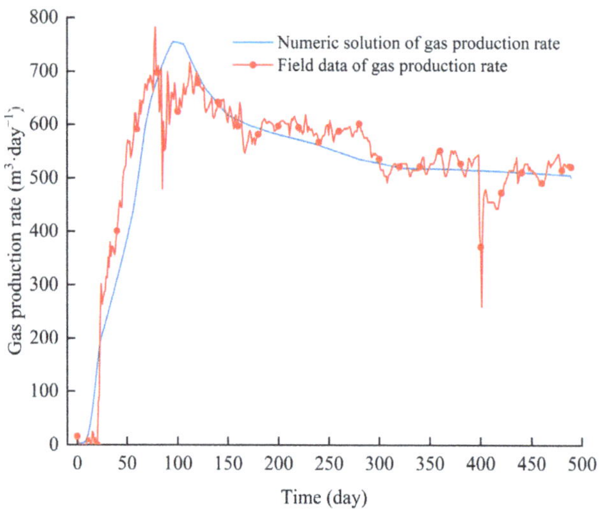

Figure 4. History fitting results of gas production.

4. Results

Compared with single-layer drainage, the cumulative gas production of full-layer drainage is increased by 33% (Figure 5a). Among them, coal seam gas production contributed the most, accounting for 75%; the shale roof was second, accounting for 21%; and the shale floor was the smallest, accounting for 4% (Figure 5b). It can be seen from Figure 5c that the cumulative gas production of the coal seam under single-layer drainage is slightly larger than that of the coal seam under full-layer drainage. This is because, during the single-layer drainage process, the methane in the roof and floor (shale) migrates to the coal seam under the influence of the vertical pore pressure difference and permeability difference among reservoirs and flows to the wellbore under the influence of the radial pore pressure difference of the reservoir. The rebound of the radial pore pressure difference and permeability of the roof and floor (shale) in the full-layer drainage process weakens the influence of the vertical pore pressure difference between reservoirs on the methane migration, so that the methane in the roof and floor (shale) flows radially to the wellbore, resulting in the cumulative gas production under single-layer drainage being slightly larger than the cumulative gas production of the coal seam under full-layer drainage.

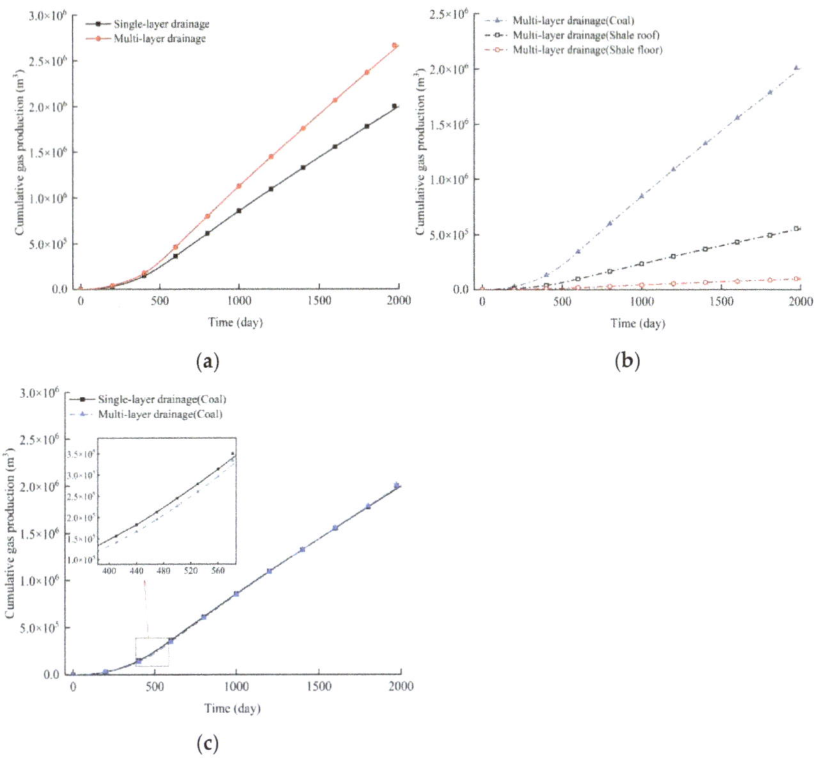

Figure 5. Cumulative gas production (**a**) under different drainage methods, (**b**) from each layer under multi-layer drainage, and (**c**) under different drainage methods (coal).

5. Discussion

5.1. Influencing Factors of Gas Production Effect of Coal Measure Gas Wells under Different Drainage Methods

5.1.1. Influence of Pressure Conduction Mode on Gas Production Effect

With the development of coal measure gas drainage, the reservoir pressure in each reservoir changes continuously. Affected by the difference in drainage methods and physical properties of different reservoirs, the reservoir pressure changes and conduction directions of coal and shale reservoirs are significantly different.

On the 30th day of drainage, the decreased range of reservoir pressure in the coal reservoir is larger than that in the shale reservoir under both drainage modes, which leads to the formation of a vertical reservoir pressure difference between the coal reservoir and the shale reservoir, and the reservoir pressure is transmitted from the high-reservoir-pressure area of the shale reservoir to the low-reservoir-pressure area of the coal reservoir (Figure 6). The pressure gradient is the main driving force for the migration of coal measure gas. Under the influence of the reservoir pressure difference between reservoirs, the migration direction of coal measures gas is also from the high-reservoir-pressure area of the shale reservoir to the low-reservoir-pressure area of the coal reservoir.

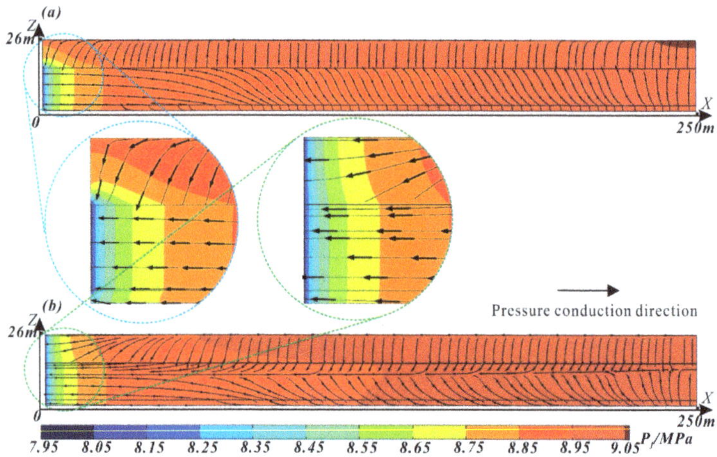

Figure 6. Direction of reservoir pressure conduction in the 30th day under different drainage mode: (**a**) single-layer drainage and (**b**) multi-layer drainage.

The pressure drop range of coal and shale reservoirs under single-layer drainage and multi-layer drainage increases gradually with the increase in drainage time. The decreased range of pore pressure in the shale reservoir near the coal seam is greater than that further away from the coal seam. Taking the reservoir pressure drop to the range of 7 MPa as an example, after 500 d, 1000 d, and 2000 d of drainage, the pressure drop range of the coal reservoir under multi-layer drainage increased by 1.5 m, 2.1 m, and 7.1 m, respectively, compared with single-layer drainage, and increased gradually with drainage time (Figure 7).

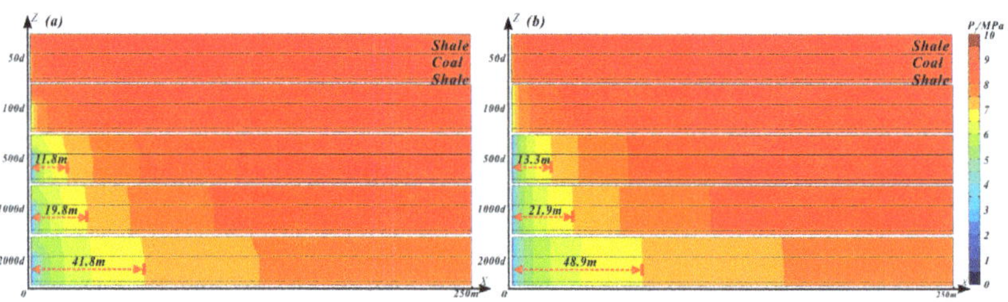

Figure 7. Dynamic distribution of reservoir pressure under different drainage methods: (**a**) single-layer drainage and (**b**) multi-layer drainage.

Combined with Figures 6 and 7, it can be seen that, compared with single-layer drainage, multi-layer drainage can effectively reduce the vertical reservoir pressure difference between layers, slow down the vertical pressure conduction between reservoirs, and promote the expansion of the radial reservoir pressure drop range of the overall reservoir.

5.1.2. The Influence of Gas Content Change on Gas Production Effect

Figure 8 shows the changes in matrix gas content with drainage time under different drainage modes at 1.4 m and 14 m from the wellbore.

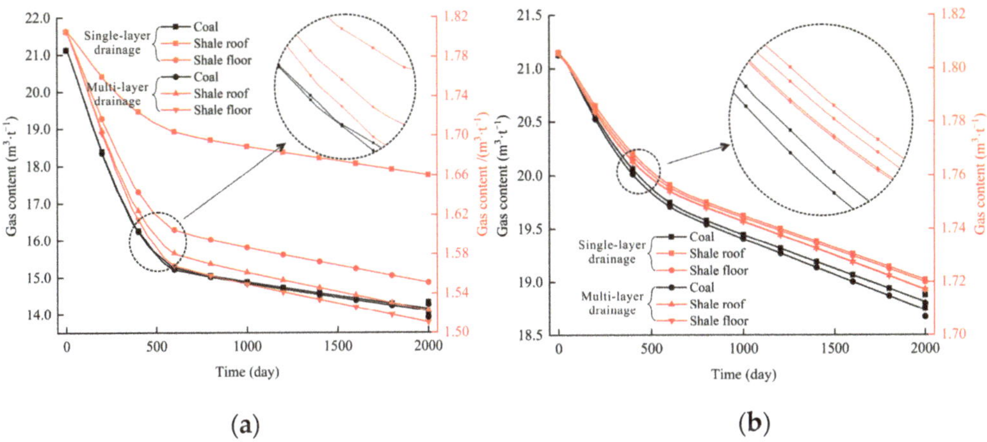

Figure 8. Change in matrix gas content under different drainage methods: (**a**) observation points (A, B, C) and (**b**) observation points (A′, B′, C′).

Under the two drainage methods, the gas content of the matrix at 1.4 m and 14 m from the wellbore showed a trend of gradual decrease with the extension of production time. Due to the differences in gas content, porosity, and permeability between coal and shale reservoirs, there are significant differences in the reduction in gas content between the two reservoirs.

At the end of single-layer drainage, the gas content of the coal reservoir decreases by 6.98 m^3/t, the gas content of the shale roof decreases by 0.15 m^3/t, and the gas content of the shale floor decreases by 0.25 m^3/t (Figure 8a); at 14 m from the wellbore, the gas content of the coal reservoir decreases by 2.31 m^3/t, the gas content of the shale roof decreases by 0.08 m^3/t, and the gas content of the shale floor decreases by 0.09 m^3/t. With the increase in distance, the drop in reservoir pressure decreases, and the desorption amount of the reservoir matrix decreases gradually (Figure 8b).

At the end of multi-layer drainage, the gas content of the coal reservoir at 1.4 m from the wellbore is reduced by 7.04 m^3/t, the gas content of the shale roof is reduced by 0.28 m^3/t, and the gas content of the shale floor is reduced by 0.30 m^3/t (Figure 8a); at a distance of 14 m from the wellbore, the gas content of the coal reservoir is reduced by 2.39 m^3/t, the gas content of the shale roof is reduced by 0.09 m^3/t, and the gas content of the shale floor is reduced by 0.09 m^3/t (Figure 8b). In addition, at a distance of 1.4 m from the wellbore, there is a significant difference in the reduction in gas content in the shale roof and floor under single-layer drainage. This is mainly due to the difference in reservoir pressure between the shale roof and floor reservoirs. The floor is thinner than the roof, and the reservoir pressure is more susceptible to the change in coal seam reservoir pressure. Under multi-layer drainage, the reservoir pressure conduction efficiency of the shale roof is improved, facilitating the desorption and diffusion of methane in the shale matrix. Compared with single-layer drainage, the reduction in gas content in each reservoir has increased, which is more conducive to the increase in production.

5.1.3. The Influence of Permeability on Gas Production Effect

Figure 9 shows the change in permeability ratio of each reservoir with drainage time under different drainage modes at different distances from the wellbore. At a distance of 1.4 m from the wellbore, under single-layer drainage, the proportion of coal seam permeability decreases first and then increases. Due to the rapid decrease in reservoir pressure of the coal reservoir at the initial stage of drainage, the amount of matrix desorption is small, and the effective stress effect is greater than the matrix shrinkage effect, resulting in an initial coal reservoir permeability decrease to the lowest value of 0.978 after 90 days

(Figure 9a). As the desorption amount of the coal matrix increases, matrix shrinkage increases, and permeability begins to increase (max = 1.153). Combined with Figure 7, it can be seen that after 500 days, the decrease in gas content of the coal matrix slows down, the shrinkage of the matrix weakens, and the increase in coal seam permeability ratio slows down.

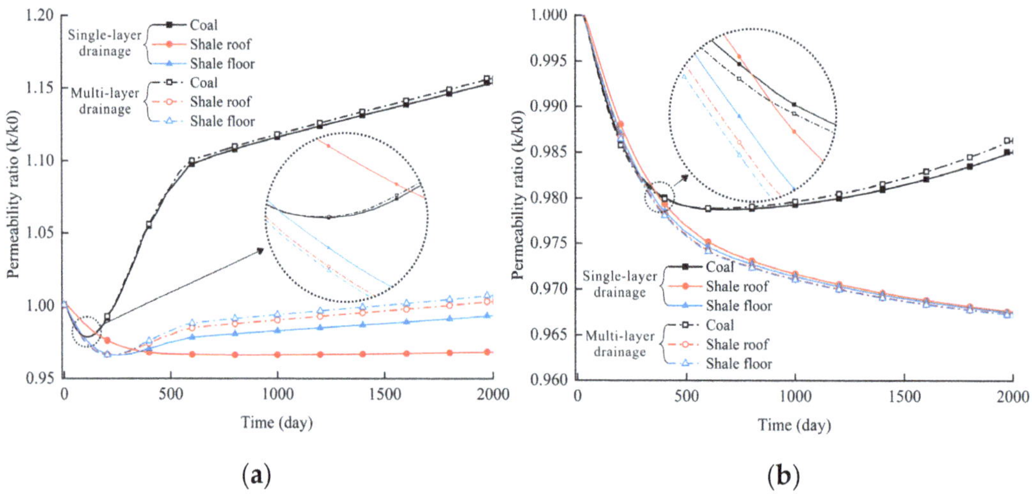

Figure 9. Permeability ratio changes under different drainage methods: (**a**) observation points (A, B, C) and (**b**) observation points (A′, B′, C′).

The permeability ratio of the shale roof shows a decreasing trend with the extension of production time. The changing trend for shale floor permeability ratio is similar to that for coal seam permeability ratio. This is mainly because the shale floor is thin, and the reservoir pressure of the reservoir is more susceptible to the influence of the coal seam, and then the shale floor matrix is more prone to desorption shrinkage, so the permeability ratio has a slight rebound. Compared with single-layer drainage, the effect of multi-layer drainage on coal reservoir permeability is relatively small, and the permeability ratio only increases by about 0.003. For the shale roof and floor, the maximum permeability ratio increases by 0.034 and 0.013, respectively.

At 14 m away from the wellbore, the permeability of each reservoir shows a decreasing trend under the two drainage methods; the permeability ratio of the coal reservoir has a slight rebound after 500 days (Figure 9b). This is mainly due to the permeability change away from the wellbore being dominated by the effective stress, resulting in a decrease in the permeability ratio.

5.2. Enlightenment of Coal Measure Gas Development Project

The simulation results show that multi-layer drainage helps increase production. Under multi-layer drainage, the shale reservoir obtains a better seepage channel, thereby improving the overall reservoir pressure transfer efficiency of the superimposed reservoir, promoting matrix desorption and permeability rebound, and achieving the purpose of increasing production. However, due to the large differences in permeability, porosity, and mechanical properties of different lithologic reservoirs, the selection of perforation and perforation intervals and the selection of fracturing technology for non-coal seam sections in multi-layer drainage need to be further studied.

6. Conclusions

A fluid–solid coupling mathematical model for typical coal-series superimposed gas reservoirs was developed and solved by the finite element method. The simulation results show that the actual gas production and the simulated gas production fit well, which verifies the accuracy of the mathematical model.

Based on the numerical simulation results, the effects of different drainage methods on the production effect were discussed from three perspectives: reservoir pressure, matrix gas content, and the permeability of superimposed reservoirs. The results show that, compared with single-layer drainage, cumulative gas production under multi-layer drainage increases by 33%; both drainage methods exhibit interlayer energy and material transfer; affected by permeability, porosity, and mechanical properties, there are obvious differences in reservoir pressure and conduction direction, matrix gas content, and permeability ratio between coal and shale reservoirs under different drainage methods. Multi-layer drainage is more conducive to the reservoir pressure conduction of shale reservoirs, promotes the desorption of shale matrix methane, enhances matrix shrinkage, promotes an increase in shale reservoir permeability, and thus increases gas production.

Author Contributions: Conceptualization, S.L. and S.S.; Data curation, W.W., Y.L. and R.D.; Formal analysis, S.L. and W.W.; Investigation, W.W., Y.L. and R.D.; Methodology, W.W.; Project administration, S.L. and S.S.; Resources, S.L. and S.S.; Writing—original draft, W.W., S.L., R.D. and Y.L.; Writing—review and editing, W.W. All authors have read and agreed to the published version of the manuscript.

Funding: This research was funded by the National Natural Science Foundation of China (No. 42030810, 41972168) and was supported by the Fundamental Research Funds for the Central Universities (No. 2023KYJD1001).

Data Availability Statement: Data are unavailable due to privacy or ethical restrictions.

Conflicts of Interest: The authors declare no conflict of interest.

eferences

Bi, C.; Hu, Z.; Tang, D.; Tao, S.; Zhang, J.; Tang, S.; Huang, H.; Tang, Y.; Yuan, Y.; Xu, Y.; et al. Research progress of coal measure gas and some important scientific problems. *Geol. China* **2021**, *48*, 402–423.

Cao, D.; Yao, Z.; Li, J. Evaluation Status and Development Trend of Unconventional Gas in Coal Measure. *Coal Sci. Technol.* **2014**, *42*, 89–92.

Yi, T.; Gao, W. Reservoir formation characteristics as well as co-exploration and co-mining orientation of Upper Permian coal-bearing gas in Liupanshui Coalfield. *J. China Coal Soc.* **2018**, *43*, 1553–1564.

Gao, W.; Yi, T.; Yan, Z.; Jin, J. Reservoir forming conditions and exploration direction of coal measure gas in Guizhou Province. *Nat. Gas Geosci.* **2022**, *33*, 799–806.

Zhang, Z.; Qin, Y.; You, Z.; Yang, Z. Distribution characteristics of in situ stress field and vertical development unit division of CBM in Western Guizhou, China. *Nat. Resour. Res.* **2021**, *30*, 3659–3671. [CrossRef]

Zhang, Z.; Qin, Y.; Yang, Z.; Li, G.; You, Z. Primary Controlling Factors of Coalbed Methane Well Productivity and High Productive Well Patterns in Eastern Yunnan and Western Guizhou, China. *Nat. Resour. Res.* **2023**, *32*, 2711–2726. [CrossRef]

Qin, Y.; Wu, J.; Li, G.; Wang, Y.; Shen, J.; Zhang, B.; Shen, Y. Patterns and pilot project demonstration of coal measures gas production. *J. China Coal Soc.* **2020**, *45*, 2513–2522.

Sang, S.; Zhou, X.; Liu, S.; Huang, H.; Wang, R.; Han, S.; Wang, J.; Pan, D.; Zhang, Z. A review of mechanical stratigraphy methodology and its application in high-efficient exploration and development of coal measure gas. *Acta Geol. Sin.* **2022**, *53*, 394–402.

Che, J.; Wang, H.; Zhang, Y.; Wang, Y.; Liu, Y.; Du, M. Field test and numerical simulation of the section mill in U-shaped wells of coalbed methane. *J. Nat. Gas Sci. Eng.* **2020**, *84*, 103681. [CrossRef]

Fan, Y.; Shu, L.; Huo, Z.; Hao, J.; Yang, L. Numerical simulation research on hydraulic fracturing promoting coalbed methane extraction. *Shock Vib.* **2021**, *2021*, 3269592.

Perera, M.S.A.; Ranjith, P.G.; Ranathunga, A.S.; Koay, A.Y.J.; Zhao, J.; Choi, S.K. Optimization of enhanced coal-bed methane recovery using numerical simulation. *J. Geophys. Eng.* **2015**, *12*, 90–107. [CrossRef]

Fang, H.; Sang, S.; Liu, S. Numerical simulation of enhancing coalbed methane recovery by injecting CO_2 with heat injection. *Petrol. Sci.* **2019**, *16*, 32–43. [CrossRef]

Yang, R.; Ma, T.; Xu, H.; Liu, W.; Hu, Y.; Sang, S. A model of fully coupled two-phase flow and coal deformation under dynamic diffusion for coalbed methane extraction. *J. Nat. Gas Sci. Eng.* **2019**, *72*, 103010. [CrossRef]

Ma, T.; Rutqvist, J.; Oldenburg, C.M.; Liu, W.; Chen, J. Fully coupled two-phase flow and poromechanics modeling of coalbed methane recovery: Impact of geomechanics on production rate. *J. Nat. Gas Sci. Eng.* **2017**, *45*, 474–486. [CrossRef]

15. Meng, S.; Li, Y.; Wang, L.; Wang, K.; Pan, Z. A mathematical model for gas and water production from overlapping fractured coalbed methane and tight gas reservoirs. *J. Pet. Sci. Eng.* **2018**, *171*, 959–973. [CrossRef]
16. Feng, Q.; Zhang, X.; Zhang, J.; Chen, D. Numerical simulation of commingling production for coalbed methane and adjoining sandstone gas reservoirs. *J. China Coal Soc.* **2014**, *39*, 169–173.
17. Li, Y.; Meng, S.; Wu, P.; Wang, Z.S.; Yu, Z.L. Numerical simulation of coal measure gases coproduction. *J. China Coal Soc.* **2018**, *43*, 1728–1737.
18. Xie, Y. Theory and Method for Compatibility Evaluation of Joint Deep Coalbed Methane and Tight Sandstone Gas Production: In a Case of Member C_{2b}~P_{1t} in Linxing Block. Ph.D. Thesis, China University of Mining and Technology, Xuhzou, China, 2022.
19. Li, L.; Kang, T.; Zhang, X.; Guo, J. Percolation Model and Numerical Simulation of Coal-based Gas in Coal-Sandstone Composite Reservoir. *Min. Res. Dev.* **2019**, *39*, 43–47.
20. Zhang, H.; Liu, J.; Elsworth, D. How sorption-induced matrix deformation affects gas flow in coal seams: A new FE model. *Int. J Rock Mech. Min. Sci.* **2008**, *45*, 1226–1236. [CrossRef]
21. Xia, T.; Zhou, F.; Gao, F.; Kang, J.; Liu, J.; Wang, J. Simulation of coal self-heating processes in underground methane-rich coal seams. *Int. J. Coal Geol.* **2015**, *141–142*, 1–12. [CrossRef]
22. Liu, Q.; Cheng, Y.; Zhou, H.; Guo, P.; An, F.; Chen, H. A mathematical model of coupled gas flow and coal deformation with gas diffusion and Klingenberg effects. *Rock Mech. Rock Eng.* **2015**, *48*, 1163–1180. [CrossRef]
23. Wang, G.; Yang, X.; Chu, X.; Shen, J.; Jiang, C. Microscale numerical simulation of Non-Darcy flow of coalbed methane. *Arab. J. Sci. Eng.* **2018**, *43*, 2547–2561. [CrossRef]
24. Kong, X.; Wang, E.; Liu, Q.; Li, Z.; Li, D.; Cao, Z.; Niu, Y. Dynamic permeability and porosity evolution of coal seam rich in CBM based on the flow-solid coupling theory. *J. Nat. Gas Sci. Eng.* **2017**, *40*, 61–71. [CrossRef]
25. Fang, H.; Li, A.; Sang, S.; Gu, C.; Yang, J.; Li, L.; Liu, H.; Xu, H.; Huang, Y. Numerical analysis of permeability rebound and recovery evolution with THM multi-physical field models during CBM extraction in crushed soft coal with low permeability and its indicative significance to CO_2 geological sequestration. *Energy* **2023**, *262*, 125395. [CrossRef]
26. Rutqvist, J.; Wu, Y.-S.; Tsang, C.-F.; Bodvarsson, G. A modeling approach for analysis of coupled multiphase fluid flow, heat transfer, and deformation in fractured porous rock. *Int. J. Rock Mech. Min. Sci.* **2002**, *39*, 429–442. [CrossRef]
27. Li, S.; Fan, C.; Han, J.; Luo, M.; Yang, Z.; Bi, H. A fully coupled thermal-hydraulic-mechanical model with two-phase flow for coalbed methane extraction. *J. Nat. Gas Sci. Eng.* **2016**, *33*, 324–336. [CrossRef]
28. Zhu, W.; Wei, C.; Liu, J.; Qu, H.; Elsworth, D. A model of coal–gas interaction under variable temperatures. *Int. J. Coal Geol.* **2011**, *86*, 213–221. [CrossRef]
29. An, F.-H.; Cheng, Y.-P.; Wang, L.; Li, W. A numerical model for outburst including the effect of adsorbed gas on coal deformation and mechanical properties. *Comput. Geotech.* **2013**, *54*, 222–231. [CrossRef]
30. Xu, H.; Tang, D.; Tang, S.; Zhao, J.; Meng, Y.; Tao, S. A dynamic prediction model for gas–water effective permeability based on coalbed methane production data. *Int. J. Coal Geol.* **2014**, *121*, 44–52. [CrossRef]
31. Naik, S.; You, Z.; Bedrikovetsky, P. Rate enhancement in unconventional gas reservoirs by wettability alteration. *J. Nat. Gas Sci. Eng.* **2015**, *26*, 1573–1584. [CrossRef]
32. Li, L. Research on Migration Mechanism and Application of Coal Measures Gas Considering Interlayer Crossflow and Dynamic Slipstream. Ph.D. Thesis, Taiyuan University of Technology, Taiyuan, China, 2019.
33. Liu, X.; Sang, S.; Zhou, X.; Wang, Z. Coupled adsorption-hydro-thermo-mechanical-chemical modeling for CO_2 sequestration and well production during CO_2-ECBM. *Energy* **2023**, *262*, 125306. [CrossRef]
34. Pan, Z.; Connell, L.D. A theoretical model for gas adsorption-induced coal swelling. *Int. J. Coal Geol.* **2007**, *69*, 243–252. [CrossRef]
35. Wu, Y.; Liu, J.; Elsworth, D.; Miao, X.; Mao, X. Development of anisotropic permeability during coalbed methane production. *J. Nat. Gas Sci. Eng.* **2010**, *2*, 197–210. [CrossRef]
36. Sang, S.; Zheng, S.; Yi, T.; Zhao, F.; Han, S.; Jia, J.; Zhou, X. Coal measures superimposed gas reservoir and its exploration and development technology modes. *Coal Geol. Explor.* **2022**, *50*, 13–21.
37. Gao, W.; Han, Z.; Jin, J.; Bai, L.; Zhou, P. Occurrence characteristics and assessment of favorable a-reas of coalbed methane exploration in Liupanshui coalfield. *Coal Geol. Explor.* **2018**, *46*, 81–89.
38. Jin, J.; Yang, Z.; Qin, Y.; Cui, Y.; Wang, G.; Yi, T.; Wu, C.; Gao, W.; Chen, J.; Li, G.; et al. Progress, potential and prospects of CBM development in Guizhou Province. *J. China Coal Soc.* **2022**, *47*, 4113–4126.
39. Zhang, Z.; Qin, Y.; Yi, T.; You, Z.; Yang, Z. Pore structure characteristics of coal and their geological controlling factors in Eastern Yunnan and Western Guizhou, China. *ACS Omega* **2020**, *5*, 19565–19578. [CrossRef]
40. Song, X.; Jiang, M.; Peng, Q.; Xiong, P. Thermal property parameters and influencing factor analysis of main rock strata in Guizhou province. *Acta Geol. Sin.* **2019**, *93*, 2092–2103.

processes

MDPI

Article

Study on Connectivity Analysis and Injection–Production Optimization of Strong Heterogeneous Sandstone Reservoir Based on Connectivity Method

Yuhui Zhou [1,2], Liang Pu [1,2,3,*], Sisi Dang [4], Jibo He [5] and Shuang Pu [6]

1 School of Petroleum Engineering, Yangtze University, Wuhan 430100, China
2 Western Research Institute, Yangtze University, Karamay 834000, China
3 Key Laboratory of Drilling and Production Engineering for Oil and Gas, Wuhan 430100, China
4 Luliang Oilfield Operation Area, Xinjiang Oilfield Company, PetroChina, Karamay 834000, China
5 The 10th Oil Production Plant, Changqing Oilfield Company, PetroChina, Qingcheng 745100, China
6 Army Logistics Academy, Chongqing 401311, China
* Correspondence: puliang_2021@163.com

Abstract: The D reservoir in the Bongor Basin, southern Chad, is highly heterogeneous. In the stage of waterflood development, the injected water is seriously channeled along the dominant channel, and the water drive effect becomes worse. At the same time, due to the strong edge and bottom water, the water flooding situation is aggravated, the water cut is increased, and the development efficiency is reduced. To accurately identify the inter-well connectivity relationship, we developed a reservoir inter-well connectivity model based on the principle of inter-well connectivity and dynamic production data and reservoir geological parameters. Thus, the plane water injection split coefficient and water injection efficiency of each reservoir layer were obtained. The results are in good agreement with the calculation results of inter-well connectivity through verification with field tracer interpretation. The practical application results show that the method can increase the annual output of oil by 1.3%, which has a good oil increase effect. In this study, a model of inter-well connectivity in multi-layer sandstone reservoirs was established for the first time. The production performance of the model injection–production well was optimized in real time by a historical fitting and production optimization algorithm and then applied to real reservoirs, so that it could effectively improve the oilfield development and optimize the injection–production structure.

Keywords: connectivity; conductivity; injection and production optimization; flushing efficiency

Citation: Zhou, Y.; Pu, L.; Dang, S.; He, J.; Pu, S. Study on Connectivity Analysis and Injection–Production Optimization of Strong Heterogeneous Sandstone Reservoir Based on Connectivity Method. Processes 2023, 11, 2816. https://doi.org/10.3390/pr11102816

Academic Editors: Tao Zhang, Dong Feng, Hung Vo Thanh, Wen Zhao, Adam Smoliński and Federica Raganati

Received: 27 August 2023
Revised: 13 September 2023
Accepted: 21 September 2023
Published: 22 September 2023

1. Introduction

Water injection development is the main development method of major oil fields at present. After a long period of water injection, major oil fields have gradually entered the middle and high water cut periods [1]. After years of waterflooding development, the oil production rate of the D block in Chad, Africa has been continuously reduced. At the same time, due to the strong heterogeneity of the reservoir, water channeling is serious in the process of water flooding development. Moreover, due to the influence of strong edge and bottom water, water flooding occurs in multiple wells, the water content rises, and the development efficiency is reduced. Thus, how to combine the existing understanding of reservoirs, quickly formulate a targeted and effective block injection–production optimization plan, adjust the injection–production structure, and improve the current situation of reservoir development is the main problem that needs to be solved urgently. Accurately identifying the injection–production connectivity between production wells and injection wells is an important factor affecting the injection–production effect. As an important research method, the inter-well connectivity method can effectively identify the connectivity between oil and water wells.

In recent years, research on inter-well connectivity has mainly been carried out from both static and dynamic aspects. The study of static connectivity is usually to evaluate the initial state before reservoir development. At present, numerous methods have been developed to describe and evaluate the static connectivity of reservoirs, including the micro-seismic analysis method, logging interpretation method, geological modeling method, and reservoir parameter analysis method. The micro-seismic analytical method mainly analyzes the structure and properties of underground rock strata by using the velocity difference of seismic waves propagating in different rock strata. Liu et al. [2] proposed a method for evaluating fracture-well connectivity in fractured sandstone reservoirs by using micro-seismic monitoring technology and tracer analysis results, and quantitatively evaluated fracture-well connectivity. Mammad Mirzayev et al. [3] analyzed the key factors affecting the development of water flooding based on the connectivity index, and combined the results of geological and seismic analysis for correlation analysis. The logging interpretation method is one of the important means to evaluate the static inter-well connectivity of reservoirs. It mainly obtains information on rock physical parameters, permeability, and porosity, and then establishes the relevant logging model to judge the connectivity between oil and water wells. Qu et al. [4] analyzed the change in reservoir plane heterogeneity by the logging interpretation static method, and found that the reservoir plane heterogeneity in the well area of the Bohai L oilfield is strong. Zhang et al. [5] studied the pore and fracture development characteristics of the deep Sinian Dengying Formation reservoir in the Sichuan Basin by logging methods, and the application of logging methods in evaluating reservoir effectiveness. The reservoir parameter analysis method infers the fluid connectivity between different wells by analyzing the differences in oil and gas composition and physical parameters of different wells. Patidar et al. [6] analyzed the relative content of alkanes such as C1–C5 by gas chromatography and liquid chromatography. Zhang et al. [7] evaluated the reservoir properties and inter-well static connectivity of different well sections by analyzing the parameters such as liquid production, gas production, and water content of different well sections. The geological modeling method divides the reservoir into different phases by analyzing factors such as reservoir rock type, pore structure, and pore connectivity. Fang et al. [8] proposed a phase model to describe the physical properties of sedimentary rocks. Through a phase analysis, the connectivity of pores in the reservoir can be judged more accurately. Ye et al. [9] proposed a well seismic four-in-one reservoir-modeling method to improve the accuracy of the geological model in the later stage of oilfield development.

Although the static method is mostly simple when evaluating inter-well connectivity it cannot accurately characterize the changes in the process of reservoir oil and gas migration. Through the study of dynamic connectivity, we can more accurately understand the flow path and speed of oil and gas in the reservoir at different times and different locations. When the reservoir changes, we can also grasp its evolution law over time. At present, numerous methods have been established to describe and evaluate the dynamic connectivity between wells. These include the tracer analysis method, numerical simulation method, and dynamic data inversion method. Tracer analysis is a method to study the transmission and mixing process of fluid in underground reservoirs. It is one of the most commonly used methods for evaluating inter-well dynamic connectivity. Cao et al. [10] quantitatively evaluated the dynamic connectivity between injection and production wells by analyzing the parameters of tracer time, concentration, continuous seeing time, and peak concentration in the benefit wells. Chen et al. [11] studied the inter-well connectivity of the injection–production well group in the Suizhong 36-1 oilfield by using the inter-well trace substance tracer technology, which provided quantitative technical parameters for the effective development of the oilfield in the later stage. Yu et al. [12] used the multi-component definition function of CMG 2017 numerical simulation software to match the history of single-well polymer production mass concentration, and realized the identification of the polymer production direction and the analysis of the output. Li et al. [13] quantitatively characterized the high-permeability channel and reservoir heterogeneity in the Changqing Oilfield by comprehensively analyzing the results of trace substance tracer monitoring and reservoir numerical simulation. Albertoni et al. [14] used the multiple linear regression model combined with the diffusion filter to clarify the main factors affecting the inter-well connectivity and quantitatively characterize the inter-well connectivity. Yousef et al. [15] proposed a capacitance–resistance model based on material balance, and Sayarpour et al. [16] proposed three improved capacitance–resistance models based on the research of the capacitance–resistance model. Based on the multi-well productivity index method, Kaviani et al. [17] evaluated the inter-well dynamic connectivity by considering the dynamic production factors such as location, skin factor, injection rate, and bottom hole flowing pressure. Based on the dynamic production data of injection–production wells, Liu et al. [18] used the combination of the Kalman filter and an artificial neural network to quantitatively characterize the dynamic connectivity between reservoirs. However, the calculation speed and accuracy of these models are slow, and they cannot consider many complex cases at the same time. Thus, Zhao et al. [19,20] proposed a connectivity model that does not depend on complex fine modeling. The reservoir injection–production system is simplified by the inter-well conductivity and connected volume. Modeling and simulation calculations are carried out by using dynamic reservoir production data and geological parameters. On this basis, the reservoir production optimization control model is established. By optimizing the injection–production structure, the injection–production optimization scheme design is completed to guide the field production application.

2. Geological Setting and Field Example

The Bongor Basin is located at the intersection of the Central African and West African rift systems, southwest of the Republic of Chad, and the northern margin of the Central African shear zone [21–23] (Figure 1). It is a meso-Cenozoic intracontinental inversion rift basin developed under the influence of the Central African shear zone. The sedimentary characteristics of the Bongor Basin reservoir mainly include five categories: gray siltstone, gray claystone, gray mudstone, gray-brown sandstone, and gray-green sandstone. These reservoirs have good porosity and permeability and are important oil and gas reserves in the Bongor Basin. The D block reservoir is located in the Bongor basin, which is an edge–bottom water reservoir controlled by structural lithology. The buried depth in the middle of the reservoir is 1270 m, and the reservoir is thick in the middle and thin at both ends. From top to bottom, the reservoir is more developed.

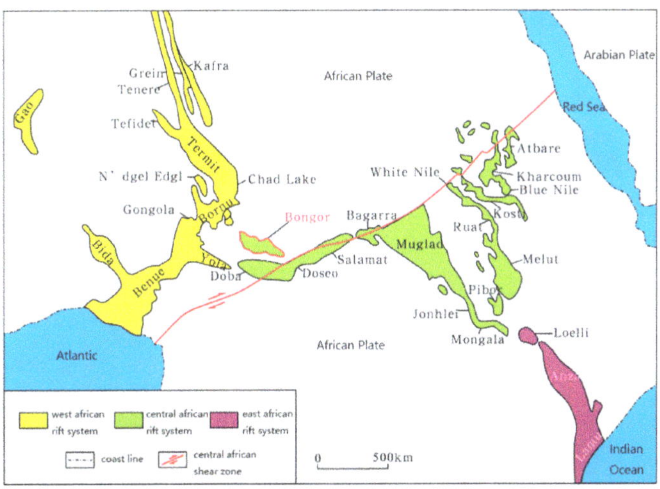

Figure 1. Regional tectonic location of Bongor basin.

3. Reservoir Inter-Well Connectivity Model

The conductivity and connected volume in the inter-well connectivity model proposed by Zhao Hui represent the flow capacity and material basis of the reservoir unit, respectively. The simulation calculation takes less time and the history matching accuracy is higher. However, at present, the application scenarios of this method are mostly single-layer sandstone reservoirs and have not been effectively extended to multi-layer sandstone reservoirs.

Therefore, based on the theory of inter-well connectivity [24], this paper will apply the method to multi-layer sandstone reservoirs. Through the characteristic production and development parameters of D reservoir in Chad, a multi-layer inter-well connectivity model is constructed to finely characterize the injection–production relationship between multi-layer injection wells and production wells in the study area. Using field tracer data to analyze and adjust the inter-well conductivity and connectivity volume values, the inter-well connectivity field is obtained in real-time. Finally, according to the average water injection efficiency of the water injection wells in the study block, each water injection well is divided into high-efficiency wells and low-efficiency wells. Through the optimization algorithm, the optimal control of the injection volume of the water injection well is realized and the purpose of controlling water-increasing oil and identifying the dominant channel of the reservoir is achieved.

3.1. Model Moisture Content Solution

Considering the oil–water two-phase flow under adiabatic conditions, ignoring the influence of gravity, the viscosity of the fluid remains constant. Based on this, the material balance equation is established:

$$\sum_{j=1}^{n_{c,i}} T_{i,j}(p_j(t) - p_i(t)) + q_i(t) = \frac{dp_i(t)}{dt} c_{t,i}(t) V_{p,i}(t) \quad (1$$

In Formula (1), $n_{c,i}$ represents the number of wells connected to well i; the unit $q_i(t$ is m^3/d. It represents the inflow and outflow liquid volume of the source and sink phas of the ith well under reservoir conditions. If well i is a production well, then the liqui volume is positive. If it is a water injection well, then the liquid volume is negative. p is the average pressure of the connected unit $V_{P,i}$. $c_{t,i}$ is the comprehensive compressio coefficient of well i. The IMPES (implicit pressure explicit saturation) scheme is used t rewrite the formula, and the difference scheme is obtained as follows:

$$p_i^n - p_i^{n-1} = \frac{\Delta t^n}{c_{t,i}^{n-1} V_{p,i}^{n-1}} \left(-\left(\sum_{j=1}^{n_{c,i}} T_{i,j}^{n-1}\right) p_i^n + \sum_{j=1}^{n_{c,i}} T_{i,j}^{n-1} p_j^n + q_i^{n-1} \right) \quad (2$$

among them,

$$T_{i,j}^{n-1} = 11.57 \frac{A_{i,j} \lambda_{i,j}^{n-1}}{L_{i,j}} = T_{i,j}^0 \frac{\lambda_{i,j}^{n-1}}{\lambda_{i,j}^0} \quad (3$$

$$V_{p,i}^{n-1} = V_{p,i}^0 (1 + C_{t,i}(p_i^{n-1} - p_i^0)) \quad (4$$

$$C_{t,i}^{n-1} = C_r + S_{w,i}^{n-1} C_w + S_{o,i}^{n-1} C_o \quad (5$$

In Formulas (3)–(5), $A_{i,j}$ and $L_{i,j}$ are the seepage cross-sectional area between wel and well j and the length of the connected unit, respectively, and the units are m^2 and $T_{i,j}^0$ and $T_{i,j}^{n-1}$ refer to the conductivity value between well i and well j at the initial tin and time $n-1$; $m^3 \cdot d^{-1} \cdot MPa^{-1}$. $V_{p,i}^0$ and $V_{p,i}^{n-1}$ refer to the control volume of well i at th initial time and the $n-1$ time point; m^3. $\lambda_{i,j}^0$ and $\lambda_{i,j}^n$ refer to the mobility value betwe well i and well j at the initial time and time $n-1$, respectively; 10^{-3} $\mu m^2 \cdot (mPa \cdot s)^{-1}$. C_w, and C_o refer to the compression coefficient of reservoir rock, reservoir crude oil, a reservoir water; MPa^{-1}. $S_{w,i}$ and $S_{o,i}$ refer to the water saturation and oil saturation at well i. The formula in [25] can be obtained by the implicit difference discrete calculation of the Formula (1). The average pressure value of the single-well control area of each well at any time is obtained, and the flow rate in the connected unit is obtained by combining the conductivity between wells.

$$q_{i,j}^n = T_{i,j}^n(p_j^n - p_i^n) \tag{6}$$

In Formula (6): $q_{i,j}^n$ is the flow rate from well j to well i; m^3/s.

Considering the influence of edge and bottom water in the reservoir, it is deduced according to the theory of front edge propulsion [26–28]. By setting an upstream node position x_u of x, the saturation of the corresponding node position is S_{wu}, and it is concluded that:

$$f'_w(S_w) = f'_w(S_{wu}) + \frac{1}{Q_{Dn}} \tag{7}$$

In Formula (7): $f'_w(S_w)$ is the derivative of water content at x_u; Q_{Dn} represents the dimensionless cumulative injection amount from x_u to x at time n;$f'_w(S_w)$ represents the water content derivative of the reservoir unit at position x.

According to the above calculation results, the water cut derivative corresponding to well j tracking to downstream well i at time n can be obtained:

$$f'_w(S_{wnij}) = \min\left\{ f'_w(S_{wnj}) + \frac{1}{Q_{Dnij}}, f'_w(S_{wki})\frac{Q_{Dkij}}{Q_{Dnji}} \right\} \tag{8}$$

In Formula (8), $f'_w(S_{wnj})$ and $f'_w(S_{wki})$ represent the derivative of water content at well j and i at time n and k, respectively; Q_{Dnij} and Q_{Dnji} denote the dimensionless cumulative flow from well $j$to well i and the dimensionless cumulative flow from well i to well j at time n, respectively. Q_{Dkij} is the dimensionless cumulative flow from well i to well j at time k.

The interpolation calculation of Formula (8) is carried out to obtain the water content of well i from the direction of well j, and then the respective water content in the upstream direction is obtained. Finally, the water content of well i at time n can be obtained:

$$f_w(S_{wni}) = \frac{\sum\limits_{j=1}^{m} q_{ij} f_w(S_{wnij})}{\sum\limits_{j=1}^{m} q_{ij}} \tag{9}$$

In Formula (9): $f_w(S_{wni})$ represents the comprehensive water cut of well i.

3.2. Connectivity Parameter Inversion

By continuously updating, adjusting, and optimizing the connected parameter values, namely the connected volume and the connected conductivity, the calculated block cumulative oil production, block water content, single-well oil production, single-well water content, and other production dynamics are historically matched. The connected characteristic parameter matrix c is solved by using the random disturbance and projection gradient algorithm [29] so that the objective function $O(c)$ can obtain the minimum value.

$$O(c) = \tfrac{1}{2}(c - c_r)^T K_B^{-1}(c - c_r) + \tfrac{1}{2}[m_{obs} - h(c)]^T K_D^{-1}[m_{obs} - h(c)] \quad (c \geq 0) \tag{10}$$

In Formula (10): c denotes the connected parameter matrix; c_r represents the prior feature parameter model estimation; K_B^{-1} represents the inverse covariance matrix of the model parameters; m_{obs} denotes the measured data; K_D^{-1} is the covariance matrix of the prior model characteristic parameters; $h(b)$ is the initial value of the results calculated by the reservoir numerical simulator. At the same time, the constraint conditions are iteratively solved:

$$c_{l+1} = c_l - \eta T * \nabla_l(c_l) \tag{11}$$

In Formula (11): c_{l+1} is the iteration variable of step $l + 1$; η represents the search step size; T represents the N_u-dimensional projection matrix; $\nabla_l(c_l)$ is the random perturbation gradient of $O(c)$.

The stochastic disturbance gradient of the objective function can be calculated by the following formula:

$$\nabla_l(c_l) = \frac{O(c_l + \varepsilon_l \Delta_l) - O(c_l)}{\varepsilon_l} \times \Delta_l \tag{12}$$

In Formula (12): ε_l denotes the gradient perturbation step; Δ_l denotes a N_u-dimensional random perturbation vector.

3.3. Calculation of Water Injection Splitting Efficiency

After the inversion of the connectivity parameters, the dynamic parameters of the oil and water wells can be obtained by calculation [30], and the dynamic parameters change with the change of the actual working system. According to the connected model after history matching, the stratified cumulative splitting coefficient and instantaneous splitting coefficient of the water well at the corresponding time, the stratified oil production and liquid production of the oil well [31], and other data are calculated.

Assuming that well i is an injection, the water distribution coefficient of the surrounding production well j is calculated to be α_{nij}, which is expressed as:

$$\alpha_{nij} = \frac{q_{ij}}{\sum\limits_{j=1}^{m} q_{ij}} \tag{13}$$

According to the above calculation results, the average water injection and oil displacement efficiency of the block at the last moment of reservoir development is calculated:

$$\varphi_w = \frac{\sum\limits_{i=1}^{m} q_i(1 - f_{wi})}{\sum\limits_{j=1}^{m} q_j} \tag{14}$$

In Formula (14): φ_w represents the average water injection splitting efficiency; f_{wi} represents the water content of well i.

The oil displacement efficiency of the water injection well indicates the amount of crude oil displaced from the oil well by the unit of water injected into the water injection well. In the case of the same amount of water being injected into the water injection well, less crude oil is displaced, indicating that the oil displacement efficiency of the water injection well to the oil well is lower, indicating that there may be channeling channels around the corresponding well, and ineffective water flooding is serious; on the contrary, if the oil displacement efficiency is higher, then the amount of crude oil displaced is more and the production situation is better. Therefore, the stratified water injection efficiency is introduced to represent the water injection efficiency of the water injection well in each layer:

$$\varphi_{wik} = \frac{\sum\limits_{t=0}^{n} \sum\limits_{j=1}^{m} q_{i,n} B_{ijk,n}(1 - f_{wjk,n})}{\sum\limits_{t-0}^{n} q_{i,n} B_{ik,n}} \tag{15}$$

In Formula (15): φ_{wik} represents the water injection efficiency of well i in the kth layer; $B_{ijk,n}$ denotes the splitting component of liquid production of well i in k layers at time n

$B_{ik,n}$ denotes the splitting coefficient of water injection from well i to well j in layer k at time n.

The average water injection efficiency of the water injection well is compared with the water injection efficiency of each layer of the single well. If the water injection efficiency of the single well is less than the average water injection efficiency of the block, then the water injection well is reduced; otherwise, the water injection is increased. At the same time, based on the updated water injection well scheme, the water injection system of the water injection well to each production well is adjusted, and the injection–production structure is adjusted to reduce the water flooding and water channeling. The specific design process is as follows.

(1) The average water injection efficiency of the block at the last moment of reservoir development is calculated and used as a criterion for judging high-efficiency wells and low-efficiency wells. The specific calculation formula is Formula (14).

(2) Efficiency evaluation and analysis of water injection wells: If the injection well $\varphi_{wik} > \varphi_w$, then it is judged that the water injection efficiency is high and the injection needs to be increased. On the contrary, if $\varphi_{wik} < \varphi_w$, then the water injection efficiency is low, and the water injection needs to be reduced. After optimization, the water injection volume of each layer of the water injection well is as follows:

$$q_{iknew} = q_{ik}(1 + W_{ik}); \tag{16}$$

among them.

$$W_{ik} = \begin{cases} W_{\max}\left(\frac{\varphi_{wik}-\varphi}{\varphi_{\max}-\varphi}\right)^{\gamma}, \varphi_{wik}<\varphi \text{ 。} \\ W_{\min}\left(\frac{\varphi-\varphi_{wik}}{\varphi-\varphi_{\min}}\right)^{\gamma}, \varphi_{wik}\geq\varphi \end{cases} \tag{17}$$

The upper limit $W_{\min}$ is set to 0.4, the lower limit is set to -0.4, and γ is set to 1.5, which is used to constrain the perturbation range W_{ik}. Through continuous iterative calculation, the dynamic analysis parameters of water injection in injection wells are optimized and updated, the injection–production structure is adjusted, and the production dynamic prediction and update are realized [32].

(3) The optimal water injection volume of the water injection wells in each layer is determined. According to the calculated water injection change of the water injection wells and the water injection splitting coefficient of water injection wells to production wells, the optimal adjustment of the production wells is realized. The calculation formula is as follows:

$$q_{inew,n} = q_{i,n} + \sum_{k=1}^{l}\sum_{i=1}^{m} q_{jk,n} w_{jk,n} A_{ijk,n} \tag{18}$$

4. Application Example

The reservoir of the D block in the Chad oilfield was put into development in 2017. At present, 46 wells have been opened, with an average porosity of 17.1%, an initial water saturation of 24%, an average permeability of 288.5 md, a comprehensive water cut of 41.8%, and a recovery degree of 6.86%. After long-term water injection development, the reservoir has experienced water flooding and water channeling, and the ineffective circulation of injected water is serious. The water content of the production wells increases rapidly, and the water content of more than half of the wells is high. The oil production rate of a single well decreases rapidly, the water flooding effect decreases, and the overall recovery rate decreases. In the current block, four production wells are shut down, nine production wells are flooded, and six production wells are insufficient in liquid supply. Therefore, accurately identifying the injection–production connectivity between injection wells and production wells, judging the dominant seepage channels of reservoirs, clarifying the distribution of remaining oil, and optimizing the injection–production structure are effective measures to reduce precipitation and increase oil production, and are also the main problems that urgently need to be solved [31,33,34].

4.1. Model Validation

Based on the geological characteristic parameters of the reservoir and the actual production dynamic data, the historical production dynamic data of the block are used as the inversion index. By appropriately modifying the inter-well-connected volume and conductivity, the automatic history matching of the reservoir is realized, and accurate inversion results are obtained as shown in Figure 2.

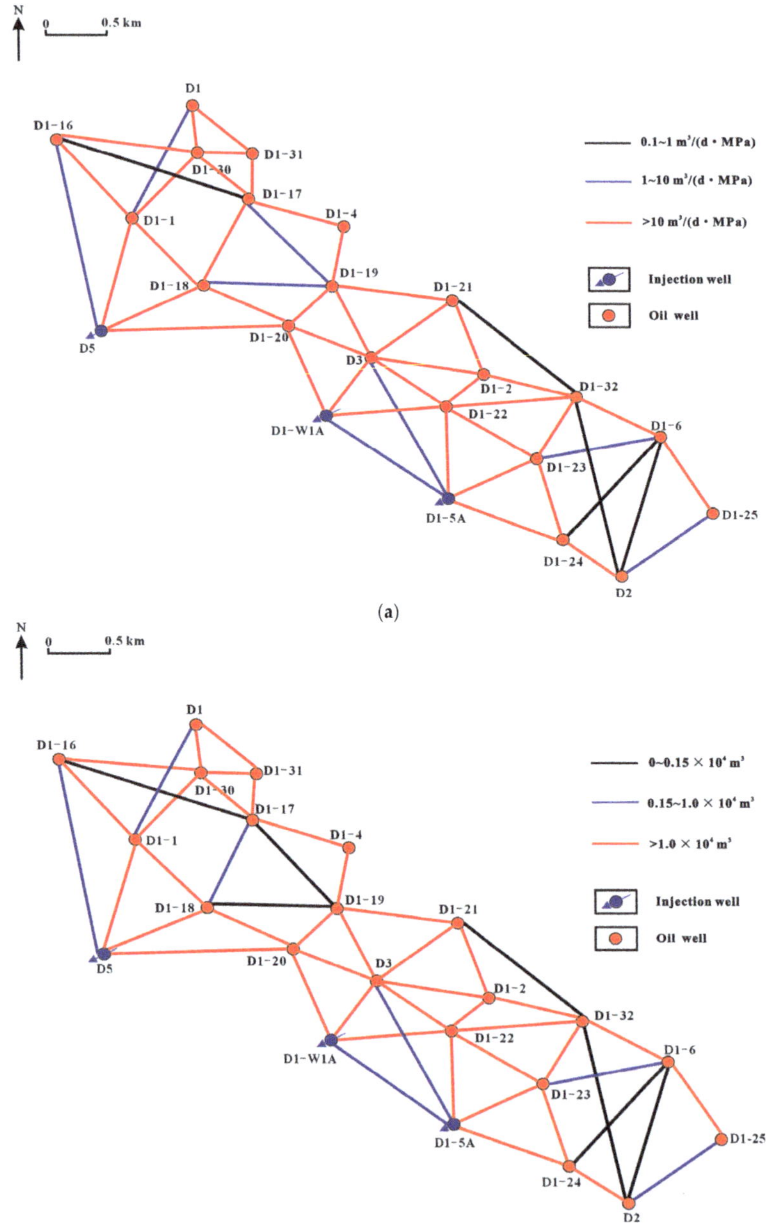

Figure 2. (**a**) Transmissibility; (**b**) Connective volume.

To verify the accuracy of the inter-well connectivity model, the D1-5A well group is selected as an example, as shown in Figure 3.

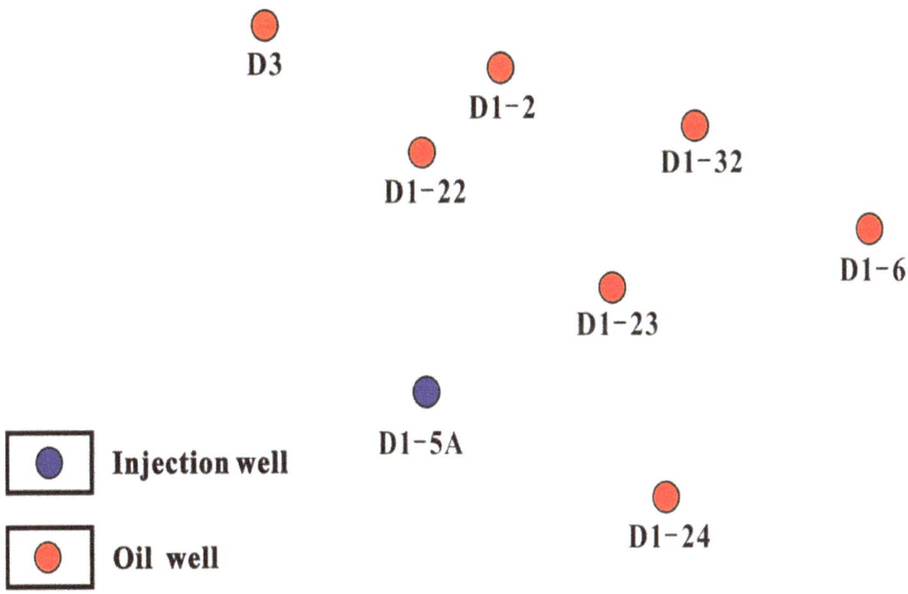

Figure 3. Schematic diagram of well group and well location.

The results of the splitting coefficient calculated by this model are compared with the actual field tracer water injection monitoring, as shown in Table 1 and Figure 4. The tracer response of the injection well D1-5A was mainly monitored in wells D1-23, D1-24, D1-22, and D3. The water flooding velocity in each direction can be inferred by the distance between the injection well and each production well and the effective time of tracer monitoring. At the same time, combined with the water injection splitting coefficient, it can be concluded that the fastest tracer monitoring response is that the water flooding speed of the D1-24 well is 84.2 m/d, and the water injection splitting coefficient is 31%, which is the highest among the four production wells. The second is well D1-23; the water flooding speed is 42.7 m/d, and the water injection splitting coefficient is 26%. The water injection splitting coefficient of well D1-22 and well D3 is consistent with the actual monitoring data of the tracer field. Among them, the tracer monitors the wells with faster water flooding speed, and the corresponding water injection splitting coefficient is larger, which fully verifies the accuracy of the model.

Table 1. Water injection splitting coefficient and tracer monitoring water flooding speed of D1-5A well group.

Well	Well Spacing /m	Water Drive Velocity/(M·d^{-1})	Water Injection Dividing Coefficient /%
D1-24	856	84.2	31
D1-23	789	62.7	26
D1-22	842	48.6	18
D3	1059	26.3	13

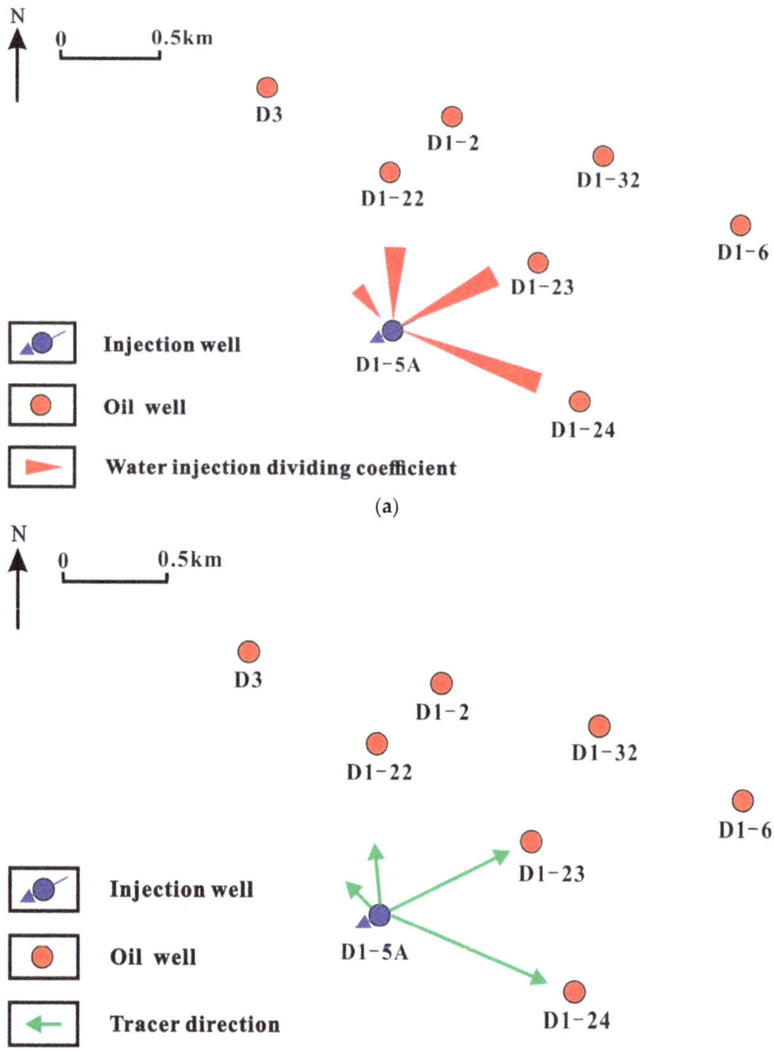

Figure 4. (**a**) Water injection dividing coefficient; (**b**) Tracer water flooding monitoring direction.

4.2. Production Optimization

The inter-well connectivity model can predict the production performance after ob taining a good historical fitting effect. By analyzing and calculating the cumulative liquic production, cumulative oil production, and cumulative water injection splitting coefficien of each oil well, the average water injection efficiency of the corresponding block is obtainec The water injection efficiency of each production well is statistically analyzed. When th water injection efficiency is lower than the average water injection efficiency of the curren block, the water injection is reduced; otherwise, the water injection is increased. It i calculated that the average water injection efficiency of the current block is 31%. Therefor when the water injection efficiency of the water injection well is lower than 31%, it is a inefficient well, and when it is higher than 31%, it is an efficient well. There are a total of si water injection wells in the study area, of which three are efficient and three are inefficien As shown in Figure 5, the water injection efficiency is low for the entire block.

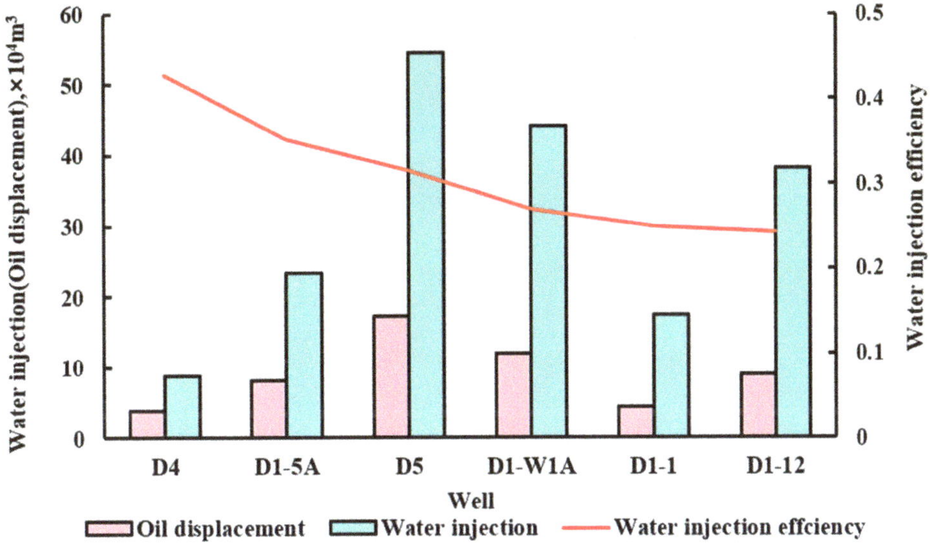

Figure 5. Water injection efficiency.

Based on the parameters of water injection efficiency of water injection wells, oil exchange rate of production wells, water cut of production wells, and injection–production efficiency of D reservoir in Chad oilfield, reasonable suggestions and measures are put forward for the adjustment of injection–production structure by using the connectivity relationship between wells and the characteristics of block development. According to the comparison of water injection efficiency, low-efficiency wells, and high-efficiency wells are judged, water injection volume is adjusted for high-efficiency and low-efficiency wells, water injection is reduced for low-efficiency wells, and water injection is increased for high-efficiency wells. At the same time, the water injection volume of water injection wells is optimized and adjusted in real time by combining the optimization algorithm. The actual water injection adjustment scheme of water injection wells is shown in Table 2.

Table 2. Adjustment strategy of injection–production structure in the study area.

Well	Optimization Measures	Daily Water Injection Rate/m^3	Water Injection Change Amount/($m^3 \cdot d^{-1}$)
D4	increase	199	22
D5	increase	97	12
D1-1	decrease	42	19
D1-12	decrease	120	40
D1-5A	increase	141	52
D1-W1A	decrease	91	31

According to the adjustment of production optimization, the cumulative oil production in the study area increased from 368.64×10^4 m^3 to 370.27×10^4 m^3, and the oil production increased by 1.63×10^4 m^3. The moisture content decreased from 53.3% to 50.7%, a decrease of 2.6%, as shown in Figures 6 and 7.

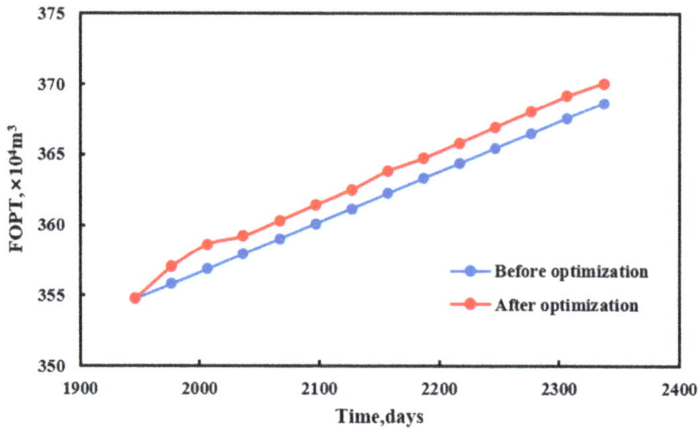

Figure 6. Before and after optimization in the study area of FOPT.

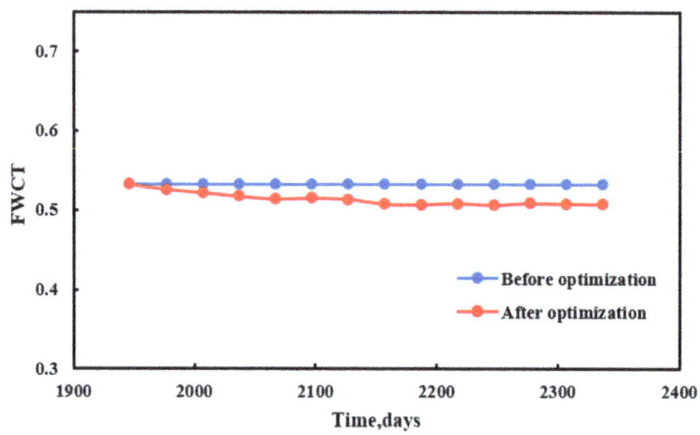

Figure 7. Before and after optimization in the study area of FWCT.

5. Conclusions

(1) Based on the theory of inter-well connectivity, a multi-layer sandstone rese voir model is established, and the reservoir dynamic parameter inversion and injectior production structure optimization were realized by using the optimization method.

(2) Based on the characteristic production and development parameters of the D rese voir in Chad, a multi-layer inter-well connectivity model was established. The inter-we conductivity and connectivity volume values were adjusted by field tracer data anal sis, and the inter-well connectivity relationship field was obtained in real time. Throug historical fitting calculation combined with tracer monitoring results, the fitting accurac exceeded 94%, which fully verified the accuracy of the model.

(3) By using the cumulative liquid production of production wells and the cumulativ splitting coefficient of water injection wells, the water injection efficiency was obtaine According to the water injection efficiency, the water injection wells were divided into in ficient wells and efficient wells. Combined with the optimization algorithm, the injectic volume of water injection wells was optimized and controlled, and the injection–productic optimization adjustment strategy was formulated. The strategy was applied to the Danie reservoir in the Chad oilfield. After one year of production, the cumulative oil productic in the study area increased by 1.3% per year, and the water cut decreased by 2.6%. Th goal of increasing oil and controlling water was realized, which provides a reference f the formulation of the oilfield development plan.

Author Contributions: Conceptualization, Y.Z. and L.P.; methodology, Y.Z., S.D. and J.H.; validation, S.P., L.P. and Y.Z.; formal analysis, L.P. and S.D.; investigation, S.P., L.P. and Y.Z.; data curation, L.P. and J.H.; writing—original draft preparation, L.P.; writing—review and editing, Y.Z. and S.D.; visualization, S.P., L.P. and J.H.; funding acquisition, Y.Z. and L.P. and S.D. All authors have read and agreed to the published version of the manuscript.

Funding: This work was financially supported by the Natural Science Foundation of Xinjiang Uygur Autonomous Region (No. 2022D01A330) and the National Natural Science Foundation of China (No. 52274030) and the "Tianchi Talent" Introduction Plan of Xinjiang Uygur Autonomous Region (2022).

Data Availability Statement: The data that support the findings of this study are available from the corresponding author upon reasonable request.

Conflicts of Interest: The authors declare that the research was conducted in the absence of any commercial or financial relationships that could be construed as a potential conflict of interest.

References

1. Wang, P.; Jiang, R.; Wang, G.; Ning, G. Numerical Simulation of Factors Affecting Water-flooding in High Water-cut Stage. *Nat. Gas Oil* **2012**, *30*, 36–38.
2. Liu, W.; Liu, T.; Ji, Y.; Zhang, L.; Chu, F.; Zhang, L. Determination of inter-well connectivity of fractured fractures in glutenite reservoirs by micro-seismic monitoring results: A case study of Mahu Oilfield in the Junggar Basin. *Oil Geophys. Prospect.* **2022**, *57*, 395–404+246.
3. Mirzayev, M.; Riazi, N.; Cronkwright, D.; Jensen, J.L.; Pedersen, P.K. Determining well-to-well connectivity using a modified capacitance model, seismic, and geology for a Bakken Waterflood. *J. Pet. Sci. Eng.* **2017**, *152*, 611–627. [CrossRef]
4. Qu, C.; Liu, B.; Song, H.; Xie, T.; Zhang, Z. Inter well connectivity of reservoir after polymer flooding based on dynamic and static data coupling judgment method—By taking Bohai L oilfield as an example. *Pet. Geol. Eng.* **2021**, *35*, 76–79.
5. Zhang, H.Y.; Zhou, X.; Wang, A.Q. Log Evaluation Method for Fractured-Vuggy Reservoir in the Dengying Formation of the Anyue Block, Sichuan Basin. *Well Logging Technol.* **2018**, *42*, 91–97.
6. Patidar, A.K.; Joshi, D.; Dristant, U.; Choudhury, T. A review of tracer testing techniques in porous media specially attributed to the oil and gas industry. *J. Pet. Explor. Prod. Technol.* **2022**, *12*, 3339–3356. [CrossRef]
7. Zhang, Y. Study on the Adjustment Strategy of the Horizontal Well Development in Bohai BZ Oilfield. Master's Thesis, China University of Petroleum, Beijing, China, 2022. [CrossRef]
8. Fang, X.; Yan, H.; Zhang, H.; Sun, X.; Peng, W. Dificulties and Technical Application of Complex Reservoir Modeling in West of South China Sea Oil and Gas Field. *Offshore Oil* **2016**, *36*, 50–55+98.
9. Ye, X.; Huo, C.; Wang, P.; Xu, J.; Yang, J. A reservoir modeling method by comprehensive utilization of various data. *Pet. Reserv. Eval. Dev.* **2017**, *7*, 1–5. [CrossRef]
10. Cao, X.; Wang, L.; Yu, Z. Application of Tracer Monitoring Technology in Reservoir Homogeneity and Reservoir Connectivity. *Unconv. Oil Gas* **2019**, *6*, 71–76.
11. Chen, W.; Huang, B.; Gao, Y. The application of inter-well trace substance tracer technology in Suizhong 36-1 oilfield. *Tianjin Sci. Technol.* **2019**, *36*, 58–61.
12. Yu, Q.; Liu, B.; Liu, C.; Zhang, X.; Zhang, X. Evaluation of inter-well dynamic connectivity based on polymer numerical simulation. *Fault-Block Oil Gas Field* **2017**, *24*, 827–830.
13. Li, N.; Yang, L.; Zheng, X.; Zhang, J.; Liu, Y.; Ma, J. Evaluation on Injection-Production Connectivity of Low-Permeability Reservoirs Based on Tracer Monitoring and Numerical Simulation. *Xinjiang Pet. Geol.* **2021**, *42*, 735–740.
14. Albertoni, A.; Lake, L.W. Inferring inter-well connectivity only from well-rate fluctuations in waterfloods. *SPE Reserv. Eval. Eng.* **2003**, *6*, 6–16. [CrossRef]
15. Yousef, A.A.; Gentil, P.; Jensen, J.L.; Lake, L.W. A Capacitance Model to Infer Inter-well Connectivity from Production and Injection Rate Fluctuations. In Proceedings of the Presented at the SPE Annual Technical Conference and Exhibition, Dallas, TX, USA, 9–12 October 2005. [CrossRef]
16. Sayarpour, M.; Kabir, C.S.; Sepehrnoori, K.; Lake, L.W. Probabilistic history matching with the capacitance-resistance model in waterfloods: A precursor to numerical modeling. In *SPE Improved Oil Recovery Symposium*; OnePetro: Richardson, TX, USA, 2010.
17. Kaviani, D.; Valkó, P.P.; Jensen, J.L. Application of the multi-well productivity index-based method to evaluate inter-well connectivity. In *SPE Improved Oil Recovery Symposium*; OnePetro: Richardson, TX, USA, 2010.
18. Liu, W.; Liu, W.; Gu, J.; Ji, C.; Sui, G. Research on inter-well connectivity of oil reservoirs based on Kalman filter and artificial neural network. *Pet. Geol. Recovery Effic.* **2020**, *27*, 118–124. [CrossRef]
19. Zhao, H.; Kang, Z.; Zhang, X.; Sun, H.; Cao, L.; Reynolds, A.C. A physics-based data-driven numerical model for reservoir history matching and prediction with a field application. *SPE J.* **2016**, *21*, 2175–2194. [CrossRef]
20. Zhao, H.; Xu, L.; Guo, Z.; Liu, W.; Zhang, Q.; Ning, X.; Li, G.; Shi, L. A new and fast waterflooding optimization workflow based on INSIM-derived injection efficiency with a field application. *J. Pet. Sci. Eng.* **2019**, *179*, 1186–1200. [CrossRef]

21. Teng, X.; Li, Y.; Yang, P.; Yang, K.; Li, N.; Xie, E. Differential structural deformation and its control factors in the eastern segment of Kuqa depression. *Pet. Geol. Recovery Effic.* **2017**, *24*, 15–21.
22. Dou, L.; Xiao, K.; Hu, Y.; Cheng, D.; Du, Y. Petroleum geology and a model of hydrocarbon accumulations in the Bongor Basin, the Republic of Chad. *Acta Pet. Sin.* **2011**, *32*, 379–386.
23. Song, H.; Dou, L.; Xiao, K.; Hu, Y.; Ren, L. An exploratory research on geological conditions of hydrocarbon pooling and distribution patterns of reservoirs in the Bongor Basin. *Oil Gas Geol.* **2009**, *30*, 762–767.
24. Zhao, H.; Zhang, X.; Wang, C.; He, H.; Xu, L.; Zhang, G.; Wang, S. Optimization of Fine water injection for reservoir stratification based on connectivity method. *J. Yangtze Univ.* **2018**, *15*, 42–51+6–7. [CrossRef]
25. Zhao, H.; Kang, Z.; Zhang, Y.; Sun, H.; Li, Y. An inter-well connectivity numerical method for geological parameter characterization and oil-water two-phase dynamic prediction. *Acta Pet. Sin.* **2014**, *35*, 922.
26. Guo, Z.; Reynolds, A.C.; Zhao, H. A physics-based data-driven model for history-matching, prediction and characterization of waterflooding performance. In *SPE Reservoir Simulation Conference*; OnePetro: Richardson, TX, USA, 2017.
27. Zhao, H.; Xie, P.; Cao, L.; Li, Y.; Zhao, Y. Reservoir production optimization method based on inter-well connectivity. *Acta Pet. Sin.* **2017**, *38*, 555.
28. Zhao, H.; Li, Y.; Cui, S.; Shang, G.; Reynolds, A.C.; Guo, Z.; Li, H.A. History matching and production optimization of water flooding based on a data-driven inter-well numerical simulation model. *J. Nat. Gas Sci. Eng.* **2016**, *31*, 48–66. [CrossRef]
29. Zhao, H.; Cao, L.; Li, Y.; Yao, J. Production optimization of oil reservoirs based on an improved simultaneous perturbation stochastic approximation algorithm. *Acta Pet. Sin.* **2011**, *32*, 1031.
30. Song, K.; Wu, Y.; Ji, B. A U-function method for estimating distribution of residual oil saturation in water drive reservoir. *Acta Pet. Sin.* **2006**, *27*, 91.
31. Shen, W.; Zhao, H.; Liu, W.; Xu, L.; Liao, M. Identification method of dominant channeling in fractured vuggy carbonate reservoir based on connectivity model. *Fault-Block Oil Gas Field* **2018**, *25*, 459–463.
32. Zhang, K.; Li, Y.; Yao, J.; Liu, J.; Yan, X. Theoretical research on production optimization of oil reservoirs. *Acta Pet. Sin.* **2010**, *31*, 78.
33. Guo, Z.; Reynolds, A.C. INSIM-FT in three-dimensions with gravity. *J. Comput. Phys.* **2019**, *380*, 143–169. [CrossRef]
34. Zhou, Y.; Lei, S.; Du, X.; Ju, S.; Li, W. Injection-production optimization of carbonate reservoir based on an inter-well connectivity model. *Energy Explor. Exploit.* **2021**, *39*, 1666–1684. [CrossRef]

processes

MDPI

Review

Development of Microbial Consortium and Its Influencing Factors for Enhanced Oil Recovery after Polymer Flooding: A Review

Hui Xiao, Zulhelmi Amir * and Mohd Usman Mohd Junaidi

Department of Chemical Engineering, Faculty of Engineering, Universiti Malaya, Kuala Lumpur 50603, Malaysia; s2178714@siswa.um.edu.my (H.X.); usmanj@um.edu.my (M.U.M.J.)
* Correspondence: zulhelmi.amir@um.edu.my

Abstract: After polymer flooding, substantial oil and residual polymers remain in reservoirs, leading to plugging and reduced recovery. MEOR (Microbial Enhanced Oil Recovery) aims to release trapped oil by utilizing microorganisms and their byproducts. The microorganisms can use residual HPAM (hydrolyzed polyacrylamide) as an energy source for polymer degradation, addressing reservoir plugging issues and improving oil recovery. However, microorganisms are sensitive to environmental conditions. This paper presents a detailed update of MEOR, including microbial products, mechanisms, and merits and demerits. The effect of the displacement fluid and conditions on microorganisms is thoroughly demonstrated to elucidate their influencing mechanism. Among these factors, HPAM and crosslinkers, which have significant biological toxicity, affect microorganisms and the efficiency of MEOR. Limited research exists on the effect of chemicals on microorganisms' properties, metabolism, and oil displacement mechanisms. The development of microbial consortium, their metabolic interaction, and oil displacement microprocesses are also discussed. In addition, prior studies lack insights into microorganisms' interaction and mechanisms using chemicals. Finally, field trials exist to examine the microbial consortium's efficiency and introduce new technologies. This review mainly explores the influencing factors on microorganisms, and confirms the credibility of MEOR after polymer flooding, providing a scientific basis for improving the theory of MEOR.

Keywords: MEOR; microbial consortium; mechanism; effect; HPAM and crosslinkers; oil displacement

tation: Xiao, H.; Amir, Z.; Mohd naidi, M.U. Development of icrobial Consortium and Its luencing Factors for Enhanced Oil covery after Polymer Flooding: A view. *Processes* **2023**, *11*, 2853. ps://doi.org/10.3390/ 11102853

ademic Editors: Zheng Sun, Zhang, Dong Feng, Wen Zhao l Hung Vo Thanh

ceived: 1 September 2023
vised: 20 September 2023
cepted: 25 September 2023
blished: 27 September 2023

1. Introduction

Crude oil is the primary energy source within the intricate capillary network in porous reservoir media. Enhanced oil recovery aims to economically extract the maximum amount of original oil in place (OOIP). It can be categorized into three stages based on the development process: primary recovery (natural energy extraction), secondary recovery (injecting water or gas to maintain reservoir pressure), and tertiary recovery (known as enhanced oil recovery or EOR) [1]. Tertiary recovery methods include chemical flooding (including polymer flooding, surfactant flooding, and alkaline flooding [2]), thermal flooding, microbial enhanced oil recovery (MEOR), and miscible flooding, which are designed to increase the amount of oil recovered by altering the reservoir conditions or fluid properties [3,4]. Figure 1 illustrates the different stages of oil recovery and their respective operating mechanisms.

In generally, primary production allows for recovery of approximately 10% of the OOIP, and secondary recovery techniques can improve the overall oil recovery by 12% [5]. Tertiary recovery techniques come into play when primary and secondary recovery methods have become ineffective and significant amounts of oil remain trapped in the reservoir. As one of the methods of tertiary oil recovery, polymer flooding technology, particularly using partially hydrolyzed polyacrylamide (HPAM), has emerged as a crucial tertiary oil recovery method and has gained widespread application [6]. This technique involves the application of HPAM to block large channels, increase the viscosity of the displacing phase, and effectively enhance the oil recovery rate by 8–15% [7]. However, polymer flooding encounters challenges related to severe plugging issues and a reduction in oil recovery. This is a consequence of the polymer solution's potential reactivity with metal ions in the original formation, resulting in the formation of compounds that are difficult to degrade [8].

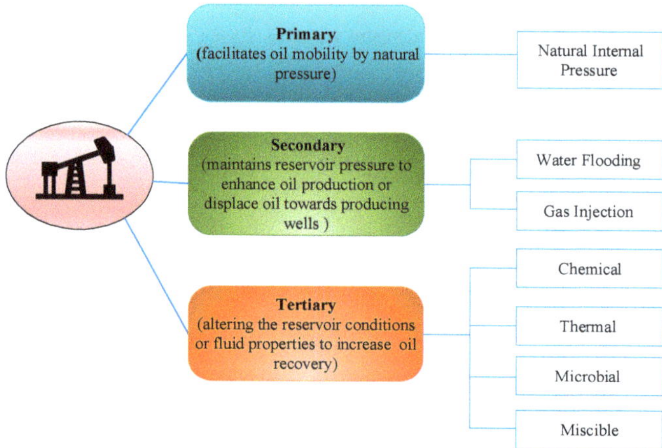

Figure 1. The distinct stages of oil recovery and their corresponding operational mechanisms.

Nowadays, the majority of oil fields have reached the stage of high-water cuts, leadin to challenges in exploitation, high costs, low recovery efficiency, and significant pressure [9 Conventional techniques often come with high costs and complex applications, resulting i a considerable volume of unrecoverable oil once the conventional technologies have reache their economic limit [10]. Hence, there is a need to explore alternative recovery methods. I recent years, microbial enhanced oil recovery (MEOR) has emerged as a prominent researc topic in energy due to its environmentally friendly nature and low operating costs [11]. Thi essential tertiary oil recovery technique utilizes microorganisms and their metabolites [12 The effectiveness of MEOR hinges on the capability of microbial communities and the metabolites to improve the fluidity of oil [13]. MEOR progress has been extensivel reviewed from both theoretical and practical perspectives [14], and it has been applied i various scenarios such as sandstone, carbonate, light oil, heavy oil, marginal reservoir post-polymer flooding reservoirs, and low/medium/high-permeability reservoirs [15]. particular approach involves conducting exploratory investigations on microorganism to enhance residual oil recovery from reservoirs following polymer flooding, potentiall improving oil recovery [7].

Following polymer flooding, a significant amount of remaining oil, approximate 50%, is retained in the reservoir, along with residual polymer that has adsorbed on the rock formations [5]. Microorganisms alter polymer molecules by breaking the amid group into carboxylic acid through hydrolysis [16]. Nuclear magnetic resonance tes have demonstrated a decrease in the content of the polymer amide group from 74.6% 60.8%, accompanied by a noticeable increase in the carboxylic group [14]. In a study b Shi et al. [17], five strains of facultative anaerobes were screened from the Daqing oil fie in MEOR after polymer flooding. These strains were found to grow and reproduce usin oil and polymer as carbon sources. The results revealed that the screened bacteria cou produce organic acids and active substances, degrade polymers and crude oil, and enhan oil recovery by 5%. Patel et al. [18] found that the microorganisms involved in MEO utilize residual HPAM as a nitrogen resource to grow, reproduce, metabolize crude o and produce surfactants in micro-pores, which can degrade polymers providing a soluti to solve the reservoir plugging problem and improve the oil recovery. Furthermore, additives used in MEOR are biodegradable and environmentally friendly [19]. Neverth less, the widely used HPAM and its crosslinkers in polymer flooding exhibit substantial biological toxicity and adverse impacts on the characteristics of subsequent oil-displacing microorganisms and oil recovery processes [20,21].

The previous research focuses on improving the oil recovery after polymer flooding, while there are few recent investigations on the effect of the chemicals on oil-displacing bacteria. This review aims to provide insights into and critically explore the influence mechanism of HPAM and crosslinkers on microorganisms and the performance of microorganisms in MEOR, which can provide a scientific basis for improving the theory of microbial oil recovery.

2. Microbial Enhanced Oil Recovery

Microbial enhanced oil recovery (MEOR) involves the utilization of microbes and their by-products, such as biosurfactants, biopolymers, biogenic acids, solvents, biogases, biomass, and emulsifiers, to stimulate the production of oil by mobilizing residual reserves [14]. According to Quraishi et al. [22], microbial flooding can be categorized into exogenous and indigenous. The category is based on the source of the strains. Exogenous microbial flooding involves the injection of microbes that have been screened under conditions similar to, but not within, the reservoir. These microbes are introduced underground to enhance oil production through their propagation and metabolites. Indigenous microbial flooding involves the utilization of remaining oil as a carbon source by microbes, making use of the active substances present in the formations. During water injection, air, inorganic salts, a phosphorus source, and a nitrogen source are introduced to facilitate the proliferation of these native microorganisms [14]. The key aspect of exogenous microbial flooding lies in developing effective production strains, and challenges include ensuring the compatibility of microorganisms, performance degradation, and high costs [3]. Indigenous microbial flooding shows excellent adaptability but lacks a subsequent procedure for developing production strains. In the process of MEOR implementation, microorganisms are commonly cultured ex situ and subsequently introduced into the reservoir by injection. As the injected water is transported, these microorganisms build up in the caprock pores, where oil is present at the interface between oil and water. After the injection of MEOR bacteria, in situ production persists, resulting in continuous alterations in the oil and reservoir properties, which facilitates the mobilization of tightly trapped oil to the surface [23].

The MEOR diagram in Figure 2 illustrates introducing a water mixture, including bacteria and/or biosurfactants, accompanied by a nutrient medium, into the reservoir as part of the MEOR process. Inside the reservoir, the bacteria facilitate the biodegradation of heavy crude oil into light components through biosurfactants. This degradation process aids in improving the movement of oil toward the production well. Furthermore, the injected bacteria experience metabolic activities that result in the production of metabolites. These metabolites contribute to multiple mechanisms, including reducing oil viscosity, lowering interfacial tension (IFT), promoting emulsification, and re-pressurizing the reservoir. These mechanisms collectively enhance the recovery of residual oil.

2.1. Microorganism

Oil reservoirs can be viewed as extensive geo-bioreactors primarily populated by sulfate-reducing bacteria (SRB), methanogens, syntrophic bacteria, and fermentative bacteria [24]. Accordingly, the dominant microbial processes observed in oil field ecosystems affected by petroleum hydrocarbons include sulfate reduction, fermentation, acetogenesis, methanogenesis, nitrate reduction, as well as iron and manganese reduction [25,26]. In addition, oil reservoirs contain a significant number of organic substances, including various inorganic ions such as sulfate and nitrate, as well as organic compounds like alkanes, alkenes, cycloalkanes, and aromatic hydrocarbons [27].

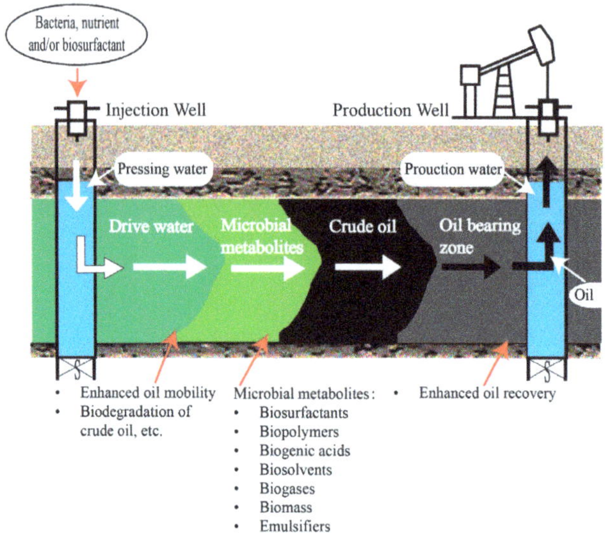

Figure 2. Diagram of MEOR.

The microbial activities in oil reservoirs significantly impact oil's chemical composition and physical–chemical properties [28]. This influence can be either positive, such as reducing viscosity in heavy oil to enhance its exploitation, or negative, leading to corrosion of drilling equipment or reservoir souring [29,30]. A large number of studies have highlighted the significance of microbial community dynamics within petroleum oilfield ecosystems [27]. The spatial distribution of microbial communities within long cores plays a role in the mechanism of oil displacement by microorganisms [31]. Generally, in the vicinity of the injection water, aerobic oil-displacement functional bacteria can decrease oil–water IFT by producing biosurfactants and emulsifying crude oil [12]. Additionally, aerobic hydrocarbon-loving microorganisms are abundant in the aerobic zone, which enhances crude oil's physical properties and residual oil fluidity through aerobic hydrocarbon metabolism. Within the middle section of the reservoir, facultative and anaerobic microorganisms coexist and produce H_2, CO_2, small molecule acids, and alcohols through anaerobic fermentation [32]. Within the anoxic conditions of the reservoir's deep environment, methanogenic microbes increase crude oil fluidity by generating CH_4 [33]. Figure 3 depicts the distribution of microorganisms in MEOR.

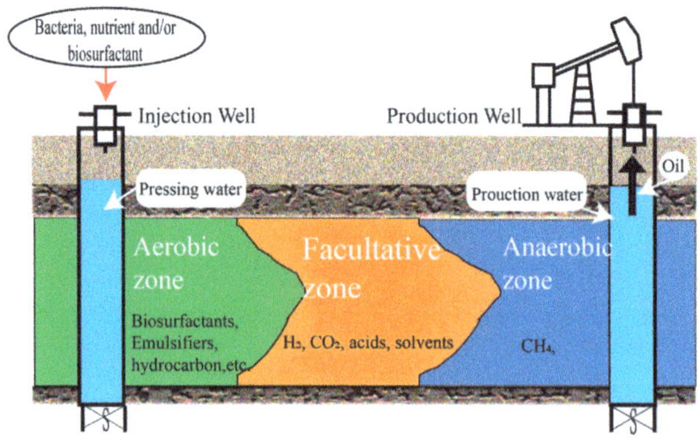

Figure 3. The spatial distribution of microorganisms in MEOR.

Oil reservoirs are known for adverse environments characterized by high temperature, high pressure, high salinity, and stringent anoxic conditions [25,34]. These environmental factors usually limit the range of microorganisms suitable for in situ processes. One adaptation of microorganisms to such harsh conditions is their ability to form biofilms. Biofilms are generally described as clusters of microorganisms surrounded by a self-produced matrix of extracellular polymeric substances (EPS), which adhere to each other or surfaces [35]. A biofilm matrix acts as a protective coating, providing resistance against external stressors in oil fields [36]. In addition, biofilms create a conducive environment for mutualistic microorganisms, promote habitat diversity through confined gradients, enable nutrient sorption for resource capture, aid nutrient transport via channel systems, facilitate the exchange of signaling molecules, and offer increased tolerance to toxic compounds through horizontal gene transfer, acting as a barrier [37]. Biofilm formation emerges as the predominant mode of microbial existence in oil reservoirs due to the advantageous protection and adaptation it provides in toxic and extreme environments.

EPS, accounting for approximately 90% of biofilm composition, are generated through the cellular lysis and hydrolysis of high molecular weight macromolecules by microorganisms [35], forming gel-like networks that confer mechanical stability to biofilms in oil environments [38]. EPS contain diverse functional groups, such as carboxyl, phosphoric, sulfhydryl, phenolic, and hydroxyl groups, contributing to cell aggregation and attachment to various surfaces [37]. They primarily consist of polysaccharides, proteins, glycoproteins, glycolipids, and nucleic acids, which can accumulate multiple substances (such as nutrients or ionic) that function as substrates in extracellular chemical reactions. Bacteria produce EPS as a survival strategy in harsh reservoir conditions characterized by extreme temperature, salinity, toxic agents, and limited nutrient availability [39]. Furthermore, EPS play a significant role in oil displacement mechanisms, influencing oil through processes such as emulsification, degradation, dispersion, aggregation, and sedimentation [40]. Previous research has predominantly focused on investigating the temporal changes in bacterial communities and genera associated with hydrocarbon degradation. As an example, Bacosa et al. [41] effectively cultured nine bacterial strains from the surface waters of the northern Gulf of Mexico that exhibited the dual ability to produce EPS and degrade hydrocarbons. EPS also act as a bio-emulsifier, facilitating efficient oil mobilization by releasing oil from rocks and creating stable oil/water emulsions [42].

To summarize, the oil reservoirs are considered harsh conditions. Microorganisms secrete EPS to protect the bacteria from adverse conditions and play an important role in MEOR. Since microorganisms are the base of MEOR, isolating and screening suitable microorganisms are vital steps in establishing an effective MEOR process. The primary selection criterion for microorganisms is their capability to degrade hydrocarbons, survive in the reservoir, and generate the desired metabolic products. Given the variations in conditions among different reservoirs, customization of the MEOR process according to specific reservoir conditions is crucial for its success.

2.2. Microbial Products

The effectiveness of the MEOR process relies on metabolite production by either exogenous or indigenous microorganisms [42]. These metabolites play a crucial role in altering the rock properties of the reservoir, including permeability, porosity, wettability, and oil viscosity. This alteration facilitates oil recovery by reducing viscosity, emulsifying the oil, pressuring the reservoir, and dissolving residual oil [3]. The bioproducts can generally be categorized into seven main groups: biosurfactants, biopolymers, biogenic acids, biosolvents, biogas, biomass, and emulsifiers.

2.2.1. Biosurfactants

Biosurfactants, a group of surface-active molecules that are synthesized by microorganisms via fermentation processes, can be categorized into five primary groups, including glycolipids, phospholipids and fatty acids, particulate surfactants, lipopeptides and lipoproteins, and polymeric surfactants [43]. These molecules exhibit great potential for improving oil recovery on account of their ability to lower IFT and surface tension, alter wettability, and create water/oil or oil/water emulsions, thereby facilitating the mobilization of oil to the surface [44,45]. The abilities mentioned above result from biosurfactants forming biofilms that interact with reservoir rocks and water/oil formations, bringing about modifications in their characteristics. *Pseudomonas aeruginosa* is an illustrative example of a biosurfactant-producing microorganism that exhibited the capability to degrade alkanes, including hexadecane and octadecane, within a 28-day incubation period [46].

As stated by Niu et al. [47], biosurfactants with lower molecular weights can reduce surface tension and IFT, whereas those with higher molecular weights display emulsifying activities, forming stable emulsions at the interface due to the close arrangement of surfactant molecules. The effectiveness of reducing IFT relies on the number of biosurfac tants required to attain the desired reduction and the adsorption of biosurfactants ontc caprocks. To achieve significant oil production, it is necessary to reduce the IFT to a range of approximately 10^{-2} to 10^{-3} mN/m, which corresponds to a capillary number of 10^{-3} tc 10^{-4} [48].

Numerous studies have reported significant oil recovery achieved through the use of biosurfactants. To produce biosurfactants of high quality with minimal losses, careful consideration of various conditions, including temperature, salinity, pH, oxygen, nutrient composition, and physical and chemical parameters, is necessary. Additionally, i is essential to investigate the characteristics of compounds such as carbon sources anc nitrogen sources, as well as the C:N ratio and the selection of bacterial strains, to ensure the production of biosurfactants that are effective [49]. Prior studies have shown that in situ production of surfactants is typically restricted by the requirement of most surfactan producing microorganisms for oxygen to support their growth. In the past few years numerous experimental studies have shown that specific microorganisms have the abilit to generate surfactants under anaerobic conditions. This ability is exceedingly advanta geous for extracting trapped oil residues situated within oxygen-depleted reservoirs' dee pores. It facilitates surfactant production and subsequently improves oil recovery withou needing oxygen during metabolic processes.

Biosurfactants offer several advantages for EOR applications, including their biodegra ability, outstanding surface and interfacial activities, low toxicity and capital cost, enoug raw materials for production, and substantial productivity. These attributes make biosu factants a cost-effective and efficient approach for achieving enhanced oil recovery.

2.2.2. Biopolymers

Injected microorganisms and their metabolites encounter significant limitations i migrating within low- and high permeability zones; they face hindrances in contactin residual oil caused by preferential fluid flow [50]. To resolve this problem, biopolyme as plugging agents are employed to selectively plug high permeability zones and rec rect injected water towards low permeability zones to enhance the recovery of residu oil [45,51].

Figure 4 visually illustrates the mechanism of selective plugging. The left-har side diagram initially portrays water flowing effortlessly through wide pore channe neglecting the narrow and low permeable zones resulting in poor sweep efficiency. contrast, the right-hand side diagram illustrates the effective plugging of the wide po channel via biopolymers. When the biopolymers adhere and grow, they deviate t water flow towards the less permeable zones, facilitating oil displacement and swe efficiency [52]. The biopolymer plugging process can be accomplished through either t injection of bacteria or the in-situ production of bacteria in the reservoir. As mention by Sen [53], microorganisms enter the reservoir via high permeability zones and establi themselves in specific laminae. The presence of the growth of biopolymers in the po throat leads to the plugging of the pore space, causing a reduction in the permeability ra As a result, plugging pore throats with biopolymer growth helps in restoring a balanc permeability across the reservoir, thereby allowing water-flooding operations to resume with conventional sweep efficiency.

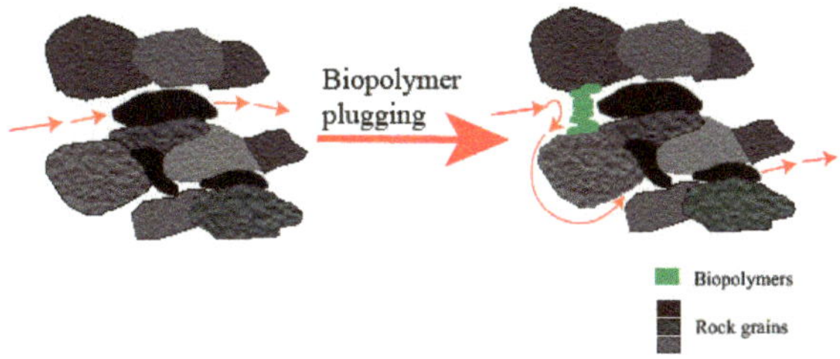

Figure 4. Illustration of the selective plugging mechanism.

Moreover, biopolymers can be employed as displacement agents to augment the viscous pressure gradient in water, thus improving the efficiency of oil sweeping. Halim et al. [52] extensively evaluated the selective plugging method using a sand pack column and the thermotolerant, anaerobic, biofilm-producing microorganism called *Bacillus licheniformis*. The study demonstrated that the biopolymer produced by the *Bacillus licheniformis* substantially augmented the biofilm thickness, resulting in the diversion of flow towards low permeability areas and subsequent oil displacement. More and more evidence indicates that microbially produced polysaccharides, including xanthan gum, have been utilized for many years and have consistently demonstrated successful outcomes in terms of achieving higher oil recovery [54]. Xanthan gum is generated through the fermentation of carbohydrates and is subsequently introduced into the water used for well flooding and wide pore channels. Microorganisms effectively plug the high permeability zones as they grow. The remarkable physical properties of xanthan gum, including shear resistance, salinity tolerance, thermal stability, and viscosity, make it an efficient option for EOR [55].

In conclusion, the application of biopolymer flooding enhances recovery effectiveness by reducing the mobility ratio between the injected water and entrapped crude oil, triggering a substantial improvement in MEOR, particularly in heterogeneous reservoirs.

2.2.3. Biogenic Acids

The production of bioacids contributes to the dissolution of carbonate rocks in the reservoir, leading to increased porosity and permeability and assisting oil displacement into the remaining reservoir [18]. Generally, low molecular weight organic acids like formic acid, acetic acid, and propionic acid are commonly employed in MEOR [19]. It has been reported that organic acids serve as the active metabolites responsible for reducing oil viscosity, thereby elevating their mobility. Laboratory experiments conducted in Daqing Oilfield demonstrated a 36% reduction in viscosity through the utilization of microbial acids [14].

2.2.4. Others

(1) Biosolvents

One significant benefit of the fermentation process is its capability of producing solvents and bioacids, which play a vital role in dissolving carbonate rocks to reveal concealed oil in the reservoir. Common examples of solvents include acetone, ethanol, and butanol. Consequently, these simultaneous processes contribute to a higher oil recovery yield [18]. Furthermore, solvents contribute to the reduction in oil viscosity and the decrease in IFT between water and oil [55].

(2) Biogases

Certain microorganisms capable of in situ gas production by fermenting carbohydrates or hydrocarbons like CH_4, CO_2, and H_2 are employed to achieve oil mobilization through re-pressurization. The reservoir containing an accumulation of gases leads to a buildup of pressure and subsequent swelling of the oil, thus promoting its mobilization to the surface [55]. Moreover, these gases can contribute to the reduction in oil viscosity by dissolving into the oil, increasing sweep efficiency.

(3) Biomass

Biomass, including microbial cells, biofilms, extracellular products, extracellular structures, water channels, and void space, plays a role in the selective plugging effect. Additionally, it contributes to the reduction in crude oil viscosity, as well as oil desulfurization and emulsification, thereby favoring the effectiveness of MEOR. Some studies have explored the approach of starving microorganisms to minimize their size and increase their penetration depth into the reservoir, followed by nutrient supply to promote the formation of biomass biofilms [10,56].

(4) Emulsifiers

Emulsifiers produced by a diverse range of microorganisms lead to the achievement of oil emulsification. This process results in the formation of stable emulsions, typically oil in water, where hydrocarbons are effectively dispersed and incorporated [18].

In summary, producing various highly advantageous metabolites by microorganisms ultimately leads to an increase in the overall oil recovery in MEOR. The widespread distribution of microbial bioproducts can significantly influence the rock characteristics of reservoirs. Table 1 comprehensively compiles the bioproducts generated by various microorganisms, including their major effects and the most suitable reservoir candidates for MEOR processes.

2.3. *Mechanisms of Microbial Enhanced Oil Recovery*

MEOR is founded on two fundamental principles. The first principle involves modifying the interfacial properties of oil–water minerals to enhance oil movement through porous media, thereby improving displacement efficiency (IFT reduction or permeability increase), driving force (reservoir pressure), fluidity (miscible flooding, viscosity reduction), and sweep efficiency (selective plugging, mobility control). The second principle focuses on microbial activity for the degradation and removal of sulfur and heavy metals from heavy oils [64]. Due to the diverse range of strains and metabolites involved, the mechanism of MEOR is intricate and comprehensive, encompassing processes such as biodegradation, emulsification, IFT reduction, and alteration of reservoir rock properties [19]. A thorough comprehension of the mechanisms of MEOR lays the groundwork for evaluating its practicability and prospective benefits.

Table 1. Microbial products.

Microbial Products	Representative Microorganisms	Effect	Limitation	Type of Formation/Reservoir	Reference
• Biosurfactant (Alasan, Surfactin, Rhamnolipid, Emulsan)	• *Acinetobacter Arthrobacter paraffineus* sp. • *Calcoaceticus* sp.	• Emulsify crude oil into an oil–water mixture. • Decrease IFT. • Modify rock wettability.	• Low displacement efficiency at the microscopic level.	• Reservoirs composed of sandstone or carbonate formations with temperatures below 50 °C and containing lighter oils with an API gravity greater than 25.	[57]
• Biopolymers (Xanthan gum, Pullulan, Levan)	• *Bacillus polymyxa* sp. • *Pseudomonas* sp. • *Brevibacterium viscogenes* sp.	• Raise the viscosity of the displacing fluid while diminishing the mobility ratio between water and oil. • Optimize the sweep regions and effectiveness with selective plugging.	• Low efficiency in sweeping fluids across the volume.	• Stratified reservoirs exhibiting distinct permeable zones.	[58]
• Bioacids (Acetic acid, formic acid, propionic acid)	• *Clostridium* sp. • *Enterobacter aerogenes* sp. • *Methanobacterium* sp.	• Improve the rock's dissolution in the pore throats to increase porosity and permeability.	• Low porosity. • Inefficient fluid drainage. • Reservoir impairment.	• Carbonate or carbonaceous reservoirs	[59]
• Biogases (CO_2, H_2, CH_4, N_2)	• *Clostridium* sp. • *Enterobacter aerogenes* sp.	• Enhance the mobility of crude oil by reducing its viscosity. • Re-establish reservoir pressure • Gas injection for partial or miscible displacement.	• Equipment corrosion. • Reservoir souring.	• Formations containing heavy crude oil with an API gravity of less than 25.	[22]
• Biosolvents (Alcohols, ketones, acetone, butanol)	• *Zymomonas mobilis* sp. • *Clostridium acetobutylicum* sp. • *Klebsiella* sp.	• Decrease the viscosity of oil through its dissolution in crude oil. • Alcohols and ketones can act as co-surfactants in the formation of micelles, leading to a decrease in critical micelle concentration (CMC) and IFT, thereby facilitating emulsification. • Optimize the porosity and permeability by dissolving heavy oil in the pore throats.	• Low displacement efficiency at the microscopic level.	• Formations containing heavy crude oil with an API gravity of less than 25. • Reservoirs with high oil wettability have undergone water flooding.	[60]

Table 1. *Cont.*

Microbial Products	Representative Microorganisms	Effect	Limitation	Type of Formation/Reservoir	Reference
• Biomass (Microbial cells, biofilms, EPS)	• *Licheniformis* • *Bacillus* • *Xanthomonas* • *Leuconostocmesenteroides* • *Campestris*	• Change rock wettability. • Employ selective plugging to optimize sweep effectiveness. • Gas-based biopolymers. • Specific decomposition of heavy oil.	• Low displacement efficiency at the microscopic level.	• Stratified reservoirs exhibiting distinct permeable zones.	[61]
• Emulsifiers	• *Candida* • *Acinetobacter* sp. • *Bacillus* sp. • *Pseudomonas* sp.	• Emulsification of oil with water.	• Sludge buildup. • Low displacement efficiency at the microscopic level.	• Formation containing high-molecular-weight waxy oil (>C_{22} alkanes), paraffinic oil, and asphaltene.	[62,63]

2.3.1. Biodegradation

Biodegradation is of paramount importance in MEOR. The main aspect of this mechanism involves the breakdown of crude oil's long-chain hydrocarbons into shorter hydrocarbons, which results in an increase of up to 30% in the light components of crude oil [14]. From one perspective, through this process, bacteria can obtain carbon for energy, growth, and reproduction by degrading hydrocarbons in crude oil [65]. From another perspective, converting heavy components into lighter ones brings about fundamental alterations in crude oil properties, leading to a reduction in viscosity and an enhancement in fluidity. Consequently, this contributes to improved oil recovery rates. This biodegradation mechanism entails direct cell contact and the involvement of biosurfactants, which decrease IFT and facilitate the emulsification of hydrocarbons [66]. Biosurfactants are commonly utilized to degrade the heavy fractions of crude oil.

Biodegradation pathways can be classified into aerobic and anaerobic biodegradation [19]. Typically, aerobic biodegradation occurs near the injection wells, while anaerobic biodegradation takes place in deep reservoirs. Aerobic biodegradation involves dissolving oxygen and injecting it into the reservoir alongside water rich in nutrients in MEOR, which has been an established and extensively employed technique for many years. Field trials have been conducted in China's Daqing Oilfield, where dissolved oxygen was injected to stimulate aerobic bacteria for degrading heavy components, thereby enhancing oil recovery [14]. During aerobic biodegradation, *Pseudomonas putida* with *alkB* genes on the OCT plasmid is capable of degrading aliphatic alkanes. This degradation process involves the enzyme alkane hydroxylase, which consists of rubredoxin reductase, oxygenase, and membrane-bound rubredoxin. These cofactors facilitate the transport of electrons from nicotinamide adenine dinucleotide phosphate to hydrocarbon substrates. The degradation of alkanols occurs through alcohol dehydrogenase, which converts them into alkanals. Subsequently, alkanals are transformed into fatty acids by aldehyde dehydrogenase and further converted to acetyl CoA by its synthetase [41]. Cell biomass biosynthesis occurs in the central precursor metabolites such as acetyl-CoA, succinate, and pyruvate [67], as depicted in Figure 5.

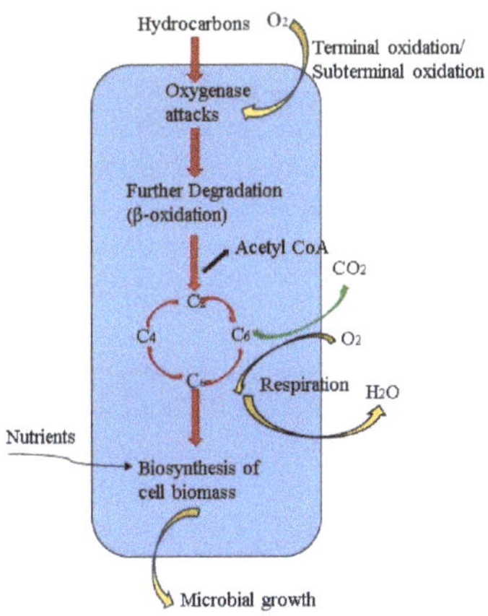

Figure 5. The fundamental concept underlying aerobic hydrocarbon degradation by microorganisms [68].

Anaerobic degradation has demonstrated the capability of biodegrading various hydrocarbons, including benzene, toluene, naphthalene, asphaltene, phenanthrene, alkane, branched alkane, and hydrocarbon mixtures [69]. Nonetheless, the application of anaerobic degradation in MEOR is restricted, possibly because of the choice of unsuitable oil-degrading microorganisms or suboptimal conditions [70]. Of all the options, the anaerobic degradation of long-chain alkanes holds the greatest promise for achieving enhanced oil recovery. Table 2 displays the commonly encountered long-chain degrading microorganisms and the scope of hydrocarbons they can degrade. Furthermore, oil-degrading microorganisms that have the ability to degrade asphaltenes are frequently regarded as highly effective, while hydrocarbon-degrading microorganisms generating biosurfactants are considered optimal for MEOR [69].

Table 2. Microorganisms that exclusively use long-chain n-alkanes as their carbon source.

Microorganisms	Range of Degradation	Reference
Rhodococcus erythropolis	C_{10}–C_{30}	[71]
Pseudomonas fluorescens	C_{12}–C_{32}	[72]
Thermophilic Bacillus	C_{15}–C_{36}	[73]
Gordonia amici	C_{18}–C_{36}	[74]
Chelatococcus daeguensis	C_{23}–C_{37}	[75]
Pseudomonas aeruginosa	C_{36}–C_{40}	[76]

Since reservoirs are generally anaerobic environments, anaerobic degradation predominantly governs the biodegradation of hydrocarbons in deep reservoirs. The thorough examination of this process has led to a clear understanding of the biochemical mechanisms involved in the anaerobic biodegradation of organisms. An illustration of this is the work by Purwasena et al. [77], who effectively isolated a thermophilic anaerobic bacterium strain belonging to the *Petrotoga* sp. from a Japanese oil reservoir. Through core flooding methods, their findings demonstrated that the bacterium could degrade long-chain n-alkanes, reduce oil viscosity, and enhance oil recovery under 80 °C. Additionally, researchers have effectively identified the key metabolites and associated genes involved in this specific biodegradation mechanism [19]. Nevertheless, a complete understanding of the biochemical processes involved in anaerobic hydrocarbon mechanisms is still being explored and investigated.

In summary, biodegradation plays a crucial role in MEOR, as research indicates that microorganisms and their products can degrade carbon atoms by utilizing them as a carbon source for metabolic processes. As a consequence, the viscosity of the oil is reduced, and the flow properties are improved; hence, promoting oil displacement towards production wells for recovery. While aerobic and anaerobic biodegradation pathways exist, anaerobic degradation predominantly controls the process. The duration of the biodegradation process varies based on factors such as hydrocarbon composition, bacterial population reservoir environment, and the amount of oxygen available or introduced into the wells.

2.3.2. Emulsification

Emulsions have an impact on increasing the oil–water interface, leading to enhance the accessibility of oil [78]. Different types of emulsions are classified based on particle size, including transparent microemulsion (nm), translucent colloid emulsion (below 1 μm) milky white emulsion (1–2 μm), fine dispersion (1 mm), and coarse emulsion (100 mm) [79] Emulsification can change the reservoir's wettability, shifting it from oil-wet to water-wet thus making trapped oil easier to recover [14].

Figure 6 provides a general visual representation of the emulsification process. On the left-hand side of the figure, water containing biosurfactant comes into contact with oil. Biosurfactants composed of both hydrophobic and hydrophilic structures, through their hydrophobic interaction properties, facilitate the formation of steady oil emulsion in water, as depicted on the right-hand side of Figure 6. These emulsions allow for th coexistence of both liquids by creating amphiphilic films at the oil–water interface [80]. T ensure a consistent dispersion of oil in water, biosurfactants are incorporated to provid stabilization. The hydrophobic interaction properties of biosurfactants enable them t attach to the interface between oil and water, improving the dispersion and facilitatin movement until the biosurfactant is either diluted or completely adsorbed on the caprocks. Emulsification of crude oil leads to the generation of various droplet sizes, exhibiting tensile, deformed, and seepage flow characteristics, which alter the permeability of the reservoir, and increase the contact area between the droplets and the reservoir medium, thereby facilitating the flow and recovery of oil [14]. Additionally, the structural alteration of bioacids contributes to changes in wettability, converting water-in-oil emulsion into oil-in-water emulsion, followed by a subsequent reduction in oil viscosity.

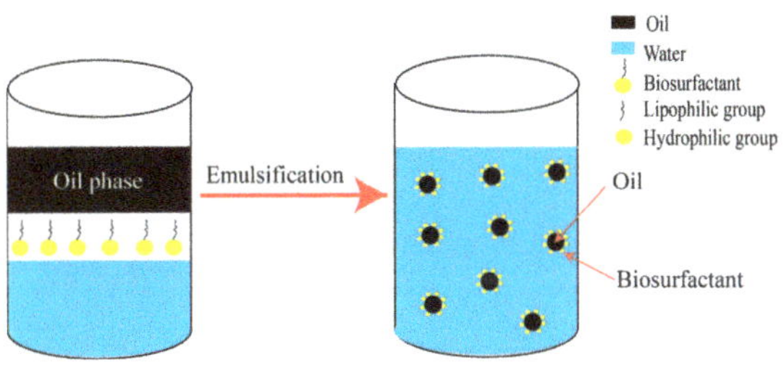

Figure 6. Schematic diagram illustrating the emulsification process.

2.3.3. Interfacial Tension Reduction

Reducing the IFT is an alternative mechanism that significantly enhances oil recovery. One of the three components contributing to the reduction in IFT is the work of adhesion, which represents the energy required to separate an oil film from an oil-wet rock surface [81]. Secondly, the elasticity of the oil/water interface is changed, enabling the two phases to mix more easily [82]. Thirdly, biosurfactants, characterized by their amphiphilic nature, have the ability to accumulate at the interface between immiscible fluids like oil and water or water and oil. This accumulation leads to a reduction in IFT, ultimately strengthening capillary forces and making oil easier to flow [83]. The formation of foams using organic materials also can lead to a reduction in IFT at the oil/water interface [23]. A visual representation of reducing IFT can be seen in Figure 7.

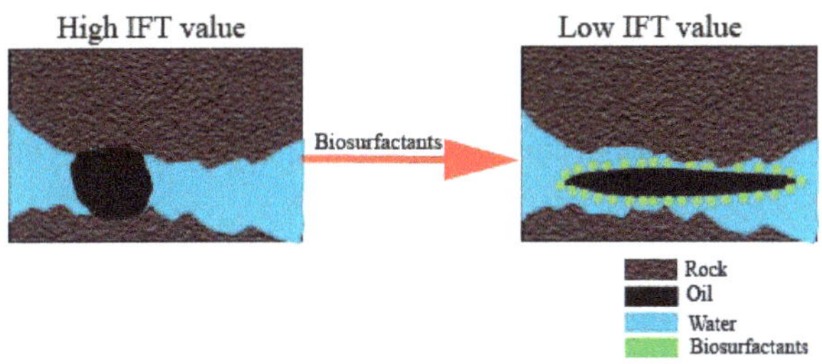

Figure 7. Graph of IFT reduction.

IFT reduction and wettability alteration are closely related mechanisms that occur concurrently and contribute significantly to EOR [84]. The reduction in IFT increases the capillary number by 3–4 orders of magnitude, resulting in improved oil flow. Simultaneously, wettability alteration promotes oil production by supporting the mobilization and displacement of oils attached to the rock surface or trapped in the pores [85,86]. The wettability condition of the rocks significantly influences the oil recovery [84]. In the case of oil-wet rocks, IFT reduction helps decrease the negative capillary force that hinders the entry of injection fluids. For intermediate-wet rocks, IFT reduction facilitates oil desorption and promotes mixing between oil and water. In the case of water-wet rocks, IFT reduction reduces positive capillary force and weakens interface elasticity, allowing trapped oil droplets to flow more easily [84]. Figure 8 depicts the process of alteration in wettability.

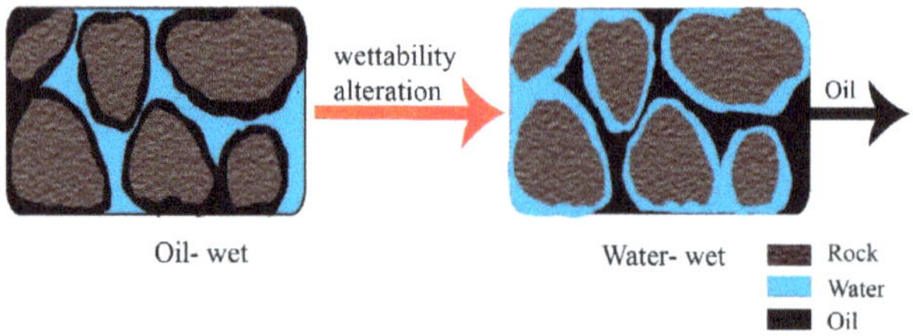

Figure 8. Illustration of alteration in wettability.

2.3.4. Altering Reservoir Rock Properties

By altering the properties of reservoir rocks, it is possible to increase their permeability and porosity. Bioacids have the potential to significantly decrease the permeability of reservoir rocks, reducing it from 284 mD to 24 mD while also diminishing the viscosity by approximately tenfold [14]. The mechanism is believed to predominantly occur in carbonate reservoirs [87]. The porosity and permeability modification can be seen in Figure 9.

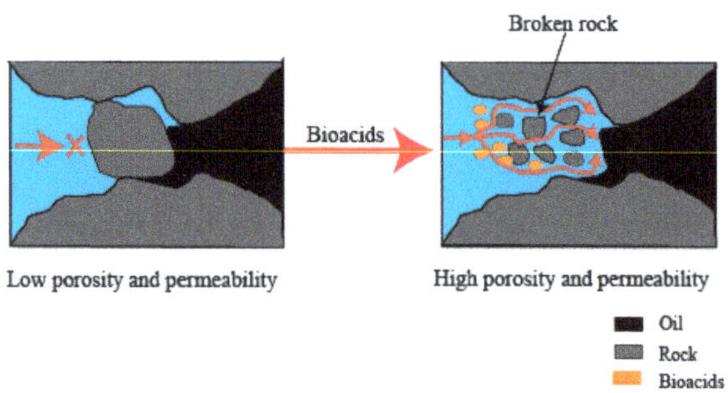

Figure 9. Diagram of the porosity and permeability modification.

Biosurfactants can undergo adsorption onto the pores of the reservoir rocks, leadin to modifications in the wetting characteristics of the porous media. Interfacial forces dicta fluid flow at the microscopic level, making pore-scale wettability a crucial factor in enhan ing oil recovery [88]. This phenomenon occurs due to the influence of the amphiprot groups in biosurfactants. Biosurfactants primarily alter the reservoir rock's wettability aid in the displacement of cells and biofilms within pore structures by adhering to them [8 The modification of wettability accelerates oil production by improving the permeabili of injected water within the reservoir matrix and boosting the capillary pressure [89]. porous media, fluids' location, distribution, and flow are governed by the wettability rock/fluid systems [90]. In non-fractured reservoirs, the transition from highly wet co ditions to neutral wettability enhances oil extraction efficiency. In fractured reservoi changing the wettability to water-wet leads to favorable capillary forces, enabling the d placement of trapped oil through water infiltration into the rock formations in the opposi direction [91]. Nevertheless, the precise mechanisms underlying the alteration of rese voir wettability during MEOR processes remain insufficiently understood and necessita additional research and exploration.

In practical terms, a mechanism must ultimately accomplish two objectives: (1) substantial enhancement in oil recovery and (2) a financially viable yield ratio that ensur the incremental yield of oil surpasses the input material for MEOR. When these criteria a fulfilled, a mechanism can be regarded as promising and deserving of further investigati The feasibility and robustness of each mechanism in the field are crucial consideratio Considering the reservoir conditions, it is essential to access fundamental reservoir da such as porosity, permeability, pressure, temperature, pH, viscosity, wettability, etc. The data are critical to determining which mechanisms are most suitable and effective for the specific reservoir conditions.

2.4. Advantages and Disadvantages of MEOR

Given the prevailing low oil prices, MEOR holds significant potential, particularly for marginal or uneconomical reservoirs. Microbial flooding has emerged as a viable substitute for other EOR techniques, showing a remarkably high success rate [92]. Compared to thermal flooding and gas flooding, microbial flooding exhibits notable advantages, primarily its eco-friendly attributes and cost-effectiveness for augmenting oil production [50]. Compared to other EOR technologies, microbial flooding possesses distinct features. Firstly, microbial products employed in the process are typically biodegradable and harmless. Secondly, implementing microbial processes is relatively straightforward in the field, as it necessitates minimal adjustments to existing facilities [93]. Thirdly, unlike thermal processes that demand significant energy consumption, MEOR is not energy-intensive, making it a cost-effective EOR method [3]. Figure 10 illustrates various cost estimates for different EOR techniques.

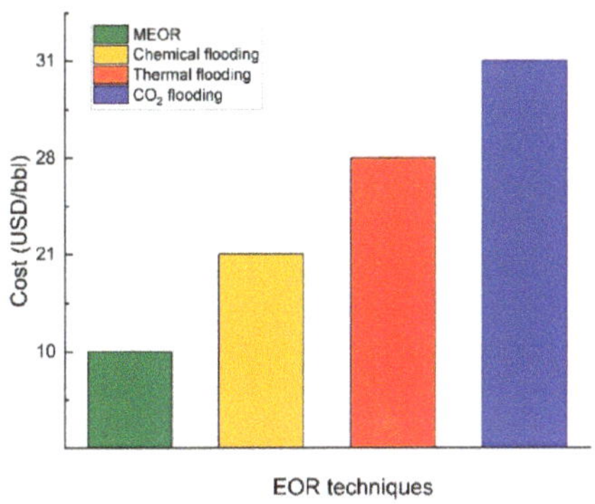

Figure 10. Estimations of cost for various EOR techniques [14].

There are also some disadvantages of MEOR. For example, the presence of SRB in MEOR is recognized for its detrimental effects [94], leading to corrosion, plugging, and reservoir souring issues in oil production, transportation, and storage [26]. Moreover, the hydrogen sulfide (H_2S) generated by SRB acts as an inhibitor, impeding the growth and metabolism of numerous bacteria, including those responsible for producing rhamnolipids [95]. Table 3 displays the current advantages and disadvantages of MEOR technology.

Table 3. Benefits and drawbacks of MEOR technology.

Benefits	Drawbacks
• Cost-effectiveness and simplicity of facility setup	• Equipment corrosion
• Affordable injection cost of materials • Minimal energy consumption associated with microbial metabolic activities • Minimal environmental pollution • Progressive enhancement of microbial metabolic activities over time • Achieving improved effects through the concurrent activation of multiple mechanisms • Significantly effective in sandstone and carbonate reservoirs	• Microorganisms' constrained ability to withstand reservoir conditions • The toxicity of microorganisms induced by the heavy metal ions • Challenges in constructing a comprehensive model that covers every aspect of MEOR • Restricted applications on offshore platforms due to the high demand for sugar in anaerobic bacteria activities
• Ability to apply to both light and heavy crude oils	

3. The Effect of a Reservoir Fluid on Microorganisms

3.1. Crude Oil

Crude oil is commonly classified into four primary categories: saturated hydrocarbons, aromatic hydrocarbons, asphaltenes, and resins, which is a mixture composed of hydrocarbons and various heteroelements such as oxygen, sulfur, nitrogen, and trace metals [96]. The properties of oil can exert a notable influence on microorganisms and subsequently affect the oil recovery process [97]. The oil's viscosity plays a crucial role in determining the concentration of metabolites needed for the dissolution, emulsification, or dispersion of oil [42]. Heavy oil's high viscosity and low flow properties pose challenges in oil extraction from the reservoir [98]. Among light components, alkanes with long carbon numbers may produce varying degrees of microbial activity [99].

In addition, heavy components in crude oil, such as bitumen, adversely affect microbial activity [100]. Moreover, volatile light components are toxic to microorganisms. Compounds such as sulfides, aromatics, and phenols are toxic and inhibit microbial growth and metabolism. Because microorganisms need a certain amount of water to carry out metabolic activities, the moisture content in crude oil also affects the activities of microorganisms [101].

3.2. Injection Water

Water injections are essential for MEOR. The quantity of water injected relies on the well maturity, reservoir pressure, or water displacement in the oil reservoir. Through water injection, exogenous microorganisms, new electron donors, and acceptors are introduced into the oil reservoir [102]. A multitude of researchers has explored microbial variabilities, metabolic activities, and the composition of microorganisms within oil reservoirs subjected to water flooding.

In certain reservoirs, chemicals or nutrients are added to the injection waters to control the native microbial structure [103]. For example, nitrite and nitrate injections are implemented to mitigate reservoir souring resulting from microbial hydrogen sulfide production. Injecting oxygen can improve aerobic hydrocarbon metabolism and facilitate oil mobilization within the well. This is accomplished by lowering the IFT between the oil and water phases by the activity of biosurfactant-producing microorganisms or changes in the oil matrix [104]. Alternatively, injecting fermentative bacteria and carbohydrates produces acids, gases, and solvents, leading to an augmentation in oil production [105].

Water injections induce a lowering in the oilfield temperature, creating a temperature gradient. Moreover, the quantity of injection impacts the chemical composition of the production water, leading to decreased levels of sulfate, nitrate, potassium, nitrite, and magnesium [106]. Production water, a byproduct of oil extraction, is transported alongside the oil phase and brought to the surface. It constitutes a combination of formation and injection water, potentially carrying particles and soluble compounds from the oil [107]. Two comparative studies have indicated that formation water exhibits higher levels of fatty acids and ammonium than production water [106]. Produced fluids with an injection water content below 10% demonstrated no notable effect on microbial community and metabolic potential.

Conversely, it was shown that fluids containing a higher proportion of injected water could affect the microbial structure and metabolic potential due to variations in the water component. A strong association was observed between the relative abundance of the genus family *Deferribacteres*, *Flexistipes*, the percentage of injection seawater, and the level of nitrite, nitrate, magnesium, sulfate, and potassium. Some wells with the highest water injection cut exhibited a higher abundance of *Epsilonproteobacteria* and *Gammaproteobacteria* isolates [106]. Numerous research studies have documented a correlation between the chemical component of oil reservoirs and the presence of operational taxonomic units (OTUs) within them. In Algerian oil fields, notable associations were observed between the relative abundance of bacterial OTUs or phyla and the concentrations of K^+ and Cl^- ions in the production waters [103].

Nonetheless, it remains uncertain whether this relevance is causative. It is plausible that the disparities in microbial community composition may arise due to factors such as variations in water content and reservoir structure, and the ion composition may merely indicate geological dissimilarities [103]. Distinct microbial structure differences between injection and production waters were evident in the offshore Halfdan and Algerian oilfields [108]. In the Algerian oilfields, the injection water exhibited higher cell richness, and bacteria predominantly populated it. In contrast, the production water had approximately one-tenth of the cell count and was mainly dominated by *Archaea* [106]. This disparity suggests a negligible relationship between oil degradation processes in the methanogenesis-dominated reservoir and the microbial structure in the production water.

3.3. The Residual Chemicals

3.3.1. Hydrolyzed Polyacrylamide (HPAM)

The concentration and composition of residues can have varying effects on microorganisms, potentially impacting negatively on the production efficiency of the field. Chemicals dissolved in formation water affect the morphology, structure, and growth metabolism of microorganisms that passively influence oil recovery. HPAM and crosslinkers are the common chemical components after polymer flooding, and this review focuses on them.

PAM (polyacrylamide) is derived from the acrylamide monomer and is commonly employed in its hydrolyzed form, known as HPAM [109]. HPAM is synthesized by either copolymerizing sodium acrylate with acrylamide or partially hydrolyzing polyacrylamide and polyacrylic acid [110], which is a water-soluble, stable, anionic, and linear polymeric compound comprising high molecular weight chains composed of acrylamide (C_3H_5NO) monomers [111]. While HPAM is not considered toxic, it is important to note that acrylamide, the starting building block used in its synthesis, is classified as a substance of very high concern [112]. The main concern revolves around the possible presence of residual acrylamide monomer in HPAM polymeric material and the associated risks linked to its degradation [113]. This degradation process can result in the conversion of acrylamide ($CH_2{=}CHCONH_2$, depicted in Figure 11). Acrylamide is a toxic chemical to many microorganisms and is included in the candidate list for authorization due to its potential health and environmental risks [114].

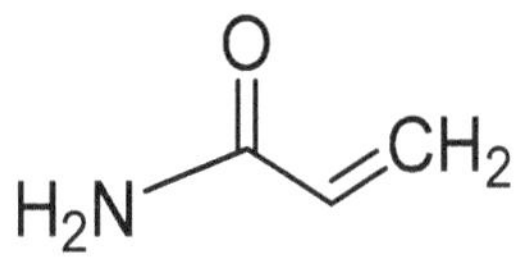

Figure 11. The chemical structure of acrylamide.

HPAM can usually be crosslinked to form a gel using a crosslinker to improve the polymer's performance, creating a highly controllable polymer network in the well. The categorization of cross-linked polymers as metallic or organic cross-linked systems is determined by the type of cross-linker employed in the process [115]. Inorganic cross-linked polymer gels that are commonly used involve the utilization of trivalent cations such as aluminum (Al^{3+}) and chromium (Cr^{3+}) as cross-linkers [116]. Metallic crosslinkers, including Cr^{3+} and Al^{3+}, have been identified for their toxicity and potential detrimental effects on microorganisms [117]. These microorganisms play a direct role as active or passive promoters of microbial metabolism [118].

Furthermore, microorganisms associated with changes in metal speciation can result in variations in metal mobility, resulting in both favorable and unfavorable consequences [119]. Polyethyleneimine (PEI) and a combination of phenol and formaldehyde are widely employed as organic cross-linkers for polymer gel formation [120]. Although the PEI crosslinker is considered non-toxic and environmentally friendly, its high cost restricts its widespread application [121]. Phenol and formaldehyde are cost-effective, but the primary concern is their toxicity, with formaldehyde mainly known for its carcinogenic properties. Research has identified alternative substitutes, such as hydroquinone (HQ) and hexamethylenetetramine (HMTA), which exhibit lower toxicity compared to phenol and formaldehyde [120].

Following polymer flooding, reservoirs contain diverse toxic components, including residual chemicals, crosslinking agents, and heavy metals [20]. Regarding residual metals, microorganisms have the capacity to induce significant alterations in metal composition amplifying or mitigating the toxicity of metals. Moreover, microorganisms can influence mineral precipitation, formation, dissolution, and deterioration due to the diverse properties exhibited on microbial-charged surfaces and metabolic activities [122]. In the context of the residual presence of HPAM/PAM in the reservoir, an observed phenomenon is the potential blockage of reservoir pores. Based on the former research, the biodegradation of PAM/HPAM takes place while microorganisms apply the carbon backbone of the polymer as a carbon source or the amide group of the polymer as a nitrogen source [123]. Observations indicate that microorganisms use acrylamide, PAM, and its derivatives as their exclusive carbon or nitrogen sources. In MEOR, certain microorganisms like *Bacillus* can utilize HPAM as a nitrogen or carbon source, enabling their growth, reproduction and metabolization of crude oil within micro pores. This microbial activity produces surfactants, which aid polymer degradation, enhance crude oil flow, and modify reservoir properties [12]. Under aerobic and anaerobic conditions, certain bacterial species within the genera *Enterobacter* sp., *Azomonas* sp., and *Clostridium* sp. have exhibited the ability to degrade PAM/HPAM, achieving degradation rates ranging from 16% to 91% [124]. Microbial degradation of the polymer results in a decrease in molecular weight, viscosity, and the conversion of amide nitrogen into ammonia, so bacterial degradation of PAM/HPAM typically leads to modifications in the structure of the polymer. The transformation of amide group (NH_2) into carboxylic acid (COOH) in PAM or HPAM has been consistently demonstrated through the application of analytical methods such as FT-IR, LC/MS, and GC/MS [125]. A proposed degradation pathway of PAM and HPAM, leading to the formation of lower molecular weight products like acrylamide, is depicted in Figure 12.

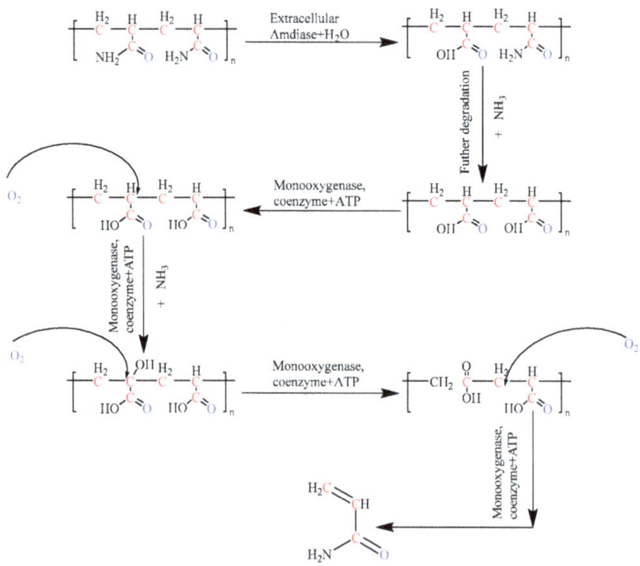

Figure 12. The pathway for the biodegradation of PAM and HPAM [124].

However, microorganisms are affected by the reservoir environment, especially HPAM and crosslinkers have enormous biological toxicity and adverse effects on the properties microorganisms and the oil displacement effect [124,126]. On the one hand, the residual agents after polymer flooding affect the growth and structure of microorganisms. According to Daqing Oilfield and Shengli Oilfield researchers, the number of microorganisms in water following polymer flooding is two orders of magnitude lower than that after water flooding [127]. Compared with water-flooded oil wells, the prevailing bacteria and archaea exhibited dissimilarity within alkali-surfactant-polyacrylamide (ASP)-flooded oil wells [128]. In both polymer-flooded and ASP-flooded production wells, *Arcobacter* and *Pseudomonas* were prevalent, while *Halomonas* was exclusively observed in the latter [21,128]. The outcomes imply that additional studies are necessary to delve into the potential effects of polymer flooding on the indigenous microbial structure in oil reservoirs.

On the other hand, after polymer flooding, there was a substantial reduction in microbial community diversity, accompanied by an increase in the prevalence of strains capable of degrading polymers. Huang et al. [129] studied microbial community structure and diversity of reservoirs after polymer flooding in the Daqing oil field at the earlier stage of microbial profile modification, which showed that the microbial community structure had undergone significant changes for the impact of exploitation by the polymer flooding. Ren et al. [21] found that the structure composition and abundance of dominant bacteria of endogenous bacteria were significantly changed after polymer flooding.

HPAM and crosslinkers in low concentrations can cause obvious negative impacts on microorganisms. Still, the primary focus of the research is centered on enhancing oil recovery following polymer flooding. At the same time, there is a lack of research on the physiological and biochemical properties, metabolic cycle, and mechanism of the influence of chemicals on the processes of MEOR.

3.3.2. Polymer Nanospheres

In the realm of petroleum field development, nanotechnology has emerged as a promising avenue for EOR applications, thanks to the benefits of nanoparticles (NPs), including their outstanding properties [130]. NPs can be categorized into inorganic NPs and organic NPs [4]. Inorganic NPs encompass metal oxide, ceramic, metal, and semiconducting NPs, whereas organic NPs include polymeric NPs and biomolecules [131]. Polymer-based NPs are colloidal entities composed of non-toxic, small, stable, water-soluble, and biodegradable polymers. Polymer NPs come in two varieties: nanospheres and nanocapsules [132].

Due to the nano size, selective water absorption, high water-absorbing capacity, tolerance to brine, low viscosity in aqueous solutions, effective dispersion in water, and other properties, nanospheres can readily penetrate interior layers to achieve comprehensive profile control, water shutoff, and flooding, with the ultimate goal of enhancing sweep efficiency [130,133]. Polymer nanospheres (PNPs) exhibit a positive impact on EOR. They have been utilized in low permeability oilfields within China and have demonstrated strong applicability in the Changqing Oilfield [133,134]. For example, polymer nanospheres and microspheres have been implemented in water mobility control, both on a pilot scale and across full fields, yielding remarkable outcomes in reducing water production, enhancing sweep efficiency, and auguring oil recovery [135]. Nonetheless, a comprehensive exploration of the mechanisms underlying the effectiveness of PNPs in enhanced oil recovery has not been conducted.

While PNPs offer numerous advantages, including cost-effectiveness, low toxicity, and exceptional durability, they also have the potential to inhibit microbial activity. Li et al. achieved the successful synthesis of antimicrobial materials through the conjugation of PNPs with Salphen Ce (IV) complexes (HSpCe) [136]. The research revealed that HSpCe@PNPs exhibited exceptional antibacterial efficacy, surpassing that of HSpCe alone. This enhancement in antibacterial activity is attributed to the synergistic antibacterial effects between HSpCe and PNPs.

PNPs are employed in EOR as an emerging technology, aiding in surmounting significant constraints associated with chemical EOR and enhancing its effectiveness. PNPs offer numerous exceptional advantages but exhibit inhibitory effects on microbial activity. Moreover, research efforts primarily focus on improving oil recovery, with a limited exploration of the mechanisms through which chemicals influence microorganisms in MEOR.

4. The Effect of Reservoir Conditions on Microorganisms

The effectiveness of MEOR is heavily influenced by the reservoir environment, including temperature, pressure, salinity, pH, permeability, porosity, crude oil viscosity, etc. [11,29]. Physicochemical conditions within oil reservoirs can profoundly affect microbial survival, growth, and metabolites [34]. Consequently, these physicochemical conditions directly impact the microbial concentration and, in turn, influence the MEOR [137].

4.1. Temperature

In the process of MEOR, temperature stands out as the foremost factor that determines the microbial structure and imposes constraints on microbial growth, metabolism, and activity [106,138]. Due to the temperature-dependent metabolic rate of bacteria, the efficiency of transport processes can be hindered at low temperatures, while excessively high temperatures can adversely affect enzymes and proteins, disrupting vital cellular activities [139]. Therefore, the microbial concentration within the reservoir, biosurfactant production, and ultimately the residual oil saturation of the reservoir varies as a function of temperature.

Petroleum oil reservoirs typically exhibit a natural temperature range from 10 to 124 °C [140]. Based on the findings of Maudgalya et al. [141], the majority of MEOR field trials have temperature limitations below 100 °C. To be specific, aerobic microorganisms are predominantly present in formations with temperatures ranging from 70 °C to 90 °C, whereas anaerobic microorganisms and facultative anaerobic microorganisms typically thrive in oil reservoirs at relatively low temperatures, usually from 30 °C to 60 °C [142]. According to Sivasankari and Govindarajan [143], microbes utilized in enhanced oil recovery processes must withstand the average temperature conditions prevalent in the reservoir.

Firstly, temperature directly influences the growth of microorganisms, which have an optimal temperature range for growth, and exceeding this range can significantly impact their growth and activity [103]. Laboratory experiments indicated that for a given microorganism, an increase in temperature from 37 °C to 73 °C resulted in a substantial decrease in bacterial concentration, and when the temperature surpassed 80 °C, the growth rate experienced a significant reduction and became extremely slow [14]. Specific temperature conditions dictate the suitability of microorganisms to inhabit the reservoir [103].

In addition, there was a notable alteration in the structure of the microbial community as the formation temperature increased [144]. Li et al. [142] employed a culture-independent approach involving CARD-FISH, qPCR, and 16S rDNA to examine the abundance, community structure, and diversity of microbes across a significant thermal gradient (50–90 °C) in the Tengchong Geothermal Field. Their findings revealed that temperature strongly influenced microbial community structure, cell growth, and activity, and the number of microorganisms exhibited a negative correlation with temperature. Moreover, at higher temperatures within the oil reservoir, there was a reduction in the diversity of the dominant bacterial species. Lin et al. [144] compared two oil reservoirs, R-A and R-G, which exhibited similar salinity and permeability but differed in temperature, with R-A being at a milder temperature of 55 °C and R-G at a higher temperature of 71 °C. The results indicated that the R-A reservoir contained a greater variety of bacterial species and displayed more intricate predominant metabolic pathways than R-G. Previous research revealed reservoirs with moderately hot temperatures around 55 °C observed the highest microbial diversity. Microorganisms inhabiting oil reservoirs exhibit varying temperature preferences, as indicated in Table 4.

… bacteria and archaea linked to temperature ranges within reservoirs.

Temperature Optimum	Phylum/Class	Order/Genus	References
• Ubiquitous	• Proteobacteria/Epsilonproteobacteria • Proteobacteria/Gammaproteobacteria • Euryarchaeota • Acinetobacter • Proteobacteria/Alphaproteobacteria	• Campylobacterales/Sulfurospirillum/Arcobacter • Pseudomonadales/Pseudomonas • Rhizobiales//Sphingomonas/Sphingomonadales/Rhizobium	[145]
• Only above 50 °C	• Atribacteria • Acidobacteria • Crenarchaeota/Thermoprotei • Euryarchaeota/Halobacteria • Thaumarchaeota/Nitrososphaeria • Euryarchaeota/Thermoplasmata • Euryarchaeota/Archaeoglobi	• Atribacter • Acidobacterium/Terriglobus • Fervidicoccales/Sulfolobales/Thermoproteales • Halobacteriales/Haloferacales • Nitrososphaerales/Nitrospira • Thermoplasmatales • Archaeoglobales	[105,146]
• Mostly above 50 °C	• Firmicutes/Clostridia • Firmicutes/Bacilli • Euryarchaeota/Thermococci • Euryarchaeota/Halobacteria • Euryarchaeota/Thermoplasmata • Proteobacteria/Hydrogenophilalia • Proteobacteria/Alphaproteobacteria • Firmicutes/Clostridia • Thermotogae • Bacteroidia • Deinococcus-Thermus/Deinococci • Nitrospirae/Nitrospira • Actinobacteria/Thermoleophilia	• Clostridiales/Thermosyntropha • Bacillales/Bacillus/Ancerobacillus • Thermococcales • Halobacteriales/Halogeometricum • Thermoplasmatales/Thermogymnomonas • Hydrogenophilales/Tepidiphilus • Rhodospirillales/Tistrella • Thermoanaerobacterales/Thermoanaerobacter • Kosmotoua • Thermals/Thermus • Nitrospirales/Thermodesulfovibrio	[106,145,147]
• Mostly below 50 °C	• Planctomycetes • Spirochaetes • Synergistetes • Chloroflflexi • Proteobacteria • Euryarchaeota • Crenarchaeota	• Planctomyces/Gemmata/Isosphaera/Pirellula/Rhodopirellula/Brocadia • Treponema/Borrelia/Leptospira • Aminobacterium/Synergistes/Thermovirga/Jonquetella • Anaerolinea/Dehalococcoides/Chloroflexus/Herpetosiphon/Caldilinea • Marinobacterium/Paracoccus/Donghicola • Methanosarcinales/Methanomicrobiales/Methanocellales/Methanocorpusculum/Methanolinea • Desulfurococcales	[105,145,146]

Furthermore, a correlation has been observed between the occurrence of functional genes of microorganisms and reservoir temperature. Cai et al. [148] examined four production wells and detected a progressive increase in the relative abundance of genes associated with the degradation of aromatic compounds, carbon cycling, and the metabolism of other organic compounds as the temperature and depth of the oil-containing stratum decreased. Correspondingly, there was a reduction in genes related to stress response, antibiotic resistance, and sulfur metabolism with diminishing temperature. For instance, optimal oil degradation rates are observed at lower temperatures, while microbial activity within reservoirs is largely absent beyond 80 °C. Therefore, general predictions can provide valuable indications for determining the in situ temperature of a reservoir. However, it is still impossible to accurately determine microbial community composition solely based on temperature.

The global demand for hydrocarbons and subsequent exploration efforts have recently shifted towards deep and ultra-deep environments, where reservoir temperatures frequently exceed 100 °C. The studies conducted by Li et al. [149] and Junzhang et al. [150] highlighted the impact of high-temperature oil reservoirs on microbial activities during oil recovery. They emphasized the necessity of identifying and isolating high thermophilic microorganisms suitable for MEOR in deep and ultra-deep reservoirs. The Jianghan Oilfield in central China documented the cultivation of thermophilic bacteria, specifically *Geobacillus kaustophilus*, capable of thriving at 100 °C, and a pilot test for paraffin removal in a well with a temperature of 117 °C demonstrated effective paraffin and plug removal [14 But the temperature criteria for MEOR reservoirs in China is still set at 80 °C. Thermophilic bacterial genera such as *Bacillus*, *Thermus*, *Thermoanaerobacter*, *Thermococcus*, and *Thermotog* have been successfully isolated from high-temperature reservoirs in regions like China, California, and the North Sea. These thermophilic bacteria represent excellent candidates for MEOR applications [144].

4.2. Salinity

The salinity in reservoirs varies widely, ranging from 100 g/L to 300 g/L depending on the corresponding depth. This significant variation in salt concentration profoundly impact the MEOR process, as microorganisms exhibit varying degrees of tolerance to differer salinity levels [151]. Using microbial flooding technology is not feasible in reservoir with high salinity in water formation [14]. An apt illustration of this phenomenon is the reduction in the metabolic generation rate of biosurfactants, gases, alcohols, and acids under high salinity conditions [152]. The high mineralization in the reservoir also tends to form precipitates when interacting with other substances, leading to blockages [92].

First and foremost, salinity directly influences microbial growth [92]. According t the report, the quantity of isolated microorganisms diminishes when the salinity of th reservoir surpasses 10% [11]. She et al. [14] isolated *Pseudomonas aeruginosa* and *Bacill subtilis* from the production fluid, and these microorganisms demonstrated successf growth within a salinity range of 1×10^5–2×10^5 ppm. But once the salinity exceeded 2 10^5 ppm, the growth rate of these microorganisms decreased.

Secondly, salinity plays a pivotal role in determining the composition of microbi communities in oil reservoirs by influencing growth and restricting bacterial activity [2 153]. The effectiveness of MEOR is contingent upon the microbial community within th reservoirs. Usually, higher levels of formation water salinity tend to suppress bacteri diversity. Reservoirs with lower salinity levels exhibit a significantly higher microbi diversity compared to reservoirs with higher salinity levels, leading to an increased c recovery [144]. For example, Xiao et al. [154] studied the community diversities of two c reservoirs in Changqing, China, exhibiting substantial salinity differences. The finding suggested that reservoirs exhibiting relatively lower salinity levels showcased increase diversity, and the efficiencies of MEOR were quantified at 8.12% (lower salinity) and 6.56 (higher salinity), correspondingly.

Additionally, salt concentrations in oil reservoirs impact the methanogenesis of microbes. The process of hydrogenotrophic methanogenesis from carbon dioxide with hydrogen was observed to occur only up to an in situ salt content of 9% [155]. Except for the above adverse effects, salinity exerts positive effects on microorganisms, reflected in the reduced CMC leading to a decrease in the required amount of ex-situ biosurfactants for surface tension reduction [156]. Table 5 displays the bacteria and archaea associated with the optimal salinity conditions in petroleum reservoirs.

Table 5. Bacteria and Archaea correlated with salinity conditions in oil reservoirs.

Phylum/Class	Order/Genus	Preferred Salinity	Reference
• Firmicutes/Clostridia	• Clostridiales/Deulfotomaculum	• higher salinity	[106]
• Proteobacteria/Deltaproteobacteria	• Desulfovibrionales/Desulfovermiculus halophilus	• higher salinity	[157]
• Firmicutes/Clostridia	• Halanaerobiales/Haloanaerabium	• higher salinity	[158]
• Euryarchaeota/Methanococci	• Methanococcales/Metanothermmeococcus	• higher salinity	[145]
• Thermotogae	• Petrotogales/Petrotoga	• higher salinity	[106]
• Euryarchaeota/Methanobacteria	• Methanobacteriales/Methanobacterium	• lower salinity	[103]
• Euryarchaeota/Methanomicrobia	• Methanomicrobia/Methanoplanus	• lower salinity	[159]

Salinity was identified as an essential screening parameter of MEOR. The salinity of the formation water directly impacts the osmotic pressure within microbial cells, thereby influencing microbial proliferation, overall abundance, and community composition. Utilizing microbial oil recovery methods is impractical in forming water with high salinity. The salinity in the reservoir varies with depth, exerting a significant influence on the MEOR process due to the varying salinity tolerance of microorganisms. Each microbe possesses its optimal salinity range, crucial in MEOR effectiveness.

4.3. pH

After polymer flooding in oil reservoirs, the variation in pH can be related to several scenarios:

Firstly, the choice of polymer for flooding can affect the pH of the injected fluid [160]. Some polymers, especially synthetic ones, may have inherent acidic or alkaline properties, leading to pH variations in the injected solution. For example, PAM/HPAM can influence the pH of the injection fluid.

Secondly, polymers injected into the reservoir can interact with reservoir rocks and fluids. These interactions can alter the pH of both the injected and reservoir fluids. On the one hand, polymers may adsorb onto rock surfaces, releasing acidic or alkaline components into the solution, affecting pH. On the other hand, chemical reactions between the injected polymer solution and reservoir fluids can result in changes in pH [161]. For instance, interactions with minerals or brines within the reservoir can lead to pH shifts due to precipitation or dissolution reactions.

Furthermore, the initial pH of the reservoir fluids and rock properties can also play a role [162]. Depending on the reservoir's geological formation and water chemistry, the pH of the injected polymer solution may need to be adjusted to optimize the recovery process.

Additionally, the reservoir's rock mineralogy and porosity variations can impact the pH response to polymer flooding [163,164]. Different regions of the reservoir may exhibit varying pH shifts based on their geological characteristics.

After polymer flooding, pH variations mainly relate to the choice of polymers, their interactions with reservoir components, chemical reactions, initial reservoir conditions, and reservoir heterogeneity. Therefore, understanding these pH effects is crucial for optimizing its impact on enhanced oil recovery.

Generally, the pH levels within oil reservoirs fall between 3 and 7 [29]. pH plays a significant role in determining the presence and distribution of microorganisms [102]. Microorganisms typically exhibit the ability to thrive across a spectrum of pH levels spanning approximately 3 to 4 units. They are classified as acidophiles ($pH < 5$) when their optimal growth occurs in acidic environments, neutrophiles ($5 < pH < 9$) in neutral conditions, and alkaliphiles ($pH > 9$) in alkaline environments.

The growth rates of microorganisms show variations depending on the pH levels, often represented by bell-shaped or triangular curves. These curves illustrate the limits of pH at which cell structural integrity is maintained and the impact of pH on cellular metabolism. In addition to influencing the metabolic activity of microorganisms, pH also impacts the performance of biosurfactants. A lower pH can result in the aggregation and sedimentation of biosurfactants, a frequently observed phenomenon in experiments aimed at the crude purification of surfactants produced by microorganisms [11].

In addition, pH can influence both the thermodynamics and kinetics of microbial respiration, ultimately shaping the composition and functionality of microbial communities [165]. A distinct relationship between pH and the compositions of microbial communities was identified at the site level for two separate pits located within the La Brea Tar Pits in Los Angeles [166]. Furthermore, variations in pH showed a positive correlation with the divergence of bacterial communities [102]. pH also impacts microorganisms' surface charge and enzymatic activity [11].

Finally, some studies found a correlation between pH value and salinity among microorganisms in oil reservoirs. An investigation conducted in twenty-two geographically distinct Chinese oil reservoirs revealed that *Alphaproteobacteria*, *Actinobacteria*, and *Deltaproteobacteria* were the dominant bacterial groups in reservoirs with neutral to alkaline conditions, characterized by pH values ranging from 7.0 to 8.2. *Pseudomonas* correlated with a decline in the pH value of formation brine within the range of 5.5 to 7.6. *Betaproteobacteria* *Gammaproteobacteria*, and *Epsilonproteobacteria* thrive in highly acidic environments and are identified in reservoirs with pH values ranging from 5.5 to 6.5 [145]. SRBs were discovered to exhibit resistance to a diverse range of salinity levels and a broad spectrum of pH values spanning from 4 to 9.5 [146]. Table 6 presents data on the bacteria and archaea found in petroleum reservoirs that thrive under optimal pH conditions.

Table 6. Bacteria and Archaea linked to pH conditions [103].

Phylum/Class	Order/Genus	Preferred pH
• Euryarchaeota/Methanomicrobia	• Methanobacteriales/Methanocorpusulum/ Methanolobus/Methanolinea/ Methanocalculus/Methanosaeta/Methanoculleus	• alkaline pH
• Actinobacteria		• alkaline pH 7.0–8.0
• Euryarchaeota/Methanococci	• Methanococcales/Methanococcs	• alkaline pH
• Proteobacteria/Alphaproteobacteria	• Rhodobacterales/paracoccus	• alkaline pH 7.0–8.2
• Proteobacteria/Archaeoglobi	• Archaeoglobales/Archaeoglobus	• alkaline pH
• Proteobacteria/Deltaproteobacteria	• Desulfuromonadales/Desulfuromonas	• alkaline pH
• Proteobacteria/Betaproteobacteria		• acidic pH 5.5–6.5
• Proteobacteria/Epsilonproteobacteria		• acidic pH 5.5–6.5
• Proteobacteria/Gammaproteobacteria	• Pseudomonadales/Pseudomonas	• acidic pH 5.5–7.6
• Euryarchaeota/Methanobacteria	• Methanobacteriales/Methanothermobacter	• acidic pH

4.4. Oxygen

Numerous microorganisms display high sensitivity to variations in oxygen concentrations [167]. Microorganisms can be classified into obligate aerobic/anaerobic bacteria, aerotolerant anaerobes, and facultative anaerobes/aerobes based on oxygen needs. Obligate aerobic bacteria rely on oxygen for their growth as their energy production and respiration mechanisms depend on the transfer of electrons to oxygen, which serves as the final electron acceptor in the electron transport reaction for these bacteria [168]. Obligate anaerobic bacteria can only survive and thrive in oxygen-free environments as their metabolic processes for energy generation are not reliant on oxygen consumption. Oxygen can be detrimental to vital enzymes in obligate anaerobic bacteria.

Nevertheless, specific bacteria are categorized as microaerophilic or aerotolerant anaerobes as they demonstrate improved growth in environments with low oxygen concentrations [152]. Oxygen frequently triggers minor metabolic processes within these bacteria, augmenting major energy production pathways. Facultative anaerobes possess the ability to adjust their metabolic processes based on the availability of oxygen. In the presence of oxygen, they employ the more efficient process of respiration, while in its absence, they resort to the less efficient process of fermentation [169]. As a result, facultative aerobes, which can adapt to aerobic and anaerobic conditions, are frequently encountered and can be effectively utilized to enhance oil recovery in oxygen-depleted reservoirs [152].

Compared to most anaerobes, aerobes possess a highly efficient metabolism and can reproduce rapidly [10]. The findings from previous studies indicate that aerobic bacterial growth and metabolic activities were hindered in the presence of anaerobic conditions [170]. Cai et al. [171] studied the impact of oxygen inoculation on oil biodegradation and the biodiversity of microorganisms in the Dagang oil field. They assessed and analyzed biodegradation through in vitro enrichment, denaturing gradient gel electrophoresis, real-time PCR, and GC-MS techniques. The results indicated that aerobic conditions could promote petroleum biodegradation and induce alterations in the microbial community structure, potentially benefiting MEOR applications. Many research studies have been conducted on injecting oxygen and nutrients to stimulate aerobic bacteria in situ for degrading heavy oil [10]. The application of this method is extensively practiced in both China and Russia [172]. In addition, as the depth and temperature of an oil environment increase, it tends to transition towards anaerobic conditions. Since oxygen concentrations in water are typically low at elevated temperatures, following the principles of Henry's law, main microorganisms exhibit anaerobic characteristics when subjected to high-temperature conditions.

In summary, petroleum oil reservoirs can be viewed as extensive geo-bioreactors that harbor diverse microorganisms. Each microorganism has specific oxygen requirements and can only achieve optimal growth under suitable conditions. Oxygen significantly impacts microbial growth, metabolism, and oil recovery.

4.5. Pressure

Pressure is also acknowledged as a critical parameter in MEOR. When examining its impact, it is observed that high hydrostatic pressures can have conflicting implications for bacterial growth [173]. Fluctuating pressure levels can induce alterations in the solubility of reservoir gas, leading to changes in crude oil viscosity and possibly influencing oil displacement efficiency. However, in comparison to other factors, pressure has a minimal impact on the growth and metabolism of oil-producing microorganisms. Simulating high pressure conditions in laboratory settings is exceptionally challenging, and there is a scarcity in the literature of studies assessing the tolerance of microorganisms to severe pressure.

4.6. Minerals

The composition of rock layers affects the activity of microorganisms. The reservoi is mainly composed of sandstone and carbonate rock, so it has a large specific surfac area, and microorganisms are readily adsorbed on the rock surface, directly affectin microorganisms' activities. In addition, the injected nutrient solution can be concentratec thus affecting the growth and reproduction of microorganisms. The clay minerals in th reservoir have a high cation exchange capacity. They are easy to hydrate, expand, an disperse under the action of formation water, resulting in a negative charge on the surfac of the particles and a positive control on the end face, which significantly impacts th activity and migration of microorganisms. Therefore, in reservoirs, the influence of roc composition, chemical properties, and formation water should be fully considered t optimize the activity environment of microorganisms.

Moreover, porosity and permeability have an essential impact on the growth an migration of microorganisms, which substantially influence the efficiency of MEOR [174 More precisely, the porosity determines the microbial penetration and the degree of growt in the pores and crevices [42]. The increase in pore size reduces permeability and adverse affects the fluid transport process [22]. When the core permeability is higher than 10 $\times 10^{-3}$ μm^2 and lower than 1500 $\times 10^{-3}$ μm^2, the reservoir conditions have adsorptio and retention effects on oil-producing bacteria [14]. There is a direct relationship betwee capillary pressure, porosity, permeability, and microbial growth rate [175]. When th capillary pressure increases, it decreases pore diameter and permeability [176]. As a resu the volume of microorganisms within the pores diminishes, causing a slowdown in the growth and metabolic rate [177].

In conclusion, the growth and activity of microorganisms in reservoirs are influence by the physical and chemical conditions of the environment, including (i) the reservo fluid: oil properties, residual chemicals, and injection water and (ii) the reservoir conditio temperature, pH, oxygen, salinity, pressure, porosity, permeability, etc. Additional the availability of essential nutrients is crucial for supporting the growth and metabo processes of the microorganisms. The specific type of microorganism introduced into th reservoir can greatly affect reservoir properties and oil recovery, and its optimal growth determined by the ecological conditions present in the reservoir. Oil displacement efficien increases as the reservoir conditions approach the optimum for microbial growth.

5. Microbial Consortium on Oil Displacement

5.1. Microbial Consortium

Various microorganisms possess the ability to degrade distinct components of crude oil. While a diverse range of microorganisms can degrade the simpler compounds found in crude oil, complex compounds such as polycyclic aromatic hydrocarbons (PAHs), asphaltenes, and resins are limited to fewer microorganisms. Mixed microbial populations are frequently employed as mixtures or formulations, along with their metabolic products, to optimize oil recovery and extend the productive lifespan of oil wells [178]. The metabolic by-products of diverse bacteria in MEOR processes possess distinct functionalities to enhance oil recovery [179]. For instance, bacteria belonging to the genera *Clostridium*, *Zymomonas*, and *Klebsiella* produce solvents like propane-2-diol acetone and butanol.

Additionally, *Enterobacter* and *Clostridium* are accountable for generating methane and hydrogen gas. These fermentation gases can contribute to re-pressurize wells, facilitating the displacement of light crude oil within the well and aiding its recovery [180]. *Clostridium* is considered one of the most suitable microorganisms for MEOR due to its highly resilient endospores, which allow them to survive in unfavorable conditions [179]. *Bacillus* is also beneficial in MEOR processes because it can produce biosurfactants that have favorable properties contributing to the effectiveness of the MEOR [181]. Thus, the concept of employing a microbial consortium was initiated. A study by Xia et al. [182] developed a microbial consortium consisting of *Serratia proteamaculans*, *Rhodococcus erythropolis*, *Rhizobium* sp., and *Alcaligenes* sp. The results showed that the microbial consortium achieved the consumption of 85.26% of the crude oil within 15 days, demonstrating a significantly higher efficiency compared to individual strains and which provided evidence to support the viability of the concept.

One of the common methods to develop the microbial consortium is cultivating a variety of microorganisms with different features to take advantage of their diverse functions and inject them into the reservoir. According to I. Lazar et al. [183], the patented Multi-strain Mixed Microbial Application consortium, which includes thermophilic, barophilic, acidophilic, and anaerobic strains isolated from oil reservoir water, has the ability to produce biosurfactants, fatty acids, alcohols, CH_4, and CO_2 at temperatures up to 90 °C, mirroring in situ conditions. The introduction of the microbial consortium into reservoirs, along with the addition of targeted nutrients, resulted in a three-fold increase in oil recovery.

Some field trials involve directly indigenous microorganisms to develop microbial consortia. Rathi et al. [184] developed the methanogenic consortium named TERIL146, consisting of indigenous microorganisms from a high-temperature oil reservoir, resulting in significant gas production (12 mmol/L) and other metabolites. The TERIL146 exhibited an incremental oil recovery of 8.3% in a sand pack assay. These findings demonstrated the successful recovery of both oil and gas by the developed indigenous methanogenic consortium, highlighting its potential application in thermophilic depleted oil wells in the Indian subcontinent.

Numerous field trials have been conducted to screen and isolate microorganisms with specific metabolites, thus enhancing oil recovery rates. In a study by Okoro et al. [185], two petroleum thermotolerant consortia were successfully isolated and characterized from reservoir sandstones. This validation confirmed that the injection of selected microorganisms to enhance oil recovery and promote the production of specific metabolites was an effective approach.

Several developments of the microbial consortium have focused on combining biodegradation with its products. Munna Bhattacharya et al. [186] researched the simultaneous output of emulsifiers and the biodegradation of crude oil using a mixed culture consisting of *Ochrobactrum pseudintermedium* and *Bacillus cereus*. Under optimized culture conditions, the microbial consortium degraded up to 70.54% of total petroleum hydrocarbons within 72 h of incubation, with crude oil as the sole carbon source (4% v/v). Additionally, the microbial consortium produced two types of EPS with exceptional emulsification activity, reduced IFT in the crude oil–water system, and improved oil recovery by 46.85%.

Integrating multiple techniques utilizing microbial consortia with diverse properties is an effective strategy for increasing oil recovery. Developing a suitable microbial consortium with good performance is crucial for successfully implementing MEOR.

5.2. Microbial Activities and Interrelations within Oil Reservoirs

Microbial populations in nature do not function independently. Specific interactions occur between spatially separated species, and they utilize soluble or volatile metabolites to communicate information. Additionally, other interactions involve close species, either forming a biofilm on the same particle or being physically associated with each other. In the environment, microbial species can form different relationships, including mutualism, competition, commensalism, and parasitism [187,188].

When utilizing a microbial consortium to degrade various components of crude oil, individual members of the microorganisms can form mutually beneficial symbiotic relationships [189]. In the process of petroleum hydrocarbon degradation, certain bacteria have demonstrated the capability of degrading metabolites that accumulate from other members, which helps mitigate the inhibitive effects of metabolites on the degradation capacity, ultimately leading to an improvement in degradation [190]. For example, the consortium consisting of *Rhodococcus* sp. and *Mycobacterium* sp. demonstrated considerable efficiency in degrading phenanthrene. This was attributed to the fact that the *Rhodococcus* sp. degraded phenanthrene, producing 1-hydroxy-2-naphthoic acid, which acted as an inhibitor for its phenanthrene-degrading activity. However, the *Mycobacterium* sp. was capable of degrading the 1-hydroxy-2-naphthoic acid, relieving the repression of phenanthrene degradation caused by the *Rhodococcus* sp. [190].

Moreover, the investigators observed that the metabolites generated by the microorganisms within the microbial consortium also facilitated the degradation of other members. As an example, the addition of biosurfactants produced by the yeast *Pseudozyma* sp. to *Pseudomonas putida* significantly improved its ability to degrade crude oil, with a substantial increase of approximately 46% in the degradation of alkanes (C_{10}–C_{24}) [191]. This phenomenon can be attributed to the modulation of cell surface proteins and alteration of the cell membrane hydrophobicity by biosurfactants, which facilitate the adhesion of microbial cells to hydrocarbons, thereby enhancing microbial crude oil degradation [192]. In MEOR, interactions among microorganisms within populations can exhibit either synergistic or detrimental effects based on the metabolic capabilities of each microorganism [193]. As an illustration, the interactions between microalgae and bacteria can encompass a broad spectrum of relationships, ranging from mutualism and commensalism to competition and parasitism [194]. Increasing evidence indicates that microbial interactions within consortia are highly species-specific and remarkably sophisticated.

Given the variety of interactions observed among microorganisms, including competition, predation, commensalism, and mutualism, numerous researchers have investigated the correlation between two microorganisms employing the Lotka–Volterra (L-V) model. Lotka [195] and Volterra [196] introduced a well-known growth model that established a novel foundation for the mathematical ecology of populations. In a study by Guerra [197], the relationship between the absolute growth rates of *Lactococcus lactis* was elucidated using the L-V model, specifically, the two predators-one prey model. In addition, Fujikawa et al. [198] applied the L-V model to illustrate microbial growth in a mixed culture containing *Escherichia coli*, *Salmonella*, and *Staphylococcus aureus* in the study. They revealed that the competition coefficients in the model remained stable. In the research by Zheng et al. [199], the growth of *Rhodotorula* sp., *R. phaseoli*, and *A. ferrooxidans* within a leaching condition was modeled utilizing the L-V model. The findings showed that the individual growth curves of microorganisms followed a logistic pattern. Additionally, the correlation between *A. ferrooxidans* and *R. phaseoli* was identified as mutualism, while the correlation between *A. ferrooxidans* and *Rhodotorula* sp. was commensalism. To investigate the contribution of microorganisms in actual reservoirs to enhanced oil recovery, Wang et al. [200] developed a mathematical model for MEOR that incorporated a competition mechanism between two bacterial species to study the impact of competitive growth and product interactions between bacteria on the efficiency of enhanced oil recovery.

Comprehensive studies conducted on the interactive growth models between single microorganisms and populations during oil recovery are comparatively scarce, while no research on the collaborative microorganisms using chemicals in oil recovery has been conducted to date. The investigations focus on oil displacement compound bacteria to establish a few bacteria with synergistic effects while there are limited studies on the relationship between microbial consortium using chemicals. Thus, based on previous works, future research is required to investigate the growth of microbial consortium using chemical agents at a specific stage during the MEOR process and verify the correlations and interactions between the microorganisms by the models. This can offer theoretical backing for the cooperative oil displacement by multiple microorganisms.

5.3. Microscopic Oil Displacing Process of Microbial Consortium

During the MEOR process, the saturation profile in a porous medium filled with oil and brine changes. However, the pore-scale physics and active mechanisms of MEOR are not entirely understood. Consequently, the flow visualization technique is employed to understand better the pore-scale behavior of MEOR mechanisms and the intricate processes involved.

The micromodel, which involves the etching of flow patterns on glass, is a valuable tool for examining various multiphase flow phenomena at the pore scale in porous media. A comprehensive literature review shows numerous researchers have extensively employed micromodels to study and visualize MEOR processes. Hongyan et al. [201] employed the dimensional microscopic visualization technique using a glass model to investigate the oil displacement mechanism and the impact of emulsified bacteria on the residual oil in a simulated reservoir environment, which revealed that the emulsified bacteria and their products exhibited a notable capability to emulsify crude oil effectively. As a result, the oil recovery was enhanced by 19%, providing valuable new insights and a theoretical foundation for applying microbial oil displacement techniques. Liu et al. [202] carried out a *Bacillus subtilis*-based MEOR process using a microfluidic device made of polymethyl methacrylate to mimic the MEOR process in reservoirs, which demonstrated that the ex situ MEOR process with *B. subtilis* achieved a recovery rate of 38.56%, while the in situ MEOR process achieved a recovery rate of 40.27%. This research provided a novel tool for comprehending the MEOR process, offering advantages in visibility and precise fluid control during the MEOR process.

While micromodels have proven to be effective in visualizing bacteria during the MEOR process, most prior investigations were performed using artificial pore-structure chips [203]. There is a consensus that artificial pore structure does not accurately represent the conditions found in natural reservoirs. Thus, Gaol et al. [204] utilized heterogeneous micromodels constructed from natural rock captured through μCT imaging to assess the MEOR process within porous media under conditions of increased pressure, temperature, and anaerobic and sterile environments, which revealed that microbial growth, along with the resulting end-products, play a crucial role in enhancing the efficiency of oil displacement during MEOR, offering novel perspectives into the pore-scale mechanisms and various factors influencing the performance of MEOR. Prior to field application, the findings may be utilized to support MEOR research on a wider scale.

Many studies describe the microscopic process of oil displacement using the flow visualization technique but rarely report the mechanism of microbial consortium on the oil displacement process using chemicals from a microscopic perspective.

6. Application of Microbial Consortium on Oil Recovery

Based on the applications and mechanisms of MEOR, its processes can be categorized as MFR (microbial flooding recovery), CMR (cycle microbial recovery), MSPR (microbial selective plugging recovery), MWR (microbial wax removal), and other methods [14]. The microbial consortium is frequently employed as a mixture or formulation to improve recovery in MEOR. Over the past decades, a considerable quantity of field trials involving microbial consortia have been conducted in various countries to assess the feasibility of MEOR. These field experiments are conducted considering reservoir permeability, temperature, salinity, pH, the specific recovery method employed, the type of reservoir, etc. Numerous field trials for MEOR have been conducted worldwide, yielding diverse outcomes. Based on global statistics from field tests, over 90% of MEOR trials have demonstrated beneficial effects [11].

Among the pioneering nations to initiate field trials for MEOR, America operated a field trial in the Lisbon oil field in 1954, which involved the activation of *Clostridium acetobutylicum* through the injection of molasses, and the mechanisms employed in this trial included biosurfactants, gases, and acids [205]. Oklahoma carried out a field experiment in the Bebee field; an in situ application of biosurfactants using the *Bacillus* strain was tested, which remarkably demonstrated a biosurfactant concentration that was one-ninth of the minimum required concentration, still enhancing oil recovery [206]. Numerous implementations and field tests of CMR and MFR have been conducted in Romania resulting in a remarkable average increase in oil production of 100% for CMR and 200% for MFR [11]. Furthermore, with the support of America, the Vizacheres oil field in Argentina underwent a 19-month injection process involving facultative anaerobic bacteria and nutrients, leading to a total increase of 20% in production, amounting to an aggregate production of 27,984 m^3 of crude oil [11]. In India, Ahmedabad carried out field trials utilizing the huff and puff technique that focused on meticulously isolating anaerobic microbial populations from the reservoir. The field trials, comprising 12 wells across 4 fields demonstrated a remarkable threefold increase in crude oil production and a substantial decrease in water cut, showcasing the approach's efficacy [53]. In Poland, Karaskiewicz conducted a series of 18 field trials from 1961 to 1969 utilizing a microbial consortium comprising *Mycobacterium*, *Arthrobacter*, *Peptococcus*, *Pseudomonas*, and *Clostridium* [183].

Indeed, China stands among the leading nations in the field of MEOR because of its practical implementation techniques in recent times. MEOR has been extensively explored through applications and field tests in various oilfields in China, including Shengli, Daqing, Changqing, Qinghai, Xinjiang, Liaohe, and Jilin. Among these applications, the predominant focus has been on microbial wax removal. Following incomplete data, 1 fields comprising 1739 wells have undergone testing for MEOR in China [19]. Specifically regarding the application of CMR in China, a combined sum of 678 wells distributed across 12 fields containing Shengli, Daqing, Zhongyuan, and Jilin oilfields have been implemented. Notably, the Shengli oil field conducted microbial huff and puff operations in 1640 wells resulting in a remarkable production increase of 2.19×10^5 tons of crude oil, while the Daqing oil field conducted polymer flooding in 518 wells across 10 blocks, with a production increase of 6.4×10^4 tons. The quantity of MFR is relatively less compared to that of CMR. Table 7 displays MEOR field experiments conducted in various countries using microbial consortia. Owing to the absence of precise data in many field trial documents, certain depictions may be lacking in detail.

Table 7. Implementation of MEOR in oil fields.

Country	Technologies	Cases	Microbial Consortium	Effect	References
USA	CMR	Single well stimulation	Mixed cultures of anaerobic bacteria	Oil production showed a substantial 230% rise over 7 months.	[207]
			Clostridium sp. and *Bacillus*	a 20 to 30% increase in the recovery of OOIP on average	[208]
	MFR	Single well stimulation waterflooding in Tulsa	*Bacillus* sp., *B. licheniformis*, *Clostridium z.*, and a Gram-negative rod	Oil extraction saw a 13% increase and a 30% decrease in the water/oil ratio over ten months.	[209]
	CMR			Oil production witnessed an improvement of up to 79%.	[209]
	CMR	Scale-up of microorganisms in a single well stimulation in Oklahoma	Facultative and anaerobic bacteria	Oil production experienced significant growth.	[11]
	MFR	Field test in Texas	Hydrocarbon-degrading bacteria	Increased oil recovery is achieved through surfactant usage, CO_2 production, pH control, and paraffin mitigation.	[210]
China	CMR, MSPR, MFR, MWR	Xinjiang, Liuzhongqu, Jilin, Huabei, Changqing, and Jing'an oil fields	*Bacillus*, *Pseudomonas*, *Bacteroides*, *Fusobacterium*, and *Eurobacterium*	Oil production demonstrated a range of increases from 2001 to 122,800 tons.	[211–213]
		Daqing, Shengli oilfield	*Bacillus cereus*, *Brevibacillusbrevis*, Hydrocarbon-degrading strains	Oil production showed a substantial boost of 165%.	
	MFR	Shengli oilfield	Indigenous microbial community	Oil increment was about 2.1×10^5 tons for 7 blocks	[214]
	MFR	Pilot tests in Baolige Oilfield	Six facultative anaerobes	Increase 2.1×10^5 tons of crude oil over 43 months for 169 oil wells	[50]

Table 7. *Cont.*

• Country	• Technologies	• Cases	• Microbial Consortium	• Effect	References
• Russia	• MFR	• Vostochno-Anzirskoe reservoir and Cheremukhovskoe heavy oil reservoir	• Hydrocarbon-oxidizing bacteria, and aerobic bacteria *Rhodococcus* and *Gordonia amicalis*	• Oil production experienced a significant rise of 1250 t.	[13]
	• MSPR	• Bashkiria reservoir	• Aerobic and anaerobic activated sludge bacteria	• Each 600-producing treated well contributed to an incremental oil recovery ranging from 1000 to 2000 tons per year.	[215]
	• MFR	• Three pilot tests in the Romashkino oilfield	• Aerobic bacteria anaerobic bacteria	• Oil recovery increased by 32.9% by generating surfactants, organic acids, carbonic acid, methane, and polysaccharides.	[216]
	• MSPR	• Oil filed		• The daily oil production escalated from 180 t to a range of 200–300 t.	[217]
	• MFR			• Oil production experienced a rise of 8% in 4 months.	[218]
• Romania	• CMR	• Oil fields	• *Bacillus*, *Clostridium*, and Gram-negative rods	• Oil production experienced a notable increase of 100–200% over a maximum of 5 months.	[183]
			• Adapted mixed enrichment cultures predominated.	• Oil production experienced an improvement of up to 200% in two wells over 1–4 years.	
• India	• MSPR, • CMR	• Oil fields	• *Clostridium*, *Thermonaero bacterium* sp., and *Thermococcus* sp.	• Improvement in oil production.	[18]
• Former East German	• ASMR, MFR	• A carbonate reservoir	• *Clostridium* and *Bacillus*	• Improved oil production and reduced the water/oil ratio from 88% to 34%.	[219]
• Former Czechoslovakia	• MFR, • ASMR, CMR	• Pilot test	• SRB, Oxidizing bacteria, Hydrocarbon, and *Pseudomonas SD.*	• Microbial growth • Oil viscosity decrease	[220]

Table 7. *Cont.*

• Country	• Technologies	• Cases	• Microbial Consortium	• Effect	References
• Poland	• MFR	• 18 field trials in Carpathian	• *Mycobacterium, Arthrobacter, Pseudomonas,* Peptococcus, and *Clostridium*	• The additional oil recovery rate varied from 20% to 200% of the original production rate over eight years.	[221]
• Hungary	• MFR	• Demjen field	• Bacteria from sewage-sludge cultures and anaerobic thermophilic bacteria	• An improvement in oil production by 12% to 60% was observed for a few weeks to 18 months.	[19]

Globally, experimental and modeling studies have been widely carried out at constant conditions to understand the effect of various microbial parameters on MEOR. In lab-scale research, most MEOR tests have shown promising results, and some field trials have demonstrated substantial oil production and greater economic effectiveness compared to chemical EOR. Despite these achievements, the predictability of MEOR's efficacy at the field scale has remained uncertain, raising doubts about the true potential of MEOR. The application of MEOR in the field is primarily constrained by the restricted understanding of microbial survival and metabolism in the oil reservoir. In particular, the residual chemicals left after polymer flooding have detrimental effects on microorganisms; however, the current research on the influencing mechanism of residual agents on microorganisms in MEOR is even less.

7. New Technologies in MEOR

Recently, innovative metagenomic methodologies have opened up new avenues for the in-depth characterization of microbial communities in oil reservoirs, allowing for a comprehensive understanding of their phylogenetic and functional profiles. These cutting-edge approaches have provided unprecedented resolution and insight into the microbial diversity and activities within oil reservoirs.

Employing metagenomic methods to explore the microbiology of oil reservoirs is anticipated to reveal previously unknown microbial strains, presenting novel discoveries in the field of science. Additionally, these approaches hold the potential to unlock opportunities for harnessing the untapped potential of these microbial strains for MEOR applications. The combination of metagenomic approaches and biotechnological methods holds tremendous promise for MEOR, and has the potential to significantly enhance MEOR processes and outcomes [18]. It is widely acknowledged that research aimed at improving MEOR necessitates the identification and isolation of new microbial strains, specifically those capable of thriving in anaerobic conditions and exhibiting rapid biomass production or the production of surface-active biopolymers [222].

The introduction of advanced DNA sequencing technology in the early 2010s enabled the precise identification and comprehensive characterization of the predominant microorganisms inhabiting oil reservoirs, thereby facilitating a deeper understanding of their metabolic pathways and functions. As a result, the MEOR technology transitioned towards directly injecting carefully selected microorganisms, typically cultivated under controlled laboratory conditions, to enable the in situ production of valuable metabolites. The underlying motivation for this approach was to substantially decrease expenses associated with procuring stimulating substrates and the need for modifying or installing additional equipment for treated water injection [178]. Due to its distinctive advantages, the utilization of sequencing technology built on 16S rDNA has gained significant traction in research related to MEOR technology [189]. Zhang et al. [223] analyzed microbial community diversity in reservoirs after polymer flooding in the Daqing oil field using amplified ribosomal DNA restriction analysis. Numerous investigations have been carried out to explore microbial consortia in oil field environments [224]. These studies have employed diverse methodologies, including culture-based approaches, culture-independent methods based on the 16S rRNA gene, and metagenomic techniques [225,226]. A previous study utilizing the PhyloChip technique based on the 16S rRNA gene investigated samples from the Alaska North Slope oil field to identify microorganisms that have the potential to play a role in methane and hydrogen sulfide production, as well as hydrocarbon degradation [227]. Despite identifying various organisms from lineages that have not been cultivated, a comprehensive understanding of these organisms' complete diversity and functional potential remains elusive.

In previous metagenomic studies of microbial community composition in different systems, the genomic DNA of coexisting organisms was extracted and sequenced [228] Ping Hu et al. [229] utilized a genome-resolved metagenomic method to analyze petroleum-reservoir-produced water samples obtained from six production wells under conditions of various depths, temperatures, and H_2S concentrations. The findings demonstrated tha the genome-resolved approach offered enhanced taxonomic resolution and enabled more detailed predictions of metabolic processes. An et al. [224] analyzed diverse hydrocarbon containing samples from a shallow oil reservoir located in Alberta, Canada, at a depth o 850 m. Interestingly, several samples exhibited a remarkable presence of genes associated with aerobic hydrocarbon metabolism, while other samples demonstrated a higher representation of the genes related to anaerobic hydrocarbon metabolism and methanogenesis In a recent study, fosmid libraries were sequenced to investigate hydrocarbon degradatior pathways in an enrichment culture [230].

Genomic analysis can identify microbial strains that possess specific genes responsibl for encoding stress resistant proteins and metabolite production, making them potentiall well-suited for MEOR applications. These strains are anticipated to exhibit resilience t elevated temperatures, pressures, salinity, and anaerobic conditions in the oil reservoi Furthermore, they are expected to possess genes encoding proteins that contribute t environmental resilience within the host organism or produce metabolites facilitating th mobilization of residual oil, thereby enhancing oil recovery [178]. Genes that encod proteins possessing characteristics such as tolerance to thermal, osmotic, and oxidativ stress and the production of desired metabolites can serve as biomarkers. Strain 101C as identified by Kapse et al. [42], contains genes encoding crucial enzymes involved i synthesizing bio-emulsifiers with mannoprotein properties that indicate strain 101C5 ca emulsify and solubilize crude oil. Utilizing these biomarkers makes it feasible to quickl identify potential strains in MEOR, thus promoting the development of an effective MEO process that can be successfully implemented in high-temperature depleted reservoirs.

Microbial proliferation and the effectiveness of microbial displacement technology a significantly influenced by the intricate conditions present in the reservoir environme Hence, prior to implementing microbial flooding, it is imperative to establish comprehe sive reservoir screening criteria, systematically analyze the reservoir characteristics, an assess the microbial diversity in each well. This process will facilitate the developme of an appropriate mathematical model for MEOR. Simultaneously, genetic engineerir technology can be utilized to construct microbial strains that exhibit exceptional tolerance the challenging conditions present in reservoir environments. Furthermore, the utilizati of genetic engineering technology and the construction of a microbial consortium m represent a groundbreaking advancement for microbial oil recovery in the future. Howev these approaches can be costly in terms of implementation. The former technique allows f the development of oil-recovery microorganisms with exceptional performance, while t latter facilitates the creation of a highly effective microbial consortium [189]. The integrati of both approaches has the potential to augment the practicality and effectiveness of MEC technology.

Currently, several metagenomic investigations focusing on oil reservoirs have provided insights into the phylogenetic profiles of indigenous microbial structures existing in these reservoirs. Nonetheless, there is a significant lack of knowledge regarding the genomic characteristics desired for MEOR. As a result, limited information is available for genetically assessing the suitability of bacteria or archaea for MEOR applications.

8. Conclusions

(1) MEOR is a critical approach for enhancing oil recovery after polymer flooding. However, the presence of residual polymers in the formation has significant biological toxicity and negatively affects microorganisms' properties and oil displacement efficiency.
(2) To improve MEOR, the key lies in screening, cultivating, and constructing a microbial consortium that can utilize HPAM and crude oil as energy sources while being compatible with indigenous bacteria, leading to the formation of dominant bacteria in the reservoir.
(3) The research focuses mainly on improving oil recovery after polymer flooding, with limited reports on the impact of chemical agents on microorganisms' physicochemical properties, metabolic cycles, and mechanisms of oil flooding.
(4) The relationships among microbial populations during the oil displacement process are highly complex, involving mutualism, benefits, competition, and potential harm. Existing studies mainly focus on creating a few bacteria with synergistic effects, but there are limited investigations into the interrelationships between microorganisms using chemicals during the MEOR process.
(5) Many reports have been published on the microprocess of oil flooding. The parameters of MEOR can be optimized based on various reservoir conditions. Building on previous research, micro-visualization model flooding experiments can further elucidate the microscopic oil displacement process. However, the mechanism of microbial consortium on the oil displacement process using chemicals is rarely reported from a microscopic perspective.

Author Contributions: Conceptualization, data curation, and writing—original draft, H.X.; validation, formal analysis, visualization, supervision, writing—review, and funding acquisition, Z.A. and M.U.M.J. All authors have read and agreed to the published version of the manuscript.

Funding: This research was funded by the Universiti Malaya under Geran Penyelidikan Fakulti, GPF014A-2023. The project was also supported by UM International Collaboration Grant: Program SATU Joint Research Scheme, ST066-2022.

Data Availability Statement: The data presented in this study are available on request from the corresponding author.

Conflicts of Interest: The authors declare no conflict of interest.

References

Lourdes, R.S.; Cheng, S.Y.; Chew, K.W.; Ma, Z.; Show, P.L. Prospects of microbial enhanced oil recovery: Mechanisms and environmental sustainability. *Sustain. Energy Technol. Assess.* **2022**, *53*, 102527. [CrossRef]

Abbas, A.H.; Ajunwa, O.M.; Mazhit, B.; Martyushev, D.A.; Bou-Hamdan, K.F.; Alsaheb, R.A.A. Evaluation of OKRA (*Abelmoschus esculentus*) Macromolecular Solution for Enhanced Oil Recovery in Kazakhstan Carbonate Reservoir. *Energies* **2022**, *15*, 6827. [CrossRef]

Guo, H.; Li, Y.; Yiran, Z.; Wang, F.; Wang, Y.; Yu, Z.; Haicheng, S.; Yuanyuan, G.; Chuyi, J.; Xian, G. Progress of Microbial Enhanced Oil Recovery in China. In Proceedings of the SPE Asia Pacific Enhanced Oil Recovery Conference, Kuala Lumpur, Malaysia, 11–13 August 2015.

El-Masry, J.F.; Bou-Hamdan, K.F.; Abbas, A.H.; Martyushev, D.A. A Comprehensive Review on Utilizing Nanomaterials in Enhanced Oil Recovery Applications. *Energies* **2023**, *16*, 691. [CrossRef]

Seright, R.S.; Wang, D. Polymer flooding: Current status and future directions. *Pet. Sci.* **2023**, *20*, 910–921. [CrossRef]

Li, X.; Zhang, F.; Liu, G. Review on Polymer Flooding Technology. In Proceedings of the IOP Conference Series: Earth and Environmental Science, Xiamen, China, 17–19 November 2021; IOP Publishing: Xiamen, China, 2021; p. 012199.

7. Gao, P.; Li, G.; Le, J.; Liu, X.; Liu, F.; Ma, T. Succession of microbial communities and changes of incremental oil in a post-polymer flooded reservoir with nutrient stimulation. *Appl. Microbiol. Biotechnol.* **2018**, *102*, 2007–2017. [CrossRef]
8. Mogollón, J.L.; Lokhandwala, T. Rejuvenating Viscous Oil Reservoirs by Polymer Injection: Lessons Learned in the Field. In Proceedings of the SPE Enhanced Oil Recovery Conference, Kuala Lumpur, Malaysia, 2–4 July 2013.
9. Song, K.; Tao, J.; Lyu, X.; Xu, Y.; Liu, S.; Wang, Z.; Liu, H.; Zhang, Y.; Fu, H.; Meng, E.; et al. Recent Advances in Polymer Flooding in China. *Molecules* **2022**, *27*, 6978. [CrossRef]
10. Bahadori, A. *Fundamentals of Enhanced Oil and Gas Recovery from Conventional and Unconventional Reservoirs*; Gulf Professional Publishing: Cambridge, MA, USA, 2018.
11. Safdel, M.; Anbaz, M.A.; Daryasafar, A.; Jamialahmadi, M. Microbial enhanced oil recovery, a critical review on worldwide implemented field trials in different countries. *Renew. Sustain. Energy Rev.* **2017**, *74*, 159–172. [CrossRef]
12. Mu, B.Z.; Nazina, T.N. Recent Advances in Petroleum Microbiology. *Microorganisms* **2022**, *10*, 1706. [CrossRef]
13. Nazina, T.; Sokolova, D.; Grouzdev, D.; Semenova, E.; Babich, T.; Bidzhieva, S.; Serdukov, D.; Volkov, D.; Bugaev, K.; Ershov, A.; et al. The Potential Application of Microorganisms for Sustainable Petroleum Recovery from Heavy Oil Reservoirs. *Sustainability* **2020**, *12*, 15. [CrossRef]
14. She, H.; Kong, D.; Li, Y.; Hu, Z.; Guo, H. Recent Advance of Microbial Enhanced Oil Recovery (MEOR) in China. *Geofluids* **2019**, *2019*, 1871392. [CrossRef]
15. Li, Y.; Yang, Y.; Sun, X.; Yang, D.; Zhang, N.; Yang, H.; Guo, H.; Zheng, J. The application of laser confocal method in microscopic oil analysis. *J. Pet. Sci. Eng.* **2014**, *120*, 52–60. [CrossRef]
16. Yuan, S.; Wang, Q. New progress and prospect of oilfields development technologies in China. *Adv. Pet. Explor. Dev.* **2018**, *45*, 698–711. [CrossRef]
17. Shi, M.; Wei, L.; Liao, G.; Han, P.; Hou, Z.; Chen, X.; Wang, Y. Laboratory Study on MEOR After Polymer Flooding. In Proceedings of the SPE International Improved Oil Recovery Conference in Asia Pacific, Kuala Lumpur, Malaysia, 20–21 October 2003.
18. Patel, J.; Borgohain, S.; Kumar, M.; Rangarajan, V.; Somasundaran, P.; Sen, R. Recent developments in microbial enhanced oil recovery. *Renew. Sustain. Energy Rev.* **2015**, *52*, 1539–1558. [CrossRef]
19. Niu, J.; Liu, Q.; Lv, J.; Peng, B. Review on microbial enhanced oil recovery: Mechanisms, modeling and field trials. *J. Pet. Sci. Eng.* **2020**, *192*, 107350. [CrossRef]
20. Al-Kindi, S.; Al-Bahry, S.; Al-Wahaibi, Y.; Taura, U.; Joshi, S. Partially hydrolyzed polyacrylamide: Enhanced oil recovery applications, oil-field produced water pollution, and possible solutions. *Environ. Monit. Assess.* **2022**, *194*, 875. [CrossRef]
21. Ren, G.; Wang, J.; Qu, L.; Li, W.; Hu, M.; Bian, L.; Zhang, Y.; Le, J.; Dou, X.; Chen, X.; et al. Compositions and Co-occurrence Patterns of Bacterial Communities Associated With Polymer- and ASP-Flooded Petroleum Reservoir Blocks. *Front. Microbiol.* **2020**, *11*, 580363. [CrossRef]
22. Quraishi, M.; Bhatia, S.K.; Pandit, S.; Gupta, P.K.; Rangarajan, V.; Lahiri, D.; Varjani, S.; Mehariya, S.; Yang, Y.-H. Exploiting microbes in the petroleum field: Analyzing the credibility of microbial enhanced oil recovery (MEOR). *Energies* **2021**, *14*, 4684 [CrossRef]
23. Yernazarova, A.; Kayirmanova, G.; Baubekova, A.; Zhubanova, A. Microbial enhanced oil recovery. In *Chemical Enhanced Oil Recovery (cEOR)—A Practical Overview*; InTech: Rijeka, Croatia, 2016; pp. 147–167.
24. Liang, B.; Wang, L.; Zhou, Z.; Mbadinga, S.M.; Zhou, L.; Liu, J.; Yang, S.-Z.; Gu, J.; Mu, B. High frequency of Thermodesulfovibrio spp. and Anaerolineaceae in association with Methanoculleus spp. in a long-term incubation of n-alkanes-degrading methanogenic enrichment culture. *Front. Microbiol.* **2016**, *7*, 1431. [CrossRef]
25. Lv, L.; Zhou, L.; Wang, L.-Y.; Liu, J.-F.; Gu, J.-D.; Mu, B.; Yang, S.Z. Selective inhibition of methanogenesis by sulfate in enrichment culture with production water from low-temperature oil reservoir. *Int. Biodeterior. Biodegrad.* **2016**, *108*, 133–141. [CrossRef]
26. Zhao, F.; Zhou, J.D.; Ma, F.; Shi, R.J.; Han, S.Q.; Zhang, J.; Zhang, Y. Simultaneous inhibition of sulfate-reducing bacteria, removal of H_2S and production of rhamnolipid by recombinant Pseudomonas stutzeri Rhl: Applications for microbial enhanced oil recovery. *Bioresour. Technol.* **2016**, *207*, 24–30. [CrossRef]
27. Yang, G.-C.; Zhou, L.; Mbadinga, S.M.; You, J.; Yang, H.-Z.; Liu, J.-F.; Yang, S.-Z.; Gu, J.-D.; Mu, B.-Z. Activation of CO_2-reducing methanogens in oil reservoir after addition of nutrient. *J. Biosci. Bioeng.* **2016**, *122*, 740–747. [CrossRef] [PubMed]
28. Varjani, S.J.; Upasani, V.N. A new look on factors affecting microbial degradation of petroleum hydrocarbon pollutants. *Int. Biodeterior. Biodegrad.* **2017**, *120*, 71–83. [CrossRef]
29. Varjani, S.J.; Gnansounou, E. Microbial dynamics in petroleum oilfields and their relationship with physiological properties of petroleum oil reservoirs. *Bioresour. Technol.* **2017**, *245 Pt. A*, 1258–1265. [CrossRef]
30. Hussain, A.; Hasan, A.; Javid, A.; Qazi, J.I. Exploited application of sulfate-reducing bacteria for concomitant treatment of metallic and non-metallic wastes: A mini review. *3 Biotech.* **2016**, *6*, 119. [CrossRef] [PubMed]
31. Song, Z.; Zhu, W.; Sun, G.; Blanckaert, K. Dynamic investigation of nutrient consumption and injection strategy in microbial enhanced oil recovery (MEOR) by means of large-scale experiments. *Appl. Microbiol. Biotechnol.* **2015**, *99*, 6551–6561. [CrossRef] [PubMed]
32. Varjani, S.J.; Upasani, V.N. Core Flood study for enhanced oil recovery through ex-situ bioaugmentation with thermo- and halo-tolerant rhamnolipid produced by Pseudomonas aeruginosa NCIM 5514. *Bioresour. Technol.* **2016**, *220*, 175–182. [CrossRef]
33. Li, J.; Liu, J.; Trefry, M.G.; Park, J.; Liu, K.; Haq, B.; Johnston, C.D.; Volk, H. Interactions of microbial-enhanced oil recovery processes. *Transp. Porous Media* **2011**, *87*, 77–104. [CrossRef]

34. Li, H.; Wang, X.-L.; Mu, B.-Z.; Gu, J.-D.; Liu, Y.-D.; Lin, K.-F.; Lu, S.-G.; Lu, Q.; Li, B.-Z.; Li, Y.-Y.; et al. Molecular detection, quantification and distribution of alkane-degrading bacteria in production water from low temperature oilfields. *Int. Biodeterior. Biodegrad.* **2013**, *76*, 49–57. [CrossRef]
35. Robles-Fernández, A.; Areias, C.; Daffonchio, D.; Vahrenkamp, V.C.; Sánchez-Román, M. The Role of Microorganisms in the Nucleation of Carbonates, Environmental Implications and Applications. *Minerals* **2022**, *12*, 1562. [CrossRef]
36. Billings, N.; Birjiniuk, A.; Samad, T.S.; Doyle, P.S.; Ribbeck, K. Material properties of biofilms-a review of methods for understanding permeability and mechanics. *Rep. Prog. Phys.* **2015**, *78*, 036601. [CrossRef]
37. Flemming, H.-C.; Wingender, J.; Szewzyk, U.; Steinberg, P.; Rice, S.A.; Kjelleberg, S. Biofilms: An emergent form of bacterial life. *Nat. Rev. Microbiol.* **2016**, *14*, 563–575. [CrossRef] [PubMed]
38. Persat, A.; Nadell, C.D.; Kim, M.K.; Ingremeau, F.; Siryaporn, A.; Drescher, K.; Wingreen, N.S.; Bassler, B.L.; Gitai, Z.; Stone, H.A. The mechanical world of bacteria. *Cell* **2015**, *161*, 988–997. [CrossRef] [PubMed]
39. Salama, Y.; Chennaoui, M.; Sylla, A.; Mountadar, M.; Rihani, M.; Assobhei, O. Characterization, structure, and function of extracellular polymeric substances (EPS) of microbial biofilm in biological wastewater treatment systems: A review. *Desalin. Water Treat.* **2016**, *57*, 16220–16237. [CrossRef]
40. Antonietta, Q.; Uta, P.; Wei-Chun, C.; Chen, X.; Shawn, D.; Laura, B.; Manoj, K.; Alicia, K.W.; Jason, B.S.; Zoe, V.F.; et al. The role of microbial exopolymers in determining the fate of oil and chemical dispersants in the ocean. *Limnol. Oceanogr. Lett.* **2016**, *1*, 3–26.
41. Bacosa, H.P.; Kamalanathan, M.; Labonté, J.M.; Hala, D.; Quigg, A.; Chiu, M.H.; Tsai, S.M.; Chin, W.C.; Sun, L.; Schwehr, K.A.; et al. Extracellular polymeric substances (EPS) producing and oil degrading bacteria isolated from the northern Gulf of Mexico. *PLoS ONE* **2018**, *13*, e0208406. [CrossRef]
42. Kapse, N.G.; Paliwal, V.; Dagar, S.S.; Rana, D.P.; Dhakephalkar, P.K. Genomics and simulated laboratory studies reveal Thermococcus sp. 101C5 as a novel hyperthermophilic archaeon possessing a specialized metabolic arsenal for enhanced oil recovery. *Antonie Van Leeuwenhoek* **2022**, *115*, 19–31. [CrossRef]
43. Aboelkhair, H.; Diaz, P.; Attia, A. Biosurfactant production using Egyptian oil fields indigenous bacteria for microbial enhanced oil recovery. *J. Pet. Sci. Eng.* **2022**, *208*, 109601. [CrossRef]
44. Geetha, S.; Banat, I.M.; Joshi, S.J. Biosurfactants: Production and potential applications in microbial enhanced oil recovery (MEOR). *Biocatal. Agric. Biotechnol.* **2018**, *14*, 23–32.
45. Wood, D.A. Microbial improved and enhanced oil recovery (MIEOR): Review of a set of technologies diversifying their applications. *Adv. Geo-Energy Res.* **2019**, *3*, 122–140. [CrossRef]
46. Karlapudi, A.P.; Venkateswarulu, T.; Tammineedi, J.; Kanumuri, L.; Ravuru, B.K.; Ramu Dirisala, V.; Kodali, V.P. Role of biosurfactants in bioremediation of oil pollution-a review. *Petroleum* **2018**, *4*, 241–249. [CrossRef]
47. Moeininia, N. Improvement of a Full Field Reservoir Model for a MEOR Application. Master's Thesis, University of Leoben, Leoben, Austria, 2018.
48. Massarweh, O.; Abushaikha, A.S. The use of surfactants in enhanced oil recovery: A review of recent advances. *Energy Rep.* **2020**, *6*, 3150–3178. [CrossRef]
49. Reis, R.; Pacheco, G.; Pereira, A.; Freire, D. Biosurfactants: Production and applications. In *Biodegradation-Life of Science*; InTech: Rijeka, Croatia, 2013; pp. 31–61.
50. Ke, C.-Y.; Lu, G.-M.; Li, Y.-B.; Sun, W.-J.; Zhang, Q.-Z.; Zhang, X.-L. A pilot study on large-scale microbial enhanced oil recovery (MEOR) in Baolige Oilfield. *Int. Biodeterior. Biodegrad.* **2018**, *127*, 247–253. [CrossRef]
51. Jeong, M.S.; Lee, Y.W.; Lee, H.S.; Lee, K.S. Simulation-based optimization of microbial enhanced oil recovery with a model integrating temperature, pressure, and salinity effects. *Energies* **2021**, *14*, 1131. [CrossRef]
52. Halim, A.Y.; Nielsen, S.M.; Lantz, A.E.; Suicmez, V.S.; Lindeloff, N.; Shapiro, A. Investigation of spore forming bacterial flooding for enhanced oil recovery in a North Sea chalk Reservoir. *J. Pet. Sci. Eng.* **2015**, *133*, 444–454. [CrossRef]
53. Sen, R. Biotechnology in petroleum recovery: The microbial EOR. *Prog. Energy Combust. Sci.* **2008**, *34*, 714–724. [CrossRef]
54. Dhanarajan, G.; Rangarajan, V.; Bandi, C.; Dixit, A.; Das, S.; Ale, K.; Sen, R. Biosurfactant-biopolymer driven microbial enhanced oil recovery (MEOR) and its optimization by an ANN-GA hybrid technique. *J. Biotechnol.* **2017**, *256*, 46–56. [CrossRef]
55. Eldeen, A.E.G.; Alhadi, A.M.A.A.; Mergani, M.A. *Microbial Enhanced Oil Recovery (MEOR)*; University of Khartoum: Khartom, Sudan, 2013.
56. Cui, K.; Zhang, Z.; Zhang, Z.; Sun, S.; Li, H.; Fu, P. Stimulation of indigenous microbes by optimizing the water cut in low permeability reservoirs for green and enhanced oil recovery. *Sci. Rep.* **2019**, *9*, 15772. [CrossRef]
57. Mujumdar, S.; Joshi, P.; Karve, N. Production, characterization, and applications of bioemulsifiers (BE) and biosurfactants (BS) produced by Acinetobacter spp.: A review. *J. Basic. Microbiol.* **2019**, *59*, 277–287. [CrossRef]
58. Elakkiya, V.T.; SureshKumar, P.; Alharbi, N.S.; Kadaikunnan, S.; Khaled, J.M.; Govindarajan, M. Swift production of rhamnolipid biosurfactant, biopolymer and synthesis of biosurfactant-wrapped silver nanoparticles and its enhanced oil recovery. *Saudi J. Biol. Sci.* **2020**, *27*, 1892–1899. [CrossRef]
59. Morais, I.; Cordeiro, A.; Teixeira, G.; Domingues, V.; Nardi, R.; Monteiro, A.; Alves, R.; Siqueira, E.; Santos, V. Biological and physicochemical properties of biosurfactants produced by Lactobacillus jensenii P_{6A} and Lactobacillus gasseri P_{65}. *Microb. Cell Fact.* **2017**, *16*, 155. [CrossRef]
60. Dowluru, S. Prospects for Industrial and Microbial Biotechnology in Oil Industry. *Biotechnology* **2014**, *3*, 37–39.

61. Jahanbani Veshareh, M.; Ganji Azad, E.; Deihimi, T.; Niazi, A.; Ayatollahi, S. Isolation and screening of Bacillus subtilis MJ01 for MEOR application: Biosurfactant characterization, production optimization and wetting effect on carbonate surfaces. *J. Pet. Explor. Prod. Technol.* **2019**, *9*, 233–245. [CrossRef]
62. Borah, D.; Chaubey, A.; Sonowal, A.; Gogoi, B.; Kumar, R. Microbial biosurfactants and their potential applications: An overview In *Microbial Biosurfactants: Preparation, Properties and Applications*; Inamuddin, A.M.I., Prasad, R., Eds.; Springer: Singapore, 2021; pp. 91–116.
63. Jadeja, N.B.; Moharir, P.; Kapley, A. Genome sequencing and analysis of strains Bacillus sp. AKBS9 and Acinetobacter sp. AKBS16 for biosurfactant production and bioremediation. *Appl. Biochem. Biotechnol.* **2019**, *187*, 518–530. [CrossRef]
64. Shibulal, B.; Al-Bahry, S.N.; Al-Wahaibi, Y.M.; Elshafie, A.E.; Al-Bemani, A.S.; Joshi, S.J. Microbial Enhanced Heavy Oil Recovery by the Aid of Inhabitant Spore-Forming Bacteria: An Insight Review. *Sci. World J.* **2014**, *2014*, 309159. [CrossRef] [PubMed]
65. Xu, X.; Liu, W.; Tian, S.; Wang, W.; Qi, Q.; Jiang, P.; Gao, X.; Li, F.; Li, H.; Yu, H. Petroleum hydrocarbon-degrading bacteria for the remediation of oil pollution under aerobic conditions: A perspective analysis. *Front. Microbiol.* **2018**, *9*, 2885. [CrossRef] [PubMed]
66. Al-Sulaimani, H.; Joshi, S.; Al-Wahaibi, Y.; Al-Bahry, S.; Elshafie, A.; Al-Bemani, A. Microbial biotechnology for enhancing oil recovery: Current developments and future prospects. *Biotechnol. Bioinf. Bioeng.* **2011**, *1*, 147–158.
67. Olajire, A.; Essien, J. Aerobic degradation of petroleum components by microbial consortia. *J. Pet. Environ. Biotechnol.* **2014**, *5*, 1 [CrossRef]
68. Bacosa, H.P.; Evans, M.M.; Wang, Q.; Liu, Z. Chapter 28—Assessing the Role of Environmental Conditions on the Degradation of Oil Following the Deepwater Horizon Oil Spill. In *Oil Spill Environmental Forensics Case Studies*; Stout, S.A., Wang, Z., Eds Butterworth-Heinemann: Oxford, UK, 2018; pp. 617–637.
69. Gudina, E.J.; Pereira, J.F.; Costa, R.; Coutinho, J.A.; Teixeira, J.A.; Rodrigues, L.R. Biosurfactant-producing and oil-degrading Bacillus subtilis strains enhance oil recovery in laboratory sand-pack columns. *J. Hazard. Mater.* **2013**, *261*, 106–113. [CrossRef]
70. Gao, H.; Zhang, J.; Lai, H.; Xue, Q. Degradation of asphaltenes by two Pseudomonas aeruginosa strains and their effects on physicochemical properties of crude oil. *Int. Biodeterior. Biodegrad.* **2017**, *122*, 12–22. [CrossRef]
71. Margesin, R.; Moertelmaier, C.; Mair, J. Low-temperature biodegradation of petroleum hydrocarbons (n-alkanes, pheno anthracene, pyrene) by four actinobacterial strains. *Int. Biodeterior. Biodegrad.* **2013**, *84*, 185–191. [CrossRef]
72. Li, Y.; Pan, J.; Ma, Y. Elucidation of multiple alkane hydroxylase systems in biodegradation of crude oil n-alkane pollution by Pseudomonas aeruginosa DN1. *J. Appl. Microbiol.* **2020**, *128*, 151–160. [CrossRef] [PubMed]
73. Sakthipriya, N.; Doble, M.; Sangwai, J.S. Influence of thermophilic Bacillus subtilis YB7 on the biodegradation of long chain paraffinic hydrocarbons ($C_{16}H_{34}$ to $C_{36}H_{74}$). *RSC Adv.* **2016**, *6*, 82541–82552. [CrossRef]
74. Zhang, J.H.; Xue, Q.H.; Gao, H.; Ma, X.; Wang, P. Degradation of crude oil by fungal enzyme preparations from Aspergillus spp for potential use in enhanced oil recovery. *J. Chem. Technol. Biotechnol.* **2016**, *91*, 865–875. [CrossRef]
75. Ke, C.-Y.; Lu, G.-M.; Wei, Y.-L.; Sun, W.-J.; Hui, J.-F.; Zheng, X.-Y.; Zhang, Q.-Z.; Zhang, X.-L. Biodegradation of crude oil by Chelatococcus daeguensis HB-4 and its potential for microbial enhanced oil recovery (MEOR) in heavy oil reservoirs. *Bioresour. Technol.* **2019**, *287*, 121442. [CrossRef]
76. Wang, Y.; Nie, M.; Wan, Y.; Tian, X.; Nie, H.; Zi, J.; Ma, X. Functional characterization of two alkane hydroxylases in a versatile Pseudomonas aeruginosa strain NY3. *Ann. Microbiol.* **2017**, *67*, 459–468. [CrossRef]
77. Purwasena, I.A.; Sugai, Y.; Sasaki, K. Estimation of the potential of an anaerobic thermophilic oil-degrading bacterium as candidate for MEOR. *J. Pet. Explor. Prod. Technol.* **2014**, *4*, 189–200. [CrossRef]
78. Vijayakumar, S.; Saravanan, V. Biosurfactants-types, sources and applications. *Res. J. Microbiol.* **2015**, *10*, 181.
79. Suthar, H.; Hingurao, K.; Desai, A.; Nerurkar, A. Evaluation of bioemulsifier mediated microbial enhanced oil recovery using sand pack column. *J. Microbiol. Methods* **2008**, *75*, 225–230. [CrossRef]
80. Reuvers, B. Emulsifying ionic apolar polymer in water: Understanding the process. *J. Coat. Technol. Res.* **2020**, *17*, 1131–114 [CrossRef]
81. Hou, B.; Wang, Y.; Cao, X.; Zhang, J.; Song, X.; Ding, M.; Chen, W. Mechanisms of enhanced oil recovery by surfactant-induced wettability alteration. *J. Dispers. Sci. Technol.* **2016**, *37*, 1259–1267. [CrossRef]
82. Tetteh, J.; Janjang, N.M.; Barati, R. Wettability alteration and enhanced oil recovery using low salinity waterflooding in limestone rocks: A mechanistic study. In Proceedings of the SPE Kingdom of Saudi Arabia Annual Technical Symposium and Exhibition Dammam, Saudi Arabia, 23–26 April 2018; OnePetro: Dammam, Saudi Arabia, 2018.
83. Najafi-Marghmaleki, A.; Kord, S.; Hashemi, A.; Motamedi, H. Experimental investigation of efficiency of MEOR process in carbonate oil reservoir using Alcaligenes faecalis: Impact of interfacial tension reduction and wettability alteration mechanism *Fuel* **2018**, *232*, 27–35. [CrossRef]
84. Deng, X.; Tariq, Z.; Murtaza, M.; Patil, S.; Mahmoud, M.; Kamal, M.S. Relative contribution of wettability Alteration and interfacial tension reduction in EOR: A critical review. *J. Mol. Liq.* **2021**, *325*, 115175. [CrossRef]
85. Joonaki, E.; Ghanaatian, S. The application of nanofluids for enhanced oil recovery: Effects on interfacial tension and coreflooding process. *Pet. Sci. Technol.* **2014**, *32*, 2599–2607. [CrossRef]
86. Kamal, M.S.; Hussein, I.A.; Sultan, A.S. Review on surfactant flooding: Phase behavior, retention, IFT, and field application *Energy Fuels* **2017**, *31*, 7701–7720. [CrossRef]

87. Kögler, F.; Mahler, E.; Dopffel, N.; Schulze-Makuch, D.; Borovina, A.; Visser, F.; Herold, A.; Alkan, H. The Microbial Enhanced Oil Recovery (MEOR) potential of Halanaerobiales under dynamic conditions in different porous media. *Pet. Sci. Eng.* **2021**, *196*, 107578. [CrossRef]
88. Khajepour, H.; Mahmoodi, M.; Biria, D.; Ayatollahi, S. Investigation of wettability alteration through relative permeability measurement during MEOR process: A micromodel study. *J. Pet. Sci. Eng.* **2014**, *120*, 10–17. [CrossRef]
89. Abolhasanzadeh, A.; Khaz'ali, A.R.; Hashemi, R.; Jazini, M. Experimental study of microbial enhanced oil recovery in oil-wet fractured porous media. *Oil Gas Sci. Technol—Rev. d'IFP Energ. Nouv.* **2020**, *75*, 10. [CrossRef]
90. Karimi, M.; Mahmoodi, M.; Niazi, A.; Al-Wahaibi, Y.; Ayatollahi, S. Investigating wettability alteration during MEOR process, a micro/macro scale analysis. *Colloid. Surf. B-Biointerfaces* **2012**, *95*, 129–136. [CrossRef]
91. Donaldson, E.C.; Alam, W. *Wettability*; Gulf Publishing Company: Houston, TX, USA, 2013.
92. Gao, C. Experiences of microbial enhanced oil recovery in Chinese oil fields. *J. Pet. Sci. Eng.* **2018**, *166*, 55–62. [CrossRef]
93. Zhang, J.; Gao, H.; Xue, Q. Potential applications of microbial enhanced oil recovery to heavy oil. *Crit. Rev. Biotechnol.* **2020**, *40*, 459–474. [CrossRef]
94. Guan, J.; Xia, L.-P.; Wang, L.-Y.; Liu, J.-F.; Gu, J.-D.; Mu, B.-Z. Diversity and distribution of sulfate-reducing bacteria in four petroleum reservoirs detected by using 16S rRNA and dsrAB genes. *Int. Biodeterior. Biodegrad.* **2013**, *76*, 58–66. [CrossRef]
95. Zhao, F.; Ma, F.; Shi, R.; Zhang, J.; Han, S.; Zhang, Y. Production of rhamnolipids by Pseudomonas aeruginosa is inhibited by H_2S but resumes in a co-culture with P. stutzeri: Applications for microbial enhanced oil recovery. *Biotechnol. Lett.* **2015**, *37*, 1803–1808. [CrossRef] [PubMed]
96. Varjani, S.J. Microbial degradation of petroleum hydrocarbons. *Bioresour. Technol.* **2017**, *223*, 277–286. [CrossRef] [PubMed]
97. Gamadi, T.; Sheng, J.; Soliman, M.; Menouar, H.; Watson, M.; Emadibaladehi, H. An experimental study of cyclic CO_2 injection to improve shale oil recovery. In Proceedings of the SPE Improved Oil Recovery Symposium, Tulsa, OK, USA, 12–16 April 2014; OnePetro: Tulsa, OK, USA, 2014.
98. Hart, A. A review of technologies for transporting heavy crude oil and bitumen via pipelines. *J. Pet. Explor. Prod. Technol.* **2014**, *4*, 327–336. [CrossRef]
99. Purwasena, I.A.; Astuti, D.I.; Syukron, M.; Amaniyah, M.; Sugai, Y. Stability test of biosurfactant produced by Bacillus licheniformis DS1 using experimental design and its application for MEOR. *J. Pet. Sci. Eng.* **2019**, *183*, 106383. [CrossRef]
100. Cameron, T.A.; Margolin, W. Construction and Characterization of Functional FtsA Sandwich Fusions for Studies of FtsA Localization and Dynamics during Escherichia coli Cell Division. *J. Bacteriol.* **2023**, *205*, 2974–2984. [CrossRef]
101. Zhao, F.; Shi, R.; Cui, Q.; Han, S.; Dong, H.; Zhang, Y. Biosurfactant production under diverse conditions by two kinds of biosurfactant-producing bacteria for microbial enhanced oil recovery. *J. Pet. Sci. Eng.* **2017**, *157*, 124–130. [CrossRef]
102. Xu, Y.; Wang, J.; Liu, Q.; Zhang, Q.; Wu, J.; Zhou, M.; Nie, Y.; Wu, X.-L. pH and Nitrate Drive Bacterial Diversity in Oil Reservoirs at a Localized Geographic Scale. *Microorganisms* **2023**, *11*, 151. [CrossRef]
103. Pannekens, M.; Kroll, L.; Müller, H.; Mbow, F.T.; Meckenstock, R.U. Oil reservoirs, an exceptional habitat for microorganisms. *New Biotech.* **2019**, *49*, 1–9. [CrossRef]
104. Kowalewski, E.; Rueslåtten, I.; Steen, K.; Bødtker, G.; Torsæter, O. Microbial improved oil recovery—Bacterial induced wettability and interfacial tension effects on oil production. *J. Pet. Sci. Eng.* **2006**, *52*, 275–286. [CrossRef]
105. Wang, L.-Y.; Duan, R.-Y.; Liu, J.-F.; Yang, S.-Z.; Gu, J.-D.; Mu, B.-Z. Molecular analysis of the microbial community structures in water-flooding petroleum reservoirs with different temperatures. *Biogeosciences* **2012**, *9*, 4645–4659. [CrossRef]
106. Vigneron, A.; Alsop, E.B.; Lomans, B.P.; Kyrpides, N.C.; Head, I.M.; Tsesmetzis, N. Succession in the petroleum reservoir microbiome through an oil field production lifecycle. *ISME J.* **2017**, *11*, 2141–2154. [CrossRef] [PubMed]
107. Dahle, H.; Garshol, F.; Madsen, M.; Birkeland, N.K. Microbial community structure analysis of produced water from a high-temperature North Sea oil-field. *Antonie Van Leeuwenhoek* **2008**, *93*, 37–49. [CrossRef] [PubMed]
108. Lenchi, N.; İnceoğlu, Ö.; Kebbouche-Gana, S.; Gana, M.L.; Llirós, M.; Servais, P.; García-Armisen, T. Diversity of Microbial Communities in Production and Injection Waters of Algerian Oilfields Revealed by 16S rRNA Gene Amplicon 454 Pyrosequencing. *PLoS ONE* **2013**, *8*, e66588. [CrossRef]
109. Braun, O.; Coquery, C.; Kieffer, J.; Blondel, F.; Favero, C.; Besset, C.; Mesnager, J.; Voelker, F.; Delorme, C.; Matioszek, D. Spotlight on the Life Cycle of Acrylamide-Based Polymers Supporting Reductions in Environmental Footprint: Review and Recent Advances. *Molecules* **2022**, *27*, 42. [CrossRef]
110. Olajire, A.A. Review of ASP EOR (alkaline surfactant polymer enhanced oil recovery) technology in the petroleum industry: Prospects and challenges. *Energy* **2014**, *77*, 963–982. [CrossRef]
111. Xiong, B.; Loss, R.D.; Shields, D.; Pawlik, T.; Hochreiter, R.; Zydney, A.L.; Kumar, M. Polyacrylamide degradation and its implications in environmental systems. *NPJ Clean Water* **2018**, *1*, 17. [CrossRef]
112. Lindeman, B.; Johansson, Y.; Andreassen, M.; Husøy, T.; Dirven, H.; Hofer, T.; Knutsen, H.K.; Caspersen, I.H.; Vejrup, K.; Paulsen, R.E. Does the food processing contaminant acrylamide cause developmental neurotoxicity? A review and identification of knowledge gaps. *Reprod. Toxicol.* **2021**, *101*, 93–114. [CrossRef]
113. Liu, J.; Ren, J.; Xu, R.; Yu, B.; Wang, J. Biodegradation of partially hydrolyzed polyacrylamide by immobilized bacteria isolated from HPAM-containing wastewater. *Environ. Prog. Sustain. Energy* **2016**, *35*, 1344–1352. [CrossRef]

114. Elshafie, A.; Joshi, S.J.; Al-Wahaibi, Y.M.; Al-Bahry, S.N.; Al-Bemani, A.S.; Al-Hashmi, A.; Al-Mandhari, M.S. Isolation and Characterization of Biopolymer Producing Omani Aureobasidium Pullulans Strains and Its Potential Applications in Microbial Enhanced Oil Recovery. In Proceedings of the SPE Oil and Gas India Conference and Exhibition, Mumbai, India, 4–6 April 2017.
115. Amir, Z.; Said, I.M.; Jan, B.M. In situ organically cross-linked polymer gel for high-temperature reservoir conformance control: A review. *Polym. Adv. Technol.* **2019**, *30*, 13–39. [CrossRef]
116. Al-Assi, A.A.; Willhite, G.P.; Green, D.W.; McCool, C.S. Formation and propagation of gel aggregates using partially hydrolyzed polyacrylamide and aluminum citrate. *SPE J.* **2009**, *14*, 450–461. [CrossRef]
117. Adewunmi, A.A.; Ismail, S.; Sultan, A.S. Study on strength and gelation time of polyacrylamide/polyethyleneimine composite gels reinforced with coal fly ash for water shut-off treatment. *J. Appl. Polym. Sci.* **2015**, *132*, 41392. [CrossRef]
118. Liu, P.; Zhang, Y.; Tang, Q.; Shi, S. Bioremediation of metal-contaminated soils by microbially-induced carbonate precipitation and its effects on ecotoxicity and long-term stability. *Biochem. Eng. J.* **2021**, *166*, 107856. [CrossRef]
119. Gadd, G.M. Metals, minerals and microbes: Geomicrobiology and bioremediation. *Microbiology* **2010**, *156*, 609–643. [CrossRef]
120. Bai, B.; Zhou, J.; Yin, M. A comprehensive review of polyacrylamide polymer gels for conformance control. *Pet. Explor. Dev.* **2015**, *42*, 525–532. [CrossRef]
121. Liu, Y.; Dai, C.; Wang, K.; Zhao, M.; Zhao, G.; Yang, S.; Yan, Z.; You, Q. New insights into the hydroquinone (HQ)–hexamethylenetetramine (HMTA) gel system for water shut-off treatment in high temperature reservoirs. *J. Ind. Eng. Chem.* **2016**, *35*, 20–28. [CrossRef]
122. Ali, H.; Khan, E.; Ilahi, I. Environmental chemistry and ecotoxicology of hazardous heavy metals: Environmental persistence, toxicity, and bioaccumulation. *J. Chem.* **2019**, *2019*, 6730305. [CrossRef]
123. Berdugo-Clavijo, C.; Sen, A.; Seyyedi, M.; Quintero, H.; O'Neil, B.; Gieg, L.M. High temperature utilization of PAM and HPAM by microbial communities enriched from oilfield produced water and activated sludge. *AMB Express* **2019**, *9*, 46. [CrossRef] [PubMed]
124. Joshi, S.J.; Abed, R.M.M. Biodegradation of Polyacrylamide and Its Derivatives. *Environ. Process.* **2017**, *4*, 463–476. [CrossRef]
125. Yu, F.; Fu, R.; Xie, Y.; Chen, W. Isolation and Characterization of Polyacrylamide-Degrading Bacteria from Dewatered Sludge. *Int. J. Environ. Res. Public Health* **2015**, *12*, 4214–4230. [CrossRef]
126. Al-Moqbali, W.; Joshi, S.J.; Al-Bahry, S.N.; Al-Wahaibi, Y.M.; Elshafie, A.E.; Al-Bemani, A.S.; Al-Hashmi, A.; Soundra Pandian, S.B. Biodegradation of Partially Hydrolyzed Polyacrylamide HPAM Using Bacteria Isolated from Omani Oil Fields. In Proceedings of the SPE EOR Conference, Oil and Gas West Asia, Muscat, Oman, 26–28 March 2018.
127. Du, C.A.; Song, Y.; Yao, Z.; Su, W.; Zhang, G.; Wu, X. Developments in in-situ microbial enhanced oil recovery in Shengli oilfield *Energy Sources Part A-Recovery Util. Environ. Eff.* **2022**, *44*, 1977–1987. [CrossRef]
128. Chen, Y.; Pan, J.; Yun, Y.; Zhi, B.; Li, G.; Li, M.; Ma, T. Halomonas plays a central role in the syntrophic community of an alkaline oil reservoir with alkali-surfactant-polymer (ASP) flooding. *Sci. Total Environ.* **2020**, *747*, 141333. [CrossRef] [PubMed]
129. Huang, Y.H.; Ren, G.L.; Yuan, H.M.; Wei, L.; Wu, X.L.; Song, K.P. Microbial Diversity Analysis of Reservoirs after Polymer Flooding in Daqing Oil Field at the Earlier Stage of Microbial Profile Modification. *Adv. Mater. Res.* **2010**, *113–116*, 835–839 [CrossRef]
130. Zhang, T.; Li, Z.; Gao, M.; Xu, Z.; Adenutsi, C.D.; You, Q. New insights into the synergism between silica nanoparticles and surfactants on interfacial properties: Implications for spontaneous imbibition in tight oil reservoirs. *J. Pet. Sci. Eng.* **2022**, *215* 110647. [CrossRef]
131. Thota, V.; C Perry, C. A review on recent patents and applications of inorganic material binding peptides. *Recent Pat. Nanotechnol* **2017**, *11*, 168–180. [CrossRef]
132. Sajid, M.; Płotka-Wasylka, J. Nanoparticles: Synthesis, characteristics, and applications in analytical and other sciences. *Microchem J.* **2020**, *154*, 104623. [CrossRef]
133. Wang, L.; Zhang, G.; Ge, J.-J.; Li, G.; Zhang, J.; Ding, B. Preparation of microgel nanospheres and their application in EOR In *SPE International Oil and Gas Conference and Exhibition in China, Beijing, China, 8–10 June 2010*; SPE: Beijing, China, 2010; p SPE-130357-MS.
134. Wang, S.; Tang, Z.; Qu, J.; Wu, T.; Liu, Y.; Wang, J.; Liu, X.; Ju, Y.; Liu, F. Research on the mechanisms of polyacrylamide nanospheres with different size distributions in enhanced oil recovery. *RSC Adv.* **2021**, *11*, 5763–5772. [CrossRef]
135. Hendraningrat, L.; Zhang, J. Polymeric nanospheres as a displacement fluid in enhanced oil recovery. *Appl. Nanosci.* **2015**, *5* 1009–1016. [CrossRef]
136. Li, T.; Huang, P.; Li, X.; Wang, R.; Lu, Z.; Song, P.; He, Y. Synthesis of polymer nanospheres conjugated Ce (IV) complexes fo constructing double antibacterial centers. *J. Inorg. Organomet. Polym. Mater.* **2022**, *32*, 883–894. [CrossRef]
137. Neha, S.; Meeta, L.; Vatsala, K.; Dhruva, P.; Nitish, K.; Amitabh, P.; Rahul, R.; Kumar, M.S.; Banwari, L. Nutrient optimization fo indigenous microbial consortia of a Bhagyam oil field: MEOR studies. *Front. Microbiol.* **2023**, *14*, 1026720.
138. Li, C.-Y.; Zhang, D.; Li, X.-X.; Mbadinga, S.M.; Yang, S.-Z.; Liu, J.-F.; Gu, J.-D.; Mu, B.-Z. The biofilm property and its correlationshi with high-molecular-weight polyacrylamide degradation in a water injection pipeline of Daqing oilfield. *J. Hazard. Mater.* **201** *304*, 388–399. [CrossRef] [PubMed]
139. Adetunji, J. Microbial Enhanced Oil Recovery. Ph.D. Thesis, Department of Chemistry, Biotechnology and Environment, Aalbor University, Esbjerg, Denmark, 2012; 181p.

140. Bachmann, R.T.; Johnson, A.C.; Edyvean, R.G.J. Biotechnology in the petroleum industry: An overview. *Int. Biodeterior. Biodegrad.* **2014**, *86*, 225–237. [CrossRef]
141. Charlier, D.; Droogmans, L. Microbial life at high temperature, the challenges, the strategies. *Cell Mol. Life Sci.* **2005**, *62*, 2974–2984. [CrossRef] [PubMed]
142. Li, H.; Yang, Q.; Li, J.; Gao, H.; Li, P.; Zhou, H. The impact of temperature on microbial diversity and AOA activity in the Tengchong Geothermal Field, China. *Sci. Rep.* **2015**, *5*, 17056. [CrossRef]
143. Olabode, O.; Ogbebor, V.; Onyeka, E.; Felix, B. The effect of chemically enhanced oil recovery on thin oil rim reservoirs. *J. Pet. Explor. Prod.* **2021**, *11*, 1461–1474. [CrossRef]
144. Lin, J.; Hao, B.; Cao, G.; Wang, J.; Feng, Y.; Tan, X.; Wang, W. A study on the microbial community structure in oil reservoirs developed by water flooding. *J. Pet. Sci. Eng.* **2014**, *122*, 354–359. [CrossRef]
145. Gao, P.; Tian, H.; Wang, Y.; Li, Y.; Li, Y.; Xie, J.; Zeng, B.; Zhou, J.; Li, G.; Ma, T. Spatial isolation and environmental factors drive distinct bacterial and archaeal communities in different types of petroleum reservoirs in China. *Sci. Rep.* **2016**, *6*, 20174. [CrossRef]
146. Li, X.-X.; Mbadinga, S.M.; Liu, J.-F.; Zhou, L.; Yang, S.-Z.; Gu, J.-D.; Mu, B.-Z. Microbiota and their affiliation with physiochemical characteristics of different subsurface petroleum reservoirs. *Int. Biodeterior. Biodegrad.* **2017**, *120*, 170–185. [CrossRef]
147. Gittel, A.; Sørensen, K.B.; Skovhus, T.L.; Ingvorsen, K.; Schramm, A. Prokaryotic Community Structure and Sulfate Reducer Activity in Water from High-Temperature Oil Reservoirs with and without Nitrate Treatment. *Appl. Environ. Microbiol.* **2009**, *75*, 7086–7096. [CrossRef]
148. Cai, M.; Nie, Y.; Chi, C.Q.; Tang, Y.Q.; Li, Y.; Wang, X.B.; Liu, Z.S.; Yang, Y.; Zhou, J.; Wu, X.L. Crude oil as a microbial seed bank with unexpected functional potentials. *Sci. Rep.* **2015**, *5*, 16057. [CrossRef]
149. Lin, Q.; He, G.; Rui, J.; Fang, X.; Tao, Y.; Li, J.; Li, X. Microorganism-regulated mechanisms of temperature effects on the performance of anaerobic digestion. *Microb. Cell Fact.* **2016**, *15*, 96. [CrossRef] [PubMed]
150. Mehta, R.; Singhal, P.; Singh, H.; Damle, D.; Sharma, A.K. Insight into thermophiles and their wide-spectrum applications. *3 Biotech.* **2016**, *6*, 81. [CrossRef] [PubMed]
151. Nmegbu, C.G.J. The effect of salt concentration on microbes during microbial enhanced oil recovery. *J. Eng. Res. Appl.* **2014**, *4*, 244–247.
152. Lee, K.S.; Kwon, T.-H.; Park, T.; Jeong, M.S. *Theory and Practice in Microbial Enhanced Oil Recovery*; Gulf Professional Publishing: Cambridge, MA, USA, 2020.
153. Fathepure, B.Z. Recent studies in microbial degradation of petroleum hydrocarbons in hypersaline environments. *Front. Microbiol.* **2014**, *5*, 173. [CrossRef]
154. Xiao, M.; Sun, S.-S.; Zhang, Z.-Z.; Wang, J.-M.; Qiu, L.-W.; Sun, H.-Y.; Song, Z.-Z.; Zhang, B.-Y.; Gao, D.-L.; Zhang, G.-Q.; et al. Analysis of bacterial diversity in two oil blocks from two low-permeability reservoirs with high salinities. *Sci. Rep.* **2016**, *6*, 19600. [CrossRef]
155. Head, I.M.; Gray, N.D.; Larter, S.R. Life in the slow lane; biogeochemistry of biodegraded petroleum containing reservoirs and implications for energy recovery and carbon management. *Front. Microbiol.* **2014**, *5*, 566. [CrossRef]
156. Amani, H.; Müller, M.M.; Syldatk, C.; Hausmann, R. Production of microbial rhamnolipid by Pseudomonas aeruginosa MM1011 for ex situ enhanced oil recovery. *Appl. Biochem. Biotechnol.* **2013**, *170*, 1080–1093. [CrossRef]
157. Kumar, S.S.; Kumar, V.; Gnaneswar Gude, V.; Malyan, S.K.; Pugazhendhi, A. Alkalinity and salinity favor bioelectricity generation potential of Clostridium, Tetrathiobacter and Desulfovibrio consortium in Microbial Fuel Cells (MFC) treating sulfate-laden wastewater. *Bioresour. Technol.* **2020**, *306*, 123110. [CrossRef]
158. Liang, R.; Davidova, I.A.; Marks, C.R.; Stamps, B.W.; Harriman, B.H.; Stevenson, B.S.; Duncan, K.E.; Suflita, J.M. Metabolic Capability of a Predominant Halanaerobium sp. in Hydraulically Fractured Gas Wells and Its Implication in Pipeline Corrosion. *Front. Microbiol.* **2016**, *7*, 988. [CrossRef]
159. Scheffer, G.; Hubert, C.R.J.; Enning, D.R.; Lahme, S.; Mand, J.; de Rezende, J.R. Metagenomic Investigation of a Low Diversity, High Salinity Offshore Oil Reservoir. *Microorganisms* **2021**, *9*, 2266. [CrossRef]
160. Choi, S.K.; Sharma, M.M.; Bryant, S.L.; Huh, C. pH sensitive polymers for novel conformance control and polymerflood applications. In Proceedings of the SPE International Conference on Oilfield Chemistry, The Woodlands, TX, USA, 20–22 April 2009; SPE Reservoir Evaluation & Engineering: Richardson, TX, USA, 2009; Volume 13, pp. 926–939.
161. Hakala, J.A.; Crandall, D.; Moore, J.; Phan, T.; Sharma, S.; Lopano, C. Laboratory-scale studies on chemical reactions between fracturing fluid and shale core from the Marcellus Shale Energy and Environmental Laboratory (MSEEL) site. In Proceedings of the SPE/AAPG/SEG Unconventional Resources Technology Conference, Austin, TX, USA, 24–26 July 2017; URTEC: Austin, TX, USA, 2017; p. 2670856.
162. Talebian, S.H.; Fahimifar, A.; Heidari, A. Review of enhanced oil recovery decision making in complex carbonate reservoirs: Fluid flow and geomechanics mechanisms. *J. Appl. Comput. Mech.* **2021**, *52*, 350–365.
163. Wilson, M.; Wilson, L.; Patey, I. The influence of individual clay minerals on formation damage of reservoir sandstones: A critical review with some new insights. *Clay Miner.* **2014**, *49*, 147–164. [CrossRef]
164. Farquhar, S.; Pearce, J.; Dawson, G.; Golab, A.; Sommacal, S.; Kirste, D.; Biddle, D.; Golding, S. A fresh approach to investigating CO_2 storage: Experimental CO_2–water–rock interactions in a low-salinity reservoir system. *Chem. Geol.* **2015**, *399*, 98–122. [CrossRef]

165. Jin, Q.; Kirk, M.F. pH as a Primary Control in Environmental Microbiology: 1. Thermodynamic Perspective. *Front. Environ. Sci.* **2018**, *6*, 21. [CrossRef]
166. Kim, J.S.; Crowley, D.E. Microbial diversity in natural asphalts of the Rancho La Brea Tar Pits. *Appl. Environ. Microbiol.* **2007**, *73*, 4579–4591. [CrossRef] [PubMed]
167. Baez, A.; Shiloach, J. Effect of elevated oxygen concentration on bacteria, yeasts, and cells propagated for production of biological compounds. *Microb. Cell Fact.* **2014**, *13*, 181. [CrossRef]
168. Berney, M.; Greening, C.; Conrad, R.; Jacobs, W.R., Jr.; Cook, G.M. An obligately aerobic soil bacterium activates fermentative hydrogen production to survive reductive stress during hypoxia. *Proc. Natl. Acad. Sci. USA* **2014**, *111*, 11479–11484. [CrossRef]
169. Khademian, M.; Imlay, J.A. How Microbes Evolved to Tolerate Oxygen. *Trends Microbiol.* **2021**, *29*, 428–440. [CrossRef]
170. Ke, C.-Y.; Sun, W.-J.; Li, Y.-B.; Lu, G.-M.; Zhang, Q.-Z.; Zhang, X.-L. Microbial enhanced oil recovery in Baolige Oilfield using an indigenous facultative anaerobic strain Luteimonas huabeiensis sp. nov. *J. Pet. Sci. Eng.* **2018**, *167*, 160–167. [CrossRef]
171. Cai, M.; Yu, C.; Wang, R.; Si, Y.; Masakorala, K.; Yuan, H.; Yao, J.; Zhang, J. Effects of oxygen injection on oil biodegradation and biodiversity of reservoir microorganisms in Dagang oil field, China. *Int. Biodeterior. Biodegrad.* **2015**, *98*, 59–65. [CrossRef]
172. Haddad, H.; Khaz'ali, A.R.; Mehrabani-Zeinabad, A.; Jazini, M. Investigating the potential of microbial enhanced oil recovery in carbonate reservoirs using Bacillus persicus. *Fuel* **2023**, *334*, 126757. [CrossRef]
173. Huang, H.-W.; Lung, H.-M.; Yang, B.B.; Wang, C.-Y. Responses of microorganisms to high hydrostatic pressure processing. *Food Control.* **2014**, *40*, 250–259. [CrossRef]
174. Dahal, S.; Poudel, S.; Thompson, R.A. Genome-scale modeling of thermophilic microorganisms. *Netw. Biol.* **2017**, *160*, 103–119.
175. Lai, B.-L.; Wei, H.-X.; Luo, Z.-N.; Zheng, T.; Lin, Y.-H.; Liu, Z.-Q.; Li, N. ZIF-8-derived Cu, N co-doped carbon as a bifunctional cathode catalyst for enhanced performance of microbial fuel cell. *Sci. Total Environ.* **2023**, *856*, 159083. [CrossRef] [PubMed]
176. Yaşar Yildiz, S.; Nikerel, E.; Toksoy Öner, E. Genome-scale metabolic model of a microbial cell factory (*Brevibacillus thermorube* 423) with multi-industry potentials for exopolysaccharide production. *OMICS* **2019**, *23*, 237–246. [CrossRef]
177. Li, J.; Wang, Y.; Fan, Z.; Tang, P.; Wu, M.; Xiao, H.; Zeng, Z. Toxicity of Tetracycline and Metronidazole in Chlorella pyrenoidosa *Int. J. Environ. Res. Public Health* **2023**, *20*, 3623. [CrossRef]
178. Nikolova, C.; Gutierrez, T. Use of Microorganisms in the Recovery of Oil From Recalcitrant Oil Reservoirs: Current State o Knowledge, Technological Advances and Future Perspectives. *Front. Microbiol.* **2020**, *10*, 2996. [CrossRef]
179. Lezorgia Nekabari, N.; Theophilus, S.; Barifcani, A.; Sarmadivaleh, M.; Iglauer, S. EOR processes, opportunities and technologica advancements. In *Chemical Enhanced Oil Recovery (cEOR)—A Practical Overview*; Romero-Zeron, L., Ed.; IntechOpen: Rijeka Croatia, 2016; pp. 3–52.
180. Hamme, J.D.V.; Singh, A.; Ward, O.P. Recent Advances in Petroleum Microbiology. *Microbiol. Mol. Biol. Rev.* **2003**, *67*, 503–54 [CrossRef]
181. Pereira, J.F.; Gudiña, E.J.; Costa, R.; Vitorino, R.; Teixeira, J.A.; Coutinho, J.A.; Rodrigues, L.R. Optimization and characterizatio of biosurfactant production by Bacillus subtilis isolates towards microbial enhanced oil recovery applications. *Fuel* **2013**, *11* 259–268. [CrossRef]
182. Xia, M.; Fu, D.; Chakraborty, R.; Singh, R.P.; Terry, N. Enhanced crude oil depletion by constructed bacterial consortiur comprising bioemulsifier producer and petroleum hydrocarbon degraders. *Bioresour. Technol.* **2019**, *282*, 456–463. [CrossRef]
183. Lazar, I.; Petrisor, I.G.; Yen, T.F. Microbial Enhanced Oil Recovery (MEOR). *Pet. Sci. Technol.* **2007**, *25*, 1353–1366. [CrossRef]
184. Rathi, R.; Lavania, M.; Kukreti, V.; Lal, B. Evaluating the potential of indigenous methanogenic consortium for enhanced oil an gas recovery from high temperature depleted oil reservoir. *J. Biotechnol.* **2018**, *283*, 43–50. [CrossRef] [PubMed]
185. Okoro, E.E.; Efajemue, E.A.; Sanni, S.E.; Olabode, O.A.; Orodu, O.D.; Ojo, T. Application of thermotolerant petroleum microbes reservoir conditions for enhanced oil recovery. *Petroleum* **2022**, *9*, 223–236. [CrossRef]
186. Bhattacharya, M.; Guchhait, S.; Biswas, D.; Singh, R. Evaluation of a microbial consortium for crude oil spill bioremediation an its potential uses in enhanced oil recovery. *Biocatal. Agric. Biotechnol.* **2019**, *18*, 101034. [CrossRef]
187. Moënne-Loccoz, Y.; Mavingui, P.; Combes, C.; Normand, P.; Steinberg, C. Microorganisms and Biotic Interactions. In *Environment Microbiology: Fundamentals and Applications*; Bertrand, J.C., Caumette, P., Lebaron, P., Matheron, R., Normand, P., Ngando, T. Eds.; Nature Publishing Group: Berlin, Germany, 2014; pp. 395–444.
188. Xu, P. Dynamics of microbial competition, commensalism, and cooperation and its implications for coculture and microbion engineering. *Biotechnol. Bioeng.* **2021**, *118*, 199–209. [CrossRef] [PubMed]
189. Wu, B.; Xiu, J.; Yu, L.; Huang, L.; Yi, L.; Ma, Y. Research advances of microbial enhanced oil recovery. *Heliyon* **2022**, *8*, e114 [CrossRef]
190. Sun, S.; Wang, H.; Yan, K.; Lou, J.; Ding, J.; Snyder, S.A.; Wu, L.; Xu, J. Metabolic interactions in a bacterial co-culture accelera phenanthrene degradation. *J. Hazard. Mater.* **2021**, *403*, 123825. [CrossRef]
191. Sajna, K.V.; Sukumaran, R.K.; Gottumukkala, L.D.; Pandey, A. Crude oil biodegradation aided by biosurfactants from Pseudozyr sp. NII 08165 or its culture broth. *Bioresour. Technol.* **2015**, *191*, 133–139. [CrossRef]
192. Sharma, S.; Pandey, L.M. Production of biosurfactant by Bacillus subtilis RSL-2 isolated from sludge and biosurfactant mediat degradation of oil. *Bioresour. Technol.* **2020**, *307*, 123261. [CrossRef]
193. Dean-Ross, D.; Moody, J.; Cerniglia, C. Utilization of mixtures of polycyclic aromatic hydrocarbons by bacteria isolated fr contaminated sediment. *FEMS Microbiol. Ecol.* **2002**, *41*, 1–7. [CrossRef]

194. Fallahi, A.; Rezvani, F.; Asgharnejad, H.; Nazloo, E.K.; Hajinajaf, N.; Higgins, B. Interactions of microalgae-bacteria consortia for nutrient removal from wastewater: A review. *Chemosphere* **2021**, *272*, 129878. [CrossRef] [PubMed]
195. Chen, L.; Lu, Z.; Wang, W. The effect of delays on the permanence for Lotka-Volterra systems. *Appl. Math. Lett.* **1995**, *8*, 71–73. [CrossRef]
196. Lu, Z.; Takeuchi, Y. Permanence and global attractivity for competitive Lotka–Volterra systems with delay. *Nonlinear Anal.-Theory Methods Appl.* **1994**, *22*, 847–856. [CrossRef]
197. Guerra, N.P. Modeling the batch bacteriocin production system by lactic acid bacteria by using modified three-dimensional Lotka–Volterra equations. *Biochem. Eng. J.* **2014**, *88*, 115–130. [CrossRef]
198. Fujikawa, H.; Munakata, K.; Sakha, M.Z. Development of a competition model for microbial growth in mixed culture. *Biocontrol Sci.* **2014**, *19*, 61–71. [CrossRef]
199. Zheng, X.; Li, D. Interaction of Acidithiobacillus ferrooxidans, Rhizobium phaseoli and Rhodotorula sp. in bioleaching process based on Lotka–Volterra model. *Electron. J. Biotechnol.* **2016**, *22*, 90–97. [CrossRef]
200. Wang, T.; Yu, L.; Xiu, L.; Huang, L.; Cui, Q.; Ma, Y. A mathematical model for microbial enhanced oil recovery considering the double-bacterial competition mechanism. *J. Pet. Sci. Eng.* **2019**, *178*, 336–343.
201. Han, H.; Zhu, W.; Song, Z.; Yue, M. Mechanisms of oil displacement by Geobacillus stearothermophilus producing bio-emulsifier for MEOR. *Pet. Sci. Technol.* **2017**, *35*, 1791–1798.
202. Liu, J.; Gao, K.; Zhang, Y.; Fan, Y. Bacillus subtilis-based microbial enhanced oil recovery (MEOR) in polymer microfluidic chip. *Can. J. Chem. Eng.* **2023**, *101*, 2307–2316. [CrossRef]
203. Gaol, C.L.; Wegner, J.; Ganzer, L. Real structure micromodels based on reservoir rocks for enhanced oil recovery (EOR) applications. *Lab Chip* **2020**, *20*, 2197–2208. [CrossRef]
204. Gaol, C.L.; Ganzer, L.; Mukherjee, S.; Alkan, H. Parameters govern microbial enhanced oil recovery (MEOR) performance in real-structure micromodels. *J. Pet. Sci. Eng.* **2021**, *205*, 108814. [CrossRef]
205. Strappa, L.; De Lucia, J.; Maure, M.; Llopiz, M.L. A novel and successful MEOR pilot project in a strong water-drive reservoir Vizcacheras Field, Argentina. In Proceedings of the SPE Improved Oil Recovery Conference, Tulsa, OK, USA, 17–21 April 2004; SPE: Tulsa, OK, USA, 2004; p. SPE-89456-MS.
206. Youssef, N.; Simpson, D.R.; McInerney, M.J.; Duncan, K.E. In-situ lipopeptide biosurfactant production by Bacillus strains correlates with improved oil recovery in two oil wells approaching their economic limit of production. *Int. Biodeterior. Biodegrad.* **2013**, *81*, 127–132. [CrossRef]
207. Bhatia, S.K.; Kim, S.-H.; Yoon, J.-J.; Yang, Y.-H. Current status and strategies for second generation biofuel production using microbial systems. *Energy Conv. Manag.* **2017**, *148*, 1142–1156. [CrossRef]
208. Johnson, A. Microbial oil release technique for enhanced oil recovery. In Proceedings of the Conference on Microbiological Processes Useful in Enhanced Oil Recovery, San Diego, CA, USA, 29 August–1 September 1979.
209. Bryant, R.; Stepp, A.; Bertus, K.; Burchfield, T.; Dennis, M. Microbial enhanced waterflooding field tests. In Proceedings of the SPE/DOE Improved Oil Recovery Symposium, Tulsa, OK, USA, 17–20 April 1994; OnePetro: Tulsa, OK, USA, 1994.
210. Oppenheimer, C.; Heibert, F. Microbial enhanced oil production field tests in Texas. In Proceedings of the DOE/NIPER Symposium on Applications of Microorganisms to Petroleum Technology, Bartlesville, OK, USA, 12 August 1987; pp. 12–13.
211. Le, J.-J.; Wu, X.-L.; Wang, R.; Zhang, J.-Y.; Bai, L.-L.; Hou, Z.-W. Progress in pilot testing of microbial-enhanced oil recovery in the Daqing oilfield of north China. *Int. Biodeterior. Biodegrad.* **2015**, *97*, 188–194. [CrossRef]
212. Qi, Y.; Wang, D.; Wu, M.; Li, G.; Lü, X.; Ma, T. Effects of resin degradation and biological emulsification on the viscosity break of heavy oils. *Acta Pet. Sin.* **2012**, *33*, 670–675.
213. Le, J.; Liu, F.; Zhang, J.; Bai, L.; Wang, R.; Liu, X.; Hou, Z.; Wu, X. A field test of activation indigenous microorganism for microbial enhanced oil recovery in reservoir after polymer flooding. *Shiyou Xuebao/Acta Pet. Sin.* **2014**, *35*, 99–106.
214. Li, C.F.; Li, Y.; Li, X.M.; Cao, Y.B.; Song, Y.T. The Application of Microbial Enhanced Oil Recovery Technology in Shengli Oilfield. *Pet. Sci. Technol.* **2015**, *33*, 556–560. [CrossRef]
215. Yulbarisov, E. Microbiological method for EOR. *Rev. L'institut Français PÉTrole* **1990**, *45*, 115–121. [CrossRef]
216. Murygina, V.; Mats, A.; Arinbasarov, M.; Salamov, Z.; Cherkasov, A. *Oil Field Experiments of Microbial Improved Oil Recovery in Vyngapour, West Siberia, Russia*; BDM Oklahoma, Inc.: Bartlesville, OK, USA, 1995.
217. Ivanov, M.; Belyaev, S.; Borzenkov, I.; Glumov, I.; Ibatullin, R. Additional oil production during field trials in Russia. *Dev. Pet. Sci.* **1993**, *39*, 373–381.
218. Yulbarisov, E. Results from analysis of petroleum gas on Introduction of biochemical processes in the oil formation. *Gazov. Delo* **1972**, *4*, 26–28.
219. Wagner, M.; Lungerhansen, D.; Nowak, U.; Ziran, B. Microbially improved oil recovery from carbonate. *Biohydrometalurgical Technol.* **1993**, *2*, 695–710.
220. Updegraff, D. Early research on microbial enhanced oil recovery. *Dev. Ind. Microbiol.* **1990**, *31*, 135–142.
221. Myszkowska-Ryciak, J.; Harton, A. Impact of nutrition education on the compliance with model food ration in 231 preschools, Poland: Results of eating healthy, growing healthy program. *Nutrients* **2018**, *10*, 1427. [CrossRef] [PubMed]
222. Nikolova, C.; Gutierrez, T. Biosurfactants and their applications in the oil and gas industry: Current state of knowledge and future perspectives. *Front. Bioeng. Biotechnol.* **2021**, *9*, 626639. [CrossRef] [PubMed]

223. Zhang, F.; She, Y.-H.; Chai, L.-J.; Banat, I.; Zhang, X.-T.; Shu, F.-C.; Wang, Z.-L.; Yu, L.-J.; Hou, D. Microbial diversity in long-term water-flooded oil reservoirs with different in situ temperatures in China. *Sci. Rep.* **2012**, *2*, 760. [CrossRef] [PubMed]
224. An, D.; Caffrey, S.M.; Soh, J.; Agrawal, A.; Brown, D.; Budwill, K.; Dong, X.; Dunfield, P.F.; Foght, J.; Gieg, L.M.; et al. Metagenomics of hydrocarbon resource environments indicates aerobic taxa and genes to be unexpectedly common. *Environ. Sci. Technol.* **2013**, *47*, 10708–10717. [CrossRef]
225. Anguita-Maeso, M.; Olivares-Garcia, C.; Haro, C.; Imperial, J.; Navas-Cortes, J.A.; Landa, B.B. Culture-Dependent and Culture-Independent Characterization of the Olive Xylem Microbiota: Effect of Sap Extraction Methods. *Front. Plant Sci.* **2020**, *10*, 1708. [CrossRef]
226. Koedooder, R.; Mackens, S.; Budding, A.; Fares, D.; Blockeel, C.; Laven, J.; Schoenmakers, S. Identification and evaluation of the microbiome in the female and male reproductive tracts. *Hum. Reprod. Update* **2019**, *25*, 298–325. [CrossRef]
227. Piceno, Y.; Reid, F.C.; Tom, L.M.; Conrad, M.E.; Bill, M.; Hubbard, C.G.; Fouke, B.W.; Graff, C.J.; Han, J.; Stringfellow, W.T.; et al. Temperature and injection water source influence microbial community structure in four Alaskan North Slope hydrocarbon reservoirs. *Front. Microbiol.* **2014**, *5*, 409. [CrossRef]
228. Lewin, A.; Johansen, J.; Wentzel, A.; Kotlar, H.K.; Drabløs, F.; Valla, S. The microbial communities in two apparently physically separated deep subsurface oil reservoirs show extensive DNA sequence similarities. *Environ. Microbiol.* **2014**, *16*, 545–558. [CrossRef]
229. Hu, P.; Tom, L.; Singh, A.; Thomas, B.C.; Baker, B.J.; Piceno, Y.M.; Andersen, G.L.; Banfield, J.F. Genome-Resolved Metagenomic Analysis Reveals Roles for Candidate Phyla and Other Microbial Community Members in Biogeochemical Transformations in Oil Reservoirs. *mBio* **2016**, *7*, e01669-15. [CrossRef] [PubMed]
230. Sierra-García, I.N.; Correa Alvarez, J.; de Vasconcellos, S.P.; Pereira de Souza, A.; dos Santos Neto, E.V.; de Oliveira, V.M. New hydrocarbon degradation pathways in the microbial metagenome from Brazilian petroleum reservoirs. *PLoS ONE* **2014**, *9*, e90087. [CrossRef] [PubMed]

processes

MDPI

Article

CO_2 Injection for Enhanced Gas Recovery and Geo-Storage in Complex Tight Sandstone Gas Reservoirs

Linqiang Zhang [1], Tongzhou Bai [2], Qibin Zhao [3], Xinghua Zhang [4], Hanlie Cheng [2,*] and Zhao Li [2]

1 China United Coalbed Methane Co., Ltd., Beijing 100011, China
2 COSL-EXPRO Testing Services Co., Ltd., Tianjin 300457, China
3 Exploration Department, CNOOC Co., Ltd., Beijing 100010, China
4 Tianjin Branch, CNOOC Co., Ltd., Tianjin 300459, China
* Correspondence: chenghl@cosl-expro.com

Abstract: With the popularization of natural gas and the requirements for environmental protection, the development and utilization of natural gas is particularly important. The status of natural gas in China's oil and gas exploration and development is constantly improving, and the country is paying more and more attention to the exploitation and utilization of natural gas. The Upper Paleozoic tight sandstone in the Ordos Basin is characterized by low porosity, low permeability and a large area of concealed gas reservoirs. By injecting carbon dioxide into the formation, the recovery rate of natural gas can be improved, and carbon neutrality can be realized by carbon sequestration. Injecting greenhouse gases into gas reservoirs for storage and improving recovery has also become a hot research issue. In order to improve the recovery efficiency of tight sandstone gas reservoirs, this paper takes the complex tight sandstone of the Upper Paleozoic in the Ordos Basin as the research object; through indoor physical simulation experiments, carried out the influence of displacement rate, fracture dip angle, core permeability, core dryness and wetness on CO_2 gas displacement efficiency and storage efficiency; and analyzed the influence of different factors on gas displacement efficiency and storage efficiency to improve the recovery and storage efficiency. The research results show that under different conditions, when the injection pore volume is less than 1 PV, the relationship between the CH_4 recovery rate and the CO_2 injection pore volume is linear, and the tilt angle is 45°. When the injection pore volume exceeds 1 PV, the CH_4 recovery rate increases slightly with the increase in displacement speed, the recovery rate of CO_2 displacement CH_4 is between 87–97% and the CO_2 breakthrough time is 0.7 PV–0.9 PV. In low-permeability and low-speed displacement cores, the diffusion of carbon dioxide molecules is more significant. The lower the displacement speed is, the earlier the breakthrough time is, and the final recovery of CH_4 slightly decreases. Gravity has a great impact on carbon sequestration and enhanced recovery. The breakthrough of high injection and low recovery is earlier, and the recovery of CH_4 is about 3.3% lower than that of low injection and high recovery. The bound water causes the displacement phase CO_2 to be partially dissolved in the formation water, and the breakthrough lags about 0.1 PV. Ultimately, the CH_4 recovery factor and CO_2 storage rate are higher than those of dry-core displacement. The research results provide theoretical data support for CO_2 injection to improve recovery and storage efficiency in complex tight sandstone gas reservoirs.

Keywords: complex tight sandstone gas reservoirs; CO_2 geo-storage; enhanced natural gas recovery; physical simulation experiment

itation: Zhang, L.; Bai, T.; Zhao, Q.; hang, X.; Cheng, H.; Li, Z. CO_2 jection for Enhanced Gas Recovery d Geo-Storage in Complex Tight ndstone Gas Reservoirs. *Processes* 23, *11*, 2059. https://doi.org/ .3390/pr11072059

ademic Editors: Tao Zhang, eng Sun, Dong Feng, Hung Thanh and Wen Zhao

ceived: 29 May 2023
vised: 5 July 2023
cepted: 7 July 2023
blished: 10 July 2023

1. Introduction

With the rapid development of the industrial society, the burning of a large amount of fossil fuels in human activities has led to the emission of a large amount of carbon dioxide into the atmosphere, resulting in global warming and a serious threat to people's living environment [1–4]. Natural gas is the cleanest fossil energy source, with high calorific value and the same amount of electricity. The carbon dioxide emitted is only 50% of that of coal and 30% less than that of oil, which is conducive to achieving low-carbon goals. Natural gas also has advantages that renewable energy does not have: for example, wind energy and light energy are unstable, and their supply reliability is not as good as natural gas, which has strong compressibility and is convenient for transportation. The demand for natural gas is increasing year by year, and the gap between supply and demand of natural gas is prominent, so it is urgent to effectively develop and utilize natural gas. The importance of natural gas as a clean energy is increasing day by day [5]. At present, most natural gas mining uses depletion mining, and the recovery rate is low. Using carbon dioxide flooding can not only improve the recovery rate of natural gas but also realize effective carbon storage and reduce carbon emissions. By sequestering carbon dioxide, reducing CO_2 emissions in the atmosphere has become an important way to slow down global warming and climate change [6]. CO_2 geological storage is a technically feasible, efficient, safe and environmentally friendly emission reduction solution. Common sites for CO_2 geological storage include deep brine formations, depleted oil and gas reservoirs, unminable coal seams and deep-sea projects. If CO_2 is simply stored, the cost is high [7]. Therefore, the large-scale utilization of CO_2 greenhouse gas by improving oil and gas recovery while in storage has become a hot research topic. At present, the Ordos Basin, Tarim Basin and Luangehai Basin in China have abundant natural gas reserves, and the storage potential of CO_2 is also huge [8]. Since the gas reservoir itself exists in the natural geological body that is most suitable for storing gaseous substances, the gas storage and sealing properties of the gas-bearing reservoir have also been fully confirmed in the natural gas occurrence and development stage [9]. The implementation of CO_2 storage has theoretical and practical advantages and practical feasibility. The existing results show that CO_2 is very unlikely to be stored in the form of gas, and the supercritical state has the greatest storage potential in the formation [10,11]. Gas reservoirs whose formation temperature and pressure exceed the critical point of CO_2 are the most suitable sites for CO_2 storage. Therefore, from the perspective of economic benefits, it is a hot research topic to achieve enhanced natural ga recovery by injecting CO_2 into gas reservoirs and using its own conditions for storage [12 There are two schemes to reduce the concentration of carbon dioxide in the atmosphere The first scheme focuses on reducing the emission of carbon dioxide into the air. The utilization of carbon dioxide mentioned in this paper can improve the oil recovery of tigh sandstone gas reservoirs and increase the supply of natural gas, so that it can replace coa oil and other resources with large carbon dioxide emissions and natural gas emission under the same power generation conditions and reduce the emission of carbon dioxid into the air. At the same time, the carbon dioxide injected into the stratum is stored i the stratum, which can effectively treat the discharged carbon dioxide. In a functiona coal-fired power plant, the 660 MWe supercritical coal-fired boiler is always based o traditional and advanced artificial intelligence, using technology to present comprehensiv data visualization, and an effective and verified artificial neural network is develope Compared with the least square support vector machine process model, the artificial neura network process model is proved to be obviously more effective and used for industrial da analysis and decision making. When the unit load is 50%, 75% and 100%, the heat inpu value can be reduced by 7.20%, 6.85% and 8.60%, respectively, without affecting the overa thermal efficiency of the power plant [13]. Negative emission technology can effective remove carbon dioxide from the atmosphere. Common technologies include afforestatic forest management, vegetation restoration and other measures. Plant photosynthesis used to absorb carbon dioxide from the atmosphere and fix it in vegetation and soil reduce the concentration of carbon dioxide in the air. Capturing carbon dioxide in iv plants growing on the external walls of houses is a quick supplementary way to achieve th global GHG emission reduction goal, which can help achieve the carbon dioxide emissic reduction goal [14].

In China, there are many studies on enhanced oil recovery by carbon dioxide g flooding in tight sandstone reservoirs in the existing literature. There are many factors th affect the effect of CO_2 flooding, which are mainly divided into three categories: reservoir parameters, formation fluid properties and gas injection methods. Among them, reservoir parameters mainly include reservoir heterogeneity, reservoir thickness and permeability, and fluid properties mainly include crude oil viscosity and crude oil density. Formation heterogeneity, oil–gas viscosity ratio, formation sedimentation rhythm, miscible–immiscible, vertical and horizontal permeability ratio and development mode are the main factors affecting CO_2 displacement efficiency and storage coefficient. The permeability of the formation has little influence on CO_2 displacement efficiency and storage coefficient, but it has great influence on injection production speed and production time, such as low-permeability reservoirs, which is not conducive to the rapid storage of CO_2 [15]. This study draws lessons from the research results of CO_2 injection in tight sandstone reservoir to understand its influencing factors and makes targeted experiments to explore the mechanism of CO_2 injection in complex tight sandstone gas reservoirs to enhance oil recovery and burial.

The research results of carbon dioxide injection into other gas reservoirs outside China show that numerical simulation is used to evaluate the influence of the duration of carbon dioxide injection period on natural gas recovery during initial gas–water contact. The research shows that when carbon dioxide is injected, an artificial barrier is formed to partially block the formation water. The final gas recovery rate is 61.98% when carbon dioxide is injected and 48.04% when the reservoir is exhausted and developed. Injecting carbon dioxide to improve ultimate oil and gas recovery [16]. The efficiency of carbon dioxide injection technology is studied by numerical simulation to control the water injection process of the V-16 reservoir in the Hadiach oil and gas condensate field and to replace the residual natural gas reserves trapped by formation water. According to the calculation results, for the control of water injection process in production formation, the number of injection wells has a considerable impact on the efficiency of the analyzed technology, which provides a more complete coverage of gas-bearing areas with carbon dioxide, making the efficiency of preventing formation water from flowing into production wells higher. The research results show that compared with depleted oil and gas field development, the predicted natural gas recovery increased by 2.95% and condensate recovery increased by 1.24% [17]. Injecting carbon dioxide into gas reservoirs can greatly improve gas reservoir recovery.

Some research results have been achieved on the implementation of CO_2 storage in natural gas reservoirs and the use of supercritical CO_2 to enhance natural gas recovery [18]. Some researchers have carried out basic research on CO_2 enhanced recovery and storage technology and simplified natural gas into pure CH_4 [19]. When the temperature and pressure of gas reservoirs with a depth of more than 800 m exceed the CO_2 critical point, CO_2 can be in a supercritical state and exists underground, and at the same time, its supercritical density is higher than that of natural gas, close to liquid, with a certain strength; the diffusion coefficient is close to that of gas, about 100 times that of liquid; and it has good fluidity. Some researchers use CO_2 storage to increase oil and gas recovery, mainly by injecting CO_2 into production wells to displace oil or gas [20]. It has been proved that CO_2 displacement technology can prolong the life of oilfields by at least 20 years. Combined with oil displacement technology, it greatly improves the profit of mining [21,22]. Some researchers have studied the potential of gas injection to enhance oil recovery in major oil areas in China. The 17 oil areas suitable for CO_2 miscible displacement have 1.6×10^8 t of geological reserves, accounting for 10.4% of the total reserves. The enhanced oil recovery rate is 16.4%, and the recoverable reserves are increased by 1.7×10^8 t. At the same time, it proves that China's oil reservoirs have considerable potential for CO_2 geological storage [23]. Some researchers take long core as the research object and, through indoor experiments and numerical simulation methods, inject supercritical CO_2 into the condensate gas reservoir circularly to obtain higher condensate oil recovery and obtain a large number of components [24]. This is mainly due to the reverse evaporation of heavy components into the CO_2 condensate gas mixture caused by thermal gradient. With the increase in temperature, the formation fluid flow rate increases, which increases the gas phase viscosity and decreases the liquid phase viscosity, promotes the thermal diffusio increases the dissolution effect of CO_2 and enhances the reverse evaporation intensity t improve its recovery [25]. Some researchers have studied the cyclic gas injection test in th fractured gas reservoir. By injecting different gases into a fractured core and conductin long core physical simulation experiments, they found that the injection timing and th composition of injected gas are two key parameters to achieve miscible gas [26–28]. Th miscible gas injected into the condensate gas reservoir has obtained greater condensat recovery. At the same time, the existence of a fracture system near the matrix is mor conducive to the depletion of pressure in the matrix and the recovery of more condensat When injecting gas under miscible and immiscible pressures, it can be observed that th higher the injection pressure and the lower the condensate content in the matrix core, th greater the condensate recovery [29].

The research outside China is mainly concentrated in North America. Using CC to enhance oil recovery is an effective way to realize CO_2 storage in depleted oil an gas reservoirs.

In 2008, the output of CO_2-EOR in the world was 27.25×10^4 barrels per day, and th in the United States was 24.79×10^4 barrels per day, accounting for 91% of the daily outp of CO_2-EOR in the world, and the number of projects accounted for 85% of the world. CC flooding includes miscible flooding, immiscible flooding and injection. CO_2 injection ca improve oil recovery by 8~15%. The most widely used EOR technology in the world located in North America, and the United States has mature CO_2-EOR technology. IEA pr dicted in 2008 that the potential of CO_2-EOR in the world is 1600×10^8~3000×10^8 barre which is equivalent to 7~14% of the current recoverable crude oil in the world [30].

At present, the research of carbon dioxide gas flooding in complex tight sandston gas reservoirs is still blank. This paper verifies the influencing factors of carbon dioxid displacement on core tight gas recovery through core displacement experiments, whic provides a reliable basis for the popularization and application of this technology.

The complex tight sandstone has the characteristics of low porosity, low permeabili and a large area of concealed gas reservoirs [31,32]. By injecting CO_2 into the formation, th recovery rate of natural gas can be improved, and the stable storage of CO_2 can be achieve at the same time [33]. In this paper, taking the complex and tight sandstone of the Upp Paleozoic in the Ordos Basin as the research object, through laboratory physical simulatio experiments, the effects of displacement rate, fissure dip, core permeability, core dry an wet on CO_2 displacement efficiency and storage efficiency were carried out [34]. Th effects of different factors on CO_2 displacing efficiency and storage efficiency are analyze which provides theoretical data support for CO_2 injection in complex tight sandstone ga reservoirs to improve recovery and storage efficiency [35].

The main purpose of this paper is to provide experimental basis for carbon dioxide di placement development of complex tight sandstone gas reservoirs in the later stage throug carbon dioxide core displacement experiments and at the same time provide necessar experimental parameters for permanent carbon sequestration of carbon-dioxide-deplete oil and gas reservoirs, so as to improve gas reservoir recovery and carbon sequestration.

2. Experimental Methods

2.1. Geological Background of the Study Area

The Ordos Basin is located in the west-central part of China. The basin is general rhombus-shaped, extending from north to south. It is adjacent to Yinshan and Daqingsha in the north, Qinling Mountains in the south, Luliang Mountain in the east and Liupa Mountain in the west. This paper selects the tight sandstone reservoirs in the Hangjin area of the Ordos Basin. The recoverable positions of natural gas in the study area are Ca boniferous and Permian, including the Lower Shihezi Formation, Upper Shihezi Formatio and Shiqianfeng Formation. Table 1 shows the specific characteristics of each stratum i the study area. The lithology of the Lower Shihezi Formation and Upper Shihezi Form tion are relatively diverse, and the coal seams and mudstones have strong hydrocarbo generation capacity, providing good source rock conditions for the formation of natural gas reservoirs in this area. Sandstone and glutenite developed in the Shiqianfeng Formation have relatively good reservoir properties, providing good reservoir conditions and good cap rock conditions for gas accumulation.

Table 1. Characteristics of each stratum in the study area.

Stratum Name	Stratigraphic Characteristics
Lower Shihezi Formation	It is widely distributed on the plane and basically distributed in the whole study area. It is composed of He1 Member, He2 Member and He3 Member longitudinally. The lithology is mainly sandstone, with a certain amount of mudstone mixed between them. The reservoir physical property of Xiahezi Formation sandstone is the best, and it is the most important reservoir in this area.
Upper Shihezi Formation	It is widely distributed on the plane. The lithology is sandstone and mudstone. The thickness of mudstone is relatively large, with a total thickness of more than 100 m. It has good capping capacity, providing good capping conditions for natural gas accumulation.
Shiqianfeng Formation	The whole area is distributed on the plane, and the lithology is dominated by sandstone and mudstone. As a regional caprock, mudstone is thicker and has stronger sealing capacity.

2.2. Experimental Scheme

In this paper, cores with different physical properties of tight sandstone gas reservoirs are used to simulate the reservoir conditions according to the temperature, pressure and formation fluid composition of tight gas reservoirs. The cores are filled with natural gas and injected with carbon dioxide at different injection rates and angles. The experiments are divided into 10 groups, and the effects of various factors on CO_2 displacement efficiency and storage efficiency are studied through the experimental results of different groups.

The tight sandstone cores in the study area are selected and divided into 10 groups of CO_2 displacement CH_4 experiments. The displacement velocities are set as 0.1 mL/min, 0.2 mL/min, 0.4 mL/min and 0.8 mL/min, respectively. The injection end dip angles are 45°, −10°, −45° and nearly horizontal. For cores with low, medium and high permeability, the permeability, respectively, is 2.18×10^{-3} μm², 9.66×10^{-3} μm² and 1.03×10^{-3} μm², and the influence group whether there is bound water is set up. The influence of core permeability, gas injection rate, the existence of formation water and gravity on the gas displacement efficiency and storage efficiency when CO_2 displaces CH_4 are studied, and the diffusion degree of CO_2-CH_4 under different influence factors is evaluated. Table 2 shows the experimental schemes for different cores.

Table 2. Experimental scheme of different cores.

xperiment Order	Influencing Factors	Experimental Content
1	Displacement Speed	0.1 mL/min horizontal displacement of medium-seepage dry core
2		0.2 mL/min horizontal displacement of dry cores with medium permeability
3		0.4 mL/min horizontal displacement of dry cores with medium permeability
4		0.8 mL/min horizontal displacement of medium-seepage dry core
5	Inclination	0.2 mL/min, inlet +45° displacement medium-seepage dry core
6		0.2 mL/min, inlet −10° displacement medium-seepage dry core
7		0.2 mL/min, inlet −45° displacement medium-seepage dry core
8	Penetration	0.2 mL/min horizontal displacement of low-seepage dry core
9		0.2 mL/min horizontal displacement of high-seepage dry core
10	Bound water	0.2 mL/min displacement of medium-seepage cores with bound water

In the experiment, the core displacement equipment of RUSKA Company in the United States was used, mainly including an injection system, a core-clamping system and a recovery system. The auxiliary equipment mainly includes a constant-pressure and constant-speed displacement pump, a gas meter and a gas chromatograph. Accordi to the displacement mechanism and similarity principle, automatic core oil displaceme device simulates formation pressure and temperature conditions and carries out vario simulation tests on cores with the help of modern science and technology. We can stu the seepage characteristics of a simulated oil displacement system in porous media (liqu permeability, oil–water relative permeability, basic water injection experiment) and t influence of oil displacement efficiency. It can be used to study oil production technolo such as fracturing, water plugging, adjusting the injection–production profile and so The device is a comprehensive device, including water flooding, chemical flooding, C gas flooding and N_2 gas flooding, which can simulate the flow of various oil displaceme systems. The annular pressure tracks the injection pressure of the gripper and maintain constant differential pressure. With manual, electric, constant pressure, tracking functio etc. The model temperature is adjustable and controllable and can be heated and controll according to the program setting. Automatic switching of pressure sensors takes place w different ranges.

Table 3 shows the composition of formation water samples in the experiment, mair including industrial high-purity CO_2 (molar concentration > 99.9%) and high-purity C (molar concentration > 99%). The formation water is prepared according to the ion conte and salinity of the on-site formation water.

Table 3. Composition of formation water samples in the experiment.

$K^+ + Na^+$	Mg^{2+}	Ca^{2+}	Cl^-	SO_4^{2-}	HCO_3^-	CO_3^{2-}	Total Salinity (mg/L
12,170	26	31	17,380	270	2431	0	32,308

All cores in the experiment are divided into three groups. Formation sampling a artificial cores are divided into relatively high permeability, medium permeability and lo permeability. At present, the drilling core technology is limited, and it is impossible directly obtain cores with a length of 1 m. During the experimental test, the temperature realized by multiple sets of heating plates in the incubator, and the pressure is monitor by the measured electronic pressure gauge. The high-pressure corrosion-resistant rubb sleeve is installed in the core holder for sealing. Table 4 shows the physical parame information table of low-permeability cores, and the cores are sorted from exit to entry a from top to bottom.

Table 4. Low-permeability core physical parameter information table.

Core Number	Core Length (cm)	Core Diameter (cm)	Pore Volume (cm^3)	Porosity (%)	Permeability (10^{-3} μm^2
1	4.31	2.51	2.35	11.2	1.8
2	4.84	2.56	2.73	11.0	1.8
3	5.11	2.49	2.85	11.3	2.3
4	5.23	2.51	2.91	11.4	2.4
5	5.32	2.49	2.83	11.0	2.3
6	4.53	2.53	2.51	11.4	1.9
7	4.64	2.48	2.83	11.2	1.9
8	5.01	2.56	2.75	11.4	2.6
9	5.32	2.56	2.82	11.1	2.1
10	5.21	2.53	2.84	11.3	2.0

2.3. Experimental Conditions and Steps

In the experiment, the core temperature is set to 80 °C, the pressure is set to 8 M (controlled back pressure) and constant speed displacement is adopted. The experimen steps are:

(1) Put the short core into the long-core holder in sequence, put it into the thermos according to the angle of the injection end and connect the displacement device;

(2) Add 10 MPa confining pressure to the core, limit the maximum pressure to 5 M at the speed of 5 mL/min through the advection pump and use the mixture of displaci

petroleum ether and anhydrous ethanol to clean the long core until the outlet fluid is free of discoloration and impurities;

(3) Use nitrogen to drive long core at 2 MPa constant pressure for 12 h, and vacuum the long core after drying for 5 h;

(4) Fill saturated high-purity CH_4 gas, increase the core confining pressure to 12 MPa, use constant pressure 8 MPa to saturate CH_4 and close the outlet valve port; when the inlet pressure reaches 8 MPa and is maintained for 4 h, CH_4 saturation can be considered as complete;

(5) Connect the high-purity CO_2 intermediate container, drive CH_4 in the core at the design speed, set the back pressure to 8 MPa, record the gas volume every 0.1 PV and test the composition of the produced gas; use a gas chromatograph to analyze CH_4 and CO_2 content to calculate CH_4 recovery rate; when the CO_2 content reaches more than 98% or the CH_4 recovery rate no longer increases, the displacement is stopped and the experiment is completed;

(6) After the experiment, vacuum the core again and saturate CH_4 to conduct the next group of displacement experiments;

(7) By changing the angle of long-core gripper, high injection low production or low injection high production with different dip angles can be realized.

3. Experimental Results and Analysis

Test Results

According to the designed experimental scheme, the experiments were carried out under the conditions of 80 °C and 8 MPa with CO_2 displacement rates of 0.1 mL/min, 0.2 mL/min, 0.4 mL/min and 0.8 mL/min on the permeability of saturated high-purity CH_4. The cores were displaced, and the effects of different displacement rates, different dip angles, different permeability cores and dry and wet cores on CO_2 storage efficiency were analyzed.

Figure 1 shows the change of core CH_4 recovery with CO_2 injection pore volume under different displacement velocities. It can be seen from Figure 1 that when the CO_2 injected pore volume is less than 1 PV, the relationship between the CH_4 recovery factor and the CO_2 injected pore volume is linearly correlated, and the inclination angle is 45°; when the CO_2 injected pore volume exceeds 1 PV, the recovery factor of CH_4 increases slightly with the increase in displacement speed; when the displacement rate is 0.8 mL/min, compared with the displacement rate of 0.1 mL/min, CH_4 recovery is about 8% higher. The results show that diffusion occurs at the front of displacement, and the mixing time of CO_2 and CH_4 in porous media increases during low-speed displacement, which is easy to form a wider miscible zone, resulting in lower CH_4 recovery when CO_2 breaks through, but the final recovery is not much different.

The CO_2 storage rate refers to the ratio of the CO_2 storage amount to the core pore volume, and the CO_2 storage ratio refers to the ratio of the CO_2 storage amount to the total amount of CO_2 injected. The CO_2 storage efficiency under different experimental conditions was studied by calculating the CO_2 storage rate and storage ratio. Figure 2 shows the relationship between the CO_2 storage rate of the core and the injected pore volume under different displacement rates. It can be seen from Figure 2 that before CO_2 breakthrough, the CO_2 storage efficiency increases in a 45° oblique line. After the CO_2 breakthrough, the smaller displacement rate makes the CO_2 storage rate increase slowly, which indicates that the mixing time of CO_2 front and CH_4 is longer, and the transition zone is wider. With the CO_2 breakthrough, the CO_2 storage rate decreases, and the final CO_2 storage rate under different displacement rates is consistent. Figure 3 shows the relationship between the CO_2 storage ratio and the injected pore volume in the core under different displacement rates. It can be seen from Figure 3 that with the continuous injection of CO_2, the initial storage ratio of CO_2 is 100%, but with the CO_2 breakthrough, the storage ratio will gradually decrease. When the CO_2 injection exceeds 1 PV, under the same PV number, the rate of CO_2 storage at different displacement speeds is basically consistent.

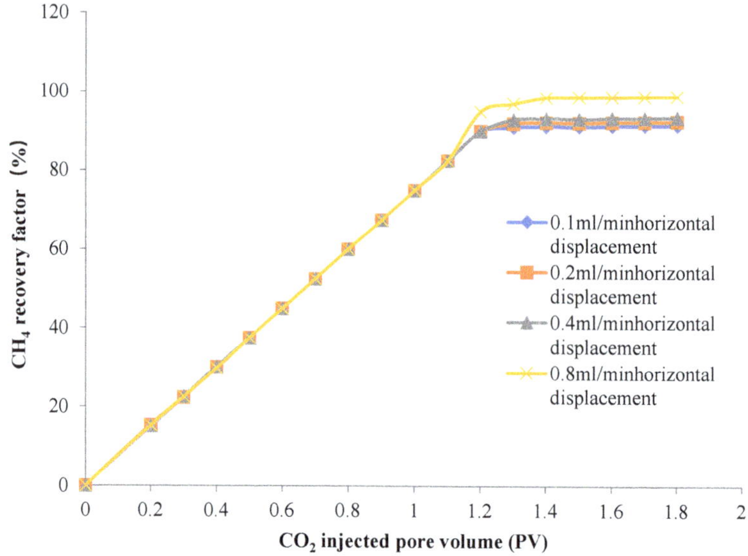

Figure 1. CH_4 recovery of cores with different displacement velocities varies with the pore volume of CO_2 injection.

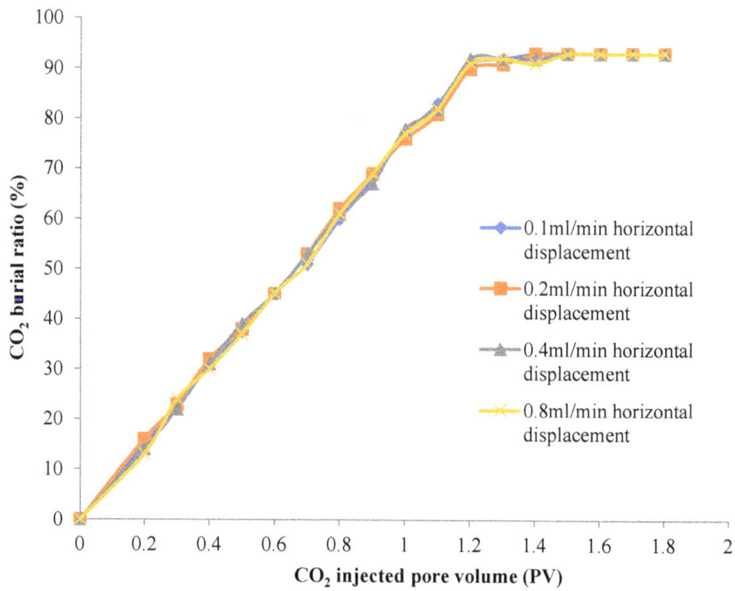

Figure 2. CO_2 storage efficiency in cores under different displacement rates.

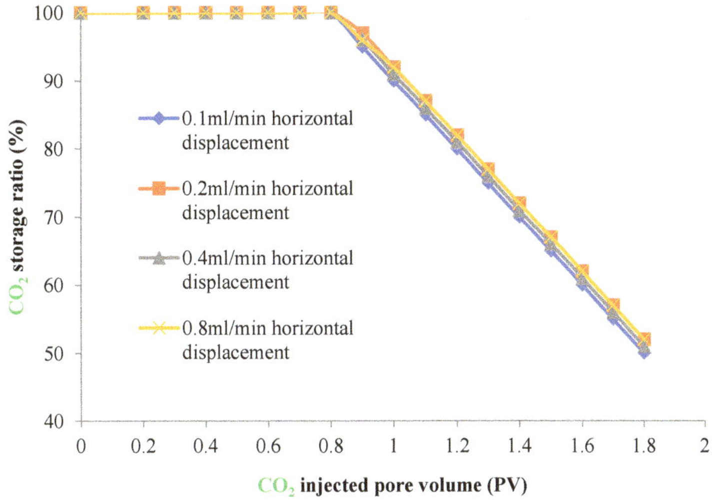

Figure 3. CO_2 storage ratio in cores under different displacement rates.

Figure 4 shows the relationship between CH_4 recovery rate of core and CO_2 injection pore volume under different displacement angles. It can be seen from Figure 4 that the CH_4 recovery factor is linearly related to the injected pore volume, and the inclination angle is 45°. At the same time, when CO_2 breakthrough occurs, the CH_4 recovery factor (76%) of high injection and low recovery is 11% lower than that of low injection and high recovery. The higher the outlet end is, the greater the recovery factor will be. The results show that under the action of gravity, gravity affects the breakthrough time of CO_2. When CO_2 is injected from a high position, the width of the transition zone is increased, the breakthrough is advanced and the recovery of CH_4 is reduced.

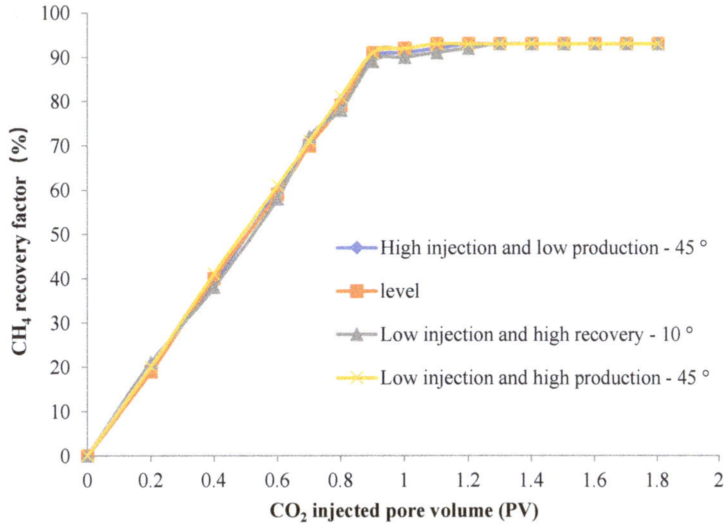

Figure 4. CH_4 recovery of cores with different displacement velocities varies with the pore volume of CO_2 injection.

Figure 5 shows the relationship between CO_2 storage rate and injected pore volume in cores under different displacement angles. It can be seen from Figure 5 that before the CO_2 breakthrough, the CO_2 storage rate increases in a 45° oblique straight line. After the CO_2

breakthrough, the growth rate of CO_2 storage rate slows down. At the same time, because the breakthrough of high injection and low production is earlier, the final CO_2 storage rate is lower than that of horizontal displacement and low injection and high production. The research results show that the mixing transition zone of CO_2 and CH_4 is wider due to gravity when displacing in high position. When CO_2 breaks through, the CO_2 will increase with CH_4 production, which will lead to a decrease in the storage rate. The higher the outlet end, the greater the CO_2 storage rate will be.

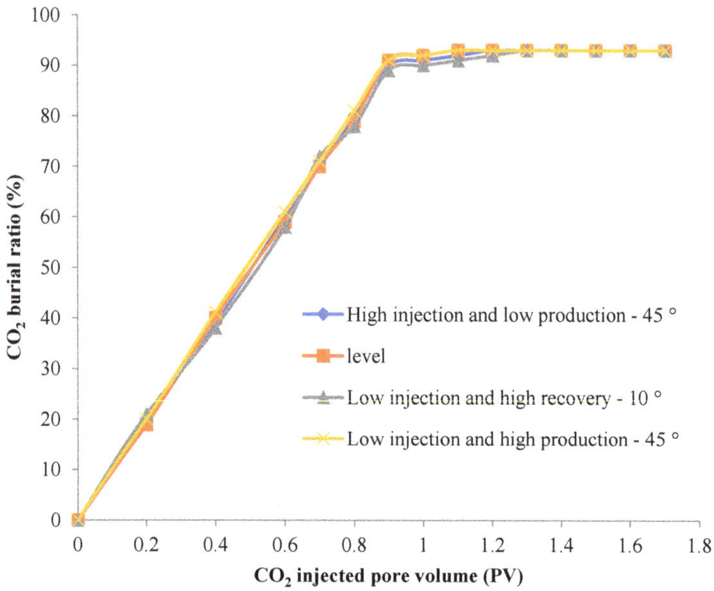

Figure 5. CO_2 storage efficiency in cores under different displacement angles.

Figure 6 shows the relationship between the buried proportion of CO_2 in the cor and the injected pore volume under different displacement velocities. It can be seen fror Figure 6 that with the continuous injection of CO_2, the storage ratio of CO_2 at the initi stage is 100%, but after the breakthrough of CO_2, the storage ratio of CO_2 will graduall decrease. After the injection exceeds 1 PV, the storage ratio of CO_2 at different displacemer angles is about 93%.

Figure 7 shows the relationship between CH_4 recovery factor and CO_2 injection por volume of cores with different permeability under the same injection conditions. It can b seen from Figure 7 that before CO_2 breakthrough, the CH_4 recovery rate is linearly relate to the injected pore volume, with an inclination of 45°; after CO_2 breakthrough, with th increase in core permeability, the recovery factor of CH_4 is higher. The research resul show that in the low-permeability core, the diffusion of CO_2 is stronger, the transitic zone between CO_2 and CH_4 is larger and the produced CH_4 is polluted, thus reducing th recovery factor.

Figure 8 shows the relationship between CO_2 storage rate and injected pore volume cores with different permeability. It can be seen from Figure 8 that before CO_2 breakthroug the CO_2 storage rate increases in a 45° oblique straight line; after CO_2 breakthrough, th CO_2 storage rate increases slowly. With the increase in core permeability, the CO_2 stora rate increases. Figure 9 shows the relationship between CO_2 storage ratio of cores wi different permeability and injected pore volume. It can be seen from Figure 9 that with t increase in permeability, the CO_2 storage effect of the core is better. The results show th the diffusion of CO_2 is stronger in low-permeability cores, and the process of displaceme of CH_4 by CO_2 tends to piston displacement with higher permeability.

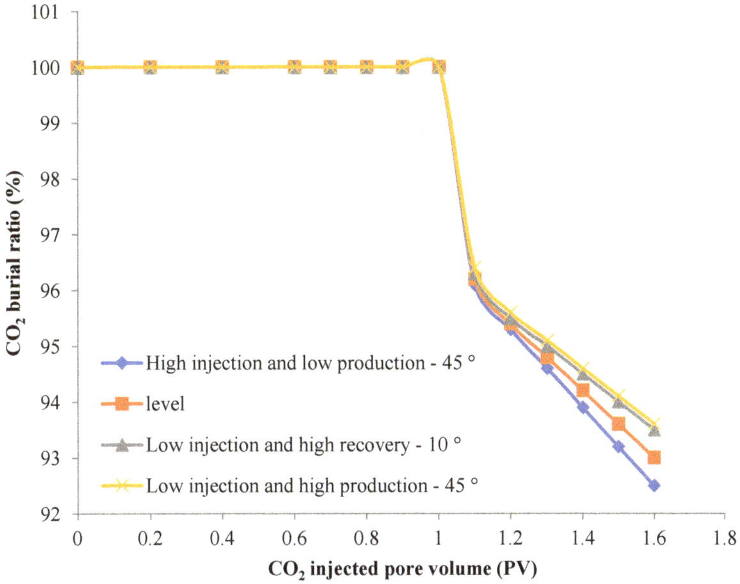

Figure 6. CO_2 storage ratio in cores under different displacement angles.

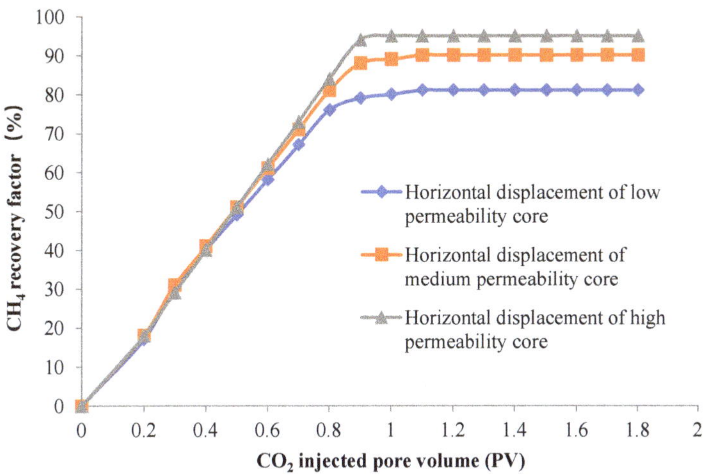

Figure 7. CH_4 recovery factor with CO_2 injection pore volume in cores with different permeability.

According to the relationship between the recovery of CH_4 in dry and wet cores and the injected pore volume, when bound water exists, there is little CO_2 effectively displaced in the initial stage of wet-core injection. When the bound water dissolves CO_2 to saturation, CH_4 is effectively displaced, so the recovery of CH_4 lags behind; at the same time, the existence of bound water makes the micropores in the core mainly filled with water, and CH_4 is more likely to be displaced by CO_2, making the CH_4 recovery of the final wet core slightly higher than that of the dry core.

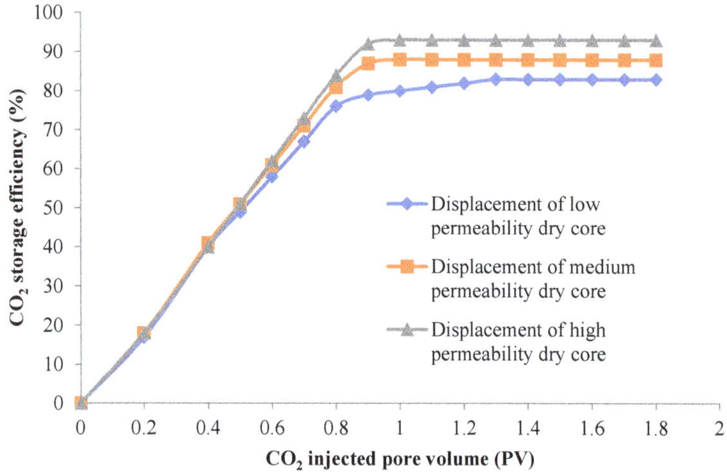

Figure 8. CO_2 storage efficiency in cores with different permeability.

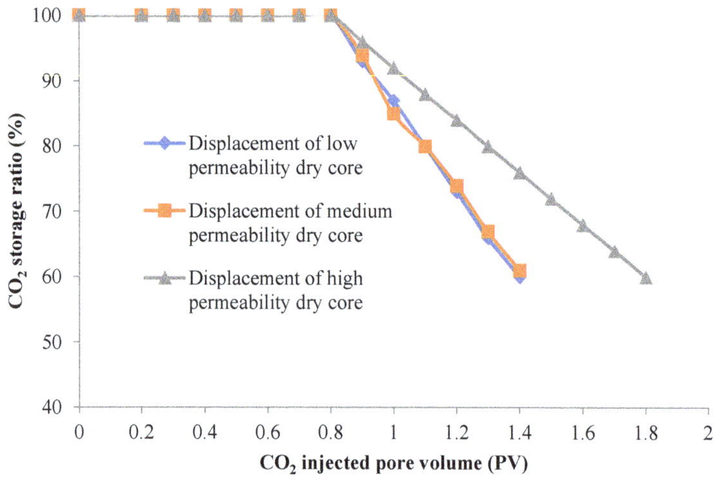

Figure 9. CO_2 storage ratio in cores with different permeability.

According to the relationship between the CO_2 storage rate and storage ratio of dr and wet cores and the pore volume of CO_2 injection, when bound water is not considere in the core, when 1.4 PV~1.6 PV of CO_2 is injected, the retention ratio is about 60%; in th presence of bound water, the CO_2 retention rate increases significantly at the same injectio of PV, which fully shows that the existence of formation water is conducive to the storag of CO_2 in the ground; at the same time, after storage, the underground saturation of CC is consistent with the degree of CH_4 production, it indicates that CO2 occupies a certai space. In addition to occupying the space of the displaced CH_4, a small part of CO_2 also dissolved in the formation water when the bound water exists. The dissolved CC accounted for about 1.89% of the pore volume under the conditions of 8 Mpa and 80 °C and the higher the pressure, the larger the dissolved amount.

CO_2 displacing CH_4 can effectively improve oil recovery and realize geological storag of CO_2 in gas reservoirs. The overall recovery ratio of CO_2 displacing CH_4 is 87–97%, an the breakthrough time of CO_2 is 0.7–0.9 PV. When CO_2 breaks through, CH_4 recover is 72–88%. When 1.5–2.4 PV is injected, the CO_2 storage rate is about 50%. When CC occupies CH_4 space in the reservoir and bound water exists, the saturation of CO_2 in the reservoir is about 55%, and the part dissolved in water accounts for 1.9% PV.

The diffusion of CO_2 molecules is more significant in low-permeability and low-velocity displacement cores. When the displacement speed is lower, the breakthrough time of CO_2 is earlier. With the increase in displacement speed, the final recovery rate of CH_4 increases slightly. When the permeability is lower, the breakthrough time of CO_2 is earlier when PV is injected at the same time, and the final CH_4 recovery rate is lower. Under the action of gravity, it has a great influence on CO_2 storage and enhanced oil recovery. The breakthrough of CO_2 with high injection and low production is earlier, the transition zone takes longer and the CH_4 recovery is about 3.3% lower than that with low injection and high production.

The dissolution of CO_2 in formation water affects the breakthrough time and recovery ratio of CO_2 displacing CH_4. When bound water exists, the displacement-phase CO_2 is partially dissolved in formation water, and the breakthrough of CO_2 lags behind about 0.1 PV, and the final CH_4 recovery ratio and CO_2 storage ratio are higher than that of dry-core displacement.

4. Conclusions

With the popularization of natural gas and the requirements of environmental protection, the development and utilization of natural gas is particularly important. The Upper Paleozoic tight sandstone in the Ordos Basin has the characteristics of low porosity, low permeability and concealed gas reservoir. By injecting CO_2 into the formation, the recovery rate of natural gas can be effectively improved, and the stable storage of CO_2 can be realized. This paper takes the complex and tight sandstone of the Upper Paleozoic in the Ordos Basin as the research object. Through laboratory experiments, the effects of displacement rate, fracture dip angle, core permeability, and core dry and wet on CO_2 displacement efficiency and storage efficiency are analyzed. The influence of these factors on CO_2 displacement and storage efficiency thus improves CH_4 recovery and CO_2 storage efficiency. The main research results are:

(1) For different cores, when the CO_2 injection pore volume is lower than 1 PV, the relationship between CH_4 recovery factor and CO_2 injection pore volume is linear, and the inclination angle is 45°. When the CO_2 injection pore volume exceeds 1 PV, the CH_4 recovery factor increases slightly with the increase in displacement speed. Before CO_2 breakthrough, the CO_2 storage efficiency increases in a 45° oblique straight line. After CO_2 breakthrough, the CO_2 storage rate increases slowly. After CO_2 breakthrough, the CO_2 storage rate decreases, and the final CO_2 storage rate is consistent. With the continuous injection of CO_2, the stored proportion of CO_2 in the initial stage is 100%, but when CO_2 breaks through, the stored proportion will gradually decrease. When CO_2 injection exceeds 1 PV, the stored proportion of CO_2 under the same PV number is basically the same.

(2) The recovery rate of CH_4 displaced by CO_2 is generally 87–97%, and the CO_2 breakthrough time is 0.7–0.9 PV. After 2.4 PV, the CO_2 storage rate is basically about 50%; when 1.5–2.4 PV is injected, the CO_2 storage rate is basically about 50%, and the CO_2 saturation in the reservoir is about 55% when there is bound water. The water part accounts for 1.9% PV; when the displacement rate is lower, the breakthrough time of CO_2 is earlier; with the increase in the displacement rate, the final recovery rate of CH_4 increases slightly; when the permeability is lower, under the same injection of PV, the CO_2 breakthrough time is earlier, and the final CH_4 recovery rate is lower; when the bound water exists, the displacement-phase CO_2 is partially dissolved into the formation water, the CO_2 breakthrough lag is about 0.1 PV, and the final CH_4 recovery rate and CO_2 storage rate are higher than that of dry-core displacement.

(3) Other conditions being unchanged, the higher the core permeability, the higher the CH_4 recovery rate and the higher the CO_2 storage rate.

Author Contributions: Conceptualization, L.Z.; methodology, H.C.; formal analysis, Z.L.; writing—original draft preparation, H.C.; writing—review and editing, Z.L.; supervision, Q.Z.; project administration, X.Z.; funding acquisition, T.B. All authors have read and agreed to the published version of the manuscript.

Funding: This research received no external funding.

Data Availability Statement: The figures and tables used to support the findings of this study are included in the article.

Acknowledgments: The authors would like to show sincere thanks to those techniques who have contributed to this research.

Conflicts of Interest: The authors declare no conflict of interest.

References

1. Wu, S.; Zou, C.; Ma, D.; Zhai, X.; Yu, H.; Yu, Z. Reservoir property changes during CO_2–brine flow-through experiments in tight sandstone: Implications for CO_2 enhanced oil recovery in the Triassic Chang 7 Member tight sandstone, Ordos Basin, China. *J. Asian Earth Sci.* **2019**, *179*, 200–210. [CrossRef]
2. Jin, L.; Hawthorne, S.; Sorensen, J.; Pekot, L.; Kurz, B.; Smith, S.; Heebink, L.; Herdegen, V.; Bosshart, N.; Torres, J.; et al. Advancing CO_2 enhanced oil recovery and storage in unconventional oil play—Experimental studies on Bakken shales. *Appl. Energy* **2017**, *208*, 171–183. [CrossRef]
3. Zhou, X.; Wang, Y.; Zhang, L.; Zhang, K.; Jiang, Q.; Pu, H.; Wang, L.; Yuan, Q. Evaluation of enhanced oil recovery potential using gas/water flooding in a tight oil reservoir. *Fuel* **2020**, *272*, 117706. [CrossRef]
4. Li, D.; Saraji, S.; Jiao, Z.; Zhang, Y. CO_2 injection strategies for enhanced oil recovery and geological storage in a tight reservoir: An experimental study. *Fuel* **2021**, *284*, 119013. [CrossRef]
5. O'Brien, G.; Gunn, R.; Raistrick, M.; Buffin, A.; Tingate, P.; Miranda, J.; Arian, N. Victorian carbon dioxide geological storage options; an engineering evaluation of storage potential in southeastern Australia. *Energy Procedia* **2011**, *4*, 4739–4746. [CrossRef]
6. Burrows, L.C.; Haeri, F.; Cvetic, P.; Sanguinito, S.; Shi, F.; Tapriyal, D.; Goodman, A.L.; Enick, R.M. A literature review of CO_2, natural gas, and water-based fluids for enhanced oil recovery in unconventional reservoirs. *Energy Fuels* **2020**, *34*, 5331–538[illegible] [CrossRef]
7. Wang, L.; Tian, Y.; Yu, X.; Wang, C.; Yao, B.; Wang, S.; Winterfeld, P.H.; Wang, X.; Yang, Z.; Wang, Y.; et al. Advances in improved/enhanced oil recovery technologies for tight and shale reservoirs. *Fuel* **2017**, *210*, 425–445. [CrossRef]
8. Zhu, C.F.; Guo, W.; Wang, Y.P.; Li, Y.J.; Gong, H.J.; Xu, L.; Dong, M.Z. Experimental study of enhanced oil recovery by CO_2 huff-n-puff in shales and tight sandstones with fractures. *Pet. Sci.* **2021**, *18*, 852–869. [CrossRef]
9. AlRassas, A.M.; Vo Thanh, H.; Ren, S.; Sun, R.; Al-Areeq, N.M.; Kolawole, O.; Hakimi, M.H. CO_2 Storage and Enhanced Oil Recovery via the Water Alternating Gas Scheme in a Mixed Transgressive Sandstone-Carbonate Reservoir: Case Study of a Large Middle East Oilfield. *Energy Fuels* **2022**, *36*, 10299–10314. [CrossRef]
10. Chaturvedi, K.R.; Sharma, T. In-situ formulation of pickering CO_2 foam for enhanced oil recovery and improved carbon storage in sandstone formation. *Chem. Eng. Sci.* **2021**, *235*, 116484. [CrossRef]
11. Wang, Y.; Shang, Q.; Zhou, L.; Jiao, Z. Utilizing macroscopic areal permeability heterogeneity to enhance the effect of CO_2 flooding in tight sandstone reservoirs in the Ordos Basin. *J. Pet. Sci. Eng.* **2021**, *196*, 107633. [CrossRef]
12. He, Y.; Qiao, Y.; Qin, J.; Tang, Y.; Wang, Y.; Chai, Z. A novel method to enhance oil recovery by inter-fracture injection and production through the same multi-fractured horizontal well. *J. Energy Resour. Technol.* **2022**, *144*, 043005. [CrossRef]
13. Muhammad Ashraf, W.; Moeen Uddin, G.; Muhammad Arafat, S.; Afghan, S.; Hassan Kamal, A.; Asim, M.; Haider Khan, M.; Waqas Rafique, M.; Naumann, U.; Niazi, S.G.; et al. Optimization of a 660 MWe Supercritical Power Plant Performance—A Case of Industry 4.0 in the Data-Driven Operational Management Part 1. Thermal Efficiency. *Energies* **2020**, *13*, 5592. [CrossRef]
14. Krzywanski, J.; Ashraf, W.M.; Czakiert, T.; Sosnowski, M.; Grabowska, K.; Zylka, A.; Kulakowska, A.; Skrobek, D.; Mistal, S.; Gao, Y. CO_2 Capture by Virgin Ivy Plants Growing Up on the External Covers of Houses as a Rapid Complementary Route to Achieve Global GHG Reduction Targets. *Energies* **2022**, *15*, 1683. [CrossRef]
15. Liu, B.; Wang, C.; Zhang, J.; Xiao, S.; Zhang, Z.; Shen, Y.; Sun, B.; He, J. Displacement mechanism of oil in shale inorganic nanopores by supercritical carbon dioxide from molecular dynamics simulations. *Energy Fuels* **2017**, *31*, 738–746. [CrossRef]
16. Matkivskyi, S.; Kondrat, O. Studying the influence of the carbon dioxide injection period duration on the gas recovery factor during the gas condensate fields development under water drive. *Min. Miner. Depos.* **2021**, *15*, 95–101. [CrossRef]
17. Matkivskyi, S. Increasing hydrocarbon recovery of Hadiach field by means of CO_2 injection as a part of the decarbonization process of the energy sector in Ukraine. *Min. Miner. Depos.* **2022**, *16*, 114–120. [CrossRef]
18. Zhang, K.; Li, S.; Liu, L. Optimized foam-assisted CO_2 enhanced oil recovery technology in tight oil reservoirs. *Fuel* **20[illegible]**, *267*, 117099. [CrossRef]
19. Sun, L.; Bai, B.; Wei, B.; Pu, W.; Wei, P.; Li, D.; Zhang, C. Recent advances of surfactant-stabilized N_2/CO_2 foams in enhanced oil recovery. *Fuel* **2019**, *241*, 83–93. [CrossRef]

20. Bai, M.; Zhang, Z.; Chen, Q.; Weifeng, S.; Du, S. Research on the Enhanced Oil Recovery Technique of Horizontal Well Volume Fracturing and CO_2 Huff-n-Puff in Tight Oil Reservoirs. *ACS Omega* **2021**, *6*, 28485–28495. [CrossRef] [PubMed]
21. Zuo, M.; Chen, H.; Qi, X.; Liu, X.; Xu, C.; Yu, H.; Brahim, M.S.; Wu, Y.; Liu, H. Effects of CO_2 injection volume and formation of in-situ new phase on oil phase behavior during CO_2 injection for enhanced oil recovery (EOR) in tight oil reservoirs. *Chem. Eng. J.* **2022**, *452*, 139454. [CrossRef]
22. Cheng, H.; Ma, P.; Dong, G.; Zhang, S.; Wei, J.; Qin, Q. Characteristics of Carboniferous Volcanic Reservoirs in Beisantai Oilfield, Junggar Basin. *Math. Probl. Eng.* **2022**, *2022*, 7800630. [CrossRef]
23. Rajkumar, P.; Pranesh, V.; Kesavakumar, R. Influence of CO_2 retention mechanism storage in Alberta tight oil and gas reservoirs at Western Canadian Sedimentary Basin, Canada: Hysteresis modeling and appraisal. *J. Pet. Explor. Prod. Technol.* **2021**, *11*, 327–345. [CrossRef]
24. Baban, A.; Keshavarz, A.; Amin, R.; Iglauer, S. Residual Trapping of CO_2 and Enhanced Oil Recovery in Oil-Wet Sandstone Core–A Three-Phase Pore-Scale Analysis Using NMR. *Fuel* **2023**, *332*, 126000. [CrossRef]
25. Lake, L.W.; Lotfollahi, M.; Bryant, S.L. CO_2 enhanced oil recovery experience and its messages for CO_2 storage. In *Science of Carbon Storage in Deep Saline Formations*; Elsevier: Amsterdam, The Netherlands, 2019; pp. 15–31.
26. Han, J.; Cheng, H.; Shi, Y.; Wang, L.; Song, Y.; Zhnag, W. Connectivity analysis and application of fracture cave carbonate reservoir in Tazhong. *Sci. Technol. Eng.* **2016**, *16*, 147–152.
27. Milad, M.; Junin, R.; Sidek, A.; Imqam, A.; Tarhuni, M. Huff-n-puff technology for enhanced oil recovery in shale/tight oil reservoirs: Progress, gaps, and perspectives. *Energy Fuels* **2021**, *35*, 17279–17333. [CrossRef]
28. Huang, X.; Li, A.; Li, X.; Liu, Y. Influence of typical core minerals on tight oil recovery during CO_2 flooding using the nuclear magnetic resonance technique. *Energy Fuels* **2019**, *33*, 7147–7154. [CrossRef]
29. Hou, Z.K.; Cheng, H.L.; Sun, S.W.; Chen, J.; Qi, D.Q.; Liu, Z.B. Crack propagation and hydraulic fracturing in different lithologies. *Appl. Geophys.* **2019**, *16*, 243–251. [CrossRef]
30. Deveci, M.; Demirel, N.Ç.; John, R.; Özcan, E. Fuzzy multi-criteria decision making for carbon dioxide geological storage in Turkey. *J. Nat. Gas Sci. Eng.* **2015**, *27 Pt 2*, 692–705. [CrossRef]
31. Bai, J.; Liu, H.; Wang, J.; Qian, G.; Peng, Y.; Gao, Y.; Yan, L.; Chen, F. CO_2, water and N_2 injection for enhanced oil recovery with spatial arrangement of fractures in tight-oil reservoirs using huff-'n-puff. *Energies* **2019**, *12*, 823. [CrossRef]
32. Alvarado, V.; Manrique, E. Enhanced oil recovery: An update review. *Energies* **2010**, *3*, 1529–1575. [CrossRef]
33. Cheng, H.; Wei, J.; Cheng, Z. Study on sedimentary facies and reservoir characteristics of Paleogene sandstone in Yingmaili block, Tarim basin. *Geofluids* **2022**, *2022*, 1445395. [CrossRef]
34. Mahzari, P.; Jones, A.P.; Oelkers, E.H. Impact of in-situ gas liberation for enhanced oil recovery and CO_2 storage in liquid-rich shale reservoirs. *Energy Sources Part A Recovery Util. Environ. Eff.* **2020**, 1–21. [CrossRef]
35. Kashkooli, S.B.; Gandomkar, A.; Riazi, M.; Tavallali, M.S. Coupled optimization of carbon dioxide storage and CO_2 enhanced oil recovery. *J. Pet. Sci. Eng.* **2022**, *208*, 109257. [CrossRef]

processes

MDPI

Article

Fracture Parameters Optimization and Field Application in Low-Permeability Sandstone Reservoirs under Fracturing Flooding Conditions

Cong Lu [1], Li Ma [1,*], Jianchun Guo [1,*], Lin Zhao [2], Shiqian Xu [1], Bugao Chen [3], Yulong Zhou [3], Haoren Yuan [3] and Zhibin Tang [3]

[1] State Key Laboratory of Oil and Gas Reservoir Geology and Exploitation, Southwest Petroleum Universit Chengdu 610500, China
[2] CNOOC Research Institute Co., Ltd., Beijing 100029, China
[3] Sinopec Shengli Oil Field Luming Oil-Gas Exploration & Development Co., Ltd., Dongying 257000, China
* Correspondence: swpumali177@163.com (L.M.); guojianchun@vip.163.com (J.G.)

Abstract: To solve engineering problems in the production process after fracturing and flooding low-permeability sandstone reservoirs, such as rapid water-cut rise and low water flooding efficienc a method for optimizing the fracture parameters of low-permeability sandstone reservoirs und fracturing flooding conditions was proposed. A rock property test experiment was first carrie out, the fracturing coefficient was defined, and an evaluation method for the brittleness index low-permeability sandstone was established to optimize the perforation location of the fracturi reservoir. A productivity numerical model for the two-phase flow of oil–water in matrix–fractu media was established to optimize the fracture morphology under fracturing flooding conditio The results showed that the quartz content, Young's modulus, and peak stress mainly affecte the fracturing coefficient of rock and are the key indicators for evaluating the brittleness of lo permeability sandstone reservoirs. For production wells in the direction of minimum horizont principal stress, the swept area of water flooding should be expanded, fracture length should b optimized to 90 m, and fracture conductivity should be 20 D·cm. For fracturing production wel in the direction of maximum horizontal principal stress, the advancing speed of the water injecti front should be slowed down to reduce the risk of water channeling in injection-production wel The optimized fracture length was 80 m, and the fracture conductivity was 25 D·cm. The applicati of these findings can markedly improve oil production and provide a reference for optimizing t fracture parameters of low-permeability sandstone reservoirs under fracturing flooding condition

Keywords: low-permeability sandstone; fracturing flooding; hydraulic fracturing; brittleness evalu tion; fracture parameters

Citation: Lu, C.; Ma, L.; Guo, J.; Zhao, L.; Xu, S.; Chen, B.; Zhou, Y.; Yuan, H.; Tang, Z. Fracture Parameters Optimization and Field Application in Low-Permeability Sandstone Reservoirs under Fracturing Flooding Conditions. *Processes* **2023**, *11*, 285. https://doi.org/10.3390/pr11010285

Academic Editor: Yidong Cai

Received: 28 December 2022
Revised: 11 January 2023
Accepted: 11 January 2023
Published: 16 January 2023

1. Introduction

With the development of oil and gas exploration and technology, oil and gas develo ment has shifted from conventional reservoirs to unconventional reservoirs that have lo permeability and are difficult to produce in China. Hydraulic fracturing is a key technolog for the effective production and economic development of unconventional reservoirs [1– Conventional water and chemical flooding can improve reservoir recovery to a certai extent; however, there are still related problems in the development of low-permeabili reservoirs [4]. The conventional fracturing process is complex, expensive, has a small scop for fracture control, and the effect of reservoir stimulation is poor. Conventional wat flooding has poor hydrodynamic connections, extensive water channeling and floodin and non-Darcy flow characteristics of fluids, which lead to problems such as "no injectio and no production" in low-permeability reservoirs. In conventional chemical flooding, t oil displacement agent adsorbs and remains in the pores of the formation rock, and t concentration decreases, which leads to a reduction in the oil displacement efficiency [5]. Therefore, "fracture-injection-production" integrated reservoir simulation technology (fracturing flooding) was proposed for the oilfield sites, which perfectly combines hydraulic fracturing technology, conventional water flooding development, and chemical flooding to form a set of continuous development technologies [6].

Since 2000, fracturing flooding technology has been applied and developed in the Daqing and Changqing oilfields, and there are some low-permeability blocks in the Shengli oilfield. The development of this technology mainly involves four stages: matrix imbibition technology for low-permeability reservoirs (2000–2009), fracture–matrix dynamic imbibition technology for fractured low-permeability reservoirs (2010–2014), energy storage fracture network fracturing technology (2015–2016), and energy storage fracturing-imbibition oil recovery technology (2017–present) [7–10]. Based on the operation mode and stimulation mechanism, the fracturing flooding process is divided into two methods: reverse fracturing flooding and forward fracturing flooding [11]. Reverse fracturing flooding involves opening the formation using a low-viscosity fluid instead of a fracturing fluid from the production well and quickly injecting liquid containing an oil-displacing agent through the fracture to increase the reservoir pressure. During the well shut-in process, the flow direction of the wetting phase that is sucked in by the reservoir matrix is opposite to the flow direction of the discharged non-wetting phase, and the reservoir recovery is improved through the enhanced oil–water imbibition displacement between the fracture and matrix. Forward fracturing flooding involves injecting a large amount of liquid containing an oil-displacing agent from an injection well at a pressure that exceeds the formation fracture pressure. Reservoir recovery is improved by increasing the reservoir pressure and oil displacement effect of the oil displacement agent. This study focuses on forward fracturing flooding technology, which has been applied to the Shengli oilfield on a large scale since 2020. More than 40 well group tests have been completed in Bonan, Niuzhuang, and other blocks, and the cumulative water injection is 110×10^4 m^3, with a cumulative increase of 33,000 tons of oil, achieving certain economic benefits [12]. However, this technique also faces two difficulties regarding the optimization of fracture parameters: poor matching of perforation horizons between injection and production wells and unreasonable fracture morphology (fracture length and fracture conductivity), resulting in engineering problems such as rapid water cut rise, low oil displacement efficiency, and short effective production time in some production wells. Therefore, optimizing the perforation location of the well and the fracture parameters of the production well are important methods for improving fracturing flooding technology.

The rock brittleness index is a key parameter for optimizing perforation location [13–16]. Hucka and Das [17] calculated rock brittleness using compressive and tensile strengths through uniaxial compressive strength and Brazilian testing. Later, many scholars used rock mechanical parameters and rock brittle mineral content to characterize rock brittleness and believed that horizon rocks with large brittleness indexes were more prone to fracture and formed a complex fracture network; therefore, the perforation location was optimized according to the brittleness index [18–20]. Presently, there are three types of methods for evaluating rock brittleness: mineral composition, mechanical parameters, and stress–strain parameters [21]. These three parameters are all related to brittleness; however, the fitting degree is not high. Therefore, it is necessary to develop a new method to characterize rock brittleness and improve the accuracy of brittleness prediction.

Optimizing and controlling the hydraulic fracture morphology is the key to the injection-production effect of the well group. Xiao and Guo [22] established a low-permeability heterogeneous geological model for the numerical simulation of hydraulic fracturing. The results show that there is an optimal value for fracture conductivity and fracture length. Xu et al. [23] used the Eclipse software to conduct orthogonal simulations for the fracture network fracturing of low-permeability oil and gas reservoirs to optimize the fracture parameters of horizontal wells. Lei et al. [24] established a non-planar three-dimensional fracture propagation model, proposed a method for the stimulation of low-permeability oil reservoirs based on the optimal design of fracture-controlled fracturing, and optimized the fracture parameters of horizontal wells. Few scholars have optimized the fracture parameters for low-permeability sandstone reservoirs under the influence of fracturing flooding [25,26]. Engineers at fracturing sites usually operate based on experience and lack theoretical guidance.

In this study, the fracture parameters were optimized from two aspects: perforation location and fracture morphology. First, considering the influence of various factors—such as the mechanical properties, mineral composition, and deformation characteristics of the rock—and using the rock fracturing coefficient as an index, a new method for brittleness evaluation of low-permeability sandstone reservoirs was proposed and the perforation location of the reservoir was optimized accordingly. Furthermore, considering the influence of the oil-displacing agent and hydraulic fractures, a productivity numerical model for the two-phase flow of oil–water in matrix–fracture dual-permeability media was established to optimize fracture morphology under fracturing flooding conditions. Finally, its application in the oil field proved that the technology has good practicability.

2. Perforation Location Optimization Based on Brittleness Evaluation

2.1. Rock Property Testing Experiment

Taking the low-permeability sandstone of the fourth member of the Shahejie formation in the Bohai Bay basin with a burial depth of 2500–3000 m as an example, 13 groups of downhole rock cores were selected for the rock property testing experiment. The stress–strain characteristics, fracture morphology, and mechanical parameters of the core were obtained through triaxial mechanical experiments.

The mineral compositions of the 13 groups of rocks were obtained using whole-rock mineral analysis. The results (Table 1) show that the mineral types of the sandstone in the fourth member of the Shahejie formation include clay minerals, quartz, feldspar and carbonate minerals, with average contents of 34, 41, 16, and 9%, respectively. The mineral composition distribution of this block was significantly different and exhibited strong heterogeneity.

Table 1. Basic parameters of the core.

Core	Mineral Composition				Mechanical Parameters		Stress–Strain Parameters			
Case	Clay (%)	Quartz (%)	Feldspar (%)	Carbonate (%)	Poisson's Ratio	Young's Modulus (MPa)	Peak Stress (MPa)	Residual Stress (MPa)	Peak Strain (%)	Residual Strain (%)
1	36.1	32.7	21	10.1	0.186	18,905.4	175	95	1.15	1.48
2	7.4	29.2	45	18.5	0.474	17,829	63	38	0.39	0.49
3	7.4	29.2	45	18.5	0.266	10,338.3	49	39	0.42	0.73
4	13.6	30.6	40.3	15.4	0.278	22,125.7	270	175	1.56	1.98
5	15.9	38.7	39.1	6.4	0.24	16,937.5	74	40	0.53	0.6
6	15.9	38.7	39.1	6.4	0.274	12,209.3	82	40	0.9	1.32
7	15.9	38.7	39.1	6.4	0.359	14,529	114	61	0.95	1.18
8	12.6	30	53.4	4	0.26	6951.2	27	16	0.58	0.82
9	12.6	30	53.4	4	0.25	9223.7	35	19	0.51	0.73
10	15.3	30	43.8	10.9	0.239	16,448.8	118	78	0.92	1.24
11	15.3	30	43.8	10.9	0.246	16,617.9	120	82	0.87	1.12
12	27.6	30.7	30.9	10.8	0.058	11,602.3	140	122	2.1	4.2
13	27.6	30.7	30.9	10.8	0.437	8629.6	38	19	0.52	0.63

The rock samples were cut into standard rock pillars (2.5 cm in diameter and 5 c in length), and triaxial mechanical experiments were conducted on the rock pillars usi the RTR-1000 rock mechanics experimenter. The Young's modulus of the sandstone in t fourth member of the Shahejie formation is between 7000 MPa and 28,000 MPa, as shown Table 1, with an average value of 14,856 MPa; the Poisson's ratio is between 0.06 and 0.4

with an average value of 0.254. Using the parameters of the triaxial mechanical experiment, the stress–strain curves of the rock pillar at different stress stages were generated, and the peak strain, residual strain, peak stress, and residual stress were calculated. The analysis shows that, under the same confining pressure, there are significant differences in the mechanical parameters, and the overall heterogeneity of the fourth member of the Shahejie formation is strong.

2.2. Fracture Morphology Analysis

The fracture morphology of the rock is important for determining the brittleness of the reservoir rock. A comparative analysis of the rock fracture modes, which was conducted after the experiment, shows that the fracture forms are divided into three types—single shear fractures, split fractures, and complex fractures—as shown in Figure 1.

(**a**) (**b**) (**c**)

Figure 1. Core fracture morphology. (**a**) Single shear fractures. (**b**) Split fractures. (**c**) Complex fractures.

Many scholars have qualitatively described the characterization of the fracture shape of the core; however, there are few quantitative characterization methods for the fracture mode of low-permeability sandstone. Through classical brittleness index analysis, it was found that there is no direct correlation among rock mechanical parameters, mineral composition, and the three fracture forms of rock [27]; therefore, it is necessary to explore a new brittleness index characterization method.

As shown in Figure 2, the fractal dimension (D) of the core end-face fracture was deduced using box theory [28], and the calculation equation is as follows:

$$D = \frac{C_p - \ln(N)}{\ln(R)} \quad (1)$$

where D is the fractal dimension of the core end-face fracture (dimensionless), R is the side length of the square boxes (cm), N is the number of square boxes containing fractures, and C_p is the proportionality constant.

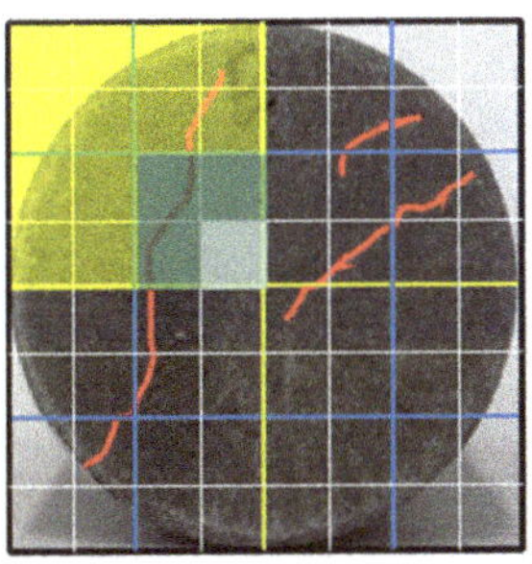

Figure 2. Schematic diagram of the box theory. The red lines represent fractures.

As shown in Figure 3, the fracture morphology on the side of the rock pillar can also characterize the fracture characteristics to a certain extent. Considering the effects of the fractal dimension and the fracture angle of the rock pillar, a rock-fracture complexity coefficient (F_c) is defined. The calculation method is shown in Equation (2), and the calculation results are listed in Table 2.

$$F_C = D \cdot \left(1 - \frac{\alpha}{90}\right) \tag{2}$$

where F_c is the fracturing coefficient (dimensionless) and α is the fracture angle (°).

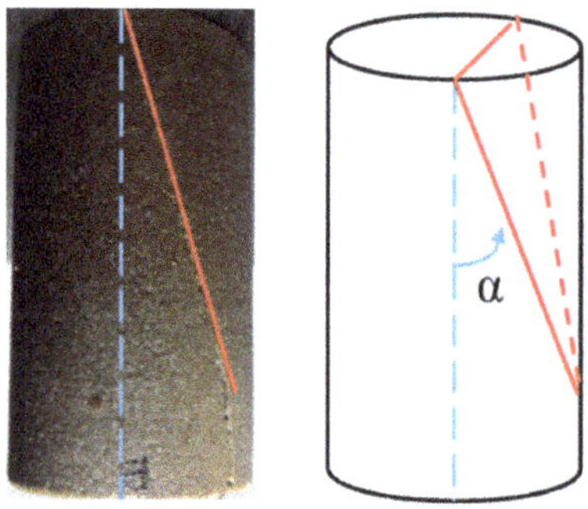

Figure 3. Schematic of the fracture angle on the side of the rock pillar.

Table 2. Fracture morphology parameters.

Core Case	Fracture Morphology	Fractal Dimension	Fracture Angle (°)	Fracturing Coefficient
1	split fractures	1.20	5	1.20
2	single shear fractures	1.50	25	1.08
3	single shear fractures	1.36	20	1.06
4	complex fractures	1.96	45	1.24
5	split fractures	1.69	12	1.17
6	split fractures	1.35	12	1.17
7	split fractures	1.94	9	1.08
8	complex fractures	1.44	35	1.20
9	split fractures	1.44	10	1.19
10	complex fractures	1.47	36	1.21
11	split fractures	1.20	4	1.20
12	single shear fractures	1.58	20	1.05
13	complex fractures	1.39	42	1.20

From Equation (2), it can be deduced that the larger the fractal dimension (D) and the smaller the fracture angle (α), the larger the fracturing coefficient (F_c) and the more complex the fracture morphology. As shown in Figure 4, the relationship between the fracturing coefficient and the fracture morphologies of the different cores was compared and analyzed. Cores with fracturing coefficients between 1 and 1.1 showed shear fracture or split fractures, cores with coefficients between 1.1 and 1.2 showed split fractures, and cores with coefficients greater than 1.2 showed complex fractures. Therefore, the core fracturing coefficient had a good correlation with the core fracture shape, which verified the reliability of the coefficient to a certain extent.

The main factors affecting the complexity of rock fractures were determined through whole-rock mineral analysis tests and triaxial mechanical experiments. As shown in Table according to single-factor correlation analysis, the factors affecting the complexity of rock fractures mainly include quartz content, Young's modulus, and peak stress [27].

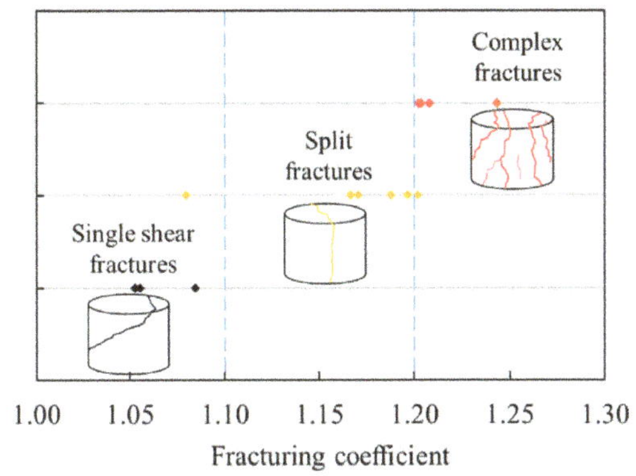

Figure 4. Core fracture morphology under different fracturing coefficients. Black dots represent single shear fractures, yellow dots represent split fractures, and red dots represent complex fractures.

Table 3. Analysis of influencing factors of rock fracture complexity.

Influencing Factor	Correlation Coefficient	Influencing Factor	Correlation Coefficient	Influencing Factor	Correlation Coefficient
Clay	0.0276	Peak stress	0.1446	Young's modulus	0.1269
Quartz	0.1071	Peak strain	0.0019	Poisson's ratio	0.0076
Feldspar	0.0004	Residual stress	0.0453	——	——
Carbonate	0.0858	Residual strain	0.0405	——	——

2.3. Evaluation of Rock Brittleness Index

Based on the above analysis, the quartz content, Young's modulus, and peak stress were selected as key indicators affecting the rock brittleness of this block, and a suitable low-permeability sandstone brittleness evaluation method for this block was established. The equation used is as follows:

$$B_{ls} = W_1 Q_n + W_2 E_n + W_3 \sigma_{pn} \tag{3}$$

where B_{ls} is the dimensionless low-permeability sandstone brittleness index; Q_n, E_n, and σ_{pn} are the normalized quartz content, Young's modulus, and peak stress, respectively; and W_1, W_2, and W_3 are the weight coefficients of each dimensionless parameter.

The weight coefficients of each influencing parameter were obtained by grey relational analysis; the quartz content, Young's modulus, and peak stress were 0.303, 0.358, and 0.339, respectively.

Young's modulus can be obtained from sonic logging and density logging data. Quartz mineral content can be obtained from elemental capture spectroscopy logging data. However, peak stress cannot be obtained from well log data; therefore, the approximated peak stress was used for the adjacent wells in the same block.

$$E = a\frac{\rho_r}{\Delta t_s^2}\left(\frac{3\Delta t_s^2 - 4\Delta t_p^2}{\Delta t_s^2 - \Delta t_p^2}\right) \tag{4}$$

where E is Young's modulus, MPa; a is unit conversion factor; ρ_r is the rock density, g/m^3; t_s is shear wave log data, μs/m; and t_p is compression wave log data, μs/m.

The brittleness index was then calculated using Equation (3). Finally, according to the calculation results, the formation with a high brittleness index was selected as the perforation location; thus, the extension position of the hydraulic fractures in the reservoir could be determined.

3. Fracture Morphology Optimization Based on Fracturing Flooding Simulation

3.1. Fracturing Flooding Mathematical Model

The following conditions were assumed in this model.

- The reservoir matrix was heterogeneous and anisotropic, with three-dimensional hexahedrons of equal horizontal thickness.
- The fracture direction was consistent with the maximum horizontal principal stress, and the fractures were homogeneous and isotropic.
- The fluid only considered the two phases of oil and water, and the oil displacement agent existed only in the water phase.
- The fluid, matrix, and fractures could be compressed to a certain extent, and the compressibility coefficient was constant.
- The reservoir was maintained at a constant temperature, and capillary imbibition was considered in the matrix.
- The bottom-hole pressure of the production well was constant; the injection rate and pressure were kept constant.
- Hydraulic fracture direction is the maximum horizontal principal stress direction, and the fracture length is constant.

The mathematical equations for the oil and water phase seepage in the matrix are as follows:

$$\frac{\partial}{\partial t}(\varphi_{\mathrm{m}}\rho_{\kappa}S_{\kappa,\mathrm{m}}) + \nabla\left(\rho_{\kappa}\vec{v}_{\kappa,\mathrm{m}}\right) = q^{\mathrm{W}}_{\kappa,\mathrm{m}} + q_{\kappa,\mathrm{fm}} \quad (5$$

$$\vec{v}_{\kappa,\mathrm{m}} = -k_{\mathrm{m}}\frac{k_{\mathrm{r}\kappa}}{\mu_{\kappa}}\nabla p_{\kappa,\mathrm{m}} \quad (6$$

where φ_{m} is the dimensionless reservoir matrix porosity, ρ_{κ} is the density of fluid κ (g/cm^3 κ denotes oil or water, $S_{\kappa,\mathrm{m}}$ is the saturation of fluid κ in the reservoir matrix (dimensionless $q^{\mathrm{W}}_{\kappa,\mathrm{m}}$ is the flow exchange term for fluid κ between the wellbore and reservoir matrix (m^3/d $q_{\kappa,\mathrm{fm}}$ is the flow exchange term for fluid κ between the wellbore and reservoir matrix (tha is, the flow of fluid κ from the fracture to the reservoir matrix (m^3/d)), k_{m} is the absolut permeability of the reservoir matrix (mD), $k_{\mathrm{r}\kappa}$ is the dimensionless relative permeability c fluid κ, μ_{κ} is the viscosity of fluid κ (mPa·s), $p_{\kappa,\mathrm{m}}$ is the pressure of fluid κ in the reservoi matrix (MPa), and t is the production time (s).

The mathematical equations for the flow of the oil and water phases in the fracture are as follows.

$$\frac{\partial}{\partial t}(\varphi_{\mathrm{f}}\rho_{\kappa}S_{\kappa,\mathrm{f}}) + \nabla\left(\rho_{\kappa}\vec{v}_{\kappa,\mathrm{f}}\right) = q^{\mathrm{W}}_{\kappa,\mathrm{f}} + q_{\kappa,\mathrm{fm}} \quad (7$$

$$\vec{v}_{\kappa,\mathrm{f}} = -k_{\mathrm{f}}\frac{k_{\mathrm{r}\kappa}}{\mu_{\kappa}}\nabla p_{\kappa,\mathrm{f}} \quad (8$$

where φ_{f} denotes the dimensionless fracture porosity, $S_{\kappa,\mathrm{f}}$ is the saturation of fluid κ i the fracture (dimensionless), $q^{\mathrm{W}}_{\kappa,\mathrm{f}}$ is the flowrate of fluid κ from the production well to th fracture (m^3/d), $q_{\kappa,\mathrm{mf}}$ is the flowrate of fluid κ from the reservoir matrix to the fractu (m^3/d), k_{f} is the absolute permeability of the fracture (mD), and $p_{\kappa,\mathrm{f}}$ is the pressure of flu κ in the fracture (MPa).

The stress-sensitive equations for porosity and permeability can be obtained by co sidering the compressibility of the fractures and matrix.

$$\begin{cases} \varphi_{\mathrm{m}} = \varphi_{\mathrm{m}0}e^{C(p_{\mathrm{m}t}-p_0)}, \varphi_{\mathrm{f}} = \varphi_{\mathrm{f}0}e^{C_{\mathrm{f}}(p_{\mathrm{ft}}-p_0)} \\ k_{\mathrm{m}} = k_{\mathrm{m}0}e^{d(p_{\mathrm{m}t}-p_0)}, k_{\mathrm{f}} = k_{\mathrm{f}0}e^{d_{\mathrm{f}}(p_{\mathrm{ft}}-p_0)} \end{cases} \quad ($$

where $\varphi_{\mathrm{m}0}$ is the dimensionless initial reservoir matrix porosity, C is the dimensionle matrix porosity compressibility, $p_{\mathrm{m}t}$ is the matrix pore pressure (MPa), p_0 is the init reservoir pressure (MPa), $\varphi_{\mathrm{f}0}$ is the dimensionless initial fracture porosity, C_{f} is the mensionless fracture porosity compressibility, p_{ft} is the fracture pressure (MPa), $k_{\mathrm{m}0}$ is t initial permeability of the reservoir matrix (mD), d is the dimensionless matrix permeability stress sensitivity coefficient, k_{f0} is the initial permeability of the fracture (mD), and d_f is the dimensionless fracture permeability stress sensitivity coefficient.

Because the permeability in the fracture was relatively large, the capillary force was regarded as zero [29]. The matrix had a capillary pressure, which is expressed as

$$p_{of} = p_{wf}, p_{om} - p_{wm} = p_c \tag{10}$$

where p_{of} is the oil pressure in the fracture (MPa), p_{wf} is the water pressure in the fracture (MPa), p_{om} is the oil pressure in the matrix (MPa), p_{wm} is the water pressure in the matrix (MPa), and p_c is the capillary pressure in the matrix (MPa).

The distribution of the oil-displacing agent was simulated by solving the conservation equation for the oil-displacing agent in the water phase.

$$\frac{\partial}{\partial t}(\varphi_m C_s \rho_s) + div(\frac{C_s k_m k_{rw} \cdot gradp}{\mu_{ws}} + \varphi S_{wm} D_w^s \cdot gradC_s) + Q_a = Q_s \tag{11}$$

where C_s is the volume fraction of the oil-displacing agent in the water phase (%), ρ_s is the density of the oil-displacing agent (kg/m^3), μ_{ws} is the solution viscosity of the oil-displacing agent and water mixture (mPa·s), D_w^s is the dimensionless diffusion coefficient of the oil-displacing agent in the water phase, Q_a is the adsorption term of the oil-displacing agent in the pore matrix, and Q_s is the injection source term of the oil-displacing agent.

The model considered the effect of the oil displacement agent concentration on the relative permeability model, capillary force, water phase viscosity, oil-displacing agent adsorption capacity, and rock wettability, and the specific influence law was obtained from the experiment [30]. The relative permeability used as the value of the miscibility function between the two endpoints was calculated in two steps. First, the endpoints of the curves were interpolated, and the immiscibility and miscibility curves were scaled to fit these points. Finally, the relative permeability values on both curves were determined, and the final relative permeability was used as an interpolation between these two values.

The production well should satisfy the bottom-hole flow pressure to maintain a constant; thus, the Peaceman model was adopted for the inner boundary condition [31]. The oil-displacing agent at the bottom of the well was given a specific initial concentration value, and the later concentration was completely updated implicitly at the end of each time step after the oil and water phase flow rates were calculated.

$$\begin{cases} q_{\kappa,f}^{w} = WI_f(p_f - p_w) \\ q_{\kappa,m}^{w} = WI_m(p_m - p_w) \\ Q_s = q_{w,m}^{w} \times C_s^{w} \end{cases} \tag{12}$$

where WI_f is the well index in the fracture grid (m^3/d·MPa), p_f is the pressure in the fracture (MPa), p_w is the bottom-hole flow pressure (MPa), WI_m is the well index in the matrix grid (m^3/d·MPa), p_m is the pressure in the matrix (MPa), $q_{w,m}^{w}$ is the water flow exchange term between the wellbore and reservoir matrix (m^3/d), and C_s^{w} is a constant that represents the volume fraction of the oil-displacing agent at the bottom of the injection well (%).

The outer boundary conditions belong to the no-flow boundary category and satisfy the Neumann model.

$$\left.\frac{\partial p}{\partial n}\right|_{\Gamma} = 0 \tag{13}$$

The initial conditions of the model include initial reservoir pressure and original oil saturation.

$$\begin{cases} p(x,y,z)|_{t=0} = C_1 \\ s(x,y,z)|_{t=0} = C_2 \end{cases} \tag{14}$$

Finally, the fracturing flooding model was solved using the finite difference method.

3.2. Numerical Simulation of Fracturing Flooding

According to the typical fracturing flooding well group (well G) in the low-permeabil sandstone reservoir of the fourth member of the Shengli oilfield, a numerical model w established, as shown in Figure 5, including one water injection well and four productic wells. The fracture direction of the fracturing well was consistent with that of the ma mum horizontal principal stress (NE45°). The half-length of the crack was 110 m, and i conductivity was 20 D·cm. The oil–water relative permeability and capillary pressure the matrix in this model were obtained from core experiments, as shown in Figure 6, ar the other reservoir geological parameters are shown in Table 4. After simulating fracturi and flooding, the water injection well injects a constant amount of liquid daily, and tl production well is produced at a constant pressure.

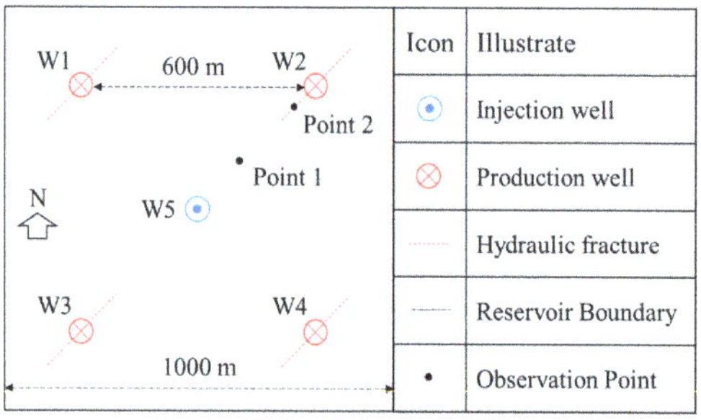

Figure 5. Schematic of fracturing flooding well group.

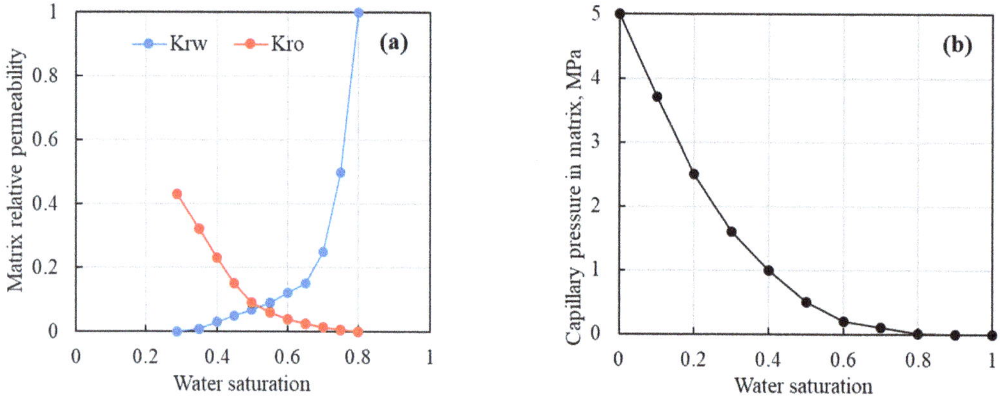

Figure 6. Matrix oil–water phase permeability (**a**) and capillary pressure curves (**b**).

Table 4. Reservoir geological parameters.

Parameters	Value	Parameters	Value
Model size (m)	1000 × 1000 × 30	Rock compression properties (MPa^{-1})	6.0×10^{-6}
Surface crude oil density ($g \cdot cm^{-3}$)	0.86	Formation water compressibility (MPa^{-1})	4.42×10^{-6}
Surface water density ($g \cdot cm^{-3}$)	1.00	Crude oil compressibility (MPa^{-1})	3.60×10^{-5}
Surface crude oil viscosity (mPa·s)	21.10	Original gas oil ratio	27.40
Formation water viscosity (mPa·s)	0.46	Formation temperature (°C)	112.00
Formation crude oil volume factor	1.30	Regional formation pressure (MPa)	30.00
Matrix porosity (%)	12	Formation pressure coefficient	1.10
Matrix permeability (mD)	5	Reservoir vertical depth (m)	2900.00

To verify the model's reliability, the calculation results of this model and CMG software results were compared under the same production and injection conditions. As shown in Figure 7, the errors in the average formation pressure and cumulative production after 3 years of well group production are within 5%, indicating that the model has good reliability.

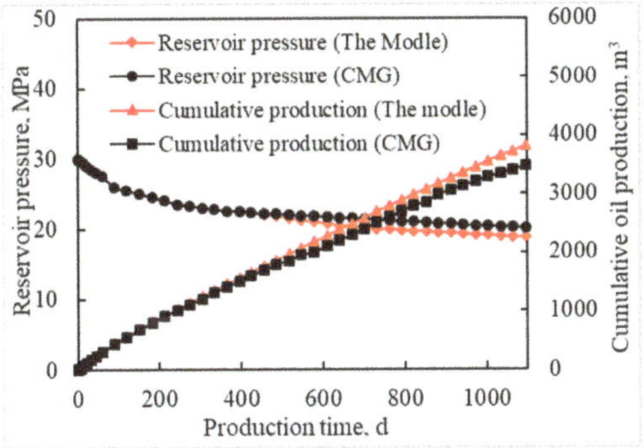

Figure 7. The calculation results of the model and CMG.

3.2.1. The Effect of Fracturing

Conventional water injection was performed in well W5 at an injection rate of 30 m^3/d. The production well was simulated under the conditions of no fracturing and fractures caused by fracturing, and the effect of fracturing on the production of the well group was analyzed. In Figure 5, point 1 represents the matrix and point 2 represents the fracture. The evolution process of reservoir physical properties in the production process is studied by analyzing the change rules of porosity and permeability at the two points. As shown in Figure 8, the stress sensitivity effect has a great impact on fracture permeability in the production process. The porosity and permeability of the reservoir matrix and fracture porosity have little change.

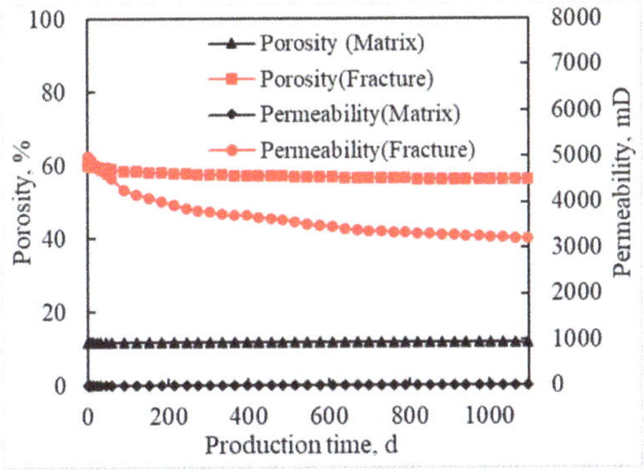

Figure 8. Evolution process of porosity and permeability.

As shown in Figure 9, using well W1 as an example, under the condition of no fracturing, the single-well productivity is low, and the initial daily oil production is only 1.58 m. After fracturing, the initial daily production of a single well increased to 8.02 m, and the multiplication ratio was 5.1. The 3-year cumulative production increased by 2066 m^3, which shows good fracturing stimulation potential.

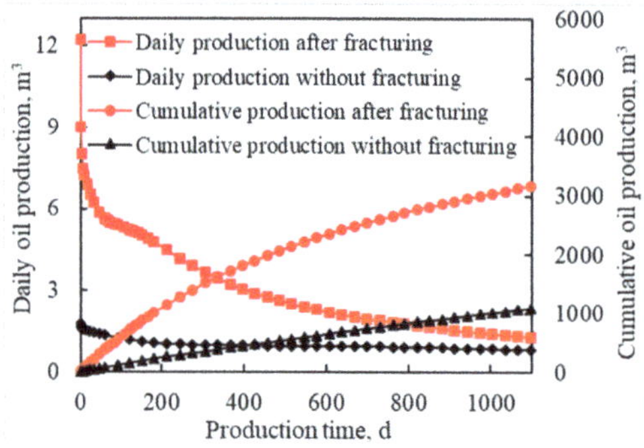

Figure 9. Production of well W1 with and without fracturing.

3.2.2. The Effect of Fracture Orientation

Well W1 was located in the minimum horizontal principal stress direction of the wate injection well (W5), and well W2 was located in the maximum horizontal principal stres direction of W5. The fracture orientations of the two wells were inconsistent with th effective direction of water injection, but the other fracture parameters were the same therefore, the production conditions were compared. As shown in Figure 10, the fractur direction of well W1 was perpendicular to the line connecting the oil injection wells. I the pressure flooding process, the water injection front advanced evenly, the water cu increased more slowly, and the initial production declined more slowly. Water injectio lasted longer. The direction of the fracture in well W2 was parallel to the line connecting th oil injection well. During the water injection process, the water flooding front protrude more considerably along the dominant fracture channel, resulting in a faster rise in th water cut and rapid decline in the production of the well. Therefore, the effect of fractur orientation on the production effect of production wells should be fully considered in th process of reservoir development and well deployment.

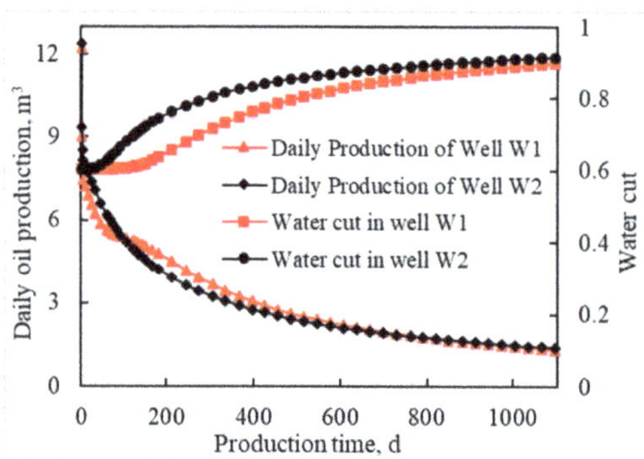

Figure 10. Production situation under different fracture orientation conditions.

3.2.3. The Effect of Oil Displacement Agent

Taking well W1 as an example, the production process of conventional water injectic and the effect of the oil displacement agent addition were simulated. The mass conce tration of the oil displacement agent was 0.5%, and the influence of the oil displaceme agent on the production effect during the water injection process was analyzed. As show in Figure 11, the production trend in the early stages of the two cases is the same; t single-well production decreases and the water cut increases. After 400 d of productic the oil-displacing agent reached the vicinity of the production well, the oil–water mobili ratio increased, and the water-cut rise rate slowed down. The daily oil production was stable at approximately 3 m^3, and the cumulative oil increase was 643.7 m^3 in 3 years.

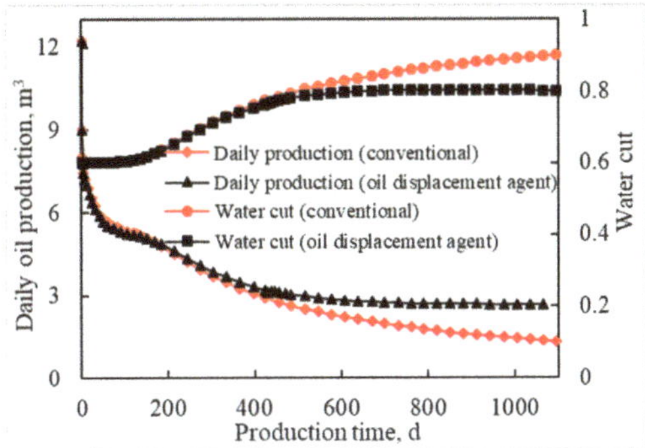

Figure 11. Influence of oil displacement agent on production efficiency.

3.2.4. Comparison between Fracturing Flooding and Conventional Fracturing

According to the fracturing flooding technology of low-permeability sandstone reservoirs in the Shengli oilfield, water was rapidly injected into the formation of well W5 at an injection rate of 0.5 m^3/min (the mass concentration of the oil displacement agent was 0.5%) to supplement the formation energy. After 2 months, a total of 43,200 m^3 was injected. Conventional water injection with an injection rate of 30 m^3/d was subsequently used, and the production well was opened for production. A comparison between the fracturing flooding process and conventional fracturing production is shown in Figure 12. The daily oil production of conventional fracturing was 8.02 m^3, and the cumulative oil production after 3 years was 3793 m^3. After the completion of fracturing flooding, the formation energy was sufficient, and the initial daily oil production was 27.3 m^3. However, the production decreased rapidly in the early stages. After 180 d of fracturing flooding, the production dropped to the same level as that of conventional fracturing (4.2 m^3). In other words, the period of fracturing flooding was valid for 6 months, and after 3 years, the cumulative oil increase was 970 m^3, and the effect of fracturing flooding was obvious.

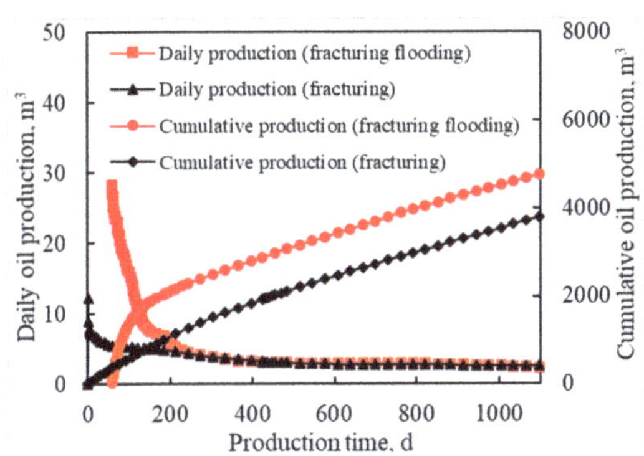

Figure 12. Comparison between fracturing flooding and conventional fracturing.

3.3. Fracture Morphology Optimization

Taking wells W1 and W2 located in different horizontal principal stress directions of injection well W5 as an example, under the condition of constant production pressure difference, the 3-year cumulative oil production was used as the index to optimize the fracture length and conductivity. As shown in Figure 13, with an increase in the fracture half-length, the 3-year cumulative oil production of well W1 continued to increase, but the rate of increase continued to decrease. Based on the increase in production, the fracture half-length of well W1 was optimized to 90 m. The production of well W2 was more significantly affected by the fracture length. When the fracture half-length was less than 80 m, the cumulative oil production increased with the fracture length. When the fracture half-length was greater than 80 m, with a further increase in the fracture length, water channeling occurred sooner in the production well owing to the influence of the water injection front, which significantly affected the production. Therefore, the optimal fracture half-length of well W2 is 80 m.

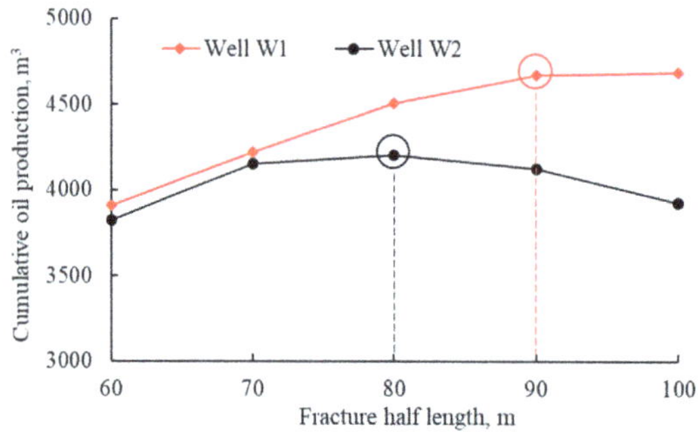

Figure 13. Fracture half-length optimization results.

Based on the optimal fracture length, the 3-year cumulative oil production under a conductivity of 15–35 D·cm was simulated. As shown in Figure 14, with an increase in conductivity, the cumulative production increases continuously but the rate of increase gradually decreases. Considering the economic benefits, the optimized fracture conductivities of wells W1 and W2 were 20 and 25 D·cm, respectively.

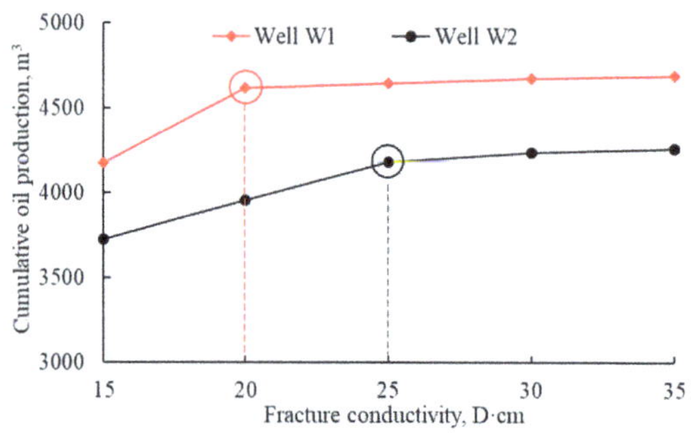

Figure 14. Fracture conductivity optimization results.

4. Field Application

Taking the low-permeability sandstone reservoir fracturing flooding well group (C well group) in the Shengli oilfield as an example, field implementation was carried out based on previous research results. According to the logging data of the well, the corresponding reservoir parameters were sorted, and a small layer with a higher brittleness index was selected as the perforation location. As shown in Figure 15, the red circle represents the preferred perforation position. According to the optimization results of the fracturing flooding numerical simulation, the half-length of the fractures in wells W1 and W4, located in the direction of the minimum horizontal principal stress of the injection well (W5) was designed to be 90 m, and their conductivity was 20 D·cm. The half-length of the fractures in wells W2 and W3, located in the direction of the maximum horizontal princip stress of the water injection well, was designed to be 80 m, and the conductivity was 25 D·cm. The construction parameters were optimized based on the fracture parameters and implemented on-site.

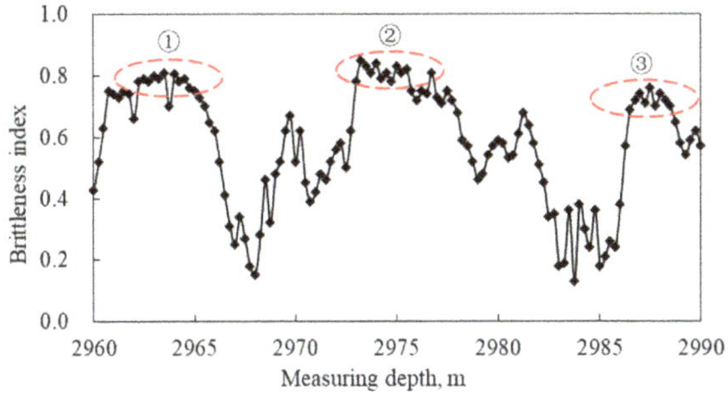

Figure 15. Perforation location optimization results. ①, ② and ③ represent three perforation positions respectively.

Considering well W1 as an example for post-fracturing analyses, well W6 is an adjacent well in the same layer as well W1, and well W6 adopts conventional methods for fracturing design and implementation. The comparison parameters are listed in Table 5. As shown in Figure 16, because the sandstone interval with a higher brittleness index is more likely to initiate fracture, the fracture pressure of well W1 at the preferred perforation location was lower, and the pumping pressure was reduced by 5–10 MPa, which effectively reduced the construction risk. According to the net pressure fitting results, the actual half-fracture length of well W1 was 87.4 m, the conductivity was 20.5 D·cm, and the errors in the design parameters were less than 3%.

Table 5. Construction parameter comparison results.

Well Name	Fracture Pressure (MPa)	Pumping Pressure (MPa)	Design Half-Fracture Length (m)	Actual Half-Fracture Length (m)	Design Conductivity (D·cm)	Actual Conductivity (D·cm)	Daily Oil Production (m^3)
W1	31.2	36~40	90	87.4	20	21.4	12.5
W6	38.5	41~50	120	116.3	30	32.8	6.9

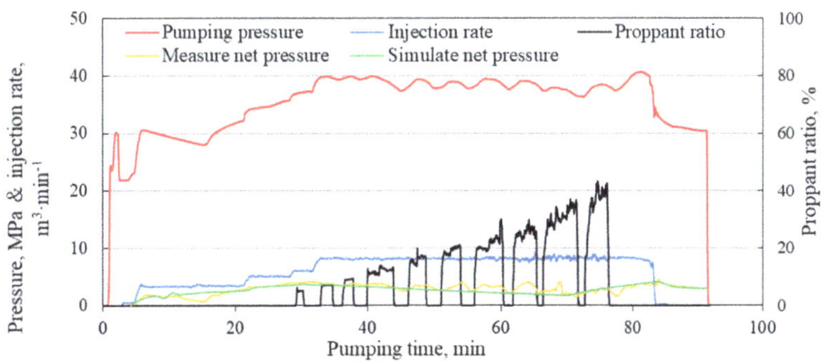

Figure 16. Construction data fitting curve.

By comparing the two wells under different fracturing parameters (Figure 17), it can be seen that the initial daily oil production of well W1 was 12.5 m^3, which is 5.6 m^3 higher than that of well W6, and the production of well W1 was stable for a longer time. This series of technologies were adopted by Shengli oilfield and applied to more than

20 low-permeability sandstone production wells in the Bohai Bay basin, and the average daily oil production per well increased by 3.5 m^3, which greatly improved the fracturing stimulation effect.

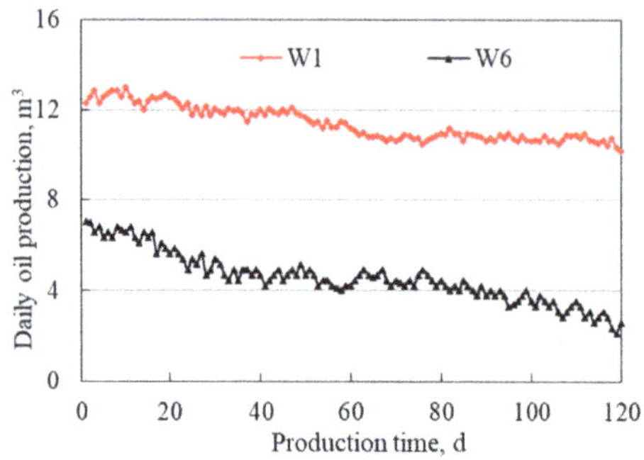

Figure 17. Daily oil production curve.

5. Conclusions

(1) Feldspar, quartz, carbonate, and clay are the main minerals of the low-permeability sandstone in the Bohai Bay basin, with contents of 41%, 34%, 9%, and 16%, respectively. The average Young's modulus and Poisson's ratio of the rocks in the fourth membe of Shahejie are 14,856 MPa and 0.254, respectively.

(2) From the three aspects of rock mineral composition, mechanical parameters, an stress–strain characteristics, it is concluded that the main controlling factors affecting the complexity of rock fractures mainly include quartz content, Young's modulu and peak stress. The brittleness index was evaluated, and the weight coefficients o the influencing factors were calculated as 0.303, 0.358, and 0.339, respectively.

(3) For production wells, in the direction of minimum horizontal principal stress, th swept area of water flooding should be expanded, the fracture length should b optimized to 90 m, and the fracture conductivity should be 20 D·cm. To fractur production wells in the direction of the maximum horizontal principal stress, th advancing speed of the water injection front should be slowed down to reduce th risk of water channeling in injection-production wells. The optimized fracture lengt was 80 m, and the fracture conductivity was 25 D·cm.

(4) The fracturing perforation location was optimized by analyzing the brittleness inde of the target interval of the reservoir, and the fracture parameters were optimized b numerical simulation of fracturing flooding. This series of technologies was adopte by the Shengli oilfield and applied to more than 20 low-permeability sandstone pr duction wells in the Bohai Bay basin, which increased the average daily oil productic per well by 3.5 m^3, prolonged the stable production time, and enhanced the fracturir stimulation effect to a certain extent.

Author Contributions: C.L., conceptualization, methodology, supervision, funding acquisitic writing—review and editing; L.M., methodology, validation, writing—original draft, investigatic writing—review and editing; J.G., supervision, funding acquisition; L.Z., supervision, fundi acquisition; S.X., methodology, validation; B.C., supervision, funding acquisition; Y.Z., supervisic funding acquisition; H.Y., supervision, funding acquisition; Z.T., supervision, funding acquisitic All authors have read and agreed to the published version of the manuscript.

Funding: This research was funded by the National Natural Science Foundation of China (gra number 52022087), National Natural Science Foundation of China Joint Fund Project Key Supp Project (grant number U21A20105), and the major project of CNOOC Beijing Research Institute: Stu on multi-scale and multi-dimensional Optimization Technology of geoengineering integration offshore low permeability reservoir (CCL2022RCPS0361KQN).

Institutional Review Board Statement: Not applicable.

Informed Consent Statement: Not applicable.

Data Availability Statement: Data are available in the article.

Acknowledgments: All individuals included in this section have consented to the acknowledgment.

Conflicts of Interest: The authors declare no conflict of interest.

Nomenclature

D	the fractal dimension of the core end-face fracture
R	the side length of the square boxes (cm)
N	the number of square boxes containing fractures
C_p	the proportionality constant
F_c	the fracturing coefficient
α	the fracture angle (°)
B_{ls}	the dimensionless low-permeability sandstone brittleness index
Q_n	the normalized quartz content
E_n	the normalized Young's modulus
σ_{pn}	the normalized peak stress
W_1, W_2, W_3	the weight coefficients of each dimensionless parameter
E	Young's modulus (MPa)
a	the unit conversion factor
ρ	the rock density (g/m^3)
t_s	the shear wave log data (μs/m)
t_p	the compression wave log data (μs/m)
$\varphi_{\rm m}$	the dimensionless reservoir matrix porosity
$\rho\kappa$	the density of fluid κ (g/cm^3)
κ	oil or water
$S_{\kappa,\rm m}$	the saturation of fluid κ in the reservoir matrix
$q^{\rm w}_{\kappa,\rm m}$	the flow exchange term for fluid κ between the wellbore and reservoir matrix (m^3/d)
$q_{\kappa,\rm fm}$	the flow exchange term for fluid κ between the wellbore and reservoir matrix—that is, the flow of fluid κ from the fracture to the reservoir matrix (m^3/d)
$k_{\rm m}$	the absolute permeability of the reservoir matrix (mD)
$k_{r\kappa}$	the dimensionless relative permeability of fluid κ
μ_κ	the viscosity of fluid κ (mPa·s)
$p_{\kappa,\rm m}$	the pressure of fluid κ in the reservoir matrix (MPa)
t	is the production time (s)
$\varphi_{\rm f}$	the dimensionless fracture porosity
$S_{\kappa,\rm f}$	the saturation of fluid κ in the fracture (dimensionless)
$q^{\rm w}_{\kappa,\rm f}$	the flowrate of fluid κ from the production well to the fracture (m^3/d)
$q_{\kappa,\rm mf}$	the flowrate of fluid κ from the reservoir matrix to the fracture (m^3/d)
$k_{\rm f}$	the absolute permeability of the fracture (mD)
$p_{\kappa,\rm f}$	the pressure of fluid κ in the fracture (MPa)
$\varphi_{\rm m0}$	the dimensionless initial reservoir matrix porosity
C	the dimensionless matrix porosity compressibility
$p_{\rm m\it t}$	the matrix pore pressure (MPa)
p_0	the initial reservoir pressure (MPa)
$\varphi_{\rm f0}$	the dimensionless initial fracture porosity
$C_{\rm f}$	the dimensionless fracture porosity compressibility
$p_{\rm ft}$	the fracture pressure (MPa)
$k_{\rm m0}$	the initial permeability of the reservoir matrix (mD)
d	the dimensionless matrix permeability stress sensitivity coefficient
$k_{\rm f0}$	the initial permeability of the fracture (mD)
$d_{\rm f}$	the dimensionless fracture permeability stress sensitivity coefficient
$p_{\rm of}$	the oil pressure in the fracture (MPa)

p_{wf}	the water pressure in the fracture (MPa)
p_{om}	the oil pressure in the matrix (MPa)
p_{wm}	the water pressure in the matrix (MPa)
p_c	the capillary pressure in the matrix (MPa)
C_s	the volume fraction of the oil-displacing agent in the water phase (%)
ρ_s	the density of the oil-displacing agent (kg/m^3)
μ_{ws}	the solution viscosity of the oil-displacing agent and water mixture (mPa·s)
D_{ws}	the dimensionless diffusion coefficient of the oil-displacing agent in the water phase
Q_a	the adsorption term of the oil-displacing agent in the pore matrix
Q_s	the injection source term of the oil-displacing agent
WI_f	the well index in the fracture grid (m^3/d·MPa)
p_f	the pressure in the fracture (MPa)
p_w	the bottom-hole flow pressure (MPa)
WI_m	the well index in the matrix grid (m^3/d·MPa)
p_m	the pressure in the matrix (MPa)
$q^{w}_{w,m}$	the water flow exchange term between the wellbore and reservoir matrix (m^3/d)
C^{w}_{s}	a constant that represents the volume fraction of the oil-displacing agent at the bottom of the injection well (%)

References

1. Wu, Q.; Xu, Y.; Wang, X.Q.; Wang, T.F.; Zhang, S.L. Volume fracturing technology of unconventional reservoirs: Connotation, optimization design and implementation. *Pet. Explor. Dev.* **2012**, *39*, 352–358. [CrossRef]
2. Xu, Y.; Lei, Q.; Chen, M.; Wu, Q.; Yang, N.Y.; Weng, D.W.; Li, D.Q.; Jiang, H. Progress and development of volume stimulation techniques. *Pet. Explor. Dev.* **2018**, *45*, 874–887. [CrossRef]
3. Lei, Q.; Guan, B.S.; Cai, B.; Wang, X.; Xu, Y.; Tong, Z.; Wang, H.Y.; Fu, H.F.; Liu, Z.; Wang, Z. Technological progress and prospects of reservoir stimulation. *Pet. Explor. Dev.* **2019**, *46*, 580–587. [CrossRef]
4. Ding, B.; Xiong, C.M.; Geng, X.F.; Guan, B.S.; Pan, J.J.; Xu, J.G.; Dong, J.F.; Zhang, C.M. Characteristics and EOR mechanisms of nanofluids permeation flooding for low-permeability oil. *Pet. Explor. Dev.* **2020**, *47*, 756–764. [CrossRef]
5. He, J.G.; Wang, H.W. Design and Effect of Fracture-flooding in Class III Oil Reservoirs. *J. Southwest Pet. Univ. Sci. Technol. Ed* **2018**, *40*, 95–104.
6. Xu, B. *Study on Mechanism and Mathematical Model of Fracturing and Percolation Enhanced Oil Recovery in Medium and Low Permeability Reservoirs: An Example in the Xingbei Development Zone of Daqing Oilfield*; Northeast Petroleum University: Daqing, China, 2019.
7. Wang, J.L.; Liu, Y.Z.; Chen, M.Q.; Liu, L.; Gao, J. Experimental study on dynamic imbibition mechanism of low permeability reservoirs. *Pet. Explor. Dev.* **2009**, *36*, 86–90.
8. Liu, H.L.; Xiong, W.; Gao, S.S.; Xue, H.; Hu, Z.M. Physical Modeling of Water-Oil Displacement in Hydraulic Fractured Horizontal Wells. *Xinjiang Pet. Geol.* **2010**, *31*, 171–173.
9. Zhang, H.N.; Chen, J.T. Insights into Energy Storage Bulk Fracturing Technology for Low-permeability Oilfields—A Case Study of Peripheral Wellblock of Jilin Oilfield. *Unconv. Oil Gas* **2015**, *2*, 55–60.
10. Wu, Z.B.; Li, L.; Yan, Y.Q. New development pattern of network fracturing and imbibition oil recovery for super-low permeability oil reservoirs. *Fault-Block Oil Gas Field* **2019**, *26*, 491–494.
11. Shi, Y.H. *Subject: Study on Fracturing Fluid's Imbibition of Chang-7 Source Based on Nuclear Magnetic Resonance*; Xi'an Shiyou University: Xi'an, China, 2018.
12. Xu, Y.H.; Meng, Y.; Shi, S.G. Design Method of Water Injection Rate for High-Pressure Water Injection in Low Permeability Reservoirs. *Inn. Mong. Petrochem. Ind.* **2021**, *47*, 118–121.
13. Lu, C.; Ma, L.; Zhang, T.; Guo, J.C.; Li, M.; Huang, B. A novel hydraulic fracturing method and case study based on proppant settlement transport model. In Proceedings of the 53rd ARMA Rock Mechanics/Geomechanics Symposium, New York, NY, USA, 23–26 June 2019.
14. Tarasov, B.; Potvin, Y. Universal criteria for rock brittleness estimation under triaxial compression. *Int. J. Rock Mech. Min. Sc* **2013**, *59*, 57–69. [CrossRef]
15. Lu, C.; Ma, L.; Li, Z.L.; Huang, F.L.; Huang, C.H.; Yuan, H.R.; Tang, Z.B.; Guo, J.C. A Novel Hydraulic Fracturing Method Based on the Coupled CFD-DEM Numerical Simulation Study. *Appl. Sci.* **2020**, *10*, 3027. [CrossRef]
16. Sun, F.Q.; Du, S.H.; Zhao, Y.P. Fluctuation of fracturing curves indicates in-situ brittleness and reservoir fracturing characteristic in unconventional energy exploitation. *Energy* **2022**, *252*, 124043. [CrossRef]
17. Hucka, V.; Das, B. Brittleness determination of rocks by different methods. *Int. J. Rock Mech. Min. Sci. Geomech. Abstr.* **197** *11*, 389–392. [CrossRef]
18. Jarvie, D.M.; Hill, R.J.; Ruble, T.E. Unconventional shale-gas systems: The Mississippian Barnett Shale of north-central Texas a one model for thermogenic shale-gas assessment. *AAPG Bull.* **2007**, *91*, 475–499. [CrossRef]

Rickman, R.; Mullen, M.J.; Petre, J.E.; Grieser, W.V.; Kundert, D. A practical use of shale petrophysics for stimulation design optimization: All shale plays are not clones of the Barnett Shale. In Proceedings of the SPE Annual Technical Conference and Exhibition, Denver, CO, USA, 21–24 September 2008.
Liu, Z.S.; Sun, Z.D. New brittleness indexes and their application in shale/clay gas reservoir prediction. *Pet. Explor. Dev.* **2015**, *42*, 117–124. [CrossRef]
Quinn, J.B.; Quinn, D.G. Indentation brittleness of ceramics: A fresh approach. *J. Mater. Sci.* **1997**, *32*, 4331–4346. [CrossRef]
Xiao, Y.; Guo, J.C. Optimization of Fracturing Parameters for Low Permeability and Heterogeneous Reservoir Production well. *Sci. Technol. Eng.* **2012**, *12*, 9019–9022.
Xu, C.C.; Chen, C.H.; Wang, B.; Wu, Y.H.; Peng, C.J. Fracture parameter optimization of network fracturing for horizontal well in low permeability and low-permeability oil reservoir. *Fault-Block Oil Gas Field* **2014**, *21*, 823–827.
Lei, Q.; Weng, D.W.; Guan, B.S.; Mu, L.J.; Xu, Y.; Wang, Z.; Guo, Y.; Li, S. A novel approach of stimulation based on fracture controlling optimization and design. *Pet. Explor. Dev.* **2020**, *47*, 1–8.
Harimi, B.; Masihi, M.; Mirzaei-Paiaman, A.; Hamidpour, E. Experimental study of dynamic imbibition during water flooding of naturally fractured reservoirs. *J. Pet. Sci. Eng.* **2018**, *174*, 1–13. [CrossRef]
Lucas, M.; Mohsen, T.; Ke, X.; Miguel, M.; Du, Y.J.; Matthew, B. Surfactant flooding in oil-wet micromodels with high permeability fractures. *Fuel* **2019**, *241*, 1117–1128.
Lu, C.; Ma, L.; Guo, J.C.; Li, X.Y.; Zheng, Y.C.; Ren, Y.; Yin, C.B.; Li, J.F.; Zhou, G.Q.; Wang, J.D.; et al. Novel Method and Case Study of a Deep Shale Fracability Evaluation Based on the Brittleness Index. *Energy Explor. Exploit.* **2021**, *40*, 1–18. [CrossRef]
He, T. *Research on Compressibility Evaluation of Low-Permeability Sandstone Reservoirs in the Second Member of Kong 2 Member in Dagang Oilfield*; Southwest Petroleum University: Chengdu, China, 2016.
Xu, S.Q.; Ren, G.T.; Younis, R.M.; Feng, Q.H. Revisiting field estimates for carbon dioxide storage in depleted shale gas reservoirs: The role of geomechanics. *Int. J. Greenh. Gas Control* **2021**, *105*, 103222. [CrossRef]
Wang, J.; Liu, H.Q.; Xia, J.; Liu, Y.T.; Hong, C.; Meng, Q.B.; Gao, Y. Mechanism simulation of oil displacement by imbibition in fractured reservoirs. *Pet. Explor. Dev.* **2017**, *44*, 761–770. [CrossRef]
Xu, S.Q.; Guo, J.C.; Feng, Q.H.; Ren, G.T.; Li, Y.Y.; Wang, S. Optimization of hydraulic fracturing treatment parameters to maximize economic benefit in tight oil. *Fuel* **2022**, *329*, 125329. [CrossRef]

processes

Article

The Development and Application of Novel Water-Based Drilling Fluid for Complex Pressure System Formation in the South China Sea

Hexing Liu [1], Yi Huang [1], Zhiqin Liu [1], Lifei Dong [2,*] and Miao Wang [2]

[1] CNOOC China Limited, Zhanjiang Branch, Zhanjiang 524057, China
[2] School of Civil Engineering, Chongqing Three Gorges University, Chongqing 404120, China
* Correspondence: 20180018@sanxiau.edu.cn

Abstract: The Y oilfield reservoir is characterized by ultra-depleted reservoir and multiple pressu systems in each well section. If conventional PLUS/KCL drilling fluid is used in directional drilli the following problems are found: the rheological stability of drilling fluid is not sufficient, resulti in increased viscosity and thickening after aging, and the single plugging material may cause lo circulation when the drilling differential pressure is more than 26 MPa, resulting in poor reserv protection effects. To solve the above problems, based on the conventional PLUS/KCL drilling flu formula, a PLUS/KCL drilling fluid system suitable for directional drilling in multi-pressure syste was formed by optimizing the addition of a treating agent to improve the rheological stability the drilling fluid, optimizing the plugging agent and compound combination to improve pluggi ability, and optimizing the particle size distribution of a temporary plugging agent to improve reservoir protection effect. The laboratory test evaluation showed that the optimized PLUS/K drilling fluid had fine plugging pressure capacity, and the intrusion depth was only 4.1 cm on natural core at 120 °C/35 MPa/12 h, effectively reducing the risk of loss-circulation. The permeabil recovery rate of the core after cutting off the polluted end was more than 90%, indicating that reservoir protection effect was good. The drilling fluid performance was stable, and the cuttir rolling recovery rate was over 90%. Field application showed that the optimized PLUS/KCL drilli fluid was used without any loss-circulation or wellbore instability, and the production of all we was over-matched, effectively solving the problem of ensuring drilling safety and reducing reserv damage under the differential pressure of multiple pressure systems.

Keywords: complex pressure system; water-based drilling fluid; loss-circulation; reservoir protecti pressure bearing capacity

Citation: Liu, H.; Huang, Y.; Liu, Z.; Dong, L.; Wang, M. The Development and Application of Novel Water-Based Drilling Fluid for Complex Pressure System Formation in the South China Sea. *Processes* **2023**, *11*, 1323. https://doi.org/10.3390/pr11051323

Academic Editors: Tao Zhang, Zheng Sun, Dong Feng, Hung Vo Thanh and Wen Zhao

Received: 20 March 2023
Revised: 18 April 2023
Accepted: 19 April 2023
Published: 25 April 2023

1. Introduction

For various reasons, the Y oilfield reservoir is characterized by ultra-depleted reserv and multiple pressure systems existing in each well interval. It is estimated that the forr tion pressure coefficient is 0.67–1.33, and the maximum drilling differential pressure c reach 35 MPa. Therefore, safe drilling and reservoir protection face great challenges [1 Conventional PLUS/KCL drilling fluid was used in the five directional wells in the ea stage of the field. Two of the directional wells were drilled into the normal pressu reservoir, and the operation process was smooth. The other three directional wells encountered the multi-pressure system reservoir, the drilling differential pressure reach 26 MPa, the reservoir protection effect was poor, and the production after perforation w lower than the production allocation. Furthermore, the conventional PLUS/KCL drilli fluid system became viscous after aging in all five directional wells, resulting in a hi density of equivalent circulating drilling fluid at the bottom hole and a high risk of lo circulation. In view of the above problems, domestic and foreign scholars have carri out relevant research. Some domestic scholars have improved the temporary pluggi technology of protective reservoirs [3–5], especially in terms of material type, particle size distribution. and proportion optimization of temporary plugging agents [6–9], which strengthened the temporary plugging performance of the system, improved the bearing capacity, and successfully solved the problem of easy loss-circulation of low-pressure reservoir. Domestic scholars chose the plugging agent and its compound combination that matched the formation of the reservoir segment to achieve a good plugging effect, form a high-quality filter cake around the shaft wall, improve the pressure bearing capacity of the formation, and reduce the risk of loss-circulation [10–16]. Many foreign drilling fluid companies have developed high-performance water-based drilling fluids that can replace the oil base, such as the ULTRADRIL system of the M-I company, the DYDRO-GUADR system of HALI-BUR-TON [17,18], and the BAKER HUGHES's PERFORMAX system [19,20]. Foreign studies on high temperature-resistant drilling fluids mainly include formate-free solid phase-resistant drilling fluids [21], bentonite-free high temperature-resistant drilling fluids [22], and high temperature-resistant silicate drilling fluids [23]. Nanotechnology has been applied to drilling fluid abroad [24,25]. The above research provides a reference for solving the drilling fluid optimization problem of multiple pressure systems. However, the drilling fluid optimization problem of the Y oilfield cannot simply copy the previous experience. Solutions should be found according to the reservoir characteristics of the Y oilfield, such as porosity, permeability, mineral composition, and content, while taking into account the economic benefits. The above research provides a reference for solving the drilling fluid optimization problem of multiple pressure systems.

2. PLUS/KCL Drilling Fluid Formula Matching Test

Y oilfield reservoir mainly includes deltaic deposits with strong heterogeneity of pore structure and inter-layers. The average porosity of the main oil group ranges from 14.3% to 18.8%, and the average permeability ranges from 7.6 to 78.3 mD. It is categorized as a medium-low porosity and low permeability reservoir, requiring a wide range distribution of temporary blocking agent particle sizes to effectively protect the reservoir [6,7]. Through quantitative analysis of the total amount of clay minerals and common non-clay minerals in the reservoir, it can be seen that the content of clay minerals in the shale is about 40%, the content of illite/smectite formation is about 20%, and the content of kaolinite is about 15%, which has certain expansion properties and enhances the inhibition of the drilling fluid system. Furthermore, in response to the problems of the existing PLUS/KCL drilling fluid in the Y oilfield, namely, that it cannot meet the safety requirements of boreholes under significant differential pressure, poor reservoir protection effects, and viscosity increases and thickening after aging, the plugging pressure-bearing performance is improved by optimizing the compounding combination of plugging agent. Furthermore, the rheological stability is improved by optimizing treating agent addition, and the reservoir protection performance of drilling fluid is improved by optimizing the particle size distribution of temporary plugging agent so that it can meet drilling requirement.

2.1. Matching Test of Particle Size Distribution of Temporary Plugging Agent

The success of temporary plugging technology for reservoir protection relies on the rapid development of a tight filter cake on the borehole to prevent filtrate and the solid phase in the drilling fluid from penetrating into the reservoir under high differential pressure and to easily remove the filter cake before production [3–5]. Therefore, this research sought to optimize the particle size distribution of the temporary plugging agent according to the D90 rule, that is, when the d_{90} value of the cumulative distribution curve of the particle size of the temporary plugging agent is equal to the maximum pore throat diameter of the reservoir section, an ideal temporary plugging effect can be obtained [8,9].

(1) Select a core from the Y oilfield and cut it into thin sections; then analyze the particle size distribution of the pore throat to obtain the maximum pore throat diameter—108.36 μm of the reservoir;

(2) Select ultra-fine calcium carbonate with high acid solubility, good compatibility with the reservoir, wide distribution, and a low price as the temporary plugging agent;
(3) According to maximum pore throat diameter of the reservoir, apply the reservoir damage assessment and protection technology system of a complex structure well to optimize the particle size distribution of ultra-fine calcium carbonate. The mass ratio of the selected 800 mesh and 400 mesh calcium carbonate was 3.35:1.

2.2. Matching Test of Plugging Agent Combination

Plugging agents and compound combinations that matched the reservoir section formation in the Y oilfield that could yield good plugging effects and a high-quality filter cake around the borehole were selected. Therefore, the formation pressure-bearing capacity could be improved, and thus the risk of loss-circulation could be reduced [10–16]. Plugging agents PF-QWY (calcium carbonate), PF-GBL (white asphalt), PF-LSF (resin), PF-LPF H (film-forming agent), PF-DYFT (sulfonated asphalt), and PF-ZP (filtration loss agent) were commonly used plugging agents in the Y oilfield, with high acid solubility, good compatibility with the formation, and good economical and applicable features. Especially with plugging agents PF-QWY (calcium carbonate) and PF-GBL (white asphalt), the acid solubility rate could go up to 90% or more directly as a basic pressure-bearing material. For acid solubility, a rate of less than 40% of PF-ZP (zero filtration loss agent) would not be considered. Therefore, it was decided to add PF-LSF (resin), PF-LPF H (film-forming agent), and PF-DYFT (sulfonated asphalt) to the base formulation after compounding—2% seawater bentonite slurry + 0.25% Na_2CO_3 (sodium carbonate) + 0.2% NaOH (sodium hydroxide) + 0.6% PF-PAC-LV (filtration loss reducer) + 3% PF-GJC (polymeric alcohol) + 0.4% PF-PLUS (coating agent) + 0.15% PF-XC (viscosifier) + 2% PF-GBL (white bitumen) + 5% KCl + 3.5% PF-QWY (calcium carbonate, 800 mesh and 400 mesh calcium carbonate in a mass ratio of 3.35:1) + barite, to determine the rheological properties, filtration loss, and plugging properties of drilling fluids under different compounding combinations (Table 1) and pressure-bearing properties (Table 2) to determine the optimal combination of plugging agents.

Table 1. Rheological properties, filtration loss, and depth of sand bed intrusion of drilling fluids under different combination of plugging agents.

Plugging Agent	Status	AV (mPa·s)	PV (mPa·s)	YP (Pa)	Φ_6/Φ_3	API Filtration/mL	HTHP Filtration (mL)	Sand Bed Intrusion Depth (cm)
Before matching test	BH	45.5	29	16.5	8/6	3.2	7.0	
PLUS/KCl	AH	77	50	27	14/10	2.6	6.2	3.3
1.5% PF-DYFT +	BH	57	34	23	9/6	4.2	8.0	
1.5% PF-LSF	AH	86	52	34	15/12	3.8	7.4	4.0
1.5% PF-DYFT +	BH	53	32	21	8.5/6	3.6	7.2	
1.5% PF-LPF-H	AH	84	51	33	15/11	3.1	6.5	3.5
1.5% PF-LSF + 1.5%	BH	46	30	16	7.5/5.5	3.0	6.8	
PF-LPF-H	AH	77	50	27	14/10	2.5	6.0	3.0

Remarks: ① The experimental conditions were aging at 120 °C for 16 h, rheological properties before and afte aging were measured at 50 °C, and HTHP filtration loss was measured at 120 °C. ② The sand particle siz of the sand filling pipe experiment was 40–60 mesh, under room temperature and 0.7 MPa. ③ AV—apparen viscosity, PV—plastic viscosity, YP—yield point, HTHP—high temperature and high pressure. Status: BH—befor hot-rolling, AH—after hot rolling.

The experiment results from Tables 2 and 3 showed that each combination of pluggin agents had a small amount of filtration, a thin filter cake, and a shallow drilling flui intrusion depth into the core. The optimal combination of blocking agent was 1.5% PF LSF (asphalt resin anti-caving agent) + 1.5% PF-LPF-H (hydroxyl film-forming agent Therefore, the PLUS/KCL drilling fluid system was initially determined as follows: 2% seawater bentonite slurry + 0.25% Na_2CO_3 (sodium carbonate) + 0.2% NaOH (sodium hydroxide) + 0.6% PF-PAC- LV (filtration loss reducer) + 3% PF-GJC (polymeric alcohol) 0.4% PF-PLUS (coating agent) + 0.15% PF-XC (viscosifier) + 2% PF-GBL (white asphalt) 5% KCl (potassium chloride) + 3.5% PF-QWY (calcium carbonate, 800 mesh and 400 mes calcium carbonate in a mass ratio of 3.35:1) + 1.5% PF-LSF (asphalt resin anti-caving agent) + 1.5% PF-LPF-H (hydroxyl film-forming agent) + barite.

Table 2. Pressure-bearing capacity of drilling fluids under different combinations of plugging agents.

Type of Base Slurry		1.5% PF-DYFT + 1.5% PF-LSF	1.5% PF-DYFT + 1.5% PF-LPF-H	1.5% PF-LSF + 1.5% PF-LPF-H
Core Number		Core G138	Core G135	Core G137
Pressure (MPa)	Pressure Time (h)	Filtration Loss (mL)		
5	0.5	3.8	3.0	2.6
10	0.5	4.2	4.2	3.8
15	0.5	5.0	4.6	4.4
20	0.5	5.8	5.6	5.0
25	12.0	12.4	9.2	9.0
Filter Cake (mm)		5.0	3.5	3.0
Intrusion Depth of Drilling Fluid (cm)		5.6	4.5	3.8

Table 3. Parameters of test core.

Core Parameters			
Core Number	Core G138	Core G135	Core G137
Length (cm)	7.70	6.80	6.90
Diameter (cm)	2.50	2.50	2.49
Gas Permeability (10^{-3} μm^2)	70	72	78

2.3. Treating Agent Matching Test

The results summarized in Table 1 showed that the viscosity of all the composite systems and the PLUS/KCl system before the matching test increased after hot rolling, indicating that temperature resistance performance needs to be optimized. Therefore, it was decided to improve the rheological properties of the system by optimizing the addition of treating agents PF-PAC-LV (filtration loss reducer), PF-PLUS (coating agent), and PF-XC (viscosifier) [26]. The test results are shown in Table 4.

Table 4. Effect of dosage variation of treatment agent on rheology.

Formula Number	Experiment Scheme	Status	AV (mPa·s)	PV (mPa·s)	YP/Pa	Φ_6/Φ_3	API Filtration Loss (mL)
1	0.6% PF-PAC-LV + 0.4% PF-PLUS + 0.15% PF-XC	BH	46.0	30	16.0	7.5/5.5	3.0
		AH	77.0	50	27.0	14/10	2.5
2	0.6% PF-PAC-LV + 0.3% PF-PLUS + 0.15% PF-XC	BH	32.0	24.5	7.5	5/3	3.5
		AH	45.0	33	12.0	6/4	3.0
3	0.6% PF-PAC-LV + 0.3% PF-PLUS + 0.10% PF-XC	BH	37.5	26	11.5	5/4	3.0
		AH	67.5	44	23.5	10/7	2.5
4	0.3% PF-PAC-LV + 0.3% PF-PLUS + 0.10% PF-XC	BH	38.0	27	11.0	6/4	3.4
		AH	45.0	30	15.0	7/4	3.0

Remarks: ① The experiment conditions were aging at 120 °C for 16 h, and the rheological properties before and after aging were measured at 50 °C. ② AV—apparent viscosity, PV—plastic viscosity, YP—yield point. Status: BH—before hot-rolling, AH—after hot rolling.

The experimental results showed that the drilling fluid of Formulation 4 was stable before and after aging, with good temperature resistance, and the addition of treating agents PF-PAC-LV (filtration loss reducer), PF-PLUS (coating agent), and PF-XC (viscosifier) were reduced to some extent, thus reducing drilling fluid cost. Therefore, the PLUS/KCL drilling fluid system after the matching test was 2% seawater bentonite slurry + 0.25% Na_2CO_3 (sodium bicarbonate) + 0.2% NaOH (sodium hydroxide) + 0.3% PF-PAC-LV (filtration loss reducer) + 3% PF-GJC (polymeric alcohol) + 0.3% PF-PLUS (coating agent) + 0.10% PF-XC (viscosifier) + 2% PF-GBL (white asphalt) + 5% KCl + 1.5% PF-LSF (asphalt resin anti-caving agent) + 1.5% PF-LPF-H (hydroxyl film-forming agent) + 3.5% PF-QWY (calcium carbonate) + barite. The calcium carbonate used was compounded by 800 mesh and 400 mesh calcium carbonate in a mass ratio of 3.35:1.

3. Performance Evaluation

3.1. Inhibition

A shale cuttings recovery experiment was used to evaluate the inhibition of drilli fluid. The cuttings from different depths of wells A1, A2, A3, and A4 in the Y oilfi were selected for rolling recovery experiments, and the results are shown in Table 5. T experiment results showed that the rolling recovery rates of cuttings from different dep in PLUS/KCL drilling fluid (before and after matching test) were above 90%, indicati that PLUS/KCL drilling fluid had better inhibition before and after the matching test, a the reduction of PF-PAC-LV, PF PLUS, and PF-XC dosages had little effects on the inhibiti performance, which met the operational requirement of strong inhibition of shale hydrati dispersion.

Table 5. Experiment results of rolling recovery of optimized PLUS/KCl drilling fluid.

Well Name	Cuttings Well Depth (m)	Status	Recovery Rate (%)
A1	2512	Before matching test	93.49
		After matching test	91.21
A2	2656	Before matching test	95.54
		After matching test	92.29
A3	3175	Before matching test	94.42
		After matching test	91.25
A4	3266	Before matching test	91.36
		After matching test	90.95

3.2. Temperature Resistance

The performance of optimized PLUS/KCL drilling fluid after aging was tested 120 °C/16 h and 130 °C/16 h. The experiment results are shown in Table 6. The experim tal results showed that the optimized PLUS/KCL drilling fluid at 120/130 °C maintain viscosity, shear force, and API&HTHP filtration loss at stable conditions, indicating that optimized PLUS/KCl drilling fluid had good temperature resistance.

Table 6. Effect of aging temperature on rheological properties of the system.

Aging Temperature (°C)	Status	AV (mPa·s)	PV (mPa·s)	YP/Pa	Φ_6/Φ_3	API Filtration Loss (mL)	HTHP Filtratio Loss (mL)
120	BH	42	30	12	7/6.5	2.0	
	AH	47	33	14	8/7	2.0	5.2
130	BH	44.5	30.5	14	8.5/6	2.6	
	AH	46	31	15	6/4	2.4	5.5

Remarks: ① AV—apparent viscosity, PV—plastic viscosity, YP—yield point, HTHP—high temperature and h pressure. Status: BH—before hot-rolling, AH—after hot rolling.

3.3. Contamination Resistance

The Y oilfield drilling operation required drilling fluid to be resistant to calciu bentonite, and sodium chloride, so the contamination resistance of optimized PLUS/K drilling fluid was tested and evaluated.

(1) Anti-calcium contamination performance

The rheology and filtration loss changes of the optimized PLUS/KCL drilling fl after the addition of different amounts of calcium chloride into the fluid were evaluat The results are shown in Table 7. The experiment results showed that when the optimiz PLUS/KCL drilling fluid was contaminated by calcium chloride, the apparent viscos of the system decreased, and the API filtration loss increased. When the calcium chlori content reached 1.5%, the apparent viscosity of the system decreased slowly, and t dynamic shear force could still reach 8 Pa, but API filtration loss did not change much. T indicated that the optimized PLUS/KCL drilling fluid had a certain ability to resist calci contamination.

Table 7. Optimized PLUS/KCL drilling fluid anti-calcium contamination performance.

Calcium Chloride Addition (%)	Status	AV (mPa·s)	PV (mPa·s)	YP/Pa	Φ_6/Φ_3	API Filtration Loss (mL)
0	BH	42.0	30	12.0	7/6.5	
	AH	47.0	33	14.0	8/7	2.0
0.5	BH	41.0	26	15.0	11/9	
	AH	42.5	39	8.5	5/4	2.8
1.0	BH	39.0	24	15.0	13/10	
	AH	34.0	24	10.0	5/4	3.8
1.5	BH	37.0	23	14.0	14/12	
	AH	30.0	22	8.0	5/3	4.0

Remarks: ① AV—apparent viscosity, PV—plastic viscosity, YP—yield point. Status: BH—before hot-rolling, AH—after hot rolling.

(2) Anti-bentonite contamination performance

The rheology and filtration loss of optimized PLUS/KCL drilling fluid after being contaminated by bentonite at room temperature and pressure were tested. The results are shown in Table 8. The experiment results showed that when the optimized PLUS/KCL drilling fluid was contaminated by bentonite, the apparent viscosity of the system increased, and the API filtration loss of the system decreased, but the variation was not significant, which met drilling requirements, indicating that the optimized PLUS/KCL drilling fluid had a certain ability to resist bentonite contamination.

Table 8. Optimized PLUS/KCL drilling fluid anti-bentonite contamination performance.

Bentonite Addition (%)	Status	AV (mPa·s)	PV (mPa·s)	YP/Pa	Φ_6/Φ_3	API Filtration Loss (mL)
0	BH	42	30	12	7/6.5	
	AH	47	33	14	8/7	2.0
5	BH	40	28	12	10/9	
	AH	46	30	16	6/4	3.2
10	BH	41	28	13	11/10	
	AH	51.5	36	15.5	7/5	3.0
15	BH	45	29	16	13/11	
	AH	52	37	15	7/5	2.8

Remarks: ① AV—apparent viscosity, PV—plastic viscosity, YP—yield point. Status: BH—before hot-rolling, AH—after hot rolling.

(3) Anti-sodium chloride contamination performance

The rheology and filtration loss values of the optimized PLUS/KCL drilling fluid after being contaminated by sodium chloride were tested. The results are shown in Table 9. The experiment results showed that when the optimized PLUS/KCL drilling fluid was contaminated with sodium chloride, the apparent viscosity of the system decreased, and the API filtration loss increased. When the sodium chloride addition reached 15%, the plastic viscosity and yield point of the system both decreased to a certain extent, but the range of variation was not significant, which met the drilling requirements, indicating that the optimized PLUS/KCL drilling fluid had a fair ability to resist sodium chloride contamination.

Table 9. Evaluation of optimized PLUS/KCL drilling fluid anti-sodium chloride contamination performance.

Sodium Chloride Addition (%)	Status	AV (mPa·s)	PV (mPa·s)	YP/Pa	Φ_6/Φ_3	API Filtration Loss (mL)
0	BH	42.0	30	12.0	7.0/6.5	
	AH	47.0	33	14.0	8.0/7.0	2.0
5	BH	36.0	23	13.0	8.5/7.0	
	AH	37.0	28	9.0	5.0/3.0	3.3
10	BH	37.5	24	13.5	9.0/7.5	
	AH	35.0	27	8.0	4.0/3.0	3.4
15	BH	37.5	24	13.5	9.0/7.5	
	AH	36.5	28	8.5	3.0/2.0	3.8

Remarks: ① AV—apparent viscosity, PV—plastic viscosity, YP—yield point. Status: BH—before hot-rolling, AH—after hot rolling.

3.4. Reservoir Protection Performance

A shale core from well A5 in the Y oilfield at a vertical depth of 3350 m was selected to evaluate the reservoir protection performance of PLUS/KCL drilling fluid (before and after matching test) by using the ZDY50-180 core dynamic contamination apparatus with reference to the oil and gas industry standard SY/T6540-2002 [27,28]. Firstly, the oil phase permeability of the core with saturated formation water was tested, and then PLUS/KCL drilling fluid was used to contaminate the core for 2 h at 120 °C and 3.5 MPa. After the core was contaminated by the high differential pressure, the permeability recovery value was tested by cutting off the contaminated end by 0.5 cm to simulate perforation. The results are shown in Table 10. The experiment results showed that the core permeability recovery value of PLUS/KCL drilling fluid before the matching test was below 80%, and the reservoir damage was great. The core permeability recovery rate of the optimized PLUS/KCL drilling fluid was above 90%, and the reservoir protection performance was greatly improved to meet the requirements of reservoir protection.

Table 10. Results of core contamination experiments before and after the PLUS/KCL drilling fluid matching test.

Core Number	Drilling Fluid	Gas Permeability (mD)	Oil Phase Original Permeability (mD)		Permeability Recovery Rate after Cut of Contaminated End (%)
			Before Contamination	After Cut of Contaminated End	
30	Before matching test	44.77	29.85	22.77	76.28
34		53.45	29.82	22.16	74.31
17	After matching test	62.58	31.29	28.52	91.14
15		67.23	42.02	38.08	90.62

3.5. Pressure Sealing Performance

A core with gas permeability of 64.87 mD was selected, and a JLX-2 dynamic plugging tester was used to evaluate the plugging capacity of the optimized PLUS/KCL drilling fluid [29]. At 120 °C, 26 MPa and 35 MPa for 12 h, the optimized PLUS/KCL drilling fluid formed filter cake with thicknesses of 2.0 mm and 3.5 mm on the core, invasion core depths were 2.5 cm and 4.2 cm, and filtration losses were 8.5 mL and 9.6 mL, respectively. The results showed that the optimized PLUS/KCL drilling fluid had a strong plugging capacity and could meet the requirements of directional drilling in the same reservoir with multiple pressure systems.

4. Field Application

The Y oilfield adopted optimized PLUS/KCL drilling fluid for directional drilling on four multi-pressure system intervals. There were three wells in this oilfield wher the conventional PLUS/KCL drilling fluid was applied before the matching test to the reservoir section of the multi-pressure system. The drilling fluid density that maintains the stability of wellbore was high, and the maximum drilling differential pressure during the drilling process was 26–28 MPa. They were caused by insufficient plugging capacity and the loss-circulation rate was about 5–15 m^3/h, and the actual drilling period was 2–3 d longer than the designed drilling period. Moreover, the particle size of the previou PLUS/KCL drilling fluid temporary plugging agent was not suitable for the reservoi in this oilfield, and the reservoir protection effect was poor, with the skin factor usuall reaching more than 15.0. The novel PLUS/KCL drilling fluid optimized the plugging agen combination, enhanced the plugging capacity, optimized the treatment agent addition improved rheological stability, and ensured stable bottom hole ECD, and no loss-circulation occurred when drilling depleted reservoir. After optimizing the particle size distribution o the temporary plugging agent, the reservoir protection performance was promising. Th skin factor of the four wells was −2.0–1.0, and the actual production of all wells exceede the production distribution. The application of the optimized PLUS/KCL drilling flui was illustrated by taking Well S of the Y oilfield as an example.

Well S in the Y oilfield was drilled into the reservoir section by using a φ215.9 mm bit with a maximum well inclination of 58° and an expected bottom pressure coefficient of 0.67–1.33. The optimized PLUS/KCL drilling fluid was used for the operation, and the following maintenance treatment measures were taken to ensure that the performance of the drilling fluid met the requirements of directional drilling in reservoir with multiple pressure systems:

(1) The drilling fluid density was strictly controlled, and the spud density was 1.26 g/cm^3. According to the actual conditions of drilling and shaker return, the drilling fluid density was adjusted to 1.30 g/cm^3 when reaching TD and 1.34 g/cm^3 before tripping to ensure wellbore safety during operation.
(2) During the operation, the concentration of plugging agent in each operation stage was strictly controlled. Before entering the reservoir section, the concentration of PF-LSF (asphalt resin anti-caving agent), PF-GBL (white asphalt), PF-LPF (hydroxyl film forming agent), and PF-QWY (calcium carbonate) were controlled to 40 kg/m^3, 30 kg/m^3, 10 kg/m^3, and 40 kg/m^3, respectively. After entering the reservoir section, the concentration of each plugging agent was adjusted in time, and the concentrations of PF-LSF (asphalt resin anti-caving agent), PF-GBL (white asphalt), PF-LPF (hydroxyl film forming agent), and PF-QWY (calcium carbonate) were controlled to 40 kg/m^3, 30 kg/m^3, 10 kg/m^3, and 60 kg/m^3, respectively.
(3) Lubricants (PF-GRA, PF-LUBE) were added in time according to the field requirements to improve the quality of the filter cake and ensure that the HTHP filtration loss was under control within 4.5–6.0 mL and the API filtration loss was 2.0–2.5 mL. Therefore, filtration into the formation could be minimized in order to protect the reservoir.
(4) An API 170 mesh/200 mesh screen was used with the shaker, and 1–2 centrifuges were run according to the situation, using solid control equipment to remove the solid phase in order to achieve the purpose of protecting the reservoir.
(5) Appropriate drilling parameters (pump rate: 2100–500 L/min, rpm: 120–140 r/min) were used to ensure hole cleaning. At the same time, operations were conducted carefully in order to reduce downhole pressure spikes and ensure that the bottom-hole ECD remained stable to prevent downhole loss-circulation.

Six formation pressure systems were encountered in Well S with formation coefficients 0.43, 0.5, 0.63, 0.71, 1.03, and 1.16, respectively. There were no downhole problems such as loss-circulation in the depleted reservoir and well control problems in the high-pressure zone. The skin factor was −1.0, and the actual production was 150% more than expected. The reservoir protection effect was good, indicating that the plugging performance and reservoir protection performance of the novel PLUS/KCL drilling fluid was significantly improved, which met the requirements of directional drilling in the reservoir section of the multi-pressure system.

5. Conclusions

(1) The PLUS/KCL drilling fluid formulation was improved by optimizing the combination of plugging agent, the addition of treating agent, and the particle size distribution of the temporary plugging agent. The results of the laboratory evaluation showed that novel PLUS/KCL drilling fluid had fine pressure bearing capacity and sealing for its invasion depth into natural core, which was only 4.2 cm at 120 °C/35 MPa for 12 h, indicating it could significantly reduce downhole loss-circulation. The laboratory performance evaluation results showed that the inhibition and temperature resistance of the optimized PLUS/KCL drilling fluid met the requirements for directional drilling of multi-pressure system reservoirs in the Y oilfield.
(2) A field application showed that the optimized PLUS/KCL drilling fluid could solve the problems of aging and thickening, downhole loss-circulation of depleted reservoir, and serious reservoir contamination under a high differential pressure under directional drilling of a multi-pressure system in the Y oilfield.

Author Contributions: Conceptualization, H.L. and Y.H.; methodology, H.L.; validation, L.D.; formal analysis, M.W.; investigation, Z.L.; resources, Y.H.; writing—original draft preparation, H.L.; writing—review and editing, H.L.; supervision, Y.H. All authors have read and agreed to the published version of the manuscript.

Funding: This work was funded by ① National Science and Technology Major Project (Project No. 2022ZX05028001-009); ② Chongqing Municipal Education Commission Science and Technology Research Program (Project No. KJQN202001203, KJZD-K202201204); ③ Natural Science Foundation of Chongqing, China (Project No. 2022NSCQ-MSX3909).

Data Availability Statement: Not applicable.

Conflicts of Interest: The authors declare no conflict of interest.

References

1. Mo, T.; Zhang, L. The optimum & fast new technique of drilling fluids in the field of WZ12-1. *Guangdong Chem. Ind.* **2015**, *42* 75–76.
2. Huang, K.; Liu, S. Technology research on sidewall stability of Weizhou12-1 in Wei2 area of north oilfield. *Oil Drill. Prod.* **2007**, *29* 7–9.
3. Li, X.; Rui, W.; Zheng, Y.; Xiao, Z. Application of multi-stage bridge temporary-plugging reservoir protection technology in the Longwangmiao Fm reservoir in the Moxi-Gaoshiti structure, Sichuan Basin. *Nat. Gas Ind.* **2015**, *35*, 76–81.
4. Zheng, L.; Wei, P. Review to shale gas formation damage for 30 years. *Oil Drill. Prod. Technol.* **2013**, *35*, 1–15.
5. Wang, S.; Zhang, J.; Zhou, S.T. Study on Reservoir Protection Drilling Fluid Technology in the Niger Oil Fields. *Sci. Technol. Eng* **2015**, *15*, 204–207.
6. Luo, J.; Yan, J.; Fang, D. Low pressure reservoirs protection in the middle block of WZ12-1 Oilfield. *Drill. Fluid Complet. Flui* **2006**, *23*, 16–21.
7. Guo, Y.; Guan, S.; Liu, Z.; Peng, W.; Cao, F. Solid-Free Organic Salt Drilling Fluid for Horizontal Wells in the Weizhou 12-1 Oilfield *Pet. Drill. Tech.* **2017**, *45*, 31–36.
8. Zhang, J.; Yan, J. Optimization of Particle Size Distribution for Temporarily Plugging/Shielding Agents in Water Base Reservoi Drilling Fluids. *Oilfield Chem.* **2005**, *1*, 1–5.
9. Zhang, J.; Yan, J. New theory and method for optimizing the particle size distribution of bridging agents in drilling fluids. *Act Pet. Sin.* **2004**, *6*, 88–91.
10. Zhang, H.; Zuo, F.; Li, H.; Li, L.; Liu, Y.; Li, H.; Zhang, Y. Method for Evaluation of Plugging of Nano-micron Fractures. *Dri Fluid Complet. Fluid* **2015**, *32*, 43–46.
11. Xu, T.; Lu, S.; He, R. Methods for Evaluating Drilling Fluid Sealing and Plugging agents and the Influential Factors. *Drill. Pro Technol.* **2009**, *26*, 60–68.
12. Zheng, L.; Zhang, M. Review of basic theory for lost circulation control. *Oil Drill. Prod. Technol.* **2012**, *34*, 1–8.
13. Kaihe, L.; Qiu, Z.; Wei, H.; Li, X.; Song, Y. Study on techniques of auto-adapting lost circulation resistance and control for drillin fluid. *Acta Pet. Sin.* **2008**, *29*, 758–760.
14. Cheng, Z.; Qiu, S.; Cao, J. Application of plugging-while-drilling technology in long open-hole in-terval of the Well Yueman-3- *Oil Drill. Prod. Technol.* **2016**, *38*, 612–616.
15. Ying, W. Mechanism Study and Field Application of Lost Circulation Resisting and Sealing for Formations with Low Pressu and Easy Lost Circulation. *Nat. Gas Ind.* **2004**, *24*, 81–83.
16. Wang, G.; Hu, G.; Li, D. Application of a new pressure-bearing and sealing technique in well L356 of Tuha Basin. *J. Yangtze Un Nat. Sci. Ed.* **2016**, *13*, 56–60.
17. Montilva, J.; Van Oort, E.; Brahim, R.; Luzardo, J.P.; McDonald, M.; Quintero, L.; Dye, B.; Trenery, J. Improved drilling performan in Lake Maracaibo using a low-salinity, high-performance water-based drilling fluid. In Proceedings of the SPE Annual Technic Conference and Exhibition, Anaheim, CA, USA, 11–14 November 2007.
18. Young, S.; Ramses, G. performance and environmental compliance-resolution of both with a unique water-based fluid. Proceedings of the SPE/IADC Indian Drilling Technology Conference and Exhibition, Mumbai, India, 16–18 October 2006.
19. Ramirez, M.A.; Clapper, D.K.; Sanchez, G.; Luna, E.; Preciado, O.; Santamaria, J. Aluminum-based HPWBM successfully replac oil-based mud to drill exploratory well in the Magellan Strait, Argentina. In Proceedings of the SPE/IADC Middle East Drilli and Technology Conference, Cairo, Egypt, 22–24 October 2007.
20. Reynolds, D.; Popplestone, A.; Hodder, M.; Gwynne, P.; Kelly, B. High-performance water-based drilling fluid helps achieve ea oil with lower capital expenditure. In Proceedings of the SPE Offshore Europe, Aberdeen, UK, 6–9 September 2005.
21. Downs, J.D.; Blaszczynski, M.; Turner, J.; Harris, M. Drilling and completing difficult HP/HT wells with the aid of cesium for-ma brines-A performance review. In Proceedings of the IADC/SPE Drilling Conference, Miami, FL, USA, 21–23 February 2006.
22. Soric, T.; Huelke, R.; Marinescu, P. Uniquely engineered water-based high-temperature drilling-in fluid increases productic cuts costs in Croatia Campaign. In Proceedings of the SPE/IADC Drilling Conference, Amsterdam, The Netherlands, 19– February 2003.

23. El Essawy, W.M.; bin Hamzah, R.; Malik, M.M.; Knox, D.; Monem, M.R.; Oswald, R.J. Novel application of sodium silicate fluids achieves significant improvement of the drilling efficiency and reduce the overall well costs by resolving borehole stability problems in East Africa shale. In Proceedings of the IADC/SPE Asia Pacific Drilling Technology Conference and Exhibition, Kuala Lumpur, Malaysia, 13–15 September 2004.
24. Sensory, T.; Chenevert, M.E.; Sharma, M. Minimizing water invasion in shale using nanoparticles. In Proceedings of the SPE Annual Technical Conference and Exhibition, New Orleans, LA, USA, 4–7 October 2009.
25. Sayyadnejad, M.A.; Ghaffarian, H.R.; Saeidi, M. Removal of hydrogen sulfide by zinc-oxide nanoparticles in drilling fluid. *Int. J. Environ. Sci. Technol.* **2008**, *5*, 565–569. [CrossRef]
26. Yue, Q.; Wang, C.; Zhang, Y. The Influence of Polyalcohols Jlx on The Properties of Polymer Drilling Fluid. *Drill. Prod. Technol.* **2001**, *24*, 67–69.
27. Zhu, H.; Xiao, P.; Xue, D. Technology of Drilling Fluids for Formation Protection. *Chem. Ind. Times* **2008**, *22*, 65–68.
28. Pang, B.; Guan, S.; Fang, M. Drilling Fluid Technology of Yacheng13-1-A12sah Extended Reach Well With High Temperature & Super Low Pore Pressure. *Drill. Prod. Technol.* **2013**, *36*, 97–100.
29. Huang, J.; Luo, P.; Li, J.; Wang, S.; Zhang, G.; Zhang, B. A Study on the Enhancement of Formation Bearing Resistance. *Drill. Prod. Technol.* **2009**, *26*, 69–71.

processes

MDPI

Article

Optimization and Performance Evaluation of Foam Acid Systems for Plugging Removal in Low Pressure Oil and Gas Reservoirs

Xiangwei Kong [1,2,*], Bing Liu [1,*], Hongxing Xu [3], Jianwen Shen [4] and Song Li [5]

1 School of Petroleum Engineering, Yangtze University, Wuhan 430100, China
2 Hubei Key Laboratory of Oil and Gas Drilling and Production Engineering, Yangtze University, Wuhan 430100, China
3 Downhole Operation Company of CNPC Chuanqing Drilling Engineering Company, Chengdu 610000, China
4 Exploration Company of SINOPEC, Chengdu 610000, China
5 Engineering Research Institute, PetroChina Southwest Oil and Gas Field Company, Chengdu 610017, China
* Correspondence: 501074@yangtzeu.edu.cn (X.K.); liubing_yangtze@163.com (B.L.)

Abstract: Foam acidization has unique advantages such as low damage, low filtration, low friction, high efficiency, excellent retardation, and fast liquid discharge rate, which is suitable for stimulation and reconstruction of low-pressure oil and gas reservoirs that have been developed over many years. It is obtained that the main chemical components of downhole plugging materials include vegetable oil, fatty acids and their esters, silicone oil, amide polymers, and additional organic components, as well as non-organic components, elemental sulfur, ferrous sulfide, iron disulfide, silicon dioxide, mineral salts, etc. The performance of foam acid was investigated by experiments, including the effective range of action of active acids, reducing filtration, increasing temperature resistance and high-temperature stability of foam acid deep wells. The new foam acid system is developed and optimized to suitable for low-pressure deep well acidification operations. Experimental evaluation optimized the acid foaming agent and foam stabilizer and developed a new foam acid formulation with foam stability, filter loss reduction, temperature resistance, and easy backflow performance. The experimental condition is that the temperature is 90 °C, the foam quality can reach more than 70% when mixed for more than 30 s, the average half-life is 38.75 min, and the liquid separation rate is 19.90 s/mL. Its suspension is better than that of conventional hydrochloric acid, its corrosion rate is 1.872 $g/m^2 \cdot h$, and the flowback rate of foam acid residue reaches 97%. Experimental evaluation has shown that the developed foam acid features high surface activity, stable foam, strong temperature resistance, significant speed and corrosion suppression, and excellent drainage assist performance. Dynamical simulation evaluation of reservoir core foam acidification demonstrated that the foam features long-life, strong suspension capacity, excellent rheology, low filtration, and significant acidization and plug removal effects, and can be used in stimulating the medium-deep high-temperature, and low-pressure oil and gas reservoirs.

Keywords: low pressure reservoir; foam acidization; plugging removal acidification; foam stability half-life; acidizing dynamic simulation

Citation: Kong, X.; Liu, B.; Xu, H.; Shen, J.; Li, S. Optimization and Performance Evaluation of Foam Acid Systems for Plugging Removal in Low Pressure Oil and Gas Reservoirs. *Processes* **2023**, *11*, 649. https://doi.org/10.3390/pr11030649

Academic Editor: Qingbang Meng

Received: 4 January 2023
Revised: 29 January 2023
Accepted: 1 February 2023
Published: 21 February 2023

1. Introduction

With the development of oil and gas fields around the world entering the middle and late stages, the long-term high-speed production oil and gas reservoirs are facing the shortage of underground oil and gas resources and the reduction of reservoir pressure leading to the gradual increase of low-pressure oil and gas wells, which makes it more difficult to stabilize the production of existing oil and gas producing blocks. Most low pressure oil and gas wells are old ones that have been in development for years. Such oil and gas wells have undergone drilling, completion, acid fracturing, workover, well kil and additional operations [1–8]. The formation energy was low and the reservoir near the well was severely damaged. Damage to such wells comes not only from non-equilibrium scale, sludge, bacteria, precipitation, particle migration, and the additional damage to the formation of gas, water, and minerals over many years caused by the production system of the well itself, but also from more prominent external operational measures. Therefore, the most prominent issue for such gas wells is to reduce formation damage. Acidification operations are a major effort to reduce or eliminate formation damage and connect oil and gas conduits near wells. Due to the significant damage to the reservoir in the vicinity of low-pressure gas wells, technicians believe that foam acid systems with limited fluid volume, strong suspension capacity, and easy flowback should be used as often as possible for repeated acidification of low-pressure oil and gas wells.

Some oil and gas reservoirs with different porosity, permeability, and rock types have been developed in damaged formations, and the extent of damage may vary after certain developments and production. In matrix acidification, it is necessary to treat all damaged formations, especially the most severely damaged ones (Zhong et al., 2020; Fang et al., 2022) [9,10]. However, fluid flows naturally into the most permeable and least damaged strata and may leave less permeable and more damaged strata untreated. To solve this problem, one of the chemical methods developed was the use of foam as a moderator, and investigations of foam acid date back to 1969 (Smith et al., 1969) [11]. Since then, numerous experimental and numerical simulation studies have been performed to investigate the influence of initial permeability, permeability contrast, foam mass, saturation conditions, injection rate, and foam rheological behavior on foam divergence. (Kennedy et al., 1992; Zerhboub et al., 1994; Rossen and Wang, 1999; Cheng et al., 2002; Nguyen et al., 2009; Letichevskiy et al., 2017; Karadkar et al., 2018; Wang et al., 2020; Wang et al., 2022a, 2022b, 2022c) [12–22]. Foam is used to partially block high-permeability and undamaged formations, diverting acid into low-permeability formations. According to the results of previous studies, foam acidization has the following main advantages. In the first one, the foam fluid has a selective effect on the formation permeability. Due to the Jamin effect, the foam has a strong plugging effect on the high permeability layer and a weak plugging effect on the low permeability reservoir. As a result, reservoirs with low permeability can be excited, enhancing the effect of acidification treatment. Second, the foaming fluid also has a selective effect on the oil and water reservoirs, that is, the acid prefers to excite the oil reservoir. In the third, the energy provided by the expansion of the gas in the foam favors the return of the residual acid flow, thus reducing the residual damage. In the last one, foam acid is a retardant acid that reduces the acid-rock reaction rate and enables deep acidification. In spite of its numerous advantages, foaming acid has some fatal defects, such as its poor resistance to temperature, its complicated formula, and its near ineffectiveness at high permeability contrasts, which severely limit its further application (Li et al., 2008) [23].

Many researchers agree that foam does not increase the viscosity of water or acid, nor does it alter the relation between the relative permeability of water and its saturation in a steady foam flow (Bernard et al., 1965; Huh et al., 1989; Sanchez and Schechter, 1989; de Vries and Wit, 1990; Friedmann et al., 1986, 1991) [24–29]. Foam directly reduces gas mobility in the rock, and the less mobile gas in turn drives down water saturation, thereby reducing the relative permeability and mobility of water. Foam partially reduces gas mobility by trapping a significant fraction of the gas in place; up to 80–99% of the gas is trapped, even if the foam flows at a high-pressure gradient (Rossen, 1990) [30]. On the other hand, the presence of foam can increase the effective viscosity of the flowing gas and thus reduce the gas mobility (Radke et al., 1990; Gdanski et al., 1993; Hill et al., 1994; Rossen, 1996) [31–34]. Both effects are related to each other. The apparent yield stress of the foam increases at high capillary pressure and a large number of bubbles in the foam are trapped in place. The presence of foam in an extremely permeable or undamaged layer can reduce the liquid saturation and relative liquid permeability, thereby reducing acid flow into the foam saturated layer. Acid can then be diverted into these layers without zonal isolation. The use of a surfactant pre-flush can improve acid diversion by helping to place additional foams in high-permeability or undamaged layers. Therefore, it is possible to preferentially block the high permeability zone during acid injection.

For oil and gas wells with low formation pressure, it is difficult for conventional acid systems to backflow residual acid after acidification due to its high density, which requires a low-density acid system with strong backflow capacity for acidification conversion. For water-sensitive formations, conventional acid systems have high filtration rates, which make the formation water sensitive and lead to reservoir pollution. The above reasons restrict the application of conventional acidization techniques in the development of oil and gas reservoirs. In order to resolve the above contradiction, the technique of acidizing formation with foam acid has been proposed. Foam acid is a foam acidizing fluid system that meets the requirements of acidizing operations with a stable performance and a good compatibility with formation using conventional acid fluids and their additives as a base fluid, salt-resistant, acid-resistant, high-temperature resistant foaming agent and foam stabilizer. Foamic acid combines the properties of a foam and an acid liquid system allowing the prepared fluid to have both the properties of a foam and the acidizing power Foamic acid has the following advantages over conventional acids:

(1) Low density (average density is generally lower than 0.5–0.8 g/cm^3, with a minimum density of 0.03–0.04 g/cm^3), low friction, suitable for low pressure oil and gas reservoirs with low energy;

(2) The lower filtration rate has less damage to the formation, so it is suitable for wate sensitive formation;

(3) Strong flowback and rock carrying capacity;

(4) The system has high viscosity and shear dilution characteristics, which can play the role of profile control and temporary plugging;

(5) It has a good retarding effect and is a retarding acid.

The above characteristics enable foam acid to be used for acidizing construction o special formations (low pressure, low permeability, water sensitive formations, multi-laye complex heterogeneous reservoirs) that cannot be involved in conventional acidizing.

Because the theoretical and experimental research work of repeated acidification o foam acid lags far behind the field requirements, the repeated acidification of foam aci lacks necessary scientific and systematic research guidance, leading to a large number c repeated acidification operations of low-pressure gas wells fail to achieve ideal result which are mainly manifested in low construction success rate, poor stimulation effec short stimulation validity period, and even ineffective repeated acidification of some lov pressure gas wells. Problems with foam stimulation manipulation techniques are main manifested by the lack of research on foam systems and application techniques. Therefor research has been carried out on the performance of foam acid in improving the effectiv action distance of active acid, reducing filtration, increasing the temperature resistance an high temperature stability of foam acid deep wells, and the foam acid system suitable fc acidizing operations in low-pressure deep wells has been developed and optimized.

2. Composition Analysis of Downhole Plug and Evaluation Conditions of Foam Acid System

2.1. Composition Analysis of Downhole Plug

The complexity of formation, additives, and interaction of various substances dete mines the complexity of plugging material composition. Different analysis methods a used for different substances in the plug: when analyzing the components of the plug, t organic substances and high molecular weight polymers in the plug are first separated ar extracted step by step with different polar solvents, and the extract is analyzed by infrar spectrum and compared with the international standard infrared spectrum for structur confirmation. For the analysis of complex organic compounds, column chromatograp or thin layer chromatography shall be adopted for further separation and purificati before identification. For liquid samples with volatile odor, use gas chromatography conduct comparative analysis and identification of components with known reagents. F the inorganic substances in the blockings, after the organic substances are separated and removed, the structure or category of the inorganic main components are first judged by infrared spectroscopy, and then the inorganic compound elements and valence states are analyzed by photoelectron spectroscopy for identification. Finally, the composition and content of the main elements are quantitatively analyzed by plasma emission spectrometer and known standard solutions. Various means are combined to confirm the composition and content of the blockage. The specific experimental results are shown in Table 1.

Table 1. Composition analysis of downhole plug.

Type of Blockage	Test Analysis Results	
	Range	Average
Organic compound (%)	0–31.9	17.57
	Hydrocarbon oil, organic acid, organic acid ester, asphalt, etc.	
Inorganic (%)	68.1–100	82.43
	Water, carbon black, metal sulfide, SiO_2, elemental sulfur, sulfate, silicate, etc.	
Water (%)	0–57.87	15.23
Iron content (%)	10.05–56.43	32.85
Elemental sulfur	Trace ~18%	

The infrared spectrogram (Figure 1) of the well scale was obtained with NICOLEt 6700 infrared spectrometer. It can be seen that the scale contains mineral oil (the characteristic peaks are at 2923.6 and 1457.9 cm^{-1}), butylene adipate (the characteristic peaks are at 1731.8, 1174.5, and 1139.7 cm^{-1}), or fatty acid esters.

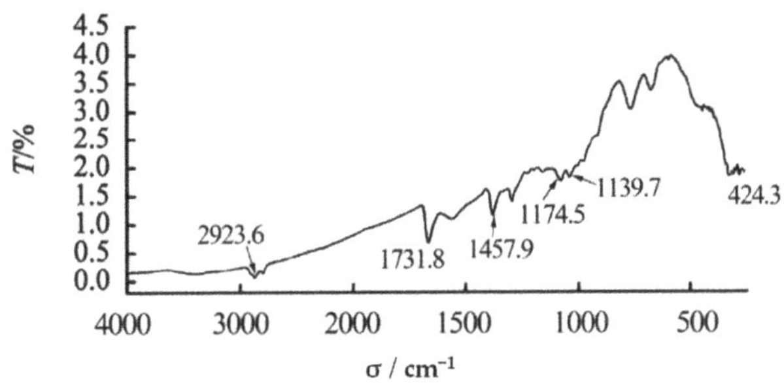

Figure 1. Infrared spectrogram of downhole plugging scale.

By XRD diffraction analysis, we can see the existence of SiO_2, FeS, ferrous disulfide, simple sulfur of rhombic crystal form, and other crystal materials by comparing with the standard card. Combined with the above crystal forms, the microscopic morphology of the plug was analyzed by scanning electron microscope to verify the binding state of each component of the plug. First, observe the overall morphology of the dirt on the plug at a lower magnification. It is found that the size of the dirt is different, the inclusion is thick, and it is dispersed in the black colloid, as shown in Figure 2a. The microscopic morphology of inclusion and colloid was observed by magnification. The former is similar to crystal cluster structure, as shown in Figure 2b, and the latter is a sub distributed inclusion and colloid morphology, in which well-defined crystal structure was also found, as shown in Figure 2c.

In summary, through infrared spectroscopy, gas chromatography, photoelectron spectroscopy, and other analysis methods, it is obtained that the main chemical components of downhole plugging materials include vegetable oil, fatty acids and their esters, silicone oil, amide polymers, and additional organic components, as well as non-organic components, elemental sulfur, ferrous sulfide, iron disulfide, silicon dioxide, mineral salts, etc.

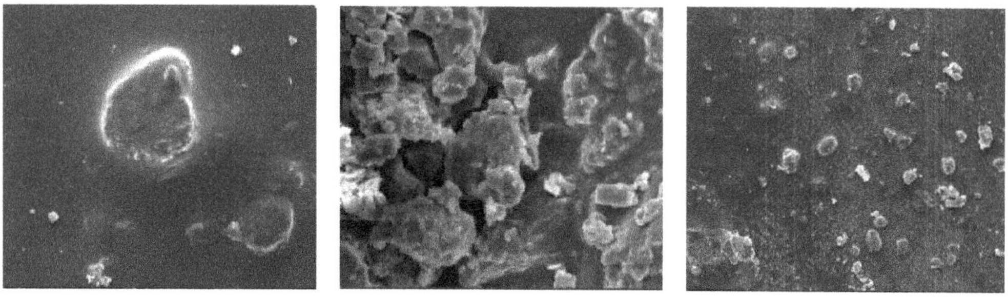

(**a**) ×500 (**b**) ×500 (**c**) ×5000

Figure 2. SEM results of downhole blocking and fouling.

2.2. Determination of Foam Quality

Acidification of foam has the unique advantages of low damage, low filtration, low friction, high efficiency, excellent retarding effect, quick liquid discharge, etc. Foam acidizing can not only effectively remove the pollution and plugging near the well, but also has a better acidizing effect for reservoirs with lower formation pressure. The indoor evaluation method of foam acid should be based on its performance characteristics. Wang et al. (2021) investigated the effects of foaming properties of ASP flooding produced liquid through a set of experiments, and the factors affecting foaming capacity and foam stabilization, including temperature, pressure and the concentrations of chemical agents, were discussed [35]. They also focused on the decay kinetic characteristics of alkaline-surfactant-polymer-strengthened foams under the ultrasonic standing wave. The performance of the diverse foams was characterized. A decay kinetic model incorporating the energy correlation was developed and validated [36].

The first evaluation condition to be determined when conducting foam acid research is foam quality r and foam half-life $t_{1/2}$. Foam quality r indicates the foaming ability of the foaming agent, and the half-life $t_{1/2}$ of foam indicates the stability of forming foam Therefore, r and $t_{1/2}$ are the two essential characteristic parameters of the foam acid system and the essential performance index that must be determined before the optimization of foam acid formula.

Under certain pressure and temperature, the quality of foam r is related to the inflated gas volume Vg, base liquid volume V_1 and foam liquid volume V_f: $r = V_g/V_f = Vg/(V_g + V_1)$ Blauer and Holcomb pointed out that the quality of the foam depends on the structure of the foam and its mobility. When the mass of the foam is 0 to 52%, the gas in the foam is spherical and dispersive and does not touch each other. In this range, the foam is a Newtonian liquid. When the mass of the foam is between 52% and 74%, the static foam is spherical or filled with diamond. The plastic viscosity and yield point of the foam shal be increased properly. When the mass of the foam is in the range of 72% to 95%, the static foam is spherical with slight contact. When flowing, the bubble deforms and the plastic viscosity and yield point increase rapidly. When the foam mass exceeds 96%, the foam becomes a useless mist. At the pressure and temperature near the bottom of the well, 60–90% can be taken. In the construction of foam acid in the United States, most foam acid liquids with a foam quality of approximately 70% are used. This acid liquid has a minute fluid consumption, moderate viscosity, and minor damage to the formation. Therefore, in this experiment, r = 70% is taken as the reference benchmark index for foam acid solution formula debugging.

2.3. Determination of the Foam Half-Life $t_{1/2}$

The half-life of the foam, $t_{1/2}$, is generally determined from the construction scal and the pumping displacement. This experiment is based on a construction time of 40 t 60 min when a 100 m^3 foam fluid is used for acidization operation under 1.5 to 3.0 m^3/mi displacement. A foaming fluid does not usually produce water, or tiny amounts of water, during its flow. $t_{1/2}$ = 20 min or so in static conditions with an indoor foam acid fluid formulation, which can meet the construction requirements of a production well within 3000 m. The half-life $t_{1/2}$ of the acid foam should reach the reference standard of $t_{1/2}$ = 30 min under indoor static testing. At the same time, given that an overly long half-life of the foam is detrimental to the control of reaction time and will increase the cost of liquid formulation, the half-life $t_{1/2}$ of the developed foam acid should be in the range of 20 min to 40 min.

3. Formula Optimization of Foam Acid System

Currently, the widely used formulation of foaming acid is: 20% HCl + foaming agent + foaming stabilizer + additive + N_2, where the additive is the CT1-3 acid corrosion inhibitor and the CT1-7 iron ion stabilizer in the CT series acid formulation, with a normal dosage of 1.0%. Thus, the key to optimizing foam acid formulations is the screening of foaming agents and foam stabilizers and the evaluation of formula compatibility.

3.1. *Foaming Agent and Its Concentration Selection*

In the formulation optimization of foaming acid solutions, foaming agents are selected based on strong foaming ability. Chemically indistinguishable from the liquid components used, when the pressure is released, the formed foam is prone to rupture low dosage. After entering the stratum, the permeability does not decrease. At present, there are numerous surfactants with different properties, but no specific acid foaming agent. It is necessary to experimentally determine the appropriate type and amount of foaming agent.

3.1.1. Test of Foaming Ability of Foaming Agent in Acid Solution

Different foaming agents are mixed into conventional acid, mixed and foamed under the same experimental conditions, and foam quality and half-life are measured. The test conditions are room temperature 20 °C and the concentration of acid solution is 5.0%. Table 2 shows that the foam stability of CT5-2 and CT5-1 in acid solution is relatively excellent.

Table 2. Foaming capacity test of foaming agent in acid solution.

Foaming Agent Type	Foam Volume mL	Foam Quality Γ, %	Half Life $t_{1/2}$, s
F-1	230	56.73	155
CT5-1	550	82.26	423
CT5-2	550	82.31	374
F-2	550	76.29	333
CT5-2H	610	80.40	344
CT5-2D	550	79.27	342

3.1.2. Critical Micelle Concentration CMC

The CMC is used to characterize the surface activity of surfactants. The CBVP-V3 surface tension meter is used to test a series of products at 25 °C. Figure 3 shows the surface tension as a function of the logarithmic concentration of surfactants for several aqueous solutions of surfactants. Under the same conditions, the critical micelle concentrations of different surfactants vary considerably, and their significance varies as well. Of these, the CMCs of CT5-1 are the most pronounced, suggesting that different surfactants have different surface activities. From the curves, the CMC values γ read from the mutation points of the values are 0.0030%, 0.0008%, and 0.0004% respectively.

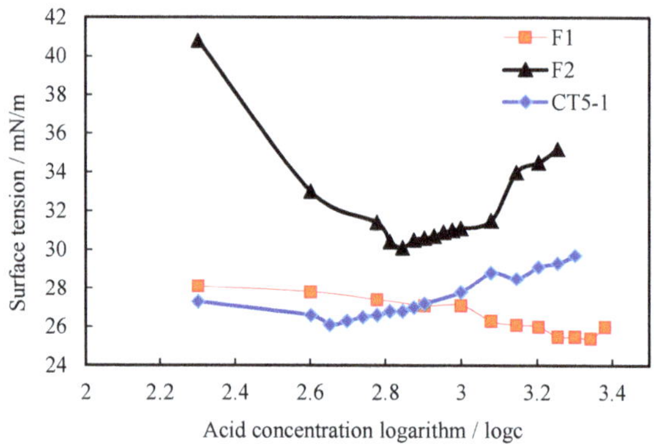

Figure 3. Curve of logarithmic relationship between surface tension and concentration of F1, F2 and CT5-1.

3.1.3. Determination of Foaming Agent Concentration

The relationship between foaming agent concentration and foam parameters is showr in Table 3. The amount of foaming agent should be as small as possible, provided that th added foaming agent is guaranteed to produce a sufficiently stable foaming solution in th regular acid. As can be seen from Table 3, the CT5-1 foams have the longest half-lives at th same foaming agent concentrations. With the exception of F-1, additional foaming agent have a foaming mass of approximately 80% and a foaming agent concentration of 1.0% a room temperature. At the same time, considering the relatively low price of the foamin agent CT5-1 and its better cost-to-performance ratio than other foaming agents, CT5-1 wa identified as the foaming agent for the foaming acid formulation with a concentration o 1.0 to 5.0%.

Table 3. Relationship between foaming agent concentration and foam parameters.

Foaming Agent Concentration %	Foam Parameters F-1		CT5-1		CT5-2		F-2		CT5-2H		CT5-2D	
	Γ %	$t_{1/2}$ s	Γ %	$t_{1/2}$ s	Γ %	$t_{1/2}$ s	Γ %	$t_{1/2}$ s	Γ %	$t_{1/2}$ s	Γ %	$t_{1/2}$ s
1.0	52.21	127	80.15	370	80.24	354	76.12	350	82.26	352	80.24	317
2.0	52.21	150	80.24	415	80.53	360	77.52	366	82.48	367	80.59	332
5.0	56.43	152	82.03	420	82.03	367	82.03	372	83.15	372	82.04	341

3.2. *Selection of Foam Stabilizer and Its Concentration*

Experiments have shown that foams composed solely of foaming agents are less stab When the temperature rises to 90 °C, the half-life of foam that characterizes the stabili of foam solution is extremely short. It is difficult to improve stability by increasing th concentration of the foaming agent. To improve the stability of the foam solution, stror molecular compounds are frequently added as foam stabilizers.

3.2.1. Selection of Foam Stabilizer

Table 4 shows the foam stabilization effect for various polymer compounds. Th stability of all foam solutions containing macromolecular compounds has been improve but the foam stability of different macromolecular compounds varies considerably. Foa parameters were tested by adding 1.0% of various foam stabilizers to the base liquid und the same conditions, with CT5-1 as the foaming agent. CT5-14 was chosen as the foa stabilizer because it had the best foam stability.

Table 4. Foam stabilization of various polymer compounds.

Foaming Agent Name	Foam Parameters with Different Foam Stabilizers											
	No Foam Stabilizer		TSY5-9		CT5-14		PAM		PAMC		PAMA	
	Γ %	$t_{1/2}$ min	Γ %	$t_{1/2}$ min	Γ %	$t_{1/2}$ min	Γ %	$t_{1/2}$ min	Γ %	$t_{1/2}$ min	Γ %	$t_{1/2}$ min
CT5-1	82.02	7.12	72.20	30	78.13	40	68.64	33	68.57	32	68.64	35
CT5-2	82.02	6.19	76.15	25	77.65	32	72.16	35	72.13	35	72.10	35
F-2	82.02	6.31	74.32	20	79.12	27	72.85	26	72.84	24	72.82	27
CT5-2H	83.24	6.28	73.27	29	77.71	34	72.12	34	72.16	35	72.13	35

3.2.2. Determination of Foam Stabilizer Concentration

The quality and stability of the foam must be considered at the same time in the commissioning of the foam acid solution formula. The concentration of the foam stabilizer has a large influence on the foam parameters. Table 5 shows the relation between CT5-14 concentration and foam parameters. It can be seen that when the concentration of CT5-14 is extremely low, the foam stabilizes poorly. When too strong, the foam stabilizes well, but the quality of the foam is extremely low. Therefore, the optimal amount of foam stabilizer has to be determined experimentally.

Table 5. Relationship between CT5-14 concentration and foam parameters.

Foaming Agent Name	Foam Parameters of CT5-14 at Different Concentrations											
	0		0.5		1.0		1.5		2.0		5.0	
	Γ %	$t_{1/2}$ min	Γ %	$t_{1/2}$ min	Γ %	$t_{1/2}$ min	Γ %	$t_{1/2}$ min	Γ %	$t_{1/2}$ min	Γ %	$t_{1/2}$ min
CT5-1	82.03	7.12	83.15	10.2	83.15	17	80.00	25	80.00	36	78.13	38

3.3. Factors Influencing the Formulation of Foam Acid and Evaluation of Compatibility of Acid Solution

3.3.1. Effect of Temperature on the Performance of Foaming Agent Foam

The effect of temperature and foaming agent on the foam parameters is shown in Table 6. Under the same experimental conditions, each foaming agent can produce foam. Foam concentrate is a thermodynamic unstable system, its stability decreases with the increase of temperature, and the stability of foam concentrate varies little at 90 °C.

Table 6. Influence of temperature and foaming agent on foam parameters.

Temperature °C	Foam Parameters											
	F-1		CT5-1		CT5-2		F-2		CT5-2H		CT5-2D	
	Γ %	$t_{1/2}$ s	Γ %	$t_{1/2}$ s	Γ %	$t_{1/2}$ s	Γ %	$t_{1/2}$ s	Γ %	$t_{1/2}$ s	Γ %	$t_{1/2}$ s
20	56.54	156	82.14	425	82.14	370	82.14	376	83.33	375	82.14	345
60	52.38	148	82.46	426	82.46	370	82.46	376	83.33	378	82.14	330
90	65.52	120	81.82	300	82.46	240	82.46	300	82.76	300	80.77	240

3.3.2. Investigation on Compatibility of Foam Acid Solution

The preparation of foaming acid requires that the foaming agent should have excellent chemical compatibility and acid resistance with the acid solution, so that it can be added to the base acid to form a transparent and homogeneous base solution, and to form a stable foaming acid after foaming. The performance of CT5-1 in acid solutions at different concentrations is shown in Table 6.

As can be seen from Table 7, CT5-1 foaming agents are closely compatible with acid solutions in the range of 10–28%.

Table 7. Influence of acid concentration on foam parameters.

Hydrochloric Acid Concentration % \ Base Fluid and Foam Performance	Foaming Agent Concentration % 2.0 5.0 10.0	Foaming Agent Concentration % 2.0		5.0		10.0	
	Compatibility of Base Solution	**r %**	**$t_{1/2}$ min**	**r %**	**$t_{1/2}$ min**	**r %**	**$t_{1/2}$ min**
10% HCl	Mutually soluble and transparent	74.42	27	79.03	22	78.04	20
15% HCl	Mutually soluble and transparent	74.86	34	79.42	26	78.63	25
20% HCl	Mutually soluble and transparent	76.65	33	76.13	32	78.12	32
28% HCl	Mutually soluble and transparent	78.14	35	76.63	33	77.18	29

3.4. Foam Acid Formula

Based on the above experimental evaluation results, assuming a foam mass of about 70% and a half-life of about 30 min, based on the characteristics of gas-liquid two-phase foams, combined with the foaming capacity of each foaming agent, the foaming stability of the foaming agent, the base fluid properties, and the cost properties, the recommended evaluation formula for foaming acid solutions was determined as follows:

20-25%HCl + 1.0%CT1-3 + 1.0%CT1-7 + 1.0-5.0%CT5-1 + 1.5-5.0%CT5-14

CT1-3 is acid corrosion inhibitor. CT1-7 is iron ion stabilizer. CT5-1 is acid foaming agent. CT5-14 is the acid liquid foam stabilizer.

In order to facilitate the performance test of the acid solution system, the typical formula for foam acid performance evaluation is:

20%HCl + 1.0%CT1-3 + 1.0%CT1-7 + 5.0%CT5-1 + 2.0%CT5-14

Based on the existing foaming acid systems, the type and addition ratios of foaming agents and foaming stabilizers are optimized for the formulation of optimized foaming acid systems. According to previous research results, the half-life of foam is 2435 s, and mass fraction of foam is 60% (Li et al., 2013; Wang et al., 2018) [37,38]. The formulation of this system uses different types of foaming agents and foam stabilizers, and increases their dosage, so that stable foam liquid can be produced in a short time, and the same grade of foam half-life can produce higher foam quality (more than 70%), thus having foam stability reducing filtration, temperature resistance, and easy backflow performance.

4. Experimental Evaluation of Foam Acid Performance

Since there is no unified evaluation standard method for foam acid, the laboratory test evaluation method for foam acid performance has been established by referring to API RP42 "Recommended Testing Principles of Surfactants for Enhanced Oil Recovery".

4.1. Factors Affecting the Performance of Foam Acid

4.1.1. Bubbling Speed

The performance parameters of CT5-1 foams produced at different foaming speeds are shown in Table 8. If the foaming speed is too low, a sufficient stable froth cannot be produced, and if the foaming speed is too extreme, there is no great benefit.

Table 8. Performance parameters of CT5-1 foam produced at different foaming speeds.

Mixing Speed	H (20,000 rpm)	M (12,000 rpm)	L (4000 rpm)
r %	72.85	74.42	75.00
$t_{1/2}$ min	36	42	40

4.1.2. Mixing Time

Table 9 shows the mixing time versus the foam parameters. At a sufficient mixing rat stable foam concentrates can be produced by mixing for more than 30 s.

Table 9. Relationship between mixing time and foam parameters.

Mixing time s	30	60	90
r %	74.30	80.00	74.27
$t_{1/2}$ min	40	41	40

It can be seen from Tables 7 and 8 that the mixing speed has a certain impact on the stability of foam acid, but the mixing time has little impact. Therefore, the mixing speed selected in the following test is medium (12,000 rpm), and the mixing time is 60 s.

4.2. Foam Parameters of Foam Acid Formula

The foam volume, foam mass, and foam half-life were measured after foaming a substrate solution consisting of typical formulations. Table 10 is shown as the experiment results.

Table 10. Foam parameters of foam acid.

Number of Measurements	1	2	3	4	Average
V_f	370	390	390	375	381
r %	72.97	74.36	74.36	73.33	73.76
$t_{1/2}$ min	36	41	41	37	38.75

4.3. Foam Stability Test

The stability of a frothy acid solution is the rate at which a certain amount of the frothy solution will burst on its own. When used for stimulation treatment of oil and gas wells, foam acid fluid requires no or small water loss during injection, so its stability is particularly critical. Holding the prepared foam solution still, observe and record the time it takes to separate the different liquid volumes. It can be seen from Table 11 that it takes 995 s to separate half of the base solution, that is, $t_{1/2} \approx 440$ min, and the liquid separation rate is 19.90 s/mL. This indicates excellent stability of the frothy acid solution.

Table 11. Stability of foam acid solution.

Volume of precipitated liquid mL	10	20	30	40	50	60	70	80	90
Time s	960	970	972	985	995	1000	1010	1015	1040
Velocity of liquid precipitation s/mL	96.0	48.5	32.50	24.60	19.90	16.70	14.45	12.74	11.55

4.4. Ability of Suspended Solid Particles of Foam Acid

The migration and deposition of acid insoluble substances (such as mineral particles and clay) released during acidizing operation may form secondary plugging and cause formation damage. In order to prevent the damage caused by the migration and deposition of acid insoluble particles, it is usually required that the acidizing residual fluid should have a certain suspension capacity to suspend the acid insoluble particles released from the formation in the residual acid, and foam acid has a significant ability to suspend solid particles.

The suspension properties of the foam acid were determined by reference to API RP42, and the test results are shown in Figure 4.

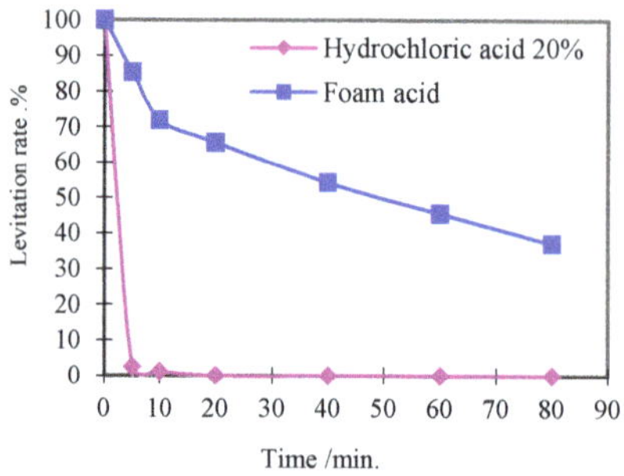

Figure 4. Foam acid suspension capacity test.

4.5. Corrosion Rate Test of Foam Acid

The corrosion rate of foam acid was tested with N80-E steel sheet under the conditio of 15 MPa, 120 °C, and 4 h by using professional evaluation method (static weight-los method). The results are shown in Table 11.

Table 12 shows that the corrosion rate of conventional acid and foam acid is consider ably inferior to the control standard (≤10 g/m^2·h), and it can be seen that the corrosio rate of foam acid is significantly lower than that of conventional acid.

Table 12. Corrosion rate test of foam acid.

Acid Type	Corrosion Rate g/m^2·h
Conventional acid	4.275
Foam acid	1.872

4.6. Drainage Performance of Foam Acid

The key performance of foam acid is its drainage assistance. The laboratory simulatio evaluation method of foam acid was established. The evaluation results are shown i Table 13, which shows that foam acid has a significant drainage effect compared wit conventional acid or gelled acid residual acid.

Table 13. Drainage performance of foam acid.

Evaluation Medium	Discharge Efficiency %	Flowback Rate %	Assisted Drainage Rate %
Conventional acid residue	32.65	95.10	200.80
Gelling acid residue	20.90	90.65	147.00
Residual acid of foam acid	39.00	97.00	360.25

5. Experimental Simulation of Foam Acid Dynamic Acidizing Plugging Removal

Based on the acidization and plug removal properties of the foam, we performe dynamic simulation experiments to evaluate the acidization and plug removal properti of the foam on the reservoir core. The tested properties mainly include the rheologic properties of the foam acid, acid rock reaction rates, retardation properties, and acid flu filtration properties.

5.1. Core Analysis of Low Pressure Gas Well Reservoir

The cores of carboniferous reservoir in eastern Sichuan Basin were selected for t experiment, and the core physical property parameters are shown in Table 14. New foa acid developed in the experiment is shown in Figure 5, and the experimental device shown in Figure 6.

Table 14. Physical parameters of core.

Coring Horizon	Coring Section (m)	Lithology	Permeability (μm^2)	Porosity (%)	Water Saturation (%)
C_2hl	4900.0–4920.5	Breccia, pinhole, and dissolved pore dolomite	1.65×10^{-2}	4.75	30

Figure 5. Appearance of foam acid.

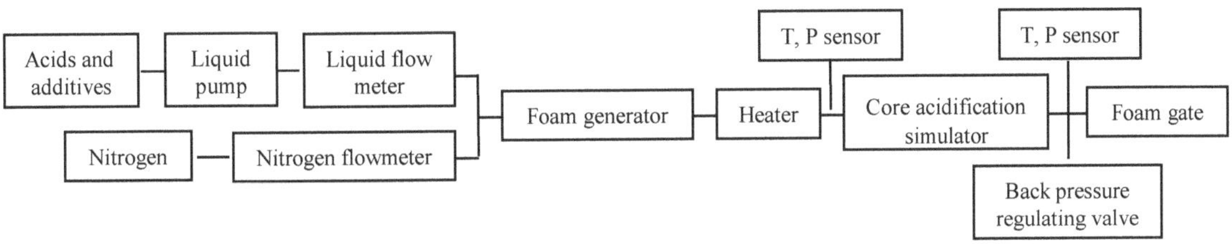

Figure 6. Flow chart of core dynamic simulation test.

5.2. Experimental Evaluation of Acid Rock Reaction Rate Retardation Performance of Foam Acid

The rate of reaction of acid with limestone and dolomite depends on the physical and chemical properties of the rock. When hydrochloric acid reacts with rock, the concentration of reaction products increases considerably. Due to the same ionic effects, the reaction rate of some residual acids will be slowed down and the reaction time of acid rocks will be prolonged. For foam acid, the delayed reaction properties of hydrochloric acid to limestone and dolomite in the presence of foam and nitrogen also affect the equilibrium of acid solution to rock reaction due to the presence of nitrogen. This leads to a delayed reaction mechanism.

$$4HCl + CaMg\,(CO_3)_2 = CaCl_2 + MgCl_2 + 2H_2O + 2CO_2$$

Assuming that this reaction equilibrium relation is dominated by pressure, the reaction rate or delayed reaction degree is a function of nitrogen, acid concentration, and reservoir pressure and temperature prior to acidification. It is believed that foams with high gas fraction have low relative filtration, and that the more unreacted acid remains in the percolation channel, the longer the acid reaction time, that is, the longer the delayed action time. Figure 7 shows a test plot of the acid rock reaction rates for conventional and frothy acids at rest. As can be seen from the figure, the reaction rate of the frothy acid is considerably lower than that of the conventional acid, and the retardation effect in the stationary state is evident. Table 15 shows the test conditions for the dynamic reaction between the foam acid and the core. Table 16 shows the test data for the dynamic response of the foam core. Figure 8 shows the reaction rate versus reaction time curves for the foam acid and the Well A trap core under dynamic conditions. As can be seen from the curves, the reaction rate of the foam acid is extremely low and the retardation performance is excellent.

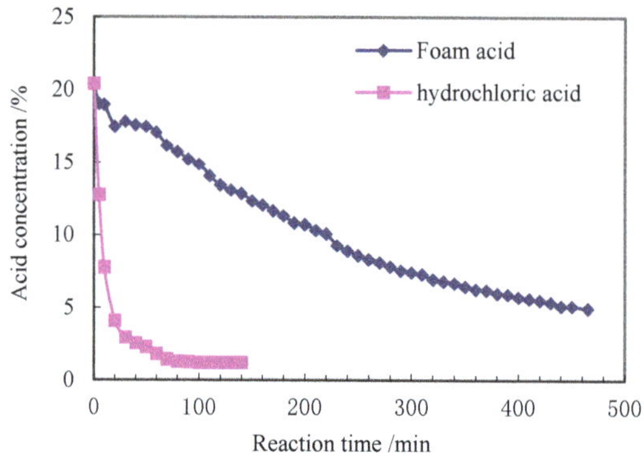

Figure 7. Retarding performance test of foam acid.

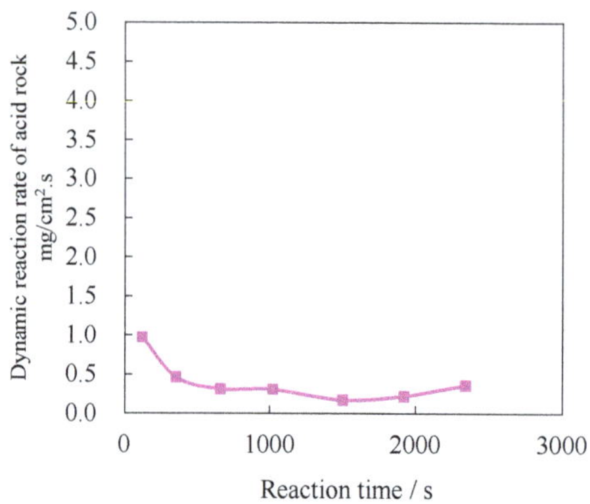

Figure 8. Dynamic reaction rate of acid rock.

5.3. Experimental Evaluation of Acid Fluid Dynamic Filtration of Foam Acid

Foam itself is an anti-filtering fluid, and the filtration coefficient of foam acid is extremely low in low permeability layers. This extremely efficient liquid produces a larger crack area than gelled or emulsified acids in the same volume without solid filtration additive. The likelihood of fluid loss during acidification is represented by the filtration coefficient, which is the flow resistance of the acid fluid filtering into the formation and is a constant. The acid liquid with low coefficient enters the production layer from the fracture with a low amount, which can produce a longer effective action distance. Howard and Fast proposed three methods to control fluid loss: first, increase viscosity to improve relative permeability. Secondly, the filtration coefficient is reduced. Third, the wall-building effect occurs. Although the foam liquid has no wall-building effect, its filtration coefficient is two orders of magnitude lower than that of the ordinary gel liquid. The core filtration test data for the developed foam acid formulation is presented in Table 17. According to the analysis in Table 16, the filtration coefficient of 75% foam acid with foam quality is 7–10 times lower than that of ordinary acid solution. Therefore, the developed foam acid formula has high apparent viscosity and low filtration, which is very conducive to enhancing the acidizing effect.

Table 15. Core dynamic reaction test conditions of foam acid.

Rock Sample Parameters			Initial Acid Concentration %	Test Temperature °C	Environmental Parameters		
Length cm	Width cm	Initial Fracture Width mm			Humidity %	Room Temperature °C	Atmospheric Pressure kPa
15.25 15.20	5.10 5.08	1.95	20	80 °C ± 1	75	20	962

Table 16. Core dynamic response test data of foam acid.

Reaction Time s	Experimental Flow mL/s	Residual Acid Concentration %	Residual Acid Density g/cm^3	Dissolution Amount g	Dissolution Volume cm^3	Fracture Width after Dissolution mm	Average Fracture Width mm
60	15.10	14.15%	1.071	63.60	23.45	3.0345	3.457
300	16.00	15.70%	1.078	50.20	18.52	2.3964	3.138
450	13.32	15.40%	1.077	22.25	8.20	1.0621	2.471
300	15.12	15.60%	1.078	23.95	8.85	1.1442	2.512
240	13.80	16.15%	1.081	19.31	7.13	0.9211	2.401
240	12.90	15.40%	1.077	21.52	7.95	1.0265	2.453
360	9.35	13.50%	1.068	21.75	8.00	1.0389	2.459

Table 17. Core dynamic filtration test results of foam acid.

Slope of Straight-Line Segment mL/min$^{0.5}$	Core Contact Area cm^2	Acid Filtration Coefficien m/min$^{0.5}$
9.776	154.65	0.0003161

6. Conclusions

Based on the experimental evaluation of foam acid single agent, a new formula foam acid solution was developed, the corresponding evaluation method was establish and the indoor performance evaluation and core dynamic performance evaluation w carried out. The following understandings were obtained:

(1) By means of infrared spectroscopy, gas chromatography, photoelectron sp troscopy, and other analytical methods, it is obtained that the main chemical composit of the plug contains organic components, such as mineral oil, fatty acids and their est silicone oil, amide polymers, and additional organic components, as well as non-orga components such as carbon black, elemental sulfur, ferrous sulfide, ferrous disulfide, sili dioxide, mineral salts, etc.

(2) The research focus of foam acidizing technology is to develop the theoretical a experimental research of foam acid, form a necessary, scientific, and systematic resea system. The foam acid properties that need urgent optimization are improved foam a stability, reduced filtration losses, enhanced temperature resistance of acid fluid in de wells, and drainage aids.

(3) A new formula of foam acid was developed, and an experimental method evaluating the acid energy of foam was established. The test evaluations have demonstra that the developed foam acid features high surface activity, stable foam, strong temperat resistance, significant speed and corrosion inhibition, and excellent drainage performa

(4) The dynamic simulation evaluation of foam acid core was completed, and the sh rheology of foam acid in casing at 80 °C was simulated and evaluated. The evaluation the dynamic delay characteristics of acid rocks and the excellent rheological propertie acid fluids shows that it has the characteristics of long foam life, strong suspension capac good rheology, and low filtration, and can be used in medium depth, high temperat and low-pressure oil wells.

Author Contributions: X.K.: Conceptualization, funding acquisition, project administration, sources, funding acquisition, writing—original draft and software. B.L.: Data curation, form analysis, methodology, project administration. S.L.: Writing—original draft and writing—revi and editing. H.X.: Project administration, resources. J.S.: Investigation, methodology, software, a visualization. All authors have read and agreed to the published version of the manuscript.

Funding: This work was supported by the Ministry of Science and Technology of Sinopec Proj Automatic linkage control technology for the drilling overflow of the fractured gas reservoir (N P22117) and Key Technology of Volume Fracturing for Medium shallow Tight Sandstone Gas Re voirs in Western Sichuan (No.: P22017).

Institutional Review Board Statement: Not applicable.

Informed Consent Statement: Not applicable.

Data Availability Statement: Not applicable.

Acknowledgments: The authors are grateful for the support of the National Natural Science Foun tion of China. Thanks to reviewers and editors for their careful review of this manuscript.

Conflicts of Interest: The authors declare no conflict of interest.

References

1. Xu, C.; Kang, Y.; You, L.; You, Z. Lost-circulation control for formation-damage prevention in naturally fractured reservoir: Mathematical model and experimental study. *SPE J.* **2017**, *22*, 1654–1670. [CrossRef]
2. Kong, X.; Liu, Z.; Jin, Y. Study on multiphase pressure wave velocity characteristics of automatic killing annulus in fractured formation in Chuanyu. *Appl. Math. Mech.* **2022**, *43*, 1370–1379. [CrossRef]
3. Xu, C.; Xie, Z.; Kang, Y.; Yu, G.; You, Z.; You, L.; Zhang, J.; Yan, X. A novel material evaluation method for lost circulation control and formation damage prevention in deep fractured tight reservoir. *Energy* **2020**, *210*, 117574. [CrossRef]
4. Chen, M.; Li, P.; Kang, Y.; Zhou, X.; You, L.; Zhang, X.; Bai, J. Effect of aqueous phase trapping in shale matrix on methane sorption and diffusion capacity. *Fuel* **2021**, *289*, 119967. [CrossRef]
5. Chen, M.; Bai, J.; Kang, Y.; Chen, Z.; You, L.; Li, X.; Liu, J.; Zhang, Y. Redistribution of fracturing fluid in shales and its impact on gas transport capacity. *J. Nat. Gas Sci. Eng.* **2021**, *86*, 103747. [CrossRef]
6. Li, S.; Fan, Y.; Gou, B.; Zhang, H.; Ye, J.; Ren, J.; Xiao, Y. Simulation of filtration fields with different completion methods in carbonate gas reservoirs. *Chem. Technol. Fuels Oils* **2021**, *57*, 698–704. [CrossRef]
7. Xu, C.; Yang, X.; Liu, C.; Kang, Y.; Bai, Y.; You, Z. Dynamic fracture width prediction for lost circulation control and formation damage prevention in ultra-deep fractured tight reservoir. *Fuel* **2022**, *307*, 121770. [CrossRef]
8. Guy, M.; Mathieu, M.; Anastopoulos, I.P.; Martínez, M.G.; Rousseau, F.; Dotto, G.L.; de Oliveira, H.P.; Lima, E.C.; Thyrel, M.; Larsson, S.H.; et al. Optimization of high temperature-resistant modified starch polyamine anti-collapse water-Based drilling fluid system for deep shale reservoir. *Molecules* **2022**, *27*, 8936. [CrossRef]
9. Zhong, H.; Yang, T.; Yin, H.; Lu, J.; Zhang, K.; Fu, C. Role of alkali type in chemical loss and ASP-flooding enhanced oil recovery in sandstone formations. *SPE Reserv. Eval. Eng.* **2020**, *23*, 431–445. [CrossRef]
10. Fang, Y.; Yang, E.; Guo, S.; Cui, C.; Zhou, C. Study on micro remaining oil distribution of polymer flooding in Class-II B oil layer of Daqing Oilfield. *Energy* **2022**, *254*, 124479. [CrossRef]
11. Smith, C.L.; Anderson, J.L.; Roberts, P. New diverting techniques for acidizing and fraction. In Proceedings of the SPE California Regional Meeting, San Francisco, CA, USA, 6–7 November 1969. [CrossRef]
12. Kennedy, D.K.; Kitziger, F.W.; Hall, B.E. Case study on the effectiveness of nitrogen foams and water-zone diverting agents in multistage matrix acid treatment. *SPE Prod. Eng.* **1992**, *7*, 203–211. [CrossRef]
13. Zerhboub, M.; Ben-Naceur, K.; Touboul, E.; Thomas, R. Matrix acidizing: A novel approach to foam diversion. *SPE Prod. Facil.* **1994**, *9*, 121–126. [CrossRef]
14. Rossen, W.; Wang, M.-W. Modeling foams for acid diversion. *SPE J.* **1999**, *4*, 92–100. [CrossRef]
15. Cheng, L.; Kam, S.; Delshad, M.; Rossen, W. Simulation of dynamic foam-acid diversion processes. *SPE J.* **2002**, *7*, 316–324. [CrossRef]
16. Nguyen, Q.P.; Zitha, P.L.; Currie, P.K.; Rossen, W.R. CT study of liquid diversion with foam. SPE Prod. *SPE Prod. Oper.* **2009**, *24*, 12–21. [CrossRef]
17. Letichevskiy, A.; Nikitin, A.; Parfenov, A.; Makarenko, V.; Lavrov, I.; Rukan, G.; Ovsyannikov, D.; Nuriakhmetov, R.; Gromovenko, A. Foam acid treatment-the key to stimulation of carbonate reservoirs in depleted oil fields of the Samara region. In Proceedings of the SPE Russian Petroleum Technology Conference, Society of Petroleum Engineers, Moscow, Russia, 16–18 October 2017. [CrossRef]
18. Karadkar, P.; Bataweel, M.; Bulekbay, A.; Alabdrabalnabi, M. Recent Advances in Foamed Acid Fracturing. Society of Petroleum Engineers. In Proceedings of the SPE Kingdom of Saudi Arabia Annual Technical Symposium and Exhibition, Dammam, Saudi Arabia, 23–26 April 2018. [CrossRef]
19. Wang, D.; Dong, Y.; Sun, D.; Yu, B. A three-dimensional numerical study of hydraulic fracturing with degradable diverting materials via CZM-based FEM. *Eng. Frac. Mech.* **2020**, *237*, 107251. [CrossRef]
20. Wang, M.; Wu, W.; Li, S.; Li, T.; Ni, G.; Fu, Y.; Zhou, W. Application and optimization for network-fracture deep acidizing technique of fractured carbonate reservoirs. *Lithosphere* **2022**, *2022*, 8685328. [CrossRef]
21. Wang, M.; Zhou, W.; Li, S.; Wu, W. Simulated investigation in wormhole expansion law of gelling acid etching and its influencing factors in deep carbonate reservoirs. *Gels* **2022**, *8*, 470. [CrossRef]
22. Wang, M.; Wu, W.; Chen, S.; Li, S.; Li, T.; Ni, G.; Fu, Y.; Zhou, W. Experimental evaluation of the rheological properties and influencing factors of gel fracturing fluid mixed with CO_2 for shale gas reservoir stimulation. *Gels* **2022**, *8*, 527. [CrossRef]
23. Li, S.; Li, Z.; Lin, R. Mathematical models for foam-diverted acidizing and their applications. *Petrol. Sci.* **2008**, *5*, 145–152. [CrossRef]
24. Bernard, G.G.; Holm, L.W.; Jacobs, W.L. Effect of foam on trapped gas saturation and on permeability of porous media to water. *SPE J.* **1965**, *5*, 295–300. [CrossRef]
25. Huh, D.G.; Handy, L.L. Comparison of steady- and unsteady-state flow of gas and foaming solution in porous media. *SPE Reserv. Eng.* **1989**, *4*, 77–84. [CrossRef]
26. San chez, J.M.; Schechter, R.S. Surfactant effects on the two-phase flow of steam-water and nitrogen-water through permeable media. *J. Pet. Sci. Eng.* **1989**, 3, 185–199. [CrossRef]
27. de Vries, A.S.; Wit, K. Rheology of gas/water foam in the quality range relevant to steam foam. *SPE Reserv. Eng.* **1990**, *5*, 185–192. [CrossRef]

28. Friedmann, F.; Jensen, J.A. Some parameters influencing the formation and propagation of foams in porous media. In Proceedings of the SPE California Regional Meeting, Oakland, CA, USA, 2–4 April 1986. [CrossRef]
29. Friedmann, F.; Chen, W.H.; Gauglitz, P.A. Experimental and simulation study of high-temperature foam displacement in porous media. *SPE Reserv. Eng.* **1991**, *6*, 37–45. [CrossRef]
30. Rossen, W.R. Theory of mobilization pressure gradient of flowing foams in porous media I: Incompressible foam. *J. Colloid Interface Sci.* **1990**, *136*, 1–53. [CrossRef]
31. Radke, C.J.; Gillis, J.V. A dual-gas tracer technique for determining trapped gas saturation during steady foam flow in porous media. In Proceedings of the SPE Annual Technical Conference and Exhibition, New Orleans, LA, USA, 23–26 September 1990. [CrossRef]
32. Gdanski, R.D. Experience and research show best designs for foam diverted acidizing. *Oil Gas J.* **1993**, *91*, 85–89.
33. Hill, A.D.; Rossen, W.R. Fluid placement and diversion in matrix acidizing. In Proceedings of the University of Tulsa Centennial Petroleum Engineering Symposium, Tulsa, OK, USA, 29–31 August 1994. [CrossRef]
34. Rossen, W.R. Foams in enhanced oil recovery. In *Foams: Fundamentals and Applications*; Prud', H.R.K., Khan, S., Eds.; Marcel Dekker: New York, NY, USA, 1996; pp. 413–464.
35. Wang, Z.; Liu, X.; Luo, H.; Peng, B.; Sun, X.; Liu, Y.; Rui, Z. Foaming properties and foam structure of produced liquid in alkali/surfactant/polymer flooding production. *J. Energy Resour. Technol.* **2021**, *143*, 103005. [CrossRef]
36. Wang, Z.-H.; Liu, X.-Y.; Zhang, H.-Q.; Wang, Y.; Xu, Y.-F.; Peng, B.-L.; Liu, Y. Modeling of kinetic characteristics of alkaline-surfactant-polymer-strengthened foams decay under ultrasonic standing wave. *Pet. Sci.* **2022**, *19*, 1825–1839. [CrossRef]
37. Li, Z.M.; Yang, L.Y.; Zhang, D.; Li, S.Y.; Mao, H. Optimization and performance evaluation on composite foam acid system. *Sci. Technol. Eng.* **2013**, *13*, 4907–4911.
38. Wang, Y.; Yuan, Q.Y.; Wu, X.; Liu, W.; Li, Q. Preparation and application of deep penetration foam acid system with higher heat resistance and salt tolerance. *Oilfield Chem.* **2018**, *35*, 406–409. [CrossRef]

processes MDPI

Article

Adsorption Behavior and Adsorption Dynamics of Micrometer-Sized Polymer Microspheres on the Surface of Quartz Sand

Jun Li [1], Taotao Luo [1,*], Wende Yan [1], Tingting Cheng [1], Keyang Cheng [1], Lu Yu [2], Jiannan Cao [1] and Zhongquan Yang [1]

1 Institute of Petroleum and Natural Gas Engineering, Chongqing University of Science and Technology, Chongqing 401331, China; 2011927@cqust.edu.cn (J.L.)
2 Exploration and Development Research Institute of Southwest Branch, SINOPEC, Chengdu 610041, China
* Correspondence: 2017039@cqust.edu.cn

Abstract: The adsorption of polymer microspheres in a stratum can directly affect its action mode and performance in the actual application process. Understanding the adsorption pattern of polymer microspheres and their adsorption mechanism can facilitate optimization of the application mode and enhance the use efficiency. Ultraviolet spectrophotometry was employed to measure the static adsorption characteristics of polymer microspheres (PMS) on the surface of quartz sand. The PMS adsorption capacity on the surface of quartz sand increased with increasing concentration. When the concentration was 1000 mg/L, the static equilibrium adsorption capacity was 402 μg/g, and monolayer adsorption was dominant. The effect of the contact time on the adsorption was investigated, and the fitting was performed using the isothermal adsorption thermodynamic equilibrium model and the adsorption kinetic model. The adsorption of 800 mg/L PMS tended to equilibrate after 0.8 h of adsorption on the surface of quartz sand, and the adsorption of 1400 mg/L PMS tended to equilibrate after 1 h of adsorption on the surface of quartz sand. Good fitting results of the kinetic adsorption process were obtained using the pseudo-first-order (PFO) model, pseudo-second-order (PSO) model, Elovich model, and mixed-order (MO) model. The effects of the temperature, particle size of the quartz sand, solid–liquid ratio, and salinity on the adsorption of PMS on the surface of quartz sand were examined. The PMS adsorption capacity on the surface of quartz sand decreased with increasing environmental temperature. The adsorption of PMS at the solid–liquid interface was an exothermic process, and the enthalpy of adsorption was negative. As the mass of the quartz sand in the solid–liquid ratio increased, the adsorption capacity decreased; a low salinity and neutral pH were conducive to the adsorption of PMS on the surface of quartz sand.

Keywords: polymer microspheres; static adsorption; isothermal adsorption thermodynamic equilibrium model; adsorption kinetic model; adsorption pattern

tation: Li, J.; Luo, T.; Yan, W.; ıeng, T.; Cheng, K.; Yu, L.; Cao, J.; ng, Z. Adsorption Behavior and lsorption Dynamics of crometer-Sized Polymer crospheres on the Surface of artz Sand. *Processes* **2023**, *11*, 1432. ps://doi.org/10.3390/pr11051432

ademic Editor: Federica Raganati

ceived: 2 April 2023
vised: 13 April 2023
cepted: 28 April 2023
blished: 8 May 2023

1. Introduction

For oilfields developed via water flooding, the water cut value of most wells increases with increasing well production time [1]. This is because, under long-term water flooding, the deposits in the stratum are gradually washed away. Large pore throats are formed in the reservoir, and the permeability increases, exhibiting the characteristics of heterogeneous reservoirs [2]. The subsequent injected water will continue to advance along the high-permeability layer and cannot effectively spread to the remaining oil area, so the reservoir cannot be efficiently exploited. To solve this problem, researchers began to focus on the theory and technique of adjusting the water injection profile in the deep layer of the oil reservoir. Polymer microspheres (PMS) are a novel plugging agent with a rapid development and wide application scope. PMS are spherical gel particles and have good water-absorbing and swelling properties. After swelling, their volume can increase from several folds to several dozen folds. In addition, since the initial size of PMS is smaller than the reservoir pore throat size, PMS can be transported to the deep layer of the reservoir. After absorbing water and swelling, relying on the retention, jamming, bridging, and sealing effects, blockage of the pore throats occurs [3], and the high permeability formation is plugged off so that the injected fluid produces an effective diversion, the sweep coefficient of the displacement agent is enhanced, and the production rate is improved.

When a fluid comes into contact with a solid, one or more components in the fluid accumulate on the surface of the solid, which is called adsorption. Adsorption also refers to the phenomenon of a substance (mainly a solid substance) with surface-absorbing molecules or ions from the surrounding medium (liquid or gas) [4]. Adsorption is a mass transfer process where molecules inside a substance and surrounding molecules have a gravitational attraction to each other, but molecules on the surface of the substance do not fully exert the external forces relative to the substance [5]. Therefore, the surface of a liquid or solid substance can adsorb other liquids or gases, especially when the surface area is large. This adsorption force can have a significant effect, so large areas of substances are often used in industry for adsorption, such as activated carbon, water film, gas, etc. Zhou et al. [6] have studied methane adsorption mechanisms on Longmaxi shale and demonstrated that methane should be not only adsorbed in micropores (<2 nm) but also in mesopores (2–50 nm) by two hypotheses. Haitao Wang et al. [7] have proposed a mathematical model describing the transient pressure behaviors of a vertical well with MRHF in a stress-sensitive CBM reservoir; the adsorption–desorption and diffusion for fluid in reservoirs are considered. Xi Zhang et al. [8] have investigated the effect of ammonium sulfate on the adsorption characteristics of low-concentration Pb(II) ions on the sulfidized hemimorphite surface and showed that ammonium sulfate could increase the maximum recovery of hemimorphite from 69.42% to 88.24% under low concentration of Pb(II) ions. Zuhao Kou et al. [9] have established a model to determine bottom-hole pressure during CO_2 injection through a hydraulic fractured multiwell-pad, considering CO_2 viscous flow, diffusion, and adsorption in shale reservoirs.

PMS have adjusting and driving functions, and their seepage process in porous media is relatively complex, involving the interaction relationship between PMS and the rocks in the reservoir. The adsorption characteristics of PMS on the surface of quartz sand need to be explored. In this study, polymer microsphere (PMS) were synthesized by suspension polymerization, and the water absorption expansion performance of PMS in deionized water and salt water was evaluated. These studies will be beneficial for analyzing the migration and sealing performance of polymer microspheres in porous media.

2. Experimental Details

2.1. Experimental Materials and Equipment

2.1.1. Experimental Materials

The inverse suspension polymerization technique is one of the methods used to prepare PMS. In this method, only a small amount of surfactant needs to be added as dispersant to ensure that the water phase is dispersed in the oil phase and the monomers are polymerized in the water phase [10]. The inverse suspension polymerization technique has the advantages of uniform polymerization, a fast reaction rate, a high yield of solution polymerization, and a simple process [11]. Additionally, in the polymerization process, the reaction system has a low apparent viscosity and good stability. In this study, the starting materials used to prepare the PMS using the inverse suspension polymerization included monomer, crosslinking agent, and initiator. The main reagents are shown in Table 1.

Table 1. Reagents used in the experiment.

	Name	Purity	Manufacturer
Reactive monomer	Acrylamide (AM)	Chemically pure	Tianjin Fucheng Chemical Reagent Factory, Tianjin, China
	2-Acrylamide-2-methylpropanesulfonic acid (AMPS)	Chemically pure	Tianjin Fucheng Chemical Reagent Factory, Tianjin, China
	Hydrophobic monomer	Purity > 80%	Self-prepared
Crosslinking agent	N,N′-methylenebisacrylamide	>98%	Chengdu Kelong Chemical Reagent Factory, Chengdu, China
Initiator	Potassium persulfate	Analytical reagent (AR)	Chengdu Kelong Chemical Reagent Factory, Chengdu, China
	Sodium sulfite	AR	Chengdu Kelong Chemical Reagent Factory, Chengdu, China
	Edetate disodium	AR	Chengdu Kelong Chemical Reagent Factory, Chengdu, China
	Quartz sand	20~30 mesh	Chengdu Kelong Chemical Reagent Factory, Chengdu, China

2.1.2. Instruments

After the PMS were prepared, the particle size distribution after the PMS absorbed water and swelled was characterized, and the ultraviolet (UV) spectrophotometry method was adopted to measure the concentration of the polymer microspheres. The instruments used are described in Table 2.

Table 2. Instruments used in the experiment.

	Instrument Name	Instrument Model	Manufacturer
Instruments for reaction	Heat-gathering constant temperature heating magnetic stirrer	DF-101S	Zhengzhou Greatwall Scientific Industrial and Trade Co., Ltd., Zhengzhou, China
	pH meter	FE20	Mettler-Toledo Instruments (Shanghai), Ltd., Shanghai, China
	Mixer	IKA	IKA-Werke GmbH & Co. KG, Staufen im Breisgau, Germany
	Electric constant temperature blast drying oven	DHG-9140A	Beijing Luxi Technology Co., Ltd., Beijing, China
Instruments for characterization	Particle Size Analyzer	Mastersizer 3000	Malvern Panalytical Ltd., Malvern, UK
	Dual-beam scanning UV-vis spectrophotometer	T9	Beijing Persee General Instrument Co., Ltd., Beijing, China
	Digital ultrasonic cleaner	KH3200DE	Kunshan Hechuang Ultrasonic Instrument Co., Ltd., Kunshan, China

2.1.3. Experiment Methodology

A Malvern Panalytical Mastersizer 3000 made by UK was used to measure the microsphere size and particle size distribution of the PMS.

Using UV spectrophotometry, the concentration and content of the substances were measured based on the Lambert–Beer law. Due to the different structures of substances, they can have absorption spectra at specific wavelengths, and the absorbance is directly proportional to the concentration. Using the strong absorption peak of amide groups in polyacrylamide at around 200 nm, Xiao et al. [12] found that there was an excellent linear relationship between the polyacrylamide concentration and UV absorbance when the polyacrylamide concentration was lower than 250 mg/L, and there were few interferences from factors such as the salinity and temperature. PMS are obtained from the polymerization of monomers, such as acrylamide, and have amide groups similar to polyacrylamide. Liu et al. [13] used UV spectrophotometry to measure the content of polymer microspheres in a solution, determined the optimal absorption wavelength of the experiment and the measurement range of the effective concentration, and calculated the concentration of the polymer microspheres in the solution according to the linear regression equation.

2.2. Experimental Contents and Steps

2.2.1. Preparation of Polymer Microspheres

Water phase: The reactive monomers, i.e., acrylamide (AM), 2-acrylamide-2-methylpropanesulfonic acid (AMPS), and N,N′-methylenebisacrylamide, were added to deionized water according to a certain ratio, and the mixture was stirred continuously. When all of the solids were dissolved, the pH of the solution was adjusted to neutral using sodium hydroxide. Subsequently, the hydrophobic monomer was added, followed by the addition of a small amount of surfactant to allow the hydrophobic monomer to be dispersed in the water. Finally, a small amount of edetate disodium was added to form the reactive water phase.

Oil phase: Certain amounts of surfactants Span80 and Tween-80 were dissolved in #3 white oil as dispersion liquids, and the reactive oil phase was formed after stirring.

Initiator: Potassium persulfate and sodium sulfite were dissolved in a small amount of deionized water for future use.

Inverse suspension polymerization process: First, the reactive oil phase was added to a three-neck flask. The reactive water phase was added to a pressure-equalizing graduated addition funnel, and then it was added dropwise to the reactive water phase while the reactive oil phase was stirred at a low speed, and oxygen was expelled by introducing nitrogen. During the dropwise addition of the reactive water phase, an initiator was added to the reaction system using a syringe. Then, the temperature was raised to the set temperature, and the reaction proceeded under continuous stirring for a certain period of time. After the reaction solution was cooled down, the upper oil phase was removed, absolute ethanol and acetone were added to perform demulsification, and washing and extraction were conducted multiple times. The PMS product was obtained after drying in an oven at 80 °C. The reaction process is shown in Figure 1.

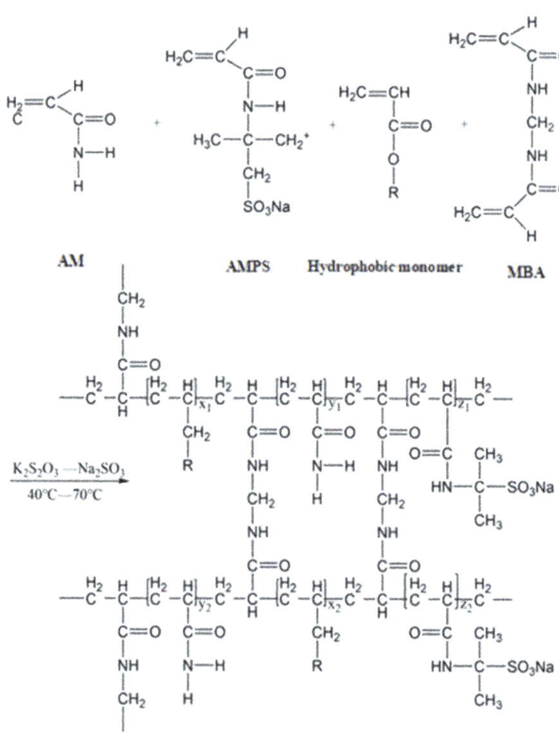

Figure 1. Reaction flowchart of PMS.

2.2.2. Particle Size Distribution Characteristics of PMS

The PMS were placed and dispersed in the deionized water to form a suspension, the initial particle size was measured, and the change in the particle size was measured after the PMS swelled completely (60 h). The simulated water of one oilfield formation with a total salinity of 14,695.89 mg/L was used as the salt solution. The PMS were placed in the simulated salt solution to swell completely (60 h), and the change in the particle size was measured.

2.2.3. Measurement of the Amount of Adsorbed PMS

(1) Plotting the standard curve

Standard aqueous solutions with concentrations of 100 mg/L and 200 mg/L were prepared, and a UV spectrophotometer was used to conduct full-range scanning. The absorption spectra are shown in Figure 2. The PMS exhibited a peak absorbance at 202 nm, and, therefore, 202 nm was selected as the wavelength to measure the PMS.

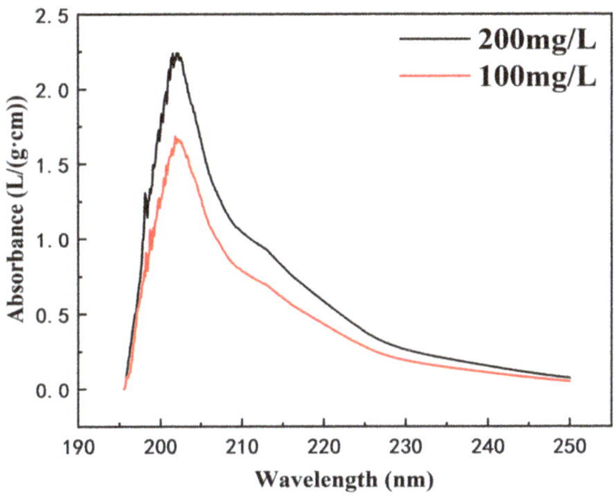

Figure 2. Ultraviolet Absorption Spectrogram of PMS.

The 50–1000 mg/L PMS standard solutions were prepared in deionized water and saline water, with a concentration gradient difference of 100 mg/L. With 202 nm as the fixed absorption wavelength, the absorbance of the various standard concentrations was measured.

(2) Calculation of the amount of adsorbed PMS

In the calculation of the amount of adsorbed PMS, the adsorption of the PMS on the wall during the liquid transfer was ignored. Using the difference in the PMS concentration before and after the adsorption and the amount of added quartz sand, the amount of PMS adsorbed by the quartz sand was calculated.

$$\Gamma = \frac{(C_0 - C)V}{\omega} \quad (1)$$

where C_0 and C are the initial concentrations of the polymer particles and the sample concentration after adsorption (mg/L), respectively; V is the volume of the PMS solution added (mL); ω is the mass of quartz sand added (g); and Γ is the amount of adsorbed PMS (μg/g).

The schematic diagram of the experimental structure of this paper is shown in Figure 3.

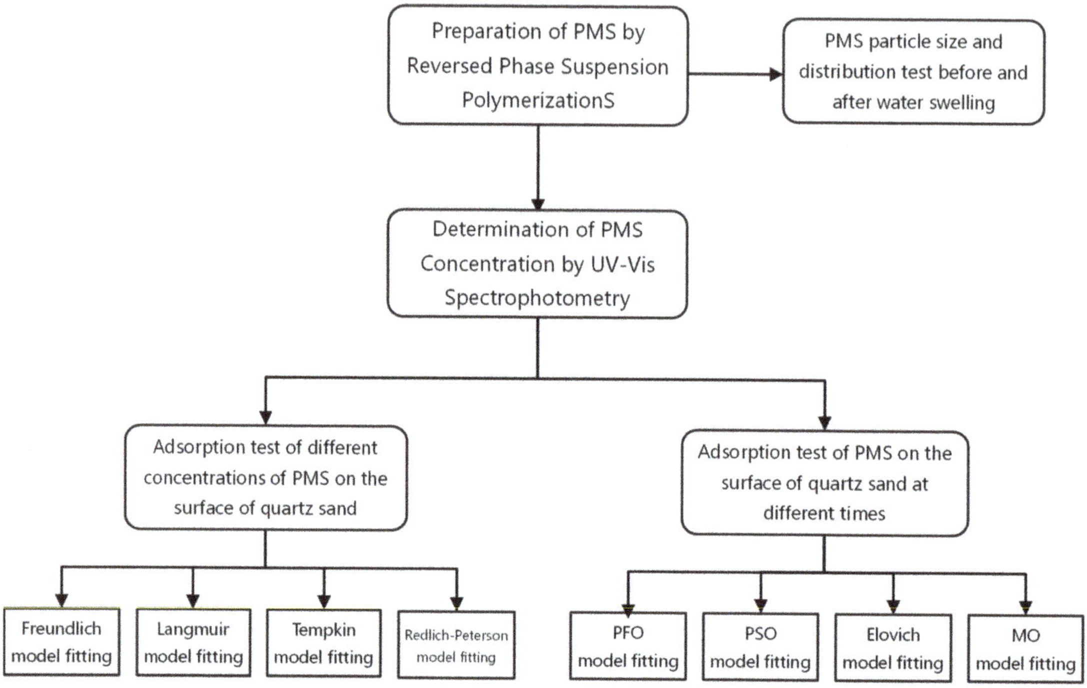

Figure 3. The Structure of all the conducted experiments.

3. Results and Discussion

3.1. Particle Size Distribution Characteristics of PMS

The initial particle size of the PMS, the particle size of the PMS after complete swelling in deionized water, and the particle size of the PMS after complete swelling in saline water are shown in Figure 4.

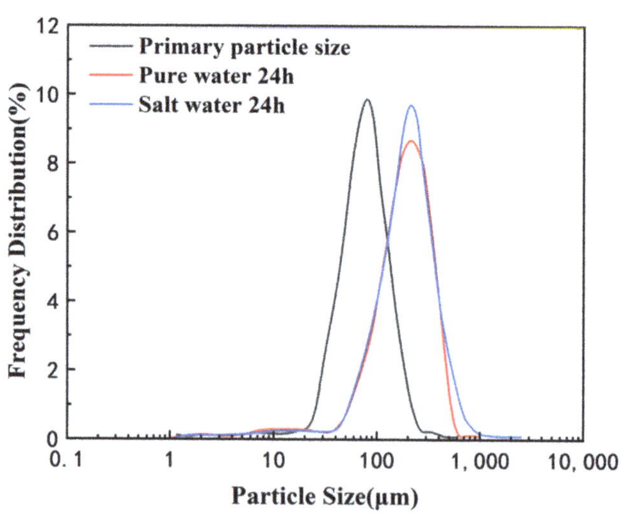

Figure 4. Particle size distribution of PMS before and after swelling.

The experimental results indicate that the PMS exhibited a unimodal normal distribution in water, and the initial average particle size D_{50} was 72 μm. After swelling in the deionized water and saline water for 60 h, the particle size of the PMS increased, and they still exhibited a unimodal normal distribution, but the distribution was wider. The average particle sizes of the PMS after swelling in the deionized water and saline water were 203 μm and 184 μm, respectively. The water entering the PMS was a gradual infiltration process. When the PMS were placed in water, the hydration of the amide and sodium sulfonate in the PMS was favorable for the stretching of the external polymer chains of the PMS, so the water was wrapped inside the PMS. Since the water infiltration and the hydration of the hydrophilic molecules inside the PMS facilitated the entrance of the water into the deep site of the PMS, the volume of the PMS gradually increased. The PMS contained a large number of crosslinks, and it was difficult for the water molecules to enter the PMS. After the hydratability of the internal molecules reached equilibrium, the water molecules could not infiltrate into the PMS, and the PMS did not swell anymore. In the saline water, the hydratability of the amide and sodium sulfonate decreased, and the ability of the water molecules to enter the microspheres decreased to a certain extent, so the swelling ability of the PMS decreased to some extent compared to that in deionized water.

3.2. Characteristics of Isothermal Adsorption of PMS on the Surface of Quartz Sand

3.2.1. Static Adsorption of Different Concentrations of PMS

First, a 5000 mg/L stock solution was prepared, and then it was diluted to 100 mg/L, 250 mg/L, 500 mg/L, 750 mg/L, 800 mg/L, 1000 mg/L, 1200 mg/L, 1400 mg/L, 1800 mg/L, 2000 mg/L, and 3000 mg/L polymer solutions. Next, 80 mL of each of the above solutions was added to a wide neck bottle, acid-washed quartz sand was added according to a solid–solution ratio of 1:10, and the mixture was mixed with a temperature heating magnetic stirrer. After sufficient contact for 24 h, the quartz sand was removed via filtration, and the PMS concentration of the filtrate was measured. In the concentration range of 100–3000 mg/L, the amount of PMS adsorbed on the surface of the quartz sand was measured twice in each experiment. The experimental results are presented in Table 3.

Table 3. Static adsorption capacity under different concentration conditions.

Adsorption Capacity (μg/g)	Concentration of Polymer Solution (mg/L)										
	100	250	500	750	800	1000	1200	1400	1800	2000	3000
#1	132	174	248	336	348	406	405	402	405	401	405
#2	124	176	246	328	347	398	406	404	407	403	401
Average	128	175	247	332	347.5	402	405.5	403	406	402	403

The experimental results indicate that, as the concentration of the PMS solution increased, the quantity of adsorbed PMS gradually increased. When the concentration was 1000 mg/L, the adsorption capacity reached equilibrium, the quantity of adsorbed PMS no longer increased, and the static equilibrium adsorption capacity was around 402 μg/g.

3.2.2. Isothermal Adsorption Thermodynamic Equilibrium Model

The four isothermal adsorption thermodynamic equilibrium models and specific parameters are described in Table 4 [14]. The monolayer adsorption on the homogeneous sites and the multilayer adsorption can be modeled using the Langmuir and Freundlich adsorptions, respectively, and the two adsorption isotherms both involve two parameters. The Temkin adsorption isotherm is also a two-parameter model that assumes multilayer adsorption. The Redlich–Peterson adsorption isotherm is the mixed form of the Langmuir and Freundlich models and has three parameters. These two models are suitable for homogeneous or heterogeneous adsorption processes. Generally, all of the adsorption models should be examined using the adsorption data to determine.

Table 4. Adsorption isotherm models.

No.	Model	Function Equation	Parameter Definition
1	Langmuir model	$q_e = \frac{q_o K_L C_e}{1+K_L C_e}$	q_0 is the maximum adsorption capacity per unit mass of adsorbent (mg/g); K_L is the Langmuir adsorption constant (L/mg); C_e is the liquid phase equilibrium concentration (mg/L).
2	Freundlich model	$qe = KFC_e^{1/n}$	q_e is the amount of adsorbed adsorbate per unit weight at adsorption equilibrium (mg/g); K_F is the adsorption constant reflecting the adsorption capacity (L/mg); the value of n is associated with the adsorption strength; C_e is the solution concentration at adsorption equilibrium (mg/L).
3	Tempkin model	$qe = B \ln KT + B \ln Ce$	K_T is the Tempkin equilibrium binding constant, which reflects the maximum binding energy (L/mg); B is the Tempkin constant related to the heat of adsorption; $B = R_T/b$, where R is the ideal gas constant (8.314 J/mol K), T is the absolute temperature (K), and b is a constant.
4	Redlich-Peterson model	$q_e = \frac{K_r C_e}{1+\alpha C_e^{\beta}}$	K_r is the Redlich–Peterson isotherm constant (L/g); β is an exponent between 0 and 1.

3.2.3. Analysis of the Fitting Results for the Isothermal Adsorption Thermodynamic Equilibrium Models

Using user-defined functions in the Origin software and linear or nonlinear models, the fitting of the adsorption data for the PMS solutions with different concentrations via the Levenberg–Marquardt optimization algorithm was evaluated using the following parameters: the sum of squared errors (SSE), mean squared error (MSE), root mean square error (RMSE), coefficient of determination (R^2), and chi-square value (X^2).

According to the Langmuir model, Freundlich model, Tempkin model, and Redlich–Peterson model, the fitting of the static adsorption data was performed. The fitting parameters and calculated results are presented in Table 5, and the model distribution curves are shown in Figure 5.

Table 5. Results for parameters of different models.

Model	Parameter Estimation Results		Model Evaluation				
	Parameter	Estimated Value	SSE	MSE	X^2	RMSE	R^2
Langmuir model	q_o K_L	482.0388 0.0031	6883.3525	625.7593	25.6762	25.0152	0.9341
Freundlich model	K_F $1/n$	45.8055 0.2916	18055.254	1641.3868	58.1445	40.5140	0.8272
Tempkin model	B KT	95.1639 0.0411	11103.728	1009.4298	33.4279	31.7715	0.8937
Redlich–Peterson model	α β K_r	0.0031 1.0000 1.4944	6883.3696	625.7609	25.6675	25.0152	0.9341

As can be seen from the fitting results presented in Table 5 and Figure 5, the paramete β of the Redlich-Peterson model was close to 1 (β is an empirical constant between 0 and 1 This β value of close to 1 indicates that the quartz sand conformed to the Langmuir mode that is, the adsorption on the surface of the quartz sand was monolayer adsorption an the adsorption action on the PMS was almost even. The fitting accuracy of the Temki model was lower than that of the Langmuir model, indicating that the adsorption of th PMS on the surface of the quartz sand was not multilayer adsorption. The three-paramete Redlich-Peterson model is a mixed form of the Langmuir model and the Freundlic model, and therefore, it could more accurately describe the isothermal adsorption proces (R^2: 0.827–0.934). The Langmuir model has the highest fitting accuracy, indicating tha

PMS has only limited adsorption sites on the surface of quartz sand, rather than unlimited adsorption sites.

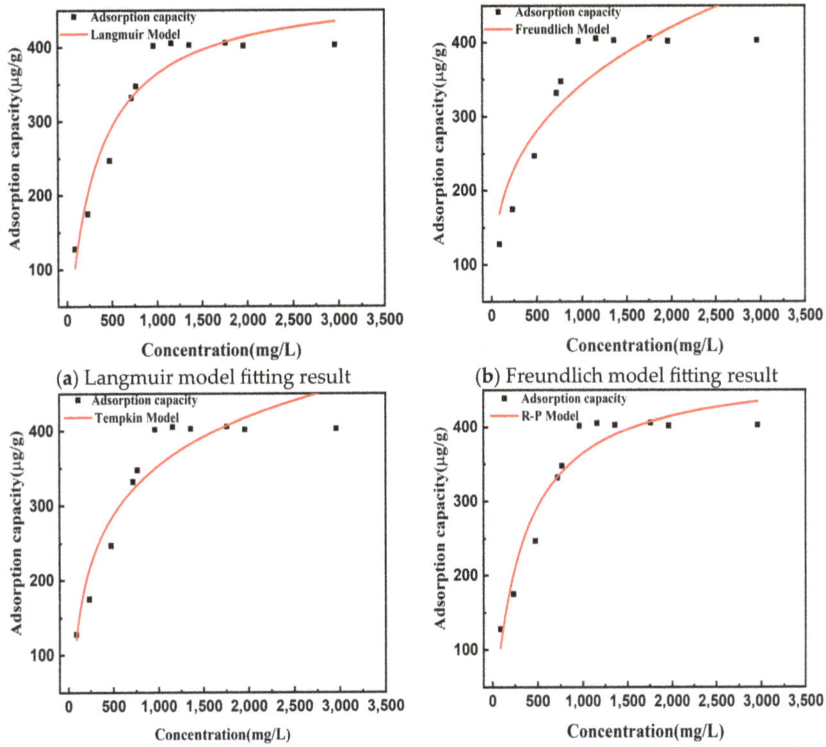

(**a**) Langmuir model fitting result (**b**) Freundlich model fitting result

(**c**) Tempkin model fitting result (**d**) Redlich–Peterson model fitting result

Figure 5. Fitting curves of different models.

3.3. Adsorption Kinetic Characteristics of PMS on the Surface of Quartz Sand

3.3.1. Static Adsorption of PMS with Different Contact Times

First, 800 mg/L and 1400 mg/L PMS solutions was prepared, 80 mL of one of the above solutions was added to a wide neck bottle, quartz sand was added according to a solid–solution ratio of 1:10, and the mixture was mixed by stirring. Sufficient contact was ensured between the polymer solution and the surface of the quartz sand, and the effect of the contact times on the adsorption of the PMS was investigated. The experimental results are shown in Figure 6.

The experimental results shown in Figure 6 reveal that, as the contact time increased, the adsorption of the PMS solution initially increased rapidly and then gradually stabilized. For the 800 mg/L PMS solution, after 0.8 h, the amount of adsorbed PMS reached 345 μg/g, and adsorption equilibrium was generally reached. For the 1400 mg/L PMS solution, after 1 h, the amount of adsorbed PMS reached 404 μg/g, and adsorption equilibrium was generally reached. This is because there were sufficient adsorption sites on the surface of the quartz sand in the initial stage of adsorption, and the PMS quickly occupied these adsorption sites. Over time, the PMS adsorbed on the surface of the quartz sand gradually accumulated, and the amount of adsorbed PMS increased rapidly. As the adsorption sites on the surface of the quartz sand were gradually occupied, there were fewer and fewer empty sites available for adsorption, and the increase in the amount of PMS adsorbed slowed down until the adsorption reached equilibrium.

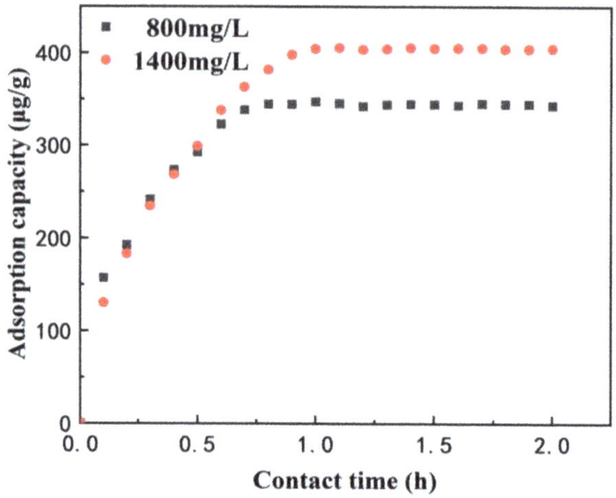

Figure 6. Adsorption of PMS with different concentrations at different contact times.

3.3.2. Kinetic Model of Adsorption Reaction

The kinetics of the adsorption process was mainly studied to describe the solut adsorption rate of the adsorbent. Investigating the adsorption kinetics can provide ar understanding of the adsorption and mass transfer process, thus elucidating the adsorptio mechanism. The commonly used adsorption kinetic models are the pseudo-first-orde (PFO) model [15], pseudo-second-order (PSO) model [16], Elovich model [17], and mixed order (MO) model [4]. The four adsorption kinetic models and their specific parameter are described in Table 6 [4]. The PFO model is based on the assumption that adsorption i controlled by the diffusion step, and the adsorption rate is proportional to the differenc between the equilibrium adsorption capacity and the adsorption capacity at time t. Th PSO is based on the assumption that the adsorption rate is limited by the chemisorptio mechanism. This chemisorption involves electron sharing and electron transfer betwee the adsorbents and adsorbates. In addition, the MO is the mixed form of the PFO an PSO. The Elovich equation is an empirical equation that describes the process of a serie of reaction mechanisms, such as the diffusion of the solute in the liquid phase or interfac and activation and deactivation of the surface. It is very suitable for processes wit high activation energy reactions. Furthermore, the Elovich equation can also reveal th irregularity in the data that the other kinetic equations ignore.

Table 6. Adsorption kinetic model.

No.	Model	Function Equation	Parameter Definition
1	PFO model	$\ln(q_e - q_t) = \ln q_e - k_1 t$	q_e and q_t are the amounts of adsorbed adsorbate at adsorption equilibrium and at time t, respectively; k_1 is the adsorption rate coefficient of the quasi-first-order model
2	PSO model	$\frac{t}{q_t} = \frac{1}{k_2 q_e^2} + \frac{t}{q_e}$	q_e is the adsorption capacity at adsorption equilibrium (μg/g); q_t is the adsorption capacity at adsorption time t (μg/g); k_2 is the rate constant of the pseudo-second-order reaction (g/mg·h).
3	Elovich model	$q_t = \frac{1}{b}\ln(ab) + \frac{1}{b}\ln t$	a is the initial adsorption rate constant; b is a parameter related to the degree of surface coverage of the adsorbent and its chemisorption activation energy.
4	MO model	$\frac{dq_t}{dt} = k_1'(q_e - q_t) + k_2'(q_e - q_t)^2$	k'_1 is the first-order rate constant of the mixed-order model (h^{-1} k'_2 is the second-order rate constant ($g{\cdot}mg^{-1}{\cdot}h^{-1}$) of the mixed-order model.

3.3.3. Analysis of the Fitting Results Using the Kinetic Models of the Adsorption Reaction

Using Visual Basic for Application (VBA), a macro language of Visual Basic, the fitting and solving of the different adsorption kinetic models were conducted using linear or nonlinear methods. The parameters used to evaluate the models included the SSE, MSE, RMSE, R^2, and X^2. The fitting parameters of each model for the adsorption of the 800 mg/L and 1400 mg/L PMS solutions under different contact times and the calculated results are presented in Table 7, and each model distribution curve is shown in Figure 6.

Table 7. Results for parameters of different models.

Model	Polymer Concentration (mg/L)	Parameter Estimation Results		Model Evaluation				
		Parameter	Estimated Value	SSE	MSE	X^2	RMSE	R^2
PFO model	800	K_1	4.329	2000.401	95.257	12.8530	9.760	0.9877
	1400	K_1	2.927	2042.978	97.285	10.3341	9.863	0.9921
PSO model	800	K_2	0.0168	3162.142	150.578	10.8220	12.271	0.9793
	1400	K_2	0.007	4965.565	236.456	14.4382	15.377	0.9800
Elovich model	800	a	10,250.5	1898.455	90.403	10.3582	9.508	0.9881
		b	0.0156					
	1400	a	3546.51	1907.757	90.846	12.3829	9.531	0.9928
		b	0.010					
MO model	800	K'_1	3.652	1806.366	86.017	9.1183	9.275	0.9883
		K'_2	0.004					
	1400	K'_1	3.098	2599.060	123.765	10.8570	11.125	0.9894
		K'_2	0.000					
		n	0.882					

The fitting results presented in Table 7 and Figure 7 show that the 4 models yielded good fitting results for the adsorption kinetic process. The results for the PFO model and PSO model indicate that the PMS were easily adsorbed on the surface of the quartz sand, and this adsorption action occurred from the beginning of the contact. With increasing contact time, the amount of PMS adsorbed gradually increased, the adsorption equilibrium was reached after 0.8 h and 1 h, respectively, and the adsorption rate was relatively fast. The Elovich model is often used to investigate the existence of a chemisorption process on the nonhomogeneous surface of the adsorbent, and the R^2 value demonstrates that the result of the Elovich model matches the adsorption kinetic process well. The Elovich model shows that the adsorption capacity of PMS on the surface of quartz sand is not uniform, which increases linearly with the increase in PMS coverage on the surface of quartz sand, and the adsorption rate of PMS on the surface of quartz sand is not uniform, which decreases exponentially with the increase in PMS adsorption on the surface of quartz sand. The MO is a mixed form of the PFO and PSO, so its fitting accuracy is relatively high, indicating that the adsorption rate of PMS on the surface of quartz sand is not only proportional to the concentration of PMS in water but also proportional to the square of the concentration of PMS in water.

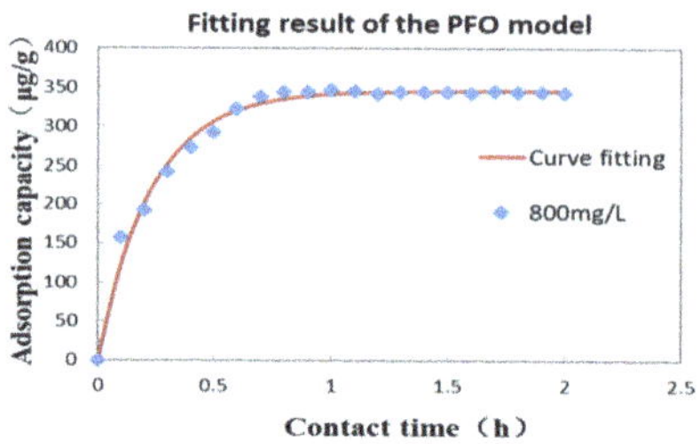

(a) PFO model 800 mg/L

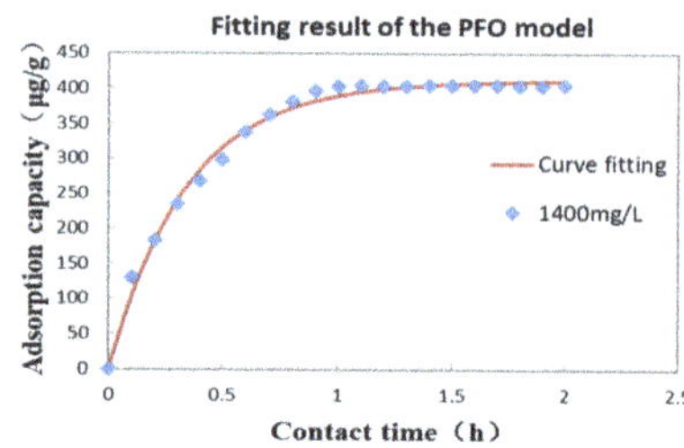

(b) PFO model 1400 mg/L

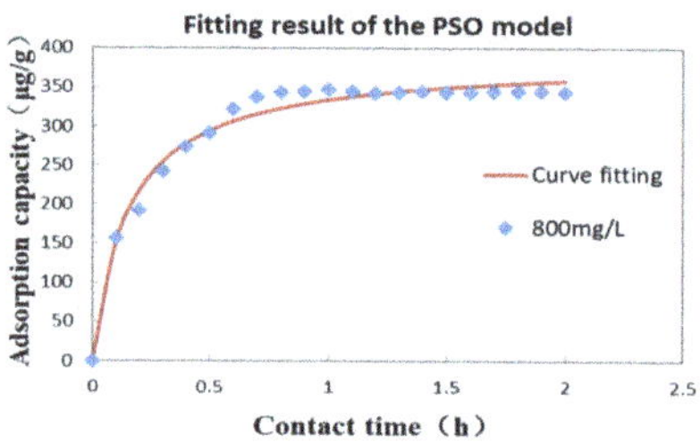

(c) PSO model 800 mg/L

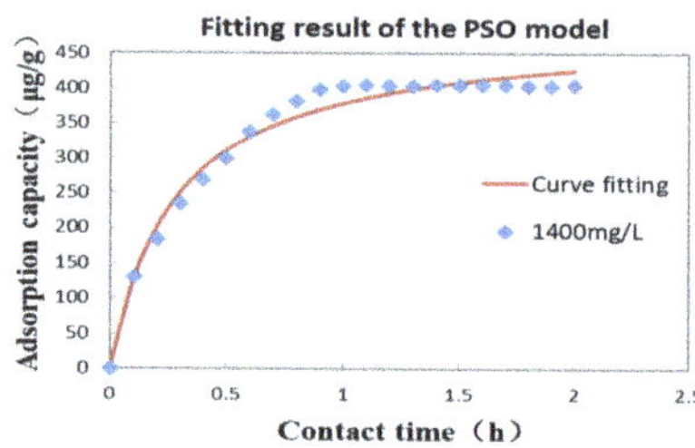

(d) PSO model 1400 mg/

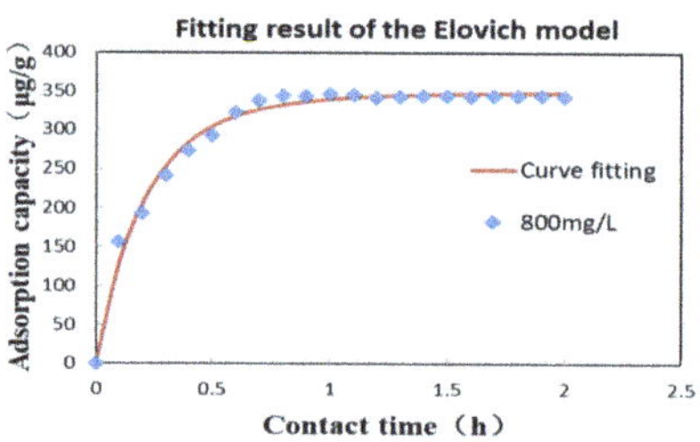

(e) Elovich model 800 mg/L

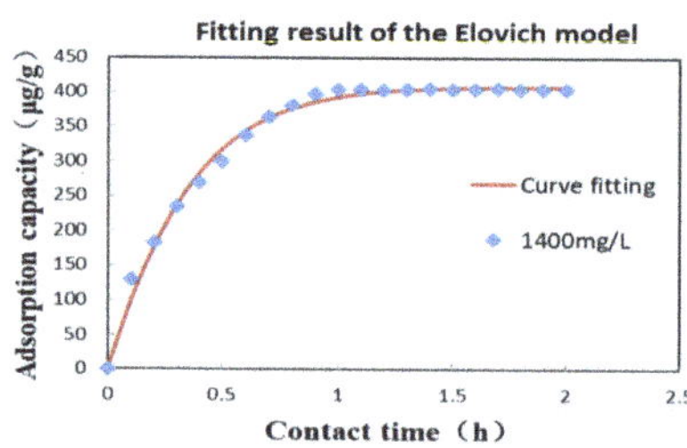

(f) Elovich model 1400 mg/L

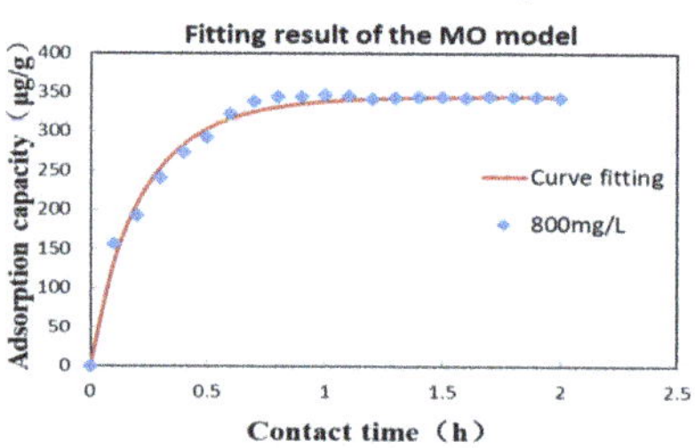

(g) MO model 800 mg/L

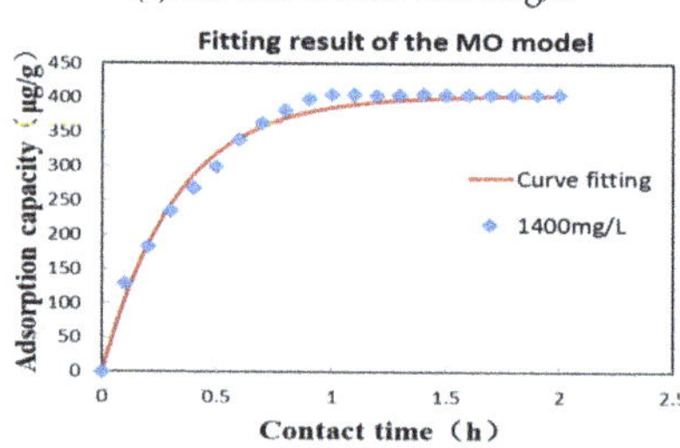

(h) MO model 1400 mg/L

Figure 7. Fitting curves of different models.

4. Conclusions

1. In this study, AM, AMPS, hydrophobic monomer, and crosslinking monomer N,N methylene bis acrylamide (MBA) were used as the reactive monomers, and the PM were prepared using an inverse suspension polymerization. The initial particle siz D_{50} was 72 μm, and the average particle sizes after swelling in deionized water an saline water were 203 μm and 184 μm, respectively. Under a fixed UV absorptio wavelength of 202 nm, the PMS exhibited a good linear relationship between th concentration and absorbance within the range of 100–500 mg/L.
2. With increasing PMS concentration, the amount of PMS adsorbed on the surface c the quartz sand gradually increased. Studies on the isothermal adsorption behavic

of PMS on the surface of quartz sand shows that the fitting accuracy of the Langmuir model for the isothermal adsorption process was relatively high, indicating that the PMS adsorption on the surface of the quartz sand was dominated by monolayer adsorption and limited adsorption positions.

3. The adsorption kinetics of PMS on the surface of quartz sand shows that good fitting of the adsorption kinetic process was obtained using the PFO model, PSO model, Elovich model, and MO model, and the PFO and PSO models indicate that PMS is easily adsorbed on the surface of quartz sand, and this adsorption has already occurred since the beginning of contact. With the increase in contact time, the adsorption amount gradually increases. The Elovich model shows that the adsorption capacity of PMS on the surface of quartz sand is not uniform. It increases linearly with the increase in PMS coverage on the surface of quartz sand. The adsorption rate of PMS on the surface of quartz sand is not uniform but decreases exponentially with the increase in PMS adsorption on the surface of quartz sand. The mixed (MO) model shows that the adsorption rate of PMS on the surface of quartz sand is not only proportional to the concentration of PMS in water, and the concentration of PMS in water is proportional to the square.

Author Contributions: Conceptualization, J.L., T.L. and T.C.; investigation, W.Y., K.C. and L.Y.; writing—original draft, J.L., T.L. and K.C.; writing—review and editing, J.C. and Z.Y. All authors have read and agreed to the published version of the manuscript.

Funding: This research was funded by the Basic and Frontier Research Programs of Chongqing Science and Technology Commission of China (Grant No. cstc2018jcyjAX0342), This research was funded by the Science and Technology Research Project of the Chongqing Municipal Education Commission of China (grant No. KJQN201901540, No. KJQN201901542, No. KJQN202101). Financial support from the Chongqing Science and Technology Bureau (Grant No. CSTB2022NSCQ-MSX1362, No. sl202100000141, Grant No. CSTB2022BSXM-JCX0152), the Key Laboratory of Ministry of Education of Enhanced Oil and Gas Recovery (Grant No NEPU-EOR-2022-02), and the Science Foundation of Chongqing University of Science and Technology, Chongqing (Grant No. 182101002) is greatly acknowledged.

Data Availability Statement: The data used to support the findings of this study are available from the corresponding author upon request.

Conflicts of Interest: The authors declare no conflict of interest.

eferences

Kang, W.; Zhou, B.; Yang, H.; Li, X.; Zhang, H.; Tang, X.; Gao, Y. Research progress of polymer microspheres for oilfield drive conditioning. *Polym. Mater. Sci. Eng.* **2020**, *36*, 173–180.

Wang, S.; Kuai, J.; Song, S.; Ju, Y.; Li, X.; Wang, C. Advances in the study of core-shell polymer microspheres for dissection. *Petrochem. Appl.* **2021**, *40*, 7–10+14.

Yang, H.; Kang, W.; Yu, Y.; Yin, X.; Wang, P.; Zhang, X. A new approach to evaluate the particle growth and sedimentation of dispersed polymer microsphere profile control system based on multiple light scattering. *Powder Technol.* **2017**, *315*, 477–485. [CrossRef]

Wang, J.; Guo, X. Adsorption kinetic models: Physical meanings, applications, and solving methods. *J. Hazard. Mater.* **2020**, *390*, 122156. [CrossRef] [PubMed]

Mozaffari, M.M.; Kermani Vahid, K.; Vahab, G.; Anis, A.; Mika, S. Adsorption isotherm models: A comprehensive and systematic review (2010–2020). *Sci. Total Environ.* **2021**, *812*, 151334. [CrossRef] [PubMed]

Zhou, S.; Ning, Y.; Wang, H.; Liu, H.; Xue, H. Investigation of methane adsorption mechanism on Longmaxi shale by combining the micropore filling and monolayer coverage theories. *Adv. Geo-Energy Res.* **2018**, *2*, 269–281. [CrossRef]

Wang, H.; Kou, Z.; Guo, J.; Chen, Z. A semi-analytical model for the transient pressure behaviors of a multiple fractured well in a coal seam gas reservoir. *J. Pet. Sci. Eng.* **2021**, *198*, 108159. [CrossRef]

Zhang, X.; Deng, J.; Wang, Y.; Li, S.; Zhao, H.; Liu, C.; Ma, Z.; Sun, Z. Adsorption characteristics of Pb (II) ions on sulfidized hemimorphite surface under ammonium sulfate system. *Int. J. Min. Sci. Technol.* **2023**, in press. [CrossRef]

Kou, Z.; Zhang, D.; Chen, Z.; Xie, Y. Quantitatively determine CO_2 geosequestration capacity in depleted shale reservoir: A model considering viscous flow, diffusion, and adsorption. *Fuel* **2022**, *309*, 122191. [CrossRef]

10. Yang, H.; Kang, W.; Liu, S.; Bai, B.; Zhao, J.; Zhang, B. Mechanism and influencing factors on the initial particle size and swelling capability of viscoelastic microspheres. *J. Dispers. Sci. Technol.* **2015**, *36*, 1673–1684. [CrossRef]
11. Yao, C.; Lei, G.; Li, L.; Gao, X. Preparation and characterization of polyacrylamide nanomicrospheres and its profile control and flooding performance. *J. Appl. Polym. Sci.* **2013**, *127*, 3910–3915. [CrossRef]
12. Xiao, Q. Adsorption performance of polyacrylamide on coal dust. *Drilling Completion Fluids* **2013**, *30*, 46–48+94,95.
13. Liu, Y.; Xu, G.; Ju, Y.; Sun, L.; Zhang, F.; Zhao, W.; Zhnag, K. UV spectrophotometric determination of polymer microsphere output solution concentration. *Sci. Technol. Eng.* **2015**, *15*, 145–149.
14. Kalam, S.; Abu-Khamsin, S.A.; Kamal, M.S.; Patil, S. Surfactant Adsorption Isotherms: A Review. *ACS Omega* **2021**, *6*, 32342–32348. [CrossRef] [PubMed]
15. Zhang, H.; Zhang, J. A review of thermodynamic equilibrium and kinetic models for biosorption. *Tianzhong J.* **2009**, *24*, 19–22.
16. Wang, Y.; Zhou, M. Structure and adsorption properties of cotton-based activated carbon fibers. *Synth. Fiber Ind.* **2009**, *32*, 11–14
17. Zhu, S.; Ye, Z.; Song, R.; He, D.; Gao, Y.; Li, J. Adsorption Behavior of Dendritic Hydrophobic Conjugated Polymers. *Fine Petrochem.* **2021**, *38*, 12–17.

processes

Article

Prediction of Single-Well Production Rate after Hydraulic Fracturing in Unconventional Gas Reservoirs Based on Ensemble Learning Model

Fan Ye [1,2], Xiaobo Li [1,2], Nan Zhang [3] and Feng Xu [1,4,*]

[1] Research Institute of Petroleum Exploration & Development, PetroChina, Beijing 100083, China
[2] Artificial Intelligence Technology R&D Center for Exploration and Development, CNPC, Beijing 100083, China
[3] Yunnan Shale Gas Exploration and Development Co., Ltd., Kunming 650000, China
[4] China National Oil and Gas Exploration and Development Co., Ltd., Beijing 100034, China
* Correspondence: jack_xf@126.com

Abstract: To address the significant challenges in determining the single-well production of tight gas and shale gas after hydraulic fracturing, artificial intelligence (AI) methods were implemented. Machine learning (ML) algorithms such as random forest (RF), extremely randomized trees (ET), lightweight gradient boosting machines (LightGBM), gradient boosting regression (GBR), and linear regression (LR) were utilized in conjunction with reservoir geology, engineering parameters, and production data to develop several foundational models for forecasting the production of unconventional gas wells. The accuracy of these models was evaluated. Based on this, improvements in the models' predictive accuracy and generalizability were achieved through the ensemble of machine learning models. Furthermore, this paper selected two representative tight and shale gas reservoirs to demonstrate the application of the ensemble model for well production forecasting, and a comparative analysis with actual production data was conducted. For tight gas reservoir A, the blending model achieved an MAE of 0.8419 and an MSE of 1.0930, with an R^2 score of 0.8812. For shale gas reservoir B, the blending model achieved an MAE of 1.4841 and an MSE of 3.1629, with an R^2 score of 0.9524. The results of the case studies indicate that the ensemble model approach employed in this study has a higher predictive accuracy and reliability than a single machine learning algorithm, and is capable of handling high-dimensional, large-scale, and imbalanced data, offering scientific validation and technical support for the assessment of the well productivity in tight and shale gas wells.

Keywords: tight gas; shale gas; hydraulic fracturing; gas rate prediction; machine learning; ensemble learning

Citation: Ye, F.; Li, X.; Zhang, N.; Xu, F. Prediction of Single-Well Production Rate after Hydraulic Fracturing in Unconventional Gas Reservoirs Based on Ensemble Learning Model. *Processes* **2024**, *12*, 1194. https://doi.org/10.3390/pr12061194

Academic Editor: Jorge Ancheyta

Received: 30 April 2024
Revised: 4 June 2024
Accepted: 7 June 2024
Published: 11 June 2024

1. Introduction

Currently, tight gas and shale gas reservoirs are increasingly recognized as significant contributors to the overall energy resources, complementing the traditional oil reservoirs. However, unconventional gas reservoirs such as tight gas and shale gas generally have characteristics such as low porosity, ultra-low permeability, and strong heterogeneity. They do not have natural productivity, and it is almost impossible to obtain an industrial gas flow without using fracturing measures. Moreover, the fracture pattern formed by multi-stage fracturing or volume fracturing is very complex, and the flow characteristics are not clear, which increases the difficulty of predicting single-well production [1–10].

In recent years, scholars from both China and other major gas-producing regions, including North America, have conducted extensive research on the prediction of production rates in tight gas and shale gas wells, using methods such as empirical formulas, analytical models, and numerical simulations [11–16]. Empirical formulas and analytical models are mainly based on theoretical models, which make it difficult to fully consider the complex flow characteristics of unconventional gas reservoirs. Furthermore, different models have different applicable conditions, resulting in significant deviations in the prediction results. Numerical models usually simplify the actual reservoir characteristics, making it challenging to accurately characterize the complex reservoir structure and dynamic fracture networks, leading to significant uncertainties in the prediction results. Therefore, it is necessary to explore integrated learning models for predicting the single-well production in unconventional gas reservoirs.

Machine learning is a branch of artificial intelligence that enables computers to learn patterns and models from data without the need for explicit programming. The fundamental principle of machine learning is the use of mathematical and statistical methods to construct mathematical models. These models are adjusted through optimization algorithms to best fit the data and generalize to new data. Common machine learning algorithms include multiple linear regression, decision trees, random forests, support vector machines, neural networks, and so on. Machine learning can delve into data to discover the underlying patterns and handle complex nonlinear relationships, and is characterized by high computational accuracy and the ease of application. It is also applicable in the field of oil and gas development.

Recently, scholars have conducted a series of studies using machine learning methods for single-well production prediction in unconventional gas reservoirs. In 2015, Ma established a sample set of influencing factors on production capacity [17]. They used BP neural network and support vector machine methods to predict the production rate of tight gas wells. The relative error of the support vector machine method ranged from 7.06% to 19.59%, while the relative error of the BP neural network ranged from 8.49% to 37.79%. In 2016, Tian et al., based on an improved BP neural network, established a shale gas production prediction model with production time, cumulative gas production, and reservoir pressure as the input layer, and gas production as the output layer [18]. The relative error prediction was 8.2%. In 2017, Zhu et al. used an adaptive threshold denoising algorithm to remove noise and then applied the BP neural network to nonlinearly fit the reservoir transformation data, thereby obtaining a production prediction model for shale gas wells [19]. Compared to the traditional BP neural network, this model showed improvements in accuracy and stability. In 2019, Lu used reservoir thickness, porosity, permeability, saturation, tubing pressure, water–gas ratio, and other constraints. They established production prediction models using recurrent neural networks and long short-term memory neural networks [20]. The average relative error of the prediction results was less than 10%. Ma et al. proposed a machine learning-based method for the uncertain prediction of shale gas production capacity using a combination of maximum information coefficient correlation analysis, a hybrid support vector machine technique, and Markov chain Monte Carlo simulation [21]. An analysis of 24 shale gas well examples showed an accuracy rate of 70.8% using this method. Fan et al. used the gray correlation method to determine the main controlling factors and weights of shale gas wells [22]. They then used a BP neural network method optimized by genetic algorithms and combined it with data mining techniques to fully utilize the geological and engineering parameters. They established a shale gas rate prediction model for horizontal wells with volume fracturing, with an average prediction error of 8.76%. In 2020, Li et al. used a NARX neural network based on a time series to establish a production rate prediction model between oil pressure and production with a prediction accuracy of up to 90% in one year of production time [23]. Chen et al. used a genetic algorithm to optimize the weight and threshold of the BP neural network and established a gas rate prediction model for shale gas horizontal wells with volume fracturing [24]. Zhao et al. used horizontal gas wells as training samples and constructed a production prediction model for horizontal wells using the BP neural network and a support vector machine [25]. The research results showed that the support vector machine had a higher prediction accuracy than the BP neural network, with a prediction accuracy of 90% for the absolute open flow rate. Yan et al., based on actual data from the Fuling shale gas wells, established a shale gas production prediction model using deep neural networks, support vector machines, and extreme gradient boosting methods, and analyzed the advantages, disadvantages, and importance of various models [26]. He et al. considered the geological factors and engineering factors comprehensively based on machine learning methods [27]. They applied the random forest method to predict the shale gas well production rate, achieving an accuracy of over 90% in the prediction results. Currently, previous studies mainly rely on a single machine learning model to predict the single-well production rate in unconventional gas reservoirs, which has certain feasibility but further improvement is needed in terms of the accuracy and stability of production prediction.

Ensemble learning is a machine learning method that leverages the strengths of multiple different learners and combines their outputs through certain strategies to achieve better results than by individual learners. The advantages of ensemble learning include reducing the risk of overfitting, improving the generalization ability, and handling high-dimensional, large-scale, and imbalanced data [28]. In this study, for the challenging task of determining the single-well production rate in tight gas and shale gas, machine learning algorithms such as random forest (RF), extremely randomized trees (ET), LightGBM (light gradient boosting machine), gradient boosting regression (GBR), and linear regression (LR) were used. We selected these algorithms due to their proven effectiveness in handling high-dimensional data and capturing the complex non-linear relationships inherent in reservoir and production data. These models are particularly adept at dealing with the imbalanced and noisy nature of the datasets typically encountered in gas production prediction tasks [29–32]. By integrating geological, reservoir, engineering parameters, and production data, several basic models for predicting the production in unconventional gas single wells was established, and the accuracy of each model was validated. Based on the predictions of multiple basic models, ensemble machine learning was employed to further improve the accuracy and generalization ability of the prediction model. This method can effectively utilize existing data resources to enhance the understanding and evaluation of tight gas reservoirs, providing a scientific basis for optimizing the development plans and improving the production efficiency.

2. Process for Predicting the Gas Rate in Unconventional Gas Reservoirs

Predicting the gas production rates in unconventional gas reservoirs involves understanding the fundamental principles of reservoir engineering and fluid dynamics. Unconventional reservoirs, such as shale gas and tight gas, have unique characteristics that influence the gas flow and production rates. Machine learning models enhance the traditional prediction methods by leveraging large datasets and identifying complex patterns that may not be apparent through conventional techniques. By incorporating features such as: (1) reservoir properties: porosity, permeability, and saturation; (2) operational parameters: pressure, temperature, and number of fracturing stages; (3) historical production data: past production rates and decline trends.

Machine learning models can provide more accurate and robust predictions. In this study, we employed models such as random forest (RF), extremely randomized trees (ET), light gradient boosting machine (LightGBM), gradient boosting regression (GBR), and linear regression (LR). These models were chosen for their ability to handle non-linearity, the interactions among variables, and high-dimensional data. By integrating these theoretical principles with advanced machine learning techniques, our approach aims to deliver precise gas production forecasts, which are crucial for optimizing field development and management strategies.

The main steps for predicting single-well production in unconventional gas reservoirs based on artificial intelligence analysis are as follows:

(1) Data collection and cleaning. Gather relevant data on the geology, reservoirs, and engineering, perform quality checks on the data, handle abnormal values and missing values, and normalize or standardize the data.

(2) Feature engineering and dimensionality reduction. Based on the characteristics of the dataset, select appropriate methods to extract the features specific to tight gas and shale gas rates, transforming high-dimensional raw data into low-dimensional feature vectors to reduce the redundant information and noise interference.
(3) Model construction and training. Choose multiple machine learning models according to the prediction objectives and problem types, utilize the historical data of known production as the training set, optimize and fit the model parameters, and evaluate the generalization ability of the models using metrics such as Mean Absolute Error (MAE) and Mean Squared Error (MSE).
(4) Production rate prediction and result analysis. Utilize the trained model to predict the production rate of the testing set and compare it with the actually observed values. Adjust or improve the above models based on the accuracy and stability of the prediction results.
(5) Ensemble of machine learning models. To further enhance the accuracy of the production rate predictions, ensemble learning can be used to integrate the prediction results of multiple base models, improving the generalization ability and accuracy of the machine learning models.

3. Production Rate Prediction in Tight Gas Field A

3.1. Overview

Gas Field A had a depth of 10,499–11,483 ft and an average effective thickness of 31 ft in the gas reservoir. The porosity ranged from 5.0% to 10.0%, and the average permeability was 0.82 mD. The pressure coefficient ranged from 0.87 to 0.94, and the temperature of the gas reservoir was between 100 and 110 °C. Starting from 2010, the field has been developed through the implementation of horizontal well development using a 3937 × 1969 ft irregular rhombus-shaped well pattern. Fracturing treatments are mainly conducted using open-hole packers for zonal isolation and supplemented with multiple fractures or volume fracturing within each section. The total injecting fluid volume per well ranged from 29.1 to 256.6 mcf, with an average of 120.8 mcf. The total sand volume per well ranged from 3.0 to 28.0 mcf, with an average of 13.7 mcf. The initial production rates of over 200 wells in the study area ranged from 0.8 to 7.4 mmcf/d. Figure 1 shows the typical production curve of a single well in Gas Field A.

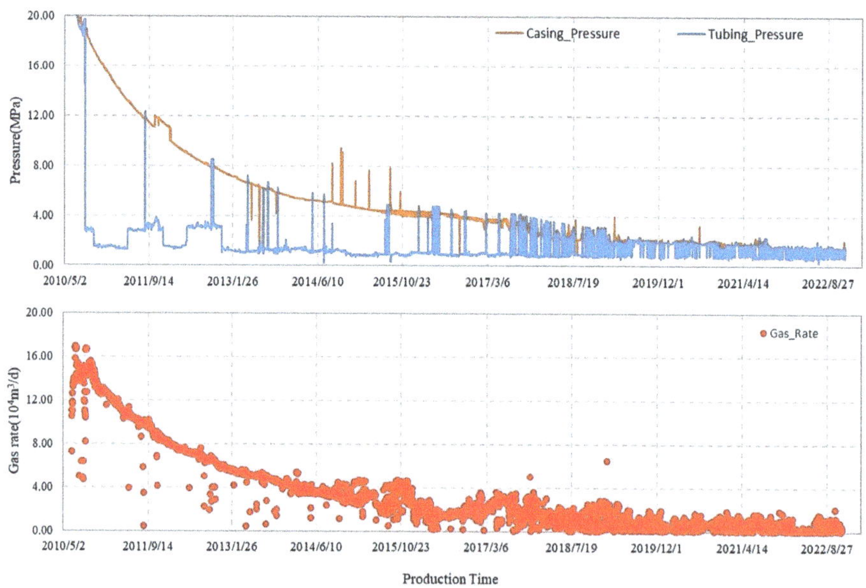

Figure 1. Typical production curve of deviated wells in Gas Field A.

3.2. Data Processing and Preparation

Data processing and preparation are complex and meticulous processes that require the selection of appropriate tools and techniques based on different data types and scenarios to achieve the optimal results. In the feature selection process, we incorporated both analytical insights and significant physical parameters critical to gas production, such as reservoir, geological parameters, and operational parameters. This holistic approach ensures that our model captures essential factors influencing the production rates, thereby enhancing the model's predictive accuracy and validity.

Geological reservoir parameters, engineering parameters, and production data were collected from over 200 wells in Gas Field A, resulting in the acquisition of the raw dataset. Some of the raw data (3.6%) were missing and were filled in using the interpolation method, while abnormal values were treated accordingly. The dataset consisted of 21 data features, and using all features for training in the model would significantly increase its complexity and computational time. This study utilized pairwise relationship graphs to visualize the relationships between two variables, assessing both the single-well production rate and the 21 features, as well as evaluating whether there were significant correlations among the 21 feature parameters. Figure 2 analyzes the pairwise relationship graph between the gas production rate and effective thickness.

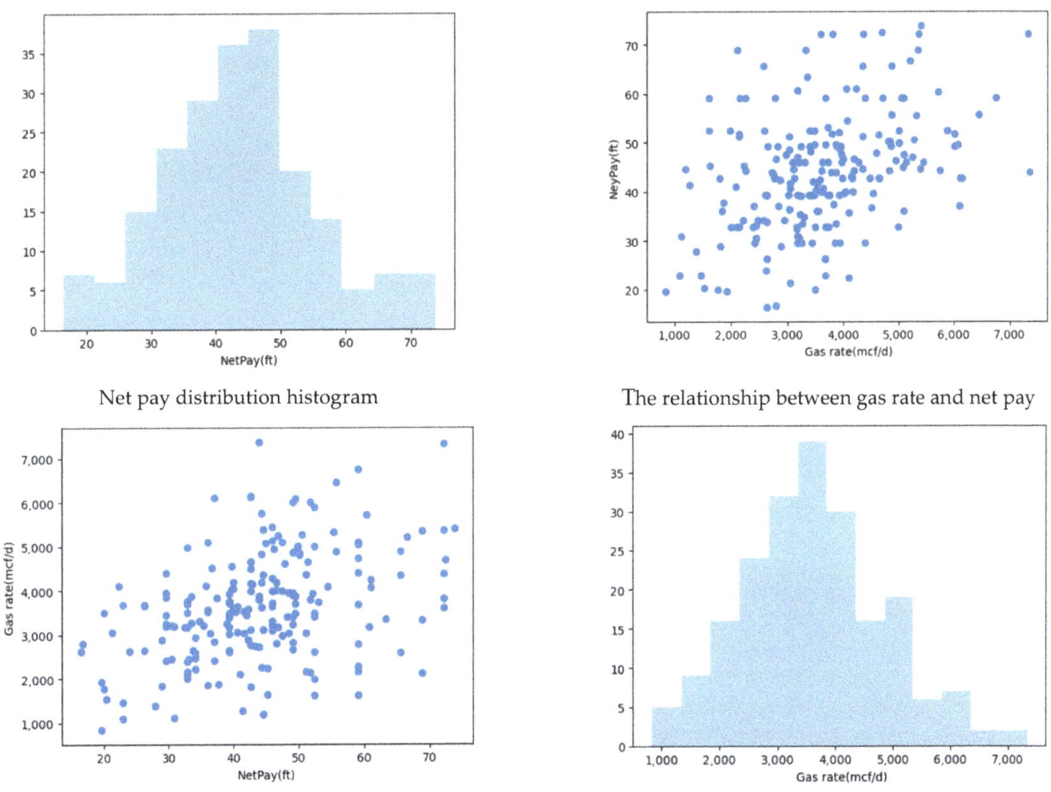

Net pay distribution histogram

The relationship between gas rate and net pay

The relationship between net pay and gas rate

Gas rate distribution histogram

Figure 2. Pair plot between gas rate and effective thickness in tight gas reservoirs.

After the analysis, the factors with a poor correlation to the production rate were removed to ensure both a sufficient data volume and improved computational efficiency. The processed dataset included 13 factors that are significantly related to the production rate, including the reservoir type, effective thickness, porosity, average gas saturation, length of the horizontal section, length of the fractured section, total fluid volume, total proppant volume, and so on. The main statistical indicators for each parameter in the dataset are detailed in Table 1.

Table 1. Main statistics for each parameter in the dataset.

Indicator \ Para. Name	Reservoir Type	Netpay(ft)	Porosity(%)	Ave. S_g (%)	Controlled Reserves/Well (mmcf)	Horizontal Length (ft)	Rate of Sand Body (%)
Min.	1.0	16.4	6.0	28.1	1448	1981.6	16.3
Max.	3.0	73.8	11.2	73.1	10,877	7217.8	100.0
Mean	1.9	44.0	8.5	55.6	4944	3568.6	85.3
Std. Dev.	0.7	12.1	0.9	9.4	1766	614.8	15.8

Indicator \ Para. Name	Effective Rate of Sand Body (%)	Total Liquid Volume (mcf)	Total Sand Volume (mcf)	Flowback Volume (mcf)	Rate of Flowback (%)	Number of Fracturing Sections
Min.	8.3	29.1	3.0	3.6	2.6	5.0
Max.	100.0	256.6	28.0	45.3	75.1	14.0
Mean	61.1	120.8	13.7	18.1	17.8	7.1
Std. Dev.	21.1	39.1	4.2	5.5	9.3	1.4

The above 13 features had different dimensions and significant differences in numerical values among the different indicators. To eliminate the influence of data dimensions, improve the data comparability, and prevent the model from being overly sensitive to features with larger values or disregarding features with smaller values, it was necessary to perform data normalization. This process restricts the preprocessed data within the range of [0, 1]. Normalization was separately applied to the training set and the test set, using the parameters obtained from the training set (such as maximum, minimum, mean, standard deviation, etc.) to normalize the test set. This ensured the consistency between the training set and the test set.

3.3. Model Selection

The collected geological, reservoir, and engineering datasets were randomly divided into the training and testing sets to avoid bias or noise in the dataset that may affect the model and to enhance the model's generalizability and accuracy. The proportion of the training and testing sets depended on the dataset size, complexity, and specific task, without a fixed standard. Considering that the dataset consisted of over 200 samples, it was important to ensure a sufficient amount of data for the model training in the training set while having adequate data in the testing set for the evaluation of the model's performance In this study, the dataset was divided into an 8:2 ratio for the training and testing sets, with 80% for the training set and 20% for the testing set.

For predicting the production of the tight sand gas rate, common algorithms for regression problems were selected: random forest (RF), extremely randomized trees (ET) LightGBM (light gradient boosting machine), gradient boosting regression (GBR), and linear regression (LR). The models' accuracy and generalization capabilities were evaluated through multiple calculations and cross-validation, resulting in the average absolute error (MAE) and mean squared error (MSE) for each model. Further model accuracy enhancement was achieved by fine-tuning the hyperparameters. The randomly generated testing set (42 wells) was used for the prediction, and the calculation results are detailed in Table 2 The accuracy of the five models for predicting the tight gas production was 0.8553, 0.8642 0.8129, 0.8017, and 0.8251, respectively. Additionally, Q-Q plots were generated for each model to observe the deviation between the calculated and predicted values, as shown in Figure 3.

Table 2. Comparison of prediction results for single-well production in tight gas using five different models.

Model Name	MAE	MSE	R^2
RF	0.9525	1.3319	0.8553
ET	0.9022	1.2497	0.8642
LightGBM	1.0335	1.7214	0.8129
GBR	1.0638	1.8242	0.8017
LR	0.9582	1.6091	0.8251

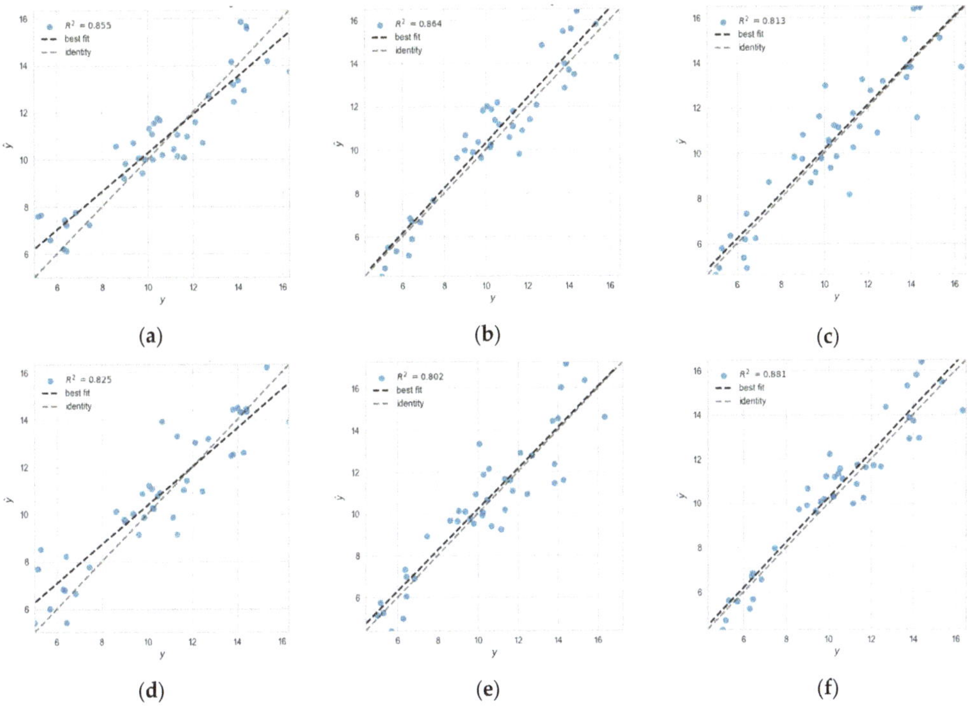

Figure 3. Q-Q plot of the prediction results for tight gas production rate. (**a**) RF model, (**b**) ET model, (**c**) LightGBM model, (**d**) LR model, (**e**) GBR model, (**f**) Blending model.

To further improve the accuracy of the model predictions, ensemble learning was used to combine the predictions from multiple base models to enhance the model's generalization ability and accuracy. Ensemble learning mainly includes two common methods: blending and stacking. Blending is a simple ensemble learning method that takes the predictions from multiple base models as features and inputs them into a meta-model to obtain the final predictions. Stacking is a more complex ensemble learning method that also takes the predictions from multiple base models as the features, but inputs them into one or more meta-models to obtain the final predictions. Comparatively, while stacking can use cross-validation to reduce overfitting and employ multiple layers of meta-models to increase the model's complexity and expressive power, its implementation is complex and requires the tuning of multiple parameters, resulting in lower computational efficiency. On the other hand, blending has the advantage of simplicity, does not require cross-validation, and has a higher computational efficiency. However, it should be noted that potential overfitting should be avoided as much as possible. See Figure 4 for the computational flow.

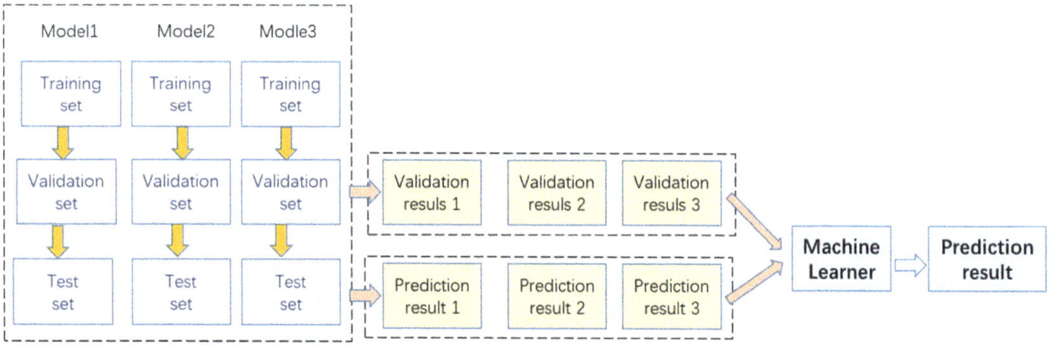

Figure 4. Blending process flowchart.

Using ensemble learning, the prediction results of three models, namely Random Forest, Extremely Randomized Trees, and Linear Regression, which demonstrated good performance in forecasting gas production, were used as the new features. These prediction results, combined with the original features, formed the new feature space of the blended training set. Then, an ensemble model was utilized to obtain the final prediction for tight sand gas production rate. The calculated results show that the blending model, by leveraging the advantages of multiple models, reduced the average absolute error (MAE) and mean squared error (MSE) compared to the five individual models mentioned above. The blending model achieved an MAE of 0.8419 and an MSE of 1.0930, with an R^2 score of 0.8812, indicating a high predictive accuracy and reliability. The detailed results can be found in Table 3. This demonstrates the applicability of ensemble learning, specifically the blending method, for tight gas well production prediction.

Table 3. Prediction results of the blending model for gas rate in tight gas reservoirs.

Model Name	MAE	MSE	R2
Blending	0.8419	1.0930	0.8812

4. Production Rate Prediction in Shale Gas Field B

4.1. Overview

Gas Field B had a depth ranging from 8202 to 13,123 ft. The average thickness of the shale reservoir was 152.2 ft, while the average thickness of the layers was 15.1 ft. The total organic carbon (TOC) content ranged from 1.5% to 3.5%, with an average of 3.2%. The porosity ranged from 3% to 8%, with an average of 6%. The gas saturation ranged from 50% to 70%, with an average of 63%. The content of brittle minerals ranged from 65% to 87%, with an average of 74%. Horizontal wells with volume fracturing were used as the development approach to achieve complex fracture networks, increase the effective stimulation volume and fracture conductivity, and reduce the distance between the shale gas and fractures. A staged technique was used, combining bridge plugs and clustered cable perforation. The length of the fracturing sections ranged from 148 m to 361 ft, with each section having 4 to 12 clusters, and cluster spacing of 16 to 33 ft. The average sand concentration was 13,228 pounds/ft, and the average fluid intensity was 3.0 mcf/ft. The pumping capacity ranged from 0.4 mcf/min to 0.6 mcf/min. The initial production rates of the 193 wells in the study area ranged from 1.2 to 1.9 mmcf/d. Figure 5 illustrates a typical single-well production curve in the study area.

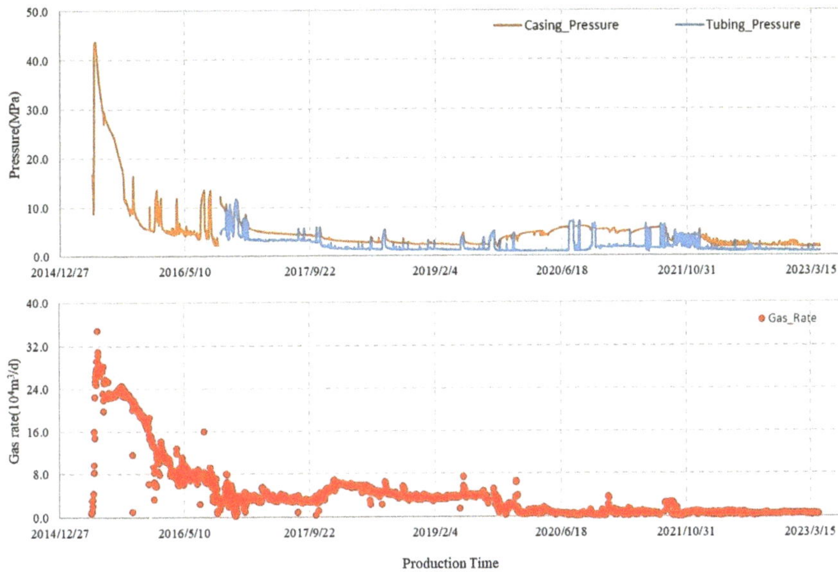

Figure 5. Typical Production curve of deviated wells in Gas Field B.

4.2. Production Forecast

Datasets were collected from 207 wells in shale gas blocks in Gas Field B, including the geological reservoir parameters, engineering parameters, and production data, to obtain the original dataset. The original data had some missing values (1.0%) which were filled in using interpolation, and the abnormal values were processed. The dataset consisted of 37 data features. Through pairwise relationship analysis, the factors with weak correlations to the production rate were eliminated. After processing, 21 factors were identified that were correlated to the production, including sweet spot zoning, the shut-in time, the length of the horizontal section, the reservoir depth, the drilling encounter rate, the layer thickness, the single-well controlled reserves, the well spacing, the length of the fracturing reconstruction, the length of the fracturing stage, the number of fracturing stages, the fluid volume per stage, the sand volume per stage, the percentage of fine sand, the reservoir classification, and the pressure. Figure 6 analyzes the pairwise relationship between the single-well production rate of shale gas and fracture length.

The above 21 features had different dimensions and significant differences in the numerical values among the indicators. To preprocess the data, normalization was performed on the training and testing sets, limiting the preprocessed data to the range of [0, 1]. Considering that the collected data samples consisted of 193 wells, to ensure sufficient data for the model training in the training set and an adequate amount of data to evaluate the model's performance in the testing set, a ratio of 8:2 was used to split the dataset into the training and testing sets, with 80% of the data allocated to the training set and 20% to the testing set. In this study, random forest (RF), extremely randomized trees (ET), light gradient boosting machine (LightGBM), gradient boosting regressor (GBR), and linear regression (LR) were used for predicting the production of tight sand gas. The models' average absolute error (MAE) and mean squared error (MSE) were obtained through multiple computations to assess the accuracy and generalization of the models using cross-validation. Further accuracy improvement was achieved by tuning the parameters of each model. Generated test data (39 wells) were fed into the models for the prediction. The accuracy rates for predicting tight gas production by applying the five models were 0.8898, 0.9468, 0.9045, 0.9109, and 0.8486, respectively. The calculation results can be found in Table 4. Similarly, Q-Q plots were created for each model to observe the deviation between the calculated values and the predicted values, as shown in Figure 7.

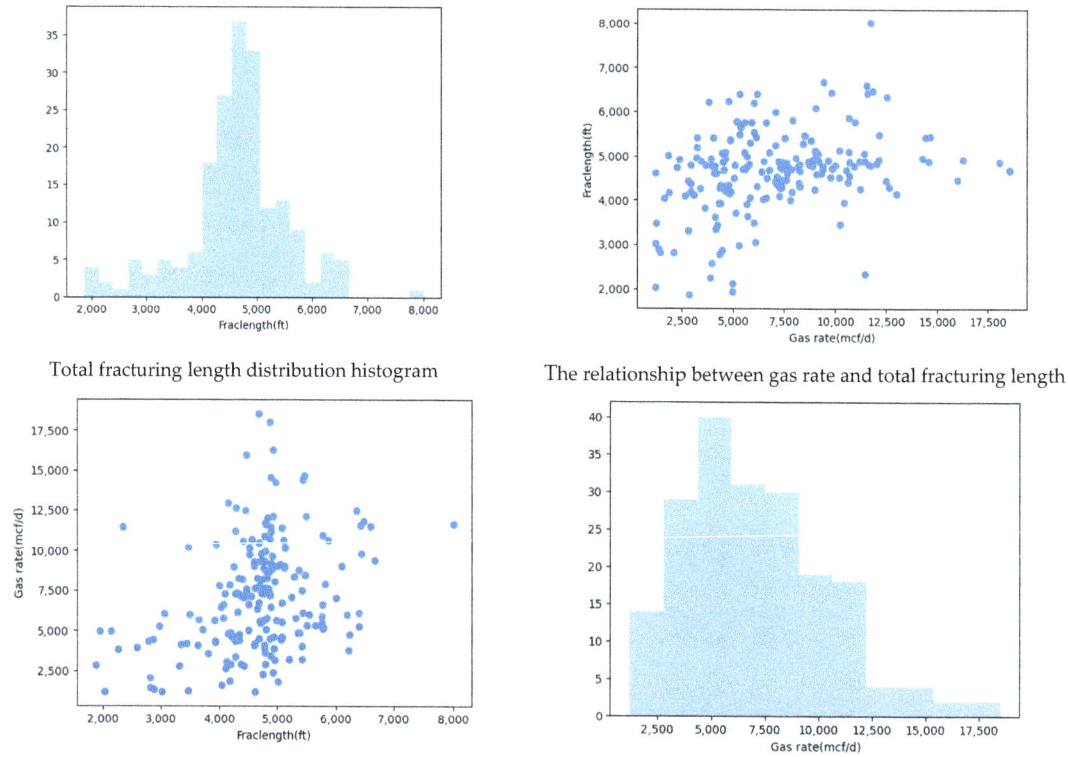

Total fracturing length distribution histogram

The relationship between gas rate and total fracturing length

The relationship between total fracturing length and gas rate

Gas rate distribution histogram

Figure 6. Pair plot between gas rate and fracture length in shale gas reservoir.

Table 4. Comparison of prediction results for single-well production in shale gas using five different models.

Model Name	MAE	MSE	R2
RF	1.9597	7.3169	0.8898
ET	1.6428	3.5326	0.9468
LightGBM	1.9183	6.3411	0.9045
GBR	1.6678	5.9207	0.9109
LR	2.6313	10.0568	0.8486

To further improve the accuracy of the model predictions, a blending ensemble learning method was adopted. The predictions of three models, namely extremely randomized trees (ET), light gradient boosting machine (LightGBM), and gradient boosting regressor (GBR), which performed well in the gas production estimation, were used as the new features. Together with the original features, these formed the new feature space of the blended training set. An ensemble model was then used to obtain the final shale gas production prediction rate. The calculation results showed that the blending model, leveraging the advantages of multiple models, achieved a reduction in the average absolute error (MAE) and mean squared error (MSE) compared to the results of the previous five models. The blending model achieved an MAE of 1.4841 and an MSE of 3.1629, with an R^2 score of 0.9524, indicating a high predictive accuracy and reliability. The detailed calculation results can be found in Table 5. This demonstrates that the blending-based ensemble learning model is more applicable for predicting the shale gas well production.

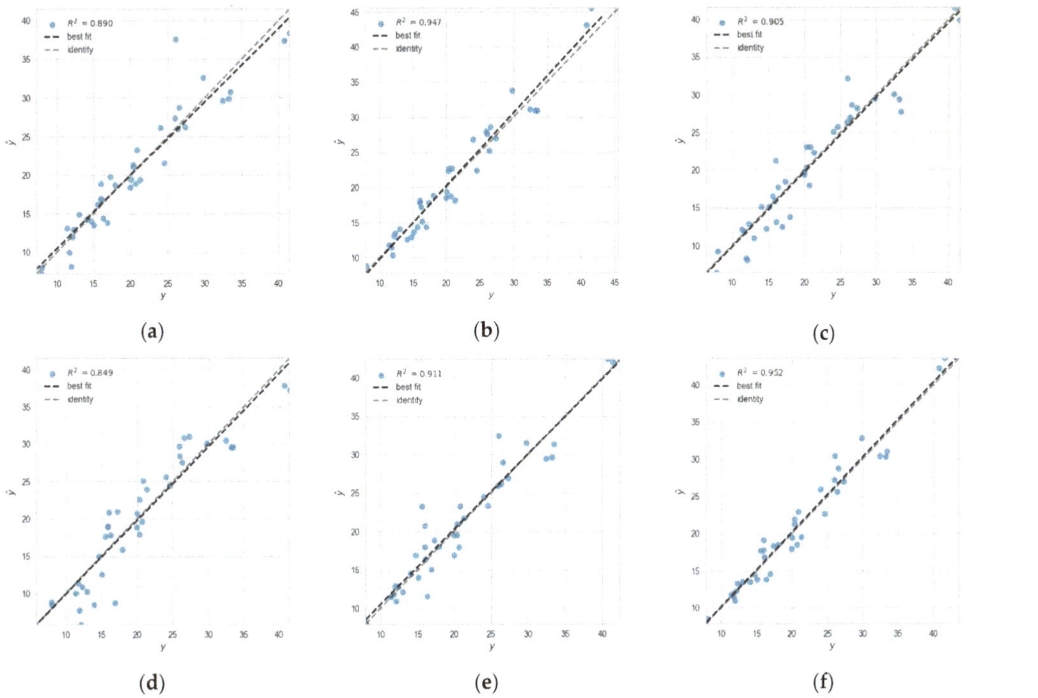

Figure 7. Q-Q plot of the prediction results for shale gas production rate. (**a**) RF model, (**b**) ET model, (**c**) LightGBM model, (**d**) LR model, (**e**) GBR model, (**f**) blending model.

Table 5. Prediction results of the blending model for gas rate in shale gas reservoirs.

Model Name	MAE	MSE	R2
Blending model	1.4841	3.1629	0.9524

5. Conclusions and Future Work

(1) This article proposes a reasonable production prediction model and method for single-well tight gas and shale gas rate based on artificial intelligence analysis. The method combines the predictions of multiple base models, such as random forest (RF), extremely randomized trees (ET), light gradient boosting machine (LightGBM), and gradient boosting regressor (GBR), and further improves the accuracy and generalization ability of the model predictions through ensemble learning.

(2) The blending-based ensemble learning model has higher accuracy and reliability for predicting the single-well gas rate of unconventional reservoirs. The blending ensemble model, utilizing the advantages of multiple basic models, predicts the production of the tight gas rate, with an accuracy rate improved to 0.8812. The blending ensemble model also improves the accuracy rate for predicting the shale gas production rate to 0.9524.

(3) Considering the characteristics of the multiple influencing factors and a limited number of effective samples in the gas production prediction rate, further research is planned to be conducted on the production prediction driven by data and knowledge graph integration. The temporal and spatial relationships of oil well production are represented using knowledge graphs, with graph embedding techniques employed to

represent the vector feature graphs. The vector features of the knowledge graph and the features extracted via deep learning are organically fused using embedding layers to achieve an integrated prediction of the gas well production rate through the fusion of data and knowledge.

Author Contributions: Research Methodology, F.Y. and X.L.; Coding, F.Y. and F.X.; Result analysis and discussion, F.X. and F.Y.; Investigation, N.Z. and F.X.; Data Collection, N.Z.; Writing, F.Y. and F.X. All authors have read and agreed to the published version of the manuscript.

Funding: This research was funded by the China National Petroleum Corporation (CNPC)'s frontier and fundamental technology research project "Research on key artificial intelligence technologies for oil and gas exploration and development—Intelligent technology of oilfield development knowledge base (2023DJ8406)".

Data Availability Statement: The data used in this research are confidential.

Acknowledgments: The authors would like to acknowledge China National Petroleum Corporation (CNPC)'s frontier and fundamental technology research project, and also thank all the researchers related to this paper.

Conflicts of Interest: The authors declare that this study received funding from "Research on key artificial intelligence technologies for oil and gas exploration and development—Intelligent technology of oilfield development knowledge base (2023DJ8406)". The funder was not involved in the study design, collection, analysis, interpretation of data, the writing of this article or the decision to submit it for publication.

References

1. Jia, A.; Wei, Y.; Guo, Z.; Wang, G.; Meng, D.; Huang, S. Development status and prospect of tight sandstone gas in China. *Na Gas Ind.* **2022**, *42*, 83–92. [CrossRef]
2. Li, L.G. Development of natural gas industry in China: Review and prospect. *Nat. Gas Ind.* **2022**, *9*, 187–196. [CrossRef]
3. Pan, J.P. New Progress and Outlook of China's Oil and Gas Exploration and Development. *Pet. Sci. Technol. Forum* **2023**, *42*, 23–3
4. Guo, J.; Lu, Q.; Liu, Z.; Zeng, F.; Guo, T.; Liu, Y.; Liu, L.; Qiu, L. Concept and key technology of "multi-scale high-density fracturing technology: A case study of tight sandstone gas reservoirs in the western Sichuan Basin. *Nat. Gas Ind.* **2023**, *43*, 67–7 [CrossRef]
5. Zhang, N. Optimization of horizontal well pattern in southeast area of Su 53 block. *Unconv. Oil Gas* **2021**, *8*, 77–83.
6. Chen, G.S.; Wu, J.F.; Liu, Y.; Huang, H.Y.; Zhao, S.J.; Chang, C.; Zhong, C.X. Geology-engineering integration key technologies fc ten billion cubic meters of shale gas productivity construction in the southern Sichuan basin. *Nat. Gas Ind.* **2021**, *41*, 72–81.
7. Gong, Y.; Gao, H.Q.; Li, X.Y. Study on the distribution characteristics of occurrence modes of shale gas in the Sichuan Basin an its periphery. *Unconv. Oil Gas* **2023**, *10*, 49–56.
8. Li, Q.; Liu, J.; Wang, S.; Guo, Y.; Han, X.; Li, Q.; Cheng, Y.; Dong, Z.; Li, X.; Zhang, X. Numerical insights into factors affectin collapse behavior of horizontal wellbore in clayey silt hydrate-bearing sediments and the accompanying control strategy. *Oce Eng.* **2024**, *297*, 117029. [CrossRef]
9. Micheal, M.; Xu, W.; Jin, J.; Yu, H.; Liu, J.; Jiang, W.; Liu, H.; Wu, H. A multi-scale quadruple-continuum model for productic evaluation of shale gas reservoirs considering complex gas transfer mechanisms and geomechanics. *J. Pet. Sci. Eng.* **202** *213*, 110419. [CrossRef]
10. Wang, Y.; Li, Q.; Dong, W.; Li, Q.; Wang, F.; Bai, H.; Zhang, R.; Owusu, A.B. Effect of different factors on the yield of epox terminated polydimethylsiloxane and evaluation of CO_2 thickening. *RSC Adv.* **2018**, *8*, 39787–39796. [CrossRef]
11. Duong, A.N. An unconventional rate decline approach for tight and fracture-dominated gas wells. In Proceedings of the Canadi Unconventional Resources and International Petroleum Conference, Calgary, AB, Canada, 19–21 October 2010.
12. Zhang, N. A new method of productivity evaluation for fractured horizontal well in tight gas reservoir. *Sci. Technol. Eng.* **20** *14*, 76–80.
13. Ye, J.W. EUR Evaluation Method for Intensive Cutting Fracturing Horizontal Wells in Tight Sandstone Reservoirs. Maste Thesis, Chengdu University of Technology, Chengdu, China, 2021.
14. Zhang, N. Classification evaluation of production dynamic for horizontal well in Su 53 block. *Unconv. Oil Gas* **2021**, *8*, 88–94.
15. Song, H.; Su, Y.; Xiong, X.; Liu, Y.; Zhong, S.C.; Wang, J.J. EUR evaluation workflow and influence factors for shale gas well. N *Gas Geosci.* **2019**, *30*, 1531–1537.
16. Zhao, Y.L.; Liang, H.B.; Jing, C.; Shang, S.F.; Li, C.Y. A new method for quick EUR evaluation of shale gas wells. *J. Southwest Univ. (Sci. Technol. Ed.)* **2019**, *41*, 124–131.
17. Ma, W.D. Productivity Regularity Research of Fractured Well in Tight Gas Reservoir. Master's Thesis, China University Petroleum, Beijing, China, 2015.

18. Tian, Y.P.; Ju, B.S. A model for predicting shale gas production decline based on the BP neural network improved by the genetic algorithm. *China Sci. Pap.* **2016**, *11*, 1710–1715.
19. Zhu, H.; Kong, D.Q.; Qian, X. Shale gas production prediction method based on adaptive threshold denoising BP neural network. *Sci. Technol. Eng.* **2017**, *17*, 128–132.
20. Lu, Z.Y. Research on Production Forecast Method of Gas Well in Tight Gas Reservoir Based on Big Data Analysis. Master's Thesis, Southwest Petroleum University, Chengdu, China, 2019.
21. Ma, W.L.; Li, Z.P.; Sun, Y.P.; Zhang, J.P.; Deng, S.Z. Non-deterministic shale gas productivity forecast based on machine learning. *Spec. Oil Gas Reserv.* **2019**, *26*, 101–105.
22. Fan, Y.; Huang, H.Y.; Wang, X.H.; Chen, J.; Deng, Y.C. Study on the production prediction technology for shale gas horizontal well by applying data mining. In Proceedings of the 31st National Natural Gas Academic Annual Conference, Hefei, China, 31 October–1 November 2019.
23. Li, Y.Z.; Bai, Y.H.; Xu, B.X. Application of NARX model in production performance prediction of tight gas wells. In Proceedings of the International Field Exploration and Development Conference, Chengdu, China, 23–25 September 2020.
24. Chen, J.; Huang, H.Y.; Liu, J.C.; Zeng, B.; Yang, X.R. Production predicting technology of shale gas fracturing Horizontal well in changning area based on the GA-BP neural network mode. *Sci. Technol. Eng.* **2020**, *20*, 1851–1858.
25. Zhao, Q.Y.; Liu, Y.; Liu, L.P.; Chen, H.; Ning, M.; Liu, J.; Gao, W.B.; Miao, R.J. Application of machine learning in the production prediction of horizontal wells in tight gas reservoirs. In Proceedings of the 17th Ningxia Young Scientists Forum Petroleum and Petrochemical Special Forum, Yinchuan, China, 15 September 2021.
26. Yan, Z.M.; Wang, T.; Liu, Z.L.; Zhuang, Z. Machine-learning-based prediction methods on shale gas recovery. *Chin. J. Solid Mech.* **2021**, *42*, 221–232.
27. He, Y.W.; He, Z.Y.; Tang, Y.; Qin, J.Z.; Song, J.J.; Wang, Y. Shale gas well production evaluation and prediction based on machine learning. *Oil Drill. Prod. Technol.* **2021**, *43*, 518–524.
28. Otchere, D.A.; Ganat, O.T.; Gholami, R.; Lawal, M. A Novel Custom Ensemble Learning Model for an Improved Reservoir Permeability and Water Saturation Prediction. *J. Nat. Gas Sci. Eng.* **2021**, *91*, 103962. [CrossRef]
29. Niu, W.; Lu, J.; Sun, Y. Development of shale gas production prediction models based on machine learning using early data. *Energy Rep.* **2022**, *8*, 1229–1237. [CrossRef]
30. Wang, H.; Qiao, L.; Lu, S.; Chen, F.; Fang, Z.; He, X.; Zhang, J.; He, T. A novel shale gas production prediction model based on machine learning and its application in optimization of multistage fractured horizontal wells. *Front. Earth Sci.* **2021**, *9*, 726537. [CrossRef]
31. Syed, F.I.; Alnaqbi, S.; Muther, T.; Dahaghi, A.K.; Negahban, S. Smart shale gas production performance analysis using machine learning applications. *Pet. Explor. Dev.* **2022**, *49*, 21–31. [CrossRef]
32. Liao, L.; Li, G.; Zhang, H.; Feng, J.; Zeng, Y.; Ke, K.; Wang, Z. Well completion optimization in Canada tight gas fields using ensemble machine learning. In Proceedings of the Abu Dhabi International Petroleum Exhibition & Conference, Abu Dhabi, United Arab Emirates, 9–12 November 2020.

processes MDPI

Article

Fast Assisted History Matching of Fractured Vertical Well in Coalbed Methane Reservoirs Using the Bayesian Adaptive Direct Searching Algorithm

Zhijun Li

Exploration and Development Research Institute, PetroChina Huabei Oilfield Company, Cangzhou 062552, China; hzb_lzj@petrochina.com.cn

Abstract: The proper understanding of reservoir properties is an important step prior to forecasting fluid productions and deploying development strategies of a coalbed methane (CBM) reservoir. The assisted history matching (AHM) technique is a powerful technique that can derive reservoir properties based on production data, which however is usually rather time-consuming because hundreds or even thousands of numerical simulation runs are required before reasonable results can be obtained. This paper proposed the use of a newly developed algorithm, namely the Bayesian adaptive direct searching (BADS) algorithm, for assisting history matching of fractured vertical CBM wells to derive reservoir property values. The proposed method was applied on representative fractured vertical wells in the low-permeable CBM reservoirs in the Qinshui Basin, China. Results showed that the proposed method is capable of deriving reasonable estimates of key reservoir properties within a number of 50 numerical simulation runs, which is far more efficient than existing methods. The superiority of the BADS algorithm in terms of matching accuracy and robustness was highlighted by comparing with two commonly used algorithms, namely particle swarm optimization (PSO) and CMA-ES. The proposed method is a perspective in laboring manual efforts and accelerating the matching process while ensuring reasonable interpretation results.

Keywords: coalbed methane; assisted history matching; numerical simulation; Bayesian adaptive direct searching (BADS); fractured vertical well

Citation: Li, Z. Fast Assisted History Matching of Fractured Vertical Well in Coalbed Methane Reservoirs Using the Bayesian Adaptive Direct Searching Algorithm. *Processes* **2023**, *11*, 2239. https://doi.org/10.3390/pr11082239

Academic Editors: Zheng Sun, Tao Zhang, Dong Feng, Wen Zhao and Hung Vo Thanh

Received: 14 May 2023
Revised: 8 July 2023
Accepted: 10 July 2023
Published: 26 July 2023

1. Introduction

Coalbed methane (CBM) has been considered to be a hazardous gas for underground coal mining activities that may potentially lead to serious disasters such as coal and/or gas outburst and gas explosion [1,2]. The development and utilization of the CBM gas is beneficial to improving underground mining safety, reduction of gas emission into the atmosphere and adding to world's energy supply [3]. To date, commercial development of CBM has been active in the countries such as the United States, China, Australia, India and Indonesia [4,5]. Although complex completion techniques such as multi-fractured horizontal well, multi-lateral horizontal and cavity have been proposed to accelerate CBM production, the hydraulic stimulated vertical wells are still the most viable option for economic development of low-permeability CBM reservoirs.

Prior to the commercial extraction of a CBM reservoir, key formation properties exerting profound effect of well productions should be reasonably estimated in order to evaluate the economic feasibility and to deploy proper development strategies (e.g., optimization of well type, placement and drainage schedule). Basically, laboratory measurements are one of the important first steps for estimating the values of the reservoir properties [6]. However, the experimental result may not fully represent the reservoir-scale characteristics due to the strong heterogeneity nature of coal seams scale. Moreover, coal samples used for laboratory experiments are typically retrieved using certain coring techniques (e.g., wireline coring and pressure coring) [7,8] that are quite expensive, and thus the experimental data ar usually available only for a very limited number of evaluation wells. Compared with the laboratory experiment method, history matching of well production data is considered to be the most economic method that gains in-depth understanding of the reservoir behavior and gives estimation of reservoir parameters exerting critical effects on well productions. A successful history match gives confidence about the estimated reservoir parameters and their distribution during production and may indicate how the reservoir may behave under different reservoir management conditions. The history matching method has been extensively used to estimate coal seam properties. Karacan [6] integrated the single-well production history matching with geostatistical proxy model to estimate the dynamic properties including gas content, gas saturation and pressure in the Black Creek coal seam of Black Warrior Basin. Feng et al. [9], Zhou [10] and Zhang et al. [11] demonstrated the use of history matching to evaluate the formation properties based on production data from vertical fractured, casing horizontal and multi-fractured horizontal wells, respectively.

Although history matching is an important and perhaps the most reliable method to estimate the formation properties, it can be quite challenging due to the large number of properties with strong uncertainties properties that lead to a wide range of reservoir responses [12]. The impacts of coalbed parameters on methane production were extensively investigated [13–15]. It was reported that coalbed parameters with the greatest impact on simulation forecasts were adsorbed gas content, desorption isotherm, water saturation, coalbed thickness, permeability, porosity, compressibility and relative permeability. These parameters may exhibit strong variations both across the plane and the layers [11]. Thus, history matching is usually quite time-consuming because a number of trials of numerical simulations may be needed before successful matching is obtained.

The assisted history matching (AHM) method has been proposed and successfully applied for estimating oil and/or gas reservoir properties [16]. The AHM method in the petroleum engineering field was first proposed for conventional sandstone reservoirs to deduce the formation properties that are difficult to measure directly from field-scale historical production data [17]. During the past two decades, AHM has been extended to solve various inverse problems such as (i) interpretation of relative permeability from laboratory core flooding data, (ii) estimation of permeability and skin factor of multilayered reservoir from injection-falloff test and (iii) characterization of hydraulic fractures in shale reservoirs from seismic and/or production data [18–20]. These previous studies indicated that the AHM is capable of deriving reasonable results with the proper experimental design and the implementation of robust algorithms. To date, various algorithms have been integrated with numerical simulation to assist history matching. Commonly used algorithms can be classified into two types. The first type is referred to as evolutionary optimization algorithms, such as the particle swarm optimization (PSO), genetic algorithm (GA), harmony search optimization (HSO), etc. The second type is referred to as the ensemble method, such as the ensemble Kalman filter (EnKF) and ensemble smoother with multiple data assimilation (ES-MDA). These algorithms have been reported to be capable of estimating the reservoir properties based on production data; however, hundreds or even thousands of numerical simulation runs are usually needed in order to obtain reasonable results [21,22]. Since numerical simulation runs regarding real well models are usually quite time-consuming, the AHM method involving these algorithms shares a common drawback of low efficiency.

To the best knowledge of the authors, the AHM method has not yet been applied for estimating the CBM reservoir properties based on well production data. This paper proposed the use of a newly developed high-efficiency algorithm, namely the Bayesian adaptive direct search (BADS) [23], to assist matching fractured vertical CBM well production in order to derive the reservoir properties. The applicability of the proposed method was tested on real CBM wells, and the superiority over typical existing algorithms was demonstrated.

2. Methods

2.1. Basics of the AHM

The AHM generally requires three components, namely a numerical or analytical simulator to simulate production behavior, an error function describing the mismatch between the simulated and real observed data and a specific algorithm to minimize the error function in order to obtain a "successful" matching.

In this study, all simulation runs were conducted using the numerical simulator GEM developed by the Computer Modelling Group [24]. The GEM is a compositional simulator that is capable of modeling CBM transport behavior including adsorption/desorption, gas diffusion, multi-phase seepage and stress- and sorption-induced permeability dynamics The GEM simulator has been widely used to model fluid production behavior [25,26], to optimize well placement [27] and to evaluate the enhanced CBM potential [28] in CBM reservoirs.

The error function to be minimized was defined as the sum of the normalized squared errors between the simulated and observed data. The error function is written as:

$$E = \frac{\sum_{i=1}^{N}\left(y_i^{obs,g} - y_i^{sim,g}\right)^2}{\sum_{i=1}^{N}\left(y_i^{obs,g} - y^{avg,g}\right)^2} + \frac{\sum_{i=1}^{N}\left(y_i^{obs,w} - y_i^{sim,w}\right)^2}{\sum_{i=1}^{N}\left(y_i^{obs,w} - y^{avg,w}\right)^2} \quad (1)$$

where y is the variable to be matched; the superscripts "*obs*" and "*sim*" denote the observed and simulated data, respectively; superscripts "*g*" and "*w*" denote the gas and water respectively; the superscript "*avg*" denotes the averaged value of the observed data; N is the total number of data points.

As stated previously, although a wide range of algorithms have been proposed to assist history matching, their efficiencies are questionable because a number of time-consuming simulation runs are needed before reasonable results can be obtained. Thus, this study proposed the usage of the BADS algorithm to accelerate the history matching process.

2.2. BADS Algorithm

The BADS algorithm is a fast, hybrid Bayesian optimization algorithm designed t solve difficult optimization problems, in particular related to fitting computational model The BADS algorithm was first proposed by [23], which was then introduced by Zhan et al. [29] to be used for estimating relative permeability of coal based on laboratory floodin test data. The BADS is a hybridization of the Bayesian optimization (BO) [30] and mes adaptive direct search (MADS) [31] algorithms. The mathematical details of the BADS a available in references [23,29]. For the completeness of this paper, the basics of the BAD are briefly introduced as follows.

Given a D_n-dimension minimization problem, BADS iteratively updates the solutio variable vector (x) in a similar manner to MADS. In other words, BADS iteratively generat and evaluates the candidate solutions in two separate steps, namely the search and po stages. In the search stage of the kth iteration, BADS generates candidate solutions b performing local optimization of the acquisition function in the neighborhood of th incumbent x_k. Commonly used acquisition functions include the lower confidence bour (LCB), upper confidence bound (UCB) and expected improvement (EI). In this study, w used the LCB acquisition function. The polling directions D_k are then constructed b scaling to the Gaussian process (GP) kernel length scales. In the polling stage, the solutic candidates $\mathbf{x}_{\text{poll}}$ can be updated according to the following equation

$$x_{poll} = \{x_k + \Delta_k^m v : v \in D_k\} \quad ($$

where Δ_k^m is the mesh size at the kth iteration.

If the poll succeeds in sufficiently improving the objective function within three consecutive steps, the incumbent is updated, and BADS switches to a new iteration with mesh and poll sizes multiplied by $\tau = 2$; otherwise, the incumbent remains unchanged, and BADS switches to a new iteration with the mesh and poll sizes divided by $\tau = 2$. These steps are repeated until a preset maximum number of iterations is met, the algorithm stalls or the poll size becomes extremely small. During the iteration process, the accuracy of the surrogate models based on the GP is updated every $2D_n$ to $5D_n$ function evaluations and whenever the accuracy of the current GP is unreliable by refitting the hyperparameters of GP.

2.3. Setup of Reservoir Model

Proper setup of the reservoir model is the first important step to ensure the success of history matching. For the primary depletion of CBM reservoirs where no injector is involved, the single-well history matching is considered to be comparable with multi-well history matching in terms of matching accuracy provided that flowing boundary of each single well is properly defined [6]. Also, a lower number of wells indicates a lower number of parameters and reduced uncertainties, which can accelerate the convergence speed and reduce the risk of being trapped in local optima during history matching. Moreover, the single-well reservoir model has a far less number of grid blocks than multi-well models does and thus runs significantly faster. Therefore, this paper is primarily concerned with the single-well history matching process.

2.3.1. Construction of the Grid Model

Two types of well patterns are commonly used in CBM reservoirs, namely the five-spot (or diamond) and rectangular patterns [27]. For either well pattern, a single-well can be considered to be located in the center of its controlled area that is defined by the flowing boundaries. In this study, 3D uniform Cartesian grid models were constructed for the well to be history matched (Figure 1). Low-permeability coalbeds generally require relatively small well spacings in order to accelerate reservoir pressure drawdown and thus improve gas production [9,32,33]. For example, CBM wells are generally deployed at spacings in the range of 200~300 m in low-permeability CBM reservoirs in the Eastern Ordos and Southern Qinshui basins, China [34]. Zhang et al.'s optimization studies on well placement showed that the optimal well spacing for coalbeds with permeability less than 1.3 mD is approximately 250 m [27]. Considering the well spacing for low-permeability coalbeds, the number of grid blocks was set to be 61 both along the *x*- and *y*-direction. The *x*- and *y*-directions represent the face-cleat and butt-cleat directions, respectively. The dimension of each grid block was calculated to be the real well spacing divided by 61 along the *x*- and *y*-directions, respectively. For single-layered CBM reservoirs, the number of grid blocks in the vertical direction is set be unity. For multi-layered CBM reservoirs, the number of grid blocks in the vertical direction is set be the total number of coal seam layers and impermeable interlayers.

For each grid model, the grid blocks are divided into two regions. The first region represents the coalbed at the in situ condition that is not affected by hydraulic fracturing stimulation. The second region represents the hydraulic stimulated area (SA) around the well. The SA is simplified to be a rectangular area with the length and width in the face-and butt-cleat directions, respectively.

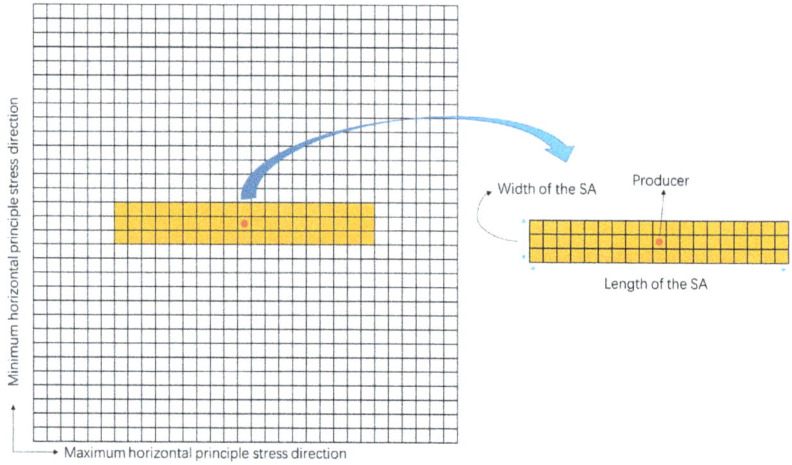

Figure 1. Illustration of the grid model.

2.3.2. Parameters to Be Tuned

Previous studies [12,15] have concluded that key parameters influencing gas and water productions in CBM reservoirs include coal seam thickness, sorption isotherm, porosity, permeability and relative permeability. For a real CBM well, the values of coal thickness, initial pressure and critical desorption pressure can be estimated with well logging and production data information. The sorption isothermal for a specific coalbed can be measured on cored samples in laboratory, which is a known parameter for the well where sample was cored. However, the sorption isothermal may vary from well to well due to differences in stress condition, water content, etc. Therefore, the sorption isothermal was also set to be tunable by adjusting the Langmuir volume and pressure within the common knowledge of the target coalbed. Porosity, permeability, permeability evolution dynamics and relative permeability are parameters associated with strong heterogeneities, which need to be tuned during history matching [11]. The permeability evolution dynamics were represented with the Palmer–Mansoori (PM) model [35], with the compressibility and sorption strain being the tunable parameters; other parameters in the PM model such as Poisson ratio, Young's Modulus and Langmuir pressure were fixed because these parameters are generally less uncertain for a given formation. The relative permeability was represented with a Corey-type model developed by Chen et al. [36] (Equation (3)). The relative permeability is varied by tuning the coefficients including the connate water saturation, endpoint water relative permeability and the exponents for the gas and water curves.

$$k_{rw} = k_{rw0} S_{wn}^{\lambda} \tag{3}$$

$$k_{rg} = (1 - S_{wn})^{\beta} \left(1 - S_{wn}^{\eta}\right) \tag{4}$$

$$S_{wn} = \frac{S_w - S_{wc}}{1 - S_{wc}} \tag{5}$$

where S_{wc} and S_{wn} are the connate and normalized water saturations, respectively; S_w is water saturation; k_{rw} and k_{rg} are relative permeability for the water and gas phases, respectively; k_{rw0} is the endpoint water relative permeability; λ, β and η are exponential coefficients that control the relative permeability curvature.

The hydraulic stimulation results in enhancement in flow capacity, which is represented by multiplying the permeability in the stimulated region by a factor in the range of one thousand to ten thousand [37]. Moreover, the embedment of proppants within stimulated zones aids the retention of the fracture aperture [38]. Therefore, the coal deformation behavior and the consequent permeability dynamics due to reservoir pressure drawdown in the stimulated zones should be different from that in the un-stimulated zones. In this regard, two respective compressibility values were assigned in the stimulated and un-stimulated zones in order to model different deformation behaviors.

As a short summary, a total number of 16 properties need to be tuned during the history matching process, which include: (1) the length and width of the stimulated area; (2) two respective permeability values in the stimulated and un-stimulated zones; (3) two respective compressibility values in the stimulated and un-stimulated zones; (4) two respective sorption strain values in the stimulated and un-stimulated zones; (5) porosity; (6) sorption isotherm represented with Langmuir volume and Langmuir pressure; (7) relative permeability represented with five characteristic coefficients shown in Equations (3)–(5), namely S_{wc}, k_{rw0}, λ, β and η.

2.4. Workflow of the AHM

The workflow of AHM in this study includes the following 5 steps.

(1) Initializing the candidate solution vector x_0 subject to boundary constraints of physical meanings. For the BADS algorithm, the candidate solution is normalized within the range of [0, 1]. The normalized vector x is transformed into the real space X according to the following equation.

$$X = X_{lb} + (X_{ub} - X_{lb}) \cdot x \tag{6}$$

where the subscripts "ub" and "lb" denote the upper and lower boundaries of the parameter in the real space.

(2) Update the reservoir grid model data file and call the numerical simulator to run the model. The well is simulated under the bottom-hole pressure constraint to derive gas and water production.

(3) Calculate the matching error between the simulated and observed production data using Equation (1).

(4) Update the solution candidate with BADS based on the candidate solution and the corresponding error value at the previous iteration.

(5) Check the number of elapsed iterations. If the number of elapsed iterations does not reach the preset tolerance of the number of iterations, go back to step 2; otherwise, cease simulation and output the optimal solution that minimizes the matching error.

3. Case Study

The proposed method was applied on a number of representative CBM wells producing in the southern Qinshui Basin in order to demonstrate its applicability. The coals in the study area are anthracite in rank, which is characterized with low porosity, low permeability and high gas content. The coal formation is considered to be initially saturated with water. The selected wells were produced from the 3# coal seam of the Shanxi Formation. Wells were placed at an average spacing of 250 m, both in the face- and butt-cleat directions. All wells were hydraulically stimulated with silkwater following similar stimulation parameters, e.g., pumping rate, total injected fracturing fluid and sand ratio.

A number of three representative wells (referred to as Well #1, #2 and #3) were selected that are typical in terms of the gas production pattern. These wells are distributed within a distance of 2 km in the same block. Well #1 showed the highest gas producibility among the selected wells, with the maximum and average daily gas rates of 5730 and 1800 m^3/d, respectively. The daily gas rates of this well exhibit the "typical" trend of a CBM well: the gas rates first climbed and then declined sharply with the production duration. The gas production curve of Well #2 also exhibits a first increasing and then a decreasing trend. However, the decline in gas productions was sharp within the initial 500 days and then became gentle after the gas rate dropped to approximately below 1500 m^3/d. The peak and average gas rates were approximately 3000 and 1206 m^3/d, respectively, which were obviously lower than that of Well #1. Compared with the aforementioned two wells, Well

#3 showed a distinctive gas production pattern: the well had been producing at a relatively steady gas rate and did not show a sharp declining trend. The peak and average gas rates of Well #3 were 3000 and 1240 m^3/d, respectively. Wells were not considered that had experienced frequent workovers and showed severe fluctuations in production rates because these wells are associated with more uncertainties such as formation damage.

Since BADS is a stochastic algorithm in nature, uncertainties exist inevitably during the iteration process, and thus the ultimate optimal solution may vary from run to run. Therefore, the AHM was run five times for each well to evaluate the uncertainties associated with the optimal solutions. For each independent run, the searching point was randomly initialized, and the maximum number of function evaluations was set to be 120. Variables that have been addressed to exert minor effects on productions of initially water-saturated CBM reservoirs were set to be fixed, including the sorption time (10 days) and fracture spacing (0.5 cm). The well radius was set to be 0.06m according to the well completion report. For each AHM run, the upper and lower boundaries for variables to be estimated are given in Table 1. The boundary conditions of the sorption isothermals were set according to laboratory tests, which suggest coals in the target area generally have Langmuir volumes and pressures in the ranges of 24 to 40 m^3/t and 1.0 to 6.0 MPa respectively. The boundary conditions for the in situ porosity, permeability and sorption strain were set empirically according to previous studies [11]. Coefficient ranges of the relative permeability model (Equation (3)) were adapted from Chen et al.'s summary on a number of relative permeability data [39]. The microseismic data monitored during the hydraulic fracturing process of CBM wells in the Qinshui basin suggest that the length of the SA is approximately 30 to 130 m, whereas the width of the SA is on the order of magnitude of tens of meters. In this study, the upper and lower boundaries were to be 50 to 100 for the length and 10 to 50 for the width of the SA. The permeability of the SA is associated with strong uncertainties and thus was varied in a relatively wide range of 1 to 50 mD. The compressibility and sorption strain were varied between 0.001 and 0.1 MPa^{-1} and between 0.005 and 0.03, respectively, in order to cover the ranges of anthracite coal according to laboratory tests.

Table 1. Boundary constraints of the variables to be estimated.

Variable	Lower Boundary	Upper Boundary
Porosity	0.001	0.05
In situ permeability, mD	0.01	1.0
In situ Compressibility, MPa^{-1}	0.001	0.1
Sorption strain	0.001	0.03
Langmuir volume, m^3/t	25	40
Langmuir pressure, MPa	1.0	6.0
S_{wc}	0	0.9
k_{rw0}	0.1	1
λ	0.1	10
β	0	5
η	1	10
Length of SA, m	50	120
Width of SA, m	10	50
Permeability of SA, mD	1	50
Compressibility of SA, MPa^{-1}	0.01	0.1

4. Results and Discussion

4.1. Performance of BADS

Figure 2 illustrates the objective function values during the iteration (minimization) process. As can be seen, the objective function decreases dramatically during the initial stage (within a number of 30 function evaluations) and then converges gradually to local optima. This suggests an alternation in mode from global exploration to local exploitation [29].

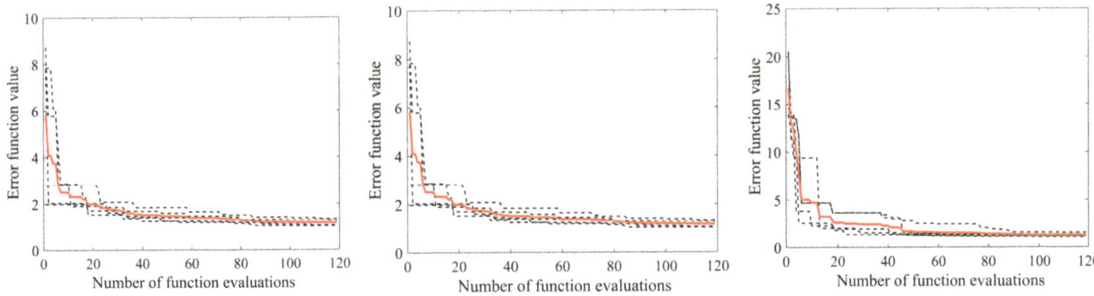

Figure 2. Evolution of the objective function value during the BADS searching process. (**a**) Well #1; (**b**) Well #2; (**c**) Well #3. The black dash lines represent five independent runs while the red solid lines are averaged values of the independent runs.

In the initial stage, BADS operated in a deterministic mode similar to the MADS algorithm, and thus the objection function values overlap with one another for the five independent runs. In the later stage (>30 iterations), BADS operated with stochasticity because the Bayesian process was involved; thus, different independent runs resulted in varied objection function values. However, very small differences were observed between different independent runs, suggesting strong robustness of BADS. It should also be noted that the number of function evaluations required for the BADS algorithm to achieve relatively stable convergence is noticeably small, which were approximately in the range of a number of 50 to 70 function evaluations. Since the numerical simulation run is usually time-consuming for a real model scenario, one may always anticipate less function evaluations in order to accelerate the AHM process. This highlights the high efficiency of the BADS algorithm in assisting achieving "successful" history matching results.

Figure 3 depicts the matching results with the minimized function values for the representative wells. As shown for each well, the independent AHM runs resulted in simulated production curves with slight deviations from one another possibly due to the stochastic nature of the BADS algorithm. Despite the distinct variation trends of the production data, the simulated production curves from all AHM runs agree well with the true ones for both gas and water phases. These observations suggested that the BADS is capable of deriving satisfactory matching results regardless of the production variation behaviors, which proved the adaptability of BADS for solving the AHM problems.

4.2. Comparison with Existing Algorithms

To demonstrate the superiority of BADS, the BADS algorithm was compared with two commonly used algorithms, namely particle swarm optimization (PSO) [40] and the covariance matrix adaptation evolution strategy (CMA-ES) [41] for history matching the aforementioned three wells. PSO and CMA-ES are stochastic algorithms with significant different underlying evolving mechanisms, which have been successfully applied for solving a variety of minimization problems including the AHM in the petroleum engineering community [42,43]. PSO and CMA-ES were independently applied for assisting the matching of the three wells. The maximum number of objective function evaluations was set to be 120, which was identical with that of the BADS. The hyperparameters of the PSO and CMA-ES were set to be default values as suggested by their original inventors. Each algorithm was independently run five times in order to evaluate their uncertainties.

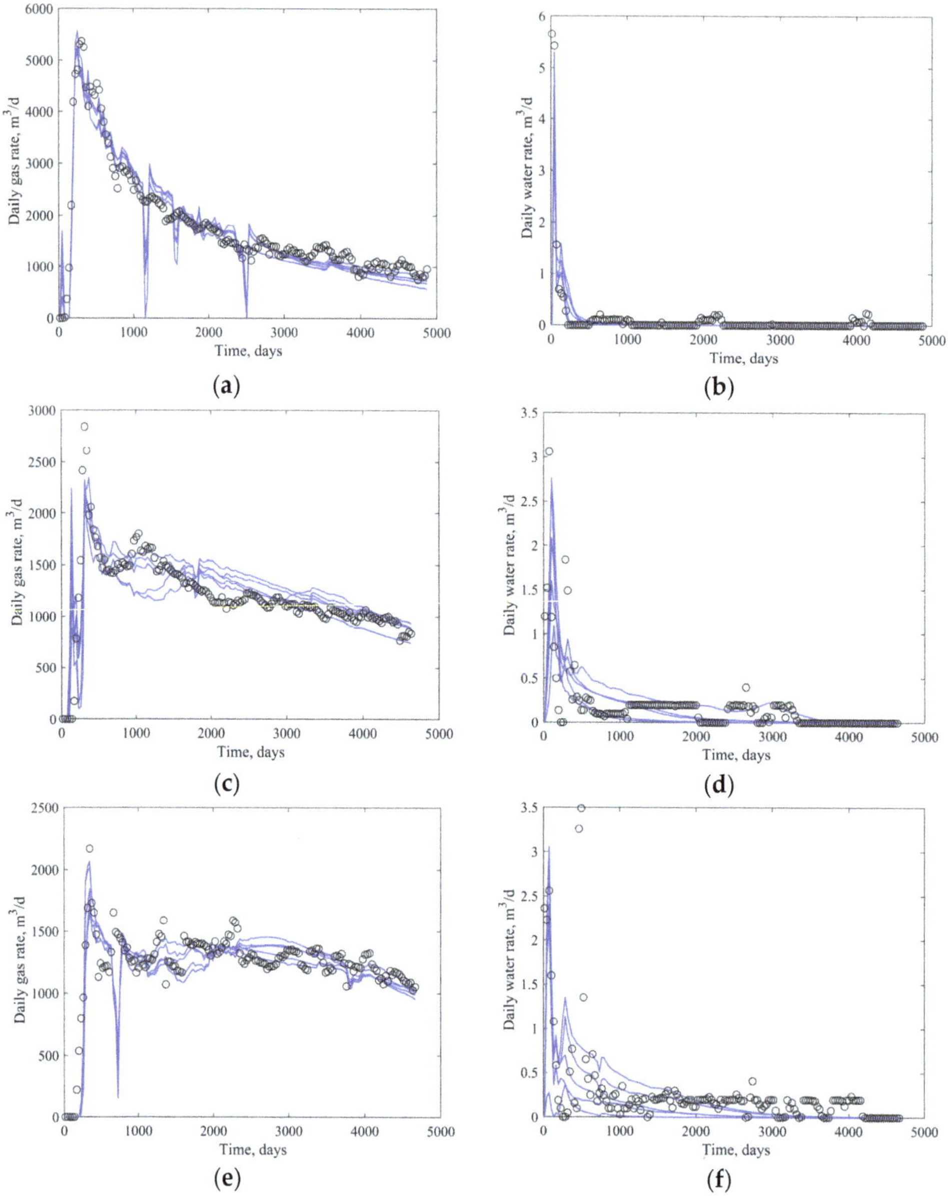

Figure 3. Comparison between the BADS-assisted simulated and true productions. (**a**,**c**,**e**) are ga productions of Well #1, #2 and #3, respectively; (**b**,**d**,**f**) are water productions of Well #1, #2 and #3 respectively. Open circles and blue lines represent the true and simulated productions, respectively

Table 2 summarizes the minimized error function values with BADS, PSO and CMA-ES for the five independent runs, with the minimum, maximum and average values show in Figure 4 for clear visualization. It can be seen from Table 2 and Figure 4 that BAD brought about smaller error function values to both the PSO and CMA-ES. The averag minimized objective function values obtained from BADS were 0.657, 1.083 and 1.186 fc Wells #1, #2 and #3, respectively. As a comparison, the CMA-ES resulted in obviousl larger averaged error function values, which are 0.838, 1.636 and 1.996 for Wells #1, #

and #3, respectively. The PSO performed better than the CMA-ES but worse than the BADS in terms of matching errors for all wells investigated. As shown in Figures 5 and 6, the resulting optimal solutions with PSO and the CMA-ES deviate severely from the true production data. The matrices and the visual figures suggest that BADS outperforms the remaining algorithms in terms of matching accuracy.

In addition to the average error, it is also important to evaluate the robustness of an algorithm considering the multiplicity nature of the history matching problems. In this study, we evaluated the robustness by comparing the deviations of the objective function values over five independent runs. As can be seen from Figure 4, very minor deviations exist with respect to the minimized errors over five independent runs with BADS. As a comparison, the minimized errors using PSO and CMA-ES are associated with obvious deviations over independent runs, especially for Wells #2 and #3 (Figure 4 and Table 2). The deviations of the minimized errors with the PSO and CMA-ES are approximately one order of magnitude of that with BADS. The large deviations among independent runs using the PSO and the CMA-ES are consistent with the resulting optimal solutions in Figures 5 and 6. As shown in Figures 5 and 6, the simulated production curves PSO and CMA-ES deviate severely from run to run, especially for Wells #2 and #3.

Table 2. Summary of the minimized objective function values using different algorithms.

Well ID	Algorithm	Minimized Objective Function Value							
		Run 1	**Run 2**	**Run 3**	**Run 4**	**Run 5**	**Max.**	**Min.**	**Avg.**
Well #1	BADS	0.729	0.626	0.644	0.661	0.624	0.729	0.624	0.657
	PSO	0.759	0.621	0.634	0.731	0.856	0.856	0.621	0.720
	CMA-ES	0.887	0.724	0.845	0.841	0.891	0.891	0.724	0.838
Well #2	BADS	1.081	1.132	1.038	1.140	1.022	1.140	1.022	1.083
	PSO	1.161	1.099	1.367	1.025	1.256	1.367	1.025	1.182
	CMA-ES	1.859	1.497	1.888	1.423	1.516	1.888	1.423	1.636
Well #3	BADS	1.102	1.041	1.128	1.396	1.265	1.396	1.041	1.186
	PSO	1.468	1.267	1.264	1.553	1.844	1.844	1.264	1.479
	CMA-ES	2.173	2.101	3.182	2.920	2.210	3.182	2.101	2.517

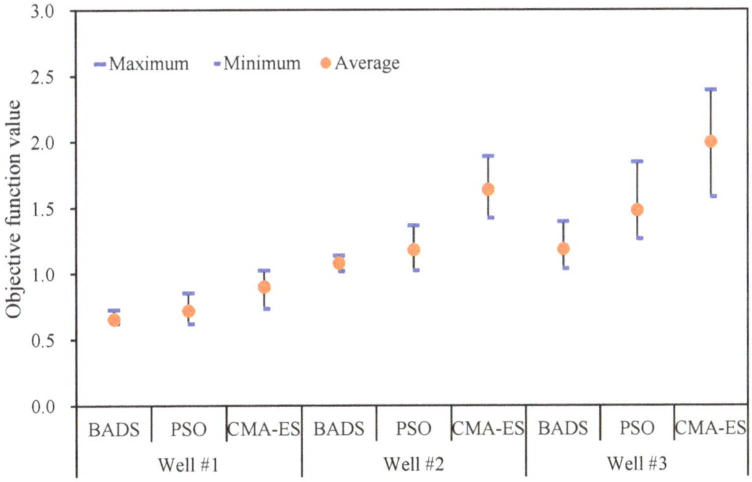

Figure 4. Comparison of BADS with PSO and CMA-ES.

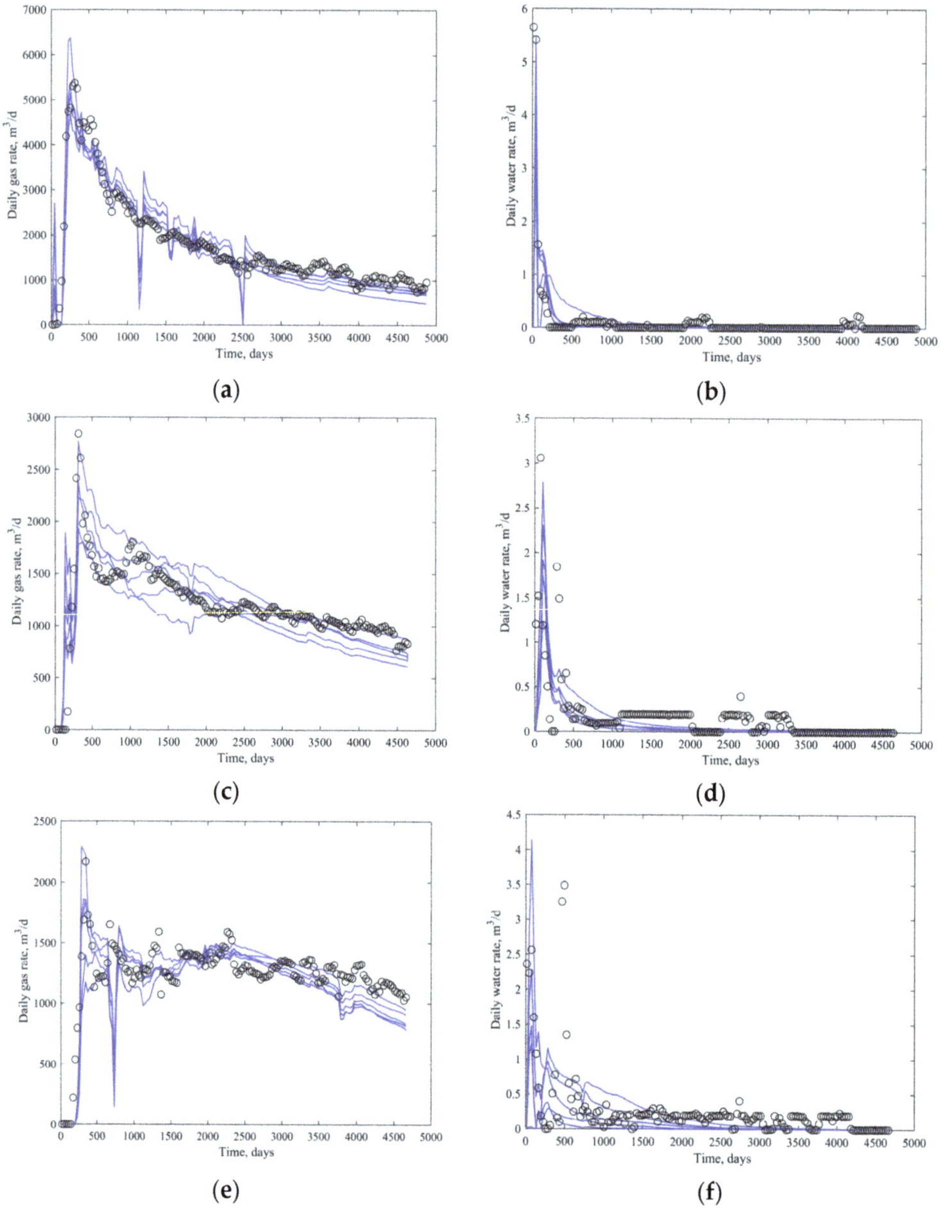

Figure 5. Comparison between the PSO-assisted simulated and true productions. (**a**,**c**,**e**) are g productions of Well #1, #2 and #3, respectively; (**b**,**d**,**f**) are water productions of Well #1, #2 and # respectively. Open circles and blue lines represent the true and simulated productions, respective

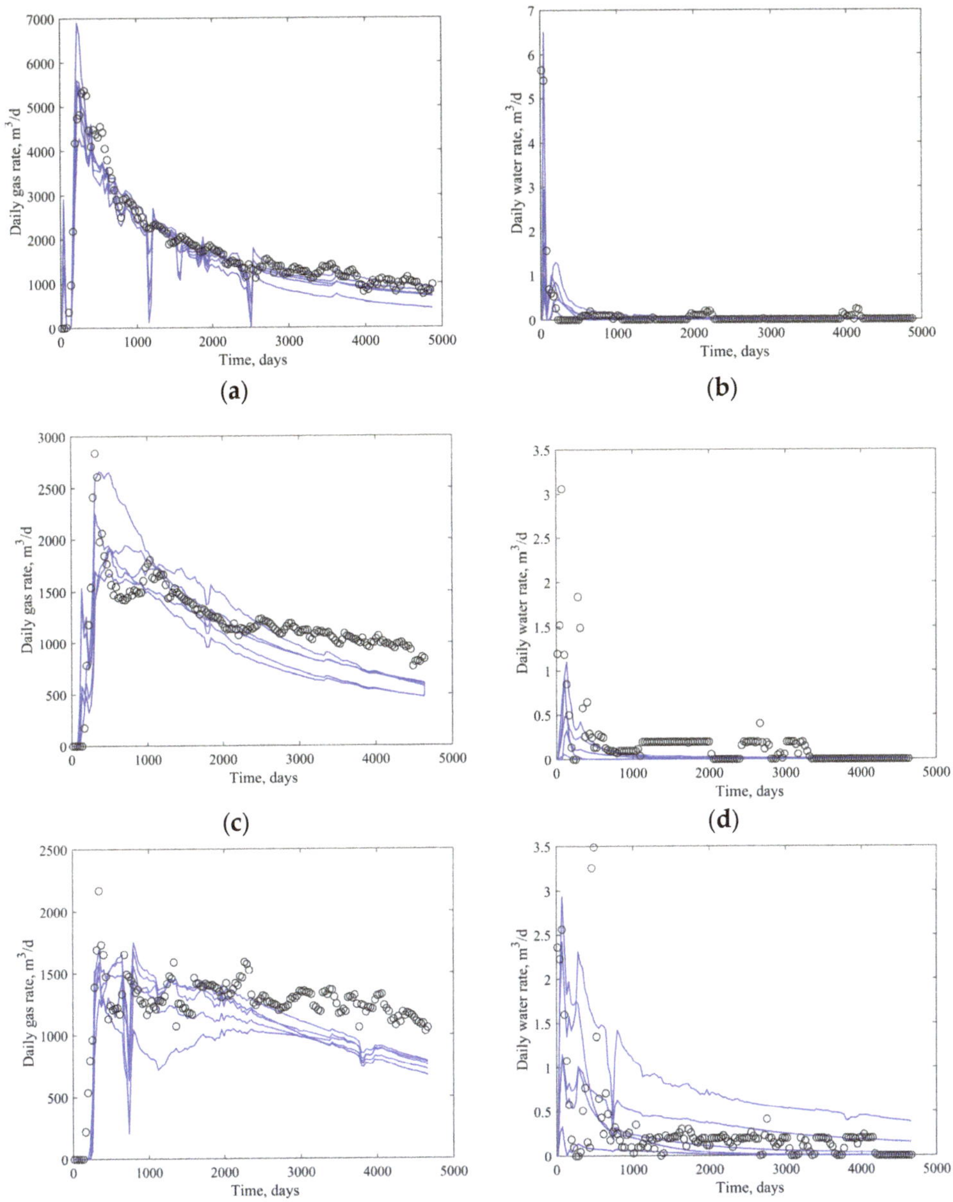

Figure 6. Comparison between the CMA-ES-assisted simulated and true productions. (**a**,**c**,**e**) are gas productions of Well #1, #2 and #3, respectively; (**b**,**d**,**f**) are water productions of Well #1, #2 and #3, respectively. Open circles and blue lines represent the true and simulated productions, respectively.

Figure 7 depicts the evolution trends of the error function values using PSO and CMA-ES as assisted algorithms. It can be seen that the error function values with the CMA-ES exhibit similar trend with the BADS; i.e., the error function drops dramatically within approximately a number of 50 function evaluations and then becomes relatively steady as more function evaluations are made. Nonetheless, CMA-ES results in remarkedly higher matching errors compared with BADS with a number of 120 function evaluations.

The error function values with the PSO exhibit similar trends with CMA-ES for Wells #1 and #3. However, the averaged error function values for Well #2 do not show an obvious convergence, which may indicate less efficiency of PSO.

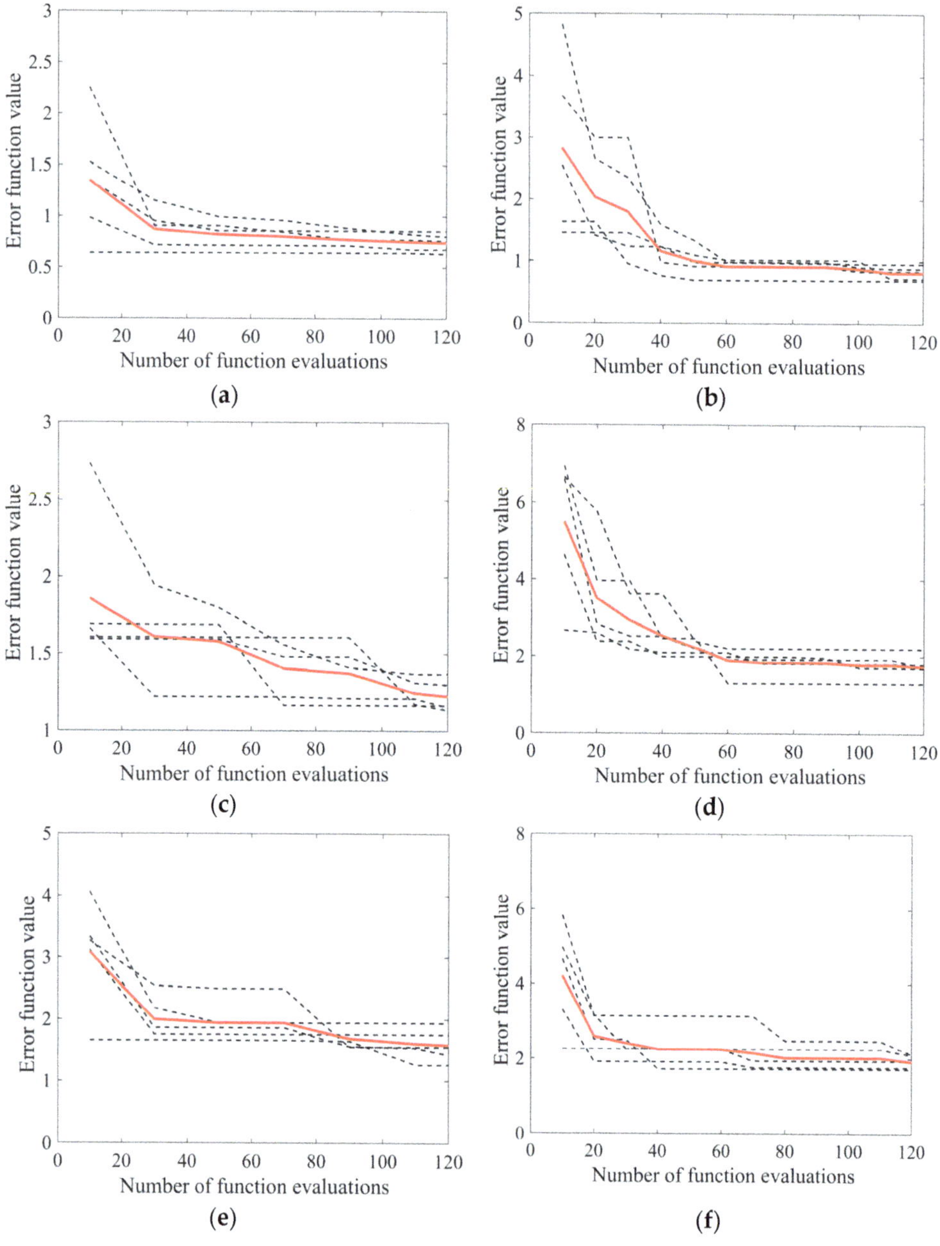

Figure 7. Evolution of the error function value during the AHM process using PSO and CAM-E (**a**,**c**,**e**) are using PSO for Well #1, #2 and #3, respectively; (**b**,**d**,**f**) are using CMA-ES for Well #1, # and #3, respectively. Dash lines represent the independent runs and red solid lines represent th averaged trend.

5. Conclusions

This paper proposed the use of the BADS algorithm to assist history matching of fractured vertical CBM wells. The proposed method was applied on CBM wells with representative production characteristics. The superiority of the BADS was highlighted by comparison with conventional algorithms including PSO and CMA-ES. Conclusions were summarized as follows.

(1) The BADS is capable of deriving reasonable results within a number of only 50–70 function evaluations, suggesting the high convergence speed and efficiency of BADS compared with existing methods. The resulting simulated gas and water productions agree well with true data.

(2) The performance of BADS in terms of matching accuracy was compared with two popular algorithms, namely PSO and CMA-ES. The resulting matching errors using BADS were significantly smaller than both the PSO and the CMA-ES algorithms. Moreover, BADS outperforms PSO and CMA-ES in terms of the robustness over repeated simulation runs.

Funding: This research was funded by PetroChina Huabei Oilfield Company, grant number HBYT-2022-JS-385.

Data Availability Statement: The data presented in this study are available on request from the corresponding author.

Acknowledgments: The author's thanks are owed to Jiyuan Zhang and Xiaodong Zhang from China University of Petroleum (East China) for their efforts in developing the code, assistance in translating the manuscript and fruitful discussions through the project.

Conflicts of Interest: The author declares no conflict of interest.

References

Díaz, M.B.; González, N.C. Control and prevention of gas out-bursts in coal mines, Riosa-Olloniego-coalfifield, Spain. *Int. J. Coal Geol.* **2007**, *69*, 235–266.

Zhang, J.; Xu, K.; Reniers, G.; You, G. Statistical analysis the characteristics of extraordinarily severe coal mine accidents (ESCMAs) in China from 1950 to 2018. *Process Saf. Environ.* **2020**, *133*, 332–340. [CrossRef]

Karacan, C.O.; Ruiz, F.A.; Cote, M.; Phipps, S. Coal mine methane: A review of capture and utilization practices with benefits to mining safety and to green-house gas reduction. *Int. J. Coal Geol.* **2011**, *86*, 121–156. [CrossRef]

Karimpouli, S.; Tahmasebi, P.; Ramandi, H.L. A review of experimental and numerical modeling of digital coalbed methane: Imaging, segmentation, fracture modeling and permeability prediction. *Int. J. Coal Geol.* **2020**, *228*, 103–552. [CrossRef]

Cong, H.R.; Wang, K.; Wang, G.D.; Yang, X.; Du, F. Underground coal seam gas displacement by injecting nitrogen: Field test and effect prediction. *Fuel* **2021**, *306*, 121–646.

Karacan, C.Ö. Single-well production history matching and geostatistical modeling as proxy to multi-well reservoir simulation for evaluating dynamic reservoir properties of coal seams. *Int. J. Coal Geol.* **2021**, *241*, 103–766. [CrossRef]

Xu, H.J.; Pan, Z.J.; Hu, B.L.; Liu, H.H.; Sun, G. A new approach to estimating coal gas content for deep core sample. *Fuel* **2020**, *277*, 118–246. [CrossRef]

Mavor, M.J.; Close, J.C.; McBane, R.A. Formation Evaluation of Exploration Coalbed Methane Wells. In Proceedings of the SPE/CIM International Technical Meeting, Calgary, AB, Canada, 10–13 June 1990.

Feng, Q.; Zhang, J.; Zhang, X.; Hu, A. Optimizing well placement in a coalbed methane reservoir using the particle swarm optimization algorithm. *Int. J. Coal Geol.* **2012**, *104*, 34–45. [CrossRef]

Zhou, F. History matching and production prediction of a horizontal coalbed methane well. *J. Petrol. Sci. Eng.* **2012**, *96–97*, 22–36. [CrossRef]

Zhang, J.; Feng, Q.; Zhang, X.; Hu, Q.; Wen, S.; Chen, D.; Zhai, Y.; Yan, X. Multi-fractured horizontal well for improved coalbed methane production in eastern Ordos basin, China: Field observations and numerical simulations. *J. Petrol. Sci. Eng.* **2020**, *194*, 107488. [CrossRef]

Akhondzadeh, H.; Keshavarz, A.; Sayyafzadeh, M.; Kalantariasl, A. Investigating the relative impact of key reservoir parameters on performance of coalbed methane reservoirs by an efficient statistical approach. *J. Nat. Gas. Sci. Eng.* **2018**, *53*, 416–428. [CrossRef]

Zuber, M.D.; Olszewski, A.J. The impact of errors in measurement of coalbed methane reservoir properties on well production forecasts. In Proceedings of the 62nd SPE Annual Technical Conference and Exhibition, Dallas, TX, USA, 27–30 September 1987.

Young, G.B.C.; Paul, G.W.; McElhiney, J.E. A parametric analysis of Fruitland Coalbed methane producibility. In Proceedings of the 67th SPE Annual Technical Conference and Exhibition, Washington, DC, USA, 4–7 October 1992.

15. Duan, L.J.; Zhou, F.D.; Xia, Z.H.; Qu, L.C.; Yi, J. New Integrated History Matching Approach for Vertical Coal Seam Gas Wells in the KN Field, Surat Basin, Australia. *Energy Fuels* **2020**, *34*, 15829–15842. [CrossRef]
16. Li, B.; Bhark, E.W.; Gross, S.J.; Billiter, T.C.; Dehghani, K. Best Practices of Assisted History Matching Using Design of Experiments. *SPE J.* **2019**, *24*, 1435–1451. [CrossRef]
17. Tripoppoom, S.; Xie, J.; Yong, R.; Wu, J.; Yu, W.; Sepehrnoori, K.; Miao, J.; Chang, C.; Li, N. Investigation of different production performances in shale gas wells using assisted history matching: Hydraulic fractures and reservoir characterization from production data. *Fuel* **2020**, *267*, 117–1097. [CrossRef]
18. Wu, H.Y.; Cheng, L.S.; Killough, J.; Huang, S.J.; Fang, S.D.; Jia, P.; Cao, R.; Xue, Y.C. Integrated characterization of the fracture network in fractured shale gas Reservoirs—Stochastic fracture modeling, simulation and assisted history matching. *J. Petrol. Sci. Eng.* **2021**, *205*, 108–886. [CrossRef]
19. Nejadi, S.; Leung, J.Y.; Trivedi, J.J.; Claudio, V. Integrated Characterization of Hydraulically Fractured Shale-Gas Reservoirs—Production History Matching. *SPE Res. Eval. Eng.* **2015**, *18*, 481–494. [CrossRef]
20. Shams, M.; El-Banbi, A.; Sayyouh, H. Harmony search optimization applied to reservoir engineering assisted history matching. *Petrol. Explor. Dev.* **2020**, *47*, 154–160. [CrossRef]
21. Zhang, Y.; Song, C.; Yang, D. A damped iterative EnKF method to estimate relative permeability and capillary pressure for tight formations from displacement experiments. *Fuel* **2016**, *167*, 306–315. [CrossRef]
22. Zhang, Y.; Fan, Z.Q.; Yang, D.Y.; Li, H.; Shirish, P. Simultaneous Estimation of Relative Permeability and Capillary Pressure for PUNQ-S3 Model With a Damped Iterative-Ensemble-Kalman-Filter Technique. *SPE J.* **2000**, *22*, 971–984. [CrossRef]
23. Acerbi, L.; Ma, W.J. Practical Bayesian Optimization for Model Fitting with Bayesian Adaptive Direct Search. *Adv. Neural Inform Process. Syst.* **2017**, *30*, 1834–1844.
24. Computer Modelling Group. *User Guide: Compositional & Unconventional Reservoir Simulator*; Computer Modelling Group: Calgary, AB, Canada, 2015.
25. Ziarani, A.S.; Aguilera, R.; Clarkson, C.R. Investigating the effect of sorption time on coalbed methane recovery through numerical simulation. *Fuel* **2011**, *90*, 2428–2444. [CrossRef]
26. Karacan, C.O.; Olea, R.A. Stochastic reservoir simulation for the modeling of uncertainty in coal seam degasification. *Fuel* **2015**, *148*, 87–97. [CrossRef]
27. Zhang, J.; Feng, Q.; Zhang, X.; Bai, J.; Karacan, C.Ö.; Wang, Y.; Elsworth, D. A two-stage step-wise framework for fast optimization of well placement in coalbed methane reservoirs. *Int. J. Coal Geol.* **2020**, *225*, 103479. [CrossRef]
28. Mohammadpoor, M.; Firouz, A.Q.; Torabi, F. Implementing Simulation and Artificial Intelligence Tools to Optimize the Performance of the CO_2 Sequestration in Coalbed Methane Reservoirs. In Proceedings of the Carbon Management Technology Conference, Orlando, FL, USA, 7–9 February 2012.
29. Zhang, J.; Zhang, B.; Xu, S.; Feng, Q.; Zhang, X.; Elsworth, D. Interpretation of Gas/Water Relative Permeability of Coal Using the Hybrid Bayesian-Assisted History Matching: New Insights. *Energies* **2021**, *14*, 626. [CrossRef]
30. Snoek, J.; Larochelle, H.; Adams, R.P. Practical Bayesian optimization of machine learning algorithms. In Proceedings of the Advances in Neural Information Processing Systems, Lake Tahoe, NV, USA, 3–6 December 2012; pp. 2951–2959.
31. Audet, C.; Dennis, J.E. Mesh adaptive direct search algorithms for constrained optimization. *SIAM J. Optimiz.* **2006**, *17*, 188–21 [CrossRef]
32. Zuber, M.D.; Kuuskraa, V.A.; Walter, K.S. Optimizing Well Spacing and Hydraulic-Fracture Design for Economic Recovery of Coalbed Methane. *SPE Form Eval.* **1990**, *5*, 98–102. [CrossRef]
33. Maricic, N.; Mohaghegh, S.D.; Artun, E. A Parametric Study on the Benefits of Drilling Horizontal and Multilateral Wells in Coalbed Methane Reservoirs. *SPE Res. Eval. Eng.* **2008**, *11*, 976–983. [CrossRef]
34. Meng, Z.; Zhang, K.; Yang, J.; Lei, J.; Wang, Y. Analysis of coal reservoir characteristics in the Qin-nan-East block and its spacing optimization of CBM development well networks. *J. China Coal Soc.* **2018**, *43*, 2525–2533, (In Chinese with English abstract).
35. Ian, P.; Mansoori, J. How Permeability Depends on Stress and Pore Pressure in Coalbeds: A New Model. *SPE Res. Eval. Eng.* **199**, *1*, 539–544.
36. Chen, D.; Pan, Z.; Liu, J.; Connell, L.D. An improved relative permeability model for coal reservoirs. *Int. J. Coal Geol.* **2013**, *1*, 45–57. [CrossRef]
37. King, G.R.; Ertekin, T. Comparative evaluation of vertical and horizontal drainage wells for the degasification of coal seams. *SPE Reserv. Eng.* **1988**, *3*, 720–734. [CrossRef]
38. Kumar, H.; Elsworth, D.; Liu, J.; Pone, D.; Mathews, J.P. Permeability evolution of propped artificial fractures in coal on injection of CO_2. *J. Petrol. Sci. Eng.* **2015**, *133*, 695–704. [CrossRef]
39. Chen, D.; Shi, J.Q.; Durucan, S.; Korre, A. Gas and water relative permeability in different coals: Model match and new insights. *Int. J. Coal Geol.* **2014**, *122*, 37–49. [CrossRef]
40. Kennedy, J. Particle swarm optimization. In Proceedings of the 1995 IEEE International Conference on Neural Networks, Perth, WA, Australia, 27 November–1 December 1995; Volume 4, pp. 1942–1948.
41. Hansen, N.; Ros, R.; Mauny, N.; Schoenauer, M.; Auger, A. Impacts of invariance in search: When CMA-ES and PSO face ill-conditioned and non-separable problems. *APPL Soft. Comput.* **2011**, *11*, 5755–5769. [CrossRef]

42. Xiang, W.; Haynes, R.D.; Feng, Q. A multilevel coordinate search algorithm for well placement, control and joint optimization. *Comput. Chem. Eng.* **2016**, *95*, 75–96.
43. Mohamed, L.; Christie, M.; Vasily, D. Comparison of Stochastic Sampling Algorithms for Uncertainty Quantification. *SPE J.* **2010**, *15*, 31–38. [CrossRef]

processes MDPI

Article

Intelligent Optimization of Gas Flooding Based on Multi-Objective Approach for Efficient Reservoir Management

Meng Gao [1], Chenji Wei [1,*], Xiangguo Zhao [2], Ruijie Huang [1,3,*], Baozhu Li [1], Jian Yang [1], Yan Gao [1], Shuangshuang Liu [1] and Lihui Xiong [1]

1 Research Institute of Petroleum Exploration and Development, PetroChina, Beijing 100083, China; gmeng@petrochina.com.cn (M.G.); lbz@petrochina.com.cn (B.L.); j.yang@petrochina.com.cn (J.Y.); gaoyan620@petrochina.com.cn (Y.G.); Liuss622@petrochina.com.cn (S.L.); xionglh@petrochina.com.cn (L.X.)
2 CNPC International Hong Kong Limited—Abu Dhabi, Abu Dhabi 999041, United Arab Emirates; zhaoxiangguo@cnpcint.com
3 Shenzhen Branch, China National Offshore Oil Corporation, Shenzhen 518000, China
* Correspondence: weichenji@petrochina.com.cn (C.W.); rogerwx@163.com (R.H.)

Abstract: The efficient development of oil reservoirs mainly depends on the comprehensive optimization of the subsurface fluid flow process. As an intelligent analysis technique, artificial intelligence provides a novel solution to multi-objective optimization (MOO) problems. In this study, an intelligent agent model based on the Transformer framework with the assistance of the multi-objective particle swarm optimization (MOPSO) algorithm has been utilized to optimize the gas flooding injection–production parameters in a well pattern in the Middle East. Firstly, 10 types of surveillance data covering 12 years from the target reservoir were gathered to provide a data foundation for model training and analysis. The prediction performance of the Transformer model reflected its higher accuracy compared to traditional reservoir numerical simulation (RNS) and other intelligent methods. The production prediction results based on the Transformer model were 21, 12, and 4 percentage points higher than those of RNS, bagging, and the bi-directional gated recurrent unit (Bi-GRU) in terms of accuracy, and it showed similar trends in the gas–oil ratio (GOR) prediction results. Secondly, the Pareto-based MOPSO algorithm was utilized to fulfil the two contradictory objectives of maximizing oil production and minimizing GOR simultaneously. After 10,000 iterations, the optimal injection–production parameters were proposed based on the generated Pareto frontier. To validate the feasibility and superiority of the developed approach, the development effects of three injection–production schemes were predicted in the intelligent agent model. In the next 400 days of production, the cumulative oil production increased by 25.3% compared to the average distribution method and 12.7% compared to the reservoir engineering method, while GOR was reduced by 27.1% and 15.3%, respectively. The results show that MOPSO results in a strategy that more appropriately optimizes oil production and GOR compared to some previous efforts published in the literature. The injection–production parameter optimization method based on the intelligent agent model and MOPSO algorithm can help decision makers to update the conservative development strategy and improve the development effect.

Keywords: injection–production optimization; artificial intelligence; multi-objective optimization; attention; gas flooding

Citation: Gao, M.; Wei, C.; Zhao, X.; Huang, R.; Li, B.; Yang, J.; Gao, Y.; Liu, S.; Xiong, L. Intelligent Optimization of Gas Flooding Based on Multi-Objective Approach for Efficient Reservoir Management. *Processes* **2023**, *11*, 2226. https://doi.org/10.3390/pr11072226

Academic Editors: Zheng Sun, Tao Zhang, Dong Feng, Wen Zhao and Hung Vo Thanh

Received: 14 June 2023
Revised: 17 July 2023
Accepted: 18 July 2023
Published: 24 July 2023

1. Introduction

Highly uncertain and disputable parameters are one of the main concerns in the process of reservoir management and decision making. Influenced by reservoir heterogeneity and unreasonable injection–production parameter combinations, the abundant remaining oil is difficult to drive and becomes more dispersed. Therefore, the original development plan is unable to achieve satisfactory performance. To improve the effectiveness of the development plan, the injectors are adjusted to reduce the inefficient or ineffective injection volume. Reasonable optimization helps to increase the oil rate and control the risk warning parameters, such as the water cut (WCT) and GOR. Therefore, a reasonable optimization strategy is an effective strategy to improve reservoir development.

To optimize the overall structure, improve the final recovery factor, and reduce development risks, the need to optimize the injection–production parameters has been proposed. In the middle and later stages of development, the existing parameters face challenges, such as unclear effectiveness and high development risks. Specifically, unreasonable parameters lead to fluid intrusion. On the one hand, this renders the remaining oil more dispersed, which restricts the development effect; on the other hand, the unreasonable adjustment of the injected fluid can easily lead to ineffective circulation and even result in high-risk producers.

Water/gas injection is an effective displacement measure that can restore the reservoir pressure and achieve a higher recovery factor. Therefore, the choice of injection fluid has a strong impact on oil recovery. Khormali [1] used two different types of formation water to simulate the effects of reservoir pressure, temperature, injection water, and pairing on inorganic salt formation, and they proposed a reasonable ratio of injection water to formation water, guiding the rational selection of optimized injection–production parameters.

One of the essential tasks in achieving efficient reservoir development is to obtain the best combination of injection–production parameters quickly and accurately. There are four commonly used methods (Table 1). The first method is to distribute the total production and injection volume to all producers and injectors evenly [2], which is simple but carries many disadvantages. Firstly, for strongly heterogeneous reservoirs, the development effectiveness varies greatly between regions. Therefore, it is difficult for a simple uniform distribution to reflect the actual conditions of the reservoir. Secondly, due to the irregular distribution of the remaining oil, it is difficult to achieve efficient potential tapping in this way.

The second method is to allocate the parameters according to the reservoir characteristics. An [3] selected the pore volume and remaining geological reserves as the distribution index. Because the index has a certain physical meaning, this method helps to achieve efficient potential tapping. Moreover, it can effectively suppress the risk of water and gas invasion, ensuring the injection effectiveness simultaneously. However, this method also has certain limitations. First, the appropriateness of the allocation has a strong relationship with the accuracy of the RNS model. Because many basic data and complicated historical fitting are required, it is difficult to reach high accuracy and ensure authenticity. Secondly, the selection of indicators does not consider specific optimization objectives, so it is difficult to determine whether the proposed strategy is the optimal solution.

The third method is based on RNS and statistical methods, represented by the response surface methodology (RSM) [4]. The RSM helps to obtain the global optimal scheme. Specifically, the RSM fits complex unknown functional relationships with relatively simple first-order or second-order polynomial models, which reduces the computational complexity. Secondly, compared with the orthogonal experimental design, the RSM can continuously analyze all levels of the experiment in the process of optimization, which greatly improves the feasibility of the model. Based on the RNS and RSM, Kang [5] designed a optimization process for the EOR and many steam and gas push (SAGP) parameters (including the injection gas type, injection time, injection gas mol, and volume). Based on the results, the best combination of all factors was effectively determined, providing a good reference for the improvement of reservoir development efficiency. Based on a higher-order polynomial equation, Wantawinj [6,7] established a simpler and more efficient reservoir agent model using the RSM. Compared with the commonly used quadratic form model, this method shows higher accuracy in the direction of shale reservoir evaluation, fracture characterization, and production prediction. However, the application of the RSM is limited due to the slow computational speed of RNS models. In addition, if the range of experimental points is not selected properly, it will be difficult to obtain good optimization results. Therefore, reasonable experimental factors and corresponding levels should determined before using the RSM.

The forth method is to optimize the parameters by defining the objective function a embedding the optimization algorithm into the RNS model. The extreme optimal va of the objective function is searched through iterations. This method can consider spec objectives, further improving the relevance of optimization. Taking the Net Present Va (NPV) as the objective function, Sarma [8] explored a new water injection optimizat strategy using the optimal control algorithm and verified it using the black oil simulat model, which showed good application results. In addition, apart from the NPV [9– different objective functions can be considered, such as GOR, WCT, and so on, depending the specific situation. However, it is still difficult to overcome some shortcomings, such the time-consuming process. On the other hand, commonly used optimization algorith such as particle swarm optimization (PSO) [12,13] and genetic algorithm (GA) [14– mainly focus on a single objective, which limits the scope of application in the oilfi Instead, two or more objective functions should be taken into consideration.

Many studies have contributed to the MOO problem [18,19] in terms of on-site quirements, which can be summarized into three aspects. The first is to assign weight factors to each objective, converting multiple objectives into a single function. In t way, a single objective optimization technique such PSO or GA can be utilized to seek optimal results. This method is easy to understand but it is unable to provide an effect solution for MOO, where the relative importance is not clear. The second is to consi the relative importance of the objective function, called lexicographic optimization. T type of method is suitable for problems where different objective functions have differ priorities. However, it cannot provide a comparative analysis of the tradeoffs betwe different functions. To address the shortcomings of the above methods, the Pareto-bas method is favored because it can provide decision makers with a tradeoff analysis [The basic idea is to iterate and plot the Pareto frontier by calculating the fitness, which suitable for problems where the importance of different objectives is unclear.

Table 1. Comparison of conventional injection–production parameter optimization methods bas on RNS.

Method	Description	Advantages	Limitations
Uniform division	Divided by uniform conditions [2]	1. Simple. 2. Suitable for weakly heterogeneous reservoirs.	1. Lack of physical basis. 2. Extremely poor applicability i highly heterogeneous reservoirs
Proportional division	Divided by reservoir properties [3]	1. Considers the actual conditions of the reservoir to a certain extent. 2. Suppresses the risk of water and gas invasion effectively.	1. Relatively dependent on the accuracy of RNS models. 2. Difficult to determine whethe the result is the optimal solutior
	Divided according to statistical methods [5–7]	1. The main factors of the reservoir could be obtained. 2. The computational complexity is reduced. 3. Can continuously analyze various experimental levels.	1. Constrained by the calculatio speed of RNS. 2. The optimization results rely heavily on the selection of experimental point ranges.
	Divided using iterative optimization [21–23]	1. Considers specific objectives, improving the relevance. 2. Obtains unique optimal result of the objective function.	1. Relatively dependent on the accuracy of RNS models. 2. Only one objective is considered mainly.

Pareto-based optimization algorithms are often embedded in population-based evo tionary algorithms, such as the non-dominated sorting genetic algorithm II (NSGA-II) a

MOPSO [19,24–28]. Wang [29] designed a fracturing optimization scheme based on the least squares support vector regression (LSSVR) and NSGA-II. They applied the proposed framework to a shale gas reservoir, shortening the optimization time and improving the accuracy, and the work provides a reference for subsequent fracturing operations. Desbordes [30] proposed a production optimization framework integrating three evolutionary algorithms, namely NSGAII, MOPSO, and the multi-objective evolutionary algorithm based on decomposition (MOEA/D). Due to the elimination of the population reinitialization step, the method can achieve relatively high fitness with a lower computational cost. Due to the limitations of RNS [31], there is an urgent need for a method that can compensate for its shortcomings in various aspects [32].

Big data and artificial intelligence (AI) technology surpass the traditional physical-driven research thinking and deeply explore the logical relationships existing within data [33]. AI has the advantages of simplicity, high speeds, and high computational accuracy because it can learn the inherent nonlinear mapping relationships of data directly. Reservoir engineers can store and manage data and conduct in-depth mining and analysis to propose development plans. Thanks to advances in computing power and the iterative updating of algorithms, AI has also become a key technology for the intelligent exploration of oilfields. These applications mainly focus on PVT prediction [34–36], missing value regression [37], well location prediction [38,39], history matching [40–42], and production prediction [43–48]. Many studies indicate that AI technology can improve the efficiency and economic benefits. Therefore, AI technology is expected to alleviate the shortcomings of RNS and play a greater role in oil exploration and development.

Transformer is a neural network that processes sequential data based on a self-attention mechanism. The core of Transformer is the attention mechanism; similar to human selective visual attention, it can learn information that is more critical to the current task goal from a variety of information, which enhances the time series modeling ability. With this ability, Transformer has achieved remarkable success in natural language processing and computer vision, demonstrating its powerful time series modeling capabilities. Guo [49] and Bai [50] predicted future traffic flows based on an attention-based graph convolutional network. The prediction results showed that the attention-based model had higher accuracy compared to traditional neural networks. Wang [51] proposed a short-term load forecasting method based on an attention mechanism and the bi-directional long short-term memory (Bi-LSTM) neural network to update data, assign weights, and perform model training and prediction, respectively. After comparing the prediction results of actual datasets, the results showed that the model with the attention mechanism had a lower calculation error, indicating the superiority of the attention mechanism. Similarly, Liu [52] combined an attention mechanism and the Bi-GRU neural network to optimize the hypertext transfer protocol's security. Niu [53] used an attention mechanism to predict wind power, achieving good results. In the petroleum industry, attention mechanisms have also been applied in production forecasting. Therefore, compared with existing deep learning methods, attention-based methods significantly improve the prediction accuracy.

This article establishes a workflow for injection–production parameters' intelligent optimization considering the multi-objective tasks of the oilfield. The basic processes include establishing an attention-based Transformer intelligent agent model, to perform multi-objective optimization tasks via the Pareto-based population evolutionary optimization algorithm. This article is divided into the following sections. In Section 2, we introduce the methodology of the Transformer model and MOPSO algorithm. Section 3 introduces the multiple objectives, the basic information of the target reservoir, the data preprocessing process, and the architecture and evaluation criteria. In Section 4, the proposed method is evaluated and compared with other methods. The discussion and conclusions are given in Sections 5 and 6.

2. Methodology

2.1. Transformer

Currently, many supervised learning algorithms can be used to build intelligent agent models. These methods can be mainly divided into classification algorithms and regression algorithms, represented by decision tree, support vector machine (SVM), random forest (RF), etc. Because of their simple structures and fast speeds, classification algorithms are also widely used in time series prediction. However, since the actual data are mostly non-stationary, the simple structures of classification algorithms do not have the ability to learn complex data. Therefore, this limits their application in time series prediction.

As another powerful tool in time series prediction, artificial neural networks (ANN) are widely used to interpret and predict times series data. The most commonly used deep learning methods for the processing of time series currently include the recurrent neural network (RNN) and its variants, such as long short-term memory (LSTM), the gated recurrent unit (GRU), Bi-LSTM, Bi-GRU, and so on. The RNN is a type of neural network structure aimed at processing time series. The unique connections between hidden layers in the RNN form a directed loop, which enables it to process time series well. As with fully connected neural networks, the simplest RNN consists of three parts: an input layer, hidden layer, and output layer. The value of the hidden layer in the RNN depends not only on the current input x_t but also the value of the previous hidden layer h_{t-1}. The value of the previous hidden layer h_{t-1} is determined by the input x_{t-2} and the previous hidden layer h_{t-2}. In other words, in an ideal situation, its hidden layer can store information for a long time.

$$o^{(t)} = g\left(Vh^{(t)} + b_o\right) \tag{1}$$

$$h^{(t)} = f\left(Ux^{(t)} + Wh^{(t-1)} + b_h\right) \tag{2}$$

where $x^{(t)}$ represents the input vector at time t; $o^{(t)}$ represents the output vector; $h^{(t-1)}$ is the hidden cell state at time $t-1$; b_o and b_h are the bias vectors; $g()$ and $f()$ are the activation functions; U, W, V are the weight matrices used for input–hidden connections, hidden–hidden connections, and hidden–output connections, respectively, with the same weight values for the same types of weight connection. L shown in Figure 1 is the loss function, and y is the label of the training set.

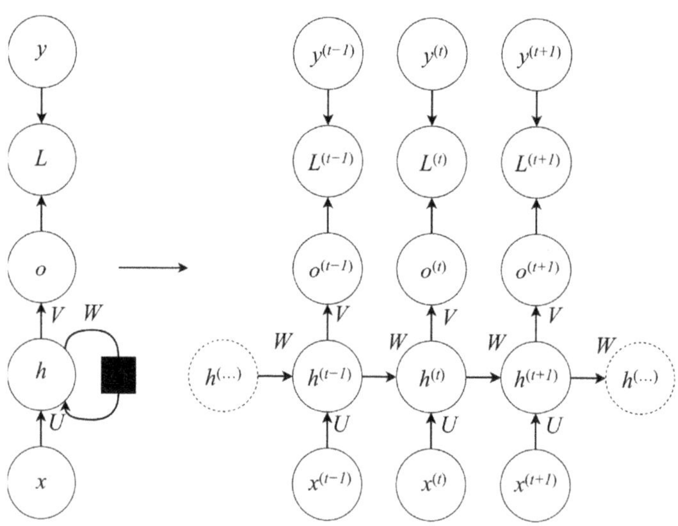

Figure 1. Schematic diagram of circulating neural network.

To update the RNN weights with the optimal U, W, V, the partial derivatives of the loss function for the three weight matrices are as shown below:

$$\frac{\partial L}{\partial V}=\sum_{t=1}^{n}\frac{\partial L^{(t)}}{\partial o^{(t)}}\cdot\frac{\partial o^{(t)}}{\partial V} \tag{3}$$

$$\frac{\partial L^{(t)}}{\partial W}=\sum_{k=0}^{t}\frac{\partial L^{(t)}}{\partial o^{(t)}}\frac{\partial o^{(t)}}{\partial h^{(t)}}\left(\prod_{j=k+1}^{t}\frac{\partial h^{(j)}}{\partial h^{(j-1)}}\right)\frac{\partial h^{(k)}}{\partial W} \tag{4}$$

$$\frac{\partial L^{(t)}}{\partial U}=\sum_{k=0}^{t}\frac{\partial L^{(t)}}{\partial o^{(t)}}\frac{\partial o^{(t)}}{\partial h^{(t)}}\left(\prod_{j=k+1}^{t}\frac{\partial h^{(j)}}{\partial h^{(j-1)}}\right)\frac{\partial h^{(k)}}{\partial U} \tag{5}$$

From Equations (3)–(5), it can be seen that there is no time dependency problem for partial derivatives of V; however, for U and W, due to the long-term dependence of the time series, $h^{(t)}$ will propagate forward with the time series. Due to the gradient vanishing and the explosive linearity of the RNN, its prediction performance is average when processing long sequence data.

To alleviate the gradient vanishing/exploding problem of RNN models, a series of variants have gradually emerged. Information in neural networks represented by LSTM [54] and GRU [55] is transmitted unidirectionally, and only past information can be utilized, without the use of future information. The emergence of bi-directional neural networks, such as Bi-LSTM [46,56] and Bi-GRU [57], has made it possible to consider both past and future information simultaneously. Bi-LSTM connects the two hidden layers of LSTM to the output layer. In this structure, both previous and future information can be utilized at the output layer and this can significantly improve the model performance. However, when using the RNN to process a vector sequence, it requires a "local encoding" of the sequence. Therefore, the RNN has a short-distance dependency and it lacks parallel processing capabilities.

Transformer consists of encoder and decoder stacks. The encoder and decoder of Transformer are both composed of n identical layers [58]. Each encoder layer has two sub-layers, consisting of a multi-head self-attention mechanism and feed forward neural network (FFNN). Each decoder layer has three sub-layers, consisting of a multi-head self-attention mechanism, multi-head attention over the output of the encoder stack, and FFNN. The attention mechanism is the core of Transformer. It indicates the importance of other tags in the input to the encoding of a given tag. The attention mechanism can solve the two problems of the RNN and its variants mentioned above. In response to the difficulty of obtaining long-term dependencies for the RNN and its variants, there is no concept of distance when calculating the attention score. The attention score of each unit and the current unit is independent of their distances, so it can avoid the problem of difficult-to-obtain long-distance dependencies. Transformer can calculate the attention score in parallel to obtain the context vector (Figure 2).

Specifically, we create three randomly initialized training parameter matrices, W^Q, W^K, and W^V first, and we multiply the encoder input vectors to obtain the corresponding matrices Q, K, and V. Then, we calculate the attention score of each input variable and other units in the data. The attention score represents the degree to which attention is focused on other parts during encoding, and it is calculated by the dot-product of Q and K. Secondly, dividing the score by the square root of the matrix K dimension helps to achieve more stable convergence in training. Subsequently, a SoftMax function is applied to obtain the weights of the values, and each SoftMax value is multiplied. Finally, we add the weighted value vectors together as the corresponding output vectors in the self-attention layer. After the vectors are correlated with each other in the self-attention layer, the corresponding vectors are transferred to the FFNN layer. In the FFNN layer, each link has no correlation, so the

FFNN layer can parallelize each input variable. The matrix of output can be computed as in the formula below:

$$\text{Attention}\ (Q, K, V) = \text{softmax}\left(\frac{QK^T}{\sqrt{d_k}}\right)V \tag{6}$$

where d_k denotes the queries and keys of the dimension.

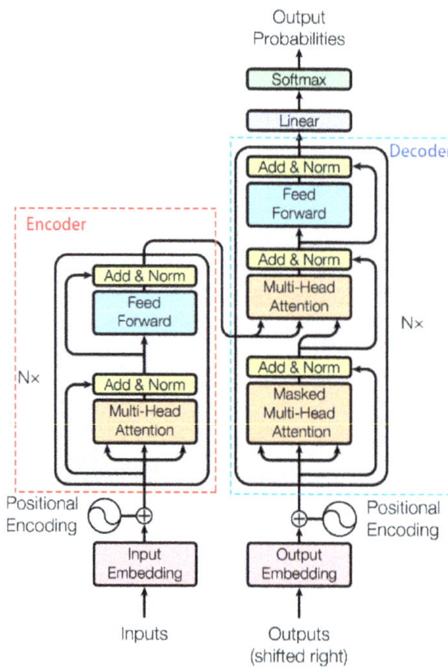

Figure 2. The Transformer model architecture [58].

Multi-head attention is an improvement of the self-attention mechanism (Figure 3). enhances the model's ability to focus on various positions by randomly initializing multipl $Q/K/V$ matrices and allows the self-attention layer to have a greater representation spac The specific method is to multiply the calculated multiple output matrices by an addition weight matrix, to obtain information that is the same size as the original and captures a attention heads.

In addition, each self-attention and FFNN layer is linked through a residual and the subjected to layer normalization operation. Furthermore, to read the position informatio of each piece of data, Transformer adds a position encoding to the input embedding an the final output vector. The calculation formula for the position encoding is as follows:

$$PE_{(pos,2i)} = \sin\left(pos/10,000^{2i/d_{\text{model}}}\right)$$

$$PE_{(pos,2i+1)} = \cos\left(pos/10,000^{2i/d_{\text{model}}}\right)$$

where *pos* is the position and *i* is the dimension.

Transformer introduces multi-head attention and self-attention modules to first "se associate" the source and target sequences, enriching the information contained in the embe ded representations of the source and target sequences themselves. Furthermore, the sub quent FFNN layers also enhance the model's expressive and parallel computing capabiliti

Therefore, it establishes a good theoretical basis for the subsequent establishment of the intelligent agent model.

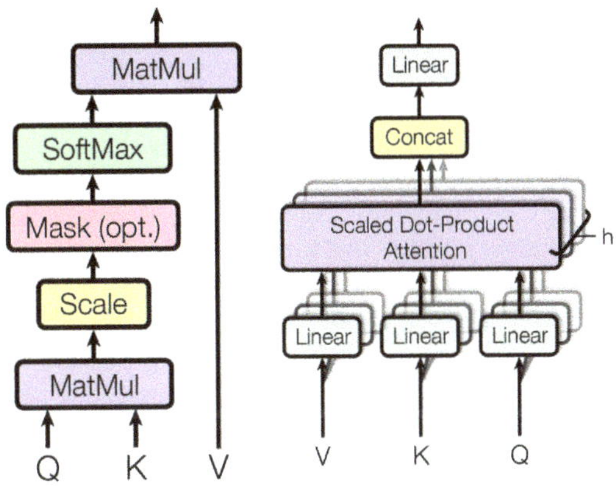

(a) (b)

Figure 3. Scaled dot-product attention (**a**); multi-head attention (**b**) [58].

2.2. MOPSO

MOO problems are mainly based on multiple conflicting objective functions. Therefore, the optimal solution obtained for a certain goal will not consider other goals in the optimal value and may even lead to degradation. The traditional MOO methods include the weighted sum method, constraint method, objective programming method, distance function method, minimax method, etc. Most of these methods decompose the problem into a single objective problem, relying on different strategies, and single objective algorithms are used to achieve optimization. Due to the reliance on prior knowledge and the inability to select the optimal solution based on different needs, the above methods are not feasible in certain contexts. For example, when MOO problems exhibit complex characteristics such as nonlinearity and high dimensionality, traditional methods are not able to achieve good results.

In recent years, evolutionary algorithms have enabled many breakthrough research achievements in the field of combinatorial and numerical optimization. PSO is an evolutionary algorithm inspired by the foraging behavior of bird populations in nature [27]. Due to its advantages, such as simple implementation, an efficient search, and fast convergence, it has been widely applied in various experimental tests and on-site practical applications.

As an evolutionary version of PSO, MOPSO is currently a popular method [19]. In MOPSO, the individual's position is treated as the solution to an optimization problem, ignoring the individual's mass and volume. The main principle is to continuously update the information exchange between individuals in the group and the optimal individual. Moreover, the entire group of individuals is guided to converge towards the optimal individual, retaining their own diversity information.

The update particles are obtained through the organic combination of population history optimal particles and individual history optimal particles. The velocity v_i^{g+1} of particle i at the next moment is determined by the current velocity v_i^g, its own optimal position $X_{i,pbest}^g$, and global optimal position $X_{i,gbest}^g$, which are jointly determined. After updating, the speed changes from the current position X_i^g to a new position X_i^{g+1}. As the iteration continues, the entire particle swarm, under the leadership of the leader, completes the search for the optimal solution. The formula is shown below:

$$X_i^{g+1} = X_i^g + v_i^{g+1}, i \subset N_p \text{ and } g \subset N_g \tag{9}$$

$$v_i^{g+1} = \omega_I v_i^g + \omega_c R_1 \left(X_{i,pbest}^g - X_i^g \right) + \omega_s R_2 \left(X_{i,gbest}^g - X_i^g \right) \tag{10}$$

The final visual solution of the MOO problem utilizes the Pareto concept. Pareto optimality refers to an ideal state of resource allocation. In the parameter space S, there is a variable Y, and if there is no variable Y in S that makes all the corresponding objective function values better than the corresponding objective function values, then Y is called the Pareto optimal solution of the objective function, also known as the non-inferior solution. Multi-objective optimization problems have more than one non-inferior solution, and the set composed of all non-inferior solutions is called the non-inferior solution set. The non-inferior solution set is mapped by the objective function to form the Pareto optimal frontier or Pareto frontier surface of the problem. For a problem with two objectives, the Pareto optimal frontier is generally a line.

3. Case Study

3.1. Basic Information of Target Reservoir

Our target is a typical carbonate reservoir in the Middle East, with an anticline reservoir morphology and an initial reservoir pressure of 4425 psi. According to the tectonics of the reservoir, it can be divided into two blocks: the crest area and the flank area. The crest area is mainly developed by injecting miscible gas, and the flank area is mainly developed by WAG. This paper focuses on the crest area. The porosity range is 5–25%, and the permeability ranges from 1.3 to 6.3 mD. Lower permeability is attributed to the blocking of the primary intergranular pore system caused by diagenesis. The specific fluid and reservoir physical parameters are shown in Table 2.

Table 2. Fluid and reservoir physical parameters.

Reservoir Physical Property	Value
Reservoir thickness	150–154 ft
Initial reservoir pressure	4425 psi
Initial bubble point pressure	2980 psi
Porosity range	5–25%
Permeability range	1.3–6.3 mD
Oil volume coefficient (at bubble point pressure)	1.44 rb/STB
Oil phase compression coefficient (at bubble point pressure)	15×10^{-6} 1/psi

Carbonate reservoirs in the Middle East are mainly composed of porous bioclastic limestone with low permeability, making them very different from fractured-cavity carbonate reservoirs. The micropore structure of a carbonate reservoir is complex, the heterogeneity is strong, and the interlayer and thief layer are generally developed. The reservoir has combinations of platform–edge and reef–shoal, with various types of porous spaces, mainly intergranular pores, while also developing intracrystalline pores, intergranular dissolved pores, intergranular pores, and cavity pores. Such reservoirs are complex and diverse differing markedly in geologic features and development modes. Therefore, efficient gas injection development is hindered by several challenges.

The crest area injects miscible gas through 11 reverse five-point well patterns, including 18 producers and 11 injectors, with a well spacing of 4100 ft. All wells are completed with 6-inch horizontal wells and open hole completion, with an average horizontal well length of approximately 4000 ft. According to the development plan, as of 2006, the sustainable production amounts to 60 Mbd/day, with a target injection rate of 100 MMscf/d.

3.2. Multi-Objective Function

Due to the presence of a gas cap, the GOR in the crest area continues to rise. While the number of wells opened remains stable, the regional production capacity continues to decline, making it difficult to achieve stable production of 60 Mbd/day. The average GOR of a single well in the reservoir is 1900–2500 scf/bbl. Since the comprehensive transition to WAG development in 2019, the GOR has not significantly decreased. Some wells have approached the shut-in targets set by resource countries, so efficient development is hindered by several challenges.

The MOPSO algorithm is utilized to mathematically define an objective function for a series of objectives in this paper. The final objective function $F(x)$ is also called the fitness function. In view of the difficulties in stabilizing the oil production and controlling the GOR in the development of the top area, this paper defines the fitness function as below:

$$F(x) = \exp\left\{\frac{Q_{|BRF(d,k|S_n)}}{GOR_{|BRF(d,k|S_n)} + 1}\right\} \tag{11}$$

Q and GOR represent the oil production and GOR calculated by the intelligent agent model. The exponential function is used to increase the sensitivity of the objective function. The larger the objective function value, the higher the gas production and the lower the GOR.

An increase in the gas injection volume will increase production, but it will also correspondingly increase the costs. Taking the P-39 well pattern as an example, we first calculate the historical average gas injection rate (GIR) of adjacent wells, as shown in Table 3, and to make the iterative simulation values more realistic, this article adds a constraint on the gas injection volume, as in Equation (12).

$$\sum_{1}^{4} GIR_n \leq 50{,}000 \text{ Mscf/d} \tag{12}$$

In this paper, GIR_n represents the gas injection rate of the n-th injector.

Table 3. Historical average gas injection volume.

Average GIR of I-14 (Mscf/d)	Average GIR of I-40 (Mscf/d)	Average GIR of I-74 (Mscf/d)	Average GIR of I-19 (Mscf/d)	Sum (Mscf/d)
12,804	8471	12,238	12,804	46,317

3.3. Model Training and Data Preprocessing

To establish a high-precision intelligent agent model, 8 types of real parameters of the reservoir were collected from 2006 onwards, including the gas injection volume and injection pressure of 4 adjacent injectors. The output characteristics are the production and GOR. It is worth mentioning that the shut-in state and choke size have a strong improvement effect on oil production and GOR, but they are strongly influenced by human factors. Therefore, to ensure the theoretical authenticity of the model, this article does not consider these two characteristics.

There are four injectors around P-39, namely I-14, I-40, I-74, and I-19. The collected dynamic data cover a range of 4500 days. The relevant characteristic curves of the well group are shown in Figure 4. The heat map between variables is shown in Figure 5.

To improve the quality of the dataset, we sequentially perform data preprocessing steps such as data standardization and data splitting. Due to the different sources of sample features and measurement units, the scales vary greatly, so it is necessary to standardize the sample and convert the features of each dimension into the same value range. Standardized processing can not only reduce the impact of manual intervention in parameter adjustment but also improve the convergence speed of the model. This article introduces a minimum–maximum scaler to normalize the historical dynamic data, and the specific formula is as follows:

$$x_{nor} = \frac{x_i - min(x_i)}{max(x_i) - min(x_i)} \tag{13}$$

where x_{nor} is the input variable after normalization.

The data used in this study are divided into a training set and a testing set, with a ratio of 9:1. The data from the first 4000 days are used as a training set to train the model. The data from the last 500 days are used as a test set to evaluate the prediction performance.

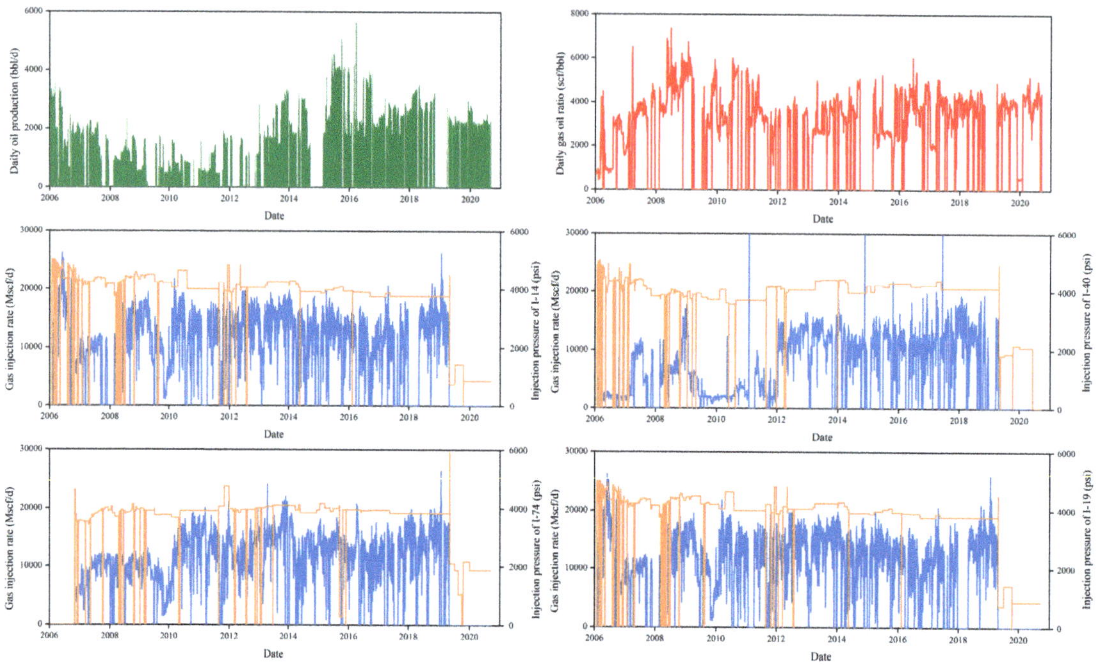

Figure 4. Parameters of producer and injector.

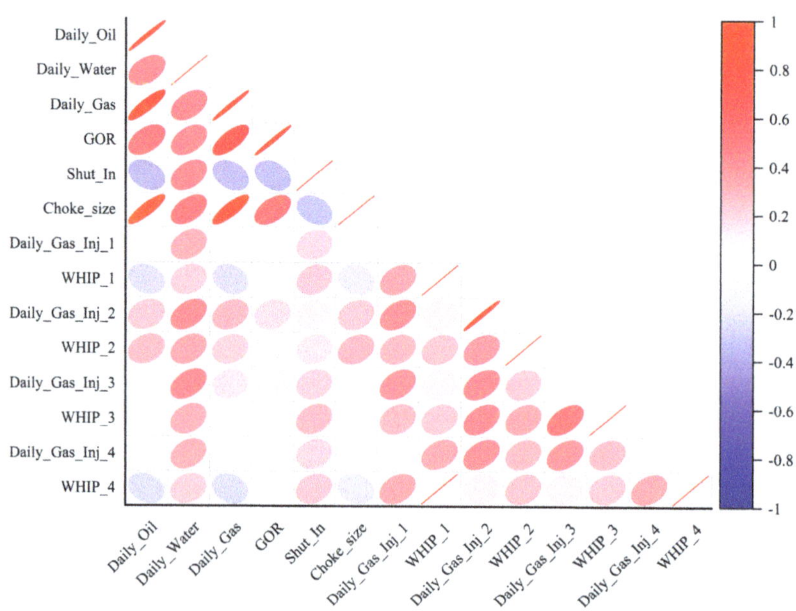

Figure 5. Heatmap of feature variables' correlation analysis.

3.4. Model Structure and Evaluation Criterion

The parameter settings involved in the process of building the intelligent agent model based on the Transformer algorithm are shown in Table 4.

Table 4. Transformer model training parameter settings.

Parameters	Value
Epoch	3000
Batch size	64
Time step	16
Learning rate	0.001
Number of hidden layers	2
Number of hidden neurons	60
Optimizer	Adam
Loss function	RMSE
Activation function	ReLU

In addition, Table 5 lists the hyperparameters of the MOPSO algorithm. The population size is 300, which means that there are 300 particles in each generation that need to be optimized. The iteration number is 10,000, and the inertia weight is 0.7. The individual confidence factor and group confidence factor are both 2.0.

Table 5. Multi-objective particle swarm optimization hyperparameters.

Parameters	Value
Number of particle groups	300
Iterations	10,000
Inertia factor	0.7
Individual confidence factor	2.0
Group confidence factor	2.0

After the prediction, the model needs to be evaluated through statistical evaluation indicators. This paper uses three commonly used error evaluation criteria to study the accuracy of the model results: the root mean square error (RMSE), mean absolute error (MAE), and median absolute error (MedAE). The formulas are as follows:

$$\text{RMSE} = \sqrt{\frac{1}{n}\sum_{i=1}^{n}(y_i - \hat{y}_i)^2} \tag{14}$$

$$\text{MAE} = \frac{1}{n}\sum_{i=1}^{n}|y_i - \hat{y}_i| \tag{15}$$

$$\text{MedAE} = \text{Median}\left(|y_1 - \hat{y}_1|, \ldots, |y_n - \hat{y}_n|\right) \tag{16}$$

where y_i is the i-th actual value, $\hat{y}_i$ is the i-th predicted value, and n is the number of samples in question. $\overline{y}_i$ is the average actual value of the sample.

To better compare the performance between algorithms, another criterion named accuracy is defined as follows:

$$\text{Accuracy} = 1 - \frac{1}{n}\sum_{i=1}^{n}\left(\left|\frac{\hat{y}_i - y_i}{y_i}\right|\right) \tag{17}$$

4. Experimental Results

4.1. Performance in Prediction of Production and GOR

Different algorithms are used to predict the production and GOR of the P-39 well, and the ability of the different algorithms to capture information in dynamic prediction is compared. The first is the temporal prediction ability of representative machine learning algorithms. From the prediction results, the machine learning algorithm-based model is not sensitive to changes in time series data, and it yields an almost smooth straight line, with only significant changes occurring at some points (Figure 6). Secondly, in the first 200 days of prediction, the bagging algorithm and RF algorithm are relatively closer to the real data in production and GOR prediction, but the error gradually increases in the later stage. Thirdly, there are fluctuations in production and GOR between 200 and 270 days; however, the machine learning algorithms are unable to capture this trend.

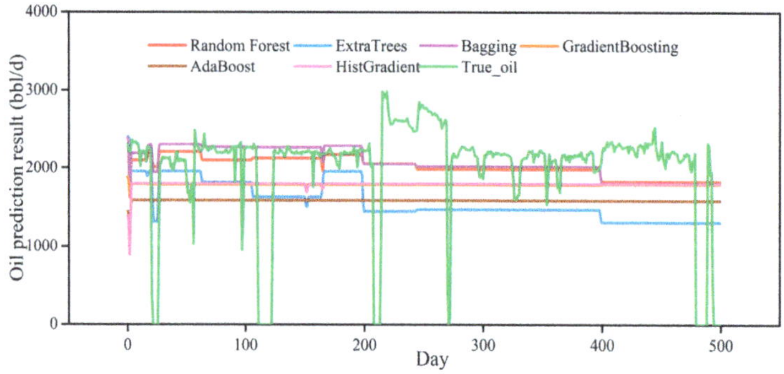

(**a**)

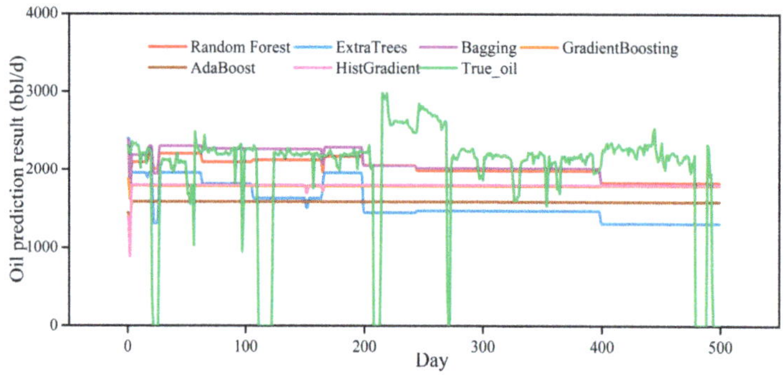

(**b**)

Figure 6. Prediction results of several machine learning algorithms. (**a**) Production prediction; (**b**) GOR prediction.

Next, we performed the temporal prediction ability comparison of the neural network algorithms represented by RNN, GRU, LSTM, Bi-LSTM, etc. From the prediction results (Figure 7), the RNN algorithm and its variants are more sensitive to time series data compared with machine learning algorithms. The prediction results have significant fluctuations, similar to the true data, indicating a preliminary trend of capturing time series information. However, due to the difficulty in obtaining long-term dependencies, in the mid- to late stages of prediction, the effectiveness decreases.

By comparing the production and GOR prediction results of common neural network algorithms, we have captured the following information. Firstly, from the overall trend the deep learning prediction results are more sensitive to time series, reflected in the large fluctuations in production and GOR. Secondly, the GOR prediction trends of Bi-GRU are more closely aligned with the true data; from the statistical error results in GOR prediction (Figure 8b), the RMSE, MAE, and MedAE of Bi-GRU are reduced by 10%, 30%, and 41% respectively, compared to the RNN. This indicates that the Bi-GRU [55] methods have alleviated the limitations of the RNN.

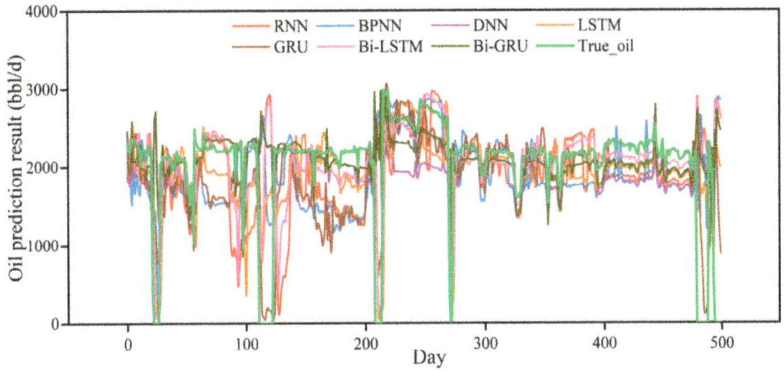

(**a**)

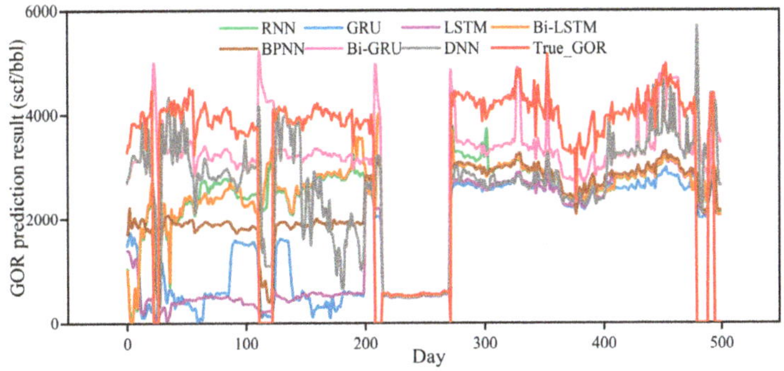

(**b**)

Figure 7. Prediction results of several deep learning algorithms. (**a**) Production prediction; (**b**) GOR prediction.

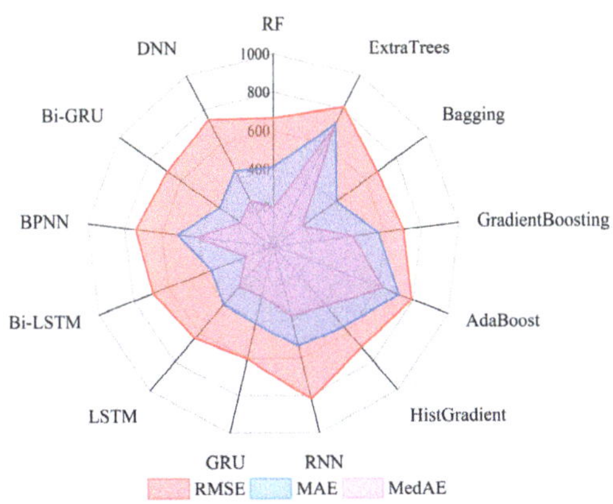

(**a**)

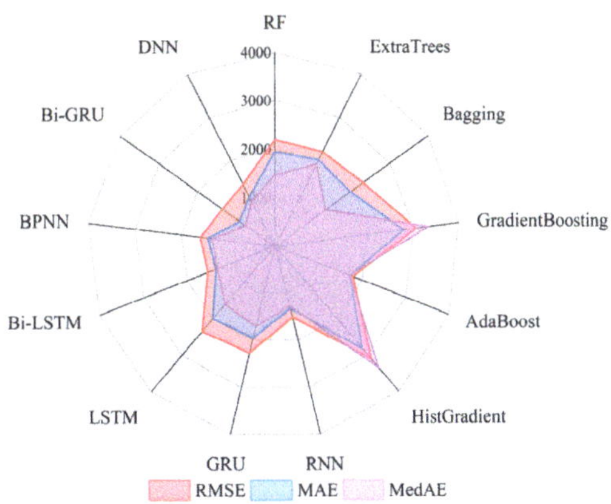

(**b**)

Figure 8. Comparison of errors of several machine learning and deep learning algorithms. (**a**) Production prediction; (**b**) GOR prediction.

The error comparison of common algorithms is shown in Figure 8. From the perspective of the production prediction performance, bagging, RF, GRU, Bi-GRU, and Bi-LSTM perform well in terms of the error criteria; the above algorithms perform better than other algorithms in MedAE especially, indicating that such algorithms have better robustness in production prediction. However, in the process of predicting GOR, the prediction error of the AI methods fluctuates greatly. At present, the bagging, AdaBoost, RNN, and Bi-GRU algorithms perform well in terms of the error criteria. Therefore, combining the prediction results of the two development indicators, the bagging and Bi-GRU algorithms have higher accuracy and will be discussed with the Transformer model further.

Finally, the prediction comparison regarding RNS, bagging, Bi-GRU, and Transformer was performed. The prediction results of the Transformer model were compared with those of the commonly used and highly accurate Bi-GRU algorithm. The comparison of the GOR prediction results showed that, compared to Bi-GRU, the RMSE, MAE, and MedAE of the Transformer algorithm were reduced by 28%, 42%, and 78%, respectively, while the accuracy increased by 9%. From the prediction results (Figure 9), the Transformer model [58] is mostly sensitive to time series data and shows better prediction performance throughout the entire prediction stage compared with bagging and Bi-GRU, which, to some extent, alleviates the problem of the traditional RNN being unable to obtain long-term dependencies (Figure 10).

Next, the differences in computational time between different methods were compared. The RNS methods and all intelligent algorithms were executed on professional workstations; the main specifications were as follows: 2 × Intel® Xeon (R) Gold 6230 CPUs @ 2.10 GHz, 256 GB DDR4 memory, and an NVIDIA RTX A5000 graphics card with 24 GB GPU memory. From Table 6, the computation time of RNS was 4600 s, while the computation time of the intelligent algorithms was significantly reduced. The Transformer algorithm has the lowest time consumption, reflecting the parallel computing ability of the attention mechanism.

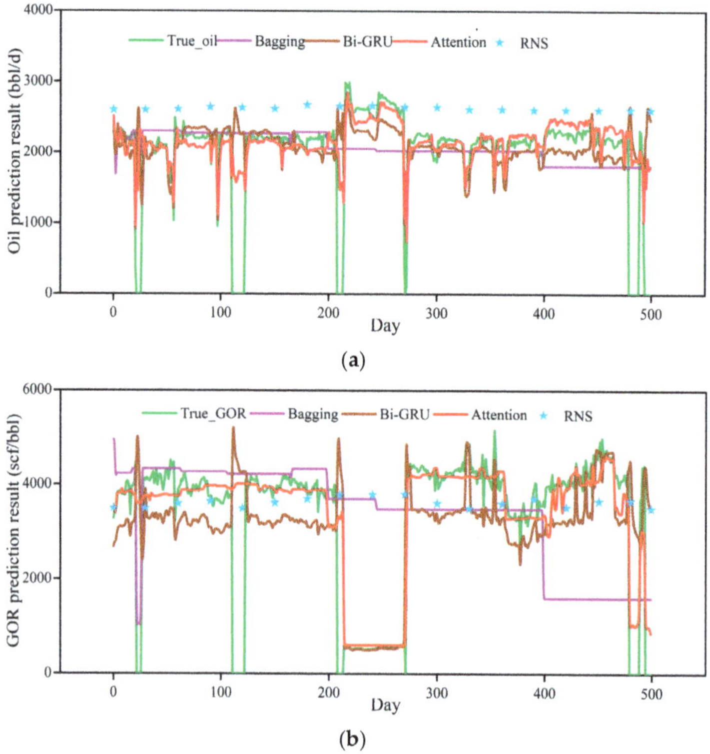

Figure 9. Prediction results of several performance prediction methods. (**a**) Production predictio (**b**) GOR prediction.

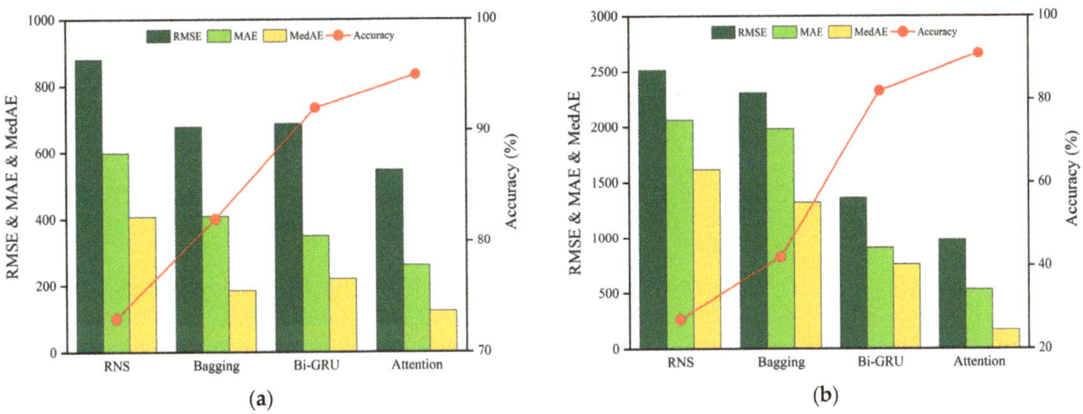

Figure 10. Comparison of errors and accuracy of performance prediction methods. (**a**) Production prediction; (**b**) GOR prediction.

Table 6. Comparison of calculation time.

Method	RNN	GRU	LSTM	Bi-LSTM	BPNN	Bi-GRU	DNN	Attention	RNS
Time (s)	960	830	890	970	1250	1080	1320	760	4600

In conclusion, compared with other methods, the intelligent agent model based on the Transformer model has higher accuracy and a faster computing speed, which can lay the foundation for further injection–production optimization.

4.2. Iteration Result of MOPSO

This section considers the integration of the MOPSO algorithm based on the agent model to find the optimization scheme for the P-39, I-14, I-40, I-74, and I-19 wells. The MOO fitness function integrating production and GOR is defined in Section 3.2, with the aim of finding the optimal solution to achieve both high production and low GOR. Moreover, the required hyperparameters in the MOPSO algorithm are given in Section 3.4.

After 10,000 iterations, the fitness curve and optimal particles of each generation are as shown in Figure 11. Each particle represents the composite pattern of the injection amount and the prediction effect. The horizontal axis represents oil production, and the vertical axis represents the GOR. From the change curve of the fitness function, the fitness gradually increased within 10,000 iterations and became stable in the later period, indicating that the update of the non-dominated solution did not occur within the chosen number of iterations.

According to the above results, the Pareto frontier can be determined. From Figure 12, the Pareto frontier can be divided into three parts, with the curves in part 2 being steeper than those in other parts, indicating that as the oil production increases, the GOR will significantly increase. In parts 1 and 3, the curve is relatively flat, indicating that the GOR is not sensitive to an increase in oil production. Therefore, based on the Pareto frontier, the injection method can be dynamically adjusted on-site. When the reservoir development target is high production, the injection method of particles corresponding to the maximum production rate can be selected. When the production well faces a high GOR, the injection mode corresponding to the minimum GOR can be selected.

Moreover, three optimization schemes are proposed for the P-39 well pattern. Firstly, the average allocation method is proposed based on the target injection volume (50,000 Mscf/d). Secondly, from the perspective of reservoir engineering, the remaining geological reserves in the plane distribution of each well in the simulation model are allocated. Finally, the iterative results proposed in this article are utilized to predict the performance. Then, based on the agent model, the development effects of different injection methods are predicted.

The comparison of the injection and production parameters for different schemes is shown in Table 7.

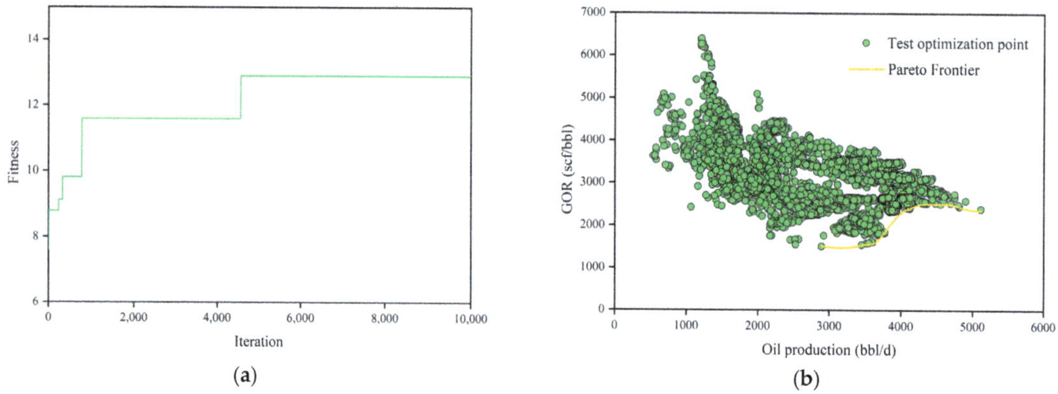

Figure 11. Schematic diagram of multi-objective particle swarm optimization results. (**a**) Fitness curve; (**b**) iteration result.

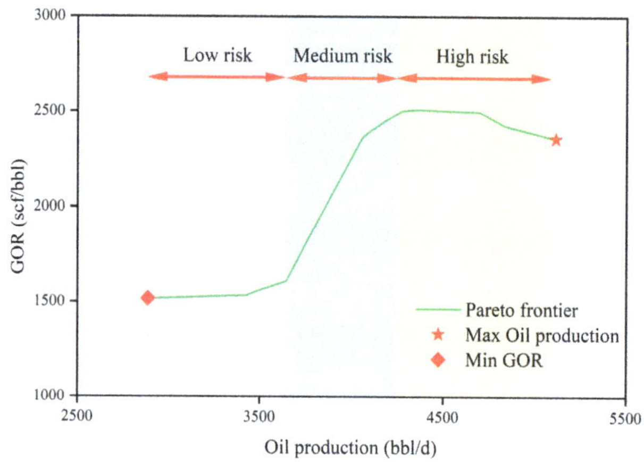

Figure 12. Pareto frontier solution for injection–production optimization based on MOPSO.

Table 7. Injection and production parameters of different schemes.

Method	GIR #1 (Mscf/d)	GIR #2 (Mscf/d)	GIR #3 (Mscf/d)	GIR #4 (Mscf/d)	Sum (Mscf/d)
Average allocation	12,500	12,500	12,500	12,500	50,000
Reservoir engineering	4457	26,360	16,027	3154	50,000
Pareto result	8198	27,260	6662	7879	50,000

The prediction results of the intelligent agent model are shown in Figure 13. The results indicate that in the next 400 days of production, the average method will face the risk of a production reduction and increasing GOR, which are caused by unreasonable injection parameters. The reservoir engineering method does not have significant production fluctuations, but the development effect needs to be further improved. The method proposed in this article results in higher production and lower GOR. In the next 400 days of production, the cumulative oil production will increase by 25.3% compared to the average distribution method and 12.7% compared to the reservoir method; GOR is reduced by

27.1% and 15.3%, respectively. The method proposed in this article helps to consider the actual situation and find the best injection production plan, which can not only ensure a production increase but also reduce the GOR and development risks.

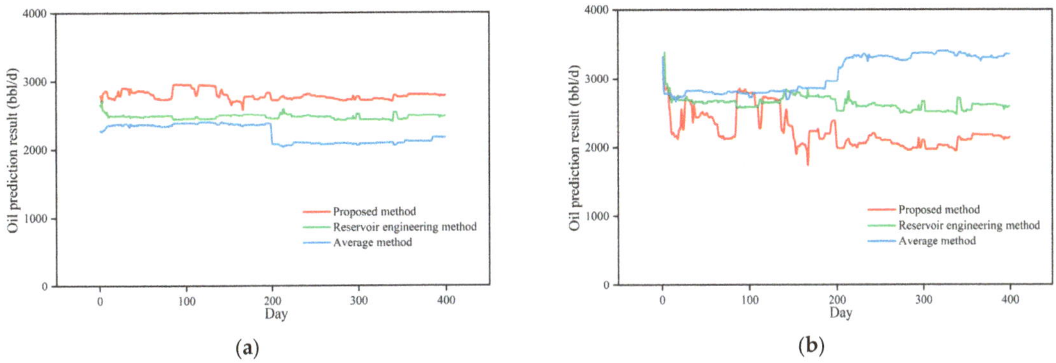

Figure 13. Comparison of prediction performance of different injection–production schemes. (**a**) Production prediction; (**b**) GOR prediction.

5. Discussion

This paper selects the Transformer algorithm from many intelligent algorithms to establish an intelligent agent model of an actual injection production well network. Through correlation analysis, this article takes 10 variables, such as oil production data, GOR, the gas injector volume, and the gas injection pressure, as characteristic variables, and the production and GOR are the output variables. This article selects common machine learning algorithms and neural network algorithms for comparison to verify the superiority of the Transformer algorithm in predicting production and GOR. Through the comparison of the evaluation criteria, including RMSE, MAE, MedAE, and accuracy, the Transformer model based on the attention mechanism has a stronger time series data capture ability, higher accuracy, and a faster calculation speed.

An intelligent multi-objective optimization method for well pattern injection–production parameters based on the Transformer model is established. Among them, the optimization objectives that are in line with the actual reservoir are defined, including simultaneously maximizing oil production and minimizing GOR. In addition, the MOPSO algorithm is used to perform 10,000 iterations to draw the Pareto frontier. Based on the Pareto frontier, it is possible to search for the optimal solution under different development requirements. The injection–production parameter results proposed in this article, and those of the traditional uniform distribution method and reservoir physical property distribution method, are input into the Transformer intelligent agent model. The model prediction results indicate that the method proposed in this article can achieve higher cumulative oil production while maintaining a low GOR. The workflow presented can provide a reference for reservoir engineers to optimize the injection–production parameters.

In future work, there will be several trends as follows. Firstly, the number of optimization goals will increase, including economic goals such as NPV and development goals such as WCT. As the number of goals increases, the representation of the Pareto boundaries will also become more complex. It is generally accepted that the Pareto frontier of two targets is a curve, while the frontier of three targets will become a camber. Therefore, in future work, we will continue to focus on the high-dimensional optimization goals of Pareto's cutting-edge visualization analysis. Secondly, we will consider more advanced AI algorithms to achieve higher prediction accuracy and faster prediction speeds. For example, the graph convolutional network (GCN) algorithm can simultaneously consider time series and spatial relationships, which means that not only the dynamic connectivity between different production wells can be considered, but also the specific flow of underground fluids can be considered. In addition, in future work, we will focus on situations that include more production wells, rather than merely a five-point well pattern. Moreover, based on a larger scope of consideration, risk prediction and real-time injection–production strategy optimization can be further achieved.

6. Conclusions

This paper proposes an intelligent optimization process for well pattern injection–production parameters based on an intelligent agent model and the MOPSO algorithm. Firstly, the ability of common machine learning, neural networks, and the Transformer algorithm in establishing agent models are compared. The Transformer algorithm has higher accuracy and faster computational speeds compared with other methods, and it improves the accuracy by 237% compared to RNS and reduces the computational time by 505%. Secondly, the fitness function of MOO is defined, which aims to maximize production while minimizing GOR. Then, the MOPSO algorithm is utilized to iterate the injection–production parameters. After 10,000 iterations, the Pareto frontier is plotted and divided into three stages according to the severity of the change in GOR during production changes. Finally, three different optimization schemes are proposed, and the development effects of different injection–production schemes are predicted in the intelligent agent model. The prediction results show that the trends are consistent with the proposed schemes. Considering the historical development capacity of the well pattern, the optimal Pareto-based particles of the gas injectors are selected as the optimization values for the development plan. After comparing the prediction performance of the different schemes, it is concluded that the Pareto-based method proposed in this article can increase the cumulative oil production by 25.3% compared to the average distribution method and 12.7% compared to the reservoir engineering method in the next 400 days; GOR is decreased by 27.1% and 15.3%, respectively, which improves the development effect. The intelligent optimization process for injection–production parameters proposed in this article can help reservoir engineers to propose adjustment strategies efficiently and accurately when facing different development scenarios.

Author Contributions: Conceptualization, M.G. and R.H.; methodology, C.W.; software, M.G.; validation, J.Y.; formal analysis, R.H.; investigation, R.H.; resources, C.W.; data curation, C.W.; writing—original draft preparation, M.G.; writing—review and editing, R.H.; visualization, X.Z. supervision, B.L. and Y.G.; project administration, L.X. and S.L. All authors have read and agreed to the published version of the manuscript.

Funding: This research was funded by Major Science and Technology Special Project of the China National Petroleum Corporation (Grant No. 2023ZZ19-03) and the National Natural Science Foundation of China (Grant No. 51974357).

Data Availability Statement: Data is unavailable due to privacy restrictions.

Conflicts of Interest: The authors declare no conflict of interest.

References

1. Khormali, A.; Petrakov, D.G.; Farmanzade, A.R. Prediction and Inhibition of Inorganic Salt Formation under Static and Dynami Conditions—Effect of Pressure, Temperature, and Mixing Ratio. *Int. J. Technol.* **2016**, *7*, 943–951. [CrossRef]
2. Chen, C.; Li, G.; Reynolds, A.C. Robust Constrained Optimization of Short- and Long-Term Net Present Value for Closed-Loo Reservoir Management. *SPE J.* **2012**, *17*, 849–864. [CrossRef]
3. An, Z.; Zhou, K.; Hou, J.; Wu, D.; Pan, Y. Accelerating Reservoir Production Optimization by Combining Reservoir Engineerin Method with Particle Swarm Optimization Algorithm. *J. Pet. Sci. Eng.* **2022**, *208*, 109692. [CrossRef]
4. Gunst, R.F.; Myers, R.H.; Montgomery, D.C. Response Surface Methodology: Process and Product Optimization Using Designe Experiments. *Technometrics* **1996**, *38*, 285. [CrossRef]
5. Kang, X.; Li, B.; Zhang, J.; Wang, X.; Yu, W. Optimization of the SAGP Process in L Oil-Sand Field with Response Surface Metho ology. In Proceedings of the Abu Dhabi International Petroleum Exhibition & Conference, Abu Dhabi, United Arab Emirat 11–14 November 2019.

6. Wantawin, M.; Dachanuwattana, S.; Sepehrnoori, K. An Iterative Response-Surface Methodology by Use of High-Degree-Polynomial Proxy Models for Integrated History Matching and Probabilistic Forecasting Applied to Shale-Gas Reservoirs. *SPE J.* **2017**, *22*, 2012–2031. [CrossRef]
7. Wantawin, M.; Sepehrnoori, K. An Iterative Work Flow for History Matching by Use of Design of Experiment, Response-Surface Methodology, and Markov Chain Monte Carlo Algorithm Applied to Tight Oil Reservoirs. *SPE J.* **2017**, *20*, 613–626. [CrossRef]
8. Sarma, P.; Aziz, K.; Durlofsky, L.J. Implementation of Adjoint Solution for Optimal Control of Smart Wells. In Proceedings of the SPE Reservoir Simulation Symposium, The Woodlands, TX, USA, 31 January–2 February 2005; p. SPE-92864-MS.
9. Mohsin Siraj, M.; Van den Hof, P.M.; Jansen, J.-D. Handling Geological and Economic Uncertainties in Balancing Short-Term and Long-Term Objectives in Waterflooding Optimization. *SPE J.* **2017**, *22*, 1313–1325. [CrossRef]
10. Da Cruz Schaefer, B.; Sampaio, M.A. Efficient Workflow for Optimizing Intelligent Well Completion Using Production Parameters in Real-Time. *Oil Gas Sci. Technol. Rev. IFP Energ. Nouv.* **2020**, *75*, 69. [CrossRef]
11. Zhao, M.; Zhang, K.; Chen, G.; Zhao, X.; Yao, J.; Yao, C.; Zhang, L.; Yang, Y. A Classification-Based Surrogate-Assisted Multiobjective Evolutionary Algorithm for Production Optimization under Geological Uncertainty. *SPE J.* **2020**, *25*, 2450–2469. [CrossRef]
12. Xue, B.; Zhang, M.; Browne, W.N. Particle Swarm Optimization for Feature Selection in Classification: A Multi-Objective Approach. *IEEE Trans. Cybern.* **2013**, *43*, 1656–1671. [CrossRef]
13. Fu, J.; Wen, X.-H. A Regularized Production-Optimization Method for Improved Reservoir Management. *SPE J.* **2018**, *23*, 467–481. [CrossRef]
14. Deb, K.; Pratap, A.; Agarwal, S.; Meyarivan, T. A Fast and Elitist Multiobjective Genetic Algorithm: NSGA-II. *IEEE Trans. Evol. Comput.* **2002**, *6*, 182–197. [CrossRef]
15. Yasari, E.; Pishvaie, M.R.; Khorasheh, F.; Salahshoor, K.; Kharrat, R. Application of Multi-Criterion Robust Optimization in Water-Flooding of Oil Reservoir. *J. Pet. Sci. Eng.* **2013**, *109*, 1–11. [CrossRef]
16. Bagherinezhad, A.; Boozarjomehry Bozorgmehry, R.; Pishvaie, M.R. Multi-Criterion Based Well Placement and Control in the Water-Flooding of Naturally Fractured Reservoir. *J. Pet. Sci. Eng.* **2017**, *149*, 675–685. [CrossRef]
17. Wang, L.; Yao, Y.; Wang, K.; Adenutsi, C.D.; Zhao, G.; Lai, F. Data-Driven Multi-Objective Optimization Design Method for Shale Gas Fracturing Parameters. *J. Nat. Gas Sci. Eng.* **2022**, *99*, 104420. [CrossRef]
18. Das, I.; Dennis, J.E. A Closer Look at Drawbacks of Minimizing Weighted Sums of Objectives for Pareto Set Generation in Multicriteria Optimization Problems. *Struct. Optim.* **1997**, *14*, 63–69. [CrossRef]
19. Coello, C.A.C.; Lechuga, M.S. MOPSO: A Proposal for Multiple Objective Particle Swarm Optimization. In Proceedings of the 2002 Congress on Evolutionary Computation. CEC′02 (Cat. No.02TH8600), Honolulu, HI, USA, 12–17 May 2002; Volume 6. [CrossRef]
20. Yasari, E.; Pishvaie, M.R. Pareto-Based Robust Optimization of Water-Flooding Using Multiple Realizations. *J. Pet. Sci. Eng.* **2015**, *132*, 18–27. [CrossRef]
21. Moshir Farahi, M.M.; Ahmadi, M.; Dabir, B. Model-Based Production Optimization under Geological and Economic Uncertainties Using Multi-Objective Particle Swarm Method. *Oil Gas Sci. Technol. Rev. IFP Energ. Nouv.* **2021**, *76*, 60. [CrossRef]
22. Farahi, M.M.M. Model-Based Water-Flooding Optimization Using Multi-Objective Approach for Efficient Reservoir Management. *J. Pet. Sci. Eng.* **2021**, *196*, 107988. [CrossRef]
23. Zhao, H.; Li, Y.; Cui, S.; Shang, G.; Reynolds, A.C.; Guo, Z.; Li, H.A. History Matching and Production Optimization of Water Flooding Based on a Data-Driven Interwell Numerical Simulation Model. *J. Nat. Gas Sci. Eng.* **2016**, *31*, 48–66. [CrossRef]
24. Coello Coello, C.A.; Reyes-Sierra, M. Multi-Objective Particle Swarm Optimizers: A Survey of the State-of-the-Art. *Int. J. Comput. Intell. Res.* **2006**, *2*, 287–308. [CrossRef]
25. Reddy, M.J.; Nagesh Kumar, D. Multi-Objective Particle Swarm Optimization for Generating Optimal Trade-Offs in Reservoir Operation. *Hydrol. Process.* **2007**, *21*, 2897–2909. [CrossRef]
26. Fallah-Mehdipour, E.; Haddad, O.B.; Mariño, M.A. MOPSO Algorithm and Its Application in Multipurpose Multireservoir Operations. *J. Hydroinform.* **2011**, *13*, 794–811. [CrossRef]
27. Zheng, Y.-J.; Ling, H.-F.; Xue, J.-Y.; Chen, S.-Y. Population Classification in Fire Evacuation: A Multiobjective Particle Swarm Optimization Approach. *IEEE Trans. Evol. Comput.* **2014**, *18*, 70–81. [CrossRef]
28. Fu, J.; Wen, X.-H. Model-Based Multiobjective Optimization Methods for Efficient Management of Subsurface Flow. *SPE J.* **2017**, *22*, 1984–1998. [CrossRef]
29. Wang, H. Optimal Well Placement Under Uncertainty Using a Retrospective Optimization Framework. *SPE J.* **2012**, *17*, 112–121. [CrossRef]
30. Desbordes, J.K.; Zhang, K.; Xue, X.; Ma, X.; Luo, Q.; Huang, Z.; Hai, S.; Jun, Y. Dynamic Production Optimization Based on Transfer Learning Algorithms. *J. Pet. Sci. Eng.* **2022**, *208*, 109278. [CrossRef]
31. Li, Y.; Jia, C.; Song, B.; Li, B.; Zhu, Y.; Qian, Q.; Wei, C. Geological Models Comparison and Selection for Multi-Layered Sandstone Reservoirs. In Proceedings of the SPE Middle East Oil & Gas Show and Conference, Manama, Bahrain, 6–9 March 2017; p. D031S026R005.
32. Gu, J.; Liu, W.; Zhang, K.; Zhai, L.; Zhang, Y.; Chen, F. Reservoir Production Optimization Based on Surrograte Model and Differential Evolution Algorithm. *J. Pet. Sci. Eng.* **2021**, *205*, 108879. [CrossRef]
33. Alkinani, H.H.; Al-Hameedi, A.T.; Dunn-Norman, S.; Flori, R.E.; Alsaba, M.T.; Amer, A.S. Applications of Artificial Neural Networks in the Petroleum Industry: A Review. In Proceedings of the SPE Middle East Oil and Gas Show and Conference, Manama, Bahrain, 18–21 March 2019; p. D032S063R002.

34. El-Sebakhy, E.A.; Sheltami, T.; Al-Bokhitan, S.Y.; Shaaban, Y.; Raharja, P.D.; Khaeruzzaman, Y. Support Vector Machines Framework for Predicting the PVT Properties of Crude Oil Systems. In Proceedings of the SPE Middle East Oil and Gas Show and Conference, Manama, Bahrain, 11–14 March 2007; p. SPE-105698-MS.
35. Madasu, S.; Rangarajan, K.P. Deep Recurrent Neural Network DRNN Model for Real-Time Multistage Pumping Data. In Proceedings of the OTC Arctic Technology Conference, Houston, TX, USA, 5–7 November 2018; p. D033S017R005.
36. Romero Quishpe, A.; Silva Alonso, K.; Alvarez Claramunt, J.I.; Barros, J.L.; Bizzotto, P.; Ferrigno, E.; Martinez, G. Innovative Artificial Intelligence Approach in Vaca Muerta Shale Oil Wells for Real Time Optimization. In Proceedings of the SPE Annual Technical Conference and Exhibition, Calgary, AB, Canada, 30 September–2 October 2019; p. D011S009R002.
37. Huang, R.; Wei, C.; Li, B.; Xiong, L.; Yang, J.; Wu, S.; Gao, Y.; Liu, S.; Zhang, C.; Lou, Y.; et al. A Data Driven Method to Predict and Restore Missing Well Head Flow Pressure. In Proceedings of the International Petroleum Technology Conference, Riyadh, Saudi Arabia, 21–23 February 2022; p. D032S154R004.
38. Ahmadi, M.-A.; Bahadori, A. A LSSVM Approach for Determining Well Placement and Conning Phenomena in Horizontal Wells. *Fuel* **2015**, *153*, 276–283. [CrossRef]
39. Nwachukwu, A.; Jeong, H.; Pyrcz, M.; Lake, L.W. Fast Evaluation of Well Placements in Heterogeneous Reservoir Models Using Machine Learning. *J. Pet. Sci. Eng.* **2018**, *163*, 463–475. [CrossRef]
40. Clarkson, C.R.; Williams-Kovacs, J.D.; Qanbari, F.; Behmanesh, H.; Heidari Sureshjani, M. History-Matching and Forecasting Tight/Shale Gas Condensate Wells Using Combined Analytical, Semi-Analytical, and Empirical Methods. *J. Nat. Gas Sci. Eng.* **2015**, *26*, 1620–1647. [CrossRef]
41. Hutahaean, J.; Demyanov, V.; Christie, M. Many-Objective Optimization Algorithm Applied to History Matching. In Proceedings of the 2016 IEEE Symposium Series on Computational Intelligence (SSCI), Athens, Greece, 6–9 December 2016; pp. 1–8.
42. Jo, S.; Jeong, H.; Min, B.; Park, C.; Kim, Y.; Kwon, S.; Sun, A. Efficient Deep-Learning-Based History Matching for Fluvial Channel Reservoirs. *J. Pet. Sci. Eng.* **2022**, *208*, 109247. [CrossRef]
43. Lee, K.; Lim, J.; Yoon, D.; Jung, H. Prediction of Shale-Gas Production at Duvernay Formation Using Deep-Learning Algorithm. *SPE J.* **2019**, *24*, 2423–2437. [CrossRef]
44. Li, X.; Ma, X.; Xiao, F.; Wang, F.; Zhang, S. Application of Gated Recurrent Unit (GRU) Neural Network for Smart Batch Production Prediction. *Energies* **2020**, *13*, 6121. [CrossRef]
45. Song, X.; Liu, Y.; Xue, L.; Wang, J.; Zhang, J.; Wang, J.; Jiang, L.; Cheng, Z. Time-Series Well Performance Prediction Based on Long Short-Term Memory (LSTM) Neural Network Model. *J. Pet. Sci. Eng.* **2020**, *186*, 106682. [CrossRef]
46. Kocoglu, Y.; Gorell, S.; McElroy, P. Application of Bayesian Optimized Deep Bi-LSTM Neural Networks for Production Forecasting of Gas Wells in Unconventional Shale Gas Reservoirs. In Proceedings of the 9th Unconventional Resources Technology Conference, Houston, TX, USA, 26–28 July 2021.
47. Huang, R.; Wei, C.; Wang, B.; Yang, J.; Xu, X.; Wu, S.; Huang, S. Well Performance Prediction Based on Long Short-Term Memory (LSTM) Neural Network. *J. Pet. Sci. Eng.* **2022**, *208*, 109686. [CrossRef]
48. Li, W.; Wang, L.; Dong, Z.; Wang, R.; Qu, B. Reservoir Production Prediction with Optimized Artificial Neural Network and Time Series Approaches. *J. Pet. Sci. Eng.* **2022**, *215*, 110586. [CrossRef]
49. Guo, S.; Lin, Y.; Feng, N.; Song, C.; Wan, H. Attention Based Spatial-Temporal Graph Convolutional Networks for Traffic Flow Forecasting. *Proc. AAAI Conf. Artif. Intell.* **2019**, *33*, 922–929. [CrossRef]
50. Bai, J.; Zhu, J.; Song, Y.; Zhao, L.; Hou, Z.; Du, R.; Li, H. A3T-GCN: Attention Temporal Graph Convolutional Network for Traffic Forecasting. *ISPRS Int. J. Geo-Inf.* **2021**, *10*, 485. [CrossRef]
51. Wang, S.; Wang, X.; Wang, S.; Wang, D. Bi-Directional Long Short-Term Memory Method Based on Attention Mechanism and Rolling Update for Short-Term Load Forecasting. *Int. J. Electr. Power Energy Syst.* **2019**, *109*, 470–479. [CrossRef]
52. Liu, Y.; Shan, L.; Yu, D.; Zeng, L.; Yang, M. An Echo State Network with Attention Mechanism for Production Prediction in Reservoirs. *J. Pet. Sci. Eng.* **2022**, *209*, 109920. [CrossRef]
53. Niu, Z.; Yu, Z.; Tang, W.; Wu, Q.; Reformat, M. Wind Power Forecasting Using Attention-Based Gated Recurrent Unit Network *Energy* **2020**, *196*, 117081. [CrossRef]
54. Hochreiter, S.; Schmidhuber, J. Long Short-Term Memory. *Neural Comput.* **1997**, *9*, 1735–1780. [CrossRef] [PubMed]
55. Cho, K.; van Merrienboer, B.; Gulcehre, C.; Bahdanau, D.; Bougares, F.; Schwenk, H.; Bengio, Y. Learning Phrase Representations Using RNN Encoder–Decoder for Statistical Machine Translation. In Proceedings of the 2014 Conference on Empirical Methods in Natural Language Processing (EMNLP), Doha, Qatar, 25–29 October 2014; pp. 1724–1734.
56. Shahid, F.; Zameer, A.; Muneeb, M. Predictions for COVID-19 with Deep Learning Models of LSTM, GRU and Bi-LSTM. *Chaos Solitons Fractals* **2020**, *140*, 110212. [CrossRef] [PubMed]
57. Li, X.; Ma, X.; Xiao, F.; Xiao, C.; Wang, F.; Zhang, S. Time-Series Production Forecasting Method Based on the Integration of Bidirectional Gated Recurrent Unit (Bi-GRU) Network and Sparrow Search Algorithm (SSA). *J. Pet. Sci. Eng.* **2022**, *208*, 109309. [CrossRef]
58. Vaswani, A.; Shazeer, N.; Parmar, N.; Uszkoreit, J.; Jones, L.; Gomez, A.N.; Kaiser, Ł.; Polosukhin, I. Attention Is All You Need *Adv. Neural Inf. Process. Syst.* **2017**, *30*.

processes

MDPI

Article

Research on the Scaling Mechanism and Countermeasures of Tight Sandstone Gas Reservoirs Based on Machine Learning

Xu Su [1], Desheng Zhou [1,*], Haiyang Wang [1,*] and Jinze Xu [1,2]

[1] College of Petroleum Engineering, Xi'an Shiyou University, Xi'an 710065, China; 22211010042@stumail.xsyu.edu.cn (X.S.); jinzxu@ucalgary.ca (J.X.)
[2] Department of Chemical and Petroleum Engineering, University of Calgary, Calgary, AB T2N 2N4, Canada
* Correspondence: dzhou@xsyu.edu.cn (D.Z.); wanghaiyang@xsyu.edu.cn (H.W.)

Abstract: The Sulige gas field is a typical "three lows" (low permeability, low pressure, and low abundance) tight sandstone gas reservoir, with formation pressures often characterized by abnormally high or low pressures. The complex geological features of the reservoir further deviate from conventional understanding, impacting the effective implementation of wellbore blockage removal measures. Therefore, it is imperative to establish the wellbore blockage mechanism, prediction model, and effective prevention measures for the target area. In this study, based on field data, we first experimentally analyzed the water quality and types of blockage in the target area. Subsequently, utilizing a BP neural network model, we established a model for predicting the risk of wellbore blockage and analyzing mitigation measures in the target reservoir. The model's prediction results, consistent with on-site actual results, demonstrate its reliability and accuracy. Experimental results show that the water quality in the target area is mainly a $CaCl_2$ type, and the predominant scales produced are $CaCO_3$ and $BaSO_4$. Model calculations reveal that temperature, pressure, and ion concentration all influence scaling, with $BaSO_4$ more influenced by pressure and $CaCO_3$ more influenced by temperature. Under the combined effect of temperature, pressure, and ion concentration, different types of scales exhibit distinct trends in scaling quantity. Combining scaling quantity calculations with wellbore contraction ratios, it was found that when the temperature, pressure, and ion concentration are within a certain range, the wellbore contraction rate can be controlled below 4%. At this point, the wellbore scaling risk is minimal, and preventive measures against wellbore scaling can be achieved by adjusting production systems, considering practical production conditions. This study investigates the mechanism of scaling in wellbores of tight sandstone gas reservoirs and proposes a cost-effective scaling prevention measure. This approach can guide the prediction of scaling risks and the implementation of scaling prevention measures for gas wells in tight sandstone reservoirs.

Keywords: tight sandstone gas reservoirs; machine learning; scaling mechanism; scale prevention measures; enhanced oil recovery

Citation: Su, X.; Zhou, D.; Wang, H.; Xu, J. Research on the Scaling Mechanism and Countermeasures of Tight Sandstone Gas Reservoirs Based on Machine Learning. *Processes* 2024, *12*, 527. https://doi.org/10.3390/pr12030527

Academic Editor: Abraham Kabutey

Received: 15 January 2024
Revised: 2 March 2024
Accepted: 5 March 2024
Published: 6 March 2024

1. Introduction

As exploration and development practices in unconventional reservoirs such as shale and tight sandstones continue to deepen, many limitations and characteristics inconsistent with actual geological conditions have been exposed in traditional geological understanding. Consequently, numerous scholars have conducted extensive research on how to improve the recovery factor of such reservoirs. This includes studying the relationship between pores and fluids from a nanoscale perspective [1,2], researching enhanced oil recovery methods [3], and reservoir modification through hydraulic fracturing [4], among other approaches. The Sulige gas field is a typical "three lows" gas field, characterized by low permeability, low pressure, and low abundance. The reservoir properties are poor, belonging to low porosity and low permeability sandstone lithology gas reservoirs. This makes the problem of wellbore plugging particularly prominent in the Sulige gas field, especially due to changes in reservoir pressure, physical properties, and lithological characteristics. These changes result in production conditions in the wellbore of gas wells in the target area being different from those of conventional gas wells. Traditional scale and corrosion control measures cannot be effectively applied to the target block.

The target block is located in the northern slope of the Shaanbei area of the Ordos Basin in China, characterized by a monocline structure trending high in the northeast and low in the southwest. The Upper Paleozoic strata in the area consist of the Permian and Carboniferous formations, including the Shi'qianfeng Formation, Shi'hezi Formation, Shanxi Formation, and Taiyuan Formation of the Permian, while the Carboniferous is predominantly represented by the Benxi Formation. The primary producing reservoir in the target block is currently the Shan 1 Member of the Shanxi Formation, with the upper reservoir being the Shi'hezi Formation and the lower reservoir being the Taiyuan Formation. Currently, the Shanxi Formation in the research block is subdivided into the Shan 1 Member and the Shan 2 Member, with the latter representing lacustrine deposits, mainly characterized by marginal shallow lake sedimentation. The Shan 1 Member primarily consists of delta-front deposits, subdivided into subaqueous distributary channels and subaqueous distributary bay microfacies. The lithology is dominated by dark gray mudstone and light gray sandstone.

The target block's formation water exhibits elevated concentrations of scale-forming ions. Additionally, the corrosion and deformation of wellbores in these production wells, coupled with frequent downhole operations, result in a significant proportion of workover fluids being retained on the inner walls of the wellbore, leading to complex wellbore conditions. Under the complex conditions in the wellbore, these scaling ions form insoluble scale deposits, leading to wellbore blockages and causing significant losses [5]. Conventional methods to address scaling issues mainly involve chemical means to delay or prevent the formation of scale deposits [6–11], or using mechanical methods to remove scaling deposits from the pipe walls [12]. The target block once employed a detoxifier primarily composed of organic sulfonic acid. Three years into production, trace amounts (0.001 mg/L) of organic sulfonic acid were detected in groundwater quality tests, posing significant health risks to the local residents and causing environmental damage. Consequently, it is imperative to grasp the intricacies of scaling and blockage mechanisms within the Sulige gas field, scrutinize the reasons behind scaling, and devise anti-scaling measures that safeguard the environment while curbing costs.

Over the years, numerous scholars have conducted in-depth research and analysis on scaling issues. The earliest study on scaling prediction was the saturation index method proposed by Langelier [13] in 1936. This method primarily relies on the principles of dissolution equilibrium theory to calculate the pH of the solution, thereby determining the tendency for scale formation. As development progresses, the underground water systems faced by oil wells become increasingly complex. Furthermore, with the deepening of the wells, the bottom temperatures continue to rise, amplifying the impact of temperature on scaling. Engineers have gradually become aware of this issue. Therefore, Davis and Stiff [14], as well as Ryznar [15], based on the saturation index method, made certain innovations and breakthroughs, mainly by introducing correction coefficients K, ionic strength, alkalinity, and other parameters to assess scaling trends. This made the method consistent with the development conditions of oil fields at that time. Skillman [16] used the thermodynamic solubility method to predict the scaling tendency of calcium sulfate. Although the prediction results were generally consistent with the actual site conditions, the method had limitations due to the lack of consideration of pressure and other equilibrium effects. By 1994, Oddo and Tomson [17] integrated thermodynamics, ionic strength, solubility product, and ion association factors based on the saturation index theory and proposed the Oddo–Tomson saturation index method. They used different equations for situations with and without a gas phase, making it one of the most widely applied models. However, as oil and gas field development progressed, and with breakthroughs in theory and technology, more scholars tended to use complex mathematical models to describe scaling issues and solve them using computers. For instance, Hasson et al. [18,19] studied the formation of $CaCO_3$ scale samples and established an ion diffusion model considering ion concentration, temperature, and pressure. However, this model only considered crystalline scale and did not account for particulate scale. Brahim et al. [20] considered the heat and mass transfer processes in the system based on the Ker–Seanton differential equation model and established a mathematical model for $CaSO_4$ scaling on the inner wall of a rectangular cross-section. Xiaoxi Yang et al. [21] established a model for predicting $CaCO_3$ scaling based on the sediment removal rate calculation model, considering fluid flow and heat transfer between the fluid and the wall surface. However, this model did not consider changes in wellhead pressure and wellbore diameter under the influence of scaling. Liang Z et al. [22] developed a mathematical model considering holdup ratio and developed corresponding software for scaling prediction. The aforementioned scholars provided excellent predictions and solutions for scaling issues, but complex theoretical models struggle to promptly adapt to increasingly complex oil field production conditions. With the continuous expansion of oil field scale and technological development, a vast amount of data is generated. Machine learning has shown great advantages in processing and utilizing data. Therefore, an increasing number of scholars are exploring how to leverage machine learning to fully utilize the massive amount of data from oil fields to address the increasingly complex on-site production conditions. Researchers such as Zhao [23], Gabetta, G. [24], Paulo A. Paz [25], Wang [26], and Bruno X. Ferreira [27] have integrated machine learning into scaling considerations and achieved significant results.

Due to the fact that BP neural networks do not require predefined mathematical equations for the mapping relationship between inputs and outputs and instead learn certain rules through training to obtain results closest to the expected output when given input values, this study is based on BP neural networks to investigate scaling problems in a block of wells in the Sulige gas field. Data from 93 wells were gathered and structured, with data from 89 wells serving as training data and the remaining four sets reserved as a validation dataset. A neural network was trained utilizing the compiled data, and an algorithm was formulated to scrutinize the distribution of major scaling anions (SO_4^{2-} and HCO_3^-) and major scaling cations (Ca^{2+} and Ba^{2+}) across various temperatures and pressures. Furthermore, the fluctuation of contraction ratio under diverse input parameters was examined, and ranges of input parameters capable of keeping the contraction ratio near or below 4% were outlined. This can aid in forecasting scaling risks and devising on-site scaling prevention strategies.

2. Data Collection

On-site deposits generally consist of sulfides, oxides, sulfates, and carbonates [28]. However, not all deposits occur in oil and gas fields; it depends on geological conditions, including parameters such as formation water quality, temperature, and pressure in the formation and wellbore. Calcium carbonate scale is the most common type of deposit and is typically widespread in production facilities such as surface pipelines and oil tubing [29].

The study area is a specific block within the Sulige gas field, with a burial depth ranging from 3500 m to 4160 m and a geothermal gradient of 3 °C/100 m. The formation water in this area is characterized by a $CaCl_2$ type, exhibiting higher production water yields in the central part and lower yields in the surrounding areas. The average salinity is 49,708.68 mg/L, with the highest salinity recorded at 93,306.86 mg/L. The entire block produces an average daily gas output of 237×10^4 m^3/d and a daily water production of 650 m^3/d. The water–gas ratio is high, measuring 2.74 $m^3/10^4$ m^3 (The typical water-to-gas ratio in the block is around 0.1 $m^3/10^4$ m^3). The pH of the formation water ranges between 6.2 and 6.5. Table 1 presents the results of the formation water quality analysis in the study block. (Data from 118 wells have been compiled for statistical analysis.)

Table 1. Research Zone 118 Wellbore Strata Water Quality Analysis Results Statistics (mg/L).

Data	Cl^-	SO_4^{2-}	HCO_3^-	Mg^{2+}	Ca^{2+}	Sr^{2+}/Ba^{2+}	Na^+/K^+	Salinity
Minimum	2753	8.23	58.79	9.024	113.5	17.83	2208	6110.5
Maximum	5464.3	269	586.33	173.98	1252.42	4525.56	32,320.8	93,306.86
Average	28,698.35	138.6	322.545	91.502	682.96	2271.7	17,264.4	49,708.68

Analysis of the blockage material composition unveiled that around 20% of the blockage comprises water and organic compounds (refer to Table 2), while the remaining 80% comprises inorganic materials. Among the organic compounds, hydrocarbons make up a minimal percentage, accounting for only 3%. Preliminary conclusions indicate that the primary culprit behind wellbore blockage is the buildup of inorganic scale deposits. Analysis of scale samples from five gas wells using a D/MAX-2400 X-ray diffractometer from Rigaku Corporation, Japan, indicates that the predominant scales in the study area are $CaCO_3$ and $BaSO_4$ (Table 3). Figure 1 displays partial scale samples from the wellbore along with their X-ray diffraction patterns (Red represents Barite (The image above) and Siderite (The image below), blue represents Magnetite, and green represents Anhydrite).

Table 2. Determination Results of Water, Oil, and Organic Matter Content.

Well Name	Well Depth, m	Water Content, %	Oil Content, %	Organic Matter, %
W1	3000	0.59	1.83	20.94
	3600	0.62	3.09	19.81
W2	3500	0.24	0.44	25.33
	4100	0.3	2.5	26.99

Table 3. Inorganic Content Determination Results.

	Inorganic Mineral Composition, %					
Well	$BaSO_4$	$FeCO_3$	$CaCO_3$	Fe_3O_4	$MgCO_3$	SiO_2
W1	38.3	4.7	40.6	5.6	7.4	3.4
W2	28.5	4.9	50.6	5.1	6.7	4.2

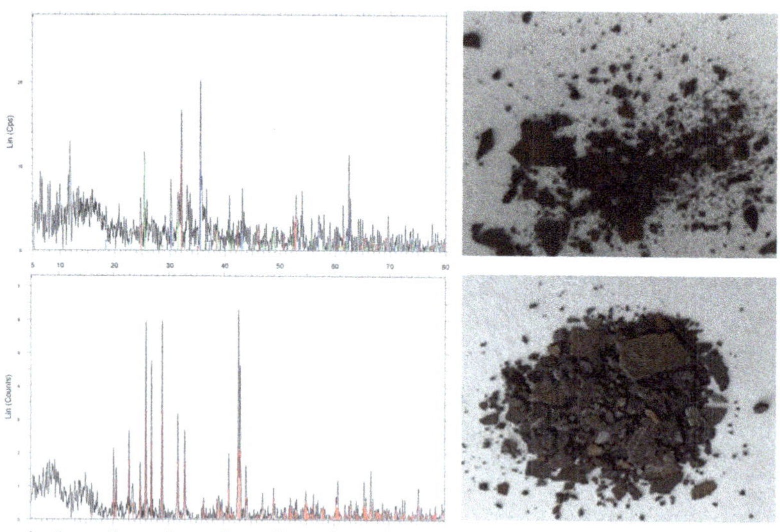

Figure 1. Scale samples from the target area's wellbores.

The scaling issue in gas wells has been extensively studied by numerous scholars. Currently, a well-established understanding of the influencing factors on scaling includes pressure, temperature, scaling ion concentration, pH, and flow velocity. However, in the study block, the gas wells exhibit high production rates (237×10^4 m^3/d) and high gas flow velocities (9.8 m/s). In fact, determining flow velocity is not easy. Considering that the basic process of scaling involves deposition and detachment of scale, we introduce the attachment coefficient to comprehensively consider both aspects. This essentially translates the factor of flow velocity into the selection of the attachment coefficient (further explained in subsequent sections), hence neglecting the influence of flow velocity. Additionally, the pH of the formation water in the study area remains stable between 6.2 and 6.5. Therefore, this study disregards the influences of flow velocity and pH on scaling.

In this study, data from 118 wells were collected, and after processing, data from 93 wells were deemed usable. The data were obtained through on-site experiments of formation water. The experiments utilized high-temperature and high-pressure reaction vessels to simulate downhole temperature and pressure, allowing for thorough scaling of the vessel's formation water. The total scale deposition was calculated by measuring the change in concentration of scaling ions in the formation water before and after titration of the scale. The specific experimental steps are as follows: (1) A certain amount of formation water was taken according to the experimental requirements and filtered to remove solid-phase impurities; (2) The concentration of scaling ions (SO_4^{2-}, HCO_3^-, Ca^{2+}, and Ba^{2+}) was determined using ion titration; (3) The processed formation water was introduced into the reaction vessel, vacuum was applied, methane was added to the designed pressure, and the reaction vessel was placed in a constant temperature chamber to maintain the design temperature for 24 h to allow for thorough scaling of the formation water; (4) The reaction vessel was opened, and the formation water was filtered. The concentration of scaling ions (SO_4^{2-}, HCO_3^-, Ca^{2+}, and Ba^{2+}) in the filtered formation water was determined using titration. Each experiment was conducted three times and the average value was taken. The total scale deposition was calculated based on the difference in ion concentration before and after the reaction. Each well's formation water sample in the target block underwent the above steps to conduct experiments and obtain the necessary data for final simulation. Table 4 presents part of the acquired data.

Table 4. Partial Training Data.

Wells	Ion Concentration, mg/L								T, °C	P, MPa	Scale Amount, mg/L	
	Cl^-	SO_4^{2-}	HCO_3^-	Mg^{2+}	Ca^{2+}	Ba^{2+}	Na^+	Salinity			$BaSO_4$	$CaCO_3$
1	40,843	198	498	77	338	2247	18,516	62,717	81	11	181.6	106.4
2	43,560	36	511	165	243	216	6742	51,473	89	11	31.5	110.9
3	33,870	83	415	125	281	1992	2643	39,409	81	10	153.2	101.4
4	41,142	193	287	113	1237	1605	8977	53,554	72	10	181.6	106.4
5	45,921	67	475	71	291	3751	3305	53,881	88	11	108.7	50.2
6	46,530	110	301	171	587	3859	23,962	75,520	85	13	182.2	107.4
7	42,149	117	422	74	177	2512	16,919	62,370	77	11	177.7	104.9
8	45,889	195	298	131	889	2883	6378	56,663	78	11	181.6	106.4
9	43,215	35	353	42	128	1977	16,679	62,429	76	11	68.8	21.1
10	53,860	92	322	100	920	1565	11,723	68,582	82	11	172.6	101.5
11	42,090	61	63	156	839	3891	2981	50,081	74	13	207.3	81.2
12	43,591	112	86	52	718	1805	31,391	77,755	73	13	176.9	102.8
13	38,875	241	262	57	503	18	18,654	58,610	85	13	180.3	106.5
14	31,300	135	345	103	1220	3198	23,176	59,477	78	8	181.7	105.7
15	40,432	116	75	76	747	3123	20,271	64,840	72	14	180.4	105.6
16	37,770	28	271	85	873	3046	22,437	64,510	70	11	17.99	66.54
17	41,357	28	285	86	949	3318	24,589	70,612	70	10	26.65	65.53
18	44,943	27	300	87	1026	3591	26,741	76,715	71	12	35.39	64.53
19	48,530	27	314	87	1103	3863	28,892	82,817	75	11	44.24	63.53
20	19,836	30	197	82	489	1684	11,679	33,997	89	13	6.32	70.64

Table 4. *Cont.*

Wells	Ion Concentration, mg/L								T, °C	P, MPa	Scale Amount, mg/L	
	Cl^-	SO_4^{2-}	HCO_3^-	Mg^{2+}	Ca^{2+}	Ba^{2+}	Na^+	Salinity			$BaSO_4$	$CaCO_3$
21	23,423	30	212	82	566	1957	13,831	40,100	89	14	15.8	69.66
22	27,010	29	227	83	642	2229	15,982	46,202	88	13	25.28	68.68
23	30,596	29	241	84	719	2501	18,134	52,305	85	13	34.76	67.69
24	34,183	28	256	85	796	2774	20,286	58,407	85	13	44.27	66.7
25	37,770	28	271	85	873	3046	22,437	64,510	85	13	53.81	65.71
26	41,357	28	285	86	949	3318	24,589	70,612	85	13	63.4	64.71
27	44,943	27	300	87	1026	3591	26,741	76,715	87	15	73.04	63.72
28	48,530	27	314	87	1103	3863	28,892	82,817	85	14	82.73	62.73
29	31,607	31	253	118	748	2657	18,645	54,060	73	12	16.28	73.26
30	35,446	31	271	127	832	2969	20,924	60,601	74	11	28.45	73.44
31	39,286	31	289	137	916	3280	23,203	67,142	74	11	40.63	73.62
32	43,125	31	306	146	1000	3592	25,483	73,683	73	11	52.83	73.79
33	46,965	31	324	155	1084	3903	27,762	80,224	74	12	65.07	73.96
34	50,804	31	341	165	1168	4214	30,041	86,766	70	10	77.35	74.13
35	20,089	31	200	90	496	1723	11,806	34,436	85	14	9.89	71.82
36	23,928	31	218	99	580	2035	14,086	40,977	90	13	22.87	72.01
37	27,768	31	236	109	664	2346	16,365	47,518	90	13	35.78	72.19
38	31,607	31	253	118	748	2657	18,645	54,060	85	14	48.64	72.37
39	35,446	31	271	127	832	2969	20,924	60,601	90	13	61.46	72.55
40	39,286	31	289	137	916	3280	23,203	67,142	85	13	74.27	72.72
41	43,125	31	306	146	1000	3592	25,483	73,683	87	14	87.08	72.89
42	46,965	31	324	155	1084	3903	27,762	80,224	85	15	99.88	73.06
43	50,804	31	341	165	1168	4214	30,041	86,766	90	13	112.69	73.23
44	52,370	27	332	97	1187	4174	31,172	89,358	57	10	38.64	73.97
45	52,622	27	335	105	1194	4213	31,300	89,797	53	9	34.49	72.78

In the end, data from 118 wells were collected on-site. After sorting, data from 93 wells were deemed suitable for training the neural network. Four wells' data were reserved as a validation dataset for the network, and the remaining data were utilized for the neural network training. Some of the training data are presented in Table 4.

3. Construction of BP Neural Network Model

3.1. Neural Network Model Theory

Artificial Neural Networks (ANNs) are systems designed to replicate the synaptic structure of biological neural networks in terms of information processing. By imitating the connections between neurons and their synaptic weights, akin to the neural pathways in the human brain, these networks possess the capability to filter, analyze, learn, and make decisions regarding information. Among them, the backpropagation (BP) neural network utilizes various gradient descent theories. It iteratively adjusts the weights and bias parameters of network nodes via backpropagation using the chain rule, aiming to minimize the network's mean square error [30]. Leveraging these principles, this paper constructs a BP neural network to simulate the evolving patterns of wellbore scale formation induced by various factors.

3.2. Model Construction

The factors influencing wellbore scale formation, including Cl^-, SO_4^{2-}, HCO_3^-, Mg^{2+}, Ca^{2+}, Ba^{2+}, Na^+, salinity, temperature, and pressure, are considered as the input layer of the neural network. The output layer consists of $CaCO_3$ and $BaSO_4$ scale quantities. Through comparative calculations, it was determined that the model achieved satisfactory accuracy with 11 hidden layers. The final constructed neural network structure is illustrated in Figure 2.

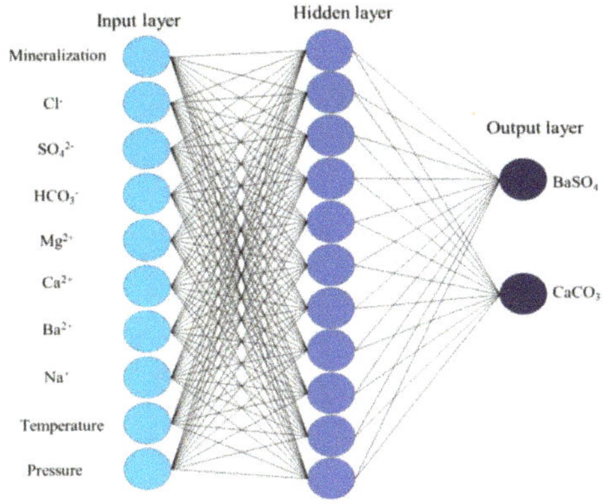

Figure 2. Neural network training diagram.

Using the data collected from the field, the neural network was trained, and four sets of test data were reserved to verify the model's reliability. The learning rate of the model was set to 0.05, with an error threshold of 0.01, and the iteration step was set to 3000. Three commonly used transfer functions in BP neural networks are logsig, tansig, and purelin. The logsig function allows input values to range arbitrarily while constraining output values between 0 and 1. Tansig permits input values of any magnitude, mapping them to an output range between −1 and +1. The linear transfer function, purelin, accommodates arbitrary input and output values. Considering the specifics of the problem domain, this study opted for purelin as the transfer function for the output layer. Furthermore, we investigated the adaptability of these three transfer functions to the training data for the output layer, contrasting the computational accuracy of the models under each function. The operational outcomes of the model are depicted in the Figure 3. Under the purelin transfer function, the training efficiency of the model was relatively poor; although the fitting coefficient reached 0.983, gradient convergence proved elusive. In the case of the logsig transfer function, the fitting degree reached 0.94, yet both gradient and error failed to converge, exhibiting divergence. However, with the tansig transfer function, the model achieved a fitting degree of 0.9834, with gradients and errors gradually converging towards their minimum values (Figure 3).

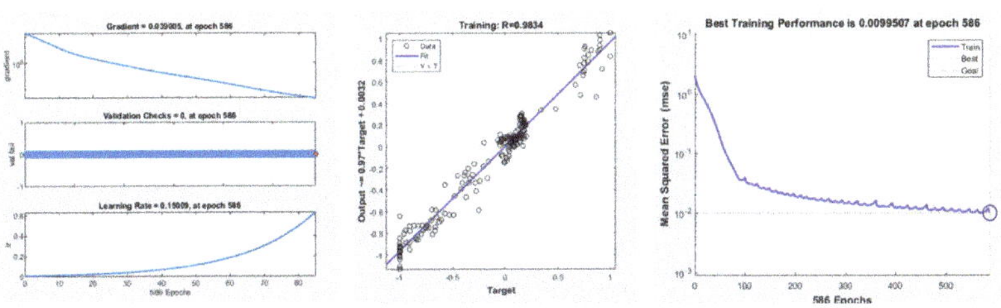

(**a**) Tansig transfer function

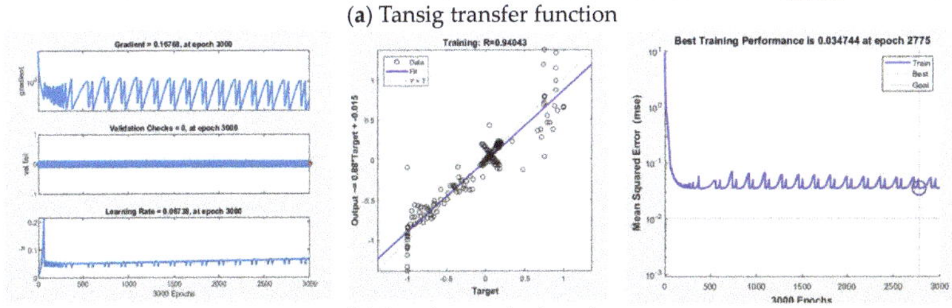

(**b**) Logsig transfer function

Figure 3. *Cont.*

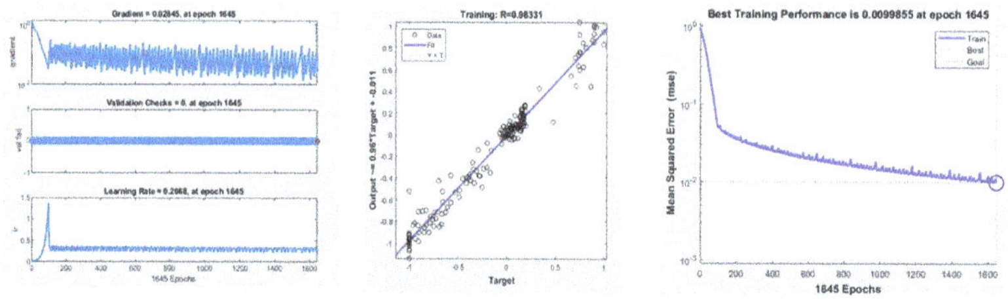

(**c**) Purelin transfer function

Figure 3. Image of the neural network training process.

3.3. Model Verification

To assess the accuracy and reliability of the model, four sets of data were set aside from the compiled dataset of 93 wells for both training and validation purposes. The input layer of the validation dataset was then supplied to the well-trained network for computations, yielding the predicted outcomes. Using neural networks trained with the three transfer functions, calculations were performed on reserved validation data. Simultaneously, under identical conditions, the scale buildup was computed using the commercial software OLI ScaleChem (V4.0.3). The final results are presented in Table 5. It is evident that under the tansig transfer function, the model's computed results exhibited the highest correlation coefficient with the experimental results, reaching 0.9982. For the logsig transfer function, the correlation coefficient was 0.7112, while for the purelin transfer function, it was 0.8405. The correlation coefficient between the results obtained from the commercial software and the experimental values was 0.9891. Hence, it can be concluded that the model based on the tansig function yielded the most accurate computational results.

Table 5. Experimental data and model prediction data.

Scale	Experimental	Tansig	Logsig	Purelin	Commercial Software
$CaCO_3$, mg/L	73.62	72.91	68.71	70.26	74.98
	73.79	73.04	53.24	65.34	70.21
	73.96	72.53	73.56	70.21	75.62
	74.13	73.61	65.41	70.53	77.69
$BaSO_4$, mg/L	40.63	38.74	30.23	48.67	38.62
	52.83	51.67	68.69	59.74	49.67
	65.07	66.1	60.21	60.29	65.01
	77.356	76.54	93.41	87.68	77.31

4. Result Analysis

The final depiction of the scale formation mechanism analysis and scale prediction algorithm process is presented in Figure 4. This algorithm initially utilizes on-site data from the research area to establish the range of various parameters for the input layer, based on which an input layer matrix is generated. Subsequently, this input layer matrix is fed into a pre-trained neural network model, and the output layer matrix is computed. At this stage, the output layer matrix reflects the scale formation under different input parameters. The analysis encompasses the diverse trends of scale formation influenced by varying factors and computes the corresponding wellbore contraction ratio. Ultimately this process furnishes preventive measures for scale formation in gas wellbores.

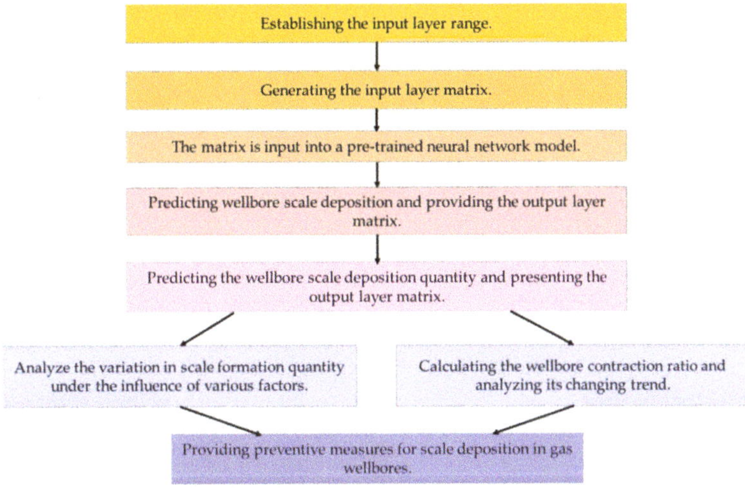

Figure 4. Illustrates the algorithmic flowchart.

Utilizing the aforementioned algorithm enables the formulation of anti-scaling measures under various conditions.

4.1. Analysis of Changes in Scaling Amount

(1) Scale anions

This study examines and forecasts scale formation patterns in the Sulige gas field wells across a range of input parameters, encompassing temperature, pressure, ion concentrations, and water salinity, through the development of algorithms. Here, temperature and pressure refer to the reservoir conditions, and scale formation quantity represents the scale concentration in unit volume of formation water at a specified pressure and temperature. The numerical ranges for temperature, pressure, and scale-forming ion concentrations are provided in Table 6. Keeping other ion concentrations constant, the study separately investigates the effects of SO_4^{2-}, HCO_3^-, Ba^{2+}, and Ca^{2+} on scale formation quantity. The simulation results are depicted in Figures 3–8, where the colors on the scale bar correspond to the scale formation quantity.

Table 6. Parameter values.

Ion	Ion Concentration, mg/L			Salinity	Temperature, °C	Pressure, MPa
	Cl^-	Mg^{2+}	Na^+			
Value	16,249	110	10,000	31,000	50–90	5–15

Figure 5 displays the distribution of $BaSO_4$ scale quantities under different SO_4^{2-} ion concentrations (with an HCO_3^- concentration at 300 mg/L, Ca^{2+} concentration at 700 mg/L, and Ba^{2+} concentration at 2000 mg/L). From Figure 5, it is evident that the $BaSO_4$ scale quantities exhibit a "diagonal" distribution under varying SO_4^{2-} ion concentrations. In the high-pressure low-temperature region (top left corner), $BaSO_4$ scale formation is relatively low, whereas in the high-temperature low-pressure region (bottom right corner), $BaSO_4$ scale formation is higher. Nonetheless, in both the high-pressure low-temperature region and the low-pressure high-temperature region, the impact of temperature and pressure on $BaSO_4$ scale formation appears to be relatively minor. On the other hand, in the low-pressure low-temperature region (bottom left corner) and the high-pressure high-temperature region (top right corner), $BaSO_4$ scale formation is significantly affected by changes in temperature and pressure. Therefore, it can be observed that in gas wells with high temperatures and low pressures, $BaSO_4$ scale formation may be more severe.

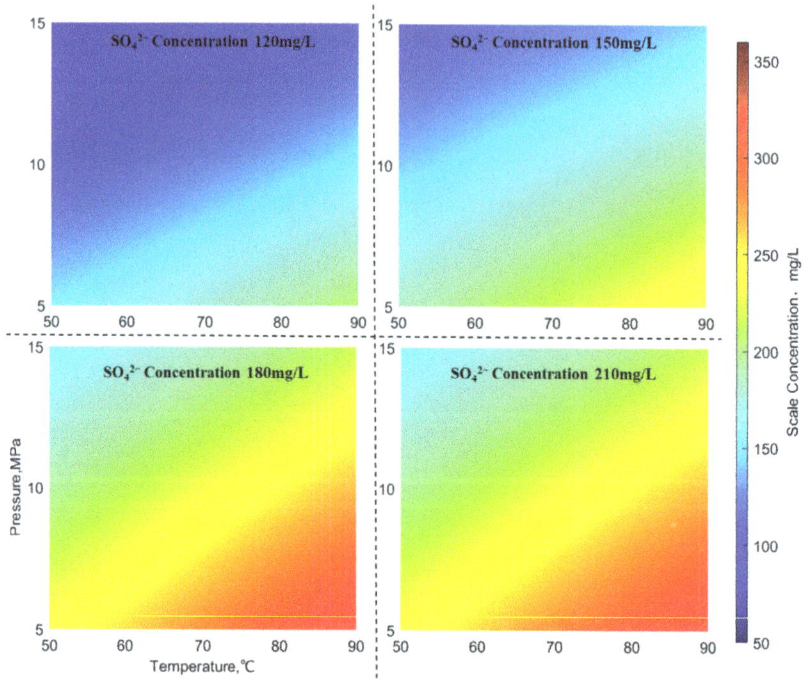

Figure 5. Distribution diagram of $BaSO_4$ scaling amounts under different SO_4^{2-} ion concentrations.

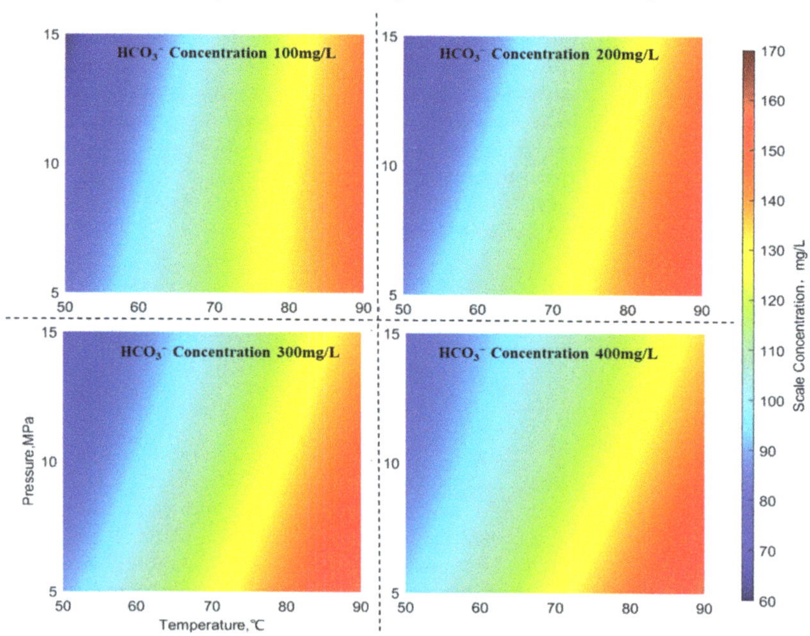

Figure 6. Distribution diagram of $CaCO_3$ scaling amounts under different HCO_3^- ion concentrations.

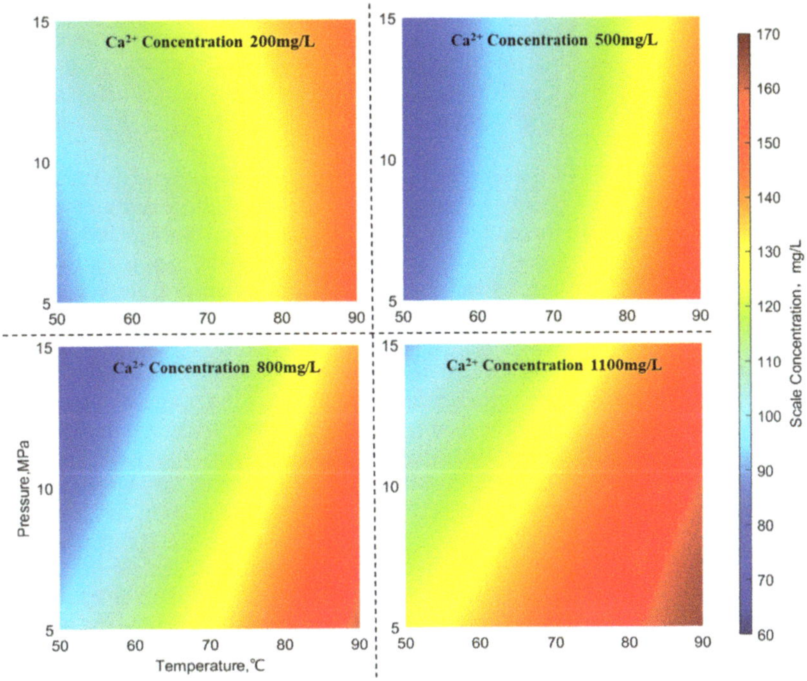

Figure 7. Distribution diagram of $CaCO_3$ scaling amounts under different Ca^{2+} ion concentrations.

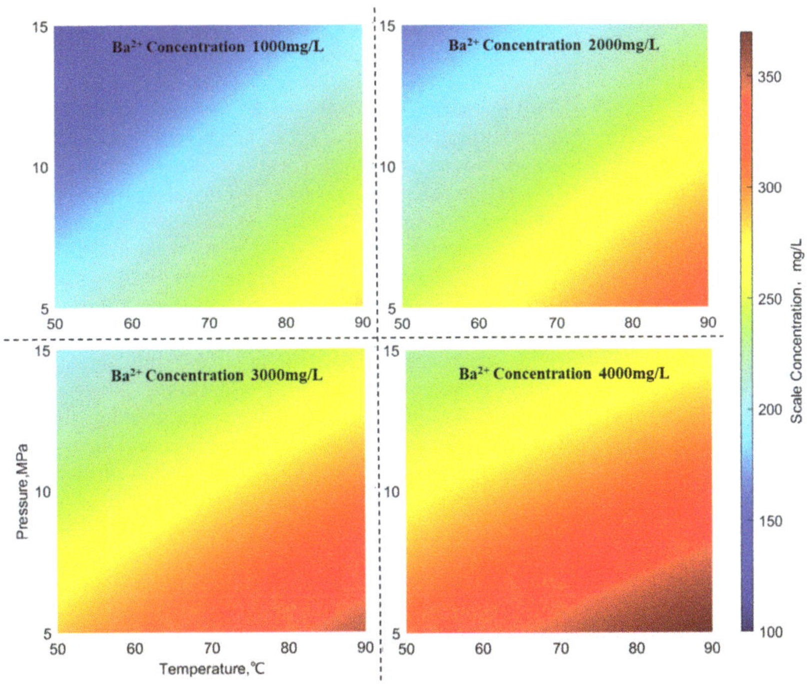

Figure 8. Distribution diagram of $BaSO_4$ scaling amounts under different Ba^{2+} ion concentrations.

Figure 6 illustrates the distribution of $CaCO_3$ scale quantities under different HCO_3^- ion concentrations (with an SO_4^{2-} concentration at 220 mg/L, Ca^{2+} concentration at 700 mg/L, and Ba^{2+} concentration at 2000 mg/L). From Figure 6, it can be observed that the $CaCO_3$ scale quantities exhibit a "wave-like" distribution under varying HCO_3- ion concentrations. In the high-temperature region (on the right side), $CaCO_3$ scale formation is higher, while in the low-temperature region (on the left side), $CaCO_3$ scale formation is lower. Both high-temperature (right side) and low-temperature (left side) regions are relatively less influenced by temperature and pressure. However, as ion concentrations rise, the effect of temperature and pressure on BaSO4 scale formation becomes markedly more significant, observed in both the high-temperature high-pressure region (top right corner) and the low-temperature low-pressure region (bottom left corner). Therefore, it can be concluded that when ion concentrations are low, $CaCO_3$ scale formation is more severe in high-temperature wells. Conversely, when ion concentrations are high, $CaCO_3$ scale formation is more serious in high-temperature low-pressure wells.

(2) Scale cations

The figure depicted in Figure 7 illustrates the distribution of $CaCO_3$ scaling amounts under different concentrations of Ca^{2+} ions (with an HCO_3^- concentration of 300 mg/L, SO_4^{2-} concentration of 220 mg/L, and Ba^{2+} concentration of 2000 mg/L). From Figure 7, it can be observed that the scaling amounts of $CaCO_3$ exhibit a "wave-like" pattern at varying concentrations of Ca^{2+} ions. In the high-temperature region (on the right side), the scaling amount of $CaCO_3$ is relatively high, whereas in the low-temperature high-pressure region (bottom left corner), the scaling amount of $CaCO_3$ is comparatively low. Both the high-temperature region (on the right side) and the low-temperature high-pressure region (bottom left corner) show minor influences of temperature and pressure on $CaCO_3$ scaling. However, with an increase in ion concentration, the impact of temperature and pressure on $CaCO_3$ scaling in the high-temperature high-pressure region (top right corner) and the low-temperature low-pressure region (bottom left corner) gradually intensifies. Additionally, the scaling of CaCO3 in the low-temperature low-pressure region gradually increases, while in the low-temperature high-pressure region (top left corner), the $CaCO_3$ scaling gradually decreases. Therefore, it can be concluded that when the ion concentration is low, $CaCO_3$ scaling is more severe in high-temperature gas wells. Conversely, when the ion concentration is high, $CaCO_3$ scaling becomes even more pronounced in high-temperature low-pressure gas wells.

The graph presented in Figure 8 illustrates the distribution of $BaSO_4$ scaling under different concentrations of Ba^{2+} ions (with an HCO_3^- concentration of 300 mg/L, SO_4^{2-} concentration of 220 mg/L, and Ca^{2+} concentration of 700 mg/L). From Figure 8, it can be observed that the $BaSO_4$ scaling amounts exhibit a "diagonal" pattern at varying Ba^{2+} ion concentrations. In the high-pressure low-temperature region (top left corner), the $BaSO_4$ scaling is relatively small, whereas in the high-temperature low-pressure region (bottom right corner), the $BaSO_4$ scaling is significantly larger. However, in both the high-pressure low-temperature area and the low-pressure high-temperature area, $BaSO_4$ scaling is relatively unaffected by changes in temperature and pressure. Conversely, in the low-pressure low-temperature region (bottom left corner) and the high-pressure high-temperature region (top left corner), $BaSO_4$ scaling is significantly affected by fluctuations in temperature and pressure. Consequently, it can be inferred that in gas wells characterized by high temperatures and low pressures, the occurrence of $BaSO_4$ scaling might be more pronounced.

In summary, the formation of $BaSO_4$ scaling under the influence of SO_4^{2-} and Ba^{2+} ions follows a "diagonal" distribution pattern. In the high-pressure low-temperature region (top left corner), the $BaSO_4$ scaling is relatively small, whereas in the high-temperature low-pressure region (bottom right corner), the $BaSO_4$ scaling is more significant. However, in both the high-pressure low-temperature area and the low-pressure high-temperature area $BaSO_4$ scaling is relatively less affected by changes in temperature and pressure. Conversely in the low-pressure low-temperature region (bottom left corner) and the high-pressure high-temperature region (top left corner), $BaSO_4$ scaling is notably influenced by variations in temperature and pressure. Under the influence of Ca^{2+} and HCO_3^- ions, $CaCO_3$ scaling exhibits a "wave-like" distribution pattern. In the high-temperature region (right side), there is a substantial buildup of $CaCO_3$ scaling, whereas in the low-temperature region (left side), the scaling is minimal. Both the high-temperature region (right side) and the low-temperature region (left side) show relatively minor variations in $CaCO_3$ scaling due to changes in temperature and pressure. However, with an increase in ion concentration, the impact of temperature and pressure on $CaCO_3$ scaling becomes increasingly significant in the high-temperature high-pressure area (top right corner) and the low-temperature low-pressure area (bottom left corner). Overall, in gas wells with high temperatures and low pressures, $BaSO_4$ scaling might be more severe. Additionally, when the ion concentration is low, there is severe $CaCO_3$ scaling in high-temperature gas wells, whereas when the ion concentration is high, the scaling becomes even more severe, especially in high-temperature low-pressure gas wells.

4.2. Analysis of Changes in Diameter Reduction Ratio

According to recent research, wellbore scale can be classified into two main types: crystalline scale and particulate scale. Crystalline scale occurs when scale ions precipitate directly onto the wellbore surface from formation water. On the other hand, particulate scale forms when scale ions aggregate into particles within the formation water and then adhere to the wellbore surface as the fluid flows. Along with the deposition of scale particles, there is a process of particle detachment. The interaction between the deposition and detachment processes [31] leads to wellbore scaling issues, gradually affecting the production of gas wells over time.

To simplify the problem, this study modeled the deposition and detachment processes of scale particles in the wellbore as the attachment of particulates from formation water, and a coefficient of attachment was assigned to describe the scaling process. The attachment of scale particles on the wellbore surface reduces the space available for fluid flow, potentially leading to wellbore blockages and impacting gas well production. Hence, to evaluate the potential risk of wellbore scaling and blockage, the Diameter Reduction Ratio (DRR) serves as a metric to gauge the extent of scaling. The DRR quantifies the ratio between the thickness of the scale layer and the internal diameter of the wellbore. Building upon the calculation of scaling quantity in the previous section, the total scaling within the predicted production lifespan is computed. This total scaling is converted into volume and averaged over the scaling intervals in the wellbore to determine the scale layer thickness. The specific calculation method is shown in Equations (1)–(3) [32].

$$m = q_w C t \times 10^{-6} \tag{1}$$

where m—the total scale mass within a certain period of time, kg;
q_w—daily water production of gas well, m^3/d;
C—scale concentration, kg/L;
t—time, d.

$$d = m/2\rho\pi r L \tag{2}$$

where d—scale layer thickness, m;
L—fouling section length, m;
ρ—density of scale, kg/m^3.

$$DRR = 2d/D \tag{3}$$

where D—wellbore inner diameter, m.

Based on the above analysis, and referring to the work of Zhang and colleagues [22], an attachment ratio of 0.73 was adopted. Based on these principles, the variations in the Diameter Reduction Ratio (DRR) over a 3-year production period were calculated under different temperatures, pressures, and ion concentrations. The calculation process overlooks changes in ion concentration, reservoir temperature and pressure, and wellbore internal diameter, and assumes scaling occurs throughout the entire wellbore segment. The selected parameters for the calculation are detailed in Table 7.

Table 7. Field data.

Daily Water Production	Geothermal Gradient	Wellbore Inner Diameter	$BaSO_4$ Density	$CaCO_3$ Density	Cl^-	Mg^{2+}	Na^+	Salinity
650 m^3/d	3 °C/100 m	0.062 m	4000 kg/m^3	2710 kg/m^3	16,249	110	10,000	31,000

Note: the unit of ion concentration and mineralization is mg/L.

Following the principles outlined above, an examination was conducted on the reduction rate of wellbore diameter concerning scale ion concentration, temperature, and pressure. In the graph, the *X*-axis and *Y*-axis denote temperature and pressure, respectively, while the *Z*-axis illustrates the wellbore's reduction rate. At any point on the surface of the graph, the height indicates the variation in the reduction rate under the current temperature, pressure, and ion concentration, with the color scale corresponding to the magnitude of the reduction rate. Practical observations on-site suggest that when the reduction rate is below 4%, there is considered to be no scaling risk, representing the position on the graph's plane. This plane divides the surface representing the reduction rate variation into two parts and is parallel to the plane of temperature and pressure. At any point on this plane, the reduction rate is 4%.

Figure 9 illustrates the variation in the wellbore reduction rate when the concentration of SO_4^{2-} changes (with an HCO_3^- concentration at 300 mg/L, Ca^{2+} concentration at 700 mg/L, and Ba^{2+} concentration at 2000 mg/L). Similarly, Figure 10 depicts the variation in the wellbore reduction rate when the concentration of HCO_3^- changes (with SO_4^{2-} concentration at 220 mg/L, Ca^{2+} concentration at 700 mg/L, and Ba^{2+} concentration at 2000 mg/L).

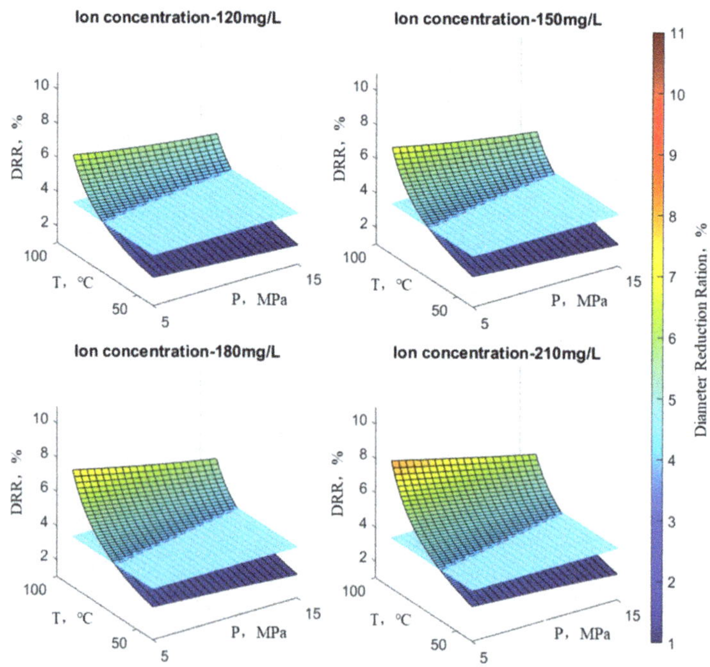

Figure 9. Distribution of wellbore diameter reduction rate under different SO_4^{2-} concentrations.

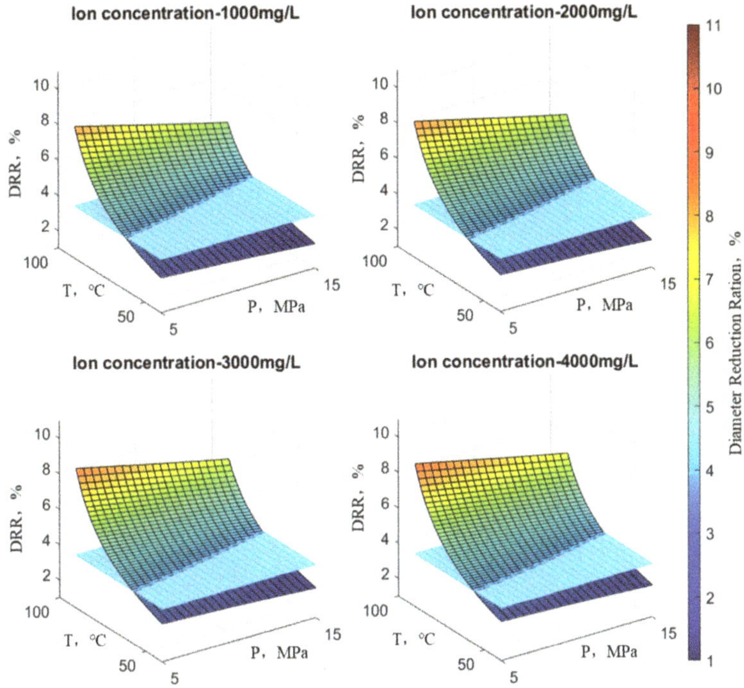

Figure 10. Distribution of wellbore diameter reduction rate under different HCO_3^- concentrations.

Figure 9 illustrates the trend in wellbore reduction rates under the influence of SO_4^{2-}. It is evident that when the SO_4^{2-} concentration remains constant, pressure has a minor impact on the reduction rate at low temperatures but a significant impact at high temperatures. Furthermore, at low pressure, temperature exerts a significant impact on the reduction rate, while at high pressure, the influence of temperature diminishes. With increasing ion concentration, the maximum reduction rate extends from the high-temperature, low-pressure region to encompass the entire surface, indicating variations in reduction rate. Overall, the reduction rate continuously increases, and the surface representing the reduction rate shifts upward. In other words, the surface above the plane where the reduction rate equals 4% becomes increasingly larger. In general, the influence of temperature on the reduction rate is greater than that of pressure.

Figure 10 illustrates the variation in wellbore reduction rates under the influence of HCO3-. It can be observed that the overall trend is similar to the reduction rate variation observed under the influence of SO_4^{2-}. Once again, temperature has a more significant impact on the reduction rate compared to pressure.

Figure 11 depicts the variation in the wellbore reduction rate when the concentration of Ca^{2+} changes (with SO_4^{2-} concentration at 220 mg/L, HCO_3^- concentration at 300 mg/L, and Ba^{2+} concentration at 2000 mg/L). Similarly, Figure 12 shows the variation in the wellbore reduction rate when the concentration of Ba^{2+} changes (with SO_4^{2-} concentration at 220 mg/L, HCO_3^- concentration at 300 mg/L, and Ca^{2+} concentration at 700 mg/L).

Figure 11 represents the variation in wellbore reduction rates under the influence of Ca^{2+}. It is evident that at an ion concentration of 200 mg/L temperature has a far greater impact on the reduction rate than pressure. This is discernible from the spatial extent of the surface's changes; the surface representing reduction rate variations is nearly perpendicular to the plane defined by reduction rate and temperature. As the ion concentration increases, the reduction rate surface starts to tilt.

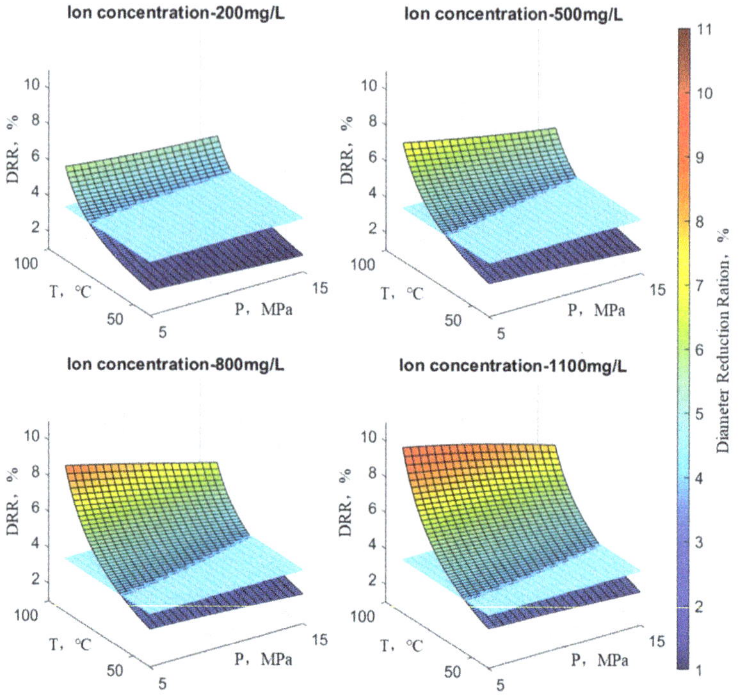

Figure 11. Distribution of wellbore diameter reduction rate under different Ca^{2+} concentrations.

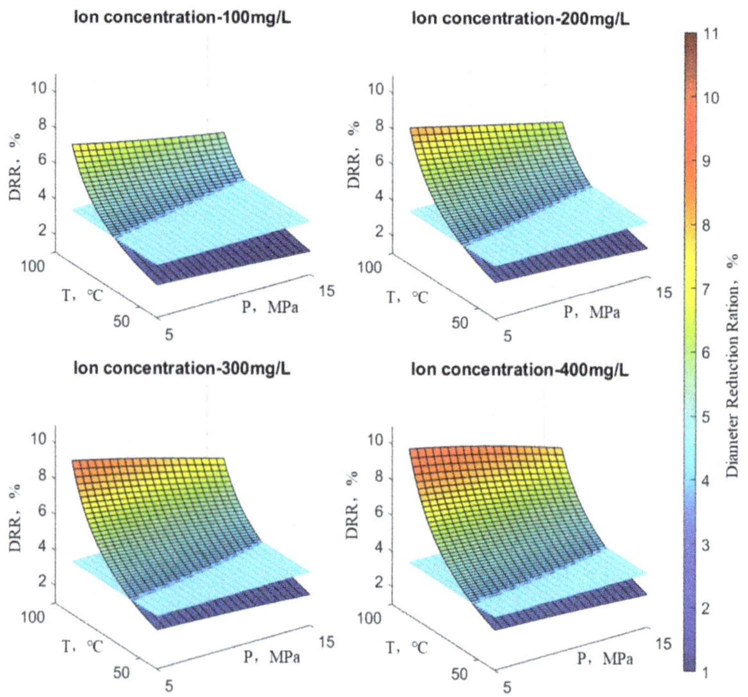

Figure 12. Distribution of wellbore diameter reduction rate under different Ba^{2+} concentrations.

This suggests that pressure exerts a lesser influence on the reduction rate at low temperatures but a more substantial impact at high temperatures. Conversely, at low pressure, temperature significantly affects the reduction rate, whereas at high pressure, the influence of temperature is comparatively minor. The wellbore reduction rate reaches its minimum at high temperatures and low pressures. In summary, the effect of temperature on the reduction rate surpasses that of pressure. Additionally, it can be observed that with increasing ion concentration, the reduction rate surface consistently shifts upward, and the portion of the surface above the 4% reduction rate plane becomes larger.

Figure 12 illustrates the variation in wellbore reduction rates under the influence of Ba^{2+}. The trend aligns with the variation observed under the influence of Ca^{2+}, emphasizing once again that temperature's impact on the reduction rate outweighs that of pressure.

Based on the above analysis, it can be observed that under certain conditions in the research area, controlling the temperature, pressure, and ion concentration of the produced water within specific ranges can maintain the borehole contraction rate below 4% over a three-year production period. Therefore, during on-site production, focusing on parameters such as temperature, pressure, and ion concentration can effectively prevent scaling issues. However, in reality, the temperature of the reservoir and the ion concentration of the produced water are uncontrollable factors determined by geological conditions. On the other hand, the bottom-hole pressure of gas wells is related to the production system, which means that the bottom-hole pressure of gas wells is controllable. Production systems for gas wells include fixed production rate systems and fixed pressure systems. The latter involves fixed wellhead (bottom-hole) pressure systems and fixed pressure difference systems. The formulation of the production system depends on geological conditions and production requirements. Thus, in the gas well production process, mitigating scaling and minimizing associated risks can be achieved by manipulating the production system to regulate bottom-hole pressure, provided certain conditions like fixed temperature and ion concentration. However, due to economic considerations and reservoir constraints, bottom-hole pressure cannot be increased indefinitely. In practice, determining the optimal magnitude of bottom-hole pressure becomes a multi-objective optimization challenge, balancing economic benefits, reservoir conditions, technical costs, and other relevant factors. In conclusion, considering the temperature and ion concentration conditions in the research area, the selected bottom-hole pressure should be adjusted to achieve a borehole contraction rate as close to 4% as possible, striking a balance between scaling prevention and economic efficiency in the given reservoir context.

5. Conclusions

The target area was a tight sandstone gas reservoir, and through the collection of on-site data and laboratory experiments, the water quality and composition of blockage materials in the target area were clarified. Combining this information with a BP neural network model, a model for predicting wellbore scaling risk and analyzing scale prevention measures in the target reservoir was established. Upon elucidating the wellbore blockage mechanism applicable to tight sandstone gas reservoirs in the target area, corresponding preventive measures were proposed. The results indicate:

(1) The water in the target area is of a $CaCl_2$ type, and the analysis of blockage material composition indicates that the predominant scales are $CaCO_3$ and $BaSO_4$. Due to significant production in the target area, it was analyzed that the fluid in the gas wells exhibits a circular flow, with a stable pH ranging between 6.2 and 6.5.

(2) Under the influence of Ba^{2+} and $SO_4{}^{2-}$, the $BaSO_4$ scaling quantity shows a "diagonal" distribution with changes in temperature and pressure. Under the influence of Ca^{2+} and $HCO_3{}^{-}$, the $CaCO_3$ scaling quantity exhibits a "wave" distribution with changes in temperature and pressure. Overall, $BaSO_4$ is more influenced by pressure, while $CaCO_3$ is more influenced by temperature.

(3) Controlling temperature, pressure, and scaling ion concentrations within a certain range can keep the contraction ratio within 4% or close to 4%, achieving the prevention

of scaling. This can be achieved through adjustments in production systems, primarily by modifying production pressure differentials. This provides guidance for predicting scaling risks and implementing scaling prevention measures in tight sandstone reservoirs.

Author Contributions: Conceptualization, X.S.; methodology, D.Z.; software, X.S.; validation, X.S.; formal analysis, H.W.; investigation, J.X.; data curation, J.X.; writing, X.S. All authors have read and agreed to the published version of the manuscript.

Funding: This research was funded by the National Natural Science Foundation of China, grant number 51934005, U23B2089; and Shaanxi Provincial Natural Science Basic Research Program Project, grant number 2024JC-YBQN-0554.

Data Availability Statement: Data are contained within the article.

Conflicts of Interest: The authors declare no conflicts of interest.

References

1. Xu, J.; Wu, K.; Yang, S.; Cao, J.; Chen, Z.; Pan, Y.; Yan, B.J.A.J. Real gas transport in tapered noncircular nanopores of shale rocks. *AICHE J.* **2017**, *63*, 3224–3242. [CrossRef]
2. Xu, J.; Wu, K.; Li, R.; Li, Z.; Li, J.; Xu, Q.; Li, L.; Chen, Z. Nano-scale Pore Size Distribution Effects on Gas Production from Fractal Shale Rocks. *Fractals* **2019**, *27*, 1950142. [CrossRef]
3. Chen, X.; Li, Y.; Liu, Z.; Trivedi, J.; Tang, Y.; Sui, M. Visualized investigation of the immiscible displacement: Influencing factors, improved method, and EOR effect. *Fuel* **2023**, *331*, 125841. [CrossRef]
4. Yang, W. Hydraulic fracture initiation for perforated wellbore coupled with the effect of fluid seepage. *Energy Rep.* **2022**, *8*, 10290–10298.
5. Quddus, A.; Al-Hadhrami, L.M. Hydrodynamically deposited $CaCO_3$ and $CaSO_4$ scales. *Desalination* **2009**, *246*, 526–533. [CrossRef]
6. Chilcott, N.P.; Phillips, D.A.; Sanders, M.G.; Collins, I.R.; Gyani, A. The Development and Application of an Accurate Assay Technique for Sulphonated Polyacrylate Co-polymer Oilfield Scale Inhibitors. In Proceedings of the International Symposium on Oilfield Scale, Aberdeen, UK, 26 January 2000.
7. Li, X.; Gao, B.; Yue, Q.; Ma, D.; Rong, H.; Zhao, P.; Teng, P. Effect of six kinds of scale inhibitors on calcium carbonate precipitation in high salinity wastewater at high temperatures. *J. Environ. Sci.* **2015**, *29*, 124–130. [CrossRef] [PubMed]
8. Johnson, A.; Eslinger, D.; Larsen, H. An Abrasive Jetting Scale Removal System. In Proceedings of the SPE/ICoTA Coiled Tubing Roundtable, Houston, TX, USA, 15 March 1998.
9. Morizot, A.P.; Neville, A. Insights into Electrodeposition of an Inhibitor Film and Its Inhibitive Effects on Calcium Carbonate Deposition. *J. Colloid Interface Sci.* **2002**, *245*, 40–49. [CrossRef] [PubMed]
10. Wanner, C.; Eichinger, F.; Jahrfeld, T.; Diamond, L.W. Unraveling the Formation of Large Amounts of Calcite Scaling in Geothermal Wells in the Bavarian Molasse Basin: A Reactive Transport Modeling Approach. *Procedia Earth Planet. Sci.* **2017**, *17*, 344–347 [CrossRef]
11. Li, Y.; Pang, Z.; Galeczka, I.M. Quantitative assessment of calcite scaling of a high temperature geothermal well in the Kangding geothermal field of Eastern Himalayan Syntax. *Geothermics* **2020**, *87*, 101844. [CrossRef]
12. Eseosa, A.; Atubokiki, A.J. Prediction and Monitoring of Oilfield Carbonate Scales Using Scale Check©. In Proceedings of the Nigeria Annual International Conference and Exhibition, Abuja, Nigeria, 30 July 2011.
13. Langelier, W.F. The analytical control of anti-corrosion water treatment. *J. AWWA* **1936**, *28*, 1500–1521. [CrossRef]
14. Stiff Jr, H.A.; Davis, L.E. A method for predicting the tendency of oil field waters to deposit calcium carbonate. *J. Pet. Techno* **1952**, *4*, 213–216. [CrossRef]
15. Ryznar, J.W. A new index for determining amount of calcium carbonate scale formed by a water. *J. Am. Water Works Assoc.* **1944** *36*, 472–483. [CrossRef]
16. Skillman, H.; McDonald, J.; Stiff, H. A simple, accurate, fast method for calculating calcium sulfate solubility in oil field brine. In Proceedings of the Spring Meeting of the Southwestern District, API, Lubbock, TX, USA, 14 March 1969.
17. Oddo, J.; Tomson, M. Why scale forms in the oil field and methods to predict it. *SPE Prod. Facil.* **1994**, *9*, 47–54. [CrossRef]
18. Hasson, D.; Avriel, M.; Resnick, W.; Rozenman, T.; Windreich, S. Mechanism of calcium carbonate scale deposition on heat-transfe surfaces. *Ind. Eng. Chem. Fundam.* **1968**, *7*, 59–65. [CrossRef]
19. Hasan, A.; Kabir, C. Heat transfer during two-Phase flow in Wellbores; Part I—Formation temperature. In Proceedings of the SP Annual Technical Conference and Exhibition, Dallas, TX, USA, 6 October 1991; p. SPE-22866-MS.
20. Brahim, F.; Augustin, W.; Bohnet, M. Numerical simulation of the fouling process. *Int. J. Therm. Sci.* **2003**, *42*, 323–334. [CrossRe
21. Yang, X.; Li, W.; Guo, L.; Liu, X.; Feng, H. Prediction of CaCO3 scaling in water injection wellbore. *Appl. Therm. Eng.* **2016**, 9 532–540. [CrossRef]
22. Zhang, L.; Yang, L.; Wang, Z.; Zhang, C.; Meng, W.; Ren, S. Experimental study on scaling and adhesion characteristics water-producing gas wellbore. *Nat. Gas Ind. B* **2021**, *8*, 252–266. [CrossRef]

23. Zhao, B.; Yang, S.; Liu, F.; Hu, Y.; Cao, S.; Men, H.; Xu, Z. Experimental study on dynamic simulation of cooling water fouling resistance prediction by support vector machine. *Proc. CSEE* **2010**, *30*, 92–97.
24. Gabetta, G.; De Masi, G.; Gentile, M.; Vichi, R.; Scapin, M. Pipeline Internal Damage Prediction by Deterministic Models and Neural Networks. In Proceedings of the Abu Dhabi International Petroleum Exhibition and Conference, Abu Dhabi, United Arab Emirates, 10 November 2014.
25. Paz, P.A.; Caprace, J.-D.; Cajaiba, J.F.; Netto, T.A. Prediction of calcium carbonate scaling in pipes using artificial neural networks. In Proceedings of the International Conference on Offshore Mechanics and Arctic Engineering, Trondheim, Norway, 25 September 2017; p. V05AT04A044.
26. Wang, J.; Lv, Z.; Liang, Y.; Deng, L.; Li, Z. Fouling resistance prediction based on GA–Elman neural network for circulating cooling water with electromagnetic anti-fouling treatment. *J. Energy Inst.* **2019**, *92*, 1519–1526. [CrossRef]
27. Ferreira, B.X.; Barbosa, C.R.H.; Cajaiba, J.o.; Kartnaller, V.; Santos, B.F. Development of Artificial Neural Network Models for the Simulation of a $CaCO_3$ Scale Formation Process in the Presence of Monoethylene Glycol (MEG) in Dynamic Tube Blocking Test Equipment. *Energy Fuels* **2022**, *36*, 2288–2299. [CrossRef]
28. Kamal, M.S.; Hussein, I.; Mahmoud, M.; Sultan, A.S.; Saad, M.A. Oilfield scale formation and chemical removal: A review. *J. Pet. Sci. Eng.* **2018**, *171*, 127–139. [CrossRef]
29. Lakshmi, D.S.; Senthilmurugan, B.; Drioli, E.; Figoli, A. Application of ionic liquid polymeric microsphere in oil field scale control process. *J. Pet. Sci. Eng.* **2013**, *112*, 69–77. [CrossRef]
30. Rumelhart, D.E.; Hinton, G.E.; Williams, R.J. Learning representations by back-propagating errors. *Nature* **1986**, *323*, 533–536. [CrossRef]
31. Pääkkönen, T.M.; Riihimäki, M.; Simonson, C.J.; Muurinen, E.; Keiski, R.L. Modeling $CaCO_3$ crystallization fouling on a heat exchanger surface—Definition of fouling layer properties and model parameters. *Int. J. Heat Mass Transf.* **2015**, *83*, 84–98. [CrossRef]
32. Mishkhes, A.T.; AlSultan, F.A.; Al-Askar, M. Method and system for predicting caliper log data for descaled wells. U.S. Patent 11,753,926, 12 September 2023.

processes

MDPI

Article

Bottomhole Pressure Prediction of Carbonate Reservoirs Using XGBoost

Hao Sun, Qiang Luo *, Zhaohui Xia, Yunbo Li and Yang Yu

Research Institute of Petroleum Exploration and Development, PetroChina, Beijing 100083, China; sunhao19982023@163.com (H.S.)
* Correspondence: riped_lq@petrochina.com.cn; Tel.: +86-158-8206-9972

Abstract: The bottomhole pressure is one of the key parameters for oilfield development and decision-making. However, due to factors such as cost and equipment failure, bottomhole pressure data is often lacking. In this paper, we established a GA-XGBoost model to predict the bottomhole pressure in carbonate reservoirs. Firstly, a total of 413 datasets, including daily oil production, daily water production, daily gas production, daily liquid production, daily gas injection rate, gas–oil ratio, and bottomhole pressure, were collected from 14 wells through numerical simulation. The production data were then subjected to standardized preprocessing and dimensionality reduction using a principal component analysis. The data were then split into training, testing, and validation sets with a ratio of 7:2:1. A prediction model for the bottomhole pressure in carbonate reservoirs based on XGBoost was developed. The model parameters were optimized using a genetic algorithm, and the average adjusted R-squared score from the cross-validation was used as the optimization metric. The model achieved an adjusted R-squared score of 0.99 and a root-mean-square error of 0.0015 on the training set, an adjusted R-squared score of 0.84 and a root-mean-square error of 0.0564 on the testing set, and an adjusted R-squared score of 0.69 and a root-mean-square error of 0.0721 on the validation set. The results demonstrated that in the case of fewer data variables, the GA-XGBoost model had a high accuracy and good generalization performance, and its performance was superior to other models. Through this method, it is possible to quickly predict the bottomhole pressure data of carbonate rocks while saving measurement costs.

Keywords: carbonate reservoirs; bottomhole pressure; machine learning; XGBoost; data driven

Citation: Sun, H.; Luo, Q.; Xia, Z.; Li, Y.; Yu, Y. Bottomhole Pressure Prediction of Carbonate Reservoirs Using XGBoost. *Processes* **2024**, *12*, 125. https://doi.org/10.3390/pr12010125

Academic Editors: Tao Zhang, Zheng Sun, Dong Feng, Hung Vo Thanh and Wen Zhao

Received: 13 November 2023
Revised: 22 December 2023
Accepted: 27 December 2023
Published: 3 January 2024

1. Introduction

The bottomhole pressure is one of the key parameters for oilfield development and decision-making. The difference between the bottomhole pressure and the formation pressure largely determines the oil production rate of a well. Traditional methods of obtaining the bottomhole pressure include installing downhole pressure sensors for measurement, a well-testing analysis, or establishing numerical simulation models. The installation of pressure gauges for all wells is costly and carries risks such as equipment lifespan and failures, leading to expensive workovers for equipment replacement and production delays. Well-testing methods also require well shutdowns, which affect economic benefits and involve complex operations. Numerical simulation methods are time-consuming, labor-intensive, and resource-demanding. Therefore, it is crucial to develop a real-time pressure prediction model that saves time, effort, and costs in order to continuously and accurately obtain the bottomhole pressure without shutting down the wells. The traditional correlation prediction methods for forecasting the bottomhole pressure rely on the establishment of empirical and semiempirical (mechanistic) models. Studies by Duns and Ros [1], Hagedorn and Brown [2], Orkiszewski [3], and Beggs and Brill [4] have focused on empirical models while Mukherjee and Brill [5], Ansari et al. [6], Chokshi et al. [7], and Gomez et al. [8] have researched semi-empirical (mechanistic) models. However, these methods are largely based on a series of specific conditions. Therefore, when applied beyond these conditions, the effectiveness of these methods deteriorates.

Artificial intelligence (AI) has a strong adaptability and flexibility in various environments and datasets, and is widely used in fields such as medicine, transportation, the Internet, and energy. In the application of petroleum engineering, AI is often used to solve problems related to data, optimization, identification, and knowledge integration [9,10]. AI technology can more accurately predict bottomhole pressure by learning patterns and correlations from a large quantity of historical data (as shown in Table 1), especially in complex reservoir conditions and dynamic oilfield development processes. Awadalla and Yousef (2016) [11] used backpropagation algorithm to find the optimal parameters of a feedforward neural network (FFNN) for predicting the bottomhole flowing pressure of vertical oil wells. Mahshid and Suren (2019) [12] predicted the bottomhole flowing pressure of coalbed methane wells using time data, linear regression, and neural-network-based methods. Ahmadi and Chen (2019) [13] compared various artificial neural network methods for predicting the bottomhole pressure of vertical wells, and the results showed that a hybrid genetic algorithm and particle swarm optimization (HGAPSO) was highly accurate. Nait Amar et al. (2018) [14] demonstrated that a hybrid model based on artificial neural network (ANN) and grey wolf optimization (GWO) (ANN-GWO) performed better than other hybrid methods such as genetic algorithm (GA), particle swarm optimization (PSO), or using only a BPNN (backpropagation neural network). Nait Amar and Zeraibi (2020) [15] proposed a hybrid model based on support vector regression (SVR) and the firefly algorithm (FFA) for predicting the bottomhole pressure (BHP) in vertical wells under multiphase flow conditions. Rathnayake et al. (2022) [16] compared multiple linear regression, linear mixed effects, and gradient boosting regression tree (XGBoost) methods for predicting the flowing bottomhole pressure (FBHP) in gas wells, and the results showed that the XGBoost method had the best prediction results.

Table 1. Summary of artificial intelligence algorithms for bottomhole pressure prediction.

Reference	Artificial Intelligence Algorithm	Application
[14]	A hybrid model based on support vector regression (SVR) and the firefly algorithm (FFA)	Vertical wells with multiphase flow
[15]	Gradient boosting regression trees (XGBoost)	Gas wells
[17]	Feedforward neural network (FFNN)	Vertical oil wells
[18]	Neural network	Coalbed methane wells
[19]	Hybrid genetic algorithm and particle swarm optimization (HGAPSO)	Vertical oil wells
[20]	Artificial neural network (ANN) and grey wolf optimization (GWO) (ANN-GWO)	Vertical wells with multiphase flow

Based on previous research, this study proposes a well bottom-pressure prediction model based on XGBoost. The XGBoost algorithm [21] is an ensemble learning algorithm based on gradient boosting trees, which has been widely applied in various fields such as healthcare [22] and telecommunications [23] and has been introduced in several articles regarding its application in the petroleum field. For example, Pan et al. applied XGBoost to predict porosity in well logging [24]; Markovic et al. applied XGBoost to predict water saturation in rock physics [25]; Zhong et al. used XGBoost to generate pseudodensity logging data [26]; Gu et al. applied it to predict permeability in geology [27]; Al Mudhafar et al. applied XGBoost to the reservoir lithology classification of carbonate rocks [28]; Zhang et al. applied it to predict hydrocarbon gases in petrochemistry [29]; Dong et al. used XGBoost for reservoir production prediction [30]; and Wang et al. applied XGBoost to predict reservoir production in highly heterogeneous sandstone [31]. These applications have demonstrated that XGBoost is an efficient, superior, and reliable algorithm. Compared with traditional machine learning algorithms, XGBoost exhibits a higher accuracy, greater flexibility, and is less prone to overfitting, demonstrating significant advantages in large-scale, efficient, and accurate problems. However, there is almost no application of XGBoost in well bottom-pressure prediction in carbonate reservoirs.

One major issue faced by machine learning models, including XGBoost, is the close relationship between the model's performance and its hyperparameters. Using default parameter values may not fully harness the powerful effects of XGBoost. Conventional grid-search algorithms require manual parameter tuning, which consumes a lot of time and effort and makes finding the global optimal value challenging. The enhanced elite strategy genetic algorithm is an improved genetic algorithm that can more easily find the global optimal solution. Therefore, this paper proposes the use of an improved genetic algorithm to optimize the hyperparameters of XGBoost in order to enhance the prediction accuracy and generalization performance of the XGBoost model. The data used in this study were production data collected from a numerical model fitted to the historical data of the X oilfield. Prior to establishing the XGBoost-based prediction model for carbonate reservoir bottomhole pressure, various preprocessing steps such as data correlation analysis and data dimensionality reduction were conducted. The optimized XGBoost model with adjusted hyperparameters was compared with ten other machine learning models, namely, support vector machines, neural networks, stochastic gradient descent, linear regression, ridge regression, decision trees, random forests, gradient boosting trees, ExtraTrees, and Adaboost, to evaluate the superior predictive performance of XGBoost. These models were selected due to their frequent utilization in machine learning problems. The novelty of this work compared to other studies in the field lies in the following aspects: (1) The X oilfield is a matrix-pore-type carbonate reservoir with localized fractures and dissolution cavities, as well as high-permeability streaks and impermeable interlayers, exhibiting an extremely strong heterogeneity. The uneven pressure distribution during the development process poses a significant challenge for pressure prediction. The use of numerical simulation for pressure prediction requires a great deal of work, and there is currently no suitable empirical formula for pressure prediction in this type of oilfield. Using XGBoost through data-driven methods for pressure prediction is a relatively feasible approach. (2) Previous research often involves collecting extensive geological, development, engineering, well logging, and seismic data to achieve optimal machine learning results. However, due to the inconvenience of a practical application caused by the need for extensive data collection, this work utilizes only six easily collectible production data variables (daily oil production daily water production, daily gas production, daily liquid production, daily gas injection and gas–oil ratio) to test the performance of the model in the case of limited data variables The utilization of a smaller set of variables in model training also facilitates the rapid implementation of real-time bottomhole pressure prediction in the field. (3) The model compares the pressure prediction performance of ten different machine learning algorithms with that of GA-XGBoost, demonstrating the superior accuracy and generalization ability of XGBoost in pressure prediction problems with limited data variables. This workflow also has certain limitations, as the data used are derived from numerical simulations Although the model has undergone historical fitting, there are still errors compared to actual production data.

2. Principles and Methods

2.1. The Principles of the XGBoost Method

XGBoost [21] is an ensemble algorithm that combines multiple weak classifiers into strong classifier based on the decision tree algorithm. It enhances the accuracy of predictive models by constructing multiple CART decision trees. The final prediction is obtained by summing the predictions of each trained decision tree in each round of training.

The principle of the XGBoost algorithm is as follows [32]:

$$\hat{y}_i = \sum_{t=1}^{n} f_t(x_i), f_t \in F \quad (1)$$

In the equation, n represents the number of trees in the model. f_t denotes a function in the function space F. $\hat{y}_i$ represents the predicted value, and x_i represents the i-th input data. F represents the set of all possible CART models.

At each iteration, the model remains unchanged, and a new function is added to the existing model. Each function corresponds to a tree, and the newly generated tree is used to fit the residual of the previous prediction. The iterative process is as follows:

$$\begin{cases} \widehat{y}_i^{(0)} = 0 \\ \widehat{y}_i^{(1)} = f_1(x_i) = \widehat{y}_i^{(0)} + f_1(x_i) \\ \widehat{y}_i^{(t)} = \widehat{y}_i^{(t-1)} + f_t(x_i) \end{cases} \tag{2}$$

The objective function of XGBoost is as follows [33]:

$$X_{\text{obj}} = \sum_{i=1}^{n} l(y, \widehat{y}) + \sum_{k=1}^{K} \Omega(f_k) \tag{3}$$

$$\Omega(f_k) = \gamma T + \lambda \frac{1}{2} \sum_{j=1}^{T} \omega_j^2 \tag{4}$$

In Equation (3), $\sum_{i=1}^{n} l(y, \widehat{y})$ is used to measure the discrepancy between predicted scores and true scores, and $\sum_{k=1}^{K} \Omega(f_k)$ is the regularization term that effectively prevents overfitting. In Equation (4), T represents the number of leaf nodes, ω represents the score of each leaf node, γ is used to control the number of leaf nodes, and λ ensures that the scores of leaf nodes are not too large.

XGBoost has several main parameters that need to be adjusted: the number of sub-models, the learning rate, the maximum depth, the minimum child weight, γ, the subsample rate, and the column sampling rate. These hyperparameters affect the final performance of the model but in different ways. Therefore, it is necessary to optimize these hyperparameters.

2.2. Genetic Algorithm

Genetic algorithms (GAs) have been proven to be superior to traditional methods when performing a global search within a complex search space [34]. Usually starting from a group of randomly generated candidates embedded with potential solutions, a GA creates a new generation in each iteration by transferring crucial information such as survival of the fittest, crossover, and mutation operations onto promising candidates selected based on the probability biased towards relatively fitter agents [35].

The enhanced elite retention strategy for genetic algorithm (SEGA) takes advantage of the best individuals (called elites) generated during evolution, which are not subjected to selection, crossover, and mutation but directly copied to the next generation. The main feature of this operation is that elites produced during evolution will not be lost or damaged due to pairing crossover and other operations, significantly improving the ability to reach global convergence.

This paper utilizes the SEGA (Sequential Evolutionary Grid Algorithm) to continuously optimize the parameters of XGBoost and achieve the maximum cross-validation adjusted R-squared score, thereby accurately predicting the bottomhole pressure. The SEGA initially generates a batch of XGBoost hyperparameter initial values within a specified range and trains them with production data from the reservoir to obtain a score for the bottomhole pressure prediction. A higher score indicates better model performance. Based on the objective of maximizing the score, the elite preservation strategy is employed to generate the next generation of superior XGBoost hyperparameter populations, which are then trained to obtain scores. This process continues until the maximum evolution generation or the optimization objective reaches a predetermined threshold. The model training process is illustrated in Figure 1. By utilizing this method, the SEGA automatically searches for the optimal parameters of the XGBoost model. Compared to conventional grid-search methods, it eliminates the need for exhaustive parameter exploration, resulting in faster optimization speed, lower computational cost, and significant time and effort savings.

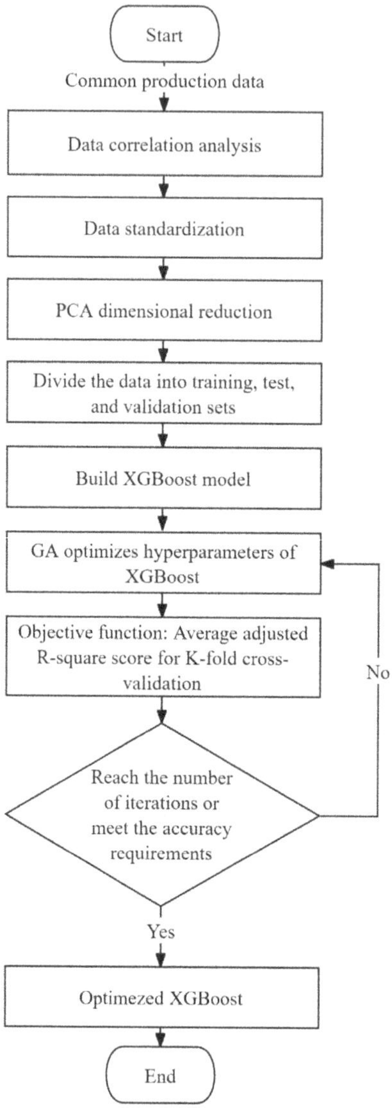

Figure 1. GA-XGBoost model flow diagram.

3. Model Verification and Application

The numerical simulation, through historical fitting, can represent, to some extent, the actual conditions of the target oilfield. The numerical model of the target oilfield had a grid size of 300 × 300 × 20 m, with a total of 239,000 effective grid cells. The boundaries were closed on all sides, and there were 8 components in the model. The model simulation parameters are shown in Table 2. Based on the actual production data and pressure test data, the fitting of a fixed oil volume and single well pressure was carried out, and the overall fitting effect was good.

Table 2. Model parameters.

Parameters	Average Reservoir Thickness per Well, m	Average Permeability, mD	Average Porosity, %	Average Oil Saturation, %	Average Production per Well, bbl
Values	225	0.19678039	12.5	78.4	30,000

Common production data (daily oil production, daily water production, daily gas production, daily liquid production, daily gas injection rate, gas–oil ratio, and bottomhole pressure) were collected from the numerical model as the dataset. The bottomhole flowing pressure was treated as the target value, while the remaining data served as input values.

A total of 413 datasets were collected from 14 wells. It can be observed from Figure 2, the distribution of the bottomhole pressure was uneven, mainly concentrated in the area of 5500 psi to 8300 psi, with the most data distributed around 7000 psi.

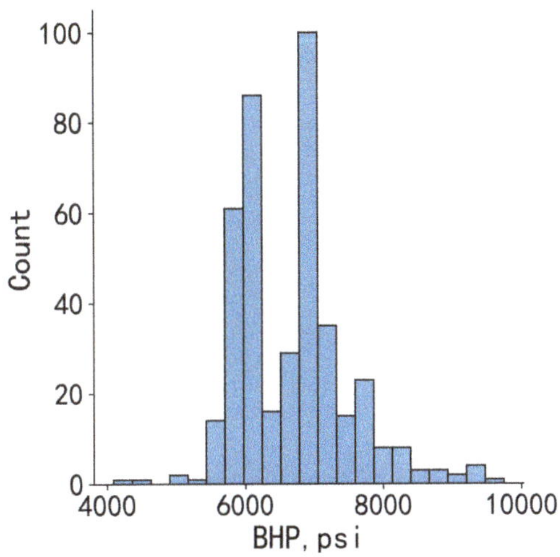

Figure 2. Data distribution histogram of bottomhole pressures.

3.1. Data Standardization

Due to the inconsistent dimensions of the original data, some of which may vary significantly, models tend to give more weight to features with larger scales during training, thus affecting the accuracy of the model. In addition, machine learning models are designed for data that follow a Gaussian distribution. If the data do not conform to a normal distribution, it will also affect the accuracy of the model. Therefore, data standardization is generally performed before training the model.

This article used the standard deviation normalization method (standardScale) to make the processed data conform to the standard normal distribution, with a mean of 0 and a standard deviation of 1. The formula is as follows:

$$x^* = \frac{x - \mu}{\sigma} \tag{5}$$

where x^* represents the standardized data, x represents the original data, μ represents the mean of the data, and σ represents the standard deviation of the data.

3.2. Dimension Reduction by Principal Component Analysis

First, the Pearson algorithm [36] was employed to calculate the correlation coefficients between each pair of features, which were then represented using a heatmap as shown in Figure 3. WOPR refers to the oil production rate of a well, WWPR refers to the water production rate of a well, WGPR refers to the gas production rate of a well, WWIR refers to the water injection rate of a well, WGIR refers to the gas injection rate of a well, GOR refers to the gas–oil ratio of a well, WWCT refers to the water cut of a well, and BHP refers to the bottomhole pressure of a well. In the figure, the values of the cells where rows intersect represent the correlation between the variables represented by rows and columns. If it is a positive number, it indicates a positive correlation between these two variables, while a negative number indicates a negative correlation between these two variables. The absolute value of the correlation coefficient ranged from 0.8 to 1.0, indicating a strong correlation between features. It can be observed that variables such as the gas–oil ratio and water cut exhibited a high correlation with the bottomhole flowing pressure (−0.91 and −0.86). Additionally, there were also high correlations among the variables themselves (e.g., the correlation between WWCT and WWPR was 0.98), which could potentially interfere with the predictive performance of the model.

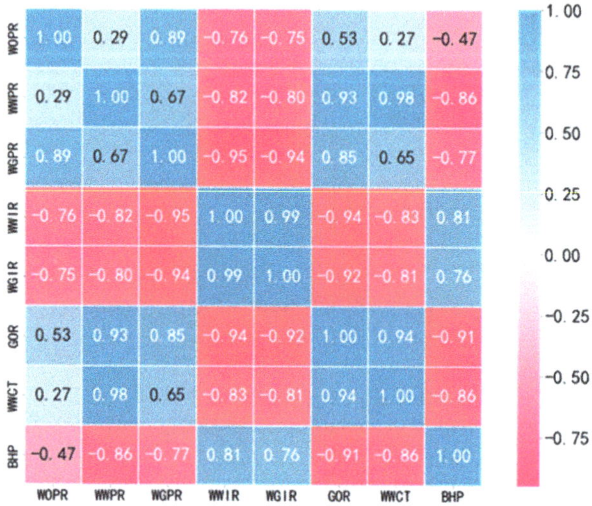

Figure 3. Heatmap for correlation analysis of data.

A principal component analysis (PCA) [37] is a multivariate statistical algorithm usec to assess the correlation between multiple variables. It employs an orthogonal transforma tion to convert a set of potentially correlated variables into a set of linearly uncorrelate variables called principal components. The PCA algorithm can transform the original dat into several principal components while preserving as much of the original informatio as possible, with each principal component being mutually independent. A PCA serve the purpose of dimensionality reduction by extracting and synthesizing relevant info mation while reducing the interference of redundant features and accelerating the mode training process.

Through the PCA algorithm, the original seven features were mapped and linearl combined to construct new features known as principal components. The importance the principal components can be evaluated based on their contribution to the variance. higher variance contribution indicates greater importance. We used Python to call the PC algorithm of the scikit learn framework and reduce the data to five dimensions based c the criterion of variance contribution rate greater than 95%. The variance contribution rat of the five principal components are shown in Table 3. From Table 3, it can be observed th the first principal component contributed the most, accounting for 52.15% of the varianc while the fifth principal component contributed the least, with a variance contribution ra of 3.65%.

Table 3. Principal component analysis data features.

Features	Feature 1	Feature 2	Feature 3	Feature 4	Feature 5
Variance contribution rate	0.52153446	0.19678039	0.11851757	0.08397276	0.03653219

The correlation heatmap of the five principal components after dimensionality reduction is presented in Figure 4. It can be seen that the absolute correlation values of the principal components after dimensionality reduction were all less than 0.5, and the principal components were independent of each other, which was conducive to the performance of the model.

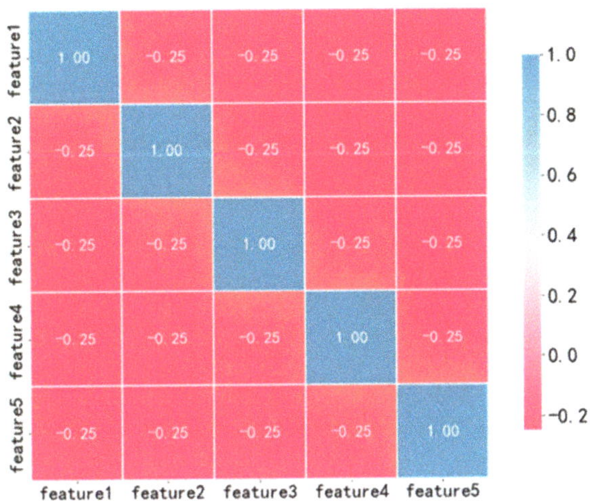

Figure 4. Correlation analysis after dimensionality reduction.

3.3. Model Training

The corresponding parameters of the genetic algorithm used in this study are presented in Table 4. The number of generations was set to 50, as it is generally believed that if convergence is not achieved by the 50th generation, it is necessary to investigate potential issues with the data or the model, as further increasing the generations may not yield significant improvements. The population size has a certain impact on the final results, with larger populations having a higher likelihood of finding the optimal solution. However, this also increases computational complexity. To ensure a balance between accuracy and efficiency, the population size was set to 20. The significance of the convergence threshold is that if the difference between the current generation and the best values from previous generations is smaller than this threshold, it indicates that the optimization of the objective has reached a stagnation state. If the evolution stagnates beyond the maximum limit set, the algorithm will stop running.

Table 4. The corresponding parameters of the genetic algorithm used in this paper.

Parameters	Parameter Range
Number of generations	50
Population size	20
Threshold for determining stagnation in single-objective optimization	1×10^{-6}
Maximum threshold value for evolution stagnation counter	10

The data were randomly split into training, testing, and validation sets with a 7:2:1 ratio, with the number of samples in each set presented in Table 5.

Table 5. Number of sample sets.

Data Set	Training Set	Test Set	Verification Set
Number	296	75	42

Using the SEGA, the parameters were iteratively adjusted to maximize the indicator of the average adjusted R-squared score obtained through cross-validation on the training set. Cross-validation [38–41] is a practical method employed in the machine learning training process, which involves dividing the data samples into smaller subsets. By performing a K-fold cross-validation, it simulates various possible datasets, enabling a thorough testing of the model's performance. This approach effectively utilizes the data to evaluate the model's performance and reduces issues such as overfitting and selection bias, thereby enhancing the model's generalization capability. Based on the empirical formula provided below, a cross-validation fold number of 6 was selected.

$$K \approx \log(n) \quad (6)$$

where K represents the number of cross-validation folds and n represents the size of the dataset. By randomly partitioning the data, each subset contains a mix of best- and worst-case scenarios. The cross-validated score obtained by averaging the scores from each subset may appear lower compared to the conventional training set score. However, it provides a more comprehensive evaluation of the model's performance across different datasets, making it a suitable objective function for optimization purposes.

The R-squared coefficient of determination indicates the goodness-of-fit of the trend line. Given a series of true values (y_i) and their corresponding predicted values, R-squared is defined as follows:

$$R^2 = 1 - \frac{\sum_i (\hat{y}_i - y_i)^2}{\sum_i (y_i - \bar{y})^2} \quad (7)$$

The R-squared value is in the range $(-\infty,1]$ and represents the proportion of variance explained by the model. It provides an intuitive understanding of the model's performance. As R-squared approaches 1, the model's performance improves. But as the number of independent variables increases, the R-squared value becomes larger, and using the adjusted R-squared value to evaluate the fitting effect of the model is more accurate. The adjusted R-squared takes into account both the sample size n and the number of independent variables k. Therefore, it reflects the modelling accuracy.

$$\text{Adjusted } R^2 = 1 - \left(1 - R^2\right)\frac{(n-1)}{(n-k-1)} \quad (8)$$

RMSE (root-mean-square Error) is simultaneously used as a supplementary validation metric to assess the performance of the model. The RMSE measures the square root of the ratio between the squared deviations of the predicted values and the observed values divided by the number of observations, n.

$$\text{RMSE} = \sqrt{\frac{1}{N}\sum_{i=1}^{n}(Y_i - f(x_i))^2} \quad (9)$$

The RMSE quantifies the deviation between predicted values and actual values, and it is particularly sensitive to outliers in the data. A smaller RMSE value indicates higher model performance.

The optimization process is illustrated in Figure 5, where the average score of the population improves from below 0.9350 to above 0.9525. The final average score of the population is comparable to that of the elites. The model optimization process terminates after the 18th generation.

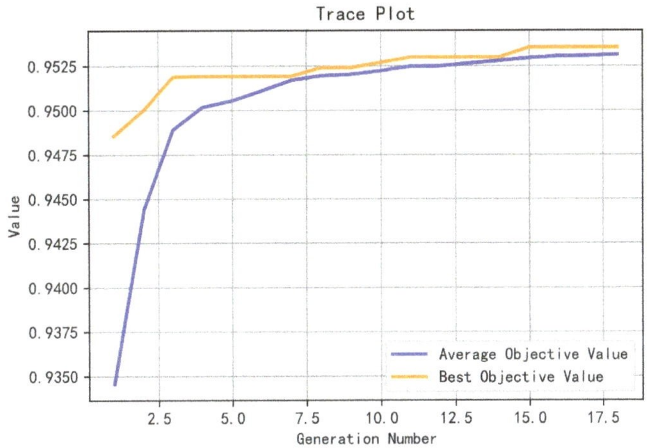

Figure 5. Genetic algorithm optimization process.

After optimization using the genetic algorithm, the optimal values of the XGBoost parameters were found and are shown in Table 6. The average adjusted R-squared score from the cross-validation was 0.9536.

Table 6. The parameters of the XGBoost model established in this paper.

Parameters	Parameter Value
n_estimators	6401
learning_rate	0.0338
max_depth	4
min_child_weight	6.5425
gamma	0.0233
subsample	0.9483
colsample_bytree	0.6000

To verify the effectiveness of XGBoost, we compared it with other common machine learning models, including support vector machines (SVM), neural networks, stochastic gradient descent (SGD), linear regression, ridge regression, decision trees, random forests, gradient boosting trees, ExtraTrees, and Adaboost. Support vector machines can handle the interaction of nonlinear features but are sensitive to missing data. Artificial neural networks can fully approximate complex nonlinear relationships but require a large number of parameters. The parameter update speed of random gradient descent is fast, but it may converge to local optima. Linear regression is easy to implement, but it cannot fit nonlinear data. Although ridge regression is less prone to overfitting compared to linear regression, it is also unable to handle nonlinear data. Decision trees are easy to understand and interpret, but they are prone to overfitting. Random forest performs well and can handle high-dimensional data, but it may overfit in some noisy problems. Gradient boosting decision trees have a good generalization ability but perform poorly on high-dimensional sparse datasets. ExtraTrees is a variant of random forests with better generalization performance at times. The Adaboost algorithm is a typical boosting algorithm that is not prone to overfitting, but imbalanced data can lead to decreased accuracy and sensitivity to outliers. Table 7 shows the parameters used by each model. These models were chosen because they have performed well in previous applications. Neural networks [13] are used for predicting bottomhole pressure in vertical multiphase flow oil wells, with a prediction error of no more than 10%. A support vector machine [42] was used for comparison in predicting bottomhole flow pressure in multiphase flow, as it represents a commonly used ML method and was therefore classified as the benchmark method in FBHP prediction. Linear regression [16] has been used to predict the bottomhole flow pressure of coalbed methane wells, with high accuracy in single-well prediction problems. Other models [43–49] have performed well in similar application fields.

Table 7. The parameters used by each model.

Model	Parameters
SVM	C: 1; epsilon: 0.1; gamma: 0.167
ANN	single hidden_layer_sizes: 100; learning_rate_init: 0.003
SGD	penalty: l2; alpha: 0.0001
Linear regression	fit_intercept: True; normalize: False
Ridge regression	alpha: 1
Decision trees	min_samples_split: 2; min_samples_leaf: 1
Random forests	n_estimators: 100; min_samples_split: 2; min_samples_leaf: 1
Gradient boosting trees	learning_rate: 0.1; n_estimators: 100; alpha: 0.9; max_depth: 3
ExtraTrees	n_estimators: 100; min_samples_split: 2; min_samples_leaf: 1
Adaboost	n_estimators: 100; learning_rate: 1

4. Results Analysis and Comparison

4.1. Results Analysis

The optimized XGBoost model had an adjusted R-squared score of 0.99 and a root-mean-square error of 0.0015 on the training set; the adjusted R-squared score on the test set was 0.84, with a root-mean-square error of 0.0564; the adjusted R-squared score of the validation set was 0.68, with a root-mean-square error of 0.0721. We drew a scatter plot with the true value as X and the predicted value as Y in Figure 6, where the comparison between the true and predicted values of the model on the training, testing, and validation sets is shown. A scattered distribution above the grey dashed line indicates that the predicted value is greater than the true value, a scattered distribution below the grey dashed line indicates that the predicted value is equal to the true value, and a scattered distribution below the grey dashed line indicates that the predicted value is less than the true value. The closer the scatter points are to the grey dashed line, the closer the predicted value is to the true value. It can be seen that the points formed by the predicted values and the true values of the model in the training set are basically distributed near the line with a slope of one, indicating that the predicted values were relatively close to the true values. For unseen data, the model still performed well in both the test and validation sets, with predicted values close to actual values, indicating good generalization performance.

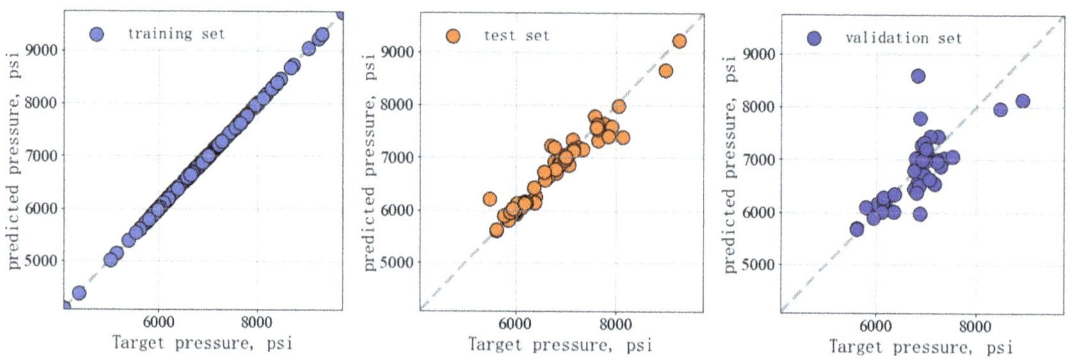

Figure 6. Predicted versus true values of the XGBoost model.

It can be seen that the validation set has significant fluctuations in the predicted values of the model at 7000 psi, with the distribution of predicted values ranging from 6000 psi to 9000 psi, which may be due to data-driven limitations. Due to the high dependence of data-driven methods on the quality and quantity of data, the prediction error of the model for unseen data such as the validation set is relatively large compared to the training set, which is inevitable. In addition, compared with traditional models, pure data-driven models do not have physical-knowledge constraints, and some predicted values may have significant deviations. In future work, improving XGBoost by incorporating physical-knowledge constraints may lead to better predictive performance.

4.2. Comparison of Models

The performance of the XGBoost model was compared with that of ten other common machine learning models, including support vector machines (SVM), neural networks, stochastic gradient descent (SGD), linear regression, ridge regression, decision trees, random forests, gradient boosting trees, ExtraTrees, and Adaboost. The comparison of adjusted R-squared scores for these models on different datasets is illustrated in Figure 7.

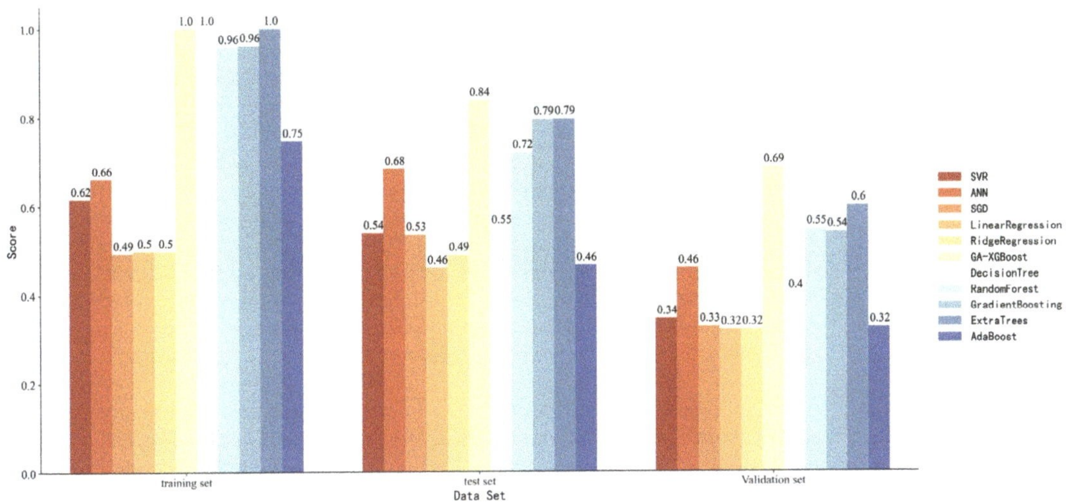

Figure 7. Comparison of Adjusted-R-squared scores for different datasets across models.

Linear regression, ridge regression, and Adaboost models performed poorly, with scores of only 0.5, 0.51, and 0.51 on the training set. This may be due to the complex nonlinear relationship between the bottomhole pressure and input variables in heterogeneous reservoir backgrounds, and these models cannot fit nonlinear data well. Previous studies used support vector regression and neural networks for predicting bottomhole pressure, and both performed well. However, with fewer data variables, support vector regression scored only 0.62 on the training set, while the neural network scored only 0.66 on the training set. It can be observed that tree-based models, such as decision trees and random forests, exhibited superior performance compared to other machine learning algorithms like support vector machines in addressing this particular problem. The training set scores of these tree-based models consistently exceeded 0.9. However, due to their tendency to overfit, the tree-based models yielded lower scores on the testing and validation sets compared to the XGBoost model. The XGBoost model optimized by GA demonstrated excellent performance across all three datasets.

5. Discussion

In this study, we selected the XGBoost algorithm, which exhibits a high accuracy and strong generalization, to establish a bottomhole pressure prediction model for carbonate reservoirs. We used a genetic algorithm with an enhanced elite strategy to optimize XGBoost's hyperparameters, with the objective function being the average adjusted R-squared score of a K-fold cross-validation, and compared its performance with other popular machine learning models. We selected six key variables, such as daily oil and water production, for the bottomhole pressure prediction, enabling a quick on-site implementation with limited data. The aim was to assess whether XGBoost could still perform well under limited data variable conditions. A correlation analysis revealed strong links between variables, potentially impacting the model training effectiveness. We used dimensionality reduction algorithms to convert the data into independent variables, minimizing interference from redundant features. We employed the cross-validation score as the genetic algorithm's optimization target, enhancing data use and the model's ability to generalize. By comparing XGBoost with commonly used machine learning algorithms, we verified the superior accuracy and generalization capabilities of XGBoost.

We established an XGBoost bottomhole pressure prediction model and optimized hyperparameters using SEGA to evaluate the pressure prediction performance of XGBoost under limited data volume. The optimized XGBoost model had an adjusted R-squared score of 0.99 and a root-mean-square error of 0.0015 on the training set; the adjusted R-squared score on the test set was 0.84, with a root-mean-square error of 0.0564; the adjusted R-squared score of the validation set was 0.69, with a root-mean-square error of 0.0721, and the superiority of GA-XGBoost was established through a comparison with other models. The poor performance of the linear regression, ridge regression, and Adaboost models may be due to the complex nonlinear relationship between the bottomhole pressure and input variables in heterogeneous reservoir backgrounds, and these models cannot fit nonlinear data well. Previous studies have used support vector regression and neural networks to predict bottomhole pressure, both of which have achieved good results. However, in the case of limited data variables, the performance of support vector regression and neural networks on the training set was also relatively average. It can be observed that tree-based models, such as decision trees and random forests, exhibited superior performance in solving this specific problem compared to other machine learning algorithms such as support vector machines. However, due to their tendency towards overfitting, tree-based models scored lower on the test and validation sets compared to XGBoost models. XGBoost, optimized by the GA model, demonstrated excellent performance on all three datasets, demonstrating the performance of the model.

6. Conclusions

The GA-XGBoost model demonstrated its ability to accurately estimate pressure values in challenging scenarios characterized by a significant heterogeneity and extreme variations in pressure distribution during the development of carbonate reservoirs. It maintained this capability even in cases with fewer data variables. This advantage of the model eliminates the need to install pressure gauges, resulting in cost savings.

Compared to numerical simulation techniques, predicting bottomhole pressure using GA-XGBoost eliminates the need for complex model building and historical fitting operations, while significantly improving computational speed. The use of genetic algorithms for automatic optimization avoids the extensive trial-and-error process required by conventional grid-search methods, saving time and effort.

Predicting the bottomhole pressure through a data-driven approach only requires commonly available production data, enabling a convenient and real-time estimation of bottomhole pressure. This method is more practical and facilitates on-site implementation of real-time bottomhole pressure prediction.

The GA-XGBoost model achieved an adjusted R-squared score of 0.99 and a root-mean-square error (RMSE) of 0.0015 on the training set. On the test set, the model obtained an adjusted R-squared score of 0.84 with an RMSE of 0.0564, while on the validation set, it achieved an adjusted R-squared score of 0.69 with an RMSE of 0.0721. Compared to other machine learning models, the GA-XGBoost model was superior in predicting bottomhole pressure with limited data features, exhibiting a high accuracy and good generalization performance. It achieved excellent results in the bottomhole pressure prediction on the training, test, and validation sets.

In future work, several trends are expected. Firstly, collecting data of various types and larger quantities will allow us to evaluate whether XGBoost's accuracy and generalization can be improved when dealing with massive test datasets, while also assessing the computational speed of XGBoost. Secondly, introducing constraints based on prior physical knowledge during the XGBoost training process may further enhance the model's accuracy, aligning the predicted results more closely with engineers' understanding.

Author Contributions: Conceptualization, H.S. and Q.L.; methodology, Z.X.; software, H.S.; validation, Z.X.; formal analysis, H.S.; investigation, Q.L.; resources, Z.X.; data curation, Y.L.; writing—original draft preparation, H.S.; writing—review and editing, Y.Y.; visualization, Q.L.; supervision, Y.L. and Y.Y.; project administration, Z.X. and Y.L. All authors have read and agreed to the published version of the manuscript.

Funding: This research was funded by Major Science and Technology Special Project of the China National Petroleum Corporation, "Research on Key Technologies for Efficient Production of Overseas Large-scale Carbonate Reservoirs" (Grant No. 2023ZZ19-04).

Data Availability Statement: Data are unavailable due to privacy restrictions.

Conflicts of Interest: Author Zhaohui Xia, Yunbo Li and Yang Yu were employed by the company Research Institute of Petroleum Exploration and Development. The remaining authors declare that the research was conducted in the absence of any commercial or financial relationships that could be construed as a potential conflict of interest. All authors have read and agreed to the published version of the manuscript.

Nomenclature

ANN	Artificial neural network
BHP	Bottomhole pressure of a well
GA	Genetic algorithm
GOR	Gas–oil ratio of a well
SGD	Stochastic gradient descent
SVR	Support vector regression
WGIR	Gas injection rate of a well
WGPR	Gas production rate of a well
WOPR	Oil production rate of a well
WWIR	Water injection rate of a well
WWCT	Water cut of a well
WWPR	Water production rate of a well
XGBoost	eXtreme Gradient Boosting

eferences

Duns, H.; Ros, N.C.J. *Vertical Flow of Gas and Liquid Mixtures in Wells*; OnePetro: Richardson, TX, USA, 1963.

Hagedorn, A.R.; Brown, K.E. Experimental Study of Pressure Gradients Occurring During Continuous Two-Phase Flow in Small-Diameter Vertical Conduits. *J. Pet. Technol.* **1965**, *17*, 475–484. [CrossRef]

Orkiszewski, J. Predicting Two-Phase Pressure Drops in Vertical Pipe. *J. Pet. Technol.* **1967**, *19*, 829–838. [CrossRef]

Beggs, D.H.; Brill, J.P. An Experimental Study of Two-Phase Flow in Inclined Pipes. *J. Pet. Technol.* **1973**, *25*, 607–617. [CrossRef]

Mukherjee, H.; Brill, J.P. Pressure Drop Correlations for Inclined Two-Phase Flow. *J. Energy Resour. Technol.* **1985**, *107*, 549–554. [CrossRef]

Ansari, A.M.; Sylvester, N.D.; Sarica, C.; Shoham, O.; Brill, J.P. A Comprehensive Mechanistic Model for Upward Two-Phase Flow in Wellbores. *SPE Prod. Facil.* **1994**, *9*, 143–151. [CrossRef]

Corsano, A. Experimental Study and the Development of a Mechanistic Model for Two-Phase Flow through Vertical Tubing. In Proceedings of the SPE Western Regional Meeting, Anchorage, AK, USA, 22–24 May 1996; SPE: Kuala Lumpur, Malaysia, 1996.

8. Gomez, L.E.; Shoham, O.; Schmidt, Z.; Chokshi, R.N.; Northug, T. Unified Mechanistic Model for Steady-State Two-Phase Flow: Horizontal to Vertical Upward Flow. *SPE J.* **2000**, *5*, 339–350. [CrossRef]
9. Wang, B. The Latest Application of Artificial Intelligence Technology in Petroleum Engineering Field. *China CIO News* **2018**, *10*, 95.
10. Zhao, X.; Chen, X.; Lan, Z.; Wang, X.; Yao, G. Pore Pressure Prediction Assisted by Machine Learning Models Combined with Interpretations: A Case Study of an HTHP Gas Field, Yinggehai Basin. *Geoenergy Sci. Eng.* **2023**, *229*, 212114. [CrossRef]
11. Awadalla, M.; Yousef, H. Neural Networks for Flow Bottom Hole Pressure Prediction. *Int. J. Electr. Comput. Eng. IJECE* **2016**, *6*, 1839. [CrossRef]
12. Firouzi, M.; Rathnayake, S. *Prediction of the Flowing Bottom-Hole Pressure Using Advanced Data Analytics*; OnePetro: Richardson, TX, USA, 2019.
13. Ahmadi, M.A.; Chen, Z. Machine Learning Models to Predict Bottom Hole Pressure in Multi-phase Flow in Vertical Oil Production Wells. *Can. J. Chem. Eng.* **2019**, *97*, 2928–2940. [CrossRef]
14. Nait Amar, M.; Zeraibi, N.; Redouane, K. Bottom Hole Pressure Estimation Using Hybridization Neural Networks and Grey Wolves Optimization. *Petroleum* **2018**, *4*, 419–429. [CrossRef]
15. Nait Amar, M.; Zeraibi, N. A Combined Support Vector Regression with Firefly Algorithm for Prediction of Bottom Hole Pressure. *SN Appl. Sci.* **2020**, *2*, 23. [CrossRef]
16. Rathnayake, S.; Rajora, A.; Firouzi, M. A Machine Learning-Based Predictive Model for Real-Time Monitoring of Flowing Bottom-Hole Pressure of Gas Wells. *Fuel* **2022**, *317*, 123524. [CrossRef]
17. Jia, D.; Liu, H.; Zhang, J.; Pei, X.; Wang, Q.; Yang, Q. Data-Driven Optimization for Fine Water Injection in a Mature Oil Field. *Pet. Explor. Dev.* **2020**, *47*, 629–636. [CrossRef]
18. Wang, H.; Mu, L.; Shi, F.; Dou, H. Production Prediction at Ultra-High Water Cut Stage via Recurrent Neural Network. *Pet. Explor. Dev.* **2020**, *47*, 1009–1015. [CrossRef]
19. Tariq, Z.; Mahmoud, M.; Abdulraheem, A. *An Artificial Intelligence Approach to Predict the Water Saturation in Carbonate Reservoir Rocks*; OnePetro: Richardson, TX, USA, 2019.
20. Artun, E.; Kulga, B. Selection of Candidate Wells for Re-Fracturing in Tight Gas Sand Reservoirs Using Fuzzy Inference. *Pet. Explor. Dev.* **2020**, *47*, 383–389. [CrossRef]
21. Chen, T.; Guestrin, C. XGBoost: A Scalable Tree Boosting System. In Proceedings of the 22nd ACM SIGKDD International Conference on Knowledge Discovery and Data Mining, San Francisco, CA, USA, 13 August 2016; pp. 785–794.
22. Ogunleye, A.; Wang, Q.-G. XGBoost Model for Chronic Kidney Disease Diagnosis. *IEEE/ACM Trans. Comput. Biol. Bioinform.* **2020**, *17*, 2131–2140. [CrossRef]
23. Dhaliwal, S.S.; Nahid, A.-A.; Abbas, R. Effective Intrusion Detection System Using XGBoost. *Information* **2018**, *9*, 149. [CrossRef]
24. Pan, S.; Zheng, Z.; Guo, Z.; Luo, H. An Optimized XGBoost Method for Predicting Reservoir Porosity Using Petrophysical Logs. *J. Pet. Sci. Eng.* **2022**, *208*, 109520. [CrossRef]
25. Markovic, S.; Bryan, J.L.; Rezaee, R.; Turakhanov, A.; Cheremisin, A.; Kantzas, A.; Koroteev, D. Application of XGBoost Model for In-Situ Water Saturation Determination in Canadian Oil-Sands by LF-NMR and Density Data. *Sci. Rep.* **2022**, *12*, 13984. [CrossRef] [PubMed]
26. Zhong, R.; Johnson, R.; Chen, Z. Generating Pseudo Density Log from Drilling and Logging-While-Drilling Data Using Extreme Gradient Boosting (XGBoost). *Int. J. Coal Geol.* **2020**, *220*, 103416. [CrossRef]
27. Gu, Y.; Zhang, D.; Bao, Z. A New Data-Driven Predictor, PSO-XGBoost, Used for Permeability of Tight Sandstone Reservoirs: A Case Study of Member of Chang 4+5, Western Jiyuan Oilfield, Ordos Basin. *J. Pet. Sci. Eng.* **2021**, *199*, 108350. [CrossRef]
28. Al-Mudhafar, W.J.; Abbas, M.A.; Wood, D.A. Performance Evaluation of Boosting Machine Learning Algorithms for Lithofacies Classification in Heterogeneous Carbonate Reservoirs. *Mar. Pet. Geol.* **2022**, *145*, 105886. [CrossRef]
29. Zhang, J.; Sun, Y.; Shang, L.; Feng, Q.; Gong, L.; Wu, K. A Unified Intelligent Model for Estimating the (Gas + n-Alkane) Interfacial Tension Based on the eXtreme Gradient Boosting (XGBoost) Trees. *Fuel* **2020**, *282*, 118783. [CrossRef]
30. Dong, Y.; Qiu, L.; Lu, C.; Song, L.; Ding, Z.; Yu, Y.; Chen, G. A Data-Driven Model for Predicting Initial Productivity of Offshore Directional Well Based on the Physical Constrained eXtreme Gradient Boosting (XGBoost) Trees. *J. Pet. Sci. Eng.* **2022**, *211*, 11017[illegible] [CrossRef]
31. Wang, Z.; Tang, H.; Cai, H.; Hou, Y.; Shi, H.; Li, J.; Yang, T.; Feng, Y. Production Prediction and Main Controlling Factors in a Highly Heterogeneous Sandstone Reservoir: Analysis on the Basis of Machine Learning. *Energy Sci. Eng.* **2022**, *10*, 4674–469[illegible] [CrossRef]
32. Zhai, L. XGBoost-Based Water Injection Profile Prediction Method and Its Application. *Pet. Geol. Recovery Effic.* **2022**, *29*, 175–18[illegible] [CrossRef]
33. Shi, J.; Xu, H.; Peng, L. Research on Complaint Management System of Manufacturing Industry Based on XGBoost. *Manu Autom.* **2023**, *45*, 76–79.
34. Chen, K.-Y. Forecasting Systems Reliability Based on Support Vector Regression with Genetic Algorithms. *Reliab. Eng. Syst. Sa* **2007**, *92*, 423–432. [CrossRef]
35. Whitley, D. A Genetic Algorithm Tutorial. *Stat. Comput.* **1994**, *4*, 65–85. [CrossRef]

36. Benesty, J.; Chen, J.; Huang, Y.; Cohen, I. Pearson Correlation Coefficient. In *Noise Reduction in Speech Processing*; Cohen, I., Huang, Y., Chen, J., Benesty, J., Eds.; Springer Topics in Signal Processing; Springer: Berlin/Heidelberg, Germany, 2009; pp. 1–4, ISBN 978-3-642-00296-0.
37. Wold, S.; Esbensen, K.; Geladi, P. Principal Component Analysis. *Chemom. Intell. Lab. Syst.* **1987**, *2*, 37–52. [CrossRef]
38. Rahimi, M.; Riahi, M.A. Reservoir Facies Classification Based on Random Forest and Geostatistics Methods in an Offshore Oilfield. *J. Appl. Geophys.* **2022**, *201*, 104640. [CrossRef]
39. Wang, G.; Ju, Y.; Carr, T.R.; Li, C.; Cheng, G. *Application of Artificial Intelligence on Black Shale Lithofacies Prediction in Marcellus Shale, Appalachian Basin*; OnePetro: Richardson, TX, USA, 2014.
40. Al-Mudhafar, W.J. *Incorporation of Bootstrapping and Cross-Validation for Efficient Multivariate Facies and Petrophysical Modeling*; OnePetro: Richardson, TX, USA, 2016.
41. Pirrone, M.; Battigelli, A.; Ruvo, L. *Lithofacies Classification of Thin Layered Reservoirs through the Integration of Core Data and Dielectric Dispersion Log Measurements*; OnePetro: Richardson, TX, USA, 2014.
42. Marfo, S.A.; Asante-Okyere, S.; Ziggah, Y.Y. A new flowing bottom hole pressure prediction model using M5 prime decision tree approach. *Model. Earth Syst. Environ.* **2022**, *8*, 2065–2073. [CrossRef]
43. Sun, P.; Huo, S.; He, T. Multiple machine learning models in estimating viscosity of crude oil: Comparisons and optimization for reservoir simulation. *J. Mol. Liq.* **2023**, *384*, 122251. [CrossRef]
44. Fan, P.; Deng, R.; Qiu, J.; Zhao, Z.; Wu, S. Well logging curve reconstruction based on kernel ridge regression. *Arab. J. Geosci.* **2021**, *14*, 1559. [CrossRef]
45. Üneş, F.; Demirci, M.; Taşar, B.; Kaya, Y.Z.; Varçin, H. Modeling of dam reservoir volume using generalized regression neural network, support vector machines and M5 decision tree models. *Appl. Ecol. Environ. Res.* **2019**, *17*, 7043–7055. [CrossRef]
46. Wang, M.; Feng, D.; Li, D.; Wang, J. Reservoir Parameter Prediction Based on the Neural Random Forest Model. *Front. Earth Sci.* **2022**, *10*, 888933. [CrossRef]
47. Qin, Y.; Ye, Z.; Zhang, C. Application of GBDT for division of petroleum reservoirs. *J. Phys. Conf. Ser.* **2020**, *1437*, 012050. [CrossRef]
48. Seyyedattar, M.; Zendehboudi, S.; Butt, S. Relative permeability modeling using extra trees, ANFIS, and hybrid LSSVM–CSA methods. *Nat. Resour. Res.* **2022**, *31*, 571–600. [CrossRef]
49. Busari, G.A.; Lim, D.H. Crude oil price prediction: A comparison between AdaBoost-LSTM and AdaBoost-GRU for improving forecasting performance. *Comput. Chem. Eng.* **2021**, *155*, 107513. [CrossRef]